AF811626

Nanotechnology in Plant Sciences

Volume 4

Series Editors

Jameel M. Al-Khayri, Department of Agricultural Biotechnology, King Faisal University, Al-Ahsa, Saudi Arabia

Shri Mohan Jain, Department of Agricultural Sciences, University of Helsinki, Helsinki, Finland

This book series provides a platform for researchers to update advances in nanotechnology research and applications relevant to various fields of plant sciences. This series is an excellent reference source of cutting-edge knowledge for agriculturists, breeders, biotechnologists, and molecular biologists interested in plant nanotechnology. It is essential for professionals, researchers, policymakers, and commercial entrepreneurs concerned with understanding the mechanisms and innovative applications of nanotechnology to achieve enhanced crop production improvement, agricultural sustainability, environment protection, and food and nutritional security.

Jameel M. Al-Khayri • T. R. Anju
Shri Mohan Jain
Editors

Green Nanotechnology Applications for Ecosystem Sustainability

Springer

Editors
Jameel M. Al-Khayri
Department of Agricultural Biotechnology
College of Agriculture and Food Sciences
King Faisal University
Al-Ahsa, Saudi Arabia

T. R. Anju
Department of Biotechnology
Newman College
Thodupuzha, Kerala, India

Shri Mohan Jain
Department of Agricultural Sciences
University of Helsinki
Helsinki, Finland

ISSN 3004-9334 ISSN 3004-9342 (electronic)
Nanotechnology in Plant Sciences
ISBN 978-3-031-90765-4 ISBN 978-3-031-90766-1 (eBook)
https://doi.org/10.1007/978-3-031-90766-1

© The Editor(s) (if applicable) and The Author(s), under exclusive license to Springer Nature Switzerland AG 2025

This work is subject to copyright. All rights are solely and exclusively licensed by the Publisher, whether the whole or part of the material is concerned, specifically the rights of translation, reprinting, reuse of illustrations, recitation, broadcasting, reproduction on microfilms or in any other physical way, and transmission or information storage and retrieval, electronic adaptation, computer software, or by similar or dissimilar methodology now known or hereafter developed.
The use of general descriptive names, registered names, trademarks, service marks, etc. in this publication does not imply, even in the absence of a specific statement, that such names are exempt from the relevant protective laws and regulations and therefore free for general use.
The publisher, the authors and the editors are safe to assume that the advice and information in this book are believed to be true and accurate at the date of publication. Neither the publisher nor the authors or the editors give a warranty, expressed or implied, with respect to the material contained herein or for any errors or omissions that may have been made. The publisher remains neutral with regard to jurisdictional claims in published maps and institutional affiliations.

This Springer imprint is published by the registered company Springer Nature Switzerland AG
The registered company address is: Gewerbestrasse 11, 6330 Cham, Switzerland

If disposing of this product, please recycle the paper.

Preface

The integration of nanotechnology with sustainability principles has revolutionized approaches to environmental and societal challenges, giving rise to the transformative field of green nanotechnology. The book, *Green Nanotechnology Applications for Ecosystem Sustainability,* provides a comprehensive exploration of how nanotechnology is being harnessed to promote ecosystem sustainability, improve agricultural productivity, and address pressing environmental issues such as water and air pollution, oil spills, and ocean contamination. It highlights innovations in manufacturing, energy harnessing, food preservation, and biomedical applications, demonstrating the vast potential of nanoscale materials to enhance efficiency and reduce environmental impact. Cutting-edge topics such as the development of advanced nanosensors for environmental and health monitoring, the use of nanomaterials in drug delivery, and the emerging field of eco-nanotoxicology are also discussed, highlighting the importance of ensuring safe and sustainable applications of these technologies.

The book consists of 18 chapters, enriched with 23 tables and 117 color figures, all meticulously designed to enhance understanding and engagement. Chapters are contributed by 67 recognized scientists representing 5 countries after a rigorous review process to maintain quality and scientific accuracy. The chapters open with an introduction that provides essential background followed by an in-depth exploration of the topic supported by high-quality color figures and tables. To guide future research, each chapter concludes with thoughtful recommendations for further study.

The book is expected to give a comprehensive exploration of green nanotechnology applications across diverse fields for ecosystem sustainability. It provides insights into cutting-edge research, methodologies, and innovations that harness the unique properties of nanomaterials for sustainable development. By presenting the latest research and advancements in green nanotechnology for sustainable ecosystem, the book aims to inspire researchers, policymakers, and industry professionals to collaborate and innovate, fostering a cleaner, healthier, and more equitable future for all.

Al-Ahsa, Saudi Arabia Jameel M. Al-Khayri
Thodupuzha, Kerala, India T. R. Anju
Helsinki, Finland Shri Mohan Jain

Acknowledgment

The editors extend their appreciation to the support of the Deanship of Scientific Research, Vice Presidency for Graduate Studies and Scientific Research, King Faisal University, Saudi Arabia (Project No. KFU252788).

Contents

About the Editors and Contributors

About the Editors

Jameel M. Al-Khayri is affiliated with the Department of Agricultural Biotechnology, King Faisal University, Saudi Arabia. He received B.S. Biology in 1984 from the University of Toledo, M.S. in Agronomy in 1988, and Ph.D. in Plant Science in 1991 from the University of Arkansas. He is the National Correspondent of the International Association of Plant Tissue Culture and Biotechnology. Al-Khayri has authored 130 research articles and 70 chapters and edited several journal special issues. In addition, he edited 30 reference books on plant biotechnology, genetic resources, breeding, genomics, and nanotechnology. In addition to teaching, advising, and research, he held administrative responsibilities as the Assistant Director of Date Palm Research Center, Head of the Department of Plant Biotechnology, and Vice Dean for Development and Quality Assurance. Al-Khayri served as a Member of Majlis Ash Shura (Saudi Legislative Council) for the 2009-2012 term. He is interested in agricultural innovations and sustainability.

T. R. Anju is an Assistant Professor in Biotechnology at Newman College, Kerala, and the Secretary of the Society for Biotechnologists, India. She completed a master's degree in Biotechnology and a doctoral degree at the Centre for Neuroscience, Cochin University of Science and Technology, Kochi, India. Anju completed her postdoctoral research on a neonatal hypoglycemia project funded by the Department of Biotechnology, Government of India. She has authored 44 publications in international peer-reviewed journals and 20 book chapters and edited 2 books. She is the recipient of Kerala State Young Scientist Award, IBS Award in Medical Biotechnology, and Prof. Edathil Vijayan Award in Neuroscience. In addition to teaching and research, Anju acts as coordinator of institutional grants funded by the Government of India. Her research is mainly focused on molecular signaling, neurotransmitter receptor functions, and nanotechnology, especially in green nanomaterials and their impact on plants, the environment and sustainable development.

Shri Mohan Jain (Adjunct) is currently associated with the Department of Agricultural Sciences, University of Helsinki, Helsinki, Finland. He has received M. Phil. (1973) and Ph.D. (1978) from Jawaharlal Nehru University, New Delhi, India. He served as a Technical Officer at Plant Breeding and Genetics, International Atomic Energy Agency (IAEA), Vienna, Austria, 1999-2005, when the agency was awarded the Nobel Peace Prize, 2005. He is on the Editorial Board of *Euphytica*, *In Vitro*, and *Propagation of Ornamental Plants*; a reviewer in *Plant Cell and Organ Culture*, *Plant Cell Reports*, and *Plants*. He has published more than 225 peer-reviewed journals, book chapters, and conference proceedings, and edited 72 books; he was an invited speaker and acted as a Chairperson in several international conferences worldwide. He is a former consultant to the IAEA, the European Union, The Government of Grenada, the Iranian Private Company, and the Egyptian Government.

Contributors

Mona A. Abdel-Fatah Chemical Engineering & Pilot Plant Department, Institute of Engineering Research and Renewable Energy, National Research Centre (NRC), Giza, Egypt

K. Aishwarya Department of Biotechnology, St. Josephs College (Autonomous), Irinjalakuda, Thrissur, Kerala, India

S. Aleena Irene Department of Biotechnology, St. Josephs College (Autonomous), Irinjalakuda, Thrissur, Kerala, India

Jameel M. Al-Khayri Department of Agricultural Biotechnology, College of Agriculture and Food Sciences, King Faisal University, Al-Ahsa, Saudi Arabia

T. R. Anju Department of Biotechnology, Newman College, Thodupuzha, Kerala, India

Sherin Antony Department of Zoology, Bharata Mata College, Ernakulam, Kerala, India

Harsh Bagdare Department of Micro-Bio-Seed Technology, SardarVallabhbhai Patel College, Mandleshwar, Madhya Pradesh, India

C. Muthu Lakshmi Bavithra Department of Agricultural Entomology, Tamil Nadu Agricultural University, Coimbatore, Tamil Nadu, India

Aviral Chaurasia Centre for Advanced Studies, Lucknow, India

Cintil Jose Chirayil Department of Chemistry, Newman College, Thodupuzha, Kerala, India

Jaiby Cyriac Department of Botany, Newman College, Thodupuzha, Kerala, India

Vinars Dawane Department of Micro-Bio-Seed Technology, SardarVallabhbhai Patel College, Mandleshwar, Madhya Pradesh, India

B. Deepa School of Nano Science and Nano Technology, Mahatma Gandhi University, Kottayam, Kerala, India

P. R. Devikrishna Department of Biotechnology, St. Josephs College (Autonomous), Irinjalakuda, Thrissur, Kerala, India

Cincy George Mahatma Gandhi University Center for Research in Chemistry, Newman College, Thodupuzha, India

Joicy George Mahatma Gandhi University Center for Research in Chemistry, Newman College, Thodupuzha, India

Naijil George Department of Biotechnology, St. Josephs College (Autonomous), Irinjalakuda, Thrissur, Kerala, India

Sumod George Department of Chemistry, Nirmala College Muvattupuzha, Ernakulam, Kerala, India

Kelvii Wei Guo Department of Mechanical and Biomedical Engineering, City University of Hong Kong, Kowloon, Hong Kong

Salwa I. Hawash Chemical Engineering & Pilot Plant Department, Institute of Engineering Research and Renewable Energy, National Research Centre (NRC), Giza, Egypt

Jayendra Kumar Himashu Department of Biotechnology, School of Life Sciences, Mahatma Gandhi Central University, Motihari, India

N. Jagathjothi Department of Agronomy, Tamil Nadu Agricultural University, Coimbatore, Tamil Nadu, India

Shri Mohan Jain Department of Agricultural Sciences, University of Helsinki, Finland

Sonia John Department of Zoology, Bharata Mata College, Ernakulam, Kerala, India

Alex Joseph Mahatma Gandhi University Center for Research in Chemistry, Newman College, Thodupuzha, India

Anu Mary Joseph Department of Chemistry, Newman College, Thodupuzha, Kerala, India

Manu M. Joseph Department of Life Sciences, School of Science, CHRIST University, Bengaluru, Karnataka, India

Simi Joseph Department of Zoology, Bharata Mata College, Ernakulam, Kerala, India

Shilpa Joy Department of Biotechnology, Sethu Institute of Technology, Kariapatti, India

Anitha Kannan Department of Biotechnology, Sethu Institute of Technology, Kariapatti, India

Vaishali Kesavan Department of Biotechnology, Sethu Institute of Technology, Kariapatti, India

Khushboo Kourani Institute of Science, Nirma University, Ahmedabad, Gujarat, India

Amee Krishnakumar Institute of Science, Nirma University, Ahmedabad, Gujarat, India

Asha Kumari Kumawat School of Environment and Sustainable Development, Central University of Gujarat, Gandhinagar, India

Livina Lazar Department of Biotechnology, Presentation College of Applied Sciences, Puthenvelikara, Kerala, India

Dilip K. Maiti Centre for Research in Nanoscience and Nanotechnology, University of Calcutta, Kolkata, India
Department of Chemistry, University of Calcutta, Kolkata, India

Kavya Manoj School of Nano Science and Nano Technology, Mahatma Gandhi University, Kottayam, Kerala, India

Saurabh Mishra Centre for Advanced Studies, Lucknow, India

Soumyadeep Mitra Centre for Research in Nanoscience and Nanotechnology, University of Calcutta, Kolkata, India
Department of Physics, Brainware University, Barasat, India

Dhannia P. Narayanan Department of Botany, Mar Athanasius College, Ernakulam, Kerala, India

Ashab Noumani Department of Biotechnology, Kalinga University, Kotni, Chhattisgarh, India

Randa M. Osman Chemical Engineering & Pilot Plant Department, Institute of Engineering Research and Renewable Energy, National Research Centre (NRC), Giza, Egypt

Bhawana Pathak School of Environment and Sustainable Development, Central University of Gujarat, Gandhinagar, Gujarat, India

Sunil Kumar Patidar Department of Chemistry, Shri Neelkantheshwar Government Post Graduate College, Khandwa, Madhya Pradesh, India

Satish Piplod Department of Chemistry, ShaheedBhagat Singh Government Post Graduate College, Pipariya, India
Govt. College, Bankhedi, Narmadapuram, Madhya Pradesh, India

L. A. Pothan Department of Chemistry, Bishop Moore College, Mavelikara, Kerala, India

P. L. Praveena Department of Biotechnology, Sethu Institute of Technology, Kariapatti, India

R. Sathya Priya Department of Agronomy, Tamil Nadu Agricultural University, Coimbatore, Tamil Nadu, India

Nageshwari Raja Department of Biotechnology, Sethu Institute of Technology, Kariapatti, India

Sudipta Ray A. K. Mitra Instiution for Girls, Dhakuria, Kolkata, India

Roshny Roy Mahatma Gandhi University Center for Research in Chemistry, Newman College, Thodupuzha, India

Salem S. Salem Botany and Microbiology Department, Faculty of Science, Al-Azhar University, Cairo, Egypt

M. Saranya Department of Agricultural Entomology, Tamil Nadu Agricultural University, Coimbatore, Tamil Nadu, India

Mohamed A. Selim Botany and Microbiology Department, Faculty of Science, Al-Azhar University, Cairo, Egypt

Sonu Sen Department of Physics, Shri Neelkantheshwar Government Post Graduate College, Khandwa, Madhya Pradesh, India

Dhriti P. Shah Institute of Science, Nirma University, Ahmedabad, Gujarat, India

R. Sharmila Department of Nematology, Tamil Nadu Agricultural University, Coimbatore, Tamil Nadu, India

Shweta Shukla Department of Chemistry, National Institute of Technology, Tiruchirappalli, India

N. Suganthi Department of Agricultural Extension, Tamil Nadu Agricultural University, Coimbatore, Tamil Nadu, India

Merin Sara Thomas Department of Chemistry, Mar Thoma College, Tiruvalla, Kerala, India

Neenu Mary Thomas Department of Physics, Morning Star Home Science College, Angamaly, Kerala, India

Sabu Thomas School of Nano Science and Nano Technology, Mahatma Gandhi University, Kottayam, Kerala, India

Emy Tomy Mahatma Gandhi University Center for Research in Chemistry, Newman College, Thodupuzha, India
Mahatma Gandhi University Center for Research in Chemistry, Nirmala College Muvattupuzha, India

Rini Thresia Varghese Department of Chemistry, Newman College, Thodupuzha, Kerala, India

Sharon Varghese Mahatma Gandhi University Center for Research in Chemistry, Newman College, Thodupuzha, India

Surya Venkatesh Department of Biotechnology, Sethu Institute of Technology, Kariapatti, India

M. Yuvaraj Soil Science and Agricultural Chemistry, Agricultural College and Research Institute, Vazhavachanur, Tamil Nadu, India

Chapter 1
Green Nanotechnology Applications for Ecosystem Sustainability

Livina Lazar, T. R. Anju, Jameel M. Al-Khayri, and Shri Mohan Jain

Abstract Green nanotechnology offers innovative solutions to halt environmental degradation and promote economic growth, positioning itself as a feasible route towards sustainable development. This chapter examines the various uses of green nanotechnology, highlighting how it may improve agricultural practices, reduce environmental pollution, optimize manufacturing processes, and promote ecosystem sustainability. These nanoparticles in agriculture develop environmentally friendly fertilizers and insecticides, increasing crop yield while reducing damage to the environment. Similar to this, green nanotechnology in manufacturing enables more effective and clean production techniques, which lower waste and resource consumption. With the use of green-synthesized nanoparticles, environmental pollutants can be effectively addressed through nano remediation, which helps to clean up contaminated sites. Green nanotechnology also protects terrestrial and aquatic environments by providing solutions for biological pollution treatment, water purification, and oil spill cleanup. It is a key player in the energy sector, helping to improve energy storage technologies and tap into renewable energy sources to create a more sustainable energy future. Additionally, it is used in the food sector to improve food quality, shelf life, and safety, as well as in medical applications for medicine delivery systems, nano sensors, and diagnostics. To guarantee the ethical and secure use of

L. Lazar
Department of Biotechnology, Presentation College of Applied Sciences, Manancherikunnu, Kerala, India

Post Graduate Department of Biosciences, M.E.S. College, Aluva, Kerala, India

T. R. Anju (✉)
Department of Biotechnology, Newman College, Thodupuzha, Kerala, India
e-mail: anju.tr@newmancollege.ac.in

J. M. Al-Khayri
Department of Agricultural Biotechnology, College of Agriculture and Food Sciences, King Faisal University, Al-Ahsa, Saudi Arabia
e-mail: jkhayri@kfu.edu.sa

S. M. Jain
Department of Agricultural Sciences, University of Helsinki, Helsinki, Finland

© The Author(s), under exclusive license to Springer Nature Switzerland AG 2025
J. M. Al-Khayri et al. (eds.), *Green Nanotechnology Applications for Ecosystem Sustainability*, Nanotechnology in Plant Sciences 4,
https://doi.org/10.1007/978-3-031-90766-1_1

nanomaterials across multiple industries, responsible research and development procedures are important to the sustainable future of nanotechnology. Through the application of green nanotechnology, interested parties can collaborate to create a future that is more robust and sustainable for the environment and society.

Keywords Agriculture · Ecosystem sustainability · Environment · Green nanotechnology · Nanomaterials · Nano remediation · Nano sensors

1.1 Introduction

Nanotechnology is being used extensively in daily life and is revolutionizing civilization as a whole (He et al. 2019). Nanotechnology is the manipulation of matter at the nanoscale (1–100 nm) across a range of fields, such as chemistry, physics, engineering, material science, medicine, and life sciences (Ciambelli et al. 2020; Mansoori et al. 2008; Mayegowda et al. 2023). Since nanotechnology has advanced so much over the last several decades, it is anticipated that we will live in a "world of nanoparticles." The small size (less than 100 nm) and enormous specific surface area of nanoparticles (NPs) provide them exceptional features in chemical, optical, magnetic electrical, and also mechanical domains (Jiang et al. 2022). Thus, a wide variety of nanoparticles and nanomaterials possess optical, magnetic, electrical, biological, and antibacterial capabilities (García-Quintero and Palencia 2021), many industrial sectors find great appeal in the demand of nanomaterials and nanotechnology, particularly in the fields of biomedicine, energy, pharmacy, food, agriculture, and environment (Ciambelli et al. 2020; Hristozov 2013). When creating nanoparticles, two general strategies are used as shown in Fig. 1.1: the top-down method employs physical, chemical, and biological techniques to break down larger

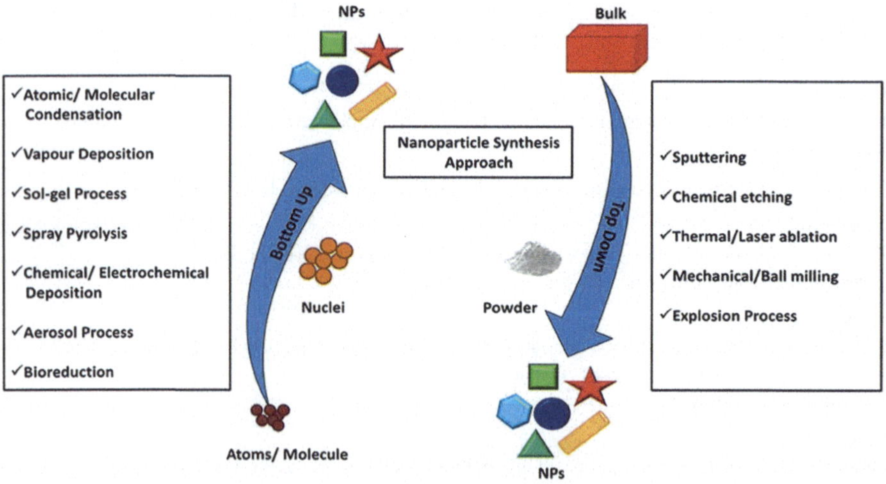

Fig. 1.1 Different strategies for synthesis of nanoparticles. (Reproduced from Bhol et al. 2020)

structures into smaller particles, while the bottom-up method synthesizes NPs at the atomic level by a variety of biological, physical, and chemical reactions to create bigger nanostructures (Mayegowda et al. 2023). A bottom-up technique called "green synthesis" creates NPs from smaller atoms. With the exception of using plant extracts in place of chemical reduction agents, the reduction process used in green synthesis is nearly identical to that of chemical reduction (Jiang et al. 2022). According to a Business Wire analysis, the worldwide nanotechnology market is look forward to reach a valuation of U.S. dollar 173.95 billion around 2025, growing at a compound annual growth rate of 18.1% during the coming few years (Ciambelli et al. 2020).

Green nanotechnology is a strategy based on materials and technologies that seeks to create safe technologies to reduce threats to the environment and human health (Ciambelli et al. 2020; Nallasamy et al. 2023; Nasrollahzadeh et al. 2019) as well economy. The scientific community has become more interested in green nanotechnology throughout the past 10 years, particularly in areas such as natural biosynthesis of nanoparticles as well as its applications (Heinemann et al. 2021) and to encourage the substitution of novel, long-lasting nanoproducts for existing ones that are more ecologically friendly (Nasrollahzadeh et al. 2019). These environmentally safe methods, sometimes known as "green nanotechnology" as well as "clean technology," are workable substitutes for chemical techniques (Mayegowda et al. 2023). Green synthesis creates biocompatible nano composites by reducing metal ions from fluids in a low-energy process using biological sources, such as bacteria, algae, fungus, plant parts, as well biological debris, as reducers in "bionanofactories." Metal ions can be reduced by biomolecules to uncharged atoms, which can then re-form into massive particles by the process of nucleation, which creates sturdy nuclei upon impact. Subsequently, stabilization takes place when various biomolecules wrap the produced particles. This stops the build-up and interference with the creation of biomolecules as well as the colour shifts brought on by the used metal components (Bhandari et al. 2023; Heinemann et al. 2021). Throughout the synthesis process, plant extracts, also known as reducing agents, donate electrons to metal ions in metal salts, reducing them to generate metals for nano-agrochemicals under carefully regulated circumstances, including pH, temperature, and reactant concentration. NPs can have functional molecules added to them to further modify their solubility, stability, or ability to target (Ali et al. 2023). Figure 1.2 schematically represents the process of green synthesis of nanomaterials from different plant parts and its potential uses.

In many ways, green nanotechnology is the logical progression of nanotechnology, motivated the need to align wide range of feasible as well as useful applications of nanotechnology along the demands that referred as sustainable development (Ciambelli et al. 2020). In order to effectively address environmental issues, green nanotechnology must be introduced as an efficient substitute for synthetic nanotechnology. This can be achieved by adhering to the principles of green science, which include creating different nanomaterials and products without endangering the environment and human health using the least amount of energy possible, and generating the least amount of waste or pollution. In order to achieve this,

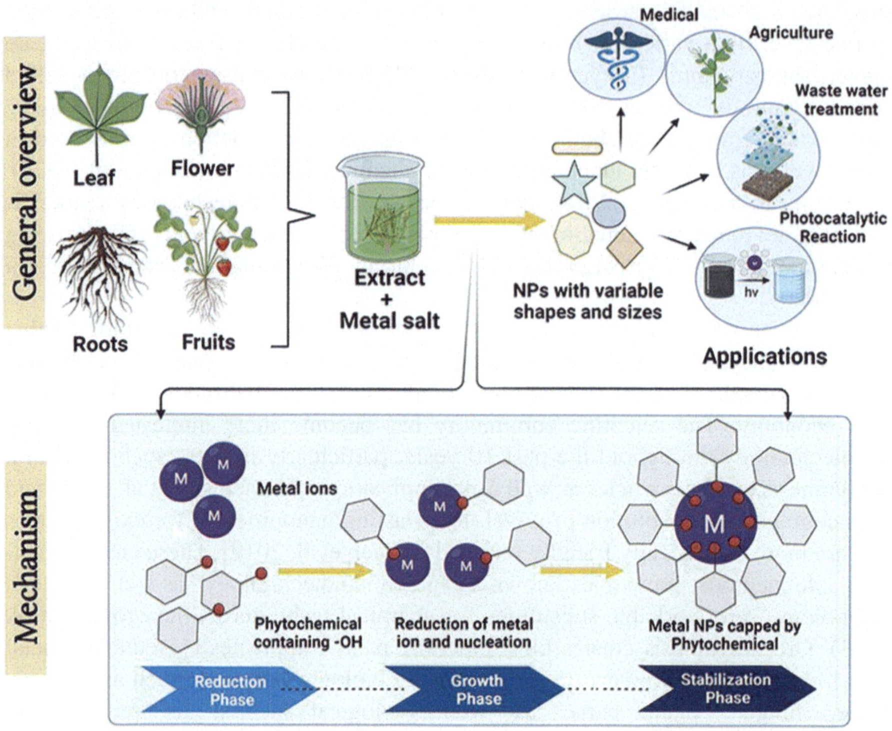

Fig. 1.2 Schematic representation of green synthesis of nanomaterials. (Reproduced from Singh et al. 2023a, b)

nanotechnological techniques and methods—such as using solid-state reactions, conserving energy, substituting water for hazardous organic solvents, modifying the structure of nanomaterials to reduce toxicity, and aspects related to recyclability and reusability—are crucial (Nasrollahzadeh et al. 2019). Green nanotechnology is a safe, low-emission, waste-reduction, energy-efficient technique. Since these products are produced using renewable resources, the environmental impact of this technology is minimal (Prasad et al. 2017). The advantages of green nanotechnology are many (Fig. 1.3). Green techniques are simple, biocompatible, easy to implement, economical, efficient, structured technique, capable of absorbing huge amounts of metal, and appropriate for large-scale production (Bhandari et al. 2023). Green nanoscience has become essential for the environmentally responsible advancement of nanotechnology, enabling a comprehensive approach to planetary safety and advancing the goal to create more secure and sustainable product (García-Quintero and Palencia 2021).

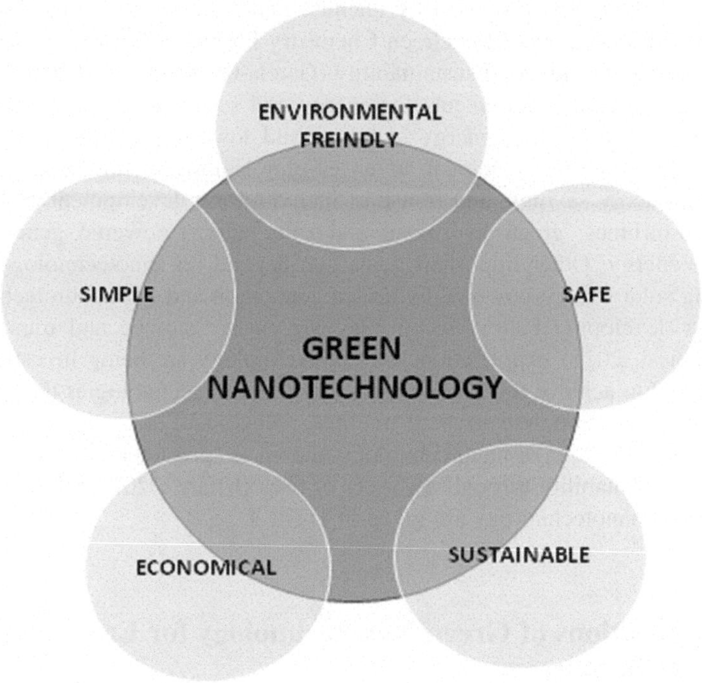

Fig. 1.3 Advantages of green nanotechnology. (Constructed by Livina Lazar)

1.2 Green Nanotechnology for Sustainable Development

"Meeting current demands without compromising future generations' ability to meet their own needs" is defined as sustainable development (Kumar et al. 2018). The application of nanotechnology is crucial to global sustainability initiatives (Pokrajac et al. 2021). The sufficient supply of resources has grown scarce due to the rise in the world's population. For the long-term, sustainable growth of human society, pollution-free technologies must be created for environmental remediation and clean energy sources. With the use of nanotechnology, "cleaner" and "greener" technologies can be developed, with significant benefits for both human health and the environment. There is a great deal of optimism in nanotechnology for the creation of new, sustainable, and environmentally friendly technologies. Green nanotechnology holds great promise to address the environmental issue and promote sustainable growth (Khan 2020). Sustainable development and economic security were the primary objectives of the United Nations Sustainable Development Goals (SDGs), which were introduced on September 15, 2015, among other topics (Ganie et al. 2021). The US Congress passed the Pollution Prevention Act in 1987, and the Life Cycle Analysis was first introduced in 1990. These events marked the beginning of the concept of sustainable development. Furthermore, the development of ecological technologies and processes would be aided by the core idea of "Green

Chemistry," which was proposed by chemist Paul Anastas to define the guiding principles and objectives of the Green Chemistry Project. This concept was developed based on environmental sustainability (García-Quintero and Palencia 2021).

Important societal requirements for energy and sustainability include reducing CO_2 emissions, stabilizing energy supplies, and lowering energy costs. Several nanotechnology and technological advances play significant functions in storage, generation, energy savings, and transport including the development of fuel cells with wind turbines, green hydrogen, and solar energy-powered generators for renewable energy. Other important goals that depend on nanotechnology include developing solar energy-powered hydrogen generation and utilization technologies as well as developing battery technology for energy storage and transportation (Pokrajac et al. 2021). Applications of nanotechnology are being investigated for their potential to achieve the functionality of traditional technologies that is used in environmental cleanup as well as to monitor, reduce, and clean up air, water, and land pollution. The area of nanotechnology known as "green nanotechnology" aims to achieve sustainability through a variety of uses (Khan 2020). Different applications of green nanotechnology are given in Fig. 1.4.

1.3 Applications of Green Nanotechnology for Ecosystem Sustainability

The sustainable use of natural resources to keep a safe environment, for the current as well as future generations, requires an ethical and responsible relationship between humans and the environment. The careful management of resources from nature, which fosters harmony among ecology and environment, can preserve the ecological and natural balance of the universe (Ganie et al. 2021). The special qualities of nanotechnology can provide technological, social, and economic problems with environmentally friendly solutions. Green nanotechnology devices purify water, remove hazardous waste, and conserve raw materials, all of which help to safeguard the environment. The large percentage of lethal chemicals used mainly in production of the nanoparticles will lead to major environmental issues. Therefore, in order to create chemical-free, ecologically safe, and less expensive nanoparticles for biological and environmental applications, biogenic oxidation of metal precursors depends critically on the purity of the resulting particles. These days, methods for producing green nanoparticles using algae, bacteria, fungus, and plants have been used (Raja et al. 2022).

The promise of nanotechnology deals the promotion of more environmentally friendly and sustainable economic growth. Between 2010 and 2035, roughly there will be a 30% increase in the world's energy demand, yet over almost 800 million individuals lack access to uncontaminated drinking water. There has never been a more pressing need for the creation of accessible and secure solutions to address these needs. This need is further reinforced by the need to lessen the impact of

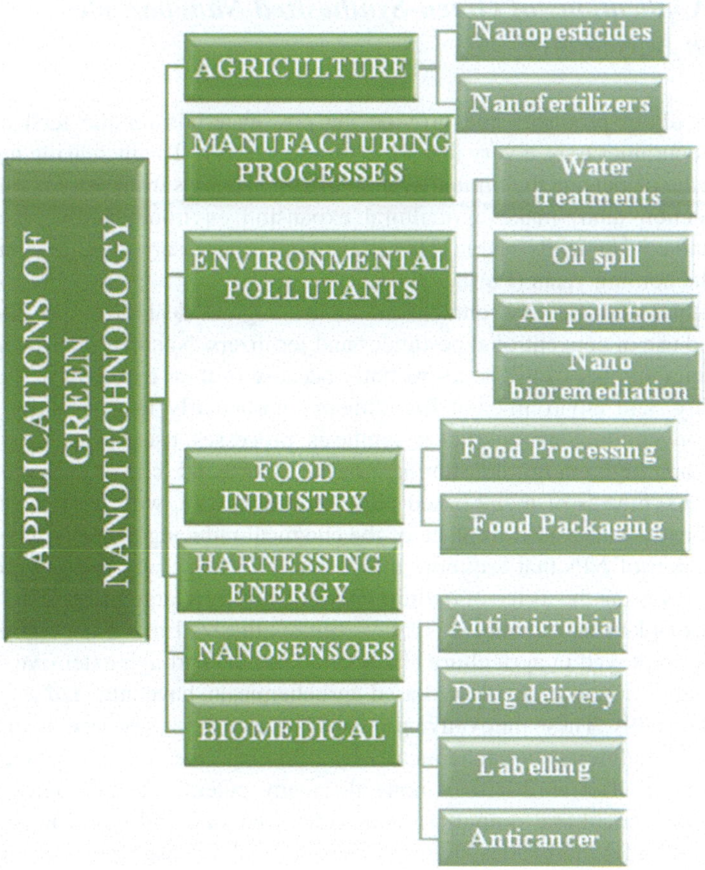

Fig. 1.4 Schematic illustration of applications of green nanotechnology. (Constructed by Livina Lazar)

carbon emissions and other environmental issues, as well as by market and supply volatility and unpredictability (Shapira and Youtie 2015). Green nanotechnology has created synthetic nanomaterials that are safe for the environment and have a variety of applications, including desalinating water, cleaning up hazardous waste sites, and treating or keeping an eye on environmental contaminants (Nasrollahzadeh et al. 2019). Sustainable alternatives in agriculture, industries and environment management can be achieved through green nanotechnology and are discussed here.

1.3.1 Applications of Green-Synthesized Nanoparticles in Agriculture

Since agriculture produces and provide the raw materials for the feed and food sectors, it continues to be a very important and stable area. The increasing number of people on the planet and the limited availability of resources from nature (soil, water, and production land) make agricultural expansion environmentally responsible, efficient and economically viable. This adjustment is necessary to achieve numerous goals in the coming years (Prasad et al. 2017).

Agriculture and the environment are under a great deal of strain due to the widespread use of conventional pesticides and fertilizers. Nanoparticles have gained attention in a number of industries recently because of their excellent performance, affordability, and environmental friendliness—particularly in agriculture sustainability. Conventional nanoparticles synthesis processes usually environmentally damaging and high energy consuming. On the other hand, chemical synthesis and plant-based synthesis of metal-based NPs are comparable, with the exception that biological extracts are used in place of the chemical reducing agent. This results in the production of NPs that are more affordable, effective, safe, and low-polluting, while also lowering the usage of traditional chemicals by a large margin. To increase the yields of plant and quality of crop, green synthesized metal nanoparticles are commonly employed in agriculture (Jiang et al. 2022). Through extensive development and studies, nanotechnology-based agrochemicals, have attracted a great deal of interest recently. These innovative products enhance the safety, sustainability, and efficacy of traditional agrochemicals by utilizing the unique properties of nanoparticles (Ali et al. 2023). Among the many potential technologies that can help increase agricultural output are nanotechnology and soil nutrition regulation, effective pesticides, herbicides, and fertilizers, as well as wastewater cleanup (Khan et al. 2022).

With the development of innovative nanotools for quick disease diagnosis and improved plant nutrient uptake, nanotechnology holds promise for advancing the food and agricultural industries. Nanotechnology use in the field of agriculture holds great promise for specific applications such as the development of nano pesticides and fertilizers that can boost productivity without contaminating soils or waterways. Furthermore, nanotechnology offers defense against a range of microbiological illnesses and insect pests. Nanotechnology can assist in preserving the wellness of agricultural plants by serving as sensors to track the quality of the soil in fields (Prasad et al. 2017).

Nanoparticles based on metals or metal oxides have been applied to the design of nanoscale coating materials, nanoscale additives, and fertilizers. Bulk materials used in conventional agricultural techniques, such as pesticides and fertilizers, are extremely toxic to the environment and ecology. These risks may be avoided by using metal- or metal oxide-based nano pesticides and fertilizers (Kshtriya et al. 2021). Nanotechnology's ability to control nutrients can have a big impact on productivity. It can also be used to monitor pesticide use and water quality for the sustainable growth of agriculture (Prasad et al. 2017).

Green synthesized NPs have the potential to completely transform crop protection, fertilization, and environmental safety through their use as plant growth regulators, fertilizers, pesticides, soil amendments, food preservatives, purification agents, biocontrol agents, biosensors, environmental remediation agents and drug delivery systems. NPs manage agricultural diseases, weeds, and insects more effectively than traditional pesticides by 31.5% thanks to their special qualities (Ali et al. 2023).

1.3.1.1 Nano Fertilizers

Fertilizers that have been designed to the nanoscale, or commonly between 1 and 100 nm in at least one dimension, are known as nano fertilizers. With a notable increase in surface area compared to volume, this nano-scale rearrangement improves reactivity and for special interactions with plant tissues that are not possible with bulkier, conventional fertilizer particles (Razauddin et al. 2023; Preetha and Balakrishnan 2017). In comparison to conventional fertilizers, nano-fertilizers (NFs) have advantages in terms of application and minimal requirement, slow-release mechanism, transportation and application cost reduction, and relatively low salt accumulation in soil (Kumar et al. 2021). Nanofertilizers uses in agriculture promises a number of benefits, including higher yields, better plant health, and better ecological outcomes. Nano fertilizers have the ability to deliver nutrients directly to the target site, reducing nutrient runoff, a major environmental concern with conventional fertilizers. They can also penetrate plant cells more effectively, releasing nutrients in a controlled manner, reducing waste, and promoting efficient nutrient use (Razauddin et al. 2023). These enhance the bioavailability of nutrients and hence effectively meet crop nutritional requirements. Through biofortification, foliar-applied nano-fertilizers improve crop nutritional quality (Kumar et al. 2021). The general characteristics of nano fertilizers is shown in Fig. 1.5.

Three types of nano fertilizers are available: (i) nanoparticulate, which include silica, carbon nanotubes, titanium dioxide, and other nanoparticles that promote plant growth; (ii) micronutrients, which include iron, copper, zinc, manganese as well nickel; and (iii) macronutrients, which include calcium, phosphorus, potassium and nitrogen. Studies have shown that when phosphatic nano fertilizers were used instead of conventional fertilizers, soybean growth increased by an astounding 32% and seed output increased by 20% (Dikshit et al. 2021; Osman et al. 2024). For soils deficient in Sulphur, fertilizers coated with nanoparticles are utilized. Sustainable kaolin and chitosan nanoparticles have demonstrated remarkable outcomes in the long-term release of N-P-K (sodium-phosphorus-potassium) fertilizers. Fertilizers that have been nano encapsulated helps in uptake of nutrients from the soil. SiO_2 and nano silica coatings protect plants from diseases and adverse weather while promoting the development of roots and seedlings (Pandey 2020). Crop fertilizer use

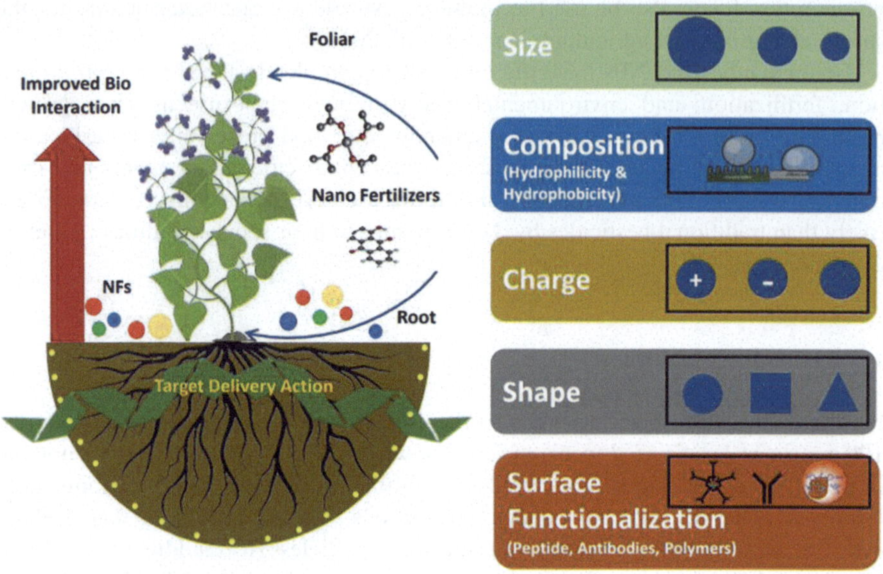

Fig. 1.5 Schematic representation of nano fertilizers (NFs) general characteristics and entry routes. (Reproduced from Zahra et al. 2022)

efficiency is increased using nano fertilizer technology. Over time, there has been a growing interest in using nanoporous zeolites in farming due to public concerns about the harmful effects of artificial fertilizers on the agro-ecosystem. The availability of low-cost natural zeolites worldwide and the increasing understanding of this phenomena have sparked significant commercial interest in the creation of zeolite-based nanofertilizers. As the phenomena becomes more well recognized and naturally occurring zeolites become more widely available, there is now a significant deal of economic interest in developing zeolite-based nano fertilizers (Preetha and Balakrishnan 2017).

1.3.1.2 Nano Pesticide

Pesticides are essential to ensure national food security and serve as the foundation for defense against significant biological disasters in agriculture. In eco-friendly green pesticide formulations, biocompatible, biodegradable, and responsive material is currently of rising interest. The development of target-oriented, controlled-release insecticides that are safe for the environment is greatly aided by nanotechnology. The most often utilized nanoparticles for nano pesticides are silica, nano-Ag, nano-TiO_2, au, and zinc (Anandhi et al. 2020).

The effectiveness of certain plant-synthesized NPs against economically significant arthropod pests, such as moths, beetles, lice, hard ticks (like *Haemaphysalis bispinosa*), louse flies (like *Hippobosca maculata*), and mosquitoes, has been studied (Athanassiou et al. 2017). The slow-releasing features of a nano encapsulated pesticide formulation with improved specificity, permeability, solubility as well as stability have currently been developed. These benefits are mostly attained by either delaying the premature degradation of the encapsulated active components or prolonging the duration of their pest control efficiency. With the advent of nano encapsulated pesticides, agricultural protection has become more environmentally friendly as human exposure and pesticide dosage have decreased (Prasad et al. 2017).

The creation of NP-based sustainable release systems may improve the efficacy and effectiveness of pesticides by lessening its harmful impacts on the environment. In general, NPs are a “nanocarrier” transport mechanism since they enter plant cells with ease. The World Health Organization (WHO) deems it safe for people to use amorphous silica as a nano biopesticide (Athanassiou et al. 2017). To keep fungi, bacteria, and insects away from plants, nanomaterials like SiO_2, TiO_2, and ZnO are becoming more common in fungicides and insecticides. To identify the pest, Au, Ag, and Cu nanoparticles are employed as electrical and biosensors. One organic insecticide that works well to address the current problem of environmental contamination is nanoparticles. Green insecticides can be encapsulated utilizing nanoparticles and stabilized active ingredients to minimize risks (Anandhi et al. 2020).

1.3.2 Green Nanotechnology in Manufacturing Processes

Environmental damage is a major consequence of industrial manufacturing. Firstly, it needs a large amount of water, energy, and ingredients. Second, it produces a large amount of harmful byproducts and wastes (gaseous, liquid, and solid) that must be disposed of or transformed into safe goods. In many industries, including chemical, pharmaceutical, semiconductor, petrochemical, materials processing, etc. an emerging platform for green chemistry and production is nanotechnology (Diallo et al. 2014). Green nanotechnology has been much more widely available commercially in recent years due to the need to reduce the risk of global warming and restrict the use of nonrenewable resources. All of this has led to the widespread application of green nanotechnology across several industries. Industrial ecology of the future will witness the replacement of current products by green nanotechnology-manufactured goods, which will lessen the negative consequences of currently employed toxic ingredients (Shah and Shah 2020).

1.3.3 Nanotechnology in Addressing Environmental Pollution

The advancement of nanotechnology has the potential to significantly enhance environmental health. Nanotechnology is used in the environment for energy storage, sensors, remediation, wastewater treatment, and polluted soil. Nanotechnology applications in the environmental sphere can be directed towards either proactive problem prevention or reactive problem solving of existing environmental issues. Remediation, water treatment and environmental sensing, are a some of the reactive uses. One of the main objectives of environmental nanotechnology is the safe design and sustainable development of nanomaterials with potential environmental benefits (Ciambelli et al. 2020). Figure 1.6 shows nanotechnology applications in combating environmental pollution.

1.3.3.1 Green Nanotechnology for Water Treatments

Even though water covers over three-quarters of Earth's surface, access to clean water remains one of the greatest problems facing humanity. In order to increase water treatment systems' efficiency, contemporary instruments that swiftly convert turning polluted water into drinkable water. The amount of toxins that come from human activity that enter the environment and water has reduced due to nanotechnology. The most important applications of nanotechnology in the environment are in water treatment, where it is used in nano filters, nano photocatalysts, carbon nanotubes, nanoparticles and nano clays. Bionanosensors are also used in this regard to quickly detect water contamination (Taran et al. 2021).

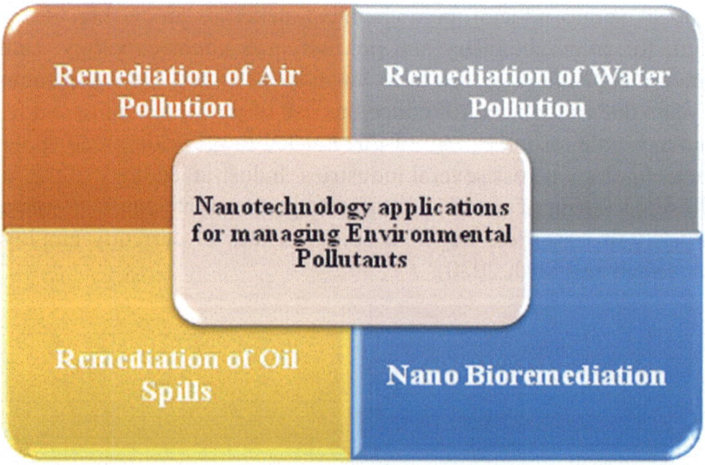

Fig. 1.6 Diverse applications of nanotechnology in managing environmental pollution. (Constructed by Livina Lazar)

One of the major applications of nanotechnology has been the supply of clean drinking water development. Desalination, treatment, and reuse are among the nanotechnology-based approaches to problems related to water scarcity. Redox active nanoparticles and nano catalysts can transform harmful organic solutions into harmless byproducts. Nano biocides can be used to destroy bacteria without producing byproducts (Shapira and Youtie 2015).

Some of the promising instruments and procedures for water treatment that nanotechnology has brought forth are: (i) Photocatalysis (ii) Nanofiltration (iii) Nano sorbents.

A light-active, nanostructured catalyst media is used in photocatalysis, a promising method of purifying water, to break down several types of contaminants found in the water. When infrared, visible, or ultraviolet radiation acts on a chemical reaction, a material called a photocatalyst is needed to affect the rate of the process or its commencement. The photocatalyst absorbs light and helps the chemical transformation of the reaction partners (Bora and Dutta 2014). Another example of nanotechnology use in this area is photocatalytic methods, which use nanoparticles like titanium dioxide as well as zinc oxide as the photocatalyst and UV radiation to irradiate contaminants (Shapira and Youtie 2015). Water containing dangerous organic pollutants can be broken down by photocatalysis into innocuous byproducts, primarily carbon dioxide and water. Using the photocatalysis technique, several alcohols, carboxylic acids, phenolic derivatives, and chlorinated aromatic pollutants have been successfully destroyed (Bhatkhande et al. 2002). Similarly, from polluted water the heavy metals and inorganic contaminants can be efficiently eliminated using this photocatalytic treatment (Bora and Dutta 2014).

Nanofiltration (NF) and other membrane technologies are becoming important components of water filtration systems. Membranes for NF are widely utilized in the treatment of drinking as well as waste water. Low pressure membrane separation is used to separate materials with diameters ranging from 0.001 to 0.1 μm. The properties of NF membranes, which are pressure-driven membranes with pore sizes ranging from 0.2 to 4 nm, fall in between those of ultra filtration and reverse osmosis membranes. NF membranes have been shown to be capable of getting rid of bacteria, turbidity, and inorganic ions like Calcium and sodium. They are employed in wastewater treatment, the elimination of organic matter that dissolved, pretreatment for saltwater desalination, groundwater softening (lower water hardness), and trace contaminants from surface level water (Sayan et al. 2013). One kind of filtration in which pressure is the force that drives is nanofiltration. Compared to conventional treatment methods, nanofiltration membranes offer a greater push or rejection of heavy metals, pesticides, and multivalent ions. As of right now, this technology is the most advanced and cutting-edge method of treating water, which can be used in house, enterprises or in industrial purpose (Yunus et al. 2012).

When purifying water, nano sorbents are frequently employed as separation media to eliminate both inorganic and organic pollutants from tainted water. As sorbents, nanoparticles are very profitable due to two essential characteristics. They are substantially larger on a mass basis than macro particles. Additionally, other reactor groups can be added to them to improve their chemical affinity for specific chemicals (Sayan et al. 2013). Polymeric adsorbents, metal oxide nanoparticles, nanoporous carbon fibers, nano clays, and zeolites are examples of non-absorbents that can remove particulates from polluted water (Shapira and Youtie 2015). Additionally, a broad range of organic compounds from the aquatic environment have been found to be highly adsorbed by carbon nanotubes. These compounds include: polynuclear aromatic hydrocarbons (PAHs), dichloro-diphenyl-trichloroethane (DDT), dioxin, as well as its metabolites, trihalomethanes, bisphenol A and nonylphenol, phthalate easters, dyes, pesticides, and herbicides like atrazine, dicamba, and sulfur on derivatives. It has been shown that nanoporous polymers copolymerized and cross-linked with functionalized carbon nanotubes have a high sorption capacity for a range of organic chemicals, including trichloroethylene and p-nitrophenol (Bhawana and Fulekar 2012).

Nanofibers are solid fibers that, when placed in a non-woven mat, have small pore sizes, a high surface area to volume ratio, and diameters in the nanoscale range. Polymer nanofibers are typically better than polymer in any other form because of their small diameter. Nanofibers can have a surface area that is up to 100 times greater than microfibers, and they are stronger, more flexible, and have superior tensile strength. Moreover, the polymer and extra no soluble particles control surface activity. Because of these characteristics, nanofibers are more efficient and adaptable than traditional polymer membranes used in liquid filtering (Yilmaz and Soylak 2021).

1.3.3.2 Nanotechnology in Air Pollution

The modern, highly developed industries have released a wide range of pollutants as a result of industrial operations or human activity into the atmosphere. These pollutants include Sulphur dioxide, particulates, hydrocarbons, nitrogen oxides, organic compounds, carbon monoxide (CO), Sulphur dioxide and heavy metals which include chromium, zinc, mercury, arsenic, cadmium, as well lead. There is a great chance that human activity, including burning petrol, coal, and oil, will alter emissions from natural sources (Yunus et al. 2012). The WHO (World Health Organization) revealed in a later study that the more than 3 million people are impacted by airborne pollution. According to studies from the World Energy Congress (WEC), if fuel use continues at its current rate, pollution will cause irreversible environmental harm by 2025. According to clinical research, breathing in particulate matter is linked to higher death rates, which are exacerbated for those with diabetes, long-term lung conditions, and inflammatory illnesses (Vaseashta 2008). Globally, pollution of air is a serious issue that has an impact on both the environment and public health. Recent developments in nanotechnology hold promise for addressing air pollution issues and greatly enhancing the efficacy and selectivity of air purification systems, which could lead to cheaper costs and improved long-term outcomes. There are several methods that nanotechnology can be applied to reduce air pollution, such as through the use of nano adsorbents, nano filters, nano catalysts, and nano sensors (Saleem et al. 2022).

Nanotechnology can be used to mitigate air pollution in a number of ways. One method is to use more surface area nano-catalysts for gaseous processes. The nano-catalysts accelerate the chemical processes that can convert toxic fumes from vehicles and factories into safe gases. Among the catalysts manganese oxide nanofiber catalyst that can cleans industrial smokestacks of volatile organic pollutants. There are still more approaches being developed (Bhawana and Fulekar 2012).

1.3.3.3 Nano Bioremediation

Bioremediations a method that transforms harmful organic components into harmless ones by detoxifying, mineralizing, and ultimately destroying them by utilizing microorganisms' enzymatic and metabolic capabilities. In essence, nano bioremediation uses NPs to boost the productivity of microorganisms and catalyze the breakdown of hazardous and waste materials by converting to harmless forms first and then allowing them to biodegrade to no risk-based levels contents (Gul et al. 2022). Thus, microorganisms may effectively and environmentally removes heavy metals from water and soil through bioremediation (Prasad et al. 2017).

A novel remediation method based on the application of nanoparticles is called nano remediation. The use of nano remediation offers novel approaches for the prompt also efficiently eliminate the pollutants present in polluted environments, enabling the pursuit of solutions to some of the most difficult problems of the twenty-first century, including the pollution problems, polluted land administration, and imbalances in the environment restoration. Nanomaterials, like nanoclusters, bimetallic nanoparticles, metal oxides, nanodots, carbon nanotubes, as well nanocomposites, which are reactive are applied in nano remediation technologies to break down and mineralize pollutants (Ganie et al. 2021).

There has been interest in the application of biologically occurring nanosized catalysts for the degradation of environmental pollutants. Because of their exceptional efficacy, unique antibacterial properties, strong photocatalytic activity, and adsorbent capacity for disinfecting and cleaning the surroundings. Nonetheless, the bulk of findings only include lab- or bench-scale research, and a thorough assessment of their effectiveness in contaminated sites is still pending. The production of tailored nanoparticles (NPs) with improved pollutant removal rates is crucial to their successful industrial application (Bhandari et al. 2023).

In the past few years, the environment has begun to contain heavy metals and some organic contaminants in a range of dangerous forms that are difficult to eradicate using conventional methods and so require the development of new, cutting-edge technologies. Because of their remarkable capacities to measure the nanoscale size impact, nanomaterials (NMs) and nanoparticles have garnered a lot of attention recently and may be the perfect option for environmental cleanup. In addition to these, environmentally acceptable and reasonably priced green synthesized nanomaterials as well as nanoparticle made using various extracts of plants can be utilized in remediation by notable results (Bratovcic et al. 2022). Figure 1.7 schematically represents the role of green nanotechnology in bioremediation.

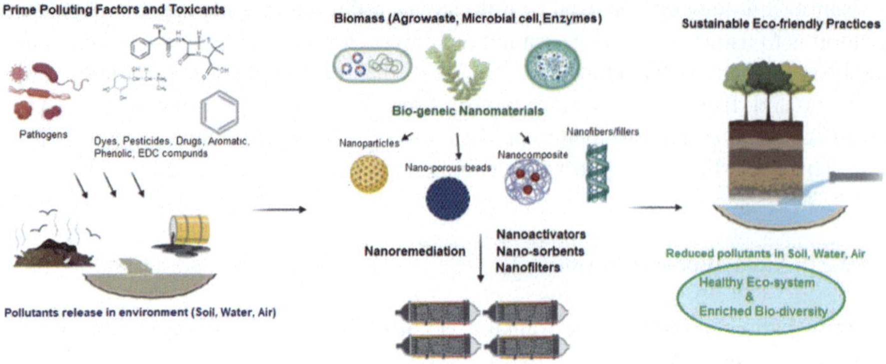

Fig. 1.7 Schematic representation of the role of green nanotechnology in bioremediation. (Reproduced from Bhandari et al. 2023)

1.3.3.4 Green Nanotechnology to Remedy Oil Spills

One of the main causes of water contamination, which has a devastating effect on the ecology and the economy, is marine oil spills. By disrupting the oceans' as well as other bodies of water's natural equilibrium, they significantly lower aquatic biodiversity. Even with current remedial measures, the damaged marine environment cannot completely recover to its natural state. It has been demonstrated that using magnetic nanomaterials, like functionalized super-paramagnetic iron oxide nanoparticles, magnetic nanocomposites and magnetic carbon nanotubes, can effectively clean up oil spills and leaks in the water. When diverse functional components, mostly polymers, polysaccharides, and inorganic and organic non-polymeric substances, are added to nanoparticles, they can be functionalized to provide them unique properties including high oil removal efficiency and increased adsorption capacity (Singh et al. 2020).

Massive oil spills are too big of a problem to be solved with traditional cleanup methods. Nanotechnology has been a promising source of new answers to many of the world's unsolved challenges in recent years. While oil spill cleaning using nanotechnology is still in its early stages, there is a lot of promise for coming years. For the last few years, there is a noticeable increase in the interest of globally investigating methods, using nanomaterials to develop appropriate solutions for cleaning up oil spills (Mishra et al. 2022).

Nanoscience may offer a viable solution for enhancing post-oil spill response technology since nanosized materials can be functionalized for highly specific uses, such as the production of magnetically sensitive sorbents or Pickering emulsions (Sadeghpour et al. 2013). Due to their ability to be removed contactlessly, based on iron oxide, magnetic nanosorbents become more and more popular since the late 1970s. However, because of their magnetic properties, iron oxide nanoparticles are an excellent approach to improve sorbents. The main disadvantage of employing particles of iron oxide for oil spill remediation is that the "naked" particles are

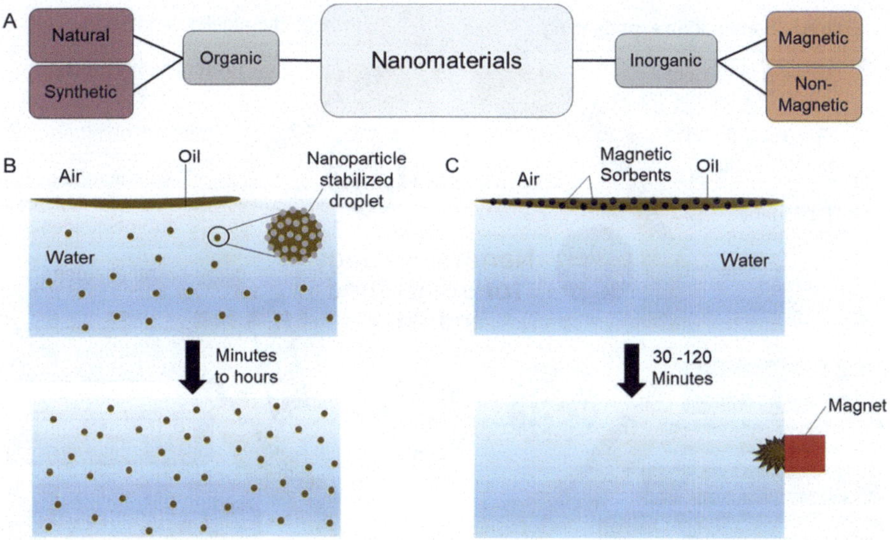

Fig. 1.8 Schematic representation of nano remediation of oil spill (**a**) Types of nanomaterials used for treating oil spills (**b**) Formation of oil-in-water Pickering emulsions (**c**) Magnetically responsive inorganic nanomaterials used for remediation. (Reproduced from Pete et al. 2021)

susceptible to aggregating and require a shell to safeguard the iron oxide core (Qiao et al. 2019).

A Pickering emulsion is one that when solid particles, such bacteria or nanoparticles, adsorb onto the oil-water interface and prevent the droplets from coalescing and the emulsion is stabilized. Halloysite is an aluminosilicate that occurs naturally in significant numbers. It may produce oil-in-water Pickering emulsions that can stay stable for several months. One potential Pickering emulsifier is Halloysite clay nanotubes. When altered, inorganic silica can also produce stable Pickering emulsions. Because native silica is so hydrophilic, it is not useful for creating Pickering emulsions. One improvement has been the addition of a surfactant coating (Pete et al. 2021). Figure 1.8 depicts different methods of nano remediation of oil spills.

1.3.4 Nanotechnology in Food Industry

The food sector uses nanotechnology in a wide range of applications, from food processing to food packaging (Fig. 1.9). Food additives, packaging and preservation, are just a few of the areas where food nanotechnology has permeated consumer goods (He et al. 2019; Khan et al. 2022). The field of food packaging and storage has been transformed by nanotechnology, ensuring food safety and hygiene. Numerous researches about the use of different compounds as food additives or in nanoscale

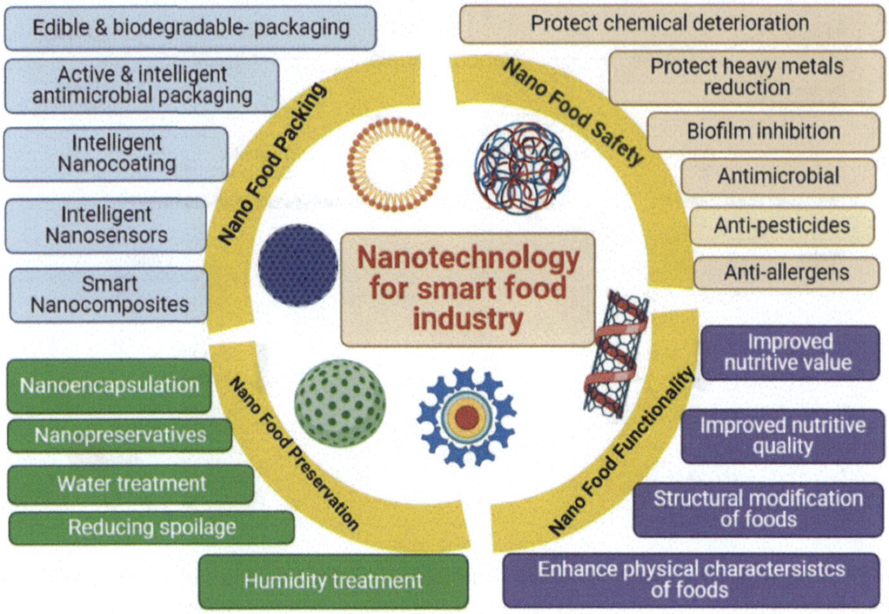

Fig. 1.9 Applications of nanotechnology in food industry. (Reproduced from Ansari 2023)

food packaging have been published. The principal regulatory agencies that supervise and formulate legislation related to food nanotechnology are the European Commission (EC) and the United States Food and Drug Administration (US FDA). Certain inorganic oxide chemicals have been approved by the US FDA for use as coloring, flavoring, and anti-caking agents in foods such puddings, sauces, and candies (Khan et al. 2022). The understanding of this cutting-edge technology has improved food safety through food processing and storage. It has also been discovered that several traditional compounds used as food additives or packaging materials partially exist at the nanoscale. For instance, up to 40% of food-grade TiO_2 NPs have recently been discovered in the nanoscale range (He et al. 2019).

Nano inside in food additives and nano outside in food packaging are the two categories of nanofood applications. For instance, the usage of nanoscale food additives can affect the shelf life, texture, flavour, and nutritional makeup of products. They can also be utilized as indicators of food quality as well as to identify food pathogens. Nanotechnologies are mostly thought to be useful in the context of food packaging to increase the products' shelf life, identify spoiled ingredients, or improve overall product quality, such as by stopping gas passage across product packaging (Ravichandran 2010).

These days, food manufacturing uses a lot of nanotechnology. Examples include animal feed, organic nano-sized additives, food supplements, and nanocarrier systems for supplement and nutrient delivery. NP is already present in many food items by nature. Casein is a milk protein that exists at the nanoscale, is found in milk, and protein filaments found in meat can also be categorized as nanomaterials. The structure and organization of the proteins within these products determine their texture and other characteristics (Prasad et al. 2017).

1.3.4.1 Food Processing

Many foods have nanoparticles added to them to extend their shelf life or to enhance their colour, flow characteristics, and stability during processing (Alfadul and Elneshwy 2010; Berekaa 2015). In fact, the creation of more wholesome foods that have less salt, sugar, and fat may be aided by nanotechnology in the fight against numerous food-related illnesses (Berekaa 2015).

For instance, anatase titanium dioxide is a typical brightener ingredient and food whitener used in sauces, some cheeses and confectionery. Aluminosilicate compounds, on the other hand, are frequently employed in powdered processed or as anticaking ingredients in granular foods. An active component is enclosed in a nanoscale capsule using nano-encapsulation. Plant sterols, probiotics, vitamins, minerals, and bioactive peptides are increasingly added to dairy products, cereals, breads, and drinks for added nutrition. These days, some of these active substances are added to meals in the form of nanoparticles or particles as small as a few hundred nanometer (Alfadul and Elneshwy 2010). Functional foods, nutritional supplement design, and nutraceuticals with nanosized ingredients and additives like vitamins, antioxidants, antimicrobials, and preservatives are among the nanotechnology's applications that currently under investigation to improved taste, absorption, and bioavailability. Some of the nutrients included in the carriers are lycopene, phytosterols, and beta-carotenes, which are employed in nutritious foods to stop cholesterol from building up (Berekaa 2015).

Food products may be changed and modified with the aid of nanotechnology to more efficiently supply nutrients like protein and antioxidants for certain nutritional and health benefits. Using an encapsulated system, nanospheres or microspheres can be produced. Many products, such as baked goods, tortillas and flat breads, processed meat products, seasonal candies, dessert mixes, chewing gum, and nutrient-dense foods, might potentially use the nano-sphere/microsphere system (Alfadul and Elneshwy 2010).

1.3.4.2 Food Packaging

In the food industry, packaging is essential because it shields the product from deterioration and harm. Therefore, it is essential to explain improved packaging methods that need to be passive, secure, affordable, reusable, and conveniently disposed of or reusable during storage and transportation. Food quality is significantly impacted by the content of the packaging materials. Increased food shelf life has been achieved by using an active packaging strategy (Jafarzadeh et al. 2022; Khan et al. 2022).

The food packaging sectors have likewise made substantial use of nanomaterials over the past 20 years. Food travels slowly from one country to another because it is sold all over the world these days. Therefore, to prevent food from spoiling and losing nutrients, packing materials must have high anti-fouling properties. NPs are used as anti-fouling agents in food packaging, which is a novel application for these metal-based particles. It is also essential to research the risk assessment of NPs when they are used as materials for food packaging (Kshtriya et al. 2021). When it comes to packaging, nanotechnology prolongs the shelf life of food by halting nutritional loss and breakdown. Besides providing an inert barrier against external conditions, food preservation appears to be greatly aided by active packaging. It usually has to do with packaging options that adjust to changing environmental conditions. They function as gas scavengers or by producing advantageous compounds like antioxidants or antimicrobials. O_2 scavengers, enzyme immobilization methods, and antimicrobials application are only a handful of the innovations in packaging that increase food stability as a result of these interactions. Controlled-release packaging is another for nanocomposites in active packaging. In this case, they can be used as delivery methods to promote the uptake of useful supplements like minerals, probiotics, and vitamins in food (Biswas et al. 2022).

The process of combining an inorganic material (calcium carbonate, for example) with an organic one (lipid, peptide, or protein) to produce a substance that is more resilient served as the model for the first nanocomposite. A packing material made of calcium carbonate and potato starch is one example. This foam can take the place of the polystyrene used in fast food since it is biodegradable and thermally stable (Rashidi and Khosravi-Darani 2011). Oxygen eliminating property of TiO_2 NPs is utilized for food packaging and processing (Pandhi et al. 2021).

Antimicrobial nanoparticles are used in active packaging to protect food from bad and rotting microorganisms, extending the product's shelf life and freshness. They are also a component of the active packaging, which increases its strength, lightness, and resistance to O_2. Effective food packaging uses a combination of polymers and nanodevices or nano sensors to detect food pathogens, pollutants, chemicals, and toxins during storage and transportation (Biswas et al. 2022). Certain metals and metal oxides, including silicon dioxide, magnesium oxide, iron, silver, and zinc oxides, as well as silicon dioxide and carbon nanoparticles, can be employed as antimicrobial substance in food packaging and occasionally as supplements in food (Berekaa 2015).

1.3.5 Green Nanotechnology for Harnessing Energy

It is well known that today's utilization of fossil fuels is unsustainable and increases greenhouse gas emissions. It is critically necessary for renewable energy technologies to offer a certain percentage of the world's primary energy supply. In light of the need to address environmental issues, research is focusing on developing "green" processing technologies of nanotechnology in relation to hydrogen and renewable energy sources like solar, wind, and ocean energy for harnessing future energy for a sustainable environment (Guo 2011). The over dependence on finite fossil resources and the steadily rising demand for energy have resulted in the severe energy problems of the twenty-first century. One approach is to build the technology and infrastructure necessary to enable a smooth transition to renewable and alternative energy sources. It is believed that nanotechnology, which combines engineering and chemistry, will be the next big thing in renewable energy applications. In order to incorporate nanoscale structures into larger material components and systems, it entails manipulating them. Nanomaterials are predicted to have higher activity due to their high surface areas relative to bulk materials. Additionally, they exhibit increased durability and stability, are more economical, and have a large potential for recycling (Ramsurn and Gupta 2013). Both naturally occurring and artificially created nanomaterials have been used to help address the energy issue and unlock the limitless potential of the bioenergy found in microorganisms. They have enabled the microbial bio electrochemical systems to produce electric current. As a unique method of extracting fatty acids from algal cultures, a variety of biocompatible, mesoporous nanoparticles that have a tendency to absorb hydrophobic molecules have been produced and studied. As a result, the use of nano farming technology to produce biofuel from algal biomass has proven to be very successful in producing biofuels without interfering with the activity of biocatalysts. It could be a significant advancement in the manufacturing of biodiesel in this way (Malik and Sangwan 2012).

The existing solar cell modules are not particularly efficient at generating the maximum amount of energy from solar radiation since solar energy is a very diluted kind of energy, and they also take up a lot of space. Solar energy harvesting can be increased using nanotechnology, occupying less physical space than existing solar energy harvesting techniques. When nanomaterials are used in solar cells, their efficiencies can be increased over a larger range, leading to substantially higher output. TiO_2 nanoparticles, graphene, and silicon nanoholes are examples of nanomaterials that can be employed in solar cells (Vatti et al. 2018).

Clean and sustainable nanotechnologies encompass the development, characterization, and application of nanomaterials and nanodevices in energy production, conversion, storage, and utilization. Fuel cells, thermoelectric materials and devices, optoelectronic devices, batteries, hydrogen production, photovoltaics, detection, and storage, as well as self-powered, flexible, and/or integrated energy systems, are just a few examples of the parts, devices, and processes that make up nanomaterials-based technology. Among all the energy storage devices, the photovoltaic device is one of the most important renewable energy technologies since it directly turns sunlight into electrical power (Singh et al. 2023a, b).

1.3.6 Biomedical Applications of Green Nanotechnology

Expanding the scope of biological applications for the synthesis of materials, environmentally friendly technology development is crucial (Nath and Banerjee 2013). Chemicals with potential risks are frequently utilized in the chemical and physical methods of NP production, including solvents that aid in preventing colloidal agglomeration, stabilizers, and reducing agents. Because of their harmful and carcinogenic effects on cells, these chemically synthesized NPs have limited applications in biomedicine and other clinical settings. This prompted scientists to develop a clean, dependable, physiologically active, and environmentally acceptable method for creating NPs (Mayegowda et al. 2023). Figure 1.10 shows various biomedical applications of green nanotechnology that are playing crucial roles in diseases diagnosis, treatment and prevention.

1.3.6.1 Anticancer

Time-release and tailored pharmaceuticals may be developed using nanoparticles. To guarantee optimal efficacy and patient safety, a powerful dosage of medication could be administered to a particular location and designed to release over a

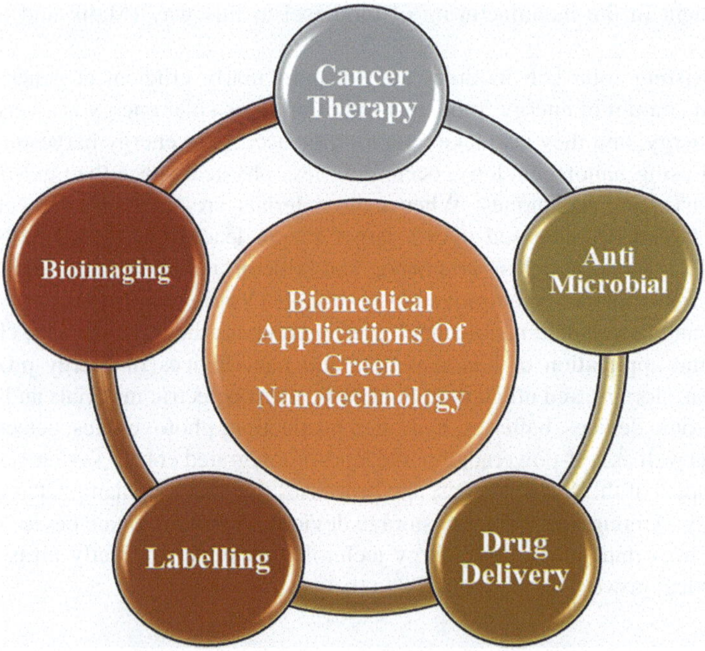

Fig. 1.10 Various biomedical applications of green nanotechnology. (Constructed by Livina Lazar)

predetermined length of time. Because of their potent light-absorbing qualities, gold nanoparticles (AuNPs) are useful as heat-mediating devices because the light energy they absorb dissipates into the area around the particles, raising the temperature there. This impact even kills malignant cells and can be employed, for example, to open polymer microcapsules for drug administration. The antibodies that are unique to the malignant cell enable the nanoparticles to function (Nath and Banerjee 2013). Cobalt oxide nanoparticles showed encouraging antitumor characteristics. *Euphorbia tiru* Calli extract-derived synthetic cobalt oxide nanoparticles shown antiproliferative efficacy against MCF-7 cell lines of breast cancer Silver nanoparticles derived from the extract of *Aerva javanica* demonstrated enhanced apoptotic potency on MCF-7 cells when combined with the anticancer drug geftinib. AgNP-mediated geftinib administration improved medication efficacy while lowering negative effects. AgNPs generated from the rhizome of *Acorus calamus* shown enhanced efficacy against A431 cancerous cells (Nallasamy et al. 2023). For the treatment of illnesses, particularly cancer, the use of Au NPs as a carrier of bioactive substances, such as anticancer medications, is highly advantageous (Ying et al. 2022). Selenium nanoparticles were applied to titanium oxide nanotubes and observed to inhibit the proliferation of MG-63 cancer cells (Mayegowda et al. 2023).

1.3.6.2 Anti-microbial

Seventeen million individuals die from infectious diseases each year globally, and because nanoparticles naturally possess antimicrobial qualities, they have significant potential to lower this death rate and manage infections. Because of their remarkable physico-chemical characteristics, nanomaterials—particularly metal nanoparticles—have tremendous biological and medicinal implications. Strong anti-microbial and antibacterial qualities are revealed by metal-based NPs. The WHO claims the antibacterial capabilities of metal nanoparticles are effectively increased when their size is reduced. There is ample evidence supporting the AgNPs' antibacterial activity. Anti-biotherapy also commonly uses NPs of different metals, like zinc oxide, gold, copper, and copper oxide nanoparticles (Kshtriya et al. 2021). NPs like silver, chitosan, copper, as well metal oxide nanoparticle mainly titanium oxide as well zinc oxide are good antimicrobial agent. NPs attach to the cell walls of microorganisms via electrostatic attraction and break their cytoplasmic membrane, which also releases the nucleic acid because of their preference for sulfur-rich amino acids. These properties make NPs potentially effective antibacterial agents (Mayegowda et al. 2023).

1.3.6.3 Drug Delivery Using Nanoparticle

Drug delivery systems are the focus of intensive research and development to improve the efficacy and distribution of active pharmaceutical molecules, such as medicines, antibodies, vaccines, enzymes, proteins, and peptides. Due to the impact of environmentally friendly and green chemistry on the nanotechnology-based drug delivery sector, the discipline known as “green nanomedicine” has emerged. Among the several green nanotechnology-driven drug delivery technologies, research indicates that polymers, nanometal particles, and biological materials have garnered the most interest. Green chemistry principles and ecologically friendly, low-side-effect synthesis techniques are being heavily prioritized in the design, synthesis, and use of these medication delivery systems (Nallasamy et al. 2023). Due of their nano size, NPs are utilized to target malignancies through drug delivery. They also help to lessen the side effects of photodynamic cancer therapy. The nontoxicity and no immunogenicity of AuNPs make them perfect for the construction of drug delivery scaffolds. Their functionalization feature further enhances their potential as drug delivery vehicles (Nath and Banerjee 2013). Due to their nanoscale size, the medications administered by the nanomedicine approach can travel through space with ease. If a particular biomarker is attached to the drug-delivering cell, the drug may also be released on the targeted cell. On the other hand, due to their large size, medications intended for delivery using conventional methods have poor biodistribution and difficulty reaching the target cells. The extended half-life of nano-encapsulated medications in blood circulation and their release’s site-specific effectiveness are two benefits over bulk medications (Kshtriya et al. 2021).

The most advanced kind of nanocarrier technology is represented by Solid Lipid Nanoparticles (SLNs), which have numerous potential applications in medication delivery, research, and therapeutic treatment. SLNs, or submicron colloidal carriers, consist of a solid lipid core matrix stabilized in an emulsifier aqueous solution. The capacity of SLNs to encapsulate pharmaceuticals into nanocarriers presents a novel prototype for drugs delivery at the secondary and tertiary stages (Kanwar et al. 2019). Lipid-based nanoparticles, which provide a different kind of nanocarrier system from emulsions, liposomes, and polymeric NPs, enable several drug delivery techniques (Nasrollahzadeh et al. 2019).

One of the most useful biomaterials in the field of nanomedicine is polysaccharides, which have been employed as carriers to create nano-based drug delivery systems. Because of their biodegradability and biocompatibility polysaccharides are excellent choices for medical applications, particularly drug delivery systems. Alginate is a polysaccharide that occurs naturally which is employed in drug-chemical immobilization systems as well as drug delivery devices to accelerate drug release. Additionally, it is a biodegradable and mucoadhesive substance that regulates drug delivery mechanisms to improve drug bioavailability (Nasrollahzadeh et al. 2019).

1.3.6.4 Bioimaging

Green produced nanoparticles offer promise in bioimaging. In imaging modalities like positron emission tomography (PET), computed tomography (CT), magnetic nanoparticles like iron oxide can be employed as contrast agents. Which are effective in the advance detection of malaria (Nallasamy et al. 2023). Iron oxide-based magnetic NPs have been applied to medication delivery, magnetic resonance imaging (NMR), cancer treatment, and magnetic fluid hyperthermia therapy. By imaging the sentinel lymph nodes (SLNs), the smallest features of the deadly malignant cells' crucial growth can be verified. Carbon NPs in SLN imaging can be synthesized using food-grade honey (Mayegowda et al. 2023).

1.3.6.5 Labelling

The metal nanoparticles' ability to absorb electrons is used for labelling with the goal to create contrast. Because AuNPs strongly absorb electrons, they are a good contrast agent for TEMs. In addition, because nanoparticles and proteins share the same size domain, nanomaterials can be used for bio tagging or labelling (Nath and Banerjee 2013).

1.3.7 Nano Sensors

Nano sensors is used for environmental monitoring. The ability of humans to safeguard the sustainability of both environmental and human health may be improved by quick and accurate sensors that can identify pollutants at the molecular level. Nanowires and nanotubes have excellent promise as building blocks for biological and chemical sensors. Single-walled carbon nanotubes have proven to have a faster response time and higher sensitivity when compared to the conventional probes currently used in the detection of gas molecules like NO_2 and NH_3 (Ciambelli et al. 2020). Nano sensors may be created to identify waterborne pollutants and hazardous gas leaks at incredibly low concentrations in air pollution (Mansoori et al. 2008).

Conventional techniques for diagnosing infectious diseases are laborious, sluggish, and inefficient. This is especially true in an emergency. Microbial biosensors have set a standard by offering a quick, affordable, and precise way to diagnose infectious diseases, hormone abnormalities, and DNA. Microbial biosensors have been used in forensic identification to assess body fluidic traces carrying DNA and miRNA as evidence (Mayegowda et al. 2023).

Furthermore, crop growth and disease, nutrient efficiency, and field conditions can all be tracked by nano sensors. One such use is wireless nano sensor technology. Notably, in food and agricultural systems, tailored nano sensors may identify infections at trace levels as well as pollutants like herbicides and pesticides (He et al. 2019). Nano sensors are designed to detect and sense a variety of pollutants, chemicals, infections in crops, pathogens, enzymes, as well soil water levels. They are also known as nano biosensors when immobilized with a variety of bio-receptors. An antibody-coated fiber-optic nano biosensor can be used to detect the dangerous bacteria *E. coli* (Pandey 2020).

1.4 Sustainable Future of Nanotechnology

One could argue that modern science's "golden child" of scientific discoveries is nanotechnology and its significant advances have made our understanding of science and paved new paths for investigation. After that, the number of nanotechnology research fields and applications increased dramatically (Chee et al. 2022). The sustainable development defined by Brundtland Commission as "development which meets the needs of the present, without compromising the ability of future generations to meet their own needs" in 1987 since "sustainability" and "futurity" are connected concepts (Basiago 1995).

When the scanning tunnelling microscope was developed, 30 years ago, scientists could now build and control novel materials and particles with sizes ranging from a few micrometers to manometers. Since then, a variety of inorganic and organic particles with unique sizes, shapes, and names—such as nanowires, nanotubes, nanorods, and nano tweezers—have been synthesized to achieve specialized goals, which may include the encapsulation of molecules, the coordination of metals, sensing, and the sieving of molecules or ions. Many of the most difficult technological, environmental, and economic issues we face today and, in the future, can be resolved by this new technology (Kumar et al. 2018).

Sustainability and future advancements are related, however from the viewpoint of the collaborator—depending on whether they are environmentalists or futurists—the timeline would differ. While future researchers will be thinking in terms of centuries, environmentalists may only be thinking in terms of a few decades. Nowadays, sustainability is critical due to the growth in various diseases and the higher death rate brought about by new technology (Shah and Shah 2020). We must take use of Internet of Things (IoT) and green innovative technologies (GTI) advancements to produce durable, biodegradable, and ecologically friendly products for a sustainable future. GTI encompasses any innovations that contribute to the development of useful products, services, or practices that minimize adverse environmental impacts and degradation while optimizing the utilization of natural resources (Bradu et al. 2023).

Products with nanotechnology capabilities will be integral to global production processes in the future. Nanotechnology has numerous uses in a variety of industries, such as energy, automotive, healthcare, and agriculture. These applications include reduced prices as well as increased productivity and efficiency. In addition to all of these benefits, sustainability is one area where nanotechnology needs to thrive. As this sector becomes more sustainable in the near future, so too will its use (Shah and Shah 2020).

Numerous advantages of nanotechnology have already been realized for the community, and there is still limitless potential to create and apply novel technologies. Yet, there are some hazards associated with nanotechnology studies and development in addition to its advantages. Unexpected effects are common during revolutions, and they occasionally give rise to responses (like the use of agricultural use of genetically modified organisms). Our understanding of the effects of nanotechnology on the environment and human health is still lacking. It is essential to be able to assess and limit these hazards in order to ensure that nanotechnology is safe and sustainable and that "nano" may rightfully be a driver of a sustainable planet and future (Pokrajac et al. 2021).

Even though it is invisible, nanotechnology is present in every aspect of our daily lives. In roughly 50 years, this new technology has evolved into complex solutions and become embedded in a wide range of industrial applications. Its applications are found in many different areas, including as electronics, food packaging, medical, cosmetics, and sensing. There are numerous benefits to each distinct application of nanotechnology. For instance, because of their extensive range of antifungal, antibacterial, and antiviral qualities, silver nanoparticles (Gaps) are an active ingredient in numerous consumer products. In the medical field, NPs and insoluble medications are mixed to help the body absorb pharmaceuticals. Nanomaterials were produced in hundreds of thousands of tones at the beginning of the 2000s, and the global market for them was projected to grow to an optimistic 5 million tons by 2025 (Chee et al. 2022).

Nanotechnology offers enormous potential in agriculture. To maximize production with minimal inputs through enhanced sensing and targeted actions, precision farming driven by nano principles would be the optimal way to raise agricultural productivity. By the use nanotechnology, crops can now use resources like water, soil, pesticides herbicides, and fertilizers more effectively. Future uses for nanotechnology include producing biofuel, using enhanced nano capsules to deliver agrochemicals, manipulating plants genetically more effectively, and utilizing nanoporous zeolites to release fertilizers and water to plants in an efficient manner—all while prioritizing the health of living things and the sustainable environment (Pandey 2020).

The energy needed cannot be produced by the existing, inefficient energy producing infrastructure. To meet energy needs, new, highly efficient renewable resources and/or sophisticated materials are needed. Examples of sustainable and environmentally friendly energy sources are solar and hydrogen energy. They encourage sustainable growth, although their full potential is still unrealized. A helpful instrument for boosting energy output and efficiency that can address the world's current energy shortage is nanotechnology. Recent developments in nanotechnology have made it possible to produce, store, and use renewable energy (Kumar et al. 2018).

1.5 Conclusion and Prospects

A new era of innovation and opportunities has been brought about by nanotechnology, especially in the areas of environmental stewardship and sustainability. Green nanotechnology applications have emerged as a practical way to ensure ecosystem sustainability and address some of the most pressing environmental issues of our time. By utilizing the unique properties of nanoparticles, such as their high surface area to volume ratio and amazing reactivity, a wide range of technology have been developed and goods helpful to the environment.

The green nanotechnology's application has a great potential to significantly reduce environmental deterioration and foster a more harmonious coexistence between humans and the natural world, ranging from pollution treatment to alternative energy production. Catalysts based on nanomaterials facilitate economical and effective wastewater treatment operations, and membranes with nanostructures provide desalination and filtering options for clean water. Furthermore, the efficiency of fuel cells, solar cells, and energy storage devices is being improved by nanoscale materials, which are propelling developments in renewable energy technology. Additionally, new paths for environmentally friendly production methods and biodegradable materials have been made possible by the fusion of nanotechnology, biotechnology, and sustainable materials design. For example, nanocellulose, which comes from sustainable resources like wood pulp, has the potential to replace conventional plastics and lessen the environmental impact of many businesses. Despite the enormous potential advantages of green nanotechnology, care must be taken to ensure that the health and environmental effects of nanomaterials are taken into account at every stage of their life cycle. Strict regulation, ethical production methods, and thorough risk analyses are essential to guarantee that the application of nanotechnology stays in line with the objectives of ecological sustainability. To sum up, the development of green nanotechnology is an important step forward in the direction of a more sustainable future. Through utilizing the distinct characteristics of nanomaterials in eco-friendly uses, we may tackle urgent environmental issues and work towards a more robust and harmonious coexistence with our planet. Adopting the tenets of green nanotechnology provides a way forward for a future generation of cleaner, healthier, and more sustainable environments.

References

Alfadul SM, Elneshwy AA (2010) Use of nanotechnology in food processing, packaging and safety—review. Afr J Food Agric Nutr Dev 10. https://doi.org/10.4314/ajfand.v10i6.58068

Ali S, Ahmad N, Dar MA et al (2023) Nano-agrochemicals as substitutes for pesticides: prospects and risks. Plan Theory 13:109. https://doi.org/10.3390/plants13010109

Anandhi S, Saminathan VR, Yasotha P et al (2020) Nano-pesticides in pest management. J Entomol Zool Stud 8:685–690

Ansari MA (2023) Nanotechnology in food and plant science: challenges and future prospects. Plan Theory. https://doi.org/10.3390/plants12132565

Athanassiou CG, Kavallieratos NG, Benelli G et al (2017) Nanoparticles for pest control: current status and future perspectives. J Pest Sci 91:1–15. https://doi.org/10.1007/s10340-017-0898-0

Basiago AD (1995) Methods of defining 'sustainability'. Sustain Dev 3:109–119. https://doi.org/10.1002/sd.3460030302

Berekaa MM (2015) Nanotechnology in food industry; advances in food processing, packaging and food safety. Int J Curr Microbiol App Sci 4:345–357

Bhandari G, Dhasmana A, Chaudhary P et al (2023) A perspective review on green nanotechnology in agro-ecosystems: opportunities for sustainable agricultural practices & environmental remediation. Agriculture 13:668. https://doi.org/10.3390/agriculture13030668

Bhatkhande DS, Pangarkar VG, Beenackers AACM (2002) Photocatalytic degradation for environmental applications—a review. J Chem Technol Biotechnol 77:102–116. https://doi.org/10.1002/jctb.532

Bhawana P, Fulekar M (2012) Nanotechnology: remediation technologies to clean up the environmental pollutants. Res J Chem Sci 2:90–96. ISSN 2231, 606X

Bhol P, Bhavya MB, Swain S et al (2020) Modern chemical routes for the controlled synthesis of anisotropic bimetallic nanostructures and their application in catalysis. Front Chem 8. https://doi.org/10.3389/fchem.2020.00357

Biswas R, Alam M, Sarkar A et al (2022) Application of nanotechnology in food: processing, preservation, packaging and safety assessment. Heliyon 8(11):e11795

Bora T, Dutta J (2014) Applications of nanotechnology in wastewater treatment—a review. J Nanosci Nanotechnol 14:613–626. https://doi.org/10.1166/jnn.2014.8898

Bradu P, Biswas A, Nair C et al (2023) Recent advances in green technology and Industrial Revolution 4.0 for a sustainable future. Environ Sci Pollut Res 30:124488–124519

Bratovcic A, Darroudi M, Sundaramanickam A, Ibrahimpasic J (2022) Application of nanotechnology in remediation of environmental pollutants. In: Sustainable management of environmental contaminants, pp 343–355. https://doi.org/10.1007/978-3-031-08446-1_12

Chee PL, Toh WL, Yew PY et al (2022) Introduction of nanotechnology and sustainability. In: Sustainable nanotechnology, pp 1–32. https://doi.org/10.1039/9781839165771-00001

Ciambelli P, La Guardia G, Vitale L (2020) Nanotechnology for green materials and processes. In: Studies in surface science and catalysis, pp 97–116. https://doi.org/10.1016/b978-0-444-64337-7.00007-0

Diallo MS, Fromer NA, Jhon MS (2014) Nanotechnology for sustainable development: retrospective and outlook. J Nanopart Res 15. https://doi.org/10.1007/s11051-013-2044-0

Dikshit P, Kumar J, Das A et al (2021) Green synthesis of metallic nanoparticles: applications and limitations. Catalysts 11:902. https://doi.org/10.3390/catal11080902

Ganie AS, Bano S, Khan N et al (2021) Nano remediation technologies for sustainable remediation of contaminated environments: recent advances and challenges. Chemosphere 275:130065. https://doi.org/10.1016/j.chemosphere.2021.130065

García-Quintero A, Palencia M (2021) A critical analysis of environmental sustainability metrics applied to green synthesis of nanomaterials and the assessment of environmental risks associated with the nanotechnology. Sci Total Environ 793:148524. https://doi.org/10.1016/j.scitotenv.2021.148524

Gul MZ, Rupula K, Rao BS (2022) Nano bioremediation: a novel application of green-nanotechnology in environmental cleanup. In: Microbes and microbial biotechnology for green remediation, pp 823–841. https://doi.org/10.1016/b978-0-323-90452-0.00040-2

Guo KW (2011) Green nanotechnology of trends in future energy: a review. Int J Energy Res 36:1–17. https://doi.org/10.1002/er.1928

He X, Deng H, Hwang H (2019) The current application of nanotechnology in food and agriculture. J Food Drug Anal 27:1–21. https://doi.org/10.1016/j.jfda.2018.12.002

Heinemann MG, Rosa CH, Rosa GR, Dias D (2021) Biogenic synthesis of gold and silver nanoparticles used in environmental applications: a review. Trend Environ Anal Chem 30:e00129. https://doi.org/10.1016/j.teac.2021.e00129

Hristozov D (2013) Benefits of nanotechnology for the environment. In: Nanotechnology and human health. Taylor & Francis, p 129

Jafarzadeh S, Hadidi M, Forough M et al (2022) The control of fungi and mycotoxins by food active packaging: a review. Crit Rev Food Sci Nutr 63:6393–6411. https://doi.org/10.1080/10408398.2022.2031099

Jiang Y, Zhou P, Zhang P et al (2022) Green synthesis of metal-based nanoparticles for sustainable agriculture. Environ Pollut 309:119755. https://doi.org/10.1016/j.envpol.2022.119755

Kanwar R, Rathee J, Salunke DB, Mehta SK (2019) Green nanotechnology-driven drug delivery assemblies. ACS Omega 4:8804–8815. https://doi.org/10.1021/acsomega.9b00304

Khan SH (2020) Green nanotechnology for the environment and sustainable development. In: Green materials for wastewater treatment, pp 13–46. https://doi.org/10.1007/978-3-030-17724-9_2

Khan F, Pandey P, Upadhyay TK (2022) Applications of nanotechnology-based agrochemicals in food security and sustainable agriculture: an overview. Agriculture 12:1672. https://doi.org/10.3390/agriculture12101672

Kshtriya V, Koshti B, Gour N (2021) Green synthesized nanoparticles: classification, synthesis, characterization, and applications. In: Biosynthesized nanomaterials, pp 173–222. https://doi.org/10.1016/bs.coac.2020.12.009

Kumar A, Sadhu SD, Singh R (2018) Nanotechnology: key for sustainable future. In: Nanotechnology in environmental science, pp 759–804. https://doi.org/10.1002/9783527808854.ch24

Kumar Y, Tiwari KN, Singh T, Raliya R (2021) Nano fertilizers and their role in sustainable agriculture. Ann Plant Soil Res 23:238–255. https://doi.org/10.47815/apsr.2021.10067

Malik P, Sangwan A (2012) Nanotechnology: a tool for improving efficiency of bio-energy. J Eng Appl Sci 1:37–49

Mansoori GA, Bastami TR, Ahmadpour A, Es'haghi Z (2008) Environmental application of nanotechnology. In: Annual review of nano research, pp 439–493. https://doi.org/10.1142/9789812790248_0010

Mayegowda SB, Sarma G, Gadilingappa MN et al (2023) Green-synthesized nanoparticles and their therapeutic applications: a review. Green Process Synth 12. https://doi.org/10.1515/gps-2023-0001

Mishra S, Chauhan G, Verma S, Singh U (2022) The emergence of nanotechnology in mitigating petroleum oil spills. Mar Pollut Bull 178:113609. https://doi.org/10.1016/j.marpolbul.2022.113609

Nallasamy L, Murugavelu GS, Ganesh S et al (2023) Green nanotechnology revolution in biomedical application and treatments. In: Nanovaccinology, pp 181–191. https://doi.org/10.1007/978-3-031-35395-6_10

Nasrollahzadeh M, Sajjadi M, Sajadi SM, Issaabadi Z (2019) Green nanotechnology. In: Interface science and technology, pp 145–198. https://doi.org/10.1016/b978-0-12-813586-0.00005-5

Nath D, Banerjee P (2013) Green nanotechnology—a new hope for medical biology. Environ Toxicol Pharmacol 36:997–1014. https://doi.org/10.1016/j.etap.2013.09.002

Osman AI, Zhang Y, Farghali M et al (2024) Synthesis of green nanoparticles for energy, biomedical, environmental, agricultural, and food applications: a review. Environ Chem Lett 22:841–887. https://doi.org/10.1007/s10311-023-01682-3

Pandey G (2020) Agri-nanotechnology for sustainable agriculture. In: Ecological and practical applications for sustainable agriculture, pp 229–249. https://doi.org/10.1007/978-981-15-3372-3_11

Pandhi S, Kumar A, Mishra S, Rai DC (2021) Potential of green nanotechnology in food processing and preservation. In: Handbook of research on food processing and preservation technologies, pp 135–161

Pete AJ, Bharti B, Benton MG (2021) Nano-enhanced bioremediation for oil spills: a review. ACS ES&T Eng 1:928–946. https://doi.org/10.1021/acsestengg.0c00217

Pokrajac L, Abbas A, Chrzanowski W et al (2021) Nanotechnology for a sustainable future: addressing global challenges with the international network 4 sustainable nanotechnology. ACS Nano 15:18608–18623. https://doi.org/10.1021/acsnano.1c10919

Prasad R, Bhattacharyya A, Nguyen QD (2017) Nanotechnology in sustainable agriculture: recent developments, challenges, and perspectives. Front Microbiol 8. https://doi.org/10.3389/fmicb.2017.01014

Preetha PS, Balakrishnan N (2017) A review of nano fertilizers and their use and functions in soil. Int J Curr Microbiol App Sci 6:3117–3133. https://doi.org/10.20546/ijcmas.2017.612.364

Qiao K, Tian W, Bai J et al (2019) Application of magnetic adsorbents based on iron oxide nanoparticles for oil spill remediation: a review. J Taiwan Inst Chem Eng 97:227–236. https://doi.org/10.1016/j.jtice.2019.01.029

Raja RK, Hazir S, Balasubramani G et al (2022) Green nanotechnology for the environment. In: Handbook of microbial nanotechnology, pp 461–478. https://doi.org/10.1016/b978-0-12-823426-6.00006-1

Ramsurn H, Gupta RB (2013) Nanotechnology in solar and biofuels. ACS Sustain Chem Eng 1: 779–797. https://doi.org/10.1021/sc400046y

Rashidi L, Khosravi-Darani K (2011) The applications of nanotechnology in food industry. Crit Rev Food Sci Nutr 51:723–730. https://doi.org/10.1080/10408391003785417

Ravichandran R (2010) Nanotechnology applications in food and food processing: innovative green approaches, opportunities and uncertainties for global market. Int J Green Nanotechnol Phys Chem 1:P72–P96. https://doi.org/10.1080/19430871003684440

Razauddin, Ninama J, Sachan K et al (2023) Effects and consequences of nano fertilizer application on plant growth and developments: a review. Int J Environ Clim Change 13:2288–2298. https://doi.org/10.9734/ijecc/2023/v13i102893

Sadeghpour A, Pirolt F, Glatter O (2013) Sub micrometer-sized pickering emulsions stabilized by silica nanoparticles with adsorbed oleic acid. Langmuir 29:6004–6012. https://doi.org/10.1021/la4008685

Saleem H, Zaidi SJ, Ismail AF, Goh PS (2022) Nanotechnology in air pollution remediation. In: Nanotechnology for environmental remediation, pp 59–76. https://doi.org/10.1002/9783527834143.ch5

Sayan B, Indranil S, Aniruddha M (2013) Role of nanotechnology in water treatment and purification: potential applications and implications. Int J Chem Sci Technol 3:59

Shah S, Shah M (2020) Nanotechnology: a scope for a sustainable future. In: Handbook of polymer and ceramic nanotechnology, pp 1–24. https://doi.org/10.1007/978-3-030-10614-0_62-1

Shapira P, Youtie J (2015) The economic contributions of nanotechnology to green and sustainable growth. In: Green processes for nanotechnology, pp 409–434. https://doi.org/10.1007/978-3-319-15461-9_15

Singh H, Bhardwaj N, Arya SK, Khatri M (2020) Environmental impacts of oil spills and their remediation by magnetic nanomaterials. Environ Nanotechnol Monit Manag 14:100305. https://doi.org/10.1016/j.enmm.2020.100305

Singh H, Desimone MF, Pandya S et al (2023a) Revisiting the green synthesis of nanoparticles: uncovering influences of plant extracts as reducing agents for enhanced synthesis efficiency and its biomedical applications. Int J Nanomedicine 18:4727–4750. https://doi.org/10.2147/ijn.s419369

Singh BP, Sharma K, Tyagi S et al (2023b) Green and sustainable technology for clean energy production: applications. In: Handbook of green and sustainable nanotechnology, pp 1–23. https://doi.org/10.1007/978-3-030-69023-6_64-1

Taran M, Safaei M, Karimi N, Almasi A (2021) Benefits and application of nanotechnology in environmental science: an overview. Biointerface Res Appl Chem 11:7860–7870. https://doi.org/10.33263/BRIAC111.78607870

Vaseashta A (2008) Green nanotechnologies for responsible manufacturing. MRS Online Proc Library 1106:305

Vatti NR, Vatti P, Vatti R (2018) High efficiency solar energy harvesting using nanotechnology. In: 2018 4th international conference for convergence in technology (I2CT). https://doi.org/10.1109/i2ct42659.2018.9058156

Yilmaz E, Soylak M (2021) Nanotechnological developments in nanofiber-based membranes used for water treatment applications. Environ Nanotechnol 5:205–259. https://doi.org/10.1007/978-3-030-73010-9_6

Ying S, Guan Z, Ofoegbu PC et al (2022) Green synthesis of nanoparticles: current developments and limitations. Environ Technol Innov 26:102336. https://doi.org/10.1016/j.eti.2022.102336

Yunus IS, Harwin, Kurniawan A et al (2012) Nanotechnologies in water and air pollution treatment. Environ Technol Rev 1:136–148. https://doi.org/10.1080/21622515.2012.733966
Zahra Z, Habib Z, Hyun H, Shahzad HMA (2022) Overview on recent developments in the design, application, and impacts of nano fertilizers in agriculture. Sustain For 14:9397. https://doi.org/10.3390/su14159397

Chapter 2
Harnessing the Silica Mesoporous Nanoparticles for Sustainable Agriculture

M. Saranya, C. Muthu Lakshmi Bavithra, M. Yuvaraj, R. Sathya Priya, N. Jagathjothi, R. Sharmila, N. Suganthi, and Jaiby Cyriac

Abstract Mesoporous nanoparticles (MSNs) are emerging as a game-changer in agriculture due to their unique properties. This chapter explores the potential of MSNs in enhancing crop improvement and protection. The high surface area and tunable pores of MSNs make them ideal for controlled delivery of nutrients, promoting efficient uptake by plants and reducing waste. Additionally, MSNs can be functionalized with various molecules to target specific pests or pathogens, offering a precision agriculture approach. Field trials demonstrated that crops treated with MSNs exhibited significant improvements in growth, stress tolerance, and yield compared to conventional treatments. The integration of MSNs into agricultural practices presents a sustainable alternative to traditional farming methods, contributing to enhanced food security and environmental conservation. Furthermore, MSNs can act as targeted carriers for proteins or genes used in crop modification, potentially leading to the development of new and improved crop varieties. However, the widespread adoption of MSNs in agriculture necessitates further research into their long-term effects on soil health, plant physiology, and environmental

---

M. Saranya (✉) · C. Muthu Lakshmi Bavithra
Department of Agricultural Entomology, Tamil Nadu Agricultural University, Coimbatore, Tamil Nadu, India

M. Yuvaraj
Soil Science and Agricultural Chemistry, Agricultural College and Research Institute, Vazhavachanur, Tamil Nadu, India

R. Sathya Priya · N. Jagathjothi
Department of Agronomy, Tamil Nadu Agricultural University, Coimbatore, Tamil Nadu, India

R. Sharmila
Department of Nematology, Tamil Nadu Agricultural University, Coimbatore, Tamil Nadu, India

N. Suganthi
Department of Agricultural Extension, Tamil Nadu Agricultural University, Coimbatore, Tamil Nadu, India

J. Cyriac
Department of Botany, Newman College, Thodupuzha, Kerala, India

© The Author(s), under exclusive license to Springer Nature Switzerland AG 2025

J. M. Al-Khayri et al. (eds.), *Green Nanotechnology Applications for Ecosystem Sustainability*, Nanotechnology in Plant Sciences 4,
https://doi.org/10.1007/978-3-031-90766-1_2

safety. This study emphasizes the need for comprehensive assessments and regulatory frameworks to guide the safe and effective use of MSNs in agriculture. By harnessing the potential of MSNs, the agricultural sector can take a significant step towards more sustainable and environmentally friendly farming practices.

Keywords Nanoparticles · Physiology · Soil health · Gene delivery · Genetic engineering

2.1 Introduction

Over the past two decades, there has been a notable global effort to apply a wide range of nanoparticles, including polymers, carbon nanotubes, quantum dots, dendrimers, and inorganic (metal and metal oxide) nanoparticles. Since the early 1990s, research on silica-based nanoparticles within the context of metal oxides has been particularly compelling. Structurally ordered mesoporous silica materials, such as MCM-4-6 and SBA-type7 mesoporous silicas, offer numerous advantages over solid silica nanoparticles, including high surface area, narrow pore size distribution, and robust chemical and physical stability. These engineered nanoparticles (ENPs) have gained prominence across various fields—such as analytical chemistry, medicine, and agriculture due to their applications as delivery vectors, biosensors, and DNA labels (Kumar et al. 2013). In agriculture, nanotechnology-based formulations like fertilizers and insecticides have significantly boosted agricultural output and paved the way for innovative approaches to sustainable farming (Rai and Ingle 2012; Saleh et al. 2019). Among all these ENPs, mesoporous silica nanoparticles (MSNs) have garnered substantial research interest due to their unique properties, including large pore volumes and surface areas, high loading capacities, remarkable chemical and thermal stabilities, and excellent biocompatibility (Setyawati and Leong 2017) (Fig. 2.1).

2.2 Synthesis of Mesoporous Nanoparticles

To create spherical, monodisperse micron-sized silica particles, Stober designed a chemical reaction system, now known as the Stober synthesis (Stober et al. 1968). Over time, numerous variations of Stober's synthesis have been developed to produce monodisperse, organized nanosized silica particles. These particles can be synthesized in basic, acidic, and neutral environments, with different shapes and sizes achieved by varying the reaction conditions. Grun et al. (1997) modified first time Stober's synthesis process using a cationic surfactant as a template, resulting in a spherical MCM-41 structure instead of a hexagonal one. This spherical MCM-41 exhibited characteristics comparable to those produced by other method.

In 1992, scientists at the Mobil Company first reported the synthesis of structurally ordered mesoporous silica materials, known as MCM-type materials (Beck et al.

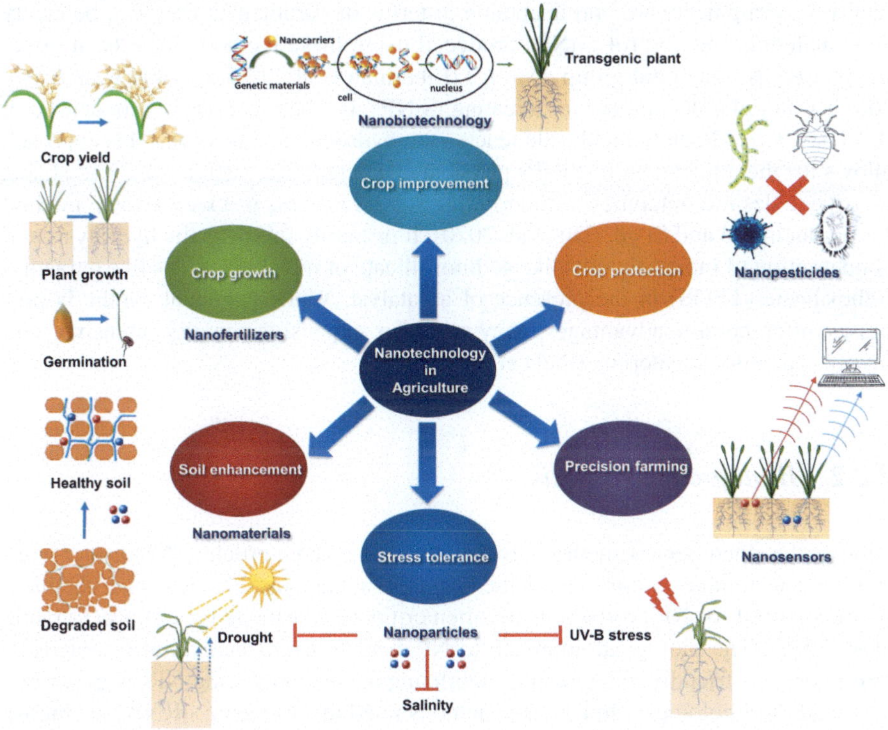

Fig. 2.1 Application of nanotechnology in agriculture. (Adopted from Shang et al. 2019)

1992). This method involved the condensation of silica precursors like sodium silicate, tetramethylammonium silicate, or tetraethyl orthosilicate under hydrothermal and basic conditions, with quaternary ammonium cationic surfactants serving as structure-directing agents. This micelle-templated approach has since led to the creation of a wide variety of mesoporous silica materials, including SBA (Zhao et al. 1998), MSU (Bagshaw et al. 1995), and FSM-types (Inagaki et al. 1993, 1996). Unlike other crystalline microporous silica materials such as zeolites, mesoporous silica materials exhibit a high degree of pore order, excellent thermal and hydrolytic stability, narrow pore size distribution, and high surface areas, despite lacking crystalline domains. However, their large size and irregular particle shapes limit their applications in various fields (Rejman et al. 2004).

2.2.1 *Chemical Synthesis*

In recent years, conventional techniques such as flame synthesis, microemulsion, Stober's approach, and the sol-gel method have been employed to create silicon nanoparticles (SiNPs) (Liou and Yang 2011). Although these chemical routes are relatively straightforward and adaptable in terms of parameters, they can be costly and challenging to control. SiNPs produced using these methods have been effectively used as functional group coatings (Liberman et al. 2014; Liu and Han 2010). Another popular technique for creating SiNPs is chemical vapor condensation (CVC), where silicon tetrachloride reacts with hydrogen and oxygen. This approach allows for the adjustment of SiNPs physical characteristics, such as size and shape, to achieve desired outcomes. Similarly, the sol-gel process is a well-known method for producing Si and Si gel (Du et al. 2020). It primarily involves the hydrolysis and condensation of inorganic salts like sodium silicate or metal alkoxides like tetraethyl orthosilicate (TEOS) in the presence of a catalyst. While chemical synthesis processes offer certain advantages, they are also expensive, energy-intensive, and involve hazardous materials (Piela et al. 2020).

2.2.2 Biological Synthesis

Among the documented methods for synthesizing nanoparticles (NPs), the green synthesis technique is considered the most advantageous. Physical and chemical processes used for NP production are often criticized for being hazardous (Karande et al. 2021). The biological approach to NP synthesis utilizes natural resources as non-toxic, eco-friendly reducing and stabilizing agents, producing NPs with various sizes and morphologies. Studies have demonstrated that biogenic siRNAs are highly effective in farming and agriculture (Snehal and Lohani 2018). Additionally, numerous studies have explored the application of biogenic SiNPs in various industries, including healthcare, with varying degrees of success. Plants and plant wastes are the most commonly used biological resources for producing SiNPs. To minimize toxicity, green approaches have been employed to manufacture NPs, which are considered less toxic and more eco-friendly (Jeelani et al. 2020).

2.3 Mesoporous Silica Nanoparticles

Silica, or silicon dioxide, is a naturally occurring substance in soil that plays a crucial physiological role in plants. Its deficiency makes plants more susceptible to lodging and infections, as it is deposited as hydrated amorphous silica in the cell wall, endoplasmic reticulum, and intercellular spaces (Rai et al. 2015). Due to its properties and natural presence in both plant and environmental systems, silica shows great promise for agricultural applications.

Mesoporous silica nanoparticles (MSNPs) offer several advantages for plants, including increased nutrient availability, improved seed germination, and enhanced growth of seedlings and roots. Silica nanoparticles can promote plant growth and development and boost resistance to abiotic stress. Consequently, the use of MSNPs in agriculture poses relatively few risks regarding their entry into the food chain and environmental toxicity. Additionally, the internalization of MSNPs into the plant system can be controlled or prevented by adjusting the size of the nanoparticles. Only nanoparticles smaller than the largest pores, typically 3–8 nm in size, can pass through the cell wall and into the plasma membrane (Carpita and Gibeaut 1993).

Microscopic methods have been used to investigate the location and fate of nanoparticles (NPs) in plants. Sun et al. (2014) localized and quantified 20 nm mesoporous silica nanoparticles (MSNPs) in lupin, wheat, maize, and Arabidopsis. They demonstrated that MSNPs entered the roots and moved to the aerial parts of the plants without adversely affecting seed germination or the transfer of MSNPs to different plant organs.

MSNPs are particularly appealing as drug delivery vehicles due to their numerous advantageous properties. Silica is inexpensive, safe, and stable both chemically and thermally (Iskandar et al. 2001). MSNPs have the potential to support sustainable agriculture by optimizing yields and ensuring the controlled application of biocides that are safe and effective for the environment and the health of humans, animals, and plants. Applications of nanoparticles for crop improvement are illustrated in (Table 2.1).

2.4 Role of Mesoporous Silica Nanoparticles (MSNs) in Physiology of Plants

Mesoporous silica nanoparticles (MSNs) are becoming increasingly popular in nanoscience and nanobiotechnology due to their large surface area, robust framework, porous and honeycomb-like structure (both ordered and random), high loading capacity, biocompatibility, and stability. MSNs have been utilized to deliver agrochemicals and DNA (Chen et al. 2011; Wanyika et al. 2012), plant hormones such as ABA (Sun et al. 2018), insoluble compounds like indole-3-butyric acid (IBA) (Ambrogi et al. 2012), pesticides (Zhao et al. 2017, 2018b; Zhu et al. 2018), proteins (Martin-Ortigosa et al. 2014), and other substances.

Sun et al. (2016) discovered that MSNs localize in chloroplasts and enhance photosynthesis in wheat and lupin. When treated with 500–1000 mg/L of MSNs, germination, plant biomass, total protein, and chlorophyll content all increased significantly. Wheat and lupin showed increased accumulation of total leaf protein and chlorophyll pigments after absorbing specific amounts of MSNs. MSNs stimulated photosynthesis in chloroplasts, leading to substantial plant growth. The study found that MSNs were absorbed by plant roots and transferred to the aerial parts of stems and leaves. Additionally, Mitra et al. (2021) found that naturally occurring MSNs derived from diatoms can enhance growth and regulate metabolic pathways in the monocot Jowar (*Sorghum bicolor* L.) and the dicot Bengal gram (*Cicer arietinum* L.).

Zinc oxide nanoparticles encased in a mesoporous SiO_2 shell ($nZnO@SiO_2$) with a primary particle diameter of around 70 nm were synthesized for the foliar delivery

Table 2.1 Nanoparticle application for the improvement of crops

Nanoparticles	Plant	Benefits	References
Nano-SiO_2	Tomato (*Solanum lycopersicum* L.)	Enhanced seed germination at low concentrations	Siddiqui and Al-Whaibi (2014)
Nano-SiO_2	Maize (*Zea mays* L.)	Increased seed germination, better plant dry weight, protein, chlorophyll, and phenol levels, and improved nutrient availability, pH, and conductivity of the growing medium	Suriyaprabha et al. (2012)
Nano-SiO_2	Tomato (*Solanum lycopersicum* L.)	Improved seed germination and increased antioxidant system under NaCl stress	Haghighi et al. (2012)
Quantum Dots (QD) and Silica coated with QDs	Rice (*Oryza sativa* L.)	Enhanced root development with QD-coated silica	Wang et al. (2014)
ZnO	*Coffea Arabica* L.	Enhanced growth, biomass accumulation and net photosynthesis	Rossi et al. (2019)
Silver nanoparticles (Ag NPs)	Cowpea (*Vigna unguiculata* L.)	Showed no phytotoxicity, but could inhibit growth of *Xanthomonas axonopodis* pv. *malvacearum* and *Xanthomonas campestris* pv. *campestris* in vitro	Vanti et al. (2019)
ZnO, CuO and Ag NPs	Plum (*Prunus domestica* L.)	Suppressed grey mold symptoms caused by *B. cinerea* and soil borne diseases	Malandrakis et al. (2019)
TiO_2 and SiO_2	Rice (*Oryza sativa* L.)	Mitigated Cadmium toxicity and improved growth by stimulating antioxidant potential and inhibiting Cadmium translocation	Rizwan et al. (2019)
Graphene NPs	Alfalfa (*Medicago sativa* L.)	Improved antioxidant enzyme activities and increased the fresh weight, dry weight and seedling root length	Chen and Wang (2021)

of zinc, a crucial micronutrient. The SiO_2 shell enhances the uptake of ZnO nanoparticles in *Solanum lycopersicum* (L.). Foliar application of micronutrients to plants reduces the consumption of fertilizers, energy, and water (Gao et al. 2021).

Various abiotic factors, such as high temperatures, droughts, flooding, salinity, and heavy metal stress, affect crops at different stages of their growth. Nanoparticles can be applied to plants through either root or foliar tissues. When nanoparticles come into contact with plants through the soil, they enter the root epidermis cell wall. This initiates a complex series of processes that transport the particles to the xylem stream and, ultimately, into the symplastic pathway, leading them to the stele before they are translocated to the leaves.

Le et al. (2014) showed that MSNPs interacted with nutrients in cotton (*Gossypium hirsutum* L.) when given to hydroponic nutrient solution, causing cultivar-dependent changes of the concentrations of numerous nutrients in both roots and shoots. Slomberg and Schoenfisch (2012) treated Arabidopsis in hydroponic culture with highly negatively charged silicon (Si) nanoparticles and observed effects on growth and chlorosis in the leaves. These effects were assumed to be a result of pH-dependent adsorption dynamics of nutrients on the nanoparticle surface.

Studies have found that mesoporous silica nanoparticles (MSNs), specifically MCM-41, which has a larger pore capacity and surface area, are more effective at higher doses compared to MCM-48 in *Vigna radiata* (L.). MSNs have been shown to increase plant growth by elevating metabolic rates and reducing stress (Mitra et al. 2022). In wheat and lupin, plant biomass, seed germination, total protein, and chlorophyll content increased after ingesting 500 and 1000 mg/L of MSNs. Even at the highest dosage of 2000 mg/L, no signs of oxidative stress or membrane damage were observed (Sun et al. 2016).

During a polyethylene glycol (PEG)-simulated drought, Arabidopsis seedlings treated with MSNs germinated more quickly and had longer primary roots. Even at the maximum recorded dosage of 5000 mg/L, reduced germination was observed in seeds exposed to drought stress. At the optimal dose of 1500 mg/L, plants treated with MSNs under non-stressed conditions showed significant increases in root length, number of lateral roots, leaf area, and shoot biomass (Tran et al. 2023). Therefore, MSNs can promote plant growth and enhance stress resistance against drought.

Additionally, applying MSNPs-Zn via foliar spray to *Cicer arietinum* (L.), especially at 60% field capacity, strengthened the plant defense mechanism, scavenged reactive oxygen species, and increased pigment accumulation and stability, which helped mitigate the effects of drought stress (Mohamadzadeh et al. 2023).

Compared to nutrients from other sources, MSNPs delivered one or more limiting nutrients to zoysia grass (*Zoysia japonica* (S.)) more efficiently, thereby accelerating growth. This suggests the potential for growing turfgrass with reduced fertilizer usage thanks to MSNPs (Adams et al. 2020).

Mesoporous nanoparticles with a 50 nm diameter and approximately 14.7 nm pore size were functionalized and tested on *Arabidopsis thaliana* (L). The seed germination rate and seedling growth were significantly higher compared to untreated controls. Plants treated with Am-MSNs showed an increase in photosynthetic pigments, enhanced photosynthetic activity, and upregulation of genes associated with carotenoid and chlorophyll production. These changes led to increased vegetative growth and seed output without causing cell or membrane damage (Lu et al. 2020).

When SiO_2 nanoparticles (NPs) were applied to maize during a drought, the production of the D2 protein increased. This protein is crucial for the plastid and forms the core of photosystem II, which is necessary for maintaining the integrity of the thylakoid membrane and ensuring efficient photosynthesis. Compared to untreated plants under drought stress, those treated with SiO_2 NPs also showed decreased expression of genes such as Aquaporins (AQPs) and osmoregulatory (OSM-34), indicating improved membrane integrity and relative water content

Table 2.2 List of drought tolerant genes regulated by nanoparticles

Nanoparticles used	Drought associated genes	Crop	Upregulation/ downregulation	References
Zinc	*Wdhn13, DREB2, P5CS, and CAT1*	Wheat	Upregulation	Sadati et al. (2022)
	WRKY1, HMA2, and ZIP1	Wheat	Upregulation	Sadati et al. (2022)
	SOD, APX, and CAT	Maize	Upregulation	Sun et al. (2020)
	MYC, MPK4	Rapeseed	Downregulation	Hezaveh et al. (2019)
Silicon	*PsbD*	Maize	Upregulation	Sharf-Eldin et al. (2023)
	OSM-34, AQPs	Maize	Downregulation	Sharf-Eldin et al. (2023)
	ABC1, Wdhn13, CHP, and EXP2	Wheat	Upregulation	Boora et al. (2023)
	DREB 2A, NAC	Rice	Upregulation	Khattab et al. (2014)
	TaSOD	Wheat	Upregulation	Ma et al. (2016)
	PC5S, PC56	Lentil	Downregulation	Biju et al. (2023)
	PsbH, PsbB, PsbD	Wheat	Upregulation	Hassan et al. (2021)

(Sharf-Eldin et al. 2023). Some of the genes regulated by nanoparticles related to drought tolerance are listed in (Table 2.2).

2.5 Role of Mesoporous Silica Nanoparticles (MSNs) in Gene Delivery

MSNs have been used to transport nucleic acids within animal cell lines and tissues, efficiently managing various diseases. Ultrasonic treatment with poly-ethyleneimine (PEI)-coated MSNs nanocarriers is an effective method for delivering genes to plant cells in suspension (Zolghadrnasab et al. 2021). GUS-encoding plasmid was transferred into suspended tobacco cells using PEI-coated MSNs (100 $\pm$ 8.7 nm) and ultrasonic treatment. The total potential of PEI-MSN for DNA adsorption was evaluated at 43.43 μg DNA mg/1 PEI-MSN. High levels of sonoporation have been demonstrated to harm cell viability. The ideal time for ultrasonic treatment is 8 min at three distinct intensities: 160, 320, and 640 W. The protein expression was monitored using a histochemical staining technique. PEI-coated MSNs effectively transmitted GUS-encoding plasmid DNA into tobacco cells. The study found that ultrasonic treatment offers a low-cost and straightforward method for gene transfer into plant cells, eliminating the need for complex equipment and safety concerns. (Zolghadrnasab et al. 2021).

Small nanoparticles (<50 nm) such as poly (amidoamine) dendrimer (Pasupathy et al. 2008), calcium phosphate (Naqvi et al. 2012), and mesoporous silica (Chang et al. 2013) have been shown to transfer genes into plant cells through ultrasonic treatment (Liu et al. 2008), CdSe quantum dots (Wang et al. 2011), and ZnS nanoparticles (Qin et al. 2012). The study found that ultrasonic treatment offers a cost-effective and simple method for transferring genes into plant cells, without the need for complex apparatus or safety concerns (Zolghadrnasab et al. 2021). The use of pDNA-loaded PEI-MSNs and ultrasonic therapy resulted in an efficient gene transfection approach in plant suspension cells. The findings demonstrate that PEI-MSNs may adsorb a substantial quantity of DNA. A study of the effect of ultrasonic therapy on tobacco plant cell viability revealed that raising the intensity from 160 to 640 W resulted in significant cell damage. However, extending the treatment duration reduced deterioration effects. PEI-MSNs as a gene carrier are essential for effective gene transfer into plant cells. Ultrasound treatment, on the other hand, has a synergistic impact that can significantly boost gene transfection efficiency (Zolghadrnasab et al. 2021).

To transfer protein into plant cells using MSNs, the site-specific recombinase "Cre" was used to cut loxP-flanked transgenic areas from the genome (Gilbertson 2003). The Cre recombinase protein was placed onto gold-coated MSNs and introduced into genetically engineered maize cells with loxP sites flanked by genes encoding selection markers and blue-fluorescent reporter proteins (Martin-Ortigosa et al. 2014). Cre-mediated recombination removes a DNA fragment in transgenic cells, activating a reporter gene and producing a red fluorescence protein.

MSNs with Cre-recombinase were administered into genetically modified *Z. mays* cells using the particle bombardment technique, resulting in successful recombination in up to 20% of the target calli. A tiny mass of red fluorescent cells was isolated and transformed into whole plants. The recombination process was validated with PCR. The results revealed that MSN-mediated transfer of CRISPR/Cas9 genome editing components in plant cells was efficient, making it a feasible DNA-free genome editing tool (Fig. 2.2). Chang et al. (2013) studied the use of MSN-mediated DNA transfer rather than mechanical force to induce transient gene expression. MSNs were supplied to *Arabidopsis* seedling roots for 1 day in MS media. The reporter transgene was detected in both the epidermal and cortical cells of the roots. Scientists assume that when the cell wall relaxes, the microfibril network loosens, enabling MSNs up to 50 nm in size to pass through unaided.

For the management of tomato pinworm, Hajiahmadi et al. (2019) sprayed and injected DNA-loaded MSNs into tomato leaves. The MSNs, which were around 40 nm in size and spherical, were developed and loaded with plasmids containing the GUS or cry1Ab genes. *Bacillus thuringiensis* produces a family of poisons known as CRY proteins. They commonly manage arthropods and nematodes (Bravo et al. 2007). MSNs can penetrate plant cells through spraying or injection. DNA transport was validated through RT-PCR detection of the transgenic (cry1Ab). Western blot analysis validated the translation. Plants with temporary cry1Ab expression were tested for *Tuta absoluta*. They were transiently expressing the MSN-delivered Cry1Ab gene protected against tomato pinworm larvae. In comparison to other

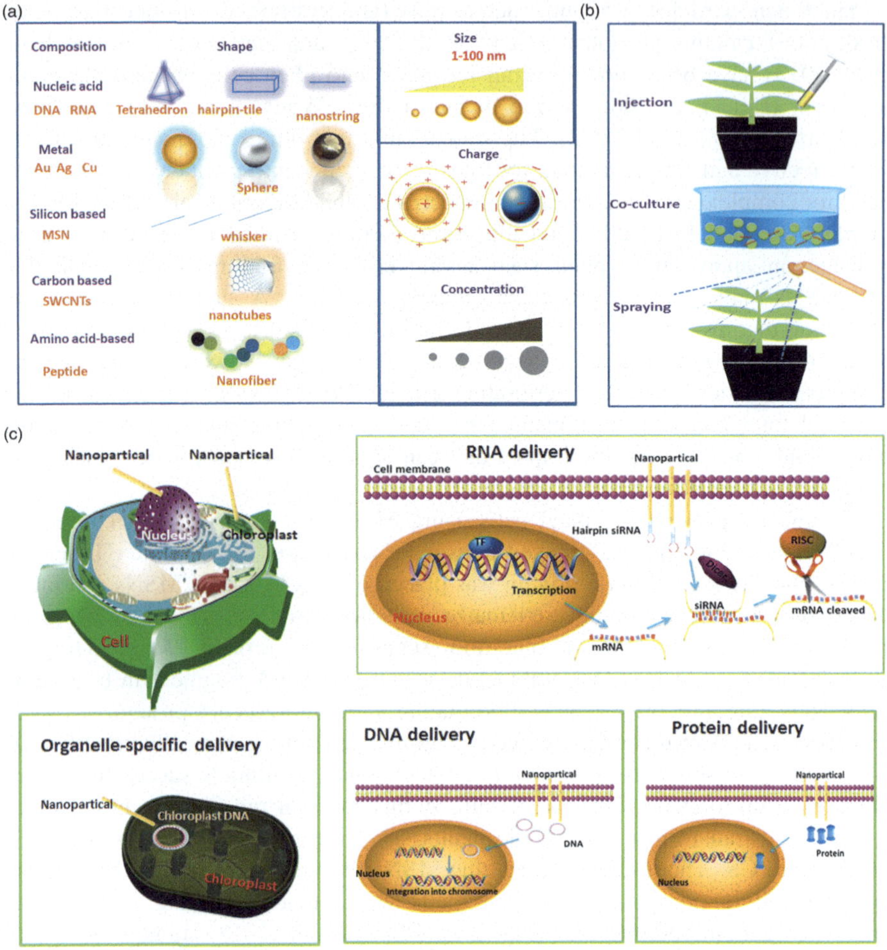

Fig. 2.2 Illustration of nanoparticles as DNA, RNA, protein carriers and delivery targets. (Adopted from Lv et al. 2020)

methods of administration, MSNs containing plasmid DNA injected through the abaxial surface of the leaves exhibited insecticidal action on tomato pinworm larvae (Hajiahmadi et al. 2019).

Torney et al. (2007) developed several MSNs and functionalized their surfaces with functional sets, such as Tri ethylene glycol, to facilitate genetic material transport in plant cells. A plasmid expressing green fluorescent protein (GFP) was introduced into surface-functionalized MSNs. After 36 h of incubation with the protoplasts, the plasmid-loaded MSNs expressed GFP. To boost MSN density and transit in plant cells, mesoporous silica NPs were coated with surface-functionalized gold NPs, inhibiting DNA from leaching from the MSNs. GFP was transiently expressed in the respective cells when these encapsulated MSNs loaded with plasmid DNA were shot into tobacco cotyledons with a particle bombardment tool. Immature maize embryos were also exposed to DNA-loaded capped MSNs and let to proliferate until a callus developed. After 10 days, the scientists discovered GFP expression in the growing callus areas.

The study discovered that employing MSNs as a delivery vehicle resulted in consistent DNA integration in callus cells. The mesoporous structure of MSNs is extremely useful since it enables the loading of enormous quantities of DNA/other materials as well as the controlled delivery of cargo to specific areas. To evaluate the simultaneous delivery of several drugs in plant cells, the authors employed a GFP-expressing gene with an inducible promoter. Plasmids encoding inducible GFP and the inducer beta-oestradiol were loaded into MSNs and then released on wild tobacco. The presence of fluorescent foci in the cell cytoplasm revealed that MSNs were capable of transmitting both the inducible GFP gene and its inducer to plant cells simultaneously. Torney and colleagues investigated the internalization process by incubating many MSNs with tobacco mesophyll protoplasts and tracking endocytosis. Their findings revealed that not only surface properties but also nanoparticle size had a major influence on cell internalization, suggesting MSNs can be employed as a viable tool for plant-cell interaction (Table 2.3).

2.6 Role of Mesoporous Silica Nanoparticles (MSNs) in Crop Protection

Mesoporous silica-based nanomaterials have gained a lot of interest for enhancing pesticide efficiency and safety because to their unique properties of low toxicity, excellent thermal and chemical stability, and, most importantly, size tunability and adaptability. Mesoporous silica nanoparticles (MSNs) serve as a nanocarrier for controlled pesticide delivery, an adsorbent for pesticide enrichment or removal in aqueous media, a catalyst for pesticide degradation, and a sensor for detection (Fig. 2.3). PYR@MSNs exhibit stable spherical shapes, uniform particle size (200 nm), high drug loading (38.9%), and enhanced UV resistance. They outperformed conventional formulations in *Fusarium graminearum* control, indicating that they may represent a novel controlled-release formulation with improved effectiveness (Feng et al. 2022).

Pesticide delivery unit made of BMMs (bimodal mesoporous silica) with a gatekeeper labelled "PMAA/Fe^{3+}". Pro@BMMs-PMAA/Fe^{3+} is a nano-pesticide with a 24.0% pesticide loading rate and good UV and thermal properties. Because of metal cooperation between -COO- and Fe^{3+}, Pro release from BMMs-PMAA/Fe^{3+} was pH dependent. Basic (pH 9.0) and mild acid (pH 5.0) environments produced greater cumulative release rates than neutral (pH 7.0). The Ritger-Peppas kinetic model anticipated the sustained release curves accurately. Biological activity tests revealed that Pro@BMMs-PMAA/Fe^{3+} nanoparticles (NPs) outperformed Pro TC in

Table 2.3 Mesoporous silica nanoparticles mediated genetic engineering in plants

S. no	Nanomaterial type	Plant source or tissue type	Cargo	Method	Effect	References
1.	Gold NP capped MSNs	*Nicotiana tabacum* L. cotyledons; *Zea mays* L. embryos	Green Fluorescent Protein (GFP) plasmid	Biolistic particle delivery	Co-delivery and controlled release of DNA and chemicals	Torney et al. (2007)
2.	Gold-coated MSNs	*Allium cepa* L. epidermis tissue	GFP gene	Particle bombardment	DNA and protein co-delivery	Martin-Ortigosa et al. (2012)
		Z. mays L. embryos	Cre protein	Passive delivery	DNA and protein co-delivery; both transient and stable expression	Martin-Ortigosa et al. (2014)
3.	Functionalized MSNs	*A. thaliana* L. roots	Cherry plasmid	Particle bombardment method	46.5% transient expression efficiency	Chang et al. (2013)
4.	Functionalized MSNs	*Solanum lycopersicum* L. fruits	pPZP122:MPI: cryIAb plasmid	Injection method	Stable transformation system	Hajiahmadi et al. (2020)
5.	Poly-ethyleneimine (PEI)-coated mesoporous silica nanoparticles (MSNs)	*Nicotiana tabacum* L.	GUS-encoding plasmid DNA	Ultrasound mediated gene delivery	Enhance the gene transfection efficiency	Zolghadrnasab et al. (2021)

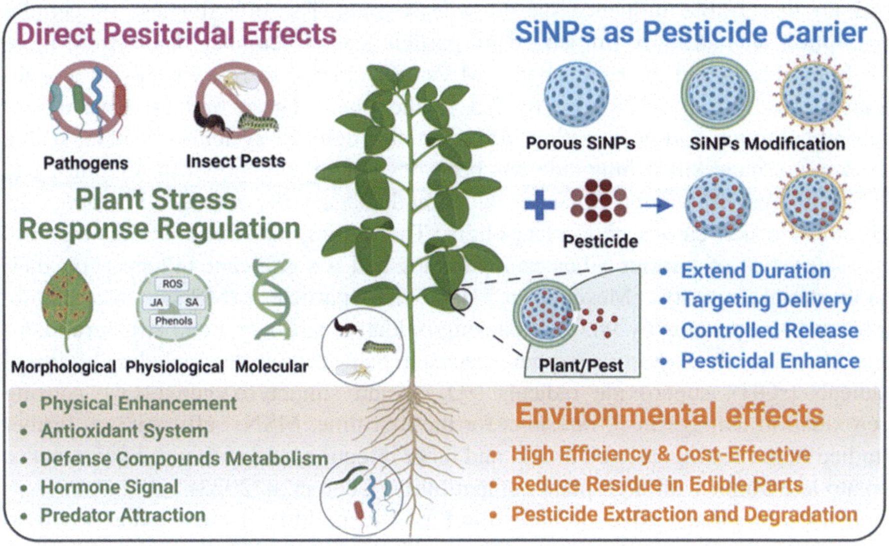

Fig. 2.3 Multiple roles of mesoporous silica in safe pesticide application. (Adopted from Yan et al. 2024)

terms of antifungal activity against *Rhizoctonia solani* at the same active component concentration, presenting a potential crop defense approach against wilt disease. Fluorescence monitoring of 5-AF-labeled NPs showed nanocarrier absorption and translocation in tomato plants and fungal mycelia, suggesting that BMMs-PMAA might be used to deliver pesticide molecules to plants and mycelia. Furthermore, Pro@BMMs-PMAA/Fe^{3+} NPs exhibited remarkable biological safety and boosted crop development as an iron supplement (Pan et al. 2022).

Pro was inserted into MSNs and coated with biodegradable pectin as a gatekeeper (Pro@MSNPec) to investigate its mobility and distribution in rice plants. The findings revealed that MSNs might spread from treated leaves and roots to other areas of the plant. Pro loaded effectively into MSNs with an average diameter of 70.89 nm. Pro@MSN-Pec outperformed Pro EC for absorption and translocation in rice plants. Pro@MSN-Pec has better fungicidal activity than Pro EC or technical Pro. This work sheds light on the translocation and dispersion of pesticide-loaded, stimuli-responsive MSNs in plants, demonstrating how these stimuli-responsive carriers might be exploited to enhance pesticide effectiveness. This study found that Pro@MSN-Pec can extend the life of the active component in rice plants. Furthermore, Pro@MSNPec has revealed a low risk of Pro accumulation in numerous rice plant sections and soils (Abdelrahman et al. 2021).

Pesticide particle size influences plant uptake and effectiveness. The size effect of MSN has an impact on the loading, release, and delivery of pyraoxystrobin (Pyr) in cucumber plants. Loading content was shown to increase with particle size. Nanoparticles as carriers enhanced the solubility of insoluble Pyr, but their size had no discernible influence on the release rate. The effectiveness of cellular absorption was strongly influenced by particle size. Cucumber plants were more likely to absorb and disperse smaller MSNs. Compared to free Pyr, Pyr's upward transit rate from Pyr@M rose by 3.5 in the plant. These findings give a new theoretical framework for creating MSN pesticide delivery systems (Xu et al. 2021).

A novel nano silica fungicide has been developed and shown to minimize late blight by producing intracellular peroxidation damage to the pathogen *Phytophthora infestans*, which causes potato late blight. The findings showed that the structural characteristics of various silica nanoparticles had a significant influence on their antibacterial activity. Mesoporous silica nanoparticles (MSNs) successfully inhibited *P. infestans* by 98.02%, causing oxidative stress and cell structural damage. MSNs produce reactive oxygen species in pathogenic cells, including hydroxyl radicals (•OH), superoxide radicals (•O_{2-}), and singlet oxygen (1O_2), causing peroxidation damage in *P. infestans* for the first time. MSNs' efficacy was further studied in pot testing, as well as leaf and tuber infection, and successful treatment of potato late blight with high plant compatibility (Chen et al. 2023).

MSN's fungicidal effect, as shown in Fig. 2.4, might be due to protein deactivation or direct contact with pathogen DNA, which induces DNA changes and inhibits replication (Petica et al. 2008; Salem et al. 2011). Furthermore, MSN's antifungal impact might be mediated by the easy disintegration of the cell wall due to its tiny size, resulting in hydrogen bonding between the cell wall's lipopolysaccharides and the surface hydroxyl groups found in MSN (Capeletti et al. 2014). Spirotetramat-loaded MSNs increased deposition, uptake, and translocation in cucumbers. MSNs may include spirotetramat in their mesoporous structure, which hinders the degradation in plants (Zhao et al. 2018a).

MSN accumulation in the membrane may cause cell lysis (Gill et al. 2005) by interfering with the transmembrane energy cycle, or it may form insoluble compounds in the fungal membrane that disrupt the electron transport chain, or it may oxidize the cell membrane due to the positive charge of MSN and the negative charge of the cell membrane, resulting in electromagnetic attraction between bot and cell (Rezaei-Zarchi et al. 2010). The surface characteristics of MSN enable this technique, since the positively charged MSN interacts with the protein thiol groups (-SH) on the fungal cell surface, causing cell lysis (Garver et al. 1998; Ma et al. 2001; Brecht et al. 2003; Kanto et al. 2004). MSN's distinguishing features, such as a high surface area (SBET) of 489 m^2/g and a small pore size (Dp) of 10.7, are expected to increase the importance of its surface morphology in cellular interaction, i.e., the number of active sites that come into contact with cells, resulting in a cytotoxic effect against early blight fungus. Furthermore, it is widely recognised that smaller particles have a higher surface area accessible for interaction, resulting in a stronger antibacterial impact than bigger particles. MSN and metalaxyl may minimise the impact of early blight fungus on the green area of tomato plants, lowering the damaged green leaf area and eventually resulting to better plant development because the leaves can perform more photosynthesis (Paveley et al. 1997).

A plant disease microenvironment-responsive nano system was built to cure Sclerotinia disease, employing disulfide-bridged mesoporous organosilica

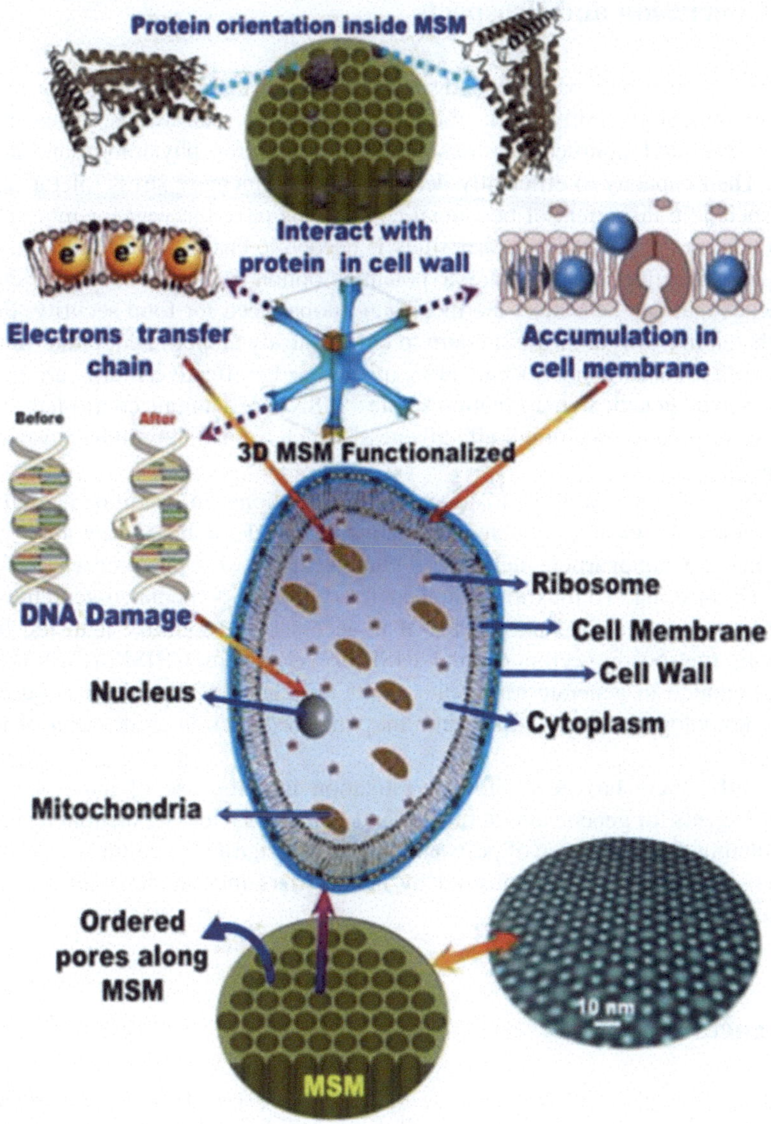

Fig. 2.4 Antifungal activity of MSN. (Adopted from Derbalah et al. 2018)

nanoparticles (MONs) as the porous support and calcium carbonate (CaC) as the capping agent. PRO-MON-CaC improves light stability and has a large loading capacity. This delivery mechanism can release PRO on its own in response to biological signals associated with Sclerotinia disease. PRO-MON-CaC protected potted rapeseed plants from Sclerotinia disease for longer than PRO emulsion in water (Gao et al. 2020).

2.7 Conclusion and Prospects

Mesoporous silica nanoparticles show more potential than other NPs due to their excellent biocompatibility and ability to penetrate cell walls. Mesoporous nanoparticles hold significant promise for improving crop physiology and performance. Their capacity to efficiently deliver nutrients, increase stress tolerance, and act as specific transporters of beneficial compounds pave the way for more robust and productive crops. While further study is needed to ensure long-term safety and environmental effects, MSNs are a potential option for sustainable agriculture approaches that can help meet the increasing global need for food security. Future research can improve the MSN system to ensure steady and efficient plant transformation. MSN-based nano systems are expected to be effective multi-cargo transporters in plant genetic transformation studies. MSNs are a unique controlled-release pesticide formulation with high effectiveness, which can assist promote sustainable agriculture.

Because of their ability to overcome restrictions in traditional delivery, nanoparticles are ideal candidates for delivering nuclease-based genome editing cargo, making nanoparticle delivery a vital technique for the progress of plant genetic engineering. To maximize the benefits of the MSN-mediated gene transformation system in many plant species, it is necessary to optimize it in the future employing transposon vectors or the CRISPR/Cas9 system. CRISPR/Cas9 may be efficient enough to generate transgenic plants utilizing MSNs since it is based on sgRNA homology, and the target gene may induce knock-in or deletion of many plant species.

Currently, there has been little investigation into the use of nanocarriers as transport agents for genetic alteration in plants, which provides a potential technique for regulating the occurrence of pests and diseases in significant cultures worldwide. The transmission of genetic material by nanocarriers must address unique plant-pathogen interactions.

References

Abdelrahman TM, Qin X, Li D, Senosy IA, Mmby M, Wan H, Li J, He S (2021) Pectinase-responsive carriers based on mesoporous silica nanoparticles for improving the translocation and fungicidal activity of prochloraz in rice plants. Chem Eng J 404:126440. https://doi.org/10.1016/j.cej.2020.126440

Adams CB, Erickson JE, Bunderson L (2020) A mesoporous silica nanoparticle technology applied in dilute nutrient solution accelerated establishment of zoysiagrass. Agrosyst Geosc Environ 3: e20006

Ambrogi V, Perioli L, Pagano C, Marmottini F, Ricci M, Sagnella A, Rossi C (2012) Use of SBA-15 for furosemide oral delivery enhancement. Eur J Pharm Sci 46(1–2):43–48. https://doi.org/10.1016/j.ejps.2012.02.004

Bagshaw SA, Prouzet E, Pinnavaia TJ (1995) Templating of mesoporous molecular sieves by nonionic polyethylene oxide surfactants. Science 269:1242–1244

Beck JS, Vartuli JC, Roth WJ, Leonowicz ME, Kresge CT, Schmitt KD (1992) A new family of mesoporous molecular sieves prepared with liquid crystal templates. J Am Chem Soc 114: 10834–10843

Biju S, Fuentes S, Gupta D (2023) Novel insights into the mechanism(s) of silicon-induced drought stress tolerance in lentil plants revealed by RNA sequencing analysis. BMC Plant Biol 23:498

Boora R, Rani N, Kumari S, Yashveer S, Kumari N, Grewal S (2023) Efficacious role of silica nanoparticles in improving growth and yield of wheat under drought stress through stress-gene upregulation. Plant Nano Biol 6:100051

Bravo A, Gill SS, Soberón M (2007) Mode of action of *Bacillus thuringiensis* Cry and Cyt toxins and their potential for insect control. Toxicon 49(4):423–435

Brecht M, Datnoff L, Nagata R, Kucharek T (2003) The role of silicon in suppressing tray leaf spot development in St. Augustine grass. Publication in University of Florida, pp 1–4

Capeletti LB, de Oliveira LF, Goncalves KD, de Oliveira JF, Saito A, Kobarg J, Santos JH, Cardoso MB (2014) Tailored silica–antibiotic nanoparticles: overcoming bacterial resistance with low cytotoxicity. Langmuir 30(25):7456–7464

Carpita NC, Gibeaut DM (1993) Structural models of primary cell walls in flowering plants: consistency of molecular structure with the physical properties of the walls during growth. Plant J 3:1–30

Chang FP, Kuang LY, Huang CA, Jane WN, Hung Y, Yue-ie CH, Mou CY (2013) A simple plant gene delivery system using mesoporous silica nanoparticles as carriers. J Mater Chem B 1(39): 5279–5287

Chen Z, Wang Q (2021) Graphene ameliorates saline-alkaline stress-induced damage and improves growth and tolerance in alfalfa (*Medicago sativa* L.). Plant Physiol Biochem 163:128–138

Chen J, Wang W, Xu Y, Zhang X (2011) Slow-release formulation of a new biological pesticide, pyoluteorin, with mesoporous silica. J Agric Food Chem 59(1):307–311. https://doi.org/10.1021/jf103640t

Chen S, Guo X, Zhang B, Nie D, Rao W, Zhang D, Lu J, Guan X, Chen Z, Pan X (2023) Mesoporous silica nanoparticles induce intracellular peroxidation damage of *Phytophthora infestans*: a new type of green fungicide for late blight control. Environ Sci Technol 57(9): 3980–3989. https://doi.org/10.1021/acs.est.2c07182

Derbalah A, Shenashen M, Hamza A, Mohamed A, El Safty S (2018) Antifungal activity of fabricated mesoporous silica nanoparticles against early blight of tomato. Egypt J Basic Appl Sci 5(2):145–150. https://doi.org/10.1016/j.ejbas.2018.05.002

Du D, Jiang Y, Feng J, Li L, Feng J (2020) Facile synthesis of silica aerogel composites via ambient-pressure drying without surface modification or solvent exchange. Vacuum 173: 109117

Feng J, Chen Z, Chen W, Sun L, Yang J, He K, Dong S, Yuan S (2022) Facile pathway to construct mesoporous silica nanoparticles loaded with pyraclostrobin: physicochemical properties, antifungal activity, and biosafety. Pest Manag Sci 78(6):2332–2341

Gao Y, Liang Y, Dong H, Niu J, Tang J, Yang J, Tang G, Zhou Z, Tang R, Shi X, Cao Y (2020) A bioresponsive system based on mesoporous organosilica nanoparticles for smart delivery of fungicide in response to pathogen presence. ACS Sustain Chem Eng 8(14):5716–5723

Gao X, Kundu A, Bueno V, Rahim AA, Ghoshal S (2021) Uptake and translocation of mesoporous SiO2-coated ZnO nanoparticles to *Solanum lycopersicum* following foliar application. Environ Sci Technol 55(20):13551–13560. https://doi.org/10.1021/acs.est.1c00447

Garver TLW, Thomas BJ, Robbins MP, Zeyen RJ (1998) Phenylalanine ammonia-lyase inhibition, auto fluorescence, and localized accumulation of silicon, calcium and manganese in oat epidermis attacked by the powdery mildew fungus *Almeria graminis* (DC) speer. Physiol Mol Plant Pathol 52:223–243

Gilbertson L (2003) Cre–lox recombination: creative tools for plant biotechnology. Trends Biotechnol 21(12):550–555. https://doi.org/10.1016/j.tibtech.2003.09.011

Gill SR, Fouts DE, Archer GL, Mongodin EF, DeBoy RT, Ravel J, Paulsen IT, Kolonay JF, Brinkac L, Beanan M, Dodson RJ (2005) Insights on evolution of virulence and resistance from the complete genome analysis of an early methicillin-resistant Staphylococcus aureus strain and a biofilm-producing methicillin-resistant Staphylococcus epidermidis strain. J Bacteriol 187(7): 2426–2438

Grun M, Lauer I, Unger KK (1997) The synthesis of micrometer-and submicrometer-size spheres of ordered mesoporous oxide MCM-41. Adv Mater 9:254–257

Haghighi M, Afifipour Z, Mozafarian M (2012) The effect of N-Si on tomato seed germination under salinity levels. J Biol Environ Sci 6:87–90

Hajiahmadi Z, Shirzadian-Khorramabad R, Kazemzad M, Sohani MM (2019) Enhancement of tomato resistance to *Tuta absoluta* using a new efficient mesoporous silica nanoparticle-mediated plant transient gene expression approach. Sci Hortic 243:367–375. https://doi.org/10.1016/j.scienta.2018.08.040

Hajiahmadi Z, Shirzadian-Khorramabad R, Kazemzad M, Sohani MM, Khajehali J (2020) A novel, simple, and stable mesoporous silica nanoparticle-based gene transformation approach in *Solanum lycopersicum*. 3 Biotech 10(8):370. https://doi.org/10.1007/s13205-020-02359-2

Hassan H, Alatawi A, Abdulmajeed A, Emam M, Khattab H (2021) Roles of Si and SiNPs in improving thermotolerance of wheat photosynthetic machinery via upregulation of PsbH, PsbB and PsbD genes encoding PSII core proteins. Horticulturae 7:16

Hezaveh TA, Pourakbar L, Rahmani F, Alipour H (2019) Interactive effects of salinity and ZnO nanoparticles on physiological and molecular parameters of rapeseed (*Brassica napus* L.). Commun Soil Sci Plant Anal 50:698–715

Inagaki S, Fukushima Y, Kuroda K (1993) Synthesis of highly ordered mesoporous materials from a layered polysilicate. J Chem Soc Chem Commun 8:680–682

Inagaki S, Koiwai A, Suzuki N, Fukushima Y, Kuroda K (1996) Syntheses of highly ordered mesoporous materials FSM-16 derived from kanemite. Bull Chem Soc Jpn 69:1449–1457

Iskandar F, Lenggoro IW, Kim TO, Nakao N, Shimada M, Okuyama K (2001) Fabrication and characterization of SiO_2 particles generated by spray method for standards aerosol. J Chem Eng Jpn 34:1285–1292

Jeelani PG, Mulay P, Venkat R, Ramalingam C (2020) Multifaceted application of silica nanoparticles. A review. SILICON 12:1337–1354

Kanto T, Miyoshi A, Ogawa T, Maekawa K, Aino M (2004) Suppressive effect of potassium silicate on powdery mildew of strawberry in hydroponics. J Gen Plant Pathol 70:207–211

Karande SD, Jadhav SA, Garud HB, Kalantre VA, Burungale SH, Patil PS (2021) Green and sustainable synthesis of silica nanoparticles. Nanotechnol Environ Eng 6:29

Khattab HI, Emam MA, Emam MM, Helal NM, Mohamed MR (2014) Effect of selenium and silicon on transcription factors NAC5 and DREB2A involved in drought-responsive gene expression in rice. Biol Plant 58:265–273

Kumar V, Guleria P, Kumar V, Yadav SK (2013) Gold nanoparticle exposure induces growth and yield enhancement in *Arabidopsis thaliana*. Sci Total Environ 461:462–468

Le VN, Rui Y, Gui X, Li X, Liu S, Han Y (2014) Uptake, transport, distribution and bio-effects of SiO2 nanoparticles in Bt-transgenic cotton. J Nanobiotechnol 12:1–15

Liberman A, Mendez N, Trogler WC, Kummel AC (2014) Synthesis and surface functionalization of silica nanoparticles for nanomedicine. Surf Sci Rep 69:132–158

Liou TH, Yang CC (2011) Synthesis and surface characteristics of nanosilica produced from alkali-extracted rice husk ash. Mater Sci Eng 176:521–529

Liu S, Han MY (2010) Silica-coated metal nanoparticles. Chem Asian J 5:36–45

Liu J, Wang FH, Wang LL, Xiao SY, Tong CY, Tang DY, Liu XM (2008) Preparation of fluorescence starch-nanoparticle and its application as plant transgenic vehicle. J Cent South Univ Technol 15:768–773

Lu X, Sun D, Zhang X, Hu H, Kong L, Rookes JE (2020) Stimulation of photosynthesis and enhancement of growth and yield in *Arabidopsis thaliana* treated with amine-functionalized mesoporous silica nanoparticles. Plant Physiol Biochem 156:566–577

Lv Z, Jiang R, Chen J, Chen W (2020) Nanoparticle-mediated gene transformation strategies for plant genetic engineering. Plant J 104(4):880–891. https://doi.org/10.1111/tpj.14973

Ma JF, Goto S, Tami K, Ichii M (2001) Role of root hairs and lateral roots in silicon uptake by rice. Plant Physiol 127:1773–1780

Ma D, Sun D, Wang C, Qin H, Ding H, Li Y, Guo T (2016) Silicon application alleviates drought stress in wheat through transcriptional regulation of multiple antioxidant defense pathways. J Plant Growth Regul 35:1–10

Malandrakis AA, Kavroulakis N, Chrysikopoulos CV (2019) Use of copper, silver and zinc nanoparticles against foliar and soil-borne plant pathogens. Sci Total Environ 670:292–299

Martin-Ortigosa S, Valenstein JS, Lin VS, Trewyn BG, Wang K (2012) Gold functionalized mesoporous silica nanoparticle mediated protein and DNA codelivery to plant cells via the biolistic method. Adv Funct Mater 22(17):3576–3582

Martin-Ortigosa S, Peterson DJ, Valenstein JS, Lin VS, Trewyn BG, Lyznik LA, Wang K (2014) Mesoporous silica nanoparticle-mediated intracellular Cre protein delivery for maize genome editing via loxP site excision. Plant Physiol 164(2):537–547. https://doi.org/10.1104/pp.113.233650

Mitra S, Mukherjee N, Das S, Sau A, Chakraborty S, Dwivedy S, Adak S, Arunava SG (2021) Regulatory role of nanoporous silica on dicot *Cicer arietinum* and monocot Sorghum bicolor. Res J Agric Sci 12:1012–1020

Mitra S, Chakraborty S, Mukherjee S, Sau A, Das S, Chakraborty B, Mitra S, Adak S, Goswami A, Hessel V (2022) A comparative study on the modulatory role of mesoporous silica nanoparticles MCM 41 and MCM 48 on growth and metabolism of dicot *Vigna radiata*. Plant Physiol Biochem 187:25–36. https://doi.org/10.1016/j.plaphy.2022.07.034

Mohamadzadeh M, Janmohammadi M, Abbasi A, Sabaghnia N, Ion V (2023) Physiochemical response of *Cicer arietinum* to zinc-containing mesoporous silica nanoparticles under water stress. Biotechnologia 104:263

Naqvi S, Maitra AN, Abdin MZ, Akmal MD, Arora I, Samim MD (2012) Calcium phosphate nanoparticle mediated genetic transformation in plants. J Mater Chem 22(8):3500–3507

Pan H, Huang W, Wu L, Hong Q, Hu Z, Wang M, Zhang F (2022) A pH dual-responsive multifunctional nanoparticle based on mesoporous silica with metal-polymethacrylic acid gatekeeper for improving plant protection and nutrition. Nano 12(4):687. https://doi.org/10.3390/nano12040687

Pasupathy K, Lin S, Hu Q, Luo H, Ke PC (2008) Direct plant gene delivery with a poly (amidoamine) dendrimer. Biotechnol J Healthc Nut Technol 3(8):1078–1082. https://doi.org/10.1002/biot.200800021

Paveley ND, Lockley KD, Sylvester-Bradley R, Thomas J (1997) Determination of fungicide spray decisions for wheat. Pestic Sci 49:379–388

Petica A, Gavriliu S, Lungu M, Buruntea N, Panzaru C (2008) Colloidal silver solutions with antimicrobial properties. Mater Sci Eng 152(1–3):22–27

Piela A, Zymanczyk-Duda E, Brzezińska-Rodak M, Duda M, Grzesiak J, Saeid A, Mironiuk M, Klimek-Ochab M (2020) Biogenic synthesis of silica nanoparticles from corn cobs husks. Dependence of the productivity on the method of raw material processing. Bioorg Chem 99: 103773

Rai M, Ingle A (2012) Role of nanotechnology in agriculture with special reference to management of insect pests. Appl Microbiol Biotechnol 94:287–293

Rai M, Ribeiro C, Mattoso L, Duran N (2015) Nanotechnologies in food and agriculture. Springer, Cham/Heidelberg/New York/Dordrecht/London, p 33

Rejman J, Oberle V, Zuhorn IS, Hoekstra D (2004) Size-dependent internalization of particles via the pathways of clathrin-and caveolae-mediated endocytosis. Biochem J 377:159–169

Rezaei ZS, Javed A, Ghani MJ, Soufian S, Barzegari FF, Bayandori MA, Mirjalili SH (2010) Comparative study of antimicrobial activities of TiO2 and CdO nanoparticles against the pathogenic strain of *Escherichia coli*. Iran J Pathol 5(2):83–89

Rizwan M, Ali S, Rehman MZ, Malik S, Adrees M, Qayyum MF, Alamri SA, Alyemeni MN, Ahmad P (2019) Effect of foliar applications of silicon and titanium dioxide nanoparticles on growth, oxidative stress, and cadmium accumulation by rice (*Oryza sativa*). Acta Physiol Plant 41:1–12

Rossi L, Fedenia LN, Sharifan H, Ma X, Lombardini L (2019) Effects of foliar application of zinc sulfate and zinc nanoparticles in coffee (*Coffea arabica* L.) plants. Plant Physiol Biochem 135: 160–166

Sadati SYR, Godehkahriz SJ, Ebadi A, Sedghi M (2022) Zinc oxide nanoparticles enhance drought tolerance in wheat via physio-biochemical changes and stress genes expression. Iran J Biotechnol 20:e3027

Saleh AM, Hassan YM, Selim S, AbdElgawad H (2019) NiO-nanoparticles induce reduced phytotoxic hazards in wheat (*Triticum aestivum* L.) grown under future climate CO_2. Chemosphere 220:1047–1057

Salem HF, Eid K, Sharaf M (2011) Formulation and evaluation of silver nanoparticles as antibacterial and antifungal agents with a minimal cytotoxic effect. Int J Drug Deliv 3:293–304

Setyawati MI, Leong DT (2017) Mesoporous silica nanoparticles as an antitumoral-angiogenesis strategy. ACS Appl Mater Interfaces 9:6690–6703

Shang Y, Hasan MK, Ahammed GJ, Li M, Yin H, Zhou J (2019) Applications of nanotechnology in plant growth and crop protection: a review. Molecules 24:2558

Sharf-Eldin AA, Alwutayd KM, El-Yazied AA, El-Beltagi HS, Alharbi BM, Eisa MA, Ibrahim MF (2023) Response of maize seedlings to silicon dioxide nanoparticles (SiO2NPs) under drought stress. Plan Theory 12:2592

Siddiqui MH, Al-Whaibi MH (2014) Role of nano-SiO_2 in germination of tomato (*Lycopersicum esculentum* seeds Mill.). Saudi J Biol Sci 21:13–17

Slomberg DL, Schoenfisch MH (2012) Silica nanoparticle phytotoxicity to *Arabidopsis thaliana*. Environ Sci Technol 46:10247–10254

Snehal S, Lohani P (2018) Silica nanoparticles: its green synthesis and importance in agriculture. J Pharmacogn Phytochem 7:3383–3393

Stober W, Fink A, Bohn DE (1968) Controlled growth of monodisperse silica spheres in the micro size range. J Colloid Interface Sci 26:62–69

Sun D, Hussain HI, Yi Z, Siegele R, Cresswell T, Kong L, Cahill DM (2014) Uptake and cellular distribution, in four plant species of fluorescently labeled mesoporous silica nanoparticles. Plant Cell Rep 33:1389–1402

Sun D, Hussain HI, Yi Z, Rookes JE, Kong L, Cahill DM (2016) Mesoporous silica nanoparticles enhance seedling growth and photosynthesis in wheat and lupin. Chemosphere 152:81–91. https://doi.org/10.1016/j.chemosphere.2016.02.096

Sun D, Hussain HI, Yi Z, Rookes JE, Kong L, Cahill DM (2018) Delivery of abscisic acid to plants using glutathione responsive mesoporous silica nanoparticles. J Nanosci Nanotechnol 18(3): 1615–1625. https://doi.org/10.1166/jnn.2018.14262

Sun L, Song F, Guo J, Zhu X, Liu S, Liu F (2020) Nano-ZnO-induced drought tolerance is associated with melatonin synthesis and metabolism in maize. Int J Mol Sci 21:782

Suriyaprabha R, Karunakaran G, Yuvakkumar R, Rajendran V, Kannan N (2012) Silica nanoparticles for increased silica availability in maize (*Zea mays* L.) seeds under hydroponic conditions. Curr Nanosci 8:902–908

Torney F, Trewyn BG, Lin VS, Wang K (2007) Mesoporous silica nanoparticles deliver DNA and chemicals into plants. Nat Nanotechnol 2(5):295–300

Tran TLC, Guirguis A, Jeyachandran T, Wang Y, Cahill DM (2023) Mesoporous silica nanoparticle-induced drought tolerance in *Arabidopsis thaliana* grown under *in vitro* conditions. Funct Plant Biol 50:889–900

Vanti GL, Nargund VB, Vanarchi R, Kurjogi M, Mulla SI, Tubaki S, Patil RR (2019) Synthesis of *Gossypium hirsutum*-derived silver nanoparticles and their antibacterial efficacy against plant pathogens. Appl Organomet Chem 33(1):e4630

Wang Q, Chen J, Zhang H, Lu M, Qiu D, Wen Y, Kong Q (2011) Synthesis of water soluble quantum dots for monitoring carrier-DNA nanoparticles in plant cells. J Nanosci Nanotechnol 11(3):2208–2214. https://doi.org/10.1166/jnn.2011.3560

Wang A, Zheng Y, Peng F (2014) Thickness-controllable silica coating of CdTe QDs by reverse microemulsion method for the application in the growth of rice. J Spectrosc 2014:1–5
Wanyika H, Gatebe E, Kioni P, Tang Z, Gao Y (2012) Mesoporous silica nanoparticles carrier for urea: potential applications in agrochemical delivery systems. J Nanosci Nanotechnol 12(3): 2221–2228. https://doi.org/10.1166/jnn.2012.5801
Xu Y, Xu C, Huang Q, Cao L, Teng F, Zhao P, Jia M (2021) Size effect of mesoporous silica nanoparticles on pesticide loading, release, and delivery in cucumber plants. Appl Sci 11(2): 575. https://doi.org/10.3390/app11020575
Yan G, Huang Q, Zhao S, Xu Y, He Y, Nikolic M, Nikolic N, Liang Y, Zhu Z (2024) Silicon nanoparticles in sustainable agriculture: synthesis, absorption, and plant stress alleviation. Front Plant Sci 15:1393458
Yu-Qin FU, Lu-Hua LI, Pi-Wu WA, Jing QU, Yong-Ping FU, Hui W, Jing-Ran SU, Chang-li L (2012) Delivering DNA into plant cell by gene carriers of ZnS nanoparticles. Chem Res Chinese Univ 28(4):672–676
Zhao D, Feng J, Huo Q, Melosh N, Fredrickson GH, Chmelka BF (1998) Triblock copolymer syntheses of mesoporous silica with periodic 50 to 300 angstrom pores. Science 279:548–552
Zhao T, Nguyen NT, Xie Y, Sun X, Li Q, Li X (2017) Inorganic nanocrystals functionalized mesoporous silica nanoparticles: fabrication and enhanced bio-applications. Front Chem 5:118. https://doi.org/10.3389/fchem.2017.00118
Zhao P, Yuan W, Xu C, Li F, Cao L, Huang Q (2018a) Enhancement of spirotetramat transfer in cucumber plant using mesoporous silica nanoparticles as carriers. J Agric Food Chem 66(44): 11592–11600
Zhao T, Chen L, Li Q, Li X (2018b) Near-infrared light triggered drug release from mesoporous silica nanoparticles. J Mater Chem B 6(44):7112–7121. https://doi.org/10.1039/C8TB01548A
Zhu F, Liu X, Cao L, Cao C, Li F, Chen C, Xu C, Huang Q, Du F (2018) Uptake and distribution of fenoxanil-loaded mesoporous silica nanoparticles in rice plants. Int J Mol Sci 19(10):2854
Zolghadrnasab M, Mousavi A, Farmany A, Arpanaei A (2021) Ultrasound-mediated gene delivery into suspended plant cells using polyethyleneimine-coated mesoporous silica nanoparticles. Ultrason Sonochem 73:105507. https://doi.org/10.1016/j.ultsonch.2021.105507

Chapter 3
Applications of Green Synthesized Nanoparticles in Agriculture

Khushboo Kourani and Amee Krishnakumar

Abstract With the advent in nanotechnology, the synthesis of green nanoparticles and its applications has been a crucial step of inception for sustainable agriculture. The synthesis of nanoparticles from biological resources paves a shift in burden from commercial and synthetic means of agriculture to economical and environment friendly methods. The application of nanomaterials from plant, micro-organisms and bio-waste based extracts has been shown to result positive with high biodegradability, enhanced functional role and less toxicity. The application of green metallic nanoparticles bio-fabricated with different biological extracts as nano-fertilizers, nano-pesticides and nano-biosensors are now widely tested in various aspects of horticulture and cultivation of crops in farming to address the problems of biotic and abiotic stresses. The lacuna to treat the pollutants by photo catalysis, disease control and efficient nutrient feed additive has been heavily tested by green metallic nanoparticles to enhance the habitat for better aquafarming and also in livestock rearing. The new age usage of bio synthesized nanoparticles in meat industry for enhanced shelf life and spoilage detection with intelligent nano-packaging opens the paradigm of areas for future. Thus, this chapter sheds light on the applications of green synthesized nanoparticles in different sectors of agriculture and their future prospects.

Keywords Aquaculture · Bio-synthesized nanoparticles · Disease control · Meat packaging · Nano sensors · Nano-fertilizer · Sustainable agriculture

K. Kourani · A. Krishnakumar (✉)
Institute of Science, Nirma University, Ahmedabad, Gujarat, India
e-mail: 20ftphds58@nirmauni.ac.in; ameenair@nirmauni.ac.in

© The Author(s), under exclusive license to Springer Nature Switzerland AG 2025

J. M. Al-Khayri et al. (eds.), *Green Nanotechnology Applications for Ecosystem Sustainability*, Nanotechnology in Plant Sciences 4,
https://doi.org/10.1007/978-3-031-90766-1_3

3.1 Introduction

In an era of modern technologies, agriculture has been a novel target for systemization and alterations to meet the increasing demand for its products and by-products. Being an important part of the ecosystem, agriculture provides habitat, resources, and sustainability to human life and their economic activities. Agriculture is majorly categorized into the domains of farming, livestock production, aquaculture, fisheries, and forestry (Muhie 2022).

With the advent of Green Revolution, agribusiness on the economic front has significantly increased by providing modernized equipment for easy and fast agriculture techniques; however, these technological developments have impacted the ecological and environmental niches leading to widespread environmental pollution (Sundararajan et al. 2024). For instance, the use of pesticides, chemical fertilizers, and herbicides has led to the degradation of soil quality, groundwater contamination, and deleterious effects on the immediate biodiversity causing disruptions in the food chain. Moreover, to meet the demands of the population and provide raw materials and resources conducive to their daily livelihood, inadvertent examples like heavy wood deforestation contribute to the loss of habitat. It is also followed by the other events such as high carbon emissions cause climate change that certainly shift the paradigm of a balanced ecosystem towards a climate change. In addition, economic activities that utilize fossil fuels that are necessity to fulfil the demands of the population have caused a up surge in air pollution leading to the massive era of global warming. Therefore, it is essential to comprehensively grasp the complexities of societal demands and their ecological ramifications, while formulating innovative and effective methodologies to enhance the global circular economy, thereby potentially benefiting the agricultural sector (Muhie 2022).

3.1.1 Nanotechnology: Inception of Nanotechnology in Agriculture

Nanotechnology, characterized as a state-of-the-art technological paradigm, demonstrates significant potential within the realm of agricultural science for transformative impact (Duhan et al. 2017; Ricardo et al. 2018). Leveraging nano-scale dimensions and offering a nano-level platform, nanoparticles exhibit considerable promise in diverse agricultural domains (Bahrulolum et al. 2021). These include augmenting crop yields, improving soil characteristics, applying therapeutic interventions in aquaculture, and facilitating bioremediation processes. For example, the technique of slow and controlled release of nanoparticles deliver efficient nutrients at the nanoscale for plants to absorb and decrease the leaching off in nearby environmental targets. Nanomaterials are also employed for soil remediation due to their large surface area where, they absorb pollutants and heavy metals leading to enhanced soil quality and fertility (Rajput et al. 2022). Further, nano-sensors and nano-devices have a multifaceted role in enhancing agriculture, where it can be either used in the detection of quality of crops, contaminants, pathogens, and spoilage indicators in food packages of meat stocks or to monitor the plant health and moisture level through nano-based seed coating which leads to enhancement of seed germination, root growth, and cultivation of crops (Lamri et al. 2021). The protection of plants under withered weather conditions with the help of nano barriers like nano-films coatings during drought, also play a significant role in increasing the shelf life of packaged food products by preventing moisture accumulation and various gas exchange. However, nanomaterials produced with physical and chemical methods that are commonly in use exhibits toxicity that restricts their applicability in agriculture, food, and health. Chemical synthesis processes often involve the use of hazardous chemicals, which may lead to the generation of harmful by-products and residues. These by-products can contaminate ecosystems, posing risks to both environmental and human health. Additionally, the production of nanoparticles through these methods may require high energy input and produce large amounts of waste, contributing to carbon emissions and resource depletion (Singh 2022). Furthermore, nanoparticles synthesized using chemical methods may exhibit lower purity and uniformity compared to those produced through more controlled techniques, impacting their effectiveness and safety in practical applications (Kurian and Elumalai 2021). Conversely, nanoparticles synthesized via biogenic means display favorable characteristics such as uniform distribution, size, and stability (Sana et al. 2020). Moreover, biological approaches enable nanoparticle synthesis under physiological conditions of pH, temperature, and pressure, offering cost-effectiveness and environmental friendliness. Hence, it is crucial to transit into the era of nanoparticle utilization employing green synthesis methodologies, aiming to enhance efficacy while mitigating challenges linked to environmental adversities (Fig. 3.1).

3.1.2 A New Era of Green Synthesized Nanoparticles

The introduction of green synthesized nanoparticles proclaims a new age in nanotechnology, defined by environmentally friendly nanoparticle production methods. Leveraging natural resources and eco-friendly synthesis pathways, this paradigm shift reflects a concentrated effort towards minimizing environmental effect and fostering sustainability in nanomaterial synthesis. Green synthesis techniques, such as plant-mediated synthesis, microbial synthesis, and bio-inspired approaches, have various benefits, including lower energy usage, elimination of harmful chemicals, and improved biocompatibility of the resulting nanoparticles (Rasmiya Begum and Jayawardana 2023). Furthermore, these green synthesis methods are consistent with green chemistry principles, emphasizing the need of reducing waste generation while increasing resource efficiency. The general approach of synthesizing green nanoparticles involves various combinations of metal ions, biological plant extracts and microbial extracts under specific conditions and reducing solutions for metal ions to form nanoparticles (Ying et al. 2022). As a result, the introduction of green

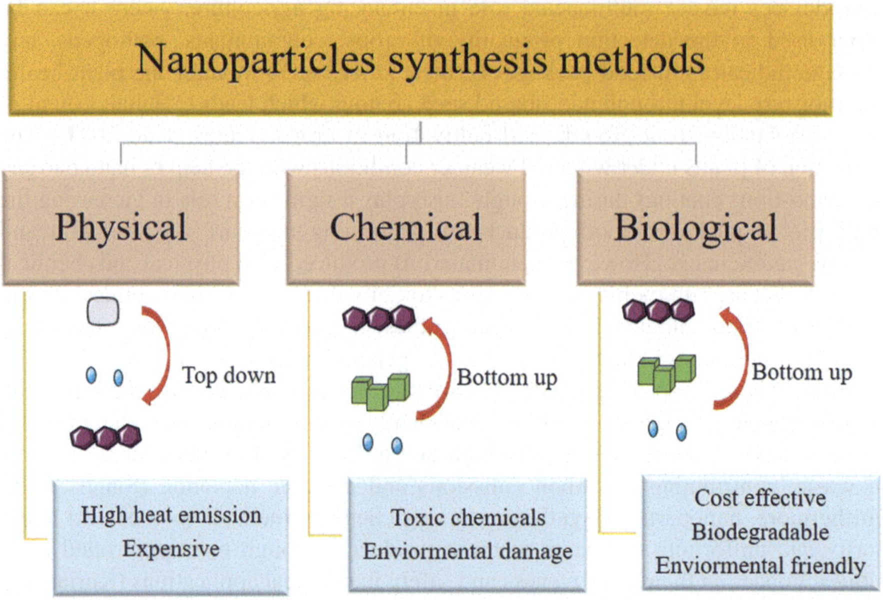

Fig. 3.1 Different methods like physical, chemical and biological used for synthesizing nanoparticles with top down or bottom up approaches. (Figure constructed by Khushboo Kourani)

synthesized nanoparticles not only holds potential for developing diverse scientific and technical applications, but also demonstrates a dedication to responsible and sustainable nanomaterial manufacturing (Javed et al. 2023). In this chapter, we have discussed the application of green synthesized nanoparticles in varied categories of agriculture. The use of bio resources such as plant, agricultural wastes and microorganism as a crucial coat to metallic nanoparticles in synthesis and their resultant application in overcoming agricultural challenges.

3.2 Green Synthesized Nanoparticles

Nanoparticles are traditionally synthesized with approaches like chemical, physical, and biological methods. The two very common methods for metallic nanoparticle synthesis are physical (uses a top-down approach) and chemical (bottom-up approach) methods where the major key disadvantages are expensive and environmentally hazardous for long-term use in agriculture (Bahrulolum et al. 2021; Salem et al. 2021). The utilization of raw materials, including toxic chemicals, and high-energy heat-generating equipment renders these methods susceptible to ecological damage. Thus, the need for a reliable environmentally friendly approach in nanotechnology has shifted to green synthesized nanoparticles as a promising tool in varied agricultural strategies. The green synthesized nanoparticles use the unique

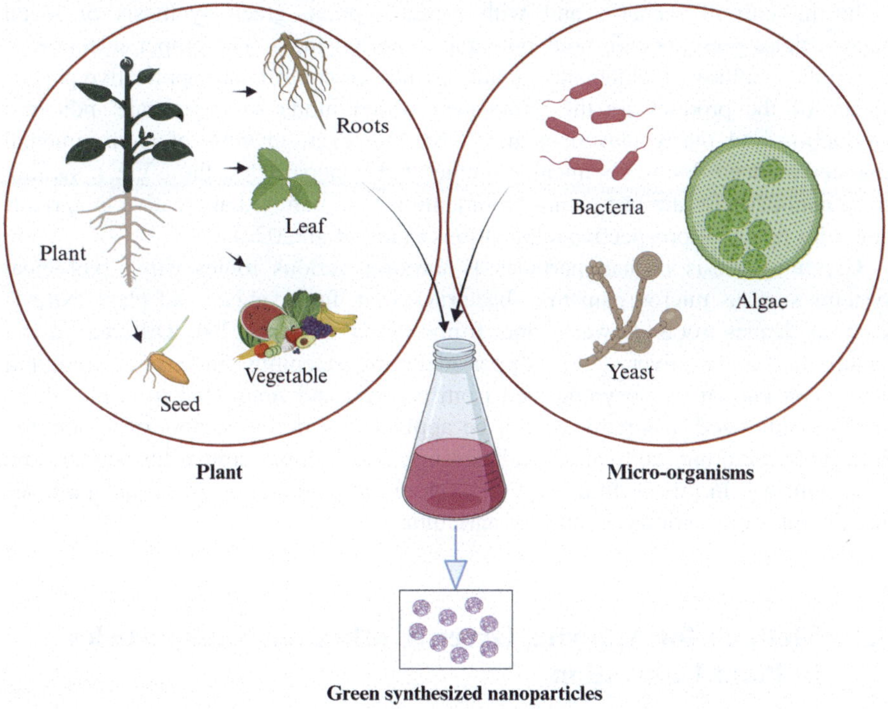

Fig. 3.2 Overview of green synthesis nanoparticles from the plant, agricultural waste, and micro-organisms. (Images created using Biorender.com)

bottom-up approach for the synthesis of metallic nanoparticles. The bottom-up approach comprises synthesizing the nanoparticles by self-assembly of single atoms and molecules scaling into desired nanostructures, shapes, and concentrations (Vijayaram et al. 2024). The 'green' symbolizes the use of environment friendly route for the synthesis of nanoparticles. The green synthesized method is also called the 'biogenic method' is more beneficial than traditional chemical and physical methods as it is cost-effective, use biodegradable and environmentally friendly raw materials, and have significant implications for balancing the ecosystem of agriculture (Bahrulolum et al. 2021). Metallic nanoparticles synthesized from plant extracts exhibit stability and homogeneous dispersion, which is achieved through exact regulation of different influencing elements such as pH, temperature, mixing ratios and retention time. Further, green synthesis, uses a variety of medicinal plants leaves such as neem, Tulsi, curry, guava and mango where it produces metallic nanoparticles with significant therapeutic properties (Huston et al. 2021). Figure 3.2 depicts the overview synthesis of green nanoparticles from bio resource such as plant, bio waste and micro-organism. These include significant anticancer, antioxidant, antibacterial, insecticidal, wound healing, hepatoprotective, immunomodulatory properties.

In the current scenario and with recent reports, green synthesis of metal nanoparticles presents enormous potential in agribusiness. The proper selection of materials, synthesis, maintenance quality assurance control, and applicative performance of the product are the parameters which limits its large-scale industrial production and purity (Javed et al. 2023). However, in view of environmental changes induced due to chemically synthesized nanoparticles, these green synthesized nanoparticles are fast gaining momentum as a preferred alternative in agriculture with potential prospective applications (Ying et al. 2022).

Green synthesis of nanoparticles is through various routes using biological systems such as microorganisms—bacteria, yeast, fungi, algae, and plant extracts such as leaves roots flowers, and fruits (Sana et al. 2020). Different metal nanoparticles like silver, gold, copper platinum, selenium silica, and zirconium have been known to biosynthesized from bacteria and fungi (Salem et al. 2022). Green synthesized nanoparticles can be applied in different sectors of agriculture such as horticulture, cultivation and farming, aquaculture, animal husbandry, and meat industry. In this section, we will explore the potential of green nanoparticles and its impact in various avenues of agriculture.

3.3 Methods for Applying Green Synthesized Nanoparticles in Plant Cultivation

Nanostructures in nanoparticles contribute a beneficial role in plant cultivation. Their easy application enhances nutrient uptake for plant growth, development and response to biotic and abiotic stress (Thabet and Alqudah 2024). Seed priming and foliar spray are two commonly used methods for applying nanoparticles in plant cultivation.

3.3.1 Seed Priming

Seed priming involves pretreatment of seeds by soaking and coating them before sowing to enhance its germination (Fig. 3.3). Priming technique increases the crop activity by preventing the crop from biotic and abiotic stress. Different priming methods are used for the betterment of crop cultivation and plant health and are applied to overcome the challenges of seed dormancy such as halo priming, hormonal priming, osmotic priming, and biogenic nano priming (Nile et al. 2022). Nano-biopriming involves the synthesis of nanoparticles within the nano-scale range utilizing biogenic raw materials. This approach represents a notable advancement with considerable potential benefits for crop cultivation. Nano-biopriming has demonstrated efficacy in modulating various aspects of seed physiology, including enhancing germination rates, overcoming seed dormancy, imparting resistance to

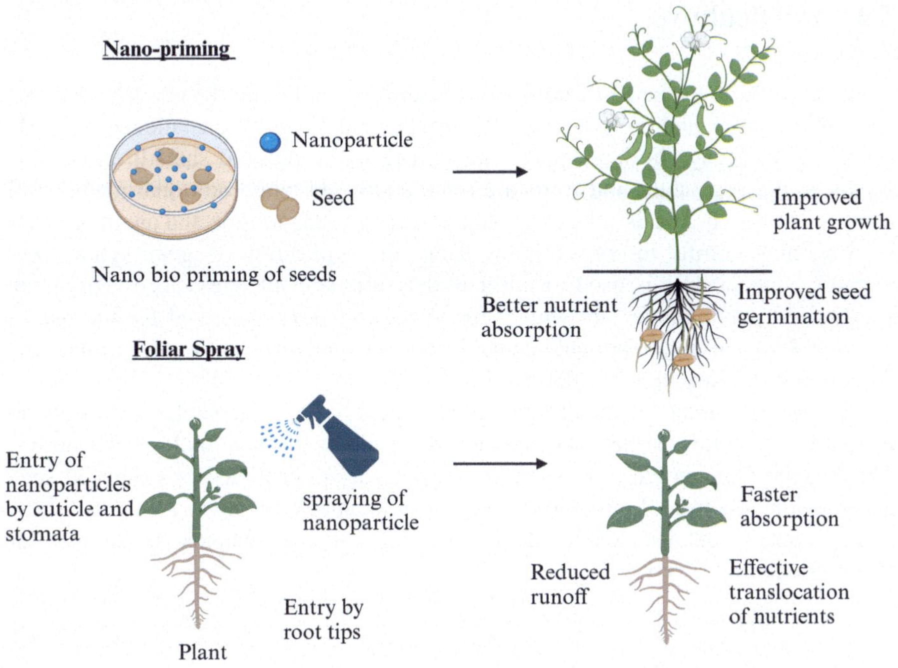

Fig. 3.3 Application of nanoparticles by nano-priming and foliar spray. (Images created using Biorender.com)

extreme environmental stresses, promoting plant growth and development, and facilitating controlled release of nutrient content within plants. Therefore, the use of nano-bio priming can be a potential tool to cultivate barren lands for forestry reclamation purposes (Nile et al. 2022).

3.3.2 Foliar Spray

Foliar sprays of nanoparticles are considered in enhancing the plant protective mechanism in comparison to traditional soil and chemical application of fertilizer (Hong et al. 2021). This approach provides a direct path for delivering nutrients, insecticides, or other compounds to the plant's foliage, where they are quickly absorbed and used by the plant (Fig. 3.3). Foliar spray techniques are widely used for enhancing the antimicrobial activity of plants against pathogens and pesticides (Nile et al. 2022). It innovates the technique for the application of nanoparticles in lesser amounts with greater impact, making them the most economical and environmentally friendly for sustainable agriculture (Ding et al. 2023).

3.4 Horticulture

Horticulture is the science of cultivating plants for edible products such as fruits, vegetables, herbs, and growing aesthetic ornamental plants. The challenges of early ripening and decay, the inadequate transportation and reduced shelf life of edible products like vegetables and fruits are often addressed with traditional methods of using synthetic chemicals. The usage of these chemicals has time and again prove to be toxic and harmful in consumption. Thus, the application of green synthesized metallic nanoparticles is used to mitigate the various challenges faced. The green synthesized nanoparticles are widely manufactured and engineered for the use of antimicrobial activity, treatment against leaf worms, increase in bio-absorption, and reduction in salinity in horticulture.

Plant-based green nanoparticles are a new and sustainable approach to nanomaterial synthesis that takes advantage of plants intrinsic biological features. These green nanoparticles are used in a variety of applications for enhanced seed germination, control of larvicidal and bacterial attack on plants, salinity stress management to enhance plant growth in saline soil that remains detrimental for plant growth and survival.

3.4.1 Seed Germination

Seed germination is the most crucial initial step for future plant development and growth. The conventional mechanism for overcoming seed dormancy entails a prolonged duration and introduces uncertainties regarding growth outcomes in agricultural settings. Thus, enhancement of seed germination and seedling growth further increases the plant growth and development. In that context, the application of biosynthesized nanoparticles was tested for enhanced seed germination and seedling growth. The use of *Cassia auriculata* leaf extract as a biomaterial in the fabrication of gold nanoparticles had been used and tested in the route of seed germination and seedling growth on the seeds of pearl millet (*Pennisetum glaucum*). The pearl millet seeds were dipped and incubated in the biosynthesized gold nanoparticle solution at different concentrations against appropriate control. The applied biosynthesized gold nanoparticles resulted in rapid and enhanced seed germination of pearl millets by 86% along with an increase in seedling length at a higher concentration of 50 μg/mL demonstrated in (Fig. 3.4). Thus, to fill in the gap in crop production of biofuel plants like pearl millet, biosynthesized nanoparticles can be a safe and efficient alternative for large-scale agricultural plant production (Parveen et al. 2016).

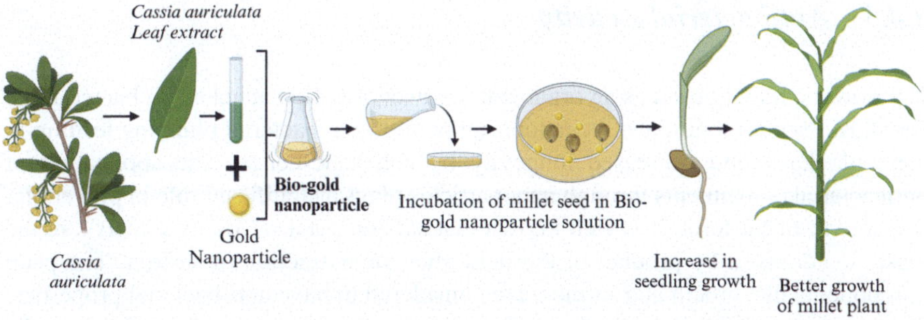

Fig. 3.4 Schematic view of nano-priming of Pearl millet seeds in green synthesized AuNPs solution for enhanced seed germination. (Image created using Biorender.com)

3.4.2 Alleviate Salinity Stress

Saline stress in the plant micro-environment has been a challenging and adverse parameter affecting plant growth, seed germination, flowering, and fruit. The excessive amount of salt in the micro-environment of plants limits the uptake of water and the absorption of essential nutrients thereby resulting in water shortage and affecting the plant development (Zhao et al. 2021). The primary stress of high salt leads to secondary stress on plant health by an increase in free radical species, osmotic stress, and alterations in the content of secondary metabolites such as flavonoids. The high content of flavonoids and anthocyanin secondary metabolites in leaf extract enables the good growth and development of plants by majorly scavenging the free radicals produced due to salinity stress. Thus, to reduce the salinity stress an eco-friendly way of synthesizing green nanoparticles using aqueous leaf extract of *Hibiscus sabdariffa* (Roselle) plant.

The *Hibiscus sabdariffa* plant has been widely used in herbal medicinal purposes with hypotensive effects. The synthesis of AgNPs with the development of aqueous leaf extract of hibiscus was used to investigate and regulate the salinity stress on the plant. Seed dipping method was used to synthesize the green synthesized Hibiscus sabdariffa silver nanoparticles (HS-AgNPs) at different concentrations, and was provided to assess the seed germination in comparison to seeds dipped in saline solution. The seedlings of the germinated plants were also subjected to HS-AgNPs with nutrients as a solution in seedling pots. The application of HS-AgNPs for seed germination and growth parameters resulted in a better concentration of 45 mg/L which suggested optimally mitigating the saline stress in the vicinity. At 60 mg/L significantly increased the plant height along with better leaf and root freshness at 120 mg/L of application with increased relative water content. The intervention also elevated the antioxidant property with increased catalase and superoxide dismutase activity as well as the increase in the content of flavonoid and anthocyanin and their gene expressions. Thus, using such plant leaf extract (HS-AgNps) can be a potential tool used as a plant defense mechanism under varied abiotic stress (Hosseini et al. 2022).

3.4.3 Antibacterial Activity

Cut flowers, hugely used as an ornamental flower are often attacked by bacteria like Bacillus, Pseudomonas, and Acinetobacter resulting in bacterial plunging leading to stem blockage and decreased longevity by cell senescence. The application of antibacterial constituents through nanoparticles plays a significant role in preventing the growth of bacteria. It is well known that saffron petals (*Crocus sativus*) are the waste materials, a by-product of the field after the extraction of saffron. The petal components like crocin and safranal are considered to have anti-bacterial properties. This petal extract has been used as a biogenic component to fabricate and reduce the metallic silver nanoparticles for its application as bactericidal treatment in cut flowers. In an agar disk diffusion method, these bio-synthesized silver nanoparticles (AgNPs) at different concentrations were tested against *Pseudomonas aeruginosa*, *Pseudomonas fluorescens* (Gram Negatives), and *Bacillus subtilus*, *Bacillus cereus*, *Acinetobacter sp.* (Gram Positives), resulted in effective antibacterial activity of pure saffron petal extract containing silver nanoparticles. The different concentrations of bio-synthesized AgNPs were also applied, increased the vase life of rose cut flowers which resulted in the lower ratio dilution. The longevity of cut flowers increased almost two times against the control (Solgi 2014). Tree peony (*Paeonia suffruticosa*) is denoted as the 'king of flowers' or 'Luoyang Hong' for its large purple colorful flowers and rich fragrance. However, it's cut flowers have short vase life due to endophytic bacterial infection in the stem end and blockage in their xylem vessel (Liu et al. 2021). Interestingly, *Eucommia ulmoides* plant has vast medicinal properties, and its leaf extract is rich in contents like chlorogenic acid, protocatechuic acid, iridoids, and flavonoids. In another study, bio-synthesized AgNPs from *E. ulmoides* leaf extract was treated against endophytic bacterial samples isolated from 'Luoyang hong' tree peony. The application of green synthesized AgNPs showed significant bacterial inhibition in *in vitro* analysis (Naing and Kim 2020). The flower diameter of the pre-treated flower with bio-synthesized AgNPs as larger during the vase period along with better water balance in comparison to than control. Further, the pre-treatment of AgNps to the tree peony significantly reduced the biofilm formation and increased the shelf life (Fig. 3.5). It could be inferred that the application of green synthesized nanoparticles has proven to be best for the cultivation of cut flowers of tree peony for better post-harvest quality and ornamental value (Ma et al. 2023).

3.4.4 Larvicidal Activity

Secondary metabolites of the plants such as polyphenols are considered as its first line defense plants against bacterial and fungal infections. This polyphenol plays a potential role as an insecticide for plant protection. Hence, diverse bio-waste materials such as fruit peels and extracts, abundant in phenolic components, present

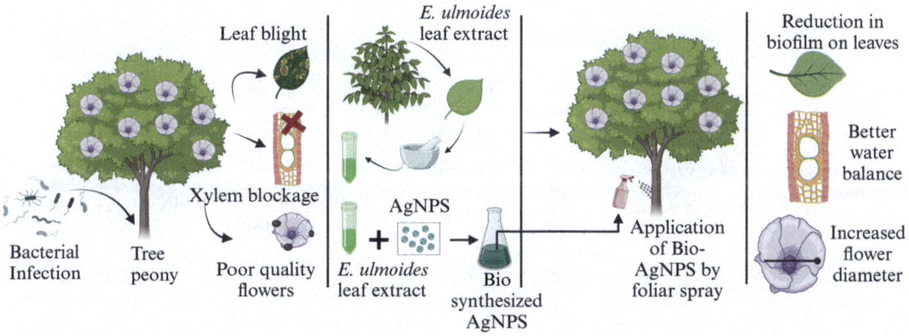

Fig. 3.5 Schematic view of application of bio-synthesized AgNPs on tree peony. (Images created using Biorender.com)

viable sources for formulating biogenic nanoparticles. Utilizing waste peels from fruits like pomegranate and watermelon, which are rich in phenolic compounds, facilitates the green synthesis of silver nanoparticles with potential applications as insecticidal agents. *Spodoptera littoralis* also known as the African cotton leaf-worm is a pest that causes leaf skeletonization, blockage of shoot and losing crops. The biosynthesized pomegranate silver nanoparticles (PPAgNPs) and watermelon silver nanoparticles (WPagNPs) resulted in increased antioxidant and antibacterial activity against bacterial and candida growth in the inhibition zone of the disc diffusion method. Moreover, the application of PPAgNPs resulted in the most effective treatment against first instar larvicidal (the stage of larvae after hatching from egg is first instar). The synergistic effect of PPAgNPs + WPAgNPs was higher with increased larvicidal activity and mortality against 1st instar of Egyptian leaf worm *Spodoptera littoralis*. Therefore, the production of green metallic nanoparticles and their utilization for combating pests and pathogens holds significant promise in the treatment and protection of horticulture plants (Saad et al. 2021).

Microbial synthesis of green nanoparticles offers scalable advantages over plant-based synthesis of green nanoparticles. These microorganisms, including bacteria, fungi, and algae, possess inherent mechanisms to reduce metal ions and facilitate the formation of nanoparticles. The microbe-based synthesis of nanoparticles provides a wider window for its differentiated benefits as high scalability, controlled and rapid synthesis, and versatility of products.

The red flour beetle (*Tribolium castaneum*) is widely found pest in stored products and grains. The pest damages both the quality and quantity of grain causing significant lack and nutritive content. The damage in crop maintenance and quality results in a reduction of the germination ratio and commercial value of grains. To overcome this load of damage at the cultivation level, the biosynthesis of nanomaterials from micro-organisms was explored for its effectivity in combination with metal materials. The biosynthesized copper nanoparticles (CuNPs) were used as a representative of nano pesticides, bio-transformed with the copper-resistant *Pseudomonas fluorescens MAL2* (El-saadony et al. 2020). It was observed that the combination of CuNPs and *Pseudomonas fluorescens* MAL2 at higher

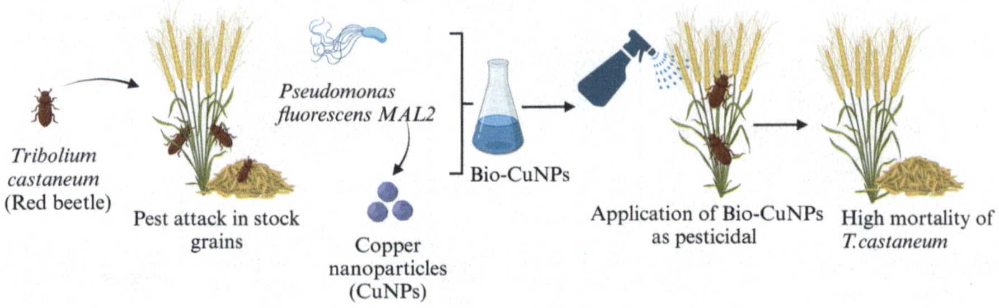

Fig. 3.6 Schematic view of pesticidal activity of microorganism based bio-synthesized CuNPs on red flour beetle. (Images created using Biorender.com)

concentrations resulted in higher mortality of *T. casteneum* in comparison to chemically synthesized CuNPs (Fig. 3.6) (El-saadony et al. 2020).

Similarly, tomato (*Solanum lycopersicum* L.), an important vegetable crop often attacked by root-nod nematodes, causing serious damage to the crop. The nematode *Meloidognye javanica* infect the tomato crop causing deleterious damage to the crop field. *Cladophora glomerata*, a freshwater green algae is an effective nematicide against root-nod nematodes. Ghareeband group evaluated the nematicide activity of AgNPs from *C. glomerata* extract as green synthesized silver nanoparticles (GSNPs). The GSNPs were tested at different concentrations against the *M. javanica* to investigate the plant response. The treatment assessed through bioassay resulted in decreased egg hatchability and increased juvenile mortality of *M. javanica* in comparison to chemically synthesized silver nanoparticles. To understand the underlying molecular mechanism of defense in plants, the defense enzyme expression of phenylalanine ammonia Lyase and polyphenol oxidase was investigated, wherein the expression levels in plants treated with *C. glomerata* extract GSNPs were higher suggesting an increase in cell wall lignification as a physical barrier to pests with no notable change of expression in the peroxidase gene levels. Thus, tomato plants infected with *M. javanica* showed significant plant growth and a better defense system against root-nod nematodes when treated with *C. glomerata* extract GSNPs (Ghareeb et al. 2020).

3.5 Cultivation and Farming

Cultivation has been the backbone of agriculture for food security and economic development. The application of nanoparticles results in the enhancement of crop traits, soil remediation, precision farming, and pest disease management. The combinatorial strategy of application of green synthesized nanoparticles contributes to increased agricultural productivity, and sustainability with environment and eco-friendly attributes. In this section, we have provided a brief overview of the utilization and potential significance of green synthesized nanoparticles in the context of cultivation and agriculture.

3.5.1 Pesticidal Activity

Bacterial leaf blight is one of the most commonly occurring diseases in rice plants affecting the growth and cultivation of the crop. The blight disease affects the output demand of the crop thereby disturbing the agribusiness chain at the large scale. Many different pesticides and pest treatment methods have been applied for the reduction of leaf blight disease. Nevertheless, synthetic pesticide treatments exacerbate the issue by inducing antimicrobial resistance, releasing toxic chemicals, and facilitating the leaching of adulterants into crops, thereby resulting in contamination and degradation of food product quality. In this regard, *Xanthomonas oryzae pv. oryzae* (Xoo), is a pathogen responsible for leaf blight disease in rice crop plants globally. To overcome the challenge of leaf blight disease, tomato extracts were used to synthesize chitosan and zinc nanoparticles to control the bacterial disease. The extract of red tomatoes higher in polyphenol content were used for the synthesis of chitosan nanoparticles and green tomato extracts rich in alkaloids and ascorbic acid were used in the synthesis of zinc oxide (ZnO) nanoparticles. The biosynthesized CSNPs and ZnONPs resulted in stronger inhibition of Xoo in diffusion assay. These nanoparticles also contributed to the reduction of ROS species, biofilm formation, and motility rate of the bacterium. Hence, the utilization of these two bio-synthesized nanoparticles holds promise as a potential intervention for Xoo infections in rice, consequently contributing to the production of high-quality agricultural yields (Fouad et al. 2020).

Wheat (*Triticum aestivum L*) crop is often infected by crown and root rot diseases (CRD), which is caused by pathogenic fungi *Fusarium graminearum* and *Fusarium culmorum*, causing death of fine roots and loss of crop. Along with varied strategies, the application of synthetic chemicals and nanoparticles has not been effective in resisting the pathogen. Thus, selenium nanoparticles biosynthesized from *lactobacillus acidophilus* M14 resulted in better antioxidant activity and antifungal ratio. The BioSeNPs at 100 μg/mL significantly increased the plant growth by increasing spike, root, and shoot length along with substantial resistance to CRD disease when compared to chemically synthesized selenium nanoparticles (El-saadony et al. 2021a, b).

3.5.2 Nano-fertilizer

Zinc is an essential nutrient for plant growth by controlling the phytohormone indole acetic acid. Zinc is used as a nano-fertilizer for widely grown, Pearl millet (*Pennisetum americanum*) cv. HHB 67 to enhance the crop production and quality.

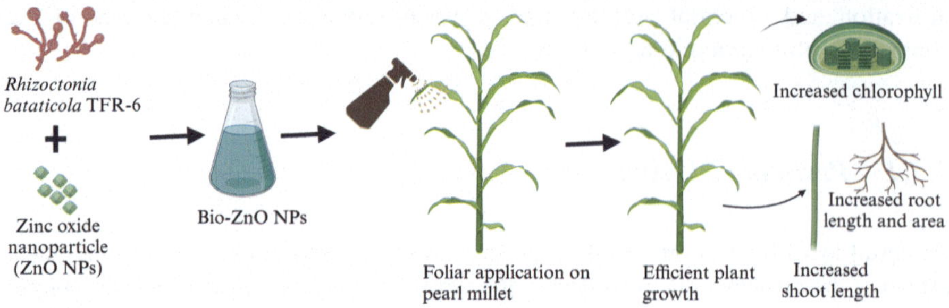

Fig. 3.7 Schematic view of fungi based ZnONPs on pearl millet for enhanced plant growth and maintenance. (Images created using Biorender.com)

The extracellular biosynthesis of Zn nano fertilizer from the fungus *Rhizoctonia bataticola TFR-6* showed higher antibacterial and antioxidant activity. The foliar application of biosynthesized Zn nano fertilizer in the crop field of pearl millet resulted in higher plant growth with increased shoot length (15.1%), root length (4.2%), and root area (24.2%), and a significant increase in chlorophyll content and soluble leaf protein as demonstrated in schematic view in (Fig. 3.7) (Tarafdar et al. 2014). Similarly in yet another study, biosynthesized selenium nanoparticles from fungi *Trichoderma spp.* as trichogenic nanoparticle were used to control downy mildew disease by *Sclerospora graminicola* in pearl millet (Nandini et al. 2017). Taken together, the employment of green synthesized nanoparticles presents encouraging outcomes as a nano-fertilizer to enhance crop productivity.

3.6 Aquafarming

The cultivation, husbandry, and harvesting of fish and aquatic plants constitute the practice of aquaculture or aquafarming. Traditionally, synthesized nanoparticles used in aquaculture have frequently led to the release of hazardous chemicals, resulting in environmental contamination, heightened pathogenic resistance, and the bio magnification of toxins released. Consequently, green synthesized nanoparticles, derived from natural sources, offer an eco-friendly alternative for aquaculture practices. The application of green synthesized nanoparticles introduces opportunities for water treatment, disease management, and nutritional supplementation within aquaculture operations as depicted in Fig. 3.9.

3.6.1 Water Treatment

Water pollution poses a significant threat to both aquatic ecosystems and human health. Nanotechnology offers a promising avenue for addressing this issue through

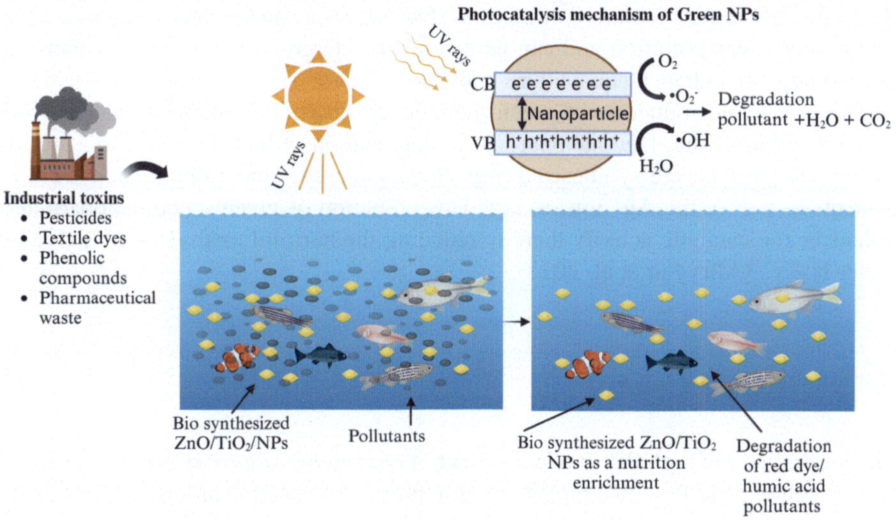

Fig. 3.8 Photocatalysis mechanism of green synthesized metal nanoparticles in water remediation for better aquaculture habitat. (Images created using Biorender.com)

the application of nanoscale remediation techniques utilizing metal nanoparticles such as zinc and titanium. Utilizing a photocatalytic mechanism in conjunction with solar irradiation, biosynthesized zinc and titanium oxide nanoparticles serve as effective nano-photocatalysts for the remediation of water pollutants. This approach has demonstrated active and efficient phytoremediation capabilities, contributing to the mitigation of water pollution. The photo generated electron-hole pairs create radicals such as hydroxyl radical and superoxide in the presence of moisture, thereby breaking down the cell walls and complex organic molecules in contaminants. The biosynthesized ZnO nanoparticles from fruit extract of *Carissa edulis* showed significant degradation of 97% Congo red a harmful textile dye; in the presence of UV light in water bodies (Fig. 3.8) (Fowsiya et al. 2016). Similarly, in another study, titanium oxide nanoparticles and ZnO nanoparticles modified by walnut shell extracts in a synergistic effect have been shown to reduce the accumulation of humic acid in water (Naghizadeh et al. 2016). More research to understand the applicative efficacy of green synthesized nanoparticles in the remediation of water to provide fresh and clean aquatic habitats is still required (Sarkar et al. 2017).

Biosynthesis of AgNPs from the leaf extract of *Biophytum sensitivum* (little tree plant) with microwave-assisted green method was carried out to degrade harmful released industrial dyes such as methyl orange and methylene blue. The microwave-assisted biosynthesis using plant leaf extract for rapid and easy synthesis of effective AgNPs was prepared by microwave irradiation. The synthesized AgNP-biorhythm efficiently acted as a nano catalyst with a combinatorial approach using sodium borohydride. The methyl orange dye colored orange red, when subjected to this treatment appeared bleached with decreased maxima absorbance in UV spectra. Similarly, the methylene blue dye faded to colorless with a decrease in absorbance.

Thus, the efficient degradation of noxious dyes serves as an applicable approach for addressing water pollution and can be extended to large-scale testing in expansive water bodies to enhance aquaculture habitats.

Enzyme based reduction of green metallic nanoparticles enhances the catalytic activity of bio synthesized nanoparticle in degradation of harmful dyes. In a study, silver nanoparticles are synthesized from *Bacillus cereus* containing bioactive components to reduce the Ag^{+} ions to Ag. This reduction of green metal nanoparticles enhances the catalytic activity thereby reducing the harmful methylene blue dye in water bodies (Alfryyan et al. 2022).

3.6.2 Disease Control

Green synthesis of AgNPs from leaf extract of *Origanum vulgare* showed significant antimicrobial activity with a high zone of inhibition to bacterial strains affecting fish culture when compared with chemically synthesized AgNPs. Further, lower hemolysis of cells was observed in green synthesized nanoparticles than in traditional nanoparticles indicating the targeted efficacy of green synthesized AgNPs (Ghetas et al. 2022). Similarly, biosynthesized AgNPs from *Acalypha indica* leaves corroborated the antimicrobial activity against *Escherichia coli* and *Vibria cholera* that act as fish pathogen (Krishnaraj et al. 2010). In another study, bio-synthesized AgNPs from biowaste peel extract of *Pomegranate granatum* was tested along with the different molar concentrations of NaOH demonstrated a better antibacterial remedy against *E.coli, Pseudomonas aeruginosa, Bacillus subtilius, Staphlococcus aureus* in fish ponds when compared to appropriate control (Jasuja et al. 2014). Many more studies reflected the use of green nanoparticles against fish pathogens in Table 3.1. Therefore, it can be concluded that the incorporation of plant based green synthesized nanoparticles plays a role in mitigating antimicrobial resistance in fish diseases and enhances the overall health of aquatic ecosystems (Reyes et al. 2021).

Shrimp aquaculture has experienced global expansion due to rising demand for dietary and various other applications like resource for fatty acids, cheese and oil. However, the presence of pathogens in shrimp cultures frequently leads to suboptimal economic outcomes for shrimp farming enterprises. The Vibrio as the most virulent causative and an opportunistic pathogen, causes a high rate of mortality in the shrimp culture. Chemical antibiotics have shown improper maintenance and outbreak of resistant bacterial strains leading to antimicrobial resistance against drugs and treatments. Hence, a combinatorial strategy has been devised to address the resistance of virulent bacterial pathogens by utilizing a variety of biosynthesized metal nanoparticles. This approach aims to reduce toxicity in aquaculture while enhancing effectiveness against bacterial strain resistance. The AgNPs are synthesized from leaf extracts of the tea plant (*Camellia sinensis*) against *Vibrio harveyi* the causative strain for infection in juvenile *Feneropenaeus indicus* (Indian prawn). The application of green AgNPs demonstrated significant decrease in the colony-forming unit of *V. harveyi* strain *in vitro* suggesting the reduction in growth of the pathogen.

Table 3.1 Green synthesized metal nanoparticles used against fish pathogens

Metal nanoparticles	Biological extract	Aquaculture pathogen	Host culture	References
Ag-NPs	*Azadirachta indica* (Neem)	*Aeromonas hydrophila*	*Cirrhinus mrigala* fingerlings	Rather et al. (2016)
Au-NPs	Aqueous extracts of *Turnera diffusa* (propanone, γ-eudesmol, hydroquinone-β-d-glu-coside (arbutin), and inositol)	*Vibrio parahaemolyticus* and *Aeromonas hydrophila*	Longfin yellowtail	Reyes et al. (2021)
Cu (II) nanoflowers	Juice and peel of blood orange	*Yersinia ruckeri* (causative for enteric red mouth disease in fish)	Fresh water fishes	Demirbas (2021)
ZnO-NPs	*Spinacia oleracea* (Spinach)	*Saprolegnia* sp.	*Labeo rohita* fingerlings	Thangapandiyan (2019)
Ag-NPs	*Gum Arabic (Acacia Arabica)*	*A. Hydrophila subsp. hydropila*	Tilapia fish	Aly et al. (2023)
Ag-NPs	*Origanum vulgare Leaves*	*S. agalactiae, Aeromonas hydrophila*, and *Vibrio alginolyticus*	Nile tilapia	Ghetas et al. (2022)

Further, the mortality rate of infected shrimps reduced to 21% in comparison to control at 91% when administered with bio-synthesized AgNPs (Vaseeharan et al. 2010).

Biosynthesis of AgNPs from red algae (*Portieria hornemannii*) was investigated against different fish pathogens such as *Vibrio harveyi*, *Vibrio parahaemolyticus*, *Vibrio alginolyitcus,* and *Vibrio anguillarum* was found effective in *in-vitro* analysis (Fatima et al. 2019). In another investigation, AgNPs synthesized from *B. subtilis* isolated from the gut of *Litopenaeus vannamei* (white leg shrimp) against *V. harveyi* and *V. parahaemolyticus* demonstrated effective treatment approach for shrimp culture at larger scale. Further, the treated shrimp culture with AgNPs also resulted in high expression of immune-related genes in the gut, gill, stomach, hepato-pancreas, and muscle (Sivaramasamy et al. 2016). Hence, the use of microbe based green metallic nanoparticles has shown an effective anti-pathogenicity in fish and shrimp aquaculture which can be a potential method for producing a large scale synthesis of green nanoparticles for disease control (Fig. 3.9).

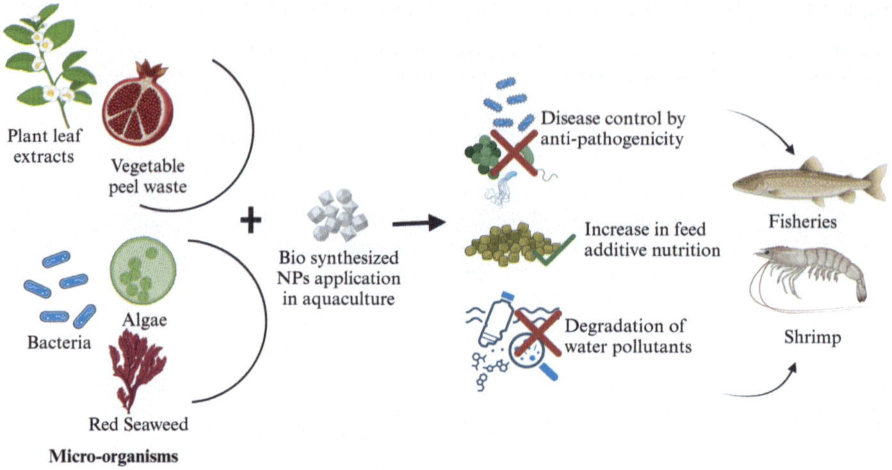

Fig. 3.9 Schematic overview of usage of green synthesized nanoparticle for disease control, water treatment and increased nutrition by feed additives in aquaculture. (Images created using Biorender.com)

3.6.3 Heavy Metal Load

Heavy metal accumulation and pathogenic bacteria contaminate aquaculture. Selenium is a vital micronutrient that enhances fish food, growth, and fertility. Selenium is a fundamental element that protects the cell membrane from enzymatic and non-enzymatic reactions due to heavy metal load. The bio-fabricated selenium particles synthesized from Se-resistant *B. subtilis* AS12 (Bio-SeNPs) was isolated from soil. Bio-SeNPs tested on Nile tilapia (*Oreochromis niloticus*) fish showed significantly increased antioxidant and antibacterial activity. Bio-SeNPs at high concentrations subjected to normal living fish showed an improvement in growth performance with an increase in weight gain of fish and high feed consumption ratio (FCR) with better feeding and swimming behavioral patterns. A similar pattern of Bio-SeNps application was observed in heavy metal-loaded (Hg and Cd) fishes with an increased survival rate of 3–4%. The accumulation of heavy metals in fish organs when assessed showed reduction of 50–87% along with reduced bacterial load in the organs in Bio-SeNPS treated fish when compared with the controls (El-saadony et al. 2022). Despite notable advancements in growth patterns and reduction of heavy metal loads, additional investigations are necessary to establish green synthesized nanoparticles as a leading method for mitigating heavy metal accumulation in aquaculture.

3.6.4 Aquatic Nutrition

In aquaculture, feed nutrition plays a pivotal role in delivering essential nutrients such as proteins, essential fatty acids, and vitamins for better health and immune functions in combating aquatic infections. Many metal nanoparticles applied in the feed of fisheries have resulted in high feed consumption ratio thereby increasing the weight gain of fishes, growth performance, and survival rate however, the accumulation of the same metals, leaves certain toxicity in traces, which later becomes a major concern for biomagnification. In a pilot study by A. Alashahi and co-workers, bio-degradable chitosan nanoparticle encapsulated with vitamin C in the fish feed resulted in an increase in the nutrition uptake in rainbow trout (*Oncorhynchus mykiss*) (Alishahi et al. 2011). The chitosan nanoparticle enhances the strategic release and adsorption efficiency in the gastrointestinal tract and vitamin C plays a pivotal role in boosting the innate immune system. In a similar study, the mixture of fish feed with chitosan nanoparticles encapsulated with vitamin C when administered orally in rainbow trout showed high release of vitamin C in the stomach, intestine, and serum for up to 48 h. Moreover, assessment of lysozyme and complement activity in the serum of fish fed with bio fabricated chitosan nanoparticles containing vitamin C exhibited the highest levels. This outcome is attributed to the synergistic impact of nanoparticles in enhancing innate immune responses in rainbow trout (Elsabee 2011). Different bio-synthesized nanoparticles are used as feed additive to increase nutrition in dietary supplement of fish are presented in Table 3.2.

Table 3.2 Different green synthesized metal nanoparticles as a feed additive in dietary supplement of fish

Metal nanoparticles	Biological extract	Host aquaculture	Application	References
ZnO-NPs	Curcumin (*Curcuma longa*)	Nile tilapia (*Oreochromis niloticus*)	Growth performance, increased anti-oxidant, and carcass quality	Yazdani et al. (2023)
ZnO-NPs	*Spinacia oleracea* (Spinach)	*Labeo rohita* fingerlings	Increase in zinc nutrition and growth	Thangapandiyan (2019)
ZnO-NPs	Leucas aspera aqueous extract and cyclodextrin complex (oxy-CD complex)	Nile tilapia (*Oreochromis niloticus*)	Improved antioxidant system, weight gain, and boost of immune system	Kurian and Elumalai (2021)

3.6.5 *Nano-biosensor*

The application of nanotechnology in aquaculture is currently being used as a new approach to detect the pollutants and pathogens affecting fish health. The industrial waste malachite green dye is highly genotoxic and carcinogenic, affecting aquaculture and evaluating its levels could be primary precautionary steps for remediation. In a study, a composite material comprising graphene oxide and gold nanoparticles was synthesized, employing tyrosine as a biological reducing agent. Graphene oxide, possessing significant electrochemical activity, was detected through Raman scattering, while gold nanoparticles served as antimicrobial agents. The amalgamation of these two metals resulted in a resilient nanoparticle structure, reduced by tyrosine, for the detection of malachite green dye in aqueous environments. The production of graphene and hybridization with gold particles was achieved by the reduction of tyrosine which resilient detection of the organic molecules of malachite green dye in water samples under surface-enhanced Raman scattering (SERs) measurement (Fu et al. 2013). Hence, there exists a necessity to pioneer novel biosensor methodologies utilizing green nanoparticles to achieve precise and accurate detection of pollutants that influence fish health.

3.7 Animal Husbandry

Animal husbandry is a branch of agriculture dealing with livestock farming, breeding, care, and maintenance of animals and supplement of food. Green nanotechnology plays a wide role in the current scenario for the improvement in animal welfare. The bio-synthesized nanomaterials are used for nutrient delivery to feed the livestock, disease management, acaricidal, insecticidal activity and immunity modulation (Fig. 3.11).

3.7.1 *Therapeutic Agents*

The green synthesized nanoparticles are used as a therapeutic effect to abiotic stress and disease management which includes antimicrobial activity, acaricidal activity and Insecticidal activity.

3.7.1.1 To Control Abiotic Stress in Domesticated Animals and Birds

Heat is one of the abiotic stresses affecting poultry production. High temperature majorly results in a reduction in feed intake, feed efficiency, drop in weight, and also results in production of free radical species in broiler chicken. To combat heat stress with eco-friendly strategies, nanotechnology is used for better catalytic efficiency and easy bioavailability. Nano-elemental selenium nanoparticles incorporated into soybean and yellow maize meal when administered to broiler chickens displayed high feed consumption ratio (FCR) compared to control group. Likewise, broiler chickens fed nano-elemental selenium in the basal diet exhibited lower abdominal fat and higher breast muscle weight compared to the control group under high ambient temperatures (El-deep et al. 2016).

In similar research, the dietary supplementation of nano-zinc photogenic nanoparticles was assessed in broiler chickens. The zinc element is an essential micronutrient often added as an additive to feeds and dietary supplementation for enhancing growth performance under high ambient temperatures. The nanoZn-photogenic (NZP) is biosynthesized from inorganic zinc salt and guava leaf extract (*Psidium guajava*) act as a bioreactor and bio stabilizer. The guava leaf extract contains active compounds of flavonoids and polyphenols improve gut health and immunity in the broiler chicken. The supplementation of NZP at a dose concentration of 90 mg/kg had beneficial effects by slightly improving performance in broiler chicken at high temperatures between 30 and 31.5 °C and high humidity ranging from 49.52% to 72.6%, however not much difference was observed in feed consumption. In addition, the antioxidant activity of broiler chicken meat and antibacterial activity in the intestine were observed to be increased in the treated group when compared with the controls (Hidayat et al. 2021). Although still in its early stages of development, there is promising potential for the utilization of green-synthesized nanoparticles as.

3.7.1.2 Antimicrobial Agents for Therapeutic Purpose

Bovine mastitis is an inflammation of mammary gland parenchyma in lactating animals arising due to physical trauma or microbial infection. The multi-drug resistant (MRD) *Staphylococcus aureus* infection in bovine results in large spread of these pathogens in raw bovine milk leading to greater economic loss. The raw milk consumption by human results in the spread of *S. aureus* infection in humans causing dermatitis, pneumonia, and septicemia. Therefore, rather than focusing solely on treatment advancement, the green synthesis of nanoparticles targeting pathogens could serve as a comprehensive solution to address multidrug resistance in animal welfare. In study by Shoeibi and co-workers, the efficacy of copper nanoparticles from the crude extract of ginger root powder and garlic root powder was evaluated. The green copper oxide nanoparticles (CuONPs) were synthesized against isolates of bovine mastitis (MDR) *Staphylococcus aureus* as Gi-doped (ginger extract) and Ga-doped (garlic extract) CuONPs. According to the diffusion method, it was found that Gi-doped CuONPs exhibited enhanced antimicrobial activity compared to Ga-doped CuONPs against MDR *S. aureus*. When subjected to dose- and time-dependent bactericidal and viability studies, biosynthesized nanoparticles exhibited the most potent effects at the highest concentration. These nanoparticles reduced proliferation and caused the formation of cavities and perforations in *S. aureus* cells via infiltration. Additionally, doped CuONPs demonstrated superior catalytic activity compared to conventional nanoparticles and crude extracts (Shoeibi et al. 2022).

3.7.1.3 Green Nanoparticles Having Acaricidal Activity

Ticks are one of the common blood-sucking eco-parasites that cause skin infections and transmit infections to the host body. Tick-borne diseases impact 80% of the cattle population causing a threat to livestock population (Banumathi et al. 2017). Many tick-borne diseases are zoonotic and highly resistant to already available insecticides and pesticides. *Babesia bovis* and *Babesia bigemina* are caused by widely spread invasive protozoa -*Rhipicephalus microplus* and *Rhipicephalus annulatus* in the cattle populations given in Table 3.3 (Zaheer et al. 2022). The AgNPs having innate antibacterial and insecticidal activity when used as a biosynthesized nanoparticle by aqueous extract of *Cissus quadrangularis* (Veld grape) resulted insignificant increase in the mortality rate of *R. microplus* in tick assay (Santhosh et al. 2012). The fabricated AgNPs might have altered the lipid layer of the tick cell membrane, thus changing the protein synthesis and increasing the oxidative stress thereby resulting in cell death (Fig. 3.10). AgNPs synthesized with various other plant extracts have performed potent acaricidal activity against different ticks such as *Hyalomma anatolicum*, *Hyalomma isaaci*, and *Haemaphysalis* (Shoeibi et al. 2022). In summary, green nanoparticles show promise as acaricidal agents within the animal husbandry industry.

Table 3.3 Different green synthesized metal nanoparticles against livestock ticks

Metal nanoparticles	Bio synthesize coating	Targeted tick	References
ZnONPs	Neem and lemon grass leaf extract	*Hyalomma, Haemophysalis,* and *Rhipicephalus*	Banumathi et al. (2017)
ZnONPs	*Lobelia Leschenaultiana*	*Rhipicephalus microplus*	Banumathi et al. (2017)
NiONPs	*Melia azedarach* leaf extract	*Hyalomma dromedarii* (cattle tick)	Ghany (2021)
AgNPs	Mimosa pudica leaf extract	Larvae of *Rhipicephalus microplus*	Banumathi et al. (2017)
AgNPs	*Manilkara zapota aqueous* leaf extract	*Rhipicephalus microplus*	Kirthi et al. (2011)
TiONPs	*Calotropis gigantean* leaf extract	Larvae of *Rhipicephalus microplus*	Periyannan et al. (2017)
TiONPs	*Mangifera Indica* leaf extract	*Rhipicephalus microplus*	Rajkumar et al. (2014)

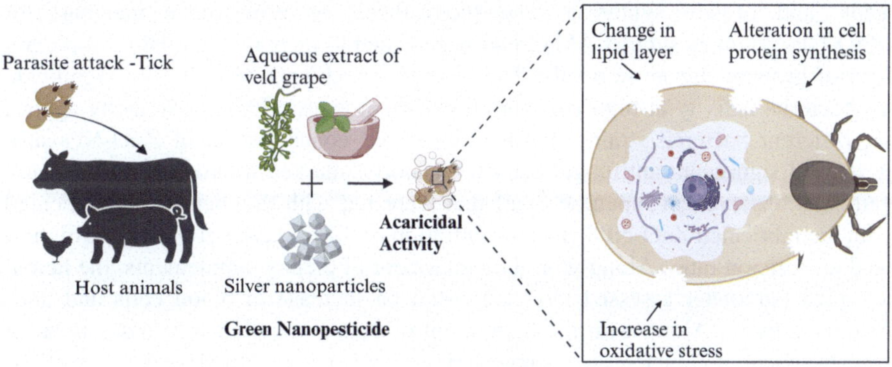

Fig. 3.10 Schematic view of application of green AgNPs as acaricidal activity and its mechanistic effect on tick

3.7.1.4 Green Nanoparticles Displaying Insecticidal Properties

Mosquitos are the primary vectors for the outbreak of major zoonotic diseases in animals. *Aedes albopictus* can affect humans and poultry animals like chickens, pigs, and cattle. Various insecticidal agents are utilized to combat disease-carrying vector; however, the development of resistance to conventional treatments is a recurring challenge. Therefore, bio-synthesized nanotechnology is being investigated and applied against these disease-spreading insects. The essential oils extracted from the secretary cavities and glandular hair cells of plants are aromatic natural products used as insect repellent and anti-microbial agents. The plant agarwood *Aquiliria sinesis* has high medicinal properties along with *Pogostemon isherba*, an aerial part of dried *Pogostemon cablin* (Blanco) is used for common cold and fever having a distinctive and intensive aromatic odor. The mixture of essential oil from both *A. sinesis* and *P. cablin* was employed to synthesize AgNPs against *A. albopictus* larvae. The bio-synthesized silver nanoparticles from *A. sinesis* (AsEO) and *P. cablin* (PcEO) larvicidal and suicidal activity resulted to be highly toxic even at lower concentration doses in comparison to monotherapy AsEO, PcEO, and AgNPS administered separately against *A. albopictus*. The small bio-fabricated AgNPS easily penetrates the epithelial gut cells of insects thereby activating an influx of change in physiological processes leading to cell death confirmed by histological studies (Ga'al et al. 2017).

3.7.2 Green Nanoparticles Coated Feed Additives

Zinc nanoparticles have been recognized as a safe metal for consumption in dietary substances. The advantageous attributes of zinc, serving as a fundamental element, lie in its involvement with metalloenzymes crucial for energy metabolism, nucleic acids, and protein synthesis. The biosynthesis of Zinc oxide nanoparticles (ZnO-NPs) from *B. subtilis* AM12 had been tested to increase the quality of dietary feed of growing Japanese quails. The bio-fabricated ZnO-NPs showed significant antibacterial activity at high concentrations when assessed with disc assay against six different bacterial strains. Further, increased concentration of ZnO-NPs also possessed significant antifungal activity. Majorly, marked improvement of weight gain was observed in Japanese quail that were fed with the dietary supplemented with bio fabricated ZnO-NPS at a concentration of 0.1 or 0.2/kg of diet as the highest feed conversion ratio. Along with the betterment of dietary supplements, the hematological parameter assessed also suggested an increase in blood cells and anti-oxidant activity. Along with the betterment of dietary supplements, assessment of hematological parameter also suggested an increase in blood cells as well as increased anti-oxidant activity (Reda et al. 2021).

Conventional poultry feed additives and synthetic treatments, employed to enhance poultry health and satisfy public demand, are often found to be environmentally unsustainable and costly. The adverse effects associated with inorganic salts and synthetic treatments contribute to diminished food quality for human consumption (Michalak et al. 2022). Thus, the use of green synthesized nanoparticles is a new niche to combat the toxicity and improve poultry health. Poor eggshell quality adversely affects the economic profitability of egg production. The quality of eggshells decreases with age in laying hens and the crack in eggs at a late stage of the laying cycle can cause economic loss. Hence, an optimal presence of calcium in hens is important for better eggshell quality (Fig. 3.11). The biosynthesis of calcium nanoparticles from biocompatible *Sargassum latifolium* brown seaweed algae extract (SL-CaNPs) was used to yield better quality eggshells. The % of egg weight and eggshell weight of hens supplemented with SL-CaNPs in their diet resulted in a significant increase compared to control. Moreover, an increase in concentration of calcium and phosphorus retention in blood was observed in the treated group along with, apical cells secreting calcium in the uterus wall of laying hens appeared high in number when analyzed in histology (El-maaty et al. 2010).

3.8 Application of Green Nanoparticles in Dairy Industry

The production of dairy products and its consumption is a large scale industry at the economic belt in animal husbandry business. The large scale production, preservation and transportation are the main challenges faced at regular basis in dairy industry (Zhao et al. 2020). With advancement in nanotechnology, the application of green synthesized nanoparticles in food preservation of dairy products is an emerging new area of explore. The plant based colloidal selenium nanoparticles bio synthesized from extracts of garlic (*Allium sativum*) has been shown to result in effective nano-biosensor for milk adulterant—salicylic acid in milk products (Aftab et al. 2023). Similarly, the iron nanoparticles bio synthesized from *Bacillus subtilis* ML6 as Bio-Fe(II) nanoparticles were used to enhance the quality, increase in shelf

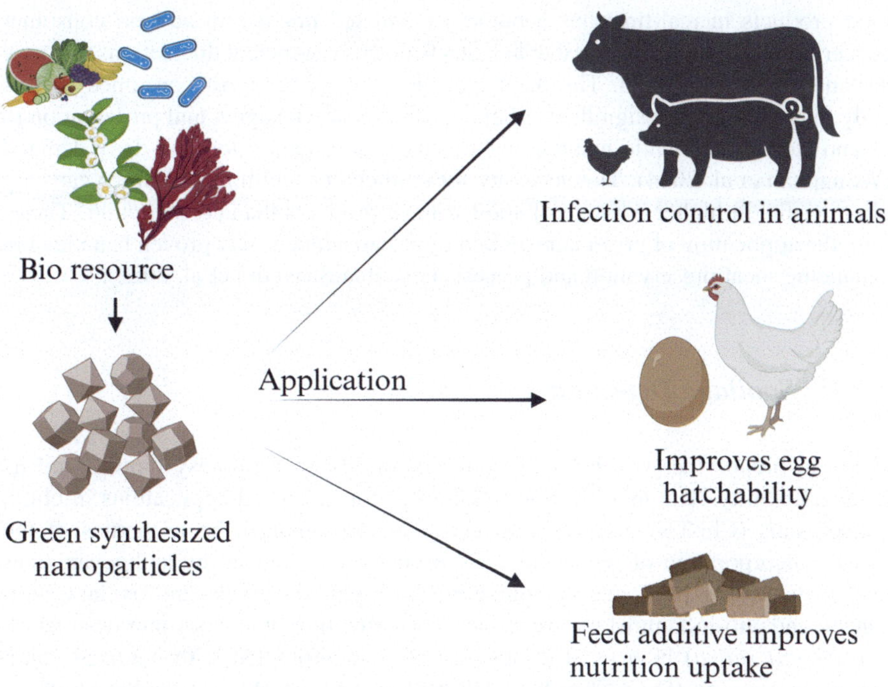

Fig. 3.11 Schematic overview of bio-synthesized nanoparticles application in disease control, enhanced egg hatchability, increase in nutrition by feed additive in livestock rearing. (Images created using Biorender.com)

life and increased iron content in yogurt (El-Saadony et al. 2021a, b). The bionanocomposites of zinc oxide nanoparticles (ZnO NPs) from biodegradable chitosan carboxymethyl cellulose (CMC) are used in smart packaging of soft white cheese from pasteurized cowmilk to prolong its shelf life. The increased shelf life of soft white cheese resulted with bio nanocomposite up to 30 days along with good anti-pathogenic properties (Youssef et al. 2016). Thus, with limited reports on application and testing of green nanoparticles in dairy products more studies are required for better understanding and potential of these nanoparticles in processing, packaging and preservation of dairy products.

3.9 Application of Green Nanoparticles in Meat Packaging and Preservation

Recent advancements in nanotechnology have highlighted its potential in smart packaging solutions aimed at optimizing food packaging and storage conditions, particularly in response to supply chain dynamics and fluctuations in environmental factors (Ramachandraiah et al. 2015). Ensuring the delivery of safe, high-quality food products in real-time has become a mandated practice to address consumer concerns and to safeguard perishable items from environmental fluctuations that may hasten their deterioration. The meat industry, one of the most consumed sectors daily, has encountered significant challenges in meat packaging and preservation to extend storage life while maintaining freshness and quality to meet daily demand (Wongphan et al. 2024). The necessity for cold chains and limited storage capacity often results in higher wastage of spoiled meat products than consumption. Therefore, the application of green nanotechnology is an added safety proven beneficial in enhancing meat preservation and processing (Ulloa-Saavedra et al. 2022).

3.9.1 Spoilage Detection

Chitosan biopolymer is often chosen for its ability to form easy biofilm and its biodegradability. The modification to chitosan for advanced applications in chitin nanocrystals (ChNCs) derived from chitin are to enhance the properties of the matrix, incorporation of physiochemical properties is attained by anthocyanins as one of the strong plant dietary containing high pH sensitivity, anti-oxidant, anticancer, and anti-inflammatory properties. The extraction of anthocyanin from Black rice (*Oryza sativa*) bran as at a larger scale is incorporated with oxidized chitin nanocrystal matrix (O-ChNCs) for intelligent nanofilm making in seafood packaging. The total volatile basic nitrogen (TVB-N) levels produced due to protein degradation were detected by changes in the color of the film duc to pH changes. The color change from purple to gray suggested increased levels of TVB-N diffused through film in packages of seafood. COB-3 film resulted in the highest color change and thus can be used as an effective indicator of spoilage in the packaging of animal-based food (Wu et al. 2019). In a similar study, halo-chromatic films were prepared by anthocyanins extracted from red cabbage (*Brassica oleracea* L.) as RCAs play a huge role in pH sensitivity incorporated with oxidized chitin nanocrystal (OCS) as COR films for determining freshness of hairtail (*Trichiurus lepturus*) and shrimp (*Penaeus vannamei*). Amongst all the tested COR films, COR-1.2 nano-film resulted in the better detector of spoilage by a change in color at three different stages. The fresh stage (reddish-purple), medium stage (brown), and spoiled stage (yellow) give clear real-time visual inspection of seafood quality and freshness (Wu et al. 2019).

Lamri et al. (2021) reported that the nanocomposite film packaging using potato starch and apple peel pectin with microencapsulation of essential oil extracted from *Zataria multiflora* and with zirconium oxide nanoparticle resulted in efficient preservation of quail meat by inhibiting the bacterial strains *E. coli* and *B. cereus* with high antioxidant activity and increased shelf life.

3.9.2 Sensory Quality of Meat Products

Nanoparticles have also been used in improving the sensory quality of meat products by the method of marination. Marinating is a common process of incorporating the flavor of spices in muscle food. The infused food with marination results in improved texture, juiciness, and taste. *Paprika oleoresin* is used as a food color which contains a high amount of carotenoid pigment from red pepper (capsicum annum L.) (Fig. 3.12). The nanoparticle paprika oleoresin is tested to evaluate the sensory quality of meat products in the marinating process for toxic-free taste in muscle-based food products. Thus, nanoparticle application of paprika showed a significant effect on the sensory quality of meat with high color intensity. However improvement of sensory quality of meat product is very new and at very early stage. Thus more investigation and strategies are required in this field to enhance the flavor of meat products by application of nanoparticles (Yusop et al. 2012).

3.9.3 To Enhance the Shelf Life of Products

Increase in shelf life of meat products are achieved by various intelligent packaging systems. Nano-emulsion-based coating from the extract of ginger (*Zingiber officinale*) essential oil (GEO) in combination with edible sodium caseinate is applied to preserve and increase the chicken breast fillet. The nano-emulsion-based coating resulted in higher antimicrobial activity in comparison to commercial gentamycin antibiotics (Abiri et al. 2023). The significant increase in shelf life was observed by a reduction in aerobic psychrophilic bacteria, molds, and yeasts suggesting active anti-bacterial activity little color change in nano-emulsion coated meat was observed in comparison to fresh chicken breast fillets. Resulted in high

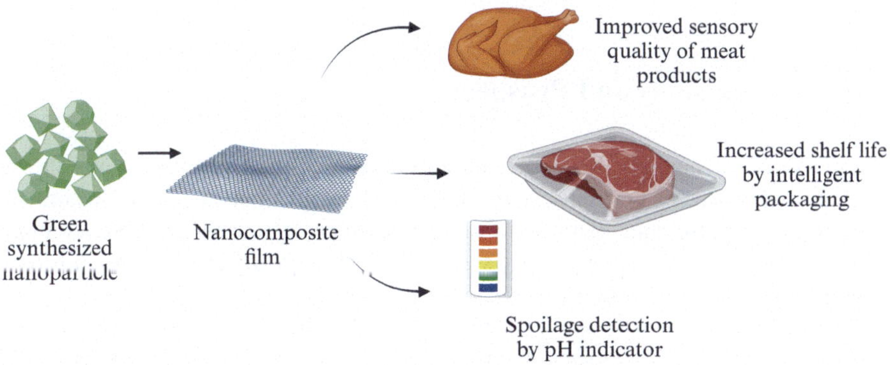

Fig. 3.12 Schematic overview of application of bio-synthesized nanoparticle as intelligent packaging, spoilage detection by pH sensitivity color change and increase in shelf life by antioxidant and antibacterial activity in meat industry. (Images created with BioRender.com)

Table 3.4 Green synthesized nanoparticle application in meat packaging and preservation

Metal nanoparticles	Biological extract	Product	Application	References
ε-Polylysine nanoparticle	Olive and green tea plant extracts	Frankfurter sausages	Nitrile-free, increased shelf life, and improved antimicrobial activity	Alirezalu et al. (2021)
Chitosan nanoparticle	Thyme phenolic	Beef burger	Improved sensory quality of meat and increased antioxidant activity	Ghaderi-ghahfarokhi et al. (2016)
Nano emulsion (AUSN-4)	Sunflower oil	Steaks	Improved shelf life	Hana et al. (2023)
Chitosan nanoparticle	*Satureja khuzestanica*	Lamb meat	Improved shelf life	Esmaeili et al. (2020)
ZnONPs	*Okra mucilage*	Chicken meat	Improve shelf life	Mohammadi et al. (2018)
Chitosan/ montmorrilontebio-nanocomposite film	Rosemary–ginger and essential oil	Chicken fillets	Increased shelf life	Piers et al. (2018)

acceptance with increasing shelf life for storage of raw poultry products (Fig. 3.12) (Noori et al. 2018).

Synthetic routes such as nitrites, beta-agonists, and gelatin material often used for meat packaging, meat preservation, and processing may cause undesirable changes in the products causing deleterious effects on the consumers. The use of natural preservatives at nanoscale formation reduces the load of toxicity level and increases the quality of products (Table 3.4): The plant extracts and essential oils provide the platform as a nano-biosensor to detect the quality of meat products at longer storage. Thus, in future advancement the nanocomposite coatings promote smart and active packaging to increase the shelf life by increasing the antimicrobial and antioxidant activity of poultry products.

3.10 Conclusions and Prospects

Green synthesis method in nanotechnology has revolutionized the synthesis of metallic nanoparticles for safer and economical application in varied sectors of agriculture. By utilizing extracts from plants, microorganisms, and agricultural waste products, these nanoparticles synthesized biologically, has shown positive outcomes in numerous studies.

The utilization of green-synthesized nanoparticles has led to showcase the economical use in nano-biosensors, nano-pesticides, and nano-fertilizers in crop cultivation. Numerous studies on the application of nanoparticles in agriculture consistently demonstrate that green synthesized nanoparticles outperform conventional nanoparticles in supporting functional roles within the agricultural sector.

However, there remains a dearth of research regarding the controlled release of these nanoparticles on a large scale in agricultural fields to ascertain permissible limits and assess environmental toxicity. Similarly, the application of green-synthesized nanoparticles in poultry, meat packaging, and aquaculture requires further investigation, as existing studies merely scratch the surface of a vast area of unknowns. More comprehensive studies are essential to understand the specific functional roles of bio-fabricated nanoparticles and to facilitate the scaling up of green-synthesized nanoparticles for effective commercialization. This necessitates deeper research into their properties and behavior in various agricultural and related applications, ensuring their safety, efficacy, and practicality on a larger scale.

References

Abiri BA, Baghaei H, Mohammadi Nafchi A (2023) Preparation and application of active bionanocomposite films based on sago starch reinforced with a combination of TiO2 nanoparticles and penganum harmala extract for preserving chicken fillets. Polymers 15(13):2889. https://doi.org/10.3390/polym15132889

Abdallah Y, Liu M, Ogunyemi SO, Ahmed T, Fouad H, Abdelazez A, Yan C, Yang Y, Chen J, Li B (2020) Bioinspired green synthesis of chitosan and zinc oxide nanoparticles with strong antibacterial activity against rice pathogen xanthomonas oryzae pv. oryzae. Molecules 25 (20):4795. https://doi.org/10.3390/molecules25204795

Aftab R, Ahsan S, Liaqat A et al (2023) Green-synthesized selenium nanoparticles using garlic extract and their application for rapid detection of salicylic acid in milk. Food Sci Technol 43. https://doi.org/10.1590/fst.67022

Alfryyan N, Kordy MGM, Abdel-Gabbar M et al (2022) Characterization of the biosynthesized intracellular and extracellular plasmonic silver nanoparticles using Bacillus cereus and their catalytic reduction of methylene blue. Sci Rep 12:1–14. https://doi.org/10.1038/s41598-022-16029-1

Alirezalu K, Hesari J, Yaghoubi M, Mousavi A (2021) Combined effects of ε-polylysine and ε-polylysine nanoparticles with plant extracts on the shelf life and quality characteristics of nitrite-free frankfurter-type sausages. Meat Sci 172:108318. https://doi.org/10.1016/j.meatsci.2020.108318

Alishahi A, Mirvaghefi A, Tehrani MR et al (2011) Shelf life and delivery enhancement of vitamin C using chitosan nanoparticles. Food Chem 126:935–940. https://doi.org/10.1016/j.foodchem.2010.11.086

Aly SM, Eissa AE, Abdel-Razek N, El-Ramlawy AO (2023) Chitosan nanoparticles and green synthesized silver nanoparticles as novel alternatives to antibiotics for preventing A. hydrophila subsp. hydrophila infection in Nile tilapia, Oreochromis niloticus. Int J Vet Sci Med 11:38–54. https://doi.org/10.1080/23144599.2023.2205338

Bahrulolum H, Nooraei S, Javanshir N et al (2021) Green synthesis of metal nanoparticles using microorganisms and their application in the agrifood sector. J Nanobiotechnol:1–26. https://doi.org/10.1186/s12951-021-00834-3

Banumathi B, Vaseeharan B, Rajasekar P, Prabhu NM, Ramasamy P, Murugan K, Canale A, Benelli G (2017) Exploitation of chemical, herbal and nanoformulated acaricides to control the cattle tick, Rhipicephalus (Boophilus) microplus - A review. Vet Parasitol 244:102–110. https://doi.org/10.1016/j.vetpar.2017.07.021

Demirbas A (2021) Comparison study of synthesized red (or blood) orange peels and juice extract-nanoflowers and their antimicrobial properties on fish pathogen (Yersinia ruckeri). Indian J Microbiol 61:324–330. https://doi.org/10.1007/s12088-021-00945-3

Ding Y, Zhao W, Zhu G et al (2023) Recent trends in foliar nanofertilizers: a review. Nanomaterials 13(21):2906

Duhan JS, Kumar R, Kumar N et al (2017) Nanotechnology: the new perspective in precision agriculture. Biotechnol Rep 15:11–23. https://doi.org/10.1016/j.btre.2017.03.002

El-deep MH, Ijiri D, Ebeid TA, Ohtsuka A (2016) Effects of dietary nano-selenium supplementation on growth performance, antioxidative status, and immunity in broiler chickens under thermoneutral and high ambient temperature conditions. J Poult Sci:1–7. https://doi.org/10.2141/jpsa.0150133

El-maaty HAA, El-khateeb AY, Al-khalaifah H et al (2010) Effects of ecofriendly synthesized calcium nanoparticles with biocompatible Sargassum latifolium algae extract supplementation on egg quality and scanning electron microscopy images of the eggshell of aged laying hens. Poult Sci 100:675–684. https://doi.org/10.1016/j.psj.2020.10.043

El-saadony MT, El-hack MEA, Taha AE (2020) Ecofriendly synthesis and insecticidal application of copper nanoparticles against the storage pest Tribolium castaneum. Nanomaterials (Basel) 10(3):587

El-saadony MT, Saad AM, Najjar AA et al (2021a) The use of biological selenium nanoparticles to suppress Triticum aestivum L. crown and root rot diseases induced by Fusarium species and improve yield under drought and heat stress. Saudi J Biol Sci 28:4461–4471. https://doi.org/10.1016/j.sjbs.2021.04.043

El-Saadony MT, Sitohy MZ, Ramadan MF, Saad AM (2021b) Green nanotechnology for preserving and enriching yogurt with biologically available iron (II). Innov Food Sci Emerg Technol 69:102645. https://doi.org/10.1016/j.ifset.2021.102645

Elsabee MZ (2011) Chitosan nanoparticle to carry vitamin C through the gastrointestinal tract and induce the non-specific immunity system of rainbow trout (Oncorhynchus mykiss). Carbohydr Polym 86:142–146. https://doi.org/10.1016/j.carbpol.2011.04.028

Esmaeili H, Cheraghi N, Khanjari A et al (2020) Incorporation of nanoencapsulated garlic essential oil into edible films: a novel approach for extending shelf life of vacuum-packed sausages. Meat Sci 166:108135. https://doi.org/10.1016/j.meatsci.2020.108135

Fatima R, Priya M, Indurthi L et al (2019) Biosynthesis of silver nanoparticles using red algae Portieria hornemannii and its antibacterial activity against fish pathogens. Microb Pathog:103780. https://doi.org/10.1016/j.micpath.2019.103780

Fowsiya J, Madhumitha G, Al-Dhabi NA, Arasu MV (2016) Photocatalytic degradation of Congo red using Carissa edulis extract capped zinc oxide nanoparticles. J Photochem Photobiol B 162:395–401. https://doi.org/10.1016/j.jphotobiol.2016.07.011

Fu WL, Zhen SJ, Huang CZ (2013) One-pot green synthesis of graphene oxide/gold nanocomposites as SERS substrates for malachite green detection. The Analyst 138(10):3075. https://doi.org/10.1039/c3an00018d

Ga'al H, Fouad H, Mao G, Tian J, Jianchu M (2017) Larvicidal and pupicidal evaluation of silver nanoparticles synthesized using Aquilaria sinensis and Pogostemon cablin essential oils against dengue and zika viruses vector Aedes albopictus mosquito and its histopathological analysis. Artif Cells Nanomed Biotechnol 46(6):1171–1179. https://doi.org/10.1080/21691401.2017.1365723

Ghaderi-ghahfarokhi M, Barzegar M, Sahari MA (2016) Nanoencapsulation approach to improve antimicrobial and antioxidant activity of thyme essential oil in beef burgers during refrigerated storage. Food Bioprocess Technol. https://doi.org/10.1007/s11947-016-1708-z

Ghany HSMA (2021) Nanoparticles against the camel tick, Hyalomma dromedarii:611–633. https://doi.org/10.1007/s10493-021-00596-5

Ghareeb RY, Alfy H, Fahmy AA, Ali HM (2020) Utilization of Cladophora glomerata extract nanoparticles as eco—nematicide and enhancing the defense responses of tomato plants infected by Meloidogyne javanica. Sci Rep:1–15. https://doi.org/10.1038/s41598-020-77005-1

Ghetas HA, Abdel-razek N, Shakweer MS et al (2022) Antimicrobial activity of chemically and biologically synthesized silver nanoparticles against some fish pathogens. Saudi J Biol Sci 29: 1298–1305. https://doi.org/10.1016/j.sjbs.2021.11.015

Hana D, Kši S, Illa R et al (2023) Agronomic investigation of spray dispersion of metal-based nanoparticles on sunflowers in real-world environments. Plants 12(9):1789

Hidayat C, Jayanegara A, Wina E (2021) Supplementation of dietary nano Zn-phytogenic on performance, antioxidant activity, and population of intestinal pathogenic bacteria in broiler chickens. Trop Anim Sci J 44:90–99

Hong J, Wang C, Wagner DC et al (2021) Nano absorption, transfer, and multiple impacts. Environ Sci. https://doi.org/10.1039/d0en01129k

Hosseini MS, Naeimi A, Boroomand N, Aalifar M (2022) Alleviating the adverse effects of salinity on Roselle plants by green synthesized nanoparticles. Sci Rep:1–12. https://doi.org/10.1038/s41598-022-22903-9

Huston M, DeBella M, DiBella M, Gupta A (2021) Green synthesis of nanomaterials. Nanomat 11 (8):2130. https://doi.org/10.3390/nano11082130

Jasuja ND, Gupta DK, Reza M, Joshi SC (2014) Green synthesis of AgNPs Stabilized with biowaste and their antimicrobial activities. Braz J Microbiol 1332:1325–1332

Javed MN, Ali S, Bangash K et al (2023) Potential and challenges in green synthesis of nanoparticles: a review. J Xi'an Shiyou Univ 19(02):1155–1165

Journal AI, Ga H, Fouad H et al (2017) Larvicidal and pupicidal evaluation of silver nanoparticles synthesized using Aquilaria sinensis and Pogostemon cablin essential oils against dengue and zika viruses vector Aedes albopictus mosquito and its histopathological analysis. Artif Cells Nanomed Biotechnol 0:1–9. https://doi.org/10.1080/21691401.2017.1365723

Kirthi AV, Rahuman AA, Rajakumar G (2011) Acaricidal, pediculocidal and larvicidal activity of synthesized ZnO nanoparticles using wet chemical route against blood feeding parasites. Parasitol Res:461–472. https://doi.org/10.1007/s00436-011-2277-8

Krishnaraj C, Jagan EG, Rajasekar S et al (2010) Synthesis of silver nanoparticles using Acalypha indica leaf extracts and its antibacterial activity against water borne pathogens. Colloids Surf B: Biointerfaces 76:50–56. https://doi.org/10.1016/j.colsurfb.2009.10.008

Kurian A, Elumalai P (2021) Study on the impacts of chemical and green synthesized (Leucas aspera and oxy-cyclodextrin complex) dietary zinc oxide nanoparticles in Nile tilapia (Oreochromis niloticus). Environ Sci Pollut Res Int 28(16):20344–20361

Lamri M, Bhattacharya T, Boukid F et al (2021) Nanotechnology as a processing and packaging tool to improve meat quality and safety. Food:1–25. https://doi.org/10.3390/foods10112633

Liu J, Lai L, Liu H et al (2021) Nano-silver treatment reduces bacterial proliferation and stem bending in cut gerbera flowers: an in vitro and in vivo evaluation. Postharvest Biol Technol 180: 111595. https://doi.org/10.1016/j.postharvbio.2021.111595

Ma Z, Zhang K, Guo W, Yu W (2023) Green synthesis of silver nanoparticles using Eucommia ulmoides leaf extract for inhibiting stem end bacteria in cut tree peony flowers. Front Plant Sci:1–14. https://doi.org/10.3389/fpls.2023.1176359

Michalak I, Dziergowska K, Alagawany M et al (2022) The effect of metal-containing nanoparticles on the health, performance and production of livestock animals and poultry. Vet Q 42:68–94. https://doi.org/10.1080/01652176.2022.2073399

Mohammadi H, Kamkar A, Misaghi A (2018) Nanocomposite films based on CMC, okra mucilage and ZnO nanoparticles: physico mechanical and antibacterial properties. Carbohydr Polym 181: 351–357. https://doi.org/10.1016/j.carbpol.2017.10.045

Muhie SH (2022) Novel approaches and practices to sustainable agriculture. J Agric Food Res 10: 100446. https://doi.org/10.1016/j.jafr.2022.100446

Naghizadeh A, Shahabi H, Ghasemi F, Zarei A (2016) Synthesis of walnut shell modified with titanium dioxide and zinc oxide nanoparticles for efficient removal of humic acid from aqueous solutions. J Water Health 14(6):989–997. https://doi.org/10.2166/wh.2016.072

Naing AH, Kim CK (2020) Application of nano-silver particles to control the postharvest biology of cut flowers: a review. Sci Hortic (Amsterdam) 270:109463. https://doi.org/10.1016/j.scienta.2020.109463

Nandini B, Hariprasad P, Prakash HS, Shetty HS, Geetha N (2017) Trichogenic-selenium nanoparticles enhance disease suppressive ability of Trichoderma against downy mildew disease caused by Sclerospora graminicola in pearl millet. 7(1). https://doi.org/10.1038/s41598-017-02737-6

Nile SH, Thiruvengadam M, Wang Y et al (2022) Nano—priming as emerging seed priming technology for sustainable agriculture—recent developments and future perspectives. J Nanobiotechnol. https://doi.org/10.1186/s12951-022-01423-8

Noori S, Zeynali F, Almasi H (2018) Antimicrobial and antioxidant efficiency of nanoemulsion-based edible coating containing ginger (Zingiber officinale) essential oil and its effect on safety and quality attributes of chicken breast fillets. Food Control 84:312–320. https://doi.org/10.1016/j.foodcont.2017.08.015

Parveen A, Bi B, Mazhari Z, Rao S (2016) Impact of bio-nanogold on seed germination and seedling growth in Pennisetum glaucum. Enzym Microb Technol:3–7. https://doi.org/10.1016/j.enzmictec.2016.04.005

Periyannan R, Prabhu NM, Ramasamy P et al (2017) Exploitation of chemical, herbal and nanoformulated acaricides to control the cattle tick, Rhipicephalus (Boophilus) microplus—a review. Vet Parasitol. https://doi.org/10.1016/j.vetpar.2017.07.021

Pires, JRA, de Souza VGL, Fernando AL (2018) Chitosan/montmorillonite bionanocomposites incorporated with rosemary and ginger essential oil as packaging for fresh poultry meat. Food Packaging and Shelf Life 17:142–149. https://doi.org/10.1016/j.fpsl.2018.06.011

Rajakumar G, Rahuman A, Roopan SM, Chung I-M, Karunanithi A, Karthikeyan, V (2014) Efficacy of larvicidal activity of green synthesized titanium dioxide nanoparticles using Mangifera indica extract against blood-feeding parasites. Parasitol Res 114(2):571–581. https://doi.org/10.1007/s00436-014-4219-8

Rajput VD, Minkina T, Upadhyay SK et al (2022) Nanotechnology in the restoration of polluted soil. Nano 12:1–16. https://doi.org/10.3390/nano12050769

Ramachandraiah K, Han SG, Chin KB (2015) Nanotechnology in meat processing and packaging: potential applications—a review. Asian Australas J Anim Sci 28:290–302. https://doi.org/10.5713/ajas.14.0607

Rasmiya Begum SL, Jayawardana NU (2023) Green synthesized metal nanoparticles as an ecofriendly measure for plant growth stimulation and disease resistance. Plant Nano Biol 3: 100028. https://doi.org/10.1016/j.plana.2023.100028

Rather MA, Bhat IA, Sharma N et al (2016) Synthesis and characterization of Azadirachta indica constructed silver nanoparticles and their immunomodulatory activity in fish. Aquac Res:1–13. https://doi.org/10.1111/are.13199

Reda FM, El-saadony MT, El-rayes TK et al (2021) Use of biological nano zinc as a feed additive in quail nutrition: biosynthesis, antimicrobial activity and its effect on growth, feed utilisation, blood metabolites and intestinal microbiota. Ital J Anim Sci 20:324–335. https://doi.org/10.1080/1828051X.2021.1886001

Reyes M, Fernando B, Veronica R et al (2021) Green synthesis of gold nanoparticles using Turnera diffusa Willd enhanced antimicrobial properties and immune response in Longfin yellowtail leukocytes. Aquac Res:1–12. https://doi.org/10.1111/are.15184

Ricardo J, Pires A, Gomes V et al (2018) Chitosan/montmorillonite bionanocomposites incorporated with rosemary and ginger essential oil as packaging for fresh poultry meat. Food Packag Shelf Life 17:142–149. https://doi.org/10.1016/j.fpsl.2018.06.011

Saad AM, Sitohy MZ, Sultan-Alolama MI, El-Tarabily KA, El-Saadony MT (2022) Green nanotechnology for controlling bacterial load and heavy metal accumulation in Nile tilapia fish using biological selenium nanoparticles biosynthesized by Bacillus subtilis AS12. Front Microbio 13:1015613. https://doi.org/10.3389/fmicb.2022.1015613

Saad AM, El-saadony MT, El-tahan AM et al (2021) Polyphenolic extracts from pomegranate and watermelon wastes as substrate to fabricate sustainable silver nanoparticles with larvicidal effect against Spodoptera littoralis. Saudi J Biol Sci 28:5674–5683. https://doi.org/10.1016/j.sjbs.2021.06.011

Salem MZM, El-hefny M, Ali HM et al (2021) Plants-derived bioactives: novel utilization as antimicrobial, antioxidant and phytoreducing agents for the biosynthesis of metallic nanoparticles. Microb Pathog 158:18–25. https://doi.org/10.1016/j.micpath.2021.105107

Salem SS, Sitohy MZ, Mohamad I (2022) Green nanotechnology for controlling bacterial load and heavy metal accumulation in Nile tilapia fish using biological selenium nanoparticles biosynthesized by Bacillus subtilis AS12. Front Microbiol:1–23. https://doi.org/10.3389/fmicb.2022.1015613

Sana SS, Kumbhakar DV, Pasha A et al (2020) Crotalaria verrucosa leaf extract mediated synthesis of zinc oxide nanoparticles: assessment of antimicrobial and anticancer activity. Molecules 25. https://doi.org/10.3390/molecules25214896

Santhoshkumar T, Rahuman AA, Bagavan A, Marimuthu S, Jayaseelan C, Kirthi AV, Kamaraj C, Rajakumar G, Zahir AA, Elango G, Velayutham K, Iyappan M, Siva C, Karthik L, Rao KV (2012) Evaluation of stem aqueous extract and synthesized silver nanoparticles using Cissus quadrangularis against Hippobosca maculata and Rhipicephalus (*Boophilus*) microplus. Exp Parasitol 132(2):156–165. https://doi.org/10.1016/j.exppara.2012.06.009

Sarkar B, Daware AV, Gupta P, Krishnani KK (2017) Nanoscale wide-band semiconductors for photocatalytic remediation of aquatic pollution. Environ Sci Pollut Res. https://doi.org/10.1007/s11356-017-0252-3

Shoeibi S, Mozdziak P, Golkar-narenji A et al (2022) The effect of metal-containing nanoparticles on the health, performance and production of livestock animals and poultry. Sci Rep 2022:1–13. https://doi.org/10.1038/s41598-022-22099-y

Singh NB (2022) Green synthesis of nanomaterials. In: Handbook of microbial nanotechnology, pp 225–254. https://doi.org/10.1016/B978-0-12-823426-6.00007-3

Sivaramasamy E, Zhiwei W, Li F, Xiang J (2016) Enhancement of vibriosis resistance in Litopenaeus vannamei by supplementation of biomastered silver nanoparticles by Bacillus subtilis. J Nanomed Nanotechnol 7:1–10. https://doi.org/10.4172/2157-7439.1000352

Solgi M (2014) Evaluation of plant-mediated silver nanoparticles synthesis and its application in postharvest physiology of cut flowers. Physiol Mol Biol Plants 20:279–285. https://doi.org/10.1007/s12298-014-0237-3

Sundararajan N, Habeebsheriff HS, Dhanabalan K et al (2024) Mitigating global challenges: harnessing green synthesized nanomaterials for sustainable crop production systems. Global Chall 8:1–15. https://doi.org/10.1002/gch2.202300187

Tarafdar JC, Raliya R, Mahawar H, Rathore I (2014) Development of zinc nanofertilizer to enhance crop production in pearl millet (*Pennisetum americanum*). Agricul Res 3(3): 257–262. https://doi.org/10.1007/s40003-014-0113-y

Thabet SG, Alqudah AM (2024) Unraveling the role of nanoparticles in improving plant resilience under environmental stress condition. Plant Soil. https://doi.org/10.1007/s11104-024-06581-2

Thangapandiyan S (2019) Green synthesized zinc oxide nanoparticles as feed additives to improve growth, biochemical, and hematological parameters in freshwater fish Labeo rohita. Biol Trace Elem Res 195(2):636–647

Ulloa-Saavedra A, García-Betanzos C, Zambrano-Zaragoza M et al (2022) Recent developments and applications of nanosystems in the preservation of meat and meat products. Food 11:1–22. https://doi.org/10.3390/foods11142150

Vaseeharan B, Ramasamy P, Chen JC (2010) Antibacterial activity of silver nanoparticles (AgNps) synthesized by tea leaf extracts against pathogenic Vibrio harveyi and its protective efficacy on juvenile Feneropenaeus indicus. Lett Appl Microbiol 50:352–356. https://doi.org/10.1111/j.1472-765X.2010.02799.x

Vijayaram S, Razafindralambo H, Sun YZ et al (2024) Applications of green synthesized metal nanoparticles—a review. Biol Trace Elem Res 202:360–386. https://doi.org/10.1007/s12011-023-03645-9

Wongphan P, Promhuad K, Srisa A et al (2024) Unveiling the future of meat packaging: functional biodegradable packaging preserving meat quality and safety. Polymers (Basel) 16. https://doi.org/10.3390/polym16091232

Wu C, Sun J, Zheng P et al (2019) Preparation of an intelligent film based on chitosan/oxidized chitin nanocrystals incorporating black rice bran anthocyanins for seafood spoilage monitoring. Carbohydr Polym 222:115006. https://doi.org/10.1016/j.carbpol.2019.115006

Yazdani Z, Shamsaie M, Khayatzadeh J et al (2023) Dietary green-synthesized curcumin-mediated zinc oxide nanoparticles promote growth performance, haemato-biochemical profile, antioxidant status, immunity, and carcass quality in Nile tilapia (Oreochromis niloticus). Aquacult Rep 32:101717. https://doi.org/10.1016/j.aqrep.2023.101717

Ying S, Guan Z, Ofoegbu PC et al (2022) Green synthesis of nanoparticles: current developments and limitations. Environ Technol Innov 26:102336. https://doi.org/10.1016/j.eti.2022.102336

Youssef AM, El-Sayed SM, El-Sayed HS et al (2016) Enhancement of Egyptian soft white cheese shelf life using a novel chitosan/carboxymethyl cellulose/zinc oxide bionanocomposite film. Carbohydr Polym 151:9–19. https://doi.org/10.1016/j.carbpol.2016.05.023

Yusop SM, Sullivan MGO, Preuß M et al (2012) Assessment of nanoparticle paprika oleoresin on marinating performance and sensory acceptance of poultry meat. LWT Food Sci Technol 46: 349–355. https://doi.org/10.1016/j.lwt.2011.08.014

Zaheer T, Ali MM, Abbas RZ et al (2022) Review article insights into nanopesticides for ticks: the superbugs of livestock. In: Oxidative medicine and cellular longevity. Wiley

Zhao L, Zhang M, Wang H, Devahastin S (2020) Effect of carbon dots in combination with aqueous chitosan solution on shelf life and stability of soy milk. Int J Food Microbiol 326:108650. https://doi.org/10.1016/j.ijfoodmicro.2020.108650

Zhao S, Zhang Q, Liu M et al (2021) Regulation of plant responses to salt stress. Int J Mol Sci 22(9): 4609

Chapter 4
Green Nanotechnology in Manufacturing Processes

Kelvii Wei Guo

Abstract To date, the greatest challenge is the foreseeable long-term impact both on human health and environment caused by nanostructures which cannot be observed even with the most sophisticated optical microscopes. Also, it is difficult to see the details of the interactions between nanostructures and the basic components of life.

With the new generation of green nanotechnology arriving to reduce the resources pressures and enhance the relevant invested profits, along with the worldwide requirements of alleviating fossil fuel usage and CO_2 emissions-energy conservation and emission reduction, the utmost rigor techniques are required to scientifically explore in an economic, efficient, energetic, environmentally friendly way. Moreover, the correlated mechanisms must be revealed in detail to sufficiently reduce the harmful effect on human health and environment. Ultimately, the green nanotechnologies are actually explored and realized so-called eco-friendly to human health and environment both.

In the twenty-first century, according to the development of clean technologies, new crosscutting kinds of green nanotechnology (nano-techniques) should be investigated to minimize or solve potential human health and environmental risks/problems, and to encourage replacement of existing products with new nano-products.

Therefore, aim to provide vital information about the growing field of green nanotechnology in manufacturing, the relevant techniques, such as chemical etching, sand blasting, shot peening/laser shock peening, physical vapor deposition (PVD), chemical vapor deposition (CVD), plasma spraying, ion implantation, high pressure torsion (HPT), equal channel angular pressing (ECAP)/equal channel angular extrusion (ECAE), cold rolling, hot rolling and wastes recycle for the friendly environment, are reviewed. Meanwhile, the future prospects for the significantly feasible green nano-techniques are also pointed out.

Keywords Green · Manufacture · Nanostructure · Nanotechnology · Process

---

K. W. Guo (✉)
Department of Mechanical and Biomedical Engineering, City University of Hong Kong, Kowloon, Hong Kong

© The Author(s), under exclusive license to Springer Nature Switzerland AG 2025
J. M. Al-Khayri et al. (eds.), *Green Nanotechnology Applications for Ecosystem Sustainability*, Nanotechnology in Plant Sciences 4,
https://doi.org/10.1007/978-3-031-90766-1_4

4.1 Introduction

In past years, huge efforts have been put into a new-booming field-nanotechnology (offering myriad new materials and methods for scientists and engineers to exploit in new applications) in order to create a desired materials with compelling properties superior to the conventional materials which ultimately save the current energy or improve the relevant efficiency or explore new sources for the practical applications, especially on the commercial scale (Pitkethly 2004). It is well known that nanomaterials (1~100 nm) have attractive properties (chemical reactivity, conductivity, electricity, mechanical strength, optical reflectivity) than the correlated bulk materials (Kostoff et al. 2007). It should be noted that for some nanomaterials, they have the captivating self-assembly capability for ordered structures (Zhang 2003). In addition, they also have vast surface areas per unit volume or weight which are favorable for chemical and electrical reactions (Bjerklie 2005; van Hove 2009). Such compelling properties of nanomaterials steer new technological direction-nanotechnology (Gruère 2012; Handford et al. 2014; Khot et al. 2012). Up to now, due to the scientists' endless heading for breakthroughs of the relevant techniques, nanotechnology has already applied in various fields-from creating new prominent materials, improving solar cell efficiency, replacing toxic chemicals, to cleaning contaminated groundwater, even applying in medical treatment-ultimately targeting and curing cancer, such as chemicals, cosmetics, electronics, energy, healthcare and materials (Andersson et al. 2003; Chandrasekaran et al. 2015; Davis 1997; Dvir et al. 2011; Guo and Tam 2017a; Leso et al. 2019; Merugu et al. 2023; Patolsky and Lieber 2005; Zäch et al. 2006).

To date, current fossil fuel usage is unsustainable and associated with greenhouse gas production. For every people on the earth, an adequate and secure energy supply is pivotal for our daily lives along with minimum adverse environmental effects. Currently, global emissions already address the severe impacts on the climate and global economy. With the growing of the world's population and the expanding of the developing world economies, the demand of energy still continues to be severe issue in the world. Ultimately, even if the potential for increasing low carbon sources of energy is realized, we should note that remaining reserves of fossil fuels will still hold a crucial role in meeting the requirements of world's energy (National Nanotechnology Initiative 2000; Umit 2009).

Furthermore, the microelectronics industry once considered clean is now known as far greater environmental impact than it first appeared. Due to the nanofabricating in the semiconductor industry, especially for the high quality of chips with longer-lasting and more energy efficient needs, nanomaterials are becoming a part of the waste stream from those semiconductor manufactures. In the meantime, more and more nanotechnology-based consumer products are pouring into the market (from computer chips to automobile parts; and from clothing to cosmetics and dietary supplements) (Hutchison 2008; Malysheva et al. 2015; Mckenzie and Hutchison 2004; Moriarty and Honnery 2008).

It should be noted that due to the lack exploration on the correlated mechanism of nanomaterials distribution and ultimate fate in the environment or about their risks to wildlife and people, the foreseeable scientists pointed out a bright future for a new field known as Green Nano-which merges green chemistry and green engineering together with nanotechnology.

In accordance with alleviating fossil fuel usage and CO_2 emissions, the development of "green" processing technologies was proposed (Guo 2012). Actually, green technologies was early addressed for clean drinking water distributed in cities for public daily routine life. Nowadays, owing to the contribution of nanotechnology (such as replacing current environmentally harmful practices with new greener ones), the green nanotechnologies-clean technologies using nanotechnology, to minimize potential environmental and human health risks associated with the manufacture, have been realized and applied in various fields. For instance, (1) adopting nanotechnologies for greening up industries; (2) directly exploring green technologies (water filters, solar cells) (Guo 2018a, 2019, 2023a, 2024).

Actually, it is hard to realize the ideal Green because no matter what we take, there are always some chemicals adopted. Therefore, the correlated rules that should be obeyed are listed as follows: (1) reaction waste prevention: Chemical reactions without waste left or to be treated; (2) safer chemicals and products: Chemicals with little or no toxicity along with full effectiveness and high efficiency; (3) less hazardous chemical syntheses: Little or no toxicity to humans and the environment; (4) renewable feedstocks usage: Raw materials and feedstocks (made from agricultural products or the wastes of other processes) are renewable rather than depleting which made from fossil fuels (petroleum, natural gas or coal) or are mined; (5) catalysts usage, not stoichiometric reagents: Stoichiometric reagents are always taken in excess and perform only once. However, catalysts can work many times in a single reaction with small amounts and recyclable; (6) chemical derivatives avoidance: Derivatives result in additional reagents and more waste generation; (7) atom economy maximizing: The less materials in, the more profit out; (8) safer solvents and reaction conditions usage: As far as possible take innocuous chemicals; (9) energy efficiency: Whenever possible at ambient temperature and pressure to make chemical reactions; (10) after use chemicals and products to degrade: After use breaking down to innocuous substances without any accumulation in the environment; (11) real-time control to prevent pollution: Real-time steer to minimize or eliminate byproducts formation; (12) the potential for accidents minimizing: Minimize the potential for chemical accidents (fires, explosions and releases to the environment) (Dey et al. 2024; Guo 2017a, b, 2018b, 2021; Mohan et al. 2024; Ningombam et al. 2024; Thamarai et al. 2024).

Also, for green nano, the ultimate goals (such as catalysis exploiting, raw materials incorporation maximizing, toxic chemicals minimum usage, waste prevention) are always the research focuses for scientists and engineers. Consequently, it is significant for green engineering to leave a harmful fingerprint on the environment. It indicates that more effort should be put into the products design (encouraging biodegradable materials usage) and processes optimization (enhancing energy efficiency). Life Cycle Assessment (LCA) evaluating a particular product impacts on the environment is a standard for the green approach abiding by. In fact, researchers must consider all the routine of materials processing from the beginning to the end, even covering the practical many years usage and their decomposition and final destination/resting place, such as raw materials mining, emissions releasing, wastes disposing, decomposing, recycling, in order to offer more opportunities to do less harm on the environment to ultimately realize our green life.

Owing to our mother nature, numerous samples give us captivating models in which all materials wisely work at high efficiency along with wastes effective recycling. For instance, the cell is indeed the quintessential green nano factory. On the basis of respecting and seeking to emulate natural processes, environmentally sustainable products and processes are capable to be investigated without harmful solvents taken. Meanwhile, energy conserves, wastes reuse, feedback loops smart controls employ and carry out at ambient temperature/at room temperature. As a result, a new paradigm, proactive rather than reactive, is finally realized without environmental problems. In the meantime, with the aid of nanotechnology, catalysts are capable of ultra-efficient properties, molecular machines self-assembly and targeted operate, wastes detoxify and sun energy generalizes. Predictably, all of them potentially and definitely contribute to long-term sustainability for future generations as more products and the relevant processing procedures successfully grafting with green nano replace the old ones which are harmful and wasteful in the conventional approaches.

Therefore, aim to provide vital information about the growing field of green nanotechnology in manufacturing, the relevant techniques (included both chemical and mechanical methods) are reviewed along with the proposed future prospects for feasible green nano-techniques.

4.2 Easy Chemical Etching for Energy

Due to the current unsustainable fossil fuel usage and the relevant greenhouse gas production, the countless efforts have been putting into exploiting significantly feasible world's eco-energy with minimal environmental impact. Among various energies, solar energy is prominent and definitely appealing as an ultimate energy source (Akbari et al. 2019). With the help of nanotechnology, the relevant efficiency is raised and the costs are lowered. During the process, silicon is inevitably utilized (single crystal, microcrystalline and amorphous). Therefore, surface engineering (SE) is crucial to improve the chemical, mechanical and physical properties of materials. For instance, in order to enhance the mechanical retention, the surfaces with engineered nanostrucutres effectively increase surface area and lower the reflectivity (Chu et al. 2009; Green 2003; Kannan and Vakeesan 2016; Sampaio and González 2017; Yerokhov et al. 2002). Up to now, the silicon wafer with the surface texturing is a popular way to enhance the correlated efficiencies because the reflection loss of surface reduces by photon trapping (Altinoluk et al. 2016; Basu et al. 2015; Herche 2017; Hernández-Callejo et al. 2019). For texturing, an easy

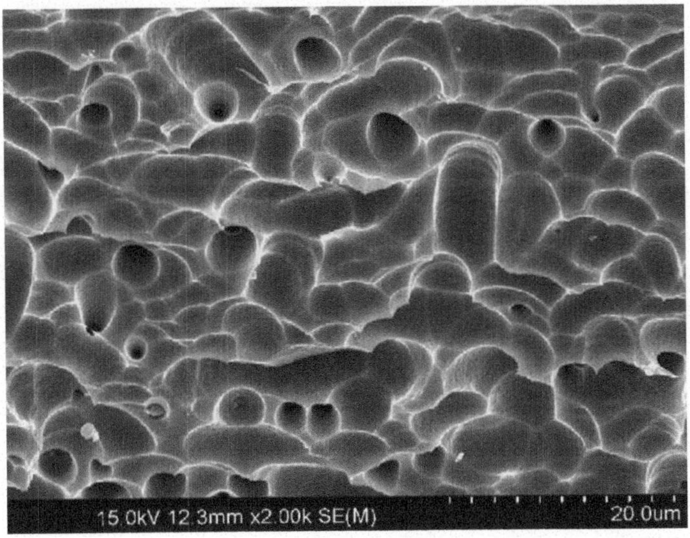

Fig. 4.1 SEM picture of *mc*-Si wafer surface after acid texturing (SDRWT). (Figure constructed by KW Guo)

chemical etching approach (saw damage removal with texturing (SDRWT)) proposed by Ju et al. (2008) is usually utilized as the step formation is avoided between grains along with nanoscale etching depth.

The details of the processed procedure are expressed as follows:

1. Immersing wafers into an ultrasonic bath at 23 °C (solution: 1(3 wt% HCl):10 deionized (DI) water) and ultrasonic cleaning for 3 min;
2. Immersing the pre-cleaned wafers into acidic etching solution (8(HF):21 (HNO_3):10(CH_3COOH):8(DI water)), temperature keeping at 26 °C;
3. After processing 3 min, out wafers from texturing bath;
4. Completely immersing into DI water and thoroughly rinsing;
5. DI water cleaning and drying by a spinning method.

The achieved texturing wafer with nanostructures is shown in Fig. 4.1 which the etching depth is about 4 μm on both sides.

By the vapor texturing, the silicon wafer can be homogeneously texturized with uniform patterns distributed. Moreover, such approach is also sound for very thin wafers owing to the minimal materials loss. Nanoscale porous formation after vapor texturing on *mc*-Si wafer surface is shown in Fig. 4.2 and the correlated high magnification is expressed inside. It indicates that the size of pores distributed on nanoscale porous surface is less than 50 nm. As a consequence, the reflectance spectra of the nanostructured silicon wafers enhance because of interference effects in such processed layer which is distinctively superior to the bulk *mc*-Si.

Nowadays, alkaline solutions are usually taken for crystalline silicon texturing with a fairly flat surface. Under such solutions, random pyramids are formed due to

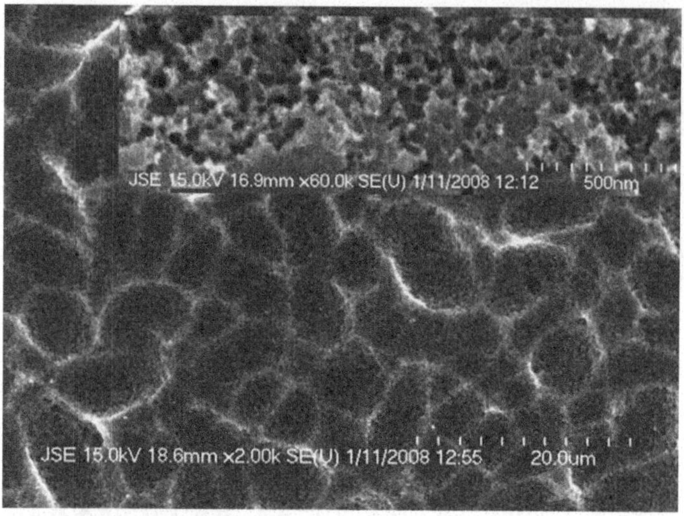

Fig. 4.2 Nano-scale porous formation after vapor texturing (inset-higher magnification). (Figure constructed by KW Guo)

the anisotropic etching. Also, saw-damage etching is always carried out to remove the surface defects and damage caused by wire sawing before the texturing. For saw-damage etching, potassium hydroxide (KOH) solution is generally utilized.

4.3 Easy Chemical Etching for Clinic

Numerous investigations on nanomaterials for biomedical applications reveal that the functions of bone cells mainly rely on the surface with the nanostructures and independent of the compared bulk materials (Manjubala et al. 2006; Price et al. 2003; Webster and Ejiofor 2004; Webster and Smith 2005). For clinic situations, the beneficial contribution of nanostructured surfaces of biomaterials is significant because (1) the only contacting part with bio-environment; (2) different properties compared with the bulk; (3) inert to host tissue without biologically active or toxic substance leaking or releasing; (4) Dominant role in implant-tissue interface (Flemming et al. 1999).

As a kind of processing method, chemical or electrochemical action (flexibility and simplicity) is often conducted on the traditional biomaterials, where acids, alkalis and hydrogen peroxide are taken for nano-structuring the surface of biomaterials (Aladjem 1973). Figure 4.3 expresses the schematic of chemical/electrochemical action. The surface of commercial pure titanium (cp-Ti) with nanostructures processed by the electrochemical approach is shown as SEM images, which the inset is higher magnification of the correlated fabricated-nanostructures. It indicates that the uniformly distributed nanostructures are advantageous to later

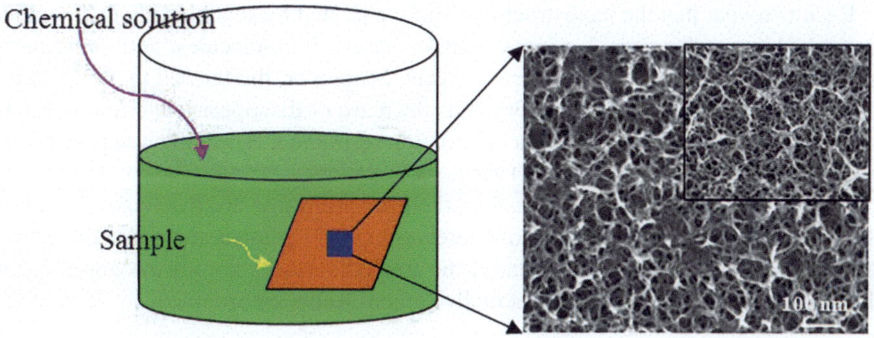

Fig. 4.3 Schematic of chemical/electrochemical action for surface modification of biomaterials. (Figure constructed by KW Guo)

clinical applications. Generally, a native protective oxide film is always formed on the surface of titanium and titanium alloys, which typically includes three different layers: (1) TiO_2 (outer layer); (2) Ti_2O_3 (intermediate layer); (3) TiO (inner layer directly contacting with the metal) (McCafferty and Wightman 1999). Once an oxidizing solution is utilized for titanium, the surface of titanium will react with the etchants through a chain of reactions. As a result, the relevant nanostructures synthesize. Simultaneously, with the etching ongoing in the oxidizing medium, Ti_2O_3 and TiO are transformed into TiO_2. Ultimately, the etching terminates at the underlying metal. As a consequence, passivation layers are naturally created along with new native layered structure (TiO_2-Ti_2O_3-TiO-Ti) reestablishing.

It should be noted that as an easy etching approach, the chemical etchants are always needed and indispensable. In other words, the chemicals are essential to silicon wafer processing for solar energy. Strictly speaking, such processes are operated under unfriendly conditions to the environment and not a 'green nanotechnology'.

4.4 Sandblasting

As a simple and effective method for nanostructuring surface, sandblasting is usually utilized without harmful chemicals involved. During blasting, in order to rough the required surface or remove contaminants from the surface, abrasive particles (SiC, Al_2O_3, TiO_2, HAp) chemically stable or biocompatible are forced onto the surface by high pressure (Hignett et al. 1987; Kern and Thompson 1994a, b). Under the effect of dynamic contact between particles and surface, the surface roughness changes and the surface activity enhances (higher surface energy-physical and chemical activities) (Wennerberg 1998). After that, the surface with desired roughness achieves, especially on (1) surface contaminants cleaning; (2) effective surface area increasing; (3) compressive residual stress producing beneficial to anti-fatigue.

Results reveal that the nanostructured surface treated by sandblasting is favorable for early better adhesion of mineral ions or atoms, biomolecules, and cells, form stronger fixation of bone or connective tissue. Moreover, the formed tissue-reaction layer is thinner and the inflammatory cells decrease or disappear led to microorganism adhesion absence and plaque accumulation restrained, especially compared with the untreated surfaces (Wen et al. 1996).

It is well known that for sandblasting, there are still some disadvantageous effects (1) various kinds of blasting media intervening; (2) blasted materials distorting (Miyakawa et al. 1996). Additionally, the treated surfaces are always rinsed with alcohol and acetone which is not actually 'green' to the environment.

4.5 Shot Peening/Laser Shock Peening

As a similar method to sandblasting, shot peening is also cold working process. In shot peening, small spherical media (shot) are always utilized for bombarding the required surface under the controllable processing parameters (direction, intensity, power). Just like a tiny hammer, each shot push onto the surface that needs to be processed. After that, small dimples or indentations form on the processed areas with high stress in compression which is beneficial to the practical applications. Under the effect of shot repeating impact on the surface, an even layer with residual compressive stress is successfully fabricated. As we all know that in a compressively stressed zone cracks are not easy to initiate or propagate (Kalainathan and Prabhakaran 2016; Soyama and Korsunsky 2022). Therefore, the compressive stresses achieved by shot peening considerably enhance the life of the processed part which are effectively prevent the failures of fatigue and stress corrosion cracks from originating at surface owing to the suppression on crack-opening by shot peening induced compressive residual stress. Results show that the shot peening induced maximum compressive residual stress is able to half the yield strength of the processed materials. In addition, under the effect of shot peening cold working, the hardness of the processed surface will also increase. Subsequently, in the practical applications, shot forming is utilized the hybrid effects of compressive stresses and cold working to form metal parts (Aparicio et al. 2003; Arulvel et al. 2023; Sealy and Guo 2010; Yocom et al. 2018).

As an advanced shot peening technology, laser shock peening was explored with the distinctive characteristics-contamination-free, no-media and non-contact. Before peening, a thin layer of water and a protective ablative layer (paint or tape) are always taken for covering the processed materials. During the shock peening, a shock wave induced by a short-lived plasma (generated by laser striking the ablative layer) travels into the processed materials, where the processing parameters of laser are 5–15 GW/cm^2 intensity, 10–30 ns nano-pulses and 3–5 mm beam width. The compressive residual stress induced by laser shock peening enhances the fatigue life of the processed materials and effectively restrain the occurrence of stress corrosion

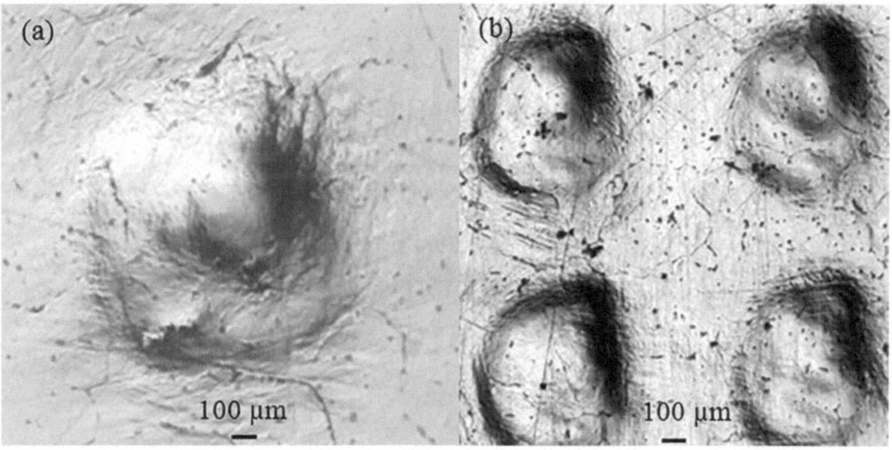

Fig. 4.4 Dent topography. (**a**) Single LSP. (**b**) Sequential LSP. (Figure constructed by KW Guo)

cracking. The optical images of dents by single laser shock peening (LSP) and sequential laser shock peening are shown in Fig. 4.4.

4.6 Physical Vapor Deposition (PVD)

As an effective surface modification process, physical vapor deposition (PVD) coating is often adopted (Reichelt and Jiang 1990). Figure 4.5 shows the correlated processing flow and SEM images included inset show the relevant TiN coating (famous for high hardness, high temperature stability, high abrasion resistance, low friction coefficient) with nanostructures. The quality of coatings relies on the precisely controllable parameters. It indicates that in cross-section the fine columnar structures closely pack without any pore. Meanwhile, on the surface of the deposited coating nanostructures uniformly distribute which are favorable for practical applications later.

Results (Vadiraj and Kamaraj 2006) reveal that TiN coatings has an outstanding ability restrain surface intrusions and extrusions formation which are the dominant sources for fatigue failures. Compared the friction coefficients of the TiN coating materials with uncoated, it is obvious that the uncoated materials are experienced with higher friction. Owing to TiN coating excellent anti-wear properties, the correlated friction distinctively reduces. It should be noted that the deposited coating cannot withstand the longer fretting process and the ruptured coating will form the networks of cracked layers at the last fretting stage as potential abrasives harmful for the surface of protected materials (Guo 2015, 2023b; Hussein et al. 2020).

As another popular coating, carbon-based coatings/films are captivating with low friction coefficient, high wear and corrosion resistance, good bearing capacity and bio compatibility. It is well known that diamond-like carbon (DLC) coatings/films,

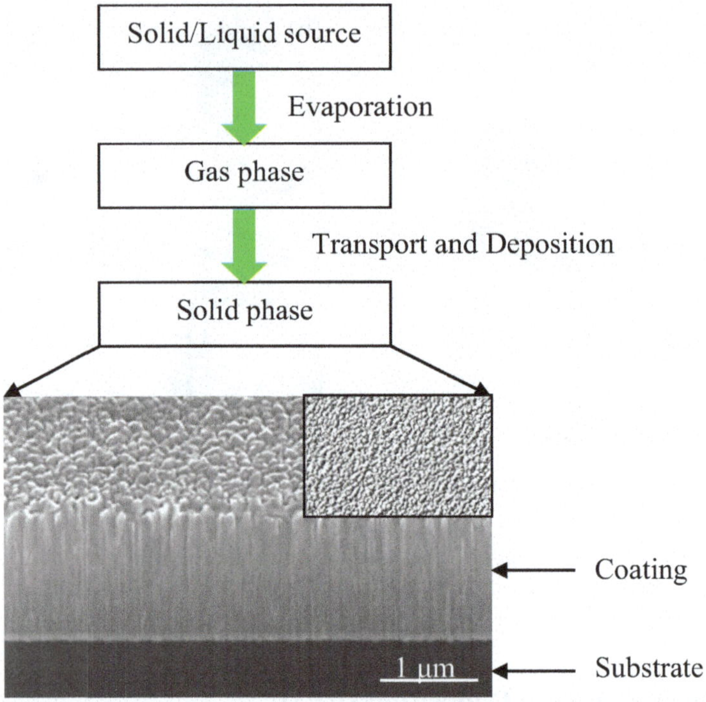

Fig. 4.5 Diagram of physical vapor deposition (PVD) process flow. (Figure constructed by KW Guo)

as an amorphous structure with a mixture of sp^2 and sp^3 C-C bonds, widely open various gates of applications. Moreover, the fraction of sp^3 C-C bonds directly affects DLC coatings/films mechanical properties. Therefore, DLC coatings/films are always doped with different suitable elements (Co, F, Mo, N, O, Si, Ti, W, V and their combinations) to restrain undesirable great intrinsic stress and increase the relevant properties to meet with the practical requirements (Anttila et al. 1997; Erdemir and Martin 2018; Martinez-Martinez and De Hosson 2014; Santavirta et al. 1999; Zia and Birkett 2021, 2023).

Zdravecká et al. (2012) utilized pulsed arc discharge PVD to dope amorphous carbon (a-C) films with lower sp^3 content and filtered pulsed arc discharge PVD to dope tetrahedral carbon films (ta-C) with a higher sp^3 (up to 85%), where the thickness of a-C coating/film was 1.25 mm and ta-C coating/film was 0.70 mm. Results indicate that the evolution characteristics of thin hard carbon layers are always an on-going theme. The correlated mechanism still needs to be revealed in detail to ultimately make coating properties hand with the practical application effectively.

4.7 Chemical Vapor Deposition (CVD)

Chemical vapor deposition (CVD), different from physical vapor deposition (PVD), is a process for depositing the relevant films onto a substrate by the chemical reaction between precursors at certain or elevated temperatures with atmospheric or lower pressures. During the deposition, volatile precursors or the correlated by-products absorb and react at the surface of the substrate to achieve the required coatings/films (Bhowmik and Rajan 2022; Choy 2003; Schalk et al. 2022; Zhang et al. 2020). The properties of deposited coatings/films (composition, crystalline structure, elements distribution, interface, residual stress) are controlled or tailored by the depositing parameters.

Nowadays, CVD is a basic manufacturing tool addressed from our daily routine life to the current cutting-edge electronics. Moreover, with the relevant theory exploring, CVD opens a door to a variety of materials and pushes the materials research in new directions (large-scale sheets of graphene production, solar cells exploit on a sheet of paper or plastic) (Hossain et al. 2020; Mbayachi et al. 2021; Naghdi et al. 2018; Sahoo et al. 2022).

Additionally, as a booming research field, biomaterials are often processed with the aid of CVD. Titanium, as most common orthopedic implant or bone substitute materials, is prone to cause inflammatory reactions (Cassar et al. 2012; Cozza et al. 2011; Cvijović-Alagić et al. 2011; Dalili et al. 2010; Hoque et al. 2022; Quinn et al. 2020) in the practical applications, even though titanium possesses outstanding mechanical properties and biocompatibility (Aparicio et al. 2011a, b; Ravelingien et al. 2010; Sunil et al. 2022; Xu et al. 2019). Titanium carbide (TiC) coating/film with high melting temperature and hardness, and excellent chemical and thermal stabilities has been successfully deposited by CVD to meet with the requirements of commercial applications. To date, titanium carbide coating/film is able to deposit on the surface of substrate at various conditions (different types of gas mixtures and gas ratio, different substrate heating temperatures, duty cycle) by CVD methods (Shanaghi et al. 2012a, b).

4.8 Plasma Spraying

As another approach for structuring the surface of workpieces at the nano scale, plasma spraying is also utilized. Figure 4.6 shows the correlated principle of plasma spraying. Similar to thermal spraying (Heimann 2006), a plasma jet taken at high temperatures projects particles or accelerates fine particles onto the surface of workpieces. After that, the projected particles condense, fuse, and cool down after hitting the surface of workpieces and form uniform coatings. Considering the effect of gravity, the required workpiece is often vertically setup.

Up to now, plasma spraying is the most versatile of the thermal spray processes because of the extremely high plasma temperatures associated with the plasma

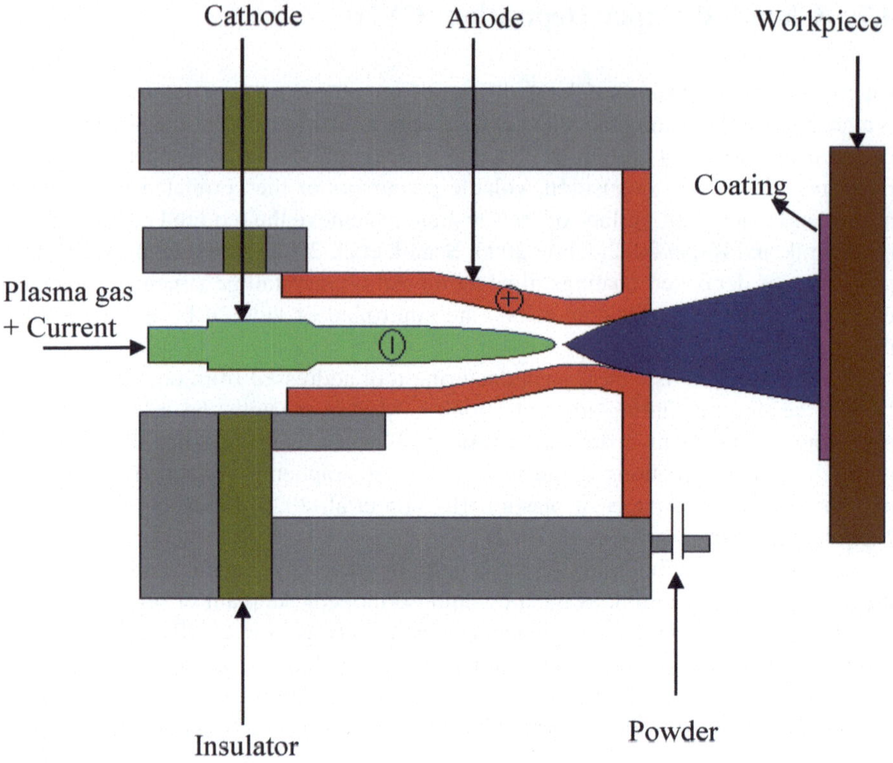

Fig. 4.6 Principle of plasma spraying. (Figure constructed by KW Guo)

plume, which makes it possible to melt or vaporize any known workpiece and commonly used in various fields, such as aerospace, oil and gas, power generation, semi-conductor industries and biomedical applications (Jacobs et al. 2012; Singh et al. 2021; Wise et al. 2012). With micrometers or several millimeters coating on the relevant workpieces produced by plasma spraying, the deposits (alumina, titania, zirconia) on the surface possess prominent mechanical properties along with the surface roughness increment and the surface chemistry modification.

It should be noted that in clinical applications, there is a controversy on the binding strength between the bone and the implant, and plasma sprayed coating's particles releasing into the host tissue (Urban et al. 2000). Also, the techniques related to plasma spraying are usually aided by other above-mentioned methods, such as chemical etching or blasting (Martin et al. 1995; Taba Junior et al. 2003). Zinger et al. (2004) and Landolt et al. (2003) reveal that with the help of sandblasting and etching, the plasma sprayed surface enhances the osteoblast activity owing to the combined effect of nanometer and micrometer structures in the coatings/films. Results elucidates that both nanoscale and microscale parameters of topography influence cell attachment *in vitro* and the bone implant interface *in vivo* along with osteoblasts prone to attach on submicron-scale structures (Anselme et al. 2002; Zhu et al. 2004). Moreover, surface modifications at the nanoscale are propitious to influence ionic or biomolecular interactions between coated surfaces with the host tissue by changing the chemical reactivity of biomedical materials resulted in a different protein adsorption or altering the mineralization of *de novo* bone formation (Flemming et al. 1999; Teixeira et al. 2003).

More attractively, due to the increasingly concerning human health, plasma spraying has been increasingly utilized in recent years for hydroxyapatite (HAp) coatings (Ong et al. 2004) with surface structures ranged from nanometers to millimeters. It is well known that as a good candidate for enhancing the biocompatibility of stainless steel, Ti and Ti-alloys, pure and crystalline hydroxyapatite ($Ca_{10}(PO_4)_6(OH)_2$) coating is extensively adopted for metallic implants due to its structure close to human bone and compatible with body fluid (Khandelwal et al. 2013; Sargin et al. 2021; Sukjai et al. 2012; Tan et al. 2012; Tucker et al. 2021; Wu et al. 2012).

For clinical applications, the common reason for coating failure is due to amorphous calcium phosphate (ACP), which is formed in a layer attributed to exceptionally high cooling rates at the interface of coating/biomaterials for semi-molten or molten calcium phosphate droplets rapid quenching. Consequently, because the formed continuous amorphous calcium phosphate layer has comparatively high solubility, preferentially dissolution *in vivo*, ACP layer acts as a fracture path to weaken the coating/biomaterials interface mechanical integrity. So far, to control ACP and other phases (such as oxyhydroxyapatite, tri-calcium phosphate) with degrading effects on HAp coatings is still a challenging theme and a more detailed understanding of the correlation between requirements for practical applications and coating properties should be systematically explored (Abdelkebir et al. 2012; Barnes et al. 2012; Cizek and Khor 2012; dos Santos et al. 2012; Karthik et al. 2023; Minati et al. 2017; Saber-Samandari and Gross 2012, 2013; Xu et al. 2012).

4.9 Ion Implantation

As a kind of plasma processing application in surface treatment, ion implantation has been utilized in a variety of fields. As a usual, an ions source along with an accelerator are required for the implantation process. Near or surrounding the workpiece that needs to be implanted, a plasma discharges and the relevant ions produce. Consequently, the ions are extracted from the discharged plasma and accelerated into the workpiece.

Actually, the aim of ion implantation is just only modify materials surface properties and the implanted workpieces range from metals and alloys to ceramics, insulators, polymers and semiconductors. Under the effect of ions bombarding and penetrating into the material surface, the lattice structure of the implanted surface changes with the surface properties (adhesion, catalysis, corrosion, dielectric, fatigue, friction, hardness, magnetic, superconductivity, resistivity, toughness, wear) simultaneously enhanced (Curtis and Wilkinson 1997; Hench and Wilson

1984; Hench and Polak 2002; Ito 1999; Kasemo 2002). The penetration depth is often in micron level. Also, the final results rely on how the beam is produced and accelerated in a multi-step process.

Due to the wide range of workpieces that can be ionized, vast species can be utilized as ions implanted into the required surface at low pressure in the vacuum chamber. Therefore, as an approach for surface treatment, the nature of the process of ion implantation is physical rather than chemical. However, for the workpieces that need to be processed, numbers of chemical reactions may occur during the implantation. Research, devoted to modifying the surface of biomaterials with C, N and O implantation, reveals that the correlated films (TiC, TiN and TiO) can be uniformly formed on the implanted surface which effectively prohibit the harmful ions releasing from the relevant biomaterials. *In vitro* biological results show that the implanted surface favors the osteoblasts proliferation (Bilek 2014; Jagielski et al. 2006; Lau et al. 2022; Mitragotri and Lahann 2009; Rautray et al. 2011; Stevens and George 2005). But, ion implantation is not only costly, but the related efficiency is also slow.

4.10 High Pressure Torsion (HPT)

For mechanical approaches (alternate options for green nanotechnology) to achieve nanocrystalline metals and alloys, severe plastic deformation (SPD) is usually utilized. In the process of HPT, high pressure and strain extremely produce with the distinctive results of exceptional grain refinement and the size of achieved nanograins is about 50–100 nm. Figure 4.7 depicts the correlated HPT setup. Generally, the workpiece that needs to be processed is a thin disk and its thickness

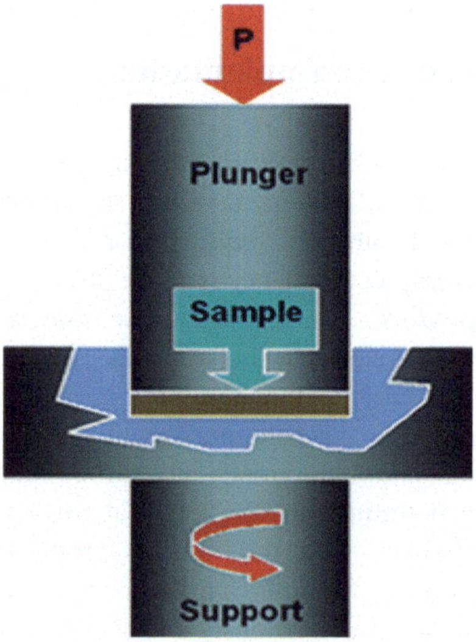

Fig. 4.7 Schematic HPT setup. (Figure constructed by KW Guo)

is about 0.8 mm. In a high pressure working area, the workpiece is placed and constrained by high pressure between two massive anvils as one keeps steady and the other one rotates. It should be noted that such approach leads to the material outward flow around the periphery of the processed workpiece (Azzeddine et al. 2022; Edalati and Horita 2016; Valiev 2004; Zhilyaev and Langdon 2008).

4.11 Equal Channel Angular Pressing (ECAP)/Equal Channel Angular Extrusion (ECAE)

For achieving metals or alloys with ultrafine grains, equal channel angular pressing (ECAP)/equal channel angular extrusion (ECAE) is often adopted. Figure 4.8 shows the correlated setup of ECAE. During the equal channel angular extrusion process, the shear plane undergoes/bears intense shear. The workpiece is forced through a channel that contains a sharp bend, resulting in large deformations at the shearing plane. Repeating the process creates very large strains.

Moreover, several passes impose to the workpiece (covering the entire cross-sectional area) to obtain exceptionally high strain. Meanwhile, an equivalent strain ε of about 1 is exerted by each pass on the workpiece. Even for hard-to-deform materials, it is feasible for ECAP/ECAE operation at elevated temperature. Under the beneficial effect of ECAP/ECAE, ultrafine grained metals or alloys can be successfully attained with nano-structural features (nanoparticles (NPs), sub-grain structures, sub-micrometer grains). Additionally, in the first one or two passes, grain

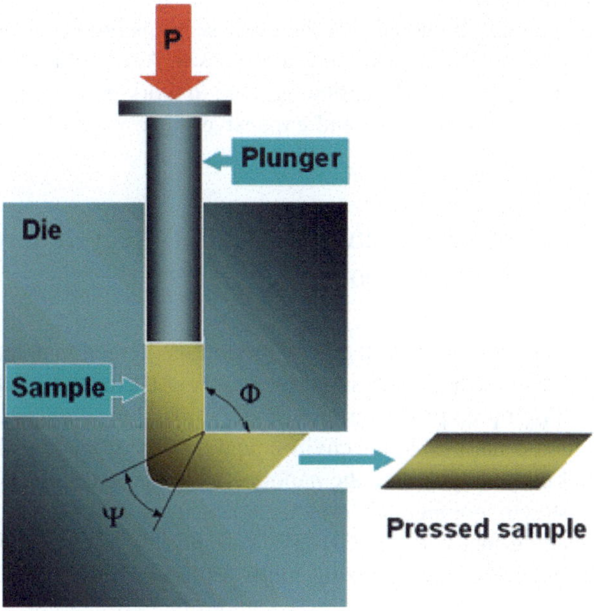

Fig. 4.8 Schematic diagram of ECAE process. (Figure constructed by KW Guo)

refinement with some characteristics of dislocation (cells, microbands, subgrains) establishes and transforms to ultrafine grains during further straining (Li et al. 2005; Valiev and Langdon 2006). To date, as a most popular severe plastic deformation (SPD) processing approach, ECAP/ECAE has been commercially conducted on various metals and alloys for grain refinement (Beyerlein and Tóth 2009; Hashemi et al. 2023; Huang and Langdon 2013; Rakshith and Seenuvasaperumal 2021). However, the size of final grain, phases and the fraction of boundaries with high angle grain strongly rely on the processing parameters as well as the properties of materials that need to be processed.

4.12 Cold Rolling and Hot Rolling

As the early and now mature mechanical processing techniques used to create sheet metal (But, cold rolling produces thinner sheets), the workpiece is processed by passes through rollers at different working temperatures, where for cold rolling, the working temperature is below the recrystallization temperature of the workpiece and for hot rolling, the working temperature is above the recrystallization temperature (which new grains are formed). But the temperature cannot be too high, otherwise the workpiece will burn and become unsuitable for the practical applications. Moreover, it is crucial to precisely control the range of temperature during the process (Aktepe and Erkal 2023; Dodd and Boddington 1980; Jeswiet et al. 2008; Laithy et al. 2019; Sakai et al. 2014; Sales et al. 2020; Song et al. 2006; Takahashi 2001; Toros et al. 2008). The difference between cold rolling and hot rolling is listed in Table 4.1.

It should be noted that the cross sections produced by hot rolling is thinner than that of cold rolling with the same passes. Also, cold rolling can reduce the grain size owing to Hall-Petch hardening. Due to working at room temperature, cold rolling is a more labor intensive and expensive process than hot rolling.

4.13 Wastes (Ethanol/Alcohol) Recycle Green Nanotechnology

As can be seen from the above-mentioned green nanotechnologies, ethanol/alcohol and acetone are always and inevitably used for the workpiece cleaning. Moreover, etching-based processes are usually utilized for the micro/nanostructures fabrication which is harmful to the environment. In order to reduce the chemical contaminations and ultimately realize the nano-technique green/ecofriendly to the environment, the waste ethanol/alcohol (directly taken from the ethanol/alcohol after the specimens cleaning-met with the GB-HW06 standard) was utilized by home-made equipment for recycling (synthesizing multi-walled carbon nanotubes (MWCNTs)). It is well

Table 4.1 Difference between cold rolling and hot rolling

	Cold rolling	Hot rolling
	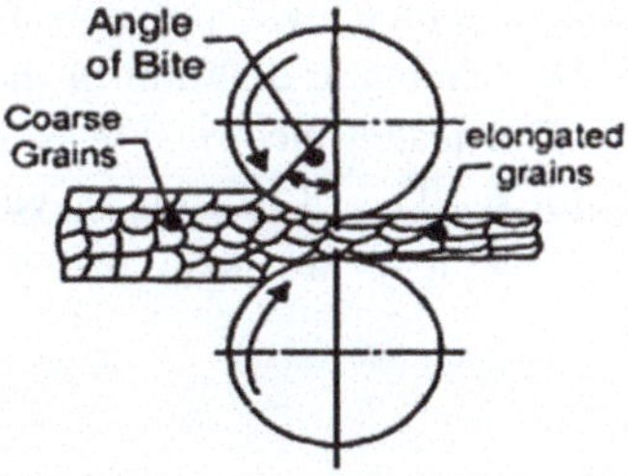	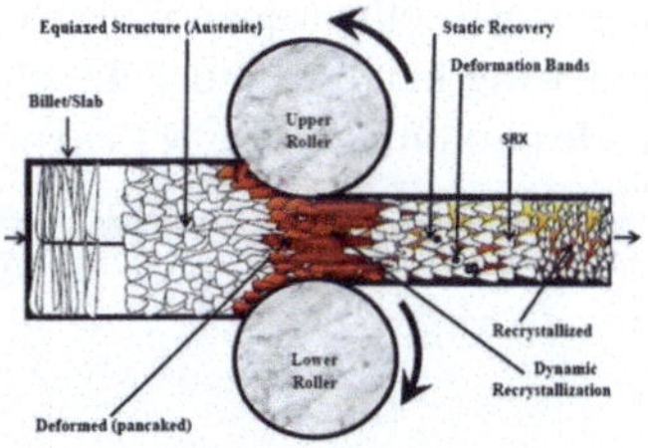
Working temperature	Below recrystallization	Above recrystallization
Heavy reduction area	Not possible	Possible
Thin section thickness	≥0.01 mm	≥1.00 mm
Surface morphology	Oxide-free & smooth surface	Grooves and metal oxides
Hardening effect	Yes	No
Property after rolling	Yield and tensile strength increased; hardness increased along with cracks occurring; ductility reduced	Toughness and strength increased; grains refined; texture achieved
Dimensional tolerance	High (~0.002 mm)	Low
Friction coefficient between rolls and workpiece	Low	High
Roll radius	Small	Larger than cold rolling
Experimental measurements	Easy	Difficult
Theoretical analysis	Easy	Difficult

Table constructed by KW Guo

known that nano-field has undergone unprecedented growth and attracted a great deal of attention (Chadar et al. 2021; Gulati et al. 2022; Mishra and Sundaram 2023; Moyer-Vanderburgh et al. 2023; Rathinavel et al. 2021; Seah et al. 2011; Zhuo et al. 2010). With the aid of nanomaterials, the electrical, magnetic, mechanical and magnetic properties are significantly enhanced. One of the most exciting classes of nanomaterials is represented by the carbon nanotubes (CNTs) (Iijima 1991, 1993). By this method, CNTs can be easily synthesized at low cost. Meanwhile, such technique is green/friendly to the environment (Guo 2020). Moreover, a new microreactor fabricated with such CNTs provides a promising potential application in μTAS (Micro total-analysis systems) for bioanalysis (Guo and Tam 2017b).

4.13.1 CNTs Synthesis

The waste ethanol/alcohol, as a carbon source, was put into a ceramic container under the heater. Figure 4.9 shows the schematic diagram of the correlated home-made equipment for CNTs synthesis. The heater (electrical resistance-T_{max} = 1680 °C) size was 50 × 30 mm along with Φ100 × 150 mm as vacuum chamber. A DC power supplier (38~40 A) was taken as the heat resource in the vacuum (1 × 10^{-2} Pa).

4.13.2 Morphology of Synthesized Vertically-Aligned CNTs

Figure 4.10 expresses the morphology of the synthesized vertically-aligned CNTs where the CNTs length is about 20~35 μm. Also, Fig. 4.11 shows the correlated TEM image. It illustrates that the diameter of the synthesized MWCNT is approximately 3~5 nm. As a kind of compelling bio-applications, the fabricated microreactor with the synthesized CNTs is shown in Fig. 4.12, where the size of microreactor unit is 100 μm × 100 μm and the gap between the reactor units is 35 μm.

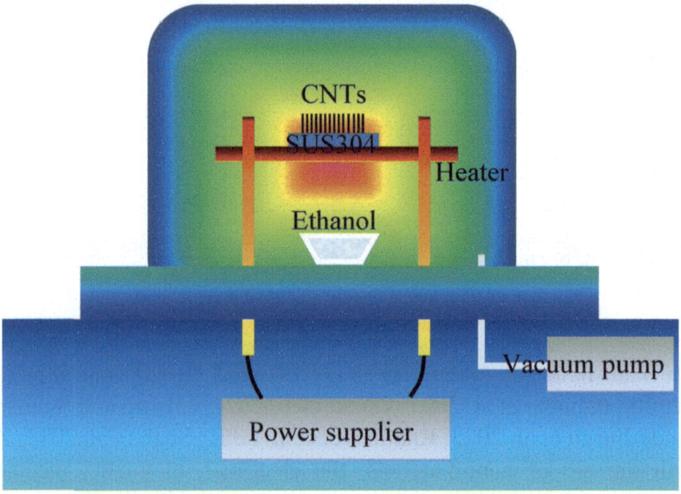

Fig. 4.9 Equipment for CNTs synthesis. (Figure constructed by KW Guo)

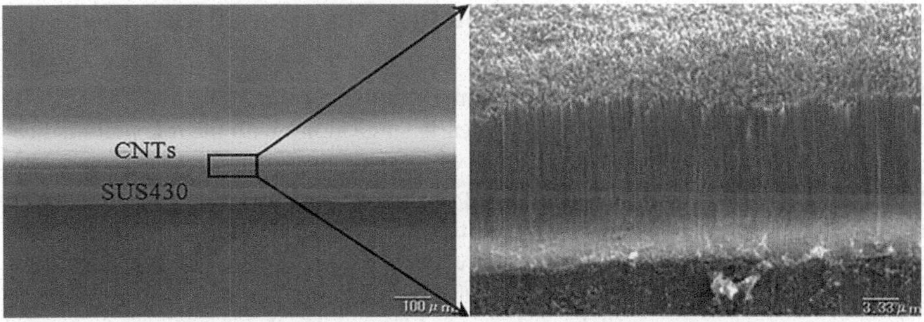

Fig. 4.10 Synthesized vertically-aligned CNTs on SUS430. (Figure constructed by KW Guo)

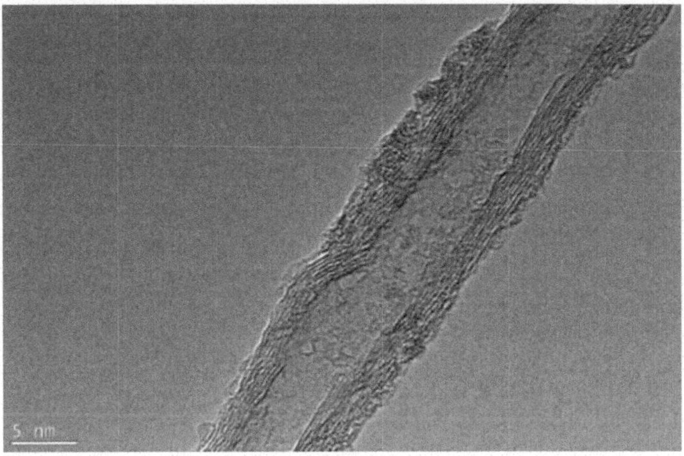

Fig. 4.11 TEM image of synthesized CNTs. (Figure constructed by KW Guo)

Fig. 4.12 Microreactors with synthesized CNTs. (Figure constructed by KW Guo)

4.14 Green Nanotechnology Potential for Human Health

Nowadays, with the increasing requirements of human health, more and more efforts are focusing on diseases diagnosing at an early stage along with the relevant medical conditions. It is a trend to realize the early warning on the health malfunctions which is well in accordance with the old saying: prevention is better than cure. Therefore, it is significant to warn or diagnose the health at earlier stage. As a consequence, man can gain a better chance to prevent a more serious conditions. With the aid of new green nanotechnology related to bio-techniques, the correlated diagnostic processes, efficiency and precision will be enhanced. Meanwhile, the initiation of targeted treatments will be precisely pinpointed along with the profound breakthroughs on successive treatment responses.

It should be noted that the tailored approaches for many diseases diagnosis, prevention and treatment with green nanotechnology should be explored by the effective integration of nanotechnology with medicine and biology. Owing to the booming growth of nanotechnology in the pasting years, it opens our eyes to new fields and attractively makes us closer vividly visualize cellular and molecular structures. As a result, we can directly tell or differentiate between normal and abnormal cells. Moreover, we can also quantify and detect minute amounts of signature molecules produced by these cells. All of these are based on well-engineered nano-tools to assist us to improve the relevant capability on medical fields, especially for clinicians.

4.14.1 Current Directions for Diagnostics

It is well known that for the developed and developing world, improved diagnosis is same required. Actually, it is crucial for individual patients to diagnose accurately and quickly. Consequently, patients can benefit from the precisely pinpointed treatments in addition to ensuring the relevant resources use efficiently. Furthermore, the spread of infectious diseases can be effectively limited. For instance, cancers (related to breast, cervix and colorectum) can be screened and early detected to finally realize mortality reduction. There are more than thirty-three million people living with HIV all over the world, but only 10% are aware about it. Also, more than eight million TB new cases are annually reported which many are undiagnosed. For developing countries, such diseases are a huge burden together with deaths from malaria in one million per year. In order to overcome such issues, better systems should be explored on the basis of resource-limited settings and ultimately realize the diseases detection as early and quickly as possible. Meanwhile, the effectiveness of treatments can be monitored or real-time controlled.

Therefore, the techniques integrated with nanotechnology (especially on green nanotechnology-such as identifying biomarkers, sensitive analytical tools) should be explored to achieve profound breakthroughs in accurate early detections with smaller samples and more precise and complete data just only from a single non-invasive detection or measurement. On such conditions, more and more rich information for precise diagnoses will ultimately achieved to reduce the correlated need for diagnosis and staying in hospital for unrequired treatments along with the associated costs. With the improvement of biomonitoring techniques, the disease risks can be individually early warned and the individual predisposition can also easily be identified along with other genetic traits. With the nano-techniques pushing forward, the personalised treatment and medication can be humanized or tailored to the specific individual needs, especially on chemical genetics and the targeted genotype rather than mass phenotype to couple with basic treatments.

Besides usual detections, chemical sciences is extraordinarily important for the medications and therapies, particularly on the effectiveness and safety. Consequently, a profound breakthrough or exploration on the chemistry of disease progression is urgent and the correlated investigations and understanding should be carried on to enable the drug treatment continuity over the life cycle of diseases. In the meantime, point-of-care diagnostics should be taken for monitoring treatment efficacy and disease progression to realize the responsive treatment, for instance, drug dose changes, hospital staying reduction. It is foreseeable that with the captivating nano-technological achievements in therapeutic devices and diagnostic techniques, more effective combined or hybrid devices will be successfully exploited.

4.14.2 Current Directions for Hygiene and Infection

In general, the infectious diseases lead to more than a fifth of human deaths and a quarter of ill-health globally. To date, more and more people died from major worldwide infectious diseases: childhood infections, diarrhoeal disease, HIV, lower respiratory tract, malaria and TB.

Recently, new infectious diseases, (for instance, COVID and SARS), antibiotic resistance challenges (Some strains of bacteria are now resistant to the antibiotics that have previously been used to treat them), current diseases worsening due to climate change, the nuclear contamination of Japan government discharging the nuclear contaminated water into the sea, will be severe in the twenty-first century.

As the main challenge, by pushing the cutting-edge techniques forward with the detailed mechanism revealed along with the technologies and practices of the anti-infective portfolio establishing, the acquired infections will be effectively minimized and even prevented.

For achieving such destination, the detailed and pinpointed understanding or mechanism of bacteria and viruses at a molecular level, even the transition of disease across species. Therefore, such progresses must be moved ahead with the profound breakthroughs and explorations in the relevant detection strategies, for instance:

(a) Exploring sensitive, effective, cheap and fast sensors for bacterial infection.
(b) Exploring new eye-catching materials with cutting-edge techniques for clinical environments, such as coatings and paints.
(c) Exploring new generation materials or enhancing materials coupled with nano-engineering tools for detecting and reducing airborne pathogens, such as air filters and the relevant sensors.

Consequently, it requires an enhanced capability to control or tailor and cope with such issues as quickly as possible on the basis of the increased incidence of new genetic and resistant strains. Moreover, it is urgent to explore novel synthetic vaccines and, antibacterial and anti-infective agents for the current situations in the twenty-first century.

4.14.3 Current Directions for Materials and Prosthetics

It is well known that organ transplants are worldwide severe issues and always required urgently. However, more people often died while on the waiting list. Moreover, a large number of patients never have a chance to appear on the waiting list. Up to now, it is estimated that the need for organ transplants is more 50% than the available. On such conditions, it is a continuing need to explore new materials with nano-techniques both for orthopedic implants (such as ankles, knees, hips) and the conventional prosthetics (artificial limbs), Moreover, the relevant new materials for prosthetics with compelling characteristics (cheap, effective, functional, sensitive) need to be explored, especially for artificial organs, eye lenses and tissues. Up to now, exploring replacement organs to enhance sustained function fully is still a main challenge for scientists.

Therefore, on the basis of detailed understanding and mechanism revealed relevant to the chemistry at the interface of biological and synthetic systems, profound breakthroughs in biocompatible material chemistry and polymer are significantly required for artificial limbs, surgical implants and equipment. Additionally, clever or smarter and/or bio-responsive drug delivery devices are also required to develop for asthma, cardiovascular disease, chronic pain relief and diabetes. Meanwhile, researching biological macromolecule materials as templates will be a hot point.

Also, it requires further investigations into cell reprogramming, stem cell transplantation, tissue types, gene therapy, soluble molecules and tissue engineering for repairing, replacing or regenerating cells, tissues, or organs. Moreover, the capability to modulate or control neural activity should be distinctively enhanced (such as brain degeneration treatment, neuron conduction modification).

Nowadays, for biomaterials and corrosion protection engineering, it is popular and attractive to explore biocompatible, biodegradable, osteo-conductive implant materials. With the relevant nano-techniques booming, as approved permanent and effective metallic implant materials, stainless steels, Cr-Co alloys, titanium and its alloys, and poly-L-Lactic acid as polymer implants are always utilized in clinical applications. However, for above-mentioned metallic materials, due to the possibility of toxic ions releasing in human environment, it often requires a second surgical operation to remove the correlated metallic implants after the patient tissue has completely healed. Furthermore, as traditional metallic implants, due to higher modulus of elasticity (stainless steel (SS): about 200 GPa; Cr-Co alloys: about 230 GPa; titanium alloys: 115 GPa), implants always carry a greater portion of the applied load. As a result, the bone beneath will experience a reduced load and will lose density in response. For polymer implants, due to their unsatisfactory mechanical strength, the applications are relatively limited.

It should be noted that magnesium based alloys are catching scientists' eyes once again with the new approaches introduced. It is promising that magnesium based alloys have possible ability to replace some metallic alloys in-used owing to the fascinating characteristics of magnesium/magnesium alloys (such as (1) lighter weight: only 1/5 of stainless steel and chrome-cobalt alloys, about 1/3 as dense as titanium based alloys; (2) more attractive and exciting elasticity modulus of magnesium based alloys, around 45 GPa, much more closely matched to that of bone beneficially preventing or reducing the likelihood of stress shielding and the associated loss of bone density (Witte 2010)). As a consequence, magnesium based alloys are ideal candidates as implant materials. Also, as the 11th most abundant element by mass in human body and basis element for bone growth and strength, nearly 50% of the ions are found within bones. Additionally, a substantial amount of Mg is daily taken into the body because magnesium ions are essential to all living cells (Song and Atrens 2003; Staiger et al. 2006; Willbold et al. 2011; Witte et al. 2005; Yun et al. 2009).

To date, there are still many obstacles for magnesium based alloys as perfect surgical materials although they (as biodegradable implants) possess many desirable properties. Due to their relatively brittle and tendency to be broken when bending magnesium alloys based implants to properly fit during the surgery, magnesium biodegradable implants cannot be utilized on a large scale. More seriously, the mechanical and physical properties magnesium alloys based implants obviously reduce due to the correlated low corrosion resistance in human body. Therefore, more efforts should be put into the relevant fields to overcome the drawbacks of magnesium alloys based implants with the final magnesium biodegradable implants applications on the commercial scale.

4.14.4 Current Directions for Drugs and Therapies

Every year, millions of deaths are worldwide caused by chronic diseases (such as cancer, cardiovascular, chronic respiratory, diabetes and others including joint problems and mental health problems). Such issue is a stubborn headache for both the developing and developed world.

More twice the number of deaths per year than infectious diseases, maternal and perinatal conditions and nutritional deficiencies combined caused by chronic diseases and the situation becomes worse and increases with high ratio. Subsequently, new generation drugs and therapies directly targeted above-mentioned diseases should be exploited to globally save human lives significantly. During these processes, the chemical sciences play a crucial role in fully understanding or revealing of the relevant basic sciences on entire drug discovery, development and healthcare landscape. After that, new kinds of the relevant therapies can be worldwide realized more effectively and efficiently. So, in order to gain such goal, some profound breakthroughs must be achieved in:

1. New biological targets identifying.
2. Chemistry of living organisms (including structural biology) detailed understanding.
3. Small molecules enhanced large molecule interactions (such as protein-DNA interactions) synthesizing and designing.
4. Chemistry and biological entities integrating for enhancing drug delivery and target.
5. Advanced chemical tools to improve clinical investigations (such as contrast agents, *in vivo* biomarkers, non-invasive monitoring) exploring.

All of these, it is a need to more detailed revealing at the molecular and cellular level with the aid or support of physical organic chemistry. Moreover, for every aspect of the drug discovery and development process, it is vital to better understand analytical sciences. Meanwhile, for validating the safety and effectiveness of the exploited drugs, it is a basic and essential part of costly, complex and lengthy drug discovery and exploring processes which communication between and within cells should be deeply studied along with external factors *in vivo* to combat disease progression. By fully revealing the overall map of drug absorption parameters within the body (for instance, the blood brain barrier), precisely targeting to particular diseased cells will improve safety and efficacy. At the same time, the relevant model systems should be developed to underpin the better understanding or detailed mechanism of extremely complex biological systems and how interventions affect these living systems over time.

At last, optimal, personalized treatment and medication designed to the individuals with specific needs. In order to realize such targets, developed breakthroughs related to deliver differentiated, specific prevention and treatment on an individual basis should be explored in advanced pharmacogenetics for tailored treatment regimes. In addition, the rapid testers should be developed by lab-on-a-chip technologies to realize the quick and effective diagnosis and treatment. Also, other techniques (such as sensitive molecular detectors, non-invasive bio-monitoring tools) should be explored in detail to reveal the information on how physical interventions work in living systems over time. For scientists, it is urgent to practically transform the relevant advanced findings into low cost and robust techniques.

4.14.5 Current Directions for Daily Nanotechnology

Any possible calcium or other deficiencies from the fingers holding the packaging can be determined by Orasensor on the outside which will be remedied by "functional food" later. As an alternative, the OLED (organic light emitting diode) label on the packaging is taken for recommending which one is the correct one. The bathroom nanoelectronics-mirror will show the personalized information for users. Moreover, if the user teeth are decayed, the attached nanotechnology-nano toothpaste (contained apatite and protein in nano scale) can be provided to prevent tooth decay and even have a beneficial aid to restore the normal condition.

For skin care, the day cream with zinc oxide nanoparticles (NPs) can effectively prevent the radiation from harmful UV. As usual, the cream is completely transparent but not white owing to the interesting characteristics of invisible nanoparticles (NPs).

4.14.6 Current Directions for Real-Time Monitoring

With the development of nanotechnology and microsystem technology, cheap analysis detectors will be easily available for the private household, even with the complicated designs. For blood analysis, a tiny jab in the finger is enough. Similarly, the data of sugar and cholesterol levels can be emailed by Internet to individuals and the related medical center. Meanwhile, an accurate analysis can be done. If it is out of the normal range, a complete medication plan for the individual will be provided at once. The patient with micro-reactors implanted in the body or coated on the surface of body will automatically remedy by the coupled nanotechnology, such as drug delivery by nanoparticles accurately and effectively target to the required area with smallest detail.

More attractively in the field of forensic science, as a permanent topic, the technology explored from the early beginning of last century relevant to fingermark has always been investigating with huge efforts put into. Up to now, the developed approaches for fingermark are effective and significant for crime investigation. With the step of nanotechnology moving forward, nanomaterials play an important role in the field of forensic science. The sensitivity of reagents is distinctively enhanced when they couple with nano-techniques along with high quality image achieved and background adsorption reduced.

4.14.7 Current Directions for Supramolecular Medication Capsules

Actually, it is extraordinarily sophisticated for the administering medications. As a future developing direction, supra-molecular hollow molecules, as nanoscale transport containers with the antennae (attached with antibodies of similar sensory proteins), should be designed and exploited to realize the effective medical treatment. For instance, when such developed capsules come into contact with structures typical of the agent responsible for the illness, they will anchor onto the targeted bacteria or cancer cells, and send a signal to the hollow molecule. Once the capsule receives the relevant signal, it opens up and releases the correlated contents. By such nano-tailored way, medications can be directly delivered to the illness source in high doses along with and minimum side-effects, without any stress burdening on the rest of organism.

4.14.8 Current Directions for Magnetic Nanoparticles for Cancer Therapy

Cancers are always headache issues for clinicians. Therefore, owing to the nanotechnology appearance, magnetic nanoparticles are developed for cancer sources. When an alternating electromagnetic field are taken for warming up such particles, they can successfully destroy the tumor. More compellingly, nanoparticles possess a captivating capability of passing through the "blood-brain barrier" filter system. Consequently, they can be utilized for brain tumors therapy.

4.14.9 Current Directions for Lab on a Chip (Micro-reactors)

With the booming semiconductor development in the twenty-first century, microtechnology coupled with nanotechnology is a hot point for scientists, especially on the field of communication (such as cell phone). As a branch application for medications, the boundaries between microsystem technology and nanotechnology are fluid. Under the beneficial effect of nano-techniques, the existing techniques should be miniaturized further and the cost should be reduced to realize the cheaper usage for individuals. The smaller and more sophisticated and sensitive microreactors should be developed for particular features of millions of cells (such as blood cells) checking at a rate of thousands per second and sort them in the living state. For instance, cell-sorters, a kind of micro-reactors, integrating the most refined electronics, micromechanics and optics into a very sophisticated and expensive device, can be utilized to sort the required cells. In general, the interest or targeted cells attach with antibodies along with a correlated dye (lighting up or fluorescing by the relevant laser light). When the cells (encased in droplets) in the cell-sorter pass through a related laser, once a fluorescent signal is spotted, the external electrical fields will make the correlated cell into a collection vessel. With the aid of nanotechnology, such cell-sorters (micro-reactors) can be scaled down to a size of a postage stamp, even ultimately making them as disposable products. Subsequently, this will significantly speed up the relevant medical progresses. It is foreseeable that for future micro-reactors or lab-on-a-chip, millions of nano-devices working together in coordination will be effectively integrated on several square centimeters or even a smaller area to achieve the relevant goals. It should be noted that the characteristics of fluids in micro-reactors are not same as the traditional, they will become viscous as honey. Labs-on-a-chip, coupled with nanotechnology, will definitely make biology revolutionize. With such sophisticated techniques, we can get the evolution of individual cells step by step by our naked eyes. Consequently, a sort of video (a video of life) will be reconstructed. After that, the mystery of life can be decoded by prodding and poking cells to see how they react.

4.14.10 Current Directions for Neuroprosthetics

Currently, some applications of microsystem technology and nanotechnology are still in the lab or trial stage. It imperatively requires intensive *in vivo* exploration on controlling inorganic electronics with organic neurons. Up to now, it is difficult to really reflect the microenvironment *in vivo*-neurons, which dramatically alter their behaviors in accordance with the change of environment. It can be seen that in the lab, cells are directly exposed to the culture solution and migrate on a flat surface. Therefore, the conditions are obviously different. In order to solve such issue, Yu et al. (2011) investigated the neuritis growth through arrays of semiconductor nanomembrane tubes. The results show that the proposed method can be utilized as a promising method for the study of interface between neuro and semiconductor.

4.14.11 Current Directions for Home Care

Nowadays, every people is eager to live to a greater age, even with permanent life. Therefore, more and more people are need to take increasingly sophisticated medical care and better nutrition. On such situations, more advanced techniques should be developed (especially on nanoelectronics). Real-time monitoring (such as, blood pressure, blood sugar, metabolism, pulse, respiration) the individual health (particularly for elderly people) should be exploited with sensitive sensor and microcomputers coated on the body or implanted in the body or woven into clothing. Once the issues occur, the correlated systems will automatically send an alert and a detailed message to the direct-contact relatives and the family doctor. Meanwhile, the patient location will be precisely pinpointed. If the problem is emergency, the precise location will be also send to the nearest point-of care hospital by GPS or others.

4.14.12 *Current Directions for Food*

To date, owing to numerous achievements on the nanoscale and a captivating window opened by nanotechnology, it is not a dream to explore more suitable food products which possess the fascinating characteristics (such as green, fresh, healthy, convenient, wonderful flavor). With the help of state-of-art nanotechniques, it will be realized with new foods with improved quality, real freshness, consumption ease and excellent taste. Mostly, such foods coupled with nanotechnology must be beneficial for human health, and even promote the health to essentially extend our life.

In order to ensure the safety of food (Esposito et al. 2005; Ligler et al. 2003; Nasongkla et al. 2006; Yih and Al-Fandi 2006), nanoparticles in food industry (cubosomes, liposomes, biopolymeric nanoparticles, nanoemulsions, micelles, nanosensor for detecting/monitoring) have been developed. Due to large surface areas per mass unit resulted in higher rates of biological activity for nanoparticles, they are ideal candidates for foods (Chau et al. 2007). However, there are still unknown gaps between nanotechnology and the explored foods although some cooking oils with nanocapsules (capsulated flavor enhancers) and nano-nutraceuticals are realized for our daily life.

There is an old saying "prevention is better than cure". For ailments treatment helping and healthy improving (such as brain injuries, cardiovascular disease, cancer, depression), even for fetal development during pregnancy, cultivating more safe foods with nanotechnology to reduce or diminish the relevant diseases is still a severe issue.

Also, sufficient results show that the growth of plant, the formation of grain, the yield of crop and the development of phenology can be altered by nanomaterials because of their unique properties. Up to now, if we want to feed the projected population of the mid-twenty-first century even at present levels (not to mention a level approaching that of developed countries), it is hard for tropical forests maintaining at the current situation unless other types of food resources are successfully developed, particularly on the issues related to human health and environment. For realizing such goals, new sources of food with enough nutrients to meet daily needs should be explored in most countries.

Moreover, the new explored foods coupled with nanotechnology must be eco-friendly to environment. Meanwhile, the correlated nano-tailored engineering techniques should be developed in a relatively safer way, particularly on side-effects or toxicity on both human health and environment.

Furthermore, according to the continue increasing on energy demand worldwide, growing of world's population, expanding of the developing world economies and even if the realization of low carbon sources of energy, for the final goal of green nanotechnology, the usage of chemicals (especially for the toxic ones) should be eliminated or minimized, the utilization of raw materials should be maximized and the wastes should be prevented. Meanwhile, for green nanotechnology, it must meet with the standards of Life Cycle Assessment (LCA) and any harms on environment should be avoided to realize the ultimate target-ecofriendly/green nano-techs which means wastes are efficiently recycled, benign materials are widely used, and energy is sufficiently used.

4.14.13 Feasible Green Techniques for Nanoparticles (NPs)

As an attractive nanomaterial, nanoparticles (NPs) possess unique characteristics in various aspects (biological, chemical, dielectric, electrical, electronic, magnetic, mechanical, optical, physical, thermal). Nanoparticles physical properties (large surface atom, large surface energy, spatial confinement and imperfections reduction) are obviously different from bulk materials due to the relevant dimension decreasing. Such pronounced size-effects lead to the surface plasmon resonance (SPR), enhanced Rayleigh scattering and surface enhanced Raman scattering (SERS) in metal nanoparticles, quantum size effect in semiconductors and supermagnetism in magnetic materials. As a result, nanoparticles have outstanding advantages in biochemical and chemical sensors, electronics, optoelectronics.

Actually, at the beginning of this century, materials and devices developing towards to nano-engineering was already a popular trend. Unlike the conventional method-a top-down process (mask application, illumination/radiation and etching on an initially unpatterned material) -for small structures generation, the bottom-up methods was utilized for smaller structures generations. At present, huge numbers of astonishing materials coupled with nanoparticles have been successfully explored (bio, electronic, inorganic, magnetic, optical, polymeric). Moreover, such nanoparticles have a pivotal capability to self-assembly or assembly into hierarchically ordered and/or arbitrarily defined structures. Consequently, the applications related to nanoparticles are booming in various fields, such as, cosmetics (Kokura et al. 2010; Müller et al. 2007), diagnostics (Jokerst et al. 2009; Kim et al. 2010; Zhang and Kataoka 2009), food technology (Dudo et al. 2011; Duncan 2011; Jafari et al. 2008), pharmacy (Kathiravan et al. 2011; Tran et al. 2011), regenerative medicine (Engel et al. 2008; Harrison and Sirivisoot 2011).

With the decades passing, nanoparticles open a fascinating window for scientists owing to their much interest in biological, chemical and physical properties. Until now, an astonishing multitude of nanoparticles have been fabricated (such as nanocages, nanohorns, nanotubes, nanospheres) along with various materials (such as carbon, gold, silver, liposomes, organic dendrimers, silicon to iron oxide, semiconductors).

In general, in order to realize the targeted usages, the surface functionalization for nanoparticles is an effective approach to tune or control particles overall properties.

It is well known that all the conventional procedures are ecologically hazardous. Endeavors are underway to explore greener avenues in the nanotechnology domain.

4.14.13.1 Green Nanoparticles-Ag Nanoparticles (Ag NPs)

As an emerging cutting-edge interdisciplinary materials, silver nanoparticles (Ag NPs) have been extensively explored and successfully applied in biosensing, catalysis, drug delivery, imaging, medicine (Ahamed et al. 2010; du Toit et al. 2010; Shegokar et al. 2011). Silver nanoparticles are also utilized in cosmetics, food industry, clothing, and sun screens because of strong antimicrobial activity (Kokura et al. 2010; Sanvicens et al. 2009).

Up to now, various techniques taken for synthesizing or produce metal nanoparticles are still expensive and sometimes related to hazardous chemicals usage. Therefore, nanoparticles green synthesis-environmentally benign technology-should be explored. Although it has reported Ag NPs green synthesis by various plants (Bhainsa and D'Souza 2006; Kathiresan et al. 2009; Mahapatra and Karak 2008; Nabikhan et al. 2010; Rajakumar and Rahuman 2011; Sanghi and Verma 2009; Shivaji et al. 2011; Zaki et al. 2011), the potential of plants as biological materials for nanoparticles synthesis is not yet to be investigated in detail. In addition, the information related to biological response of human cells to green synthesized Ag NPs is also very limited.

In early time, Ahamed et al. (2011) reported a simple, cost-effective and environmentally benign synthesis of Ag NPs at ambient conditions using garlic clove extract as a reducing and stabilizing agent in order to apply to biological response of Ag NPs in human lunge pithelial (A549) cells. The results show that the synthesized nanoparticles is about 12 ± 4.8 nm in spherical shape with smooth surfaces. Also, the achieved silver nanoparticles are potentially biocompatible to human lung epithelial cells even at a higher concentration (50 mg/ml). In the same year, Guidelli et al. (2011) explored a totally green synthesis of colloidal Ag NPs using the NRL extracted from *H. brasiliensis*. The correlated procedure is cheap, fast, simple and occurs at a relatively low temperature (water boiling temperature). Meanwhile, it is eco-friendly to environment and devoid of electrochemical, photochemical, or irradiation processes. The results show that the size of obtained silver nanoparticles (highly crystalline) ranges from 20 to 86 nm with a mean diameter size of around 50 nm.

4.14.13.2 Green Nanoparticles-Gold Nanoparticles (Au NPs)

As a kind of promising materials for electronics and medical imaging, gold nanoparticles (Au NPs) as well as silver nanoparticles attracts a cadre of scientists and engineers to put tremendous efforts into the relevant techniques exploring with environmentally benign and friendly to more people/planet ultimate goals. For Au NPs, one captivating characteristic-surface plasmon resonance (SPR) can be tailored or tuned for biomedical applications in photothermal therapy, tissue/tumor imaging, drug delivery and immunochromatographic identification of pathogens in clinical specimens from the initial development (Cao et al. 2011; Chen et al. 2010; Delfino and Cannistraro 2009). Due to the potential therapeutic applications (as anti-arthritic agent, as anti-angiogenesis, as anti-malarial agent) of Au NPs along with biohydrogen applications, various breakthroughs have been exploring with the profound theory revealed in detail. However, for the current approaches taken for synthesizing Au NPs, the standard way is utilizing large amounts of toxic solvents which can be flammable and explosive. At the beginning of the twenty-first century, Hutchison's group (Mckenzie and Hutchison 2004; Sweeney et al. 2006) developed an innovative synthesis method with a new catalyst, non-toxic solvents involved and purification by nanoporous filtration. Such proposed method is not only proved to be faster, safer and greener but also much cheaper than the old methods. Meanwhile, it illustrates that nanoparticles synthesized by green techniques can both save money and boost efficiency (Mckenzie and Hutchison 2004; Sweeney et al. 2006; Hutchison 2008).

At the same time, Philip (2009) developed a greener Au NPs synthesis method with natural honey acting as both reducing and protecting agents in water. Moreover, the synthesis was carried out at room temperature. Also, the achieved Au NPs (highly crystalline) is spherical with the average size of about 15 nm. In addition, Narayanan and Sakthivel (2011) explored a facile greener method which is eco-friendly to environment with the cell-free filtrate of fungus-*Sclerotium rolfsii*. The obtained Au NPs are of isotrophic spherical shape with anisotrophic nanostructures. Moreover, the shape and size of the achieved particles can be tunable or controllable by the cell-free filtrate of fungus-*S. rolfsii* and the ratio of gold salt and such synthesis process is quite rapid and completes in 10–15 min. Furthermore, the synthesized Au NPs are quite stable in aqueous solution for 2 months.

4.14.13.3 Green Nanostructures

As a greener approach to make a broad range of organic nanoparticles, particle replication in non-wetting templates (PRINT) was successfully developed at early stage of this century. Relatively large quantities of organic nanostructures can be fabricated with the aid of organic chemistry and microelectronics manufacturing (Gratton et al. 2007). By such technique, any shape nanoparticles can be rapidly realized with a low waste and error-free occurring. Also, the required functional groups can be added or doped for biomedical or clinical applications with tunable properties.

Until now, variable strategies are taken for realizing a combination of analysis capability, optimized optical, electrical and/or magnetic properties and targeting specificity. Meanwhile, facing the severe burden stressed on human health, an exciting avenues are heading for new green nanoparticle with innovative multifunctions for biomedical applications, especially on drug delivery and biomedical imaging, which are beneficial to the clinical therapies (diagnostics, neuropathologies, cancer).

Although huge numbers of nanoparticles are rapidly explored, they can still be classified into two major types from the analytical standpoint. (1) Containing inorganic elements (usually metals and metal oxides, *e.g.* Ag, Au, SiO_2, TiO_2, Fe_3O_4, semiconductors); (2) Basing on organic molecules (liposomes, dendrimers, carbon nanotubes (CNTs)). Moreover, the applications to new analytical methodologies and/or devices with the aim of determining species in solution are really scarce even great effort also devoted to characterization, understanding, and improvement of nanostructures' properties.

Therefore, the safety of nanoparticles and nanostructures are still key issues although they can promise endless opportunities. Noteworthy semiconductor metals (colloidal gold and iron oxide crystals) are commonly utilized and have demonstrated toxicity. It is well known that the reactivity of such tiny particles may be due to the correlated large surface area in comparison to their overall mass. Moreover, such tiny particles can easily permeate the skin and blood-brain barrier, which are harmful to human health and will lead to several potential toxicities. Consequently, detailed investigations on any toxicity issues associated with these nanoparticle and nanostructures should be carried out with the full mechanism fully revealed.

4.15 Conclusions and Prospects

With the help of abundance knowledge from various research fields, the green nanotechnologies will be explored and utilized in sufficient detail to harness nanostructures for our use. Although many relevant techniques currently are still in the lab or utilized at small scale and have yet to reach their envisioned performance, with nano-techniques octopus-into clinic applications (besides the industrial applications), green nanotechnology will be captivatingly dominant focus on our daily routine lives. Moreover, due to unsustainable current fossil fuel usage and greenhouse gas production, it requires society moving from the conventional style economy to a more green/eco-friendly way. Therefore, sustainable solutions must be explored with more efficient solutions to produce, refine and use during such transition. Moreover, the approaches to sufficiently enhance the efficiency of conventional techniques and the relevant recycles should be deeply investigated. Meanwhile, with the aid of green nanotechnologies, renewable resources should be efficiently used to diminish or extinguish the side-effect of conventional method.

Furthermore, new green nanotechnologies or hybrid eco-friendly nanotechniques should be explored along with the detailed mechanism revealed. By the relevant processing parameters precisely controlling, the width and depth of the nano-textured features can be tailored, sized and shaped over a wide range resulted in higher quality. It should be noted that new green nanotechnology exploring will require new technologies with the breakthrough and profound theory establishing. In the coming future, research on green nanotechnologies will be realized to prevent contaminations and remedy the already existed environmental problems and ultimately make the new clean eco-science and technology practical on a commercial scale in our daily life.

References

Abdelkebir K, Morin-Grognet S, Gaudière F et al (2012) Biomimetic layer-by-layer templates for calcium phosphate biomineralization. Acta Biomater 8(9):3419–3428

Ahamed M, AlSalhi MS, Siddiqui MKJ (2010) Silver nanoparticle applications and human health. Clin Chim Acta 411:1841–1848

Ahamed M, Khan MAM, Siddiqui MKJ et al (2011) Green synthesis, characterization and evaluation of biocompatibility of silver nanoparticles. Phys E 43:1266–1271

Akbari H, Browne MC, Ortega A et al (2019) Efficient energy storage technologies for photovoltaic systems. Sol Energy 192:144–168

Aktepe R, Erkal BG (2023) State-of-the-art review on measurement techniques and numerical modeling of geometric imperfections in cold-formed steel members. J Constr Steel Res 207: 107942

Aladjem A (1973) Anodic oxidation of titanium and its alloys. J Mater Sci 8(5):688–704

Altinoluk SH, Ciftpinar HE, Demircioglu O et al (2016) Light trapping by micro and nano-hole texturing of single crystalline silicon solar cells. Energy Procedia 92:291–296

Andersson AS, Bäckhed F, von Euler A et al (2003) Nanoscale features influence epithelial cell morphology and cytokine production. Biomaterials 24(20):3427–3436

Anselme K, Bigerelle M, Noel B et al (2002) Effect of grooved titanium substratum on human osteoblastic cell growth. J Biomed Mater Res 60:529–540

Anttila A, Lappalainen R, Tiainen VM, Hakovirta M (1997) Superior attachment of high-quality hydrogen-free amorphous diamond films to solid materials. Adv Mater 9:1161–1164

Aparicio C, Gil FJ, Fonseca C et al (2003) Corrosion behavior of commercially pure titanium shot blasted with different materials and sizes of shot particles for dental implant applications. Biomaterials 24(2):263–273

Aparicio C, Padrós A, Gil FJ (2011a) *In vivo* evaluation of micro-rough and bioactive titanium dental implants using histometry and pull-out tests. J Mech Behav Biomed Mater 4(8): 1672–1682

Aparicio C, Rodriguez D, Gil FJ (2011b) Variation of roughness and adhesion strength of deposited apatite layers on titanium dental implants. Mater Sci Eng C 31(2):320–324

Arulvel S, Rufuss DDW, Jain A et al (2023) Laser processing techniques for surface property enhancement: focus on material advancement. Surf Interfaces Part A 42:103293

Azzeddine H, Bradai D, Baudin T, Langdon TG (2022) Texture evolution in high-pressure torsion processing. Prog Mater Sci 125:100886

Barnes D, Johnson S, Snell R, Best S (2012) Using scratch testing to measure the adhesion strength of calcium phosphate coatings applied to poly(carbonate urethane) substrates. J Mech Behav Biomed Mater 6:128–138

Basu PK, Khanna A, Hameiri Z (2015) The effect of front pyramid heights on the efficiency of homogeneously textured inline-diffused screen-printed monocrystalline silicon wafer solar cells. Renew Energy 78:590–598

Beyerlein IJ, Tóth LS (2009) Texture evolution in equal-channel angular extrusion. Prog Mater Sci 54(4):427–510

Bhainsa KC, D'Souza SF (2006) Extracellular biosynthesis of silver nanoparticles using the fungus Aspergillus fumigatus. Colloids Surf B: Biointerfaces 47:160–164

Bhowmik S, Rajan AG (2022) Chemical vapor deposition of 2D materials: a review of modeling, simulation, and machine learning studies. iScience 25(3):103832

Bilek MM (2014) Biofunctionalization of surfaces by energetic ion implantation: review of progress on applications in implantable biomedical devices and antibody microarrays. Appl Surf Sci 310:3103–3110
Bjerklie S (2005) Thinking big with nanotechnology: nano-coatings expected to revolutionize surface finishing. Met Finish 103:46–47
Cao XD, Ye YK, Liu SQ (2011) Gold nanoparticle-based signal amplification for biosensing. Anal Biochem Method Biol Sci 417:1–16
Cassar G, Banfield S, Avelar-Batista WJC et al (2012) Micro-abrasion wear testing of triode plasma diffusion and duplex treated Ti-6Al-4V alloy. Wear 274–275:377–387
Chadar R, Afzal O, Alqahtani SM, Kesharwani P (2021) Carbon nanotubes as an emerging nanocarrier for the delivery of doxorubicin for improved chemotherapy. Colloids Surf B: Biointerfaces 208:112044
Chandrasekaran S, Nann T, Voelcker NH (2015) Nanostructured silicon photoelectrodes for solar water electrolysis. Nano Energy 17:308–322
Chau CF, Wu SH, Yen GC (2007) The development of regulations for food nanotechnology. Trends Food Sci Technol 18:269–280
Chen Y, Flowers K, Calizo M, Bishnoi SW (2010) The role of protein binding in the poisoning of gold nanoparticle catalysts. Colloids Surf B: Biointerfaces 76:241–247
Choy KL (2003) Chemical vapor deposition of coatings. Prog Mater Sci 48(2):57–170
Chu AK, Wang JS, Tsai ZY, Lee CK (2009) A simple and cost-effective approach for fabricating pyramids on crystalline silicon wafers. Sol Energy Mater Sol Cells 93:1276–1280
Cizek J, Khor KA (2012) Role of in-flight temperature and velocity of powder particles on plasma sprayed hydroxyapatite coating characteristics. Surf Coat Technol 206(8–9):2181–2191
Cozza RC, Tanaka DK, Souza RM (2011) Friction coefficient and wear mode transition in micro-scale abrasion tests. Tribol Int 44(12):1878–1889
Curtis A, Wilkinson C (1997) Topographical control of cells. Biomaterials 18(24):1573–1583
Cvijović-Alagić I, Cvijović Z, Mitrović S et al (2011) Wear and corrosion behavior of Ti-13Nb-13Zr and Ti-6Al-4V alloys in simulated physiological solution. Corros Sci 53(2):796–808
Dalili N, Edrisy A, Farokhzadeh K et al (2010) Improving the wear resistance of Ti-6Al-4V/TiC composites through thermal oxidation (TO). Wear 269(7–8):590–601
Davis SS (1997) Biomédical applications of nanotechnology—implications for drug targeting and gene therapy. Trends Biotechnol 15(6):217–224
Delfino I, Cannistraro S (2009) Optical investigation of the electron transfer protein azurin-gold nanoparticle system. Biophys Chem 139:1–7
Dey P, Raza MJ, Khera A et al (2024) Recent progress of functionalized nanomaterials-based biosensing for monitoring of food- and water-borne pathogens. Environ Nanotechnol Monit Manag 21:100914
Dodd B, Boddington P (1980) The causes of edge cracking in cold rolling. J Mech Work Technol 3(3–4):239–252
dos Santos EA, Moldovan S, Mateescu M et al (2012) Physical-chemical and biological behavior of an amorphous calcium phosphate thin film produced by RF-magnetron sputtering. Mater Sci Eng C 32(7):2086–2095
du Toit LC, Pillay V, Choonara YE (2010) Nano-microbicides: challenges in drug delivery, patient ethics and intellectual property in the war against HIV/AIDS. Adv Drug Deliv Rev 62:532–546
Dudo A, Choi DH, Scheufele DA (2011) Food nanotechnology in the news. Coverage patterns and thematic emphases during the last decade. Appetite 56:78–89
Duncan TV (2011) Applications of nanotechnology in food packaging and food safety: barrier materials, antimicrobials and sensors. J Colloid Interface Sci 363:1–24
Dvir T, Timko BP, Kohane DS, Langer R (2011) Nanotechnological strategies for engineering complex tissues. Nat Nanotechnol 6(1):13–22
Edalati K, Horita Z (2016) A review on high-pressure torsion (HPT) from 1935 to 1988. Mater Sci Eng A 652:325–352

Engel E, Michiardi A, Navarro M et al (2008) Nanotechnology in regenerative medicine: the materials side. Trends Biotechnol 26:39–47

Erdemir A, Martin JM (2018) Superior wear resistance of diamond and DLC coatings. Curr Opin Solid State Mater Sci 22(6):243–254

Esposito E, Cortesi R, Drechsler M et al (2005) Cubosome dispersions as delivery systems for percutaneuos administration of indomethacin. Pharm Res 22(12):2163–2173

Flemming RG, Murphy CJ, Abrams GA et al (1999) Effects of synthetic micro-and nano-structured surfaces on cell behavior. Biomaterials 20(6):573–588

Gratton SEA, Pohlhausa PD, Lee J et al (2007) Nanofabricated particles for engineered drug therapies: a preliminary biodistribution study of PRINT™ nanoparticles. J Control Release 121:10–18

Green MA (2003) Crystalline and thin film silicon solar cells: state of the art and future potential. Sol Energy 74(3):181–192

Gruère GP (2012) Implications of nanotechnology growth in food and agriculture in OECD countries. Food Policy 37(2):191–198

Guidelli EJ, Ramos AP, Zaniquelli MED, Baffa O (2011) Green synthesis of colloidal silver nanoparticles using natural rubber latex extracted from Hevea brasiliensis. Spectrochim Acta Part A: Mol Biomol Spectrosc 82(1):140–145

Gulati S, Lingam BHN, Kumar S et al (2022) Improving the air quality with functionalized carbon nanotubes: sensing and remediation applications in the real world. Chemosphere 299:134468

Guo KW (2012) Green nanotechnology of trends in future energy: a review. Int J Energy Res 36:1–17

Guo KW (2015) Surface engineered nanostructures on metallic biomedical materials for anti-abrasion. In: Aliofkhazrae M (ed) Anti-abrasive nanocoatings-current future applications. Woodhead Publishing, pp 349–383

Guo KW (2017a) Surface texturing for silicon solar energy by wet acid. J Nanosci Adv Technol 2(1):24–29

Guo KW (2017b) Porous carbon materials for clean and sustainable energy. Rev Energy Technol Policy Res 3(2):51–64

Guo KW (2018a) Strategies of porous carbon materials for future energy. In: Srivastava N (ed) Green nanotechnology for biofuel production. Springer, Cham, pp 79–92

Guo KW (2018b) Biomaterials degradation and bioabsorbability. In: Martínez LMT (ed) Biomedical potentials of marine enzymes handbook of ecomaterials. Springer, Cham, pp 1–28

Guo KW (2019) Biofuel cells with enzymes as a catalyst. In: Srivastava N (ed) New and future developments in microbial biotechnology and bioengineering-from cellulose to cellulase: strategies to improve biofuel production. Elsevier, pp 261–282

Guo KW (2020) Green/ecofriendly micropunching techniques for energy application. In: Kharissova OV (ed) Handbook of nanomaterials and nanocomposites for energy and environmental applications. Springer, Cham, pp 1–18

Guo KW (2021) Simple surface texturing for green energy-silicon solar energy. Nanotechnol Adv Mater Sci 3(2):1–13

Guo KW (2023a) Nanoporous silicon materials for bioapplications by electrochemical approach. In: Gomathi T (ed) Nanobiomaterials: research trends and applications. CRC Press/Taylor & Francis Group, Boca Raton, pp 188–196

Guo KW (2023b) NiTi shape memory alloys surface treatment for biomedical applications. MOJ Appl Bionic Biomech 7(1):149–153

Guo KW (2024) Nanoporous silicon materials for solar energy by electrochemical approach. In: Kulkarni NV (ed) Handbook of emerging materials for sustainable energy. Elsevier, pp 119–128

Guo KW, Tam HY (2017a) Ti microholes potential for biomedical application punched by 75WC/25Co micropunch. J Appl Biotechnol Bioeng 4(3):572–576

Guo KW, Tam HY (2017b) Simple potential bioanalysis micro reactor fabricated by vertically-aligned multi-walled carbon nanotubes (MWCNTs). J Appl Biotechnol Bioeng 3(2):305–308

Handford CE, Dean M, Henchion M et al (2014) Implications of nanotechnology for the agri-food industry: opportunities, benefits and risks. Trends Food Sci Technol 40(2):226–241

Harrison BS, Sirivisoot S (2011) Applications of nanotechnology for regenerative medicine. In: Principles of regenerative medicine, pp 529–540. https://doi.org/10.1016/B978-0-12-381422-7.10030-6

Hashemi M, Alizadeh R, Langdon TG (2023) Recent advances using equal-channel angular pressing to improve the properties of biodegradable Mg-Zn alloys. J Magnes Alloys 11(7): 2260–2284

Heimann RB (2006) Thermal spraying of biomaterials. Surf Coat Technol 201(5):2012–2019

Hench LL, Polak JM (2002) Third-generation biomedical materials. Science 295(5557):1014–1017

Hench LL, Wilson J (1984) Surface-active biomaterials. Science 226(4675):226630–226636

Herche W (2017) Solar energy strategies in the U.S. utility market. Renew Sustain Energy Rev 77: 590–595

Hernández-Callejo L, Gallardo-Saavedra S, Alonso-Gómez V (2019) A review of photovoltaic systems: design, operation and maintenance. Sol Energy 188:426–440

Hignett B, Andrew TC, Downing W et al (1987) Surface cleaning, finishing and coating. In: Wood WG (ed) Metals handbook, vol 5. American Society for Metals, Metals Park, pp 107–127

Hoque ME, Showwa NN, Ahmed M et al (2022) Titanium and titanium alloys in dentistry: current trends, recent developments, and future prospects. Heliyon 8(11):e11300

Hossain S, Abdalla AM, Suhaili SBH et al (2020) Nanostructured graphene materials utilization in fuel cells and batteries: a review. J Energ Storage 29:101386

Huang Y, Langdon TG (2013) Advances in ultrafine-grained materials. Mater Today 16(3):85–93

Hussein MA, Adesina A, Kumar AM et al (2020) Mechanical, *in vitro* corrosion, and tribological characteristics of TiN coating produced by cathodic arc physical vapor deposition on Ti20Nb13Zr alloy for biomedical applications. Thin Solid Films 709:138183

Hutchison JE (2008) Greener nanoscience: a proactive approach to advancing applications and reducing implications of nanotechnology. ACSNano 2:395–402

Iijima S (1991) Helical microtubules of graphitic carbon. Nature 354:56–58

Iijima S (1993) Single-shell carbon nanotubes of 1-nm diameter. Nature 363:603–605

Ito Y (1999) Surface micropatterning to regulate cell functions. Biomaterials 20(23):2333–2342

Jacobs T, Morent R, De Geyter N et al (2012) Plasma surface modification of biomedical polymers: influence on cell-material interaction. Plasma Chem Plasma Process 32(5):1039–1073

Jafari SM, Assadpoor E, Bhandari B, He YH (2008) Nano-particle encapsulation of fish oil by spray drying. Food Res Int 41:172–183

Jagielski J, Piatkowska A, Aubert P et al (2006) Ion implantation for surface modification of biomaterials. Surf Coat Technol 200(22–23):6355–6361

Jeswiet J, Geiger M, Engel U et al (2008) Metal forming progress since 2000. CIRP J Manuf Sci Technol 1(1):2–17

Jokerst JV, Raamanathan A, Christodoulides N et al (2009) Nano-bio-chips for high performance multiplexed protein detection: determinations of cancer biomarkers in serum and saliva using quantum dot bioconjugate labels. Biosens Bioelectron 24:3622–3629

Ju M, Gunasekaran M, Kim K et al (2008) A new vapor texturing method for multicrystalline silicon solar cell applications. Mater Sci Eng B 153:66–69

Kalainathan S, Prabhakaran S (2016) Recent development and future perspectives of low energy laser shock peening. Opt Laser Technol 81:137–144

Kannan N, Vakeesan D (2016) Solar energy for future world: a review. Renew Sust Energ Rev 62: 1092–1105

Karthik C, Rajalakshmi S, Thomas S, Thomas V (2023) Intelligent polymeric biomaterials surface driven by plasma processing. Curr Opin Biomed Eng 26:100440

Kasemo B (2002) Biological surface science. Surf Sci 500(1):656–677

Kathiravan A, Jhonsi MA, Renganathan R (2011) Photoinduced interaction of colloidal TiO_2 nanoparticles with lysozyme: evidences from spectroscopic studies. J Lumin 131:1975–1981

Kathiresan K, Manivannan S, Nabeel MA, Dhivya B (2009) Studies on silver nanoparticles synthesized by a marine fungus, Penicillium fellutanum isolated from coastal mangrove sediment. Colloids Surf B: Biointerfaces 71:133–137

Kern M, Thompson VP (1994a) Sandblasting and silica coating of a glass-infiltrated alumina ceramic-volume loss, morphology, and changes in the surface-composition. J Prosthet Dent 71:453–461

Kern M, Thompson VP (1994b) Effects of sandblasting and silica-coating procedures on pure titanium. J Dent 22:300–306

Khandelwal H, Singh G, Agrawal K et al (2013) Characterization of hydroxyapatite coating by pulse laser deposition technique on stainless steel 316 L by varying laser energy. Appl Surf Sci 265:30–35

Khot LR, Sankaran S, Maja JM et al (2012) Applications of nanomaterials in agricultural production and crop protection: a review. Crop Prot 35:64–70

Kim JY, Choi WI, Kim YH et al (2010) *In vivo* tumor targeting of pluronic-based nano-carriers. J Control Release 147:109–117

Kokura S, Handa O, Takagi T et al (2010) Silver nanoparticles as a safe preservative for use in cosmetics. Nanomed Nanotechnol Biol Med 6:570–574

Kostoff RN, Koytcheff RG, Lau GY (2007) Global nanotechnology research literature overview. Technol Forecast Soc Change 74:1733–1747

Laithy ME, Wang L, Harvey TJ et al (2019) Further understanding of rolling contact fatigue in rolling element bearings—a review. Tribol Int 140:105849

Landolt D, Chauvy PF, Zinger O (2003) Electrochemical micromachining, polishing and surface structuring of metals: fundamental aspects and new developments. Electrochim Acta 48:3185–3201

Lau K, Heu C, Moore MJ et al (2022) Effect of plasma ion immersion implantation on physiochemical and biological properties of silk towards creating a versatile biomaterial platform. Mater Today Adv 13:100212

Leso V, Fontana L, Iavicoli I (2019) Biomedical nanotechnology: occupational views. Nano Today 24:10–14

Li SY, Beyerlein IJ, Bourke MAM (2005) Texture formation during equal channel angular extrusion of fcc and bcc materials: comparison with simple shear. Mater Sci Eng A 394(1–2):66–77

Ligler FS, Taitt CR, Shriver-Lake LC et al (2003) Array biosensor for detection of toxins. Anal Bioanal Chem 377:469–477

Mahapatra SS, Karak N (2008) Silver nanoparticle in hyperbranched polyamine: synthesis, characterization and antibacterial activity. Mater Chem Phys 112:1114–1119

Malysheva A, Lombi E, Voelcker NH (2015) Bridging the divide between human and environmental nanotoxicology. Nat Nanotechnol 10:835–844

Manjubala I, Scheler S, Bossert J, Jandt KD (2006) Mineralisation of chitosan scaffolds with nanoapatite formation by double diffusion technique. Acta Biomater 2(1):75–84

Martin JY, Schwartz Z, Hummert TW et al (1995) Effect of titanium surface roughness on proliferation, differentiation, and protein synthesis of human osteoblast-like cells (MG63). J Biomed Mater Res 29(3):389–401

Martinez-Martinez D, De Hosson JTM (2014) On the deposition and properties of DLC protective coatings on elastomers: a critical review. Surf Coat Technol 258:677–690

Mbayachi VB, Ndayiragije E, Sammani T et al (2021) Graphene synthesis, characterization and its applications: a review. Results Chem 3.100163

McCafferty E, Wightman JP (1999) An X-ray photoelectron spectroscopy sputter profile study of the native air-formed oxide film on titanium. Appl Surf Sci 143(1):92–100

Mckenzie LC, Hutchison JE (2004) Green nanoscience: an integrated approach to greener products, processes, and applications. Chem Today 25:30–33

Merugu R, Sharma S, Rastogi M (2023) An overview of nanotechnological applications in biomedical engineering. Mater Today Proc Part 2 76:412–417

Minati L, Migliaresi C, Lunelli L et al (2017) Plasma assisted surface treatments of biomaterials. Biophys Chem 229:151–164

Mishra S, Sundaram B (2023) Efficacy and challenges of carbon nanotube in wastewater and water treatment. Environ Nanotechnol Monit Manag 19:100764

Mitragotri S, Lahann J (2009) Physical approaches to biomaterial design. Nat Mater 8(1):15–23

Miyakawa O, Watanabe K, Okawa S et al (1996) Surface contamination of titanium by abrading treatment. Dent Mater 15:11–21

Mohan NS, Bhuvaneswari S, Anitha R, Vijayalakshmi V (2024) Improved photocatalytic degradation of organic pollutants using green synthesized ZnO and Ni@ZnO NPs for environmental applications. Environ Nanotechnol Monit Manag 21:100922

Moriarty P, Honnery D (2008) The prospects for global green car mobility. J Clean Prod 16:1717–1726

Moyer-Vanderburgh K, Park SJ, Fornasiero F (2023) Growth of carbon nanotube forests on flexible metal substrates: advances, challenges, and applications. Carbon 206:402–421

Müller RH, Petersen RD, Hommoss A, Pardeike J (2007) Nanostructured lipid carriers (NLC) in cosmetic dermal products. Adv Drug Deliv Rev 59:522–530

Nabikhan A, Kandasamy K, Raj A, Alikunhi NM (2010) Synthesis of antimicrobial silver nanoparticles by callus and leaf extracts from saltmarsh plant, *Sesuvium portulacastrum* L. Colloids Surf B: Biointerfaces 79:488–493

Naghdi S, Rhee KY, Park SJ (2018) A catalytic, catalyst-free, and roll-to-roll production of graphene via chemical vapor deposition: low temperature growth. Carbon 127:1–12

Narayanan KB, Sakthivel N (2011) Facile green synthesis of gold nanostructures by NADPH-dependent enzyme from the extract of *Sclerotium rolfsii*. Colloids Surf A Physicochem Eng Asp 380:156–161

Nasongkla N, Bey E, Ren JM et al (2006) Multifunctional polymeric micelles as cancer targeted MRI ultrasensitive drug delivery systems. Nano Lett 6:2427–2430

National Nanotechnology Initiative (2000) Leading to the next industrial revolution. Committee on Technology, National Science, and Technology Council, Washington, DC

Ningombam L, Mana T, Apum G et al (2024) Nano-bioremediation: a prospective approach for environmental decontamination in focus to soil, water and heavy metals. Environ Nanotechnol Monit Manag 21:100931

Ong JL, Carnes DL, Bessho K (2004) Evaluation of titanium plasma-sprayed and plasma-sprayed hydroxyapatite implants *in vivo*. Biomaterials 25:4601–4606

Patolsky F, Lieber CM (2005) Nanowire nanosensors. Mater Today 8:20–28

Philip D (2009) Honey mediated green synthesis of gold nanoparticles. Spectrochim Acta Part A 73:650–653

Pitkethly MJ (2004) Nanomaterials-the driving force. Mater Today 7:20–29

Price RL, Waid MC, Haberstroh KM, Webster TJ (2003) Selective bone cell adhesion on formulations containing carbon nanofibers. Biomaterials 24:1877–1887

Quinn J, McFadden R, Chan CW, Carson L (2020) Titanium for orthopedic applications: an overview of surface modification to improve biocompatibility and prevent bacterial biofilm formation. iScience 23(11):101745

Rajakumar G, Rahuman AA (2011) Larvicidal activity of synthesized silver nanoparticles using *Eclipta prostrata* leaf extract against filariasis and malaria vectors. Acta Trop 118:196–203

Rakshith M, Seenuvasaperumal P (2021) Review on the effect of different processing techniques on the microstructure and mechanical behavior of AZ31 Magnesium alloy. J Magnes Alloy 9(5): 1692–1714

Rathinavel S, Priyadharshini K, Panda D (2021) A review on carbon nanotube: an overview of synthesis, properties, functionalization, characterization, and the application. Mater Sci Eng B 268:115095

Rautray TR, Narayanan R, Kim KH (2011) Ion implantation of titanium based biomaterials. Prog Mater Sci 56(8):1137–1177

Ravelingien M, Hervent AS, Mullens S et al (2010) Influence of surface topography and pore architecture of alkali-treated titanium on *in vitro* apatite deposition. Appl Surf Sci 256(11): 3693–3697

Reichelt K, Jiang X (1990) The preparation of thin films by physical vapour deposition methods. Thin Solid Films 191(1):91–126

Saber-Samandari S, Gross KA (2012) Nano-indentation on amorphous calcium phosphate splats: effect of droplet size on mechanical properties. J Mech Behav Biomed Mater 16:29–37

Saber-Samandari S, Gross KA (2013) Contact nanofatigue shows crack growth in amorphous calcium phosphate on Ti, Co-Cr and Stainless steel. Acta Biomater 9(3):5788–5794

Sahoo S, Sahoo G, Jeong SM, Rout CS (2022) A review on supercapacitors based on plasma enhanced chemical vapor deposited vertical graphene arrays. J Energ Storage 53:105212

Sakai T, Belyakov A, Kaibyshev R et al (2014) Dynamic and post-dynamic recrystallization under hot, cold and severe plastic deformation conditions. Prog Mater Sci 60:130–207

Sales WF, Schoop J, da Silva LRR et al (2020) A review of surface integrity in machining of hardened steels. J Manuf Process 58:136–162

Sampaio PGV, González MOA (2017) Photovoltaic solar energy: conceptual framework. Renew Sustain Energy Rev 74:590–601

Sanghi R, Verma P (2009) Biomimetic synthesis and characterization of protein capped silver nanoparticles. Bioresour Technol 100:501–504

Santavirta SS, Lappalainen R, Anttila A et al (1999) The counterface, surface smoothness, tolerances, and coatings in total joint prostheses. Clin Orthop Relat Res 369:92–102

Sanvicens N, Pastells C, Pascual N, Marco MP (2009) Nanoparticle-based biosensors for detection of pathogenic bacteria. Trends Anal Chem 28:1243–1252

Sargin F, Erdogan G, Kanbur K, Turk A (2021) Investigation of *in vitro* behavior of plasma sprayed Ti, TiO_2 and HA coatings on PEEK. Surf Coat Technol 411:126965

Schalk N, Tkadletz M, Mitterer C (2022) Hard coatings for cutting applications: physical *vs.* chemical vapor deposition and future challenges for the coatings community. Surf Coat Technol 429:127949

Seah CM, Chai SP, Mohamed AR (2011) Synthesis of aligned carbon nanotubes. Carbon 49:4613–4635

Sealy MP, Guo YB (2010) Surface integrity and process mechanics of laser shock peening of novel biodegradable magnesium-calcium (Mg-Ca) alloy. J Mech Behav Biomed Mater 3:488–496

Shanaghi A, Rouhaghdam ARS, Ahangarani S, Chu PK (2012a) Effect of plasma CVD operating temperature on nanomechanical properties of TiC nanostructured coating investigated by atomic force microscopy. Mater Res Bull 47(9):2200–2205

Shanaghi A, Rouhaghdam ARS, Ahangarani S et al (2012b) Effects of duty cycle on microstructure and corrosion behavior of TiC coatings prepared by DC pulsed plasma CVD. Appl Surf Sci 258(7):3051–3057

Shegokar R, Jansch M, Singh KK, Müller RH (2011) *In vitro* protein adsorption studies on nevirapine nanosuspensions for HIV/AIDS chemotherapy. Nanomed Nanotechnol Biol Med 7:333–340

Shivaji S, Madhu S, Singh S (2011) Extracellular synthesis of antibacterial silver nanoparticles using psychrophilic bacteria. Process Biochem 46:1800–1807

Singh J, Chatha SS, Singh H (2021) Synthesis and characterization of plasma sprayed functional gradient bioceramic coating for medical implant applications. Ceram Int Part A 47(7): 9143–9155

Song GL, Atrens A (2003) Understanding magnesium corrosion mechanism: a framework for improved alloy performance. Adv Eng Mater 5:837–858

Song R, Ponge D, Raabe D et al (2006) Overview of processing, microstructure and mechanical properties of ultrafine grained bcc steels. Mater Sci Eng A 441(1–2):1–17

Soyama H, Korsunsky AM (2022) A critical comparative review of cavitation peening and other surface peening methods. J Mater Process Technol 305:117586

Staiger MP, Pietak AM, Huadmai J, Dias G (2006) Magnesium and its alloys as orthopedic biomaterials: a review. Biomaterials 27:1728–1734
Stevens MM, George JH (2005) Exploring and engineering the cell surface interface. Science 310(5751):1135–1138
Sukjai O, Asanithi P, Limsuwan P, Limsuwan S (2012) Growth of hydroxyapatite on silk fibers using simulated body fluid. Procedia Eng 32:1087–1093
Sunil BR, Kiran ASK, Ramakrishna S (2022) Surface functionalized titanium with enhanced bioactivity and antimicrobial properties through surface engineering strategies for bone implant applications. Curr Opin Biomed Eng 23:100398
Sweeney SF, Woehrle GH, Hutchison JE (2006) Rapid purification and size separation of gold nanoparticles via diafiltration. J Am Chem Soc 128:3190–3197
Taba Junior M, Novaes AB Jr, Souza SL et al (2003) Radiographic evaluation of dental implants with different surface treatments: an experimental study in dogs. Implant Dent 12:252–258
Takahashi R (2001) State of the art in hot rolling process control. Control Eng Pract 9(9):987–993
Tan F, Naciri M, Dowling D, Al-Rubeai M (2012) *In vitro* and *in vivo* bioactivity of CoBlast hydroxyapatite coating and the effect of impaction on its osteoconductivity. Biotechnol Adv 30(1):352–362
Teixeira AI, Abrams GA, Bertics PJ et al (2003) Epithelial contact guidance on well-defined micro- and nanostructured substrates. J Cell Sci 116:1881–1892
Thamarai P, Kamalesh R, Saravanan A et al (2024) Emerging trends and promising prospects in nanotechnology for improved remediation of wastewater contaminants: present and future outlooks. Environ Nanotechnol Monit Manag 21:100913
Toros S, Ozturk F, Kacar I (2008) Review of warm forming of aluminum-magnesium alloys. J Mater Process Technol 207(1–3):1–12
Tran VT, Benoît JP, Venier-Julienne MC (2011) Why and how to prepare biodegradable, monodispersed, polymeric microparticles in the field of pharmacy? Int J Pharm 407:1–11
Tucker BS, Aliakbarshirazi S, Vijayan VM et al (2021) Nonthermal plasma processing for nanostructured biomaterials and tissue engineering scaffolds: a mini review. Curr Opin Biomed Eng 17:100259
Umit BD (2009) How green are the chemicals used as liquid fuels in direct liquid-feed fuel cells? Environ Int 35:626–631
Urban RM, Jacobs JJ, Tomlinson MJ et al (2000) Dissemination of wear particles to the liver, spleen, and abdominal lymph nodes of patients with hip or knee replacement. J Bone Joint Surg Am 82:457–476
Vadiraj A, Kamaraj M (2006) Characterization of fretting fatigue damage of PVD TiN coated biomedical titanium alloys. Surf Coat Technol 200:4538–4542
Valiev RZ (2004) Nanostructuring of metals by severe plastic deformation for advanced properties. Nat Mater 3:511–516
Valiev RZ, Langdon TG (2006) Principles of equal-channel angular pressing as a processing tool for grain refinement. Prog Mater Sci 51(7):881–981
Van Hove MA (2009) Atomic-scale structure: from surface to nanomaterials. Surf Sci 603:1301–1305
Webster TJ, Ejiofor JU (2004) Increased osteoblast adhesion on nanophase metals: Ti, Ti6Al4V, and CoCrMo. Biomaterials 25:4731–4739
Webster TJ, Smith TA (2005) Increased osteoblast function on PLGA composites containing nanophase titania. J Biomed Mater Res Part A 74:677–686
Wen X, Wang X, Zhang N (1996) Microrough surface of metallic biomaterials: a literature review. J Biomed Mater Eng 6:173–189
Wennerberg A (1998) The importance of surface roughness for implant incorporation. Int J Mach Tools Manuf 38:657–662
Willbold E, Kaya AA, Kaya RA et al (2011) Corrosion of magnesium alloy AZ31 screws is dependent on the implantation site. Mater Sci Eng B 176(20):1835–1840

Wise SG, Waterhouse A, Kondyurin A et al (2012) Plasma-based biofunctionalization of vascular implants. Nanomedicine (London) 7(12):1907–1916

Witte F (2010) The history of biodegradable magnesium implants: a review. Acta Biomater 6:1680–1692

Witte F, Kaese V, Haferkamp H et al (2005) *In vivo* corrosion of four magnesium alloys and the associated bone response. Biomaterials 26:3557–3563

Wu XD, Song XF, Li DS et al (2012) Preparation of mesoporous nano-hydroxyapatite using a surfactant template method for protein delivery. J Bionic Eng 9(2):224–233

Xu JL, Joguet D, Cizek J et al (2012) Synthesis and characterization on atomphospheric plasma sprayed amorphous silica doped hydroxyapatite coatings. Surf Coat Technol 206(22): 4659–4665

Xu YB, Chen CX, Hellwarth PB, Bao XP (2019) Biomaterials for stem cell engineering and biomanufacturing. Bioactive Mater 4:366–379

Yerokhov VY, Hezel R, Lipinski M et al (2002) Cost-effective methods of texturing for silicon solar cells. Sol Energy Mater Sol Cells 72:291–298

Yih TC, Al-Fandi M (2006) Engineered nanoparticles as precise drug delivery systems. J Cell Biochem 97:1184–1190

Yocom CJ, Zhang X, Liao YL (2018) Research and development status of laser peen forming: a review. Opt Laser Technol 108:32–45

Yu MR, Huang Y, Ballweg J et al (2011) Semiconductor nanomembrane tubes: three-dimensional confinement for controlled neurite outgrowth. ACS Nano 5:2447–2457

Yun YH, Dong ZY, Yang D et al (2009) Biodegradable Mg corrosion and osteoblast cell culture studies. Mater Sci Eng C 29:1814–1821

Zäch M, Hägglund C, Chakarov D, Kasemo B (2006) Nanoscience and nanotechnology for advanced energy systems. Curr Opin Solid State Mater Sci 10:132–143

Zaki S, Kady MFE, Abd-El-Haleem D (2011) Biosynthesis and structural characterization of silver nanoparticles from bacterial isolates. Mater Res Bull 46:1571–1576

Zdravecká E, Tiainen VM, Konttinen YT et al (2012) Relationships between the fretting wear behavior and mechanical properties of thin carbon films. Vacuum 86:675–680

Zhang S (2003) Fabrication of novel biomaterials through molecular self-assembly. Nat Biotechnol 21:1171–1178

Zhang MZ, Kataoka K (2009) Nano-structured composites based on calcium phosphate for cellular delivery of therapeutic and diagnostic agents. Nano Today 4:508–517

Zhang J, Wang F, Shenoy VB et al (2020) Towards controlled synthesis of 2D crystals by chemical vapor deposition (CVD). Mater Today 40:132–139

Zhilyaev AP, Langdon TG (2008) Using high-pressure torsion for metal processing: fundamentals and applications. Prog Mater Sci 53(6):893–979

Zhu X, Chen J, Scheideler L et al (2004) Cellular reactions of osteoblasts to micron- and submicron-scale porous structures of titanium surfaces. Cells Tissues Organs 178:13–22

Zhuo CW, Hall B, Richter H, Levendis Y (2010) Synthesis of carbon nanotubes by sequential pyrolysis and combustion of polyethylene. Carbon 48:4024–4034

Zia AW, Birkett M (2021) Deposition of diamond-like carbon coatings: conventional to non-conventional approaches for emerging markets. Ceram Int 47(20):28075–28085

Zia AW, Hussain SA, Rasul S et al (2023) Progress in diamond-like carbon coatings for lithium-based batteries. J Energy Storage Part E 72:108803

Zinger O, Anselme K, Denzer A et al (2004) Time-dependent morphology and adhesion of osteoblastic cells on titanium model surfaces featuring scale-resolved topography. Biomaterials 25:2695–2711

Chapter 5
Sensors Containing Hybrid Nanomaterials to Access Environmental Pollution Level

Ashab Noumani, Harsh Bagdare, Jayendra Kumar Himashu, Sonu Sen, Sunil Kumar Patidar, Satish Piplod, Vinars Dawane, and Bhawana Pathak

Abstract The sensing technology escalating a remarkable progress due to their incomparable selectivity and sensitivity-based applicability. The advancement in technology owes much to the integration of material sciences with nanotechnology, resulting in the emergence of highly effective sensing technologies. The integration of advanced nanomaterials into sensing platforms has enabled the development of more cost-effective, faster and consistent evaluation methods. This advancement holds great promise for applications across scientific fields, particularly in environmental monitoring, pollution detection, and complex environmental forensic

---

A. Noumani
Department of Biotechnology, Kalinga University, Naya Raipur, Chhattisgarh, India

H. Bagdare · V. Dawane
Department of Micro-Bio-Seed Technology, Sardar Vallabhbhai Patel College, Mandleshwar, Madhya Pradesh, India

J. K. Himashu
Department of Biotechnology, School of Life Sciences, Mahatma Gandhi Central University, Motihari, Bihar, India

S. Sen
Department of Physics, Shri Neelkantheshwar Government Post Graduate College, Khandwa, Madhya Pradesh, India

S. K. Patidar
Department of Chemistry, Government Holkar (Model, Autonomous) Science College, Indore, Madhya Pradesh, India

S. Piplod
Department of Chemistry, Shaheed Bhagat Singh Government Post Graduate College, Narmadapuuram, Madhya Pradesh, India

Govt. College, Narmadapuuram, Madhya Pradesh, India
e-mail: dr.satishpiplode@mp.gov.in

B. Pathak (✉)
School of Environment and Sustainable Development, Central University of Gujarat, Gandhinagar, Gujarat, India
e-mail: Bhawna.pathak@cug.ac.in

© The Author(s), under exclusive license to Springer Nature Switzerland AG 2025
J. M. Al-Khayri et al. (eds.), *Green Nanotechnology Applications for Ecosystem Sustainability*, Nanotechnology in Plant Sciences 4,
https://doi.org/10.1007/978-3-031-90766-1_5

investigations, making nano sensors a focal area for ongoing research and development efforts. In this context, nano sensors incorporating hybrid materials, including non-carbon containing materials such as metals, metal oxides, their composites, and non-metals; alongside carbon-containing materials like carbon nanoparticles, carbon dots, and graphene-based nanomaterials, have garnered significant attention for their promising applications in environmental science and research. Therefore, the current chapter emphasizes the significance and function of sensors based on hybrid nanomaterials, encompassing both carbon-based and non-carbon-based variations. It also explores their recent progressions and the primary factors that have elevated them to a favored choice in sensing technologies. The chapter extensively explores the environmental uses of sensors based on hybrid nanomaterials, as well as the primary challenges and anticipated future challenges linked to this technology.

Keywords Environmental applications · Hybrid nanomaterials · Nano sensors · Pollution · Sensing technology · Sustainable development

5.1 Introduction

The growing worries about environmental pollution in the past few years have resulted to an increase in the advancement of sensing technologies that can monitor and identify different pollutants with high sensitivity and selectivity (Zhang et al. 2021a, b). The heightened emphasis on environmental monitoring arises from the acknowledgment of the substantial influence of pollution on ecosystems and human well-being. With the increasing pollution levels worldwide (McCrink-Goode 2014), there is an escalating need to create inventive methods for identifying and reducing pollution. On-carbon containing nanohybrid materials (Dao et al. 2023) have become intriguing options for environmental sensing applications due to their distinctive features and versatile functionalities as shown in Fig. 5.1. These materials have been extensively studied among the wide range of materials explored for this purpose. Nanohybrids amalgamate the beneficial characteristics of several nanomaterials, including metal oxides, metal nanoparticles, polymers, ceramics, and composites, in order to improve sensing capabilities (Deshmukh et al. 2018; Nunes et al. 2019). These materials provide a flexible framework for customizing sensor qualities to meet specific requirements for environmental monitoring.

To appreciate the crucial significance of non-carbon-containing nanohybrid materials in environmental sensing, it is imperative to first delve into the pressing issue of pollution. Pollution, in its various manifestations, is a significant menace to ecosystems and human well-being on a global scale (Corvalan et al. 2005). From industrial effluents to automotive emissions, contaminants infiltrate the air, water, and soil, cascading severe implications on biodiversity and public well-being. As a result, the pressing requirement for strong monitoring and mitigation techniques has sparked the search for advanced sensing technologies that can accurately detect and measure pollutants with exceptional precision.

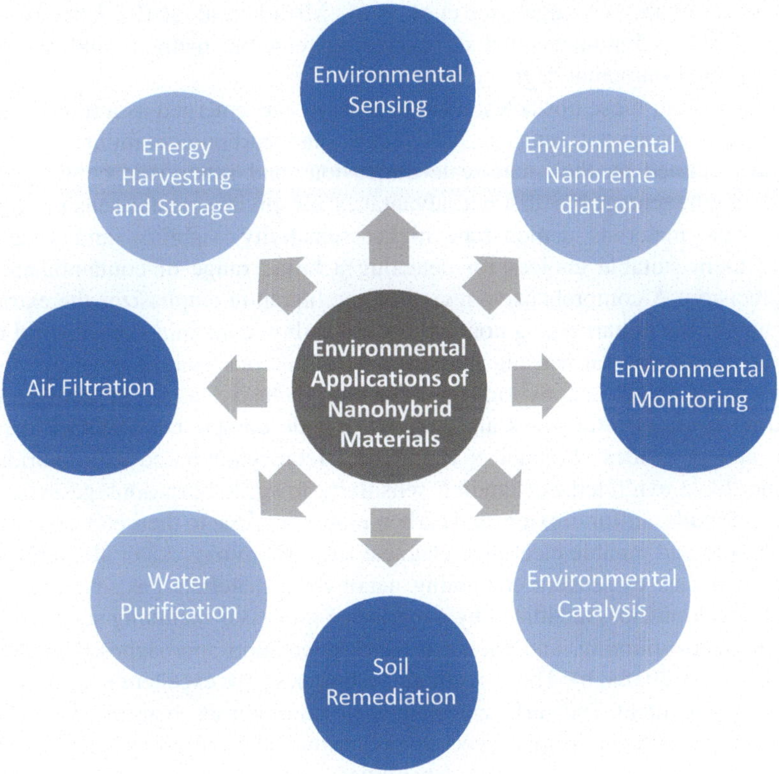

Fig. 5.1 Nanohybrid materials and their various applications. (Figure created by Jayendra Kumar Himanshu)

A prominent environmental issue is air pollution, which arises from the release of dangerous pollutants into the atmosphere from industrial activity, traffic, and agricultural practices. These pollutants include sulfur dioxide, nitrogen oxides, particulate matter, ozone, and volatile organic compounds, which pose substantial health concerns to individuals and ecosystems (Manisalidis et al. 2020). Effective monitoring of air quality is essential for reducing these dangers and adopting focused measures to lower pollution levels.

Water pollution is a significant problem that endangers both human health and aquatic ecosystems by contaminating freshwater supplies. Industrial discharges, agricultural runoff, and inappropriate waste disposal lead to the accumulation of heavy metals, pesticides, and organic contaminants in water bodies (Madhav et al. 2020). Monitoring water quality is crucial for assuring the safety of drinking water sources, preserving aquatic habitats, and safeguarding public health.

Soil pollution, stemming from industrial activities, mining operations, and inappropriate waste management, poses considerable difficulties to agricultural production and ecosystem health. Heavy metals, pesticides, and other contaminants can accumulate in soil, creating concerns to plant development, food safety, and groundwater quality (Alengebawy et al. 2021; Mishra et al. 2019). Effective monitoring of soil pollution is vital for executing remedial methods and preventing additional environmental degradation.

In response to these obstacles, nanotechnology has emerged as a transformative force in environmental sensing. Non-carbon and carbon-containing nanohybrid materials, defined by their nanoscale dimensions and synergistic combination of various components, offer distinct advantages for environmental sensing applications. These materials demonstrate higher sensitivity, stability, and selectivity, making them suitable choices for detecting a broad range of contaminants with high precision. A comprehensive review of the literature emphasizes the extraordinary gains made in harnessing non-carbon and carbon containing nanohybrid materials for environmental sensing applications. Numerous studies have proved the usefulness of these nanohybrids in detecting inclusive array of contaminants, extending from harmful gases and volatile organic compounds (VOCs) to heavy metals and pesticides (Mirzaei et al. 2016). Metal oxide-based nanohybrids, for example, have exhibited outstanding sensitivity towards hazardous gases such as nitrogen dioxide, sulfur dioxide, and carbon monoxide, due to their elevated surface-volume ratio and tunable electronic characteristics (Dimitrievska et al. 2023).

The synthesis of sensors containing nanohybrid materials has undergone substantial developments, facilitated by a myriad of processes including sol-gel, hydrothermal, electrospinning, and chemical vapor deposition approaches (Harish et al. 2022; Liu et al. 2018a, b). These synthetic techniques give excellent control over the morphology, content, and surface chemistry of nanohybrids, enabling the tuning of their sensing capability to meet specific environmental monitoring requirements. In addition to their enhanced sensing capabilities, non-carbon containing nanohybrid materials offer various intrinsic advantages that render them well-suited for environmental sensing applications. Their nanoscale dimensions endow them with large surface areas and enhanced reactivity, facilitating rapid and selective detection of target analytes. Moreover, the integration of varied components inside a single nanohybrid structure imparts synergistic functions, enabling simultaneous detection of several contaminants with better accuracy and reliability. Table 5.1 shows the main techniques used for characterization of hybrid nanomaterials.

The utilization of nanomaterials containing nanohybrid composition offers enormous potential for overcoming the difficulties associated with traditional sensing platforms, such as low selectivity, susceptibility to interference, and excessive power consumption (Sun et al. 2021). By leveraging the unique features of nanohybrids, researchers can construct compact, portable, and cost-effective sensing devices capable of real-time monitoring in varied environmental circumstances. Non-carbon and carbon-containing nanohybrid materials represent a paradigm shift in environmental sensing, giving new prospects for the creation of sophisticated sensing systems with enhanced performance and usefulness. Through synergistic integration of various components at the nanoscale, these materials hold the key to opening new horizons in pollution monitoring and mitigation, paving the road towards a sustainable future.

Table 5.1 Various methods and their applications used in the characterization of nanoparticles (Mourdioudis et al. 2018)

Technique	Main information derived
XRD	Crystal structure, composition, crystalline grain size determination
XPS	Electronic structure, elemental composition, oxidation states, ligand binding (surface-sensitive)
FTIR	Surface composition, ligand binding. Very important in the understanding of specific class of biomolecules responsible for the biofibrication.
TEM	NP size, size monodispersity, shape, aggregation state, detect and localize/ quantify NPs in matrices, study growth kinetics
HRTEM	All information by conventional TEM but also on the crystal structure of single particles. Distinguish monocrystalline, polycrystalline and amorphous NPs. Study defects
SEM-HRSEM, T-SEM-EDX	Morphology, dispersion of NPs in cells and other matrices/supports, precision in lateral dimensions of NPs, quick examination–elemental composition

This chapter explores the synthesis processes, characterization, and applications of non-carbon and carbon-containing nanohybrid materials such as silica-titania, zinc oxide-gold, cerium oxide-zirconia, silver-titanate, copper oxide-montmorillonite across various environmental sensing domains, illuminating its transformational potential in combatting environment pollution. After the discussion on selective and preferential case studies the chapter also critically examines the key bottle neck challenges and the prospects associated with the sensing technology.

5.2 Non-Carbon Containing Nanohybrid Materials for Sensing—Selective Case Studies

Non-carbon containing nanohybrid materials offer versatile applications in environmental remediation due to their unique properties and functionalities. This chapter explores the synthesis, characterization, and applications of various non-carbon containing nanohybrids, mainly silica-titania, zinc oxide-gold, cerium oxide-zirconia, silver-titanate, copper oxide-montmorillonite, cadmium sulfide-zinc oxide, and nickel-doped cadmium sulfide/zinc telluride, and few others. These nanohybrids demonstrate enhanced performance in applications such as photocatalysis, energy storage, sensing, and wastewater treatment. The synthesis methods, structural characterization, and performance evaluation of these nanohybrids are discussed, highlighting their potential for addressing environmental challenges and advancing sustainable technologies (Fig. 5.2).

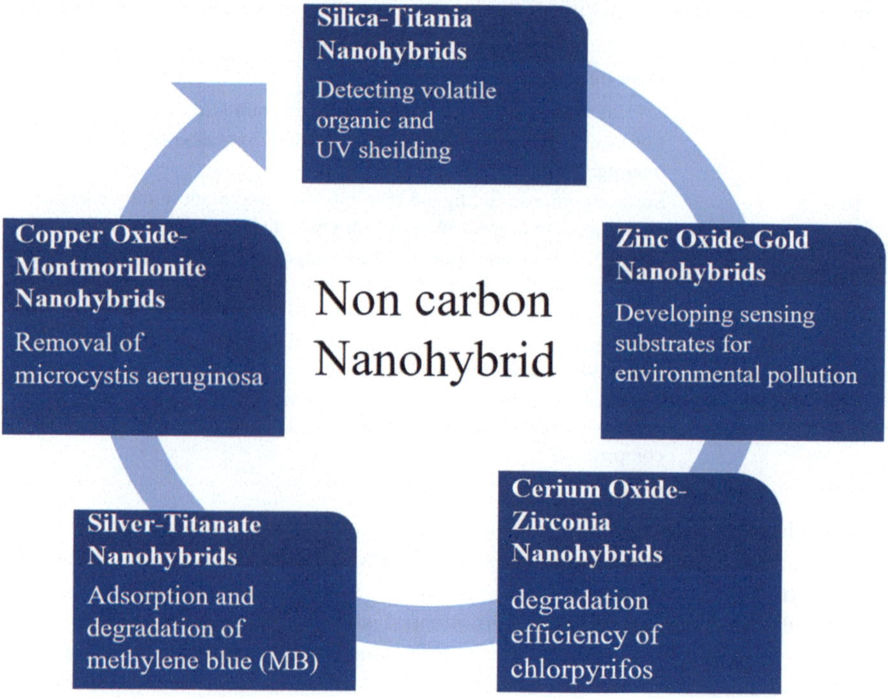

Fig. 5.2 Selective nanohybrids having non-carbon-containing materials and applications. (Figure created by Jayendra Kumar Himanshu)

5.2.1 Silica-Titania Nanohybrids Materials

Silica (SiO_2) and titania (TiO_2) nanoparticles can be combined to form nanohybrids with enhanced photocatalytic properties (Iman Amir et al. 2015). Silica provides a stable matrix for dispersing TiO_2 nanoparticles, leading to improved photocatalytic efficiency for applications such as water purification and air pollution control (Amlouk et al. 2006). The synthesis typically begins with the selection of a silica precursor, such as tetraethyl orthosilicate (TEOS) or sodium silicate. The silica precursor is hydrolyzed in the presence of water and a catalyst, such as ammonia or an acid, to form silanol groups (Si-OH). This is followed by condensation reactions, leading to the formation of silica nanoparticle. Stabilizing agents or surfactants, such as polyvinylpyrrolidone (PVP) or cetyltrimethylammonium bromide (CTAB), may be added to regulate the stability, shape, and size of silica nanoparticles. Titanium precursor, typically titanium tetraisopropoxide (TTIP) or titanium tetrachloride ($TiCl_4$), is selected for deposition of titania on silica nanoparticles. The surface of silica nanoparticles may be modified to enhance the deposition of titania. This can be achieved through functionalization with silane coupling-agents, such as vinyltriethoxysilane (VTES) or 3-aminopropyltriethoxysilane (APTES). The deposition of titania onto silica nanoparticles can be achieved through various methods, including atomic layer deposition, chemical vapor deposition or sol-gel. In the sol-gel method, the titania precursor is hydrolyzed and condensed in the presence of silica nanoparticles under controlled conditions to form a micro titania layer on surface of silica. Annealing and Drying: The resulting Silica-Titania Nanohybrids are typically annealed at elevated temperatures to improve crystallinity and remove any residual solvent or organic species. The annealing process may be followed by drying under vacuum or ambient conditions to obtain the final nanohybrid material (Wang et al. 2005; Jin et al. 2009; Liaw et al. 2011). Characterization methods like X-ray diffraction (XRD) are employed for analyzing the crystalline structure and phase composition of the Silica-Titania Nanohybrids. Transmission electron microscopy (TEM) and Scanning electron microscopy (SEM) are utilized to analyses the distribution, size and morphology of nanoparticles within the nanohybrid material. Energy-dispersive X-ray spectroscopy (EDS) and Fourier-transform infrared spectroscopy (FTIR) and can be employed to analyses the chemical composition and surface functional groups of the nanohybrid material (Kang et al. 2013; Abdullah et al. 2019).

5.2.2 *Zinc Oxide-Gold Nanohybrids*

Zinc oxide (ZnO) nanoparticles decorated with gold (Au) nanoparticles exhibit enhanced catalytic activity for various environmental applications. These nanohybrids are utilized in photocatalysis, gas sensing, and wastewater treatment due to their synergistic effects and improved performance compared to individual components. Such as zinc acetate or zinc nitrate, is dissolved in a suitable solvent, often ethanol or water. A base, typically sodium hydroxide (NaOH) or ammonia (NH_3), is added to the zinc precursor solution to initiate the hydrolysis and precipitation of zinc hydroxide. The resulting zinc hydroxide precipitate is then heated at elevated temperatures, typically around 100–200 °C, to convert it into zinc oxide nanoparticles through a dehydration process. The synthesized ZnO nanoparticles are washed with a solvent, such as ethanol or deionized water, to remove any residual impurities or by-products. They are then dried under vacuum or at ambient conditions to obtain dry ZnO nano powder. Gold precursor, often chloroauric acid ($HAuCl_4$), is dissolved in a suitable solvent, typically ethanol or water. The gold precursor solution is subsequently combined with a reducing agent, such as citric acid or sodium borohydride ($NaBH_4$), to catalyze the reduction of gold ions into gold nanoparticles. The freshly prepared gold nanoparticle solution is added dropwise to the ZnO nanoparticle suspension under stirring. The mixture is allowed to incubate and age for a certain period, typically several hours to overnight, to facilitate the adsorption and deposition of gold nanoparticles on surface of Zinc Oxide nanoparticles (Perumal et al. 2015, 2016; Rahman and Ghosh 2016).

As far as their selective applications are concerned, In one experiment it used in nano biosensor for early exposure of HPV-16 infected cervical cancer cells. The sensor, based on interdigitated electrode (IDE) chips, utilizes nanorods of gold doped zinc oxide (Au-ZnO) coated with HPV-16 viral DNA bioreceptors. This nanohybrid film enhances sensitivity and biocompatibility, enabling excellent recognition of the HPV-16 E6 oncogene, a key biomarker for HPV-induced cervical cancer. The sensor exhibits the capability to detect concentrations as low as 1 femtomolar of the viral E6 gene target and remains stable for over 5 weeks with excellent reproducibility (Ramesh et al. 2019). In another experiment it is synthesized "spiked nano rosette" nanostructures via aqueous hydrothermal route using Au and ZnO, exhibiting ultra-crystalline rose-like texture with nanowires. Biorecognition studies showed high selectivity and detection limit of 1×10^{-12} M, making these nanostructures promising for sensitive nucleic acid detection in nano biosensors and disease diagnostics (Haarindraprasad et al. 2023).

5.2.3 Cerium Oxide-Zirconia Nanohybrids

Cerium oxide (CeO_2) nanoparticles combined with zirconia (ZrO_2) nanoparticles form nanohybrids with enhanced redox properties and catalytic activity. These nanohybrids are used in catalytic converters for automotive exhaust treatment, as well as in water treatment applications for pollutant degradation and detoxification (Verma et al. 2020). Cerium precursor, such as cerium nitrate or cerium chloride, is dissolved in a suitable solvent, typically ethanol or water. A base, such as ammonium hydroxide (NH_4OH) or sodium hydroxide (NaOH), is added to the cerium precursor solution to initiate the precipitation of cerium hydroxide. Heat Treatment: The resulting cerium hydroxide precipitate is then heated at elevated temperatures, typically around 300–600 °C, to convert it into cerium oxide nanoparticles through a dehydration process. Washing and Drying: The synthesized CeO_2 nanoparticles are washed with a solvent, such as ethanol or deionized water, to remove any residual impurities or by-products. They are then dried under vacuum or at ambient conditions to obtain dry CeO_2 nanopowder. Zirconium precursor, such as zirconium nitrate or zirconium chloride, is typically dissolved in an appropriate solvent, such as ethanol or water. A base, similar to the one used for cerium oxide synthesis, is added to the zirconium precursor solution to initiate the precipitation of zirconium hydroxide. The resulting zirconium hydroxide precipitate is then heated at elevated temperatures, typically around 400–800 °C, to convert it into zirconia nanoparticles through a dehydration process. Washing and Drying: The synthesized ZrO_2 nanoparticles are washed with a solvent, similar to the one used for cerium oxide, to remove any residual impurities or by-products. They are then dried under vacuum or at ambient conditions to obtain dry ZrO_2 nanopowder. A solution containing cerium precursor, such as cerium nitrate, is prepared in a suitable solvent. The cerium oxide precursor solution is then added dropwise to the zirconia nanoparticle suspension under stirring. The mixture is allowed to incubate and age for a certain period, typically several hours to overnight, to facilitate the adsorption and deposition of cerium oxide onto the surface of zirconia nanoparticles (Oladejo et al. 2022).

As far as their applications are concerned, in a recent work Zirconia and cerium-doped zirconia nanoparticles (ZCO) were synthesized. Binary nanohybrids ZrO_2@rGO and ZCO-2@rGO, and ternary nanohybrids ZrO_2@rGO@MoS_2 and ZCO-2@rGO@MoS_2 were prepared. Tribo-activity was assessed in paraffin oil (PO), with ZCO-2@rGO@MoS_2 exhibiting the highest performance. Nanoparticles and nanosheets prevented agglomeration and controlled restacking, enhancing activity. SEM, AFM, and EDX analysis confirmed the role of elements in wear-track formation, supported by XPS revealing a tribofilm composition of SO_4^{2-}, ceria, Fe_2O_3, zirconia, MoS_2, MoO_3, and rGO (Rathee et al. 2021).

5.2.4 *Silver-Titanate Nanohybrids*

Silver (Ag) nanoparticles supported on titanate nanotubes form nanohybrids with antibacterial and catalytic properties. These nanohybrids are applied in water disinfection, wound healing, and air purification due to their strong antimicrobial activity and catalytic degradation of organic pollutants (Ali et al. 2022).

Titanium precursor, such as titanium isopropoxide or titanium sulfate, is typically dissolved in an appropriate solvent, such as ethanol or water. A strong alkali, such as potassium hydroxide (KOH) or sodium hydroxide (NaOH), is added to the titanium precursor solution to initiate the development of titanate nanotubes. The mixture is exposed to hydrothermal action at elevated temperatures, typically around 100–200 °C, for several hours. This promotes the growth and self-assembly of titanate nanotubes. The synthesized titanate nanotubes are washed with a solvent, such as deionized water, to remove any residual impurities or by-products. They are then dried under vacuum or at ambient conditions to obtain dry titanate nanotubes. Silver precursor, often silver nitrate ($AgNO_3$), is typically dissolved in an appropriate solvent, such as ethanol or water. The silver precursor solution is mixed with a reducing agent, such as glucose or sodium borohydride ($NaBH_4$), to initiate the reduction of silver ions to form silver nanoparticles. The newly prepared silver nanoparticle solution is added dropwise to the titanate nanotube suspension under stirring. The mixture is allowed to incubate and age for a certain period, typically several hours to overnight, to facilitate the adsorption and deposition of silver nanoparticles onto the surface of titanate nanotubes (Alghuthaymi et al. 2021). Characterization method like X-ray diffraction (XRD) are employed to analyses the structure of crystal and phase composition of the Ag-TiO_2 nanohybrids. Transmission electron microscopy (TEM) and Scanning electron microscopy (SEM) are employed to inspect the size, distribution and morphology of nanoparticles within the nanohybrid material. Energy-dispersive X-ray spectroscopy (EDS) is a useful tool for examining the chemical composition and distribution of elements within the Ag-TiO_2 nanohybrids (Ali et al. 2022).

As far as their applications are concerned, in a study Ag/TNTs-P400 demonstrated superior photocatalytic degradation of MB compared to other samples. In vitro cytotoxicity tests against breast cancer cell lines showed that Ag/TNTs-P400

(IC50 = 8.58 μg/mL) exhibited the highest anti-cancer activity among all samples. Geometry optimization was performed using DFT at the B3LYP/6-311G* level. Docking studies revealed that both TNTs-400 and Ag/TNTs-P400 demonstrated robust affinities as inhibitors against multiple bacterial kinase active sites. Furthermore A nanohybrid plasmonic photocatalyst was developed, consisting of Ag-AgX (X = Cl, Br, I) nanoparticles anchored on titanate nanotubes (TNTs). This hybrid, termed Ag-AgX-TNT, was synthesized through a process involving silver ion incorporation, light irradiation and hydrogen halide injection. The high surface area of Ag-AgX-TNT facilitated the adsorption and degradation of methylene blue (MB) below visible light. Titanate nanotubes adsorb MB onto their surfaces, thereby improving degradation facilitated by Ag-AgX nanoparticles. The superior performance of Ag-AgX-TNT below visible light is attributed to its localized surface plasmonic resonance effect. Thus, this combined titanate hybrid/silver halides holds promise for wastewater treatment using solar irradiation (Tang et al. 2012).

5.2.5 Copper Oxide-Montmorillonite Nanohybrids

Copper oxide (CuO) nanoparticles dispersed within montmorillonite clay matrices exhibit enhanced adsorption and catalytic properties. These nanohybrids are utilized for elimination of organic pollutants, heavy metals, and dyes from wastewater, as well as for catalytic oxidation reactions in environmental remediation processes. Montmorillonite clay, a type of layered silicate mineral, is selected as the host material for the nanohybrid. Montmorillonite is exfoliated and activated to increase its surface area and enhance its reactivity. This can be achieved by treatment with cationic surfactants or organic modifiers to intercalate and expand the interlayer spacing of the clay. The activated montmorillonite is washed with a solvent, such as water or ethanol, to remove any residual surfactants or impurities. It is then dried under vacuum or at ambient conditions to obtain dry montmorillonite nano clay. Copper precursor, such as copper nitrate or copper chloride, is typically dissolved in an appropriate solvent, such as ethanol or water. The copper oxide precursor solution is added dropwise to the montmorillonite nano clay suspension under stirring. The mixture is subjected to hydrothermal treatment at elevated temperatures, typically around 100–200 °C, for several hours. This promotes the growth and nucleation of copper oxide nanoparticles onto the surface of montmorillonite nano clay. The synthesized CuO-MMT nanohybrids are washed with a solvent, such as deionized water, to remove any residual impurities or by-products. They are then dried under vacuum or at ambient conditions to obtain dry CuO-MMT nanohybrid material (Khaorapapong et al. 2015; Hossain et al. 2023; Bhat et al. 2022; Khan et al. 2016).

As far as their applications are concerned, one study shows that CuO/MMT nanohybrids have likely to serve as an antimicrobial agent, with copper oxide nanoparticles offering antimicrobial properties and MMT support helping to mitigate toxicity concerns. Incorporating copper oxide nanoparticles into the interlayers of MMT through thermal decomposition demonstrates strong antibacterial activity against *E. coli*, attributed to the release of Cu ions upon interaction with an aqueous phase. Recent work, highlights the alkaline ion exchange method as an efficient approach to fabricate Cu/MMT nanocomposites, exhibiting effective antimicrobial activity against S. aureus and E. coli in broth media. The nanocomposite's sustained antibacterial efficacy is attributed to the diffusion of Cu into the MMT, resulting in lower Cu^{2+} ion release compared to Cu NPs attached to the MMT surface (Hossain et al. 2023).

5.2.6 Cadmium Sulfide-Zinc Oxide Nanohybrids

The combination of cadmium sulfide (CdS) nanoparticles with zinc oxide (ZnO) nanoparticles results in nanohybrids exhibiting improved photocatalytic activity suitable for solar-driven water splitting and degradation of pollutants. These nanohybrids are studied for their possible applications in environmental remediation and renewable energy generation. Zinc precursor, such as zinc acetate or zinc nitrate, is dissolved in a suitable solvent, typically ethanol or water. A base, typically sodium hydroxide (NaOH) or ammonia (NH_3), is added to the zinc precursor solution to initiate the precipitation of zinc hydroxide. The resulting zinc hydroxide precipitate is then heated at elevated temperatures, typically around 100–200 °C, to convert it into zinc oxide nanoparticles through a dehydration process. The synthesized ZnO nanoparticles are washed with a solvent, such as ethanol or deionized water, to remove any residual impurities or by-products. They are then dried under vacuum or at ambient conditions to obtain dry ZnO nanopowder. Cadmium precursor, such as cadmium acetate or cadmium chloride, is dissolved in a suitable solvent, typically ethanol or water. A sulfide source, such as thiourea or sodium sulfide, is added to the cadmium precursor solution under stirring to initiate the precipitation of cadmium sulfide nanoparticles. The mixture is heated at elevated temperatures, typically around 100–200 °C, for several hours to promote the growth and nucleation of cadmium sulfide nanoparticles. The synthesized CdS nanoparticles are washed with a solvent, such as ethanol or deionized water, to remove any residual impurities or by-products. They are then dried under vacuum or at ambient conditions to obtain dry CdS nanopowder. A solution containing cadmium precursor and a sulfide source is prepared in a suitable solvent. Addition to ZnO Nanoparticles: The CdS precursor solution is added dropwise to the ZnO nanoparticle suspension under stirring. The mixture is allowed to incubate and age for a certain period, typically several hours to overnight, to facilitate the adsorption and deposition of cadmium sulfide onto the surface of zinc oxide nanoparticles (Kamran et al. 2024). TEM, SEM and XRD are used to analyses the crystal structure and phase composition of the CdS-ZnO nanohybrids and examine the morphology, size, and distribution of nanoparticles within the nanohybrid material respectively. Energy-dispersive X-ray spectroscopy (EDS) can be utilized to analyze the chemical composition and elemental distribution of the CdS-ZnO nanohybrids (Devi and Tharmaraj 2020).

As far as their applications are concerned, the multifunctional Ni-CdS/ZnTe nanocomposites were synthesized via so no chemical method for enhanced photocatalytic activity and energy storage. The optimized sample (5% Ni-CdS/ ZnTe) exhibited excellent photocatalytic degradation of MB, MO, and RhB dyes under visible light. Electrochemical investigation revealed impressive specific capacitance (517.6 F/g), energy density (17.97 Wh/kg), and power density (250 W/kg) at 1 A/g current density. These results suggest promising applications of Ni-CdS/ZnTe nanocomposites in environmental remediation and energy storage (Kamran et al. 2024). In other experiment A novel CdS–ZnS/Zirconium–titanium phosphate (ZTP) composite photocatalyst was synthesized via a two-step sulfidation technique, exhibiting improved visible light photocatalytic activity for hydrogen production. Characterization techniques including XRD, N_2 adsorption–desorption, DRUV-vis, PL, SEM/EDX, XPS, and TEM were employed. The 5CdS–ZnS/ZTP catalyst demonstrated the highest hydrogen production (2142.7 μmol) and apparent quantum efficiency (9.6%) among all tested catalysts under visible light illumination (Biswal et al. 2011).

5.3 Carbon Containing Nanohybrid Materials for Sensing—Selective Case Studies

Carbon containing nanohybrid materials have been making a big noise in the arena of sensing technology and environmental forensics. They have been considered tube preferential in various environmental applications. Thus present part of this chapter discusses the synthesis methods and characterization techniques employed in the development of these materials to understand this nano-engineering method logy well. Subsequently, the properties and functionalities carbon containing nanohybrid materials have been critically examined to compare and highlighting their advantages over traditional materials. The applications include but are not limited to water purification, air filtration, soil remediation, and renewable energy generation have been discussed briefly in the form of various case studies and examples to illustrate the effectiveness of these materials in addressing specific environmental challenges.

To explore the potential of carbon based nanohybrids, Brahma et al. (2016) worked on CuNPs/fullerene-C60/multi-walled carbon nanotubes (MWCNTs) nanocomposite film for paracetamol detection. They used lectrochemical method based on square wave voltammetry (SWV) for paracetamol detection. Chronocoulometric, electrochemical impedance spectroscopy (EIS) and cyclic voltammetry (CV), have been used for characterization of sensor. They reported linear range from 4.0×10^{-9} to 4.0×10^{-7} M and detection limit of 7.3×10^{-11} M for paracetamol detection. In an attempt by Wang et al. (2016) investigated cadmium telluride (CdTe) quantum dots (QDs) and carbon nanodots (CDs) hybrid sensor for Cu^{+2} detection. Radiometric fluorescence nano sensor exhibits the dual-emissions peaked at 437 and 654 nm. They reported 0.36 nM detection limit for Cu^{+2} in real sample. One work by Sun et al. (2015) developed graphene quantum dots and CdTe QDs nano sensor for Cu^{+2} ion. They reported very simple nano sensor for Cu^{+2} detection by visual image. It can be easily seen by naked eye. They also reported 5.3×10^{-8} M detection limit for target.

Another work by Jin et al. (2017) examined metal-organic frameworks and CdTe quantum dots Radiometric fluorescence (RF) sensors for 6-mercaptopurine detection. It is a type of dual-emission nano sensor. They reported that hybrid nano sensor exhibit higher activity than CdTe QDs and MOFs. They reported linear range and detection limit of 0–50 μM and 0.15 μM respectively. Recently, Nguyen et al. (2024) developed gold nanoparticles and carbon quantum dots composite sensor for amikacin detection. It is based on "on-off" fluorescence nano sensor mechanism. They reported linear range and low detection limit of 10^{-10}–10^{-8} M and 3.5×10^{-11} M respectively. They also reported introduction of Cu^{+2} ion could enhance the efficiency of the nano sensor.

In a faithful attempts by, Zhu et al. (2022) worked on cadmium telluride quantum dots (CdTe QDs), carbon dots and metal–organic framework (ZIF-8) nano sensor for Ag^{+} detection. It is a type of radiometric fluorescent nano sensor. They reported linear range and limit of detection of 0.1–20 μM and 1.49 nM respectively. Similarly, Ma et al. (2023) developed mesoporous silica spheres-coated nitrogen-doped carbon dots (N-CDs@SiO_2) and gold nanoclusters (AuNCs) nano sensor for glyphosate detection. This nano sensor works on "signal on–off–on" radiometric fluorescent method. They reported 5–100 ng/mL and 3.4 ng/mL linear range and limit of detection respectively.

In a dense work by Shafiei and Hassaninejad-Darzi (2023) investigated Ag nanoparticle, polyoxometalate, reduced graphene oxide (Ag@POM@rGO), and ionic liquid (IL) modified carbon electrode nano sensor for dopamine (DA) and acetaminophen (AC) detection. Response surface methodology (RSM) has been used for investigation. They reported 0.05–115.04 μM and 17.0 nM linear dynamic ranges and limit of detection respectively for dopamine. They also reported 0.1–137.90 μM and 37.0 nM linear dynamic ranges and limit of detection respectively for acetaminophen. Another case by Xiang et al. (2018) worked on CDs-coated dual-emission silica nanoparticle for chromium detection. They reported that this nanosor easily detect Cr (III) and Cr (VI) ion in water sample. Rhodamine B dye also used for coupling mechanism with silica nanoparticles. They reported 0–500 ng mL^{-1} and .3 ng mL^{-1} linear range and detection limit respectively for Cr (VI) ion.

In a genuine work by Yuan et al. (2014) developed bis(dithiocarbamato) copper (II) functionalized carbon nanodots (CuDTC2-CDs) nanosensor for Hg^{+2} detection. It is based on "turn on" fluorescence nano sensor mechanism. Both electron transfer and energy transfer mechanism have been involved in the process. They reported 4 ppb limit for Hg^{+2} detection in water sample under optimum conditions. Similarly, Liu et al. (2014) examined carbon dots (CDs) and Rhodamine B-doped silica nanoparticles based nano sensor for Cu^{2+} detection. It is a radiometric fluorescent nano sensor for environmental as well biomedical applications. They reported

0–3 × 10^{-6} M and 35.2 nM linear range and limit of detection respectively under optimum conditions.

An authentic work by Zhang et al. (2018) investigated Sulfur–nitrogen co-doped carbon dots (SNCDs) Nao sensor for hypochlorous acid and Zn^{+2} detection. The nano sensor was prepared using Facile one-step solvothermal strategy. It is a "off–on" nano sensor for hypochlorous acid and Zn^{+2} detection in water sample detection. They reported linear range of 8.41–84.12 μM with a detection limit of 0.30 μM based on turn on process under suitable conditions. Similarly, Panda et al. (2018) developed azide (N_3)-functionalized CQD nano sensor for pesticide flumioxazin detection. It is a turn-on fluorescence nano sensor. Cu(II) and ascorbic acid plays an important role in the method. Combination of Alkynyl group, azide group and Cu (II) makes this sensor specific and selective toward flumioxazin. They reported 0–35.4 μg/l and 0.027 μg/l linear range and detection limit respectively.

In a selective work by Tumay et al. (2021) developed ferrocene-thiophene modified by carbon nanotube (FT@CNT) nano sensor for simultaneous determination of parathion and Chlorantraniliprole detection. They reported linear rang and detection limit 0.02–6.50 μmol/L and 5.3 nmol/L respectively for parathion. They also reported linear rang and detection limit 0.01–7.00 μmol/L and 8.1 nmol/L, respectively for Chlorantraniliprole. They easily detected these pesticides in food and soil sample. In another work by Liu et al. (2018a, b) developed N-doped carbon dots nano sensor for H^+ and Hg^{++} detection. They synthesized this sensor using ammonium citrate and betaine hydrochloride by hydrothermal method. They reported that H^+ and Hg^+ both can easily be detected by as prepared sensor. 1.69 nM is the detection limit has been observed for Hg^+ under suitable conditions.

A significant work by Li et al. (2020) worked on Copper nanoclusters @ nitrogen-doped CQDs (CuNCs-CNQDs) for lead (II) ions detection. It is based on radiometric fluorescence probe. They reported 0.010–2.5 mg/L and 0.0031 mg/L linear range and detection limit respectively. Similarly, Yin et al. (2022) worked on self-assembled copper nanoclusters (CuNCs) and carbon dots driven by Al^{3+} ions for sulfur ions detection. They reported linear range and low detection limit from 0.5 to 40 μM, and 0.16 μM respectively using fluorescence spectrometer. They also detect sulfur ion using smartphone-integrated test strip. They reported linear range and low detection limit from 1–40 μM to 0.42 μM respectively. Similarly, Hao et al. (2022) developed nitrogen-doped carbon dots (N-CDs) and morin-Al^{3+} based nano sensor for fluoride detection. This Radiometric fluorescence sensor based on inner filter effect. They reported wide linear range and a low detection limit 0.5–150 μM and 55.8 nM respectively using this method. They also reported detection limit of 2.09 μM using smartphone-assisted visual sensing platform.

In an eye-catching work Liu et al. (2019) worked on gold nanoclusters and carbon dots nano sensor for Cu^{2+} and alkaline phosphatase (ALP) detection. Both can be easily detected simultaneously using this method. They reported 0.12–15 U/L and 0.05 U/L linear range and low detection respectively for Cu^{+2} ion in water sample. Similarly, Tang et al. (2023) investigated nitrogen-doped carbon dots for curcumin and Fluoride ion sensing. Sensor was synthesized using one step hydrothermal method. It is based on "OFF-ON-OFF" mechanism for on-site visual quantitative detection of target. They reported 0–40 μM and 42 nM linear range and low detection limit respectively for Fluoride ion. Alqahtani et al. (2023) developed boron and nitrogen co-doped carbon dots sensor for Al^{3+} and F^{-} ions in water samples. It is based on "ON-OFF-ON" switches. They reported linear range of 0.03–90 μM and detection limit 9.0 nM for Al^{+3}. They also reported range of 0.18–80 μM with a detection limit of 50.0 nM for Fluoride ion.

In a remarkable attempt by Zhang et al. (2021a, b) developed carbon dots/ curcumin system for fluoride ion sensing. It is based on turn on mechanism. Sensor was synthesized using hydrothermal method. They reported limits of detection and quantitation 0.39 and 1.30 μM, respectively in tap water sample. Similarly, Mohandoss et al. (2023) investigated phosphorus, sulfur, and Nitrogen Co-doped carbon dots for Al^{3+} and Fe^{3+} ions. Characterization have been done using various spectroscopic and microscopic methods. Size of as prepared sensor has been 5.2 nm. This photoluminescent is based on "off-on-off" switches. They reported low detection limit 10.8 nM and 50.7 nM for Al^{3+} and Fe^{3+} respectively.

In a reproducible experiment by Pang et al. (2019) worked on nitrogen and boron co-doped carbon dots for mercury(II) and bio thiols detection. Sensor has been synthesized by hydrothermal method. It is based on based on "on-off-on" fluorescence response. They reported 2.8 μM detection limit for Hg^{+2}ion in liquid solution. Similarly, Huang et al. (2019) developed phosphorus, sulfur, and Nitrogen tri-doped carbon dots for chromium(VI) detection. The sensor was synthesized by hydrothermal method. They reported detection limit and linear range 0.23 μM and 1–500 μM respectively (Table 5.2).

Thus, these above case studies have been encouraged research and development domains to use carbon containing materials for making the nanohybrid materials and develop various environmental sensors.

5.4 Key Challenges Involved with Nanohybrid Sensors Technology

Nanotechnology has emerged as a revolutionary field with the potential to address various environmental challenges. One promising avenue is the development and utilization of non-carbon containing nanohybrid materials, which offer unique properties and functionalities for environmental applications (Köhler 2021). Recent advancements in nanohybrid sensing technology have yielded a wide array of nanohybrid sensors, showcasing their diverse applications. Consequently, the sensing domain has expanded significantly in recent times, with increased exploration of nanohybrid sensors to address significant constraints in fabricating trustworthy and cost-effective hybrid sensors (Nethravathi et al. 2022). Thus, the primary challenge in advanced nanohybrid sensors technology is directly associated with cost, response to sensitivity, Linearly, multifunctionality, sensitivity, response time, detection mechanism, longevity, and toxicity issues are key considerations (Fig. 5.3) Another

Table 5.2 shows the latest case studies of hybrid nanomaterials

S. N	Material	Target molecules	Condition	Result	References
1	WO_3/g-C_3N_4/ V_2O_5 nanocomposite	Methylene blue (MB) dye	UV–visible-light irradiation	Complete degradation occurs over 120 min	Parasuraman et al. (2024)
2	Pr-V_2O_5/rGO nanocomposite	Pb^{2+}	Electrocatalytic degradation	Obtained lower limit of detection 2.919 nM and sensitivity 0.8113 μA/nM cm^2	Harshitha et al. (2024)
3	Nanomagnetite/ graphene oxide composite	Diazinon pesticide	Adsorption and photo-Fenton degradation	99% degradation was reached after 50 min	Khoj et al. (2024)
4	Rosin-TiO_2	Linear low-density polyethylene	Photodegradation	Impressive degradation of LLDPE is observed after 1 month	Amato et al. (2024)
5	Ag/TiO/ polyvinyl alcohol	Cr, Ni, Zn, and Cu heavy metal ions	UV lamp	More than 70% of the heavy metal ions were removed after 180 min	Dehghani and Delnavaz (2024)

challenging aspect is managing the nanohybrid sensors to detect tiny amounts of substances in various samples, such as soil, biofluids, water bodies and air. Additionally, the detection limit and response recovery time can fluctuate greatly, impacting the efficiency of the sensor (Martín-Yerga et al. 2012). Furthermore, when targeting a specific substance, the data obtained from a nanohybrid sensors must be precise enough to ensure consistency within a certain range. It's crucial for sensors to avoid systematic errors to maintain the accuracy and reliability of the results. Therefore, proper calibration using adequate standard samples is essential. Since many nanohybrid sensors applications are utilized in both clinical and non-clinical settings, it would be prudent to assess their toxicity, which may vary based on their size. Conversely, given the well-established fact that the toxicity of nanohybrid sensors or nano sensors is highly contingent on their size, their applications have been subject to heightened scrutiny (Gupta et al. 2022). Consequently, numerous questions have arisen, such as why nanocarbon materials are chosen for use, what fabrication methods are necessary to obtain them, what inherent characteristics enable them to fulfil their intended purpose, what level of engineering is requisite to manage and evaluate all influencing parameters involved in sensing technology, how we can effectively evaluate, manipulate, and design the processes involved, the longevity of materials or products, whether they maintain the same quality or desired catalytic activities when transitioning to another medium or being isolated from their original medium, as well as an evaluation of their further applicability and fate with respect to toxicities (Saleh et al. 2014).

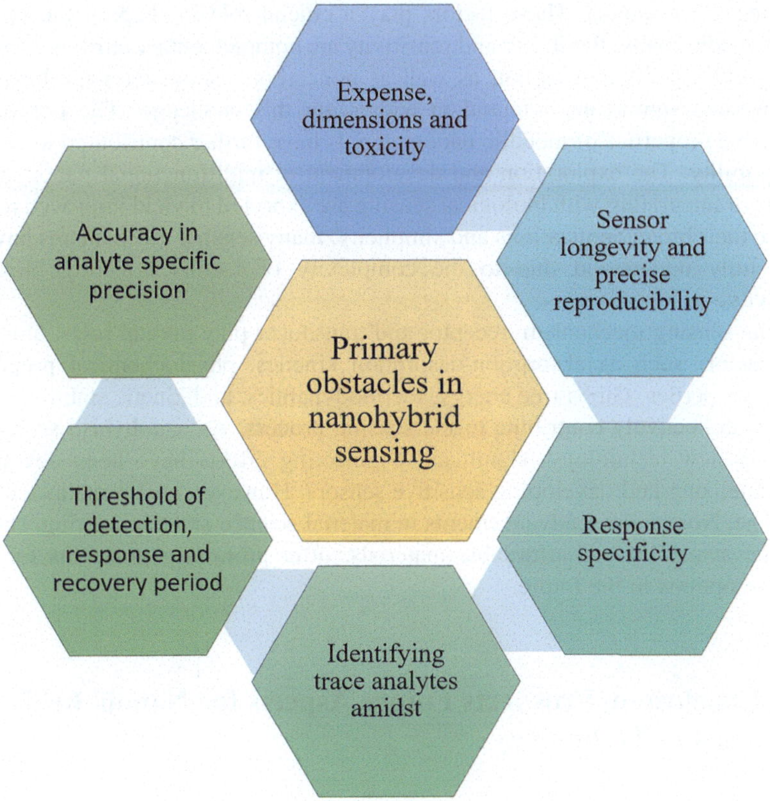

Fig. 5.3 Key challenges for nanohybrid materials as sensing materials. (Figure created by Jayendra Kumar Himanshu)

While non-carbon containing nanohybrid materials offer significant promise for environmental applications, there are various challenges that must be addressed to enable their widespread adoption. These challenges include the scalability and cost-effectiveness of synthesis methods, ensuring the long-term stability and recyclability of materials, and mitigating potential environmental and health risks associated with their use. To advance this field, future research efforts should concentrate on developing innovative synthesis approaches, optimizing material properties, and assessing environmental performance in real-world scenarios. Collaboration among scientists, engineers, policymakers, and stakeholders is crucial to surmount these obstacles and fully leverage the potential of non-carbon containing nanohybrid materials for environmental sustainability.

Another significant challenge associated with nanohybrid sensors is the composition and arrangement of the sensor mechanism, as they directly influence the sensor's performance. The material structure of the receptor, or bioreceptor, varies depending on factors such as the type of sample being analyzed, the detection mechanism used, the type of transducer for signal conversion, and the detector for capturing the response. These factors play a crucial role in shaping the sensor's design. Additionally, flexibility and sensitivity are complementary attributes for both nanohybrid sensors and carbon as well as noncarbon nanomaterials, which have demonstrated significant potential in addressing this challenge. The outstanding conductive properties of metallic nanomaterials have further contributed to enhancing flexibility. The exploration and development of multifunctional nanomaterials capable of integrating with biological sensing are expected to yield improved results. Despite their broad applications and simplicity, many sensing mechanisms have yet to be fully understood due to the complexity of factors affecting biosensor sensitivity.

In the sensing mechanism, receptor and transducer play pivotal roles, alongside other factors such as absorption-desorption kinetics, physiochemical properties, surface properties, Gibbs free energy, thermodynamics, and kinetic stability. These domains collectively contribute to the sensing process. As nanohybrid sensor is a relatively new technology, significant engineering efforts have been devoted to conceptualizing and developing sensitive sensors. However, cost remains a critical challenge. Nonetheless, advancements in material science and engineering, coupled with the availability of affordable materials, offer promising prospects for cost-effective options in the future.

5.5 Conclusion, Prospects Future Aspects for Nanohybrid Sensors Technology

Nanohybrid materials offer versatility and effectiveness in addressing diverse environmental issues. Their distinct features and capabilities render them invaluable for tasks like water purification, air filtration, soil remediation, and renewable energy endeavors. Through ongoing research and advancements, these materials stand poised to greatly contribute to the creation of a more sustainable and robust future sensors and positively enhance our understanding about detection technology. The discussed key challenges including sensitivity, selectivity, price and their toxicity are among most influential parameters that are needed to be recognized carefully. The future of this technology belongs to the scientific advancements that can overcome the key challenges and offer more sustainable approach to the nano sensing domain associated with the domains.

The advent of nanohybrid technology has ushered in a new era for highly sensitive and selective onsite analyte analysis across various domains. With the utilization of nanohybrid materials as solid substrate platforms, several functional nanohybrid materials have emerged, offering promising prospects for advanced hybrid sensing technology. However, the application of hybrid sensors or noncarbon-based hybrid sensor technology for environmental analysis is still in its nascent stage and faces significant challenges due to the inherent complexities of environmental analysis. Materials and noncarbon-based materials are being evaluated for the development of hybrid sensors aimed at environmental applications, particularly pollution mitigation. Engineered nanohybrid composites, in combination with versatile nanohybrid structures such as particle rods, wires, and tubes, enhance the sensitivity and flexibility of modified biosensors for analyzing analytes and pollutants. Paper-based devices offer a potential platform for miniaturized, cost-effective, and user-friendly biosensors or nanohybrid composites, meeting the need for onsite detection of environmental samples or pollutants (e.g., soil, rivers, water bodies, seawater, air). The recent incorporation of nanohybrid materials and nanohybrid composites into these biosensors introduces new strategies for enhancing their analytical performance. While numerous biosensors have been developed and tested in laboratory settings against a wide range of environmental contaminants, only a few are currently deployed. Thus, further advances and efforts are necessary to bridge innovation and facilitate the development of efficient, automated, real-time hybrid biosensors with high-throughput analysis capabilities for environmental pollutants. Undoubtedly, it is imperative to prioritize efforts towards adopting essential technologies to mitigate environmental pollutants. Various tools and techniques are currently available or in development to enhance nanohybrid material-based nanohybrid sensors by refining specific dimensions, including their electrochemical, optical, and magnetic properties. These modifications facilitate the development of single molecule hybrid biosensors and high throughput biosensor arrays. Furthermore, an important future prospect lies in modulating the surface of electrodes, directly influencing their interfaces, material preparation, characterization, and control mechanisms, particularly for noncarbon-based sensing technologies. Enhancing the signal-to-noise ratio, improving the transduction process, and strengthening the signal are essential future requirements. Biological molecules and nanohybrid materials possess unique structures and functions, highlighting the need for further exploration in the field of noncarbon-based sensing technology. Achieving better scientific understanding of the mechanisms and interactions between biomolecules on the surface of prepared biosensors and nanohybrid materials is crucial for maximizing current utilities and fabricating next-generation biosensors/nanohybrid sensors, thereby advancing robust and versatile noncarbon-based sensing. Both carbon-based and noncarbon-based nanohybrid sensors have found applications across various domains, including biomedical, agriculture, and environmental sectors. Nanohybrid material-based biosensors, encompassing both carbon and noncarbon materials, hold promising prospects for widespread use in medical diagnostics, food analysis, process control, and environmental assessments in the near future.

References

Abdullah M, Bakhtiar H, Islam S, Aziz MSA, Krishnan G (2019) Third-order nonlinearity and optical limiting properties of sol–gel-based bromophenol blue dye immobilized in silica–titania nanohybrid. J Sol-Gel Sci Technol 89:361–369

Alengebawy A, Abdelkhalek ST, Qureshi SR, Wang MQ (2021) Heavy metals and pesticides toxicity in agricultural soil and plants: ecological risks and human health implications. Toxics 9(3):42

Alghuthaymi MA, Rajkuberan C, Kalia A, Bhardwaj K, Bhardwaj P, Abd-Elsalam KA, Valis M, Kuca K (2021) Nanohybrid antifungals for control of plant diseases: current status and future perspectives. J Fungi 7(1):48

Ali IO, Salama TM, Gawad AA, El-Henawy AA, Ghazy MB, Bakr MF (2022) Silver nanoparticles@ titanate nanotubes composite: synthesis, characterization, applications and docking. Inorg Chem Commun 137:109187

Alqahtani YS, Mahmoud AM, Mahnashi MH, Ali R, Shahin RY, El-Wekil MM, Batakoushy HA (2023) Facile fabrication of boron and nitrogen co-doped carbon dots for "ON–OFF–ON" fluorescence sensing of Al^{3+} and F^{-} ions in water samples. RSC Adv 13(34):23736–23744

Amato P, Fantauzzi M, Sannino F, Ritacco I, Santoriello G, Camellone MF, Imparato C, Bifulco A, Vitiello G, Caporaso L, Rossi A (2024) Indirect daylight oxidative degradation of polyethylene microplastics by a bio-waste modified TiO_2-based material. J Hazard Mater 463:132907

Amlouk A, El Mir L, Kraiem S, Alaya S (2006) Elaboration and characterization of TiO_2 nanoparticles incorporated in SiO_2 host matrix. J Phys Chem Solids 67(7):1464–1468

Bhat AH, Rangreez TA, Chisti H-T-N (2022) Wastewater treatment and biomedical applications of montmorillonite based nanocomposites: a review. Curr Anal Chem 18(3):269–287

Biswal N, Das DP, Martha S, Parida KM (2011) Efficient hydrogen production by composite photocatalyst CdS–ZnS/Zirconium–titanium phosphate (ZTP) under visible light illumination. Int J Hydrog Energy 36(21):13452–11346

Brahma PK, Suresh L, Lokesh V, Nizamuddin S (2016) Fabrication of highly sensitive and selective nanocomposite film based on CuNPs/fullerene-C60/MWCNTs: an electrochemical nanosensor for trace recognition of paracetamol. Anal Chim Acta 917:107–116

Corvalan C, Hales S, McMichael AJ (2005) Ecosystems and human well-being: health synthesis. World Health Organization

Dao AQ, Nguyen DM, Toan TTT (2023) A review of applying modified/functionalized non-carbon materials to remove emergent heavy ions pollutants. Clean Techn Environ Policy 26:1–15

Dehghani MT, Delnavaz M (2024) UV-light-responsive Ag/TiO_2/PVA nanocomposite for photocatalytic degradation of Cr, Ni, Zn, and Cu heavy metal ions. Sci Rep 14(1):5195

Deshmukh MA, Shirsat MD, Ramanaviciene A, Ramanavicius A (2018) Composites based on conducting polymers and carbon nanomaterials for heavy metal ion sensing. Crit Rev Anal Chem 48(4):293–304

Devi S, Tharmaraj V (2020) Quantum dots in green photocatalytic applications for degradation of environmental pollutants and hydrogen evolution. In: Green photocatalysts for energy and environmental process. Springer, pp 87–108

Dimitrievska I, Paunovic P, Grozdanov A (2023) Recent advancements in nano sensors for air and water pollution control. Mater Sci Eng 7:113–128

Gupta R, Chauhan H, Garg VK, Kataria N (2022) Chemical and physical properties of nanoparticles and hybrid materials. In: Sustainable nanotechnology for environmental remediation. Elsevier, pp 199–220

Haarindraprasad RP, Thurga Devi N, Thevendran R, Maheswaran S (2023) Spiked nanorosette: a novel zinc oxide-gold textured nanostructure for impedance biosensing of target DNA signatures. Biotechnol J 18(8):2300092

Hao Y, Dong W, Liu Y, Wen X, Shuang S, Hu Q, Dong C, Gong X (2022) Nitrogen-doped carbon dots coupled with morin-Al^{3+}: cleverly design an integrated sensing platform for ratiometric optical dual-mode and smartphone-assisted visual detection of fluoride ion. J Hazard Mater 439: 129596

Harish V, Ansari MM, Tewari D, Gaur M, Yadav AB, García-Betancourt ML, Barhoum A (2022) Nanoparticle and nanostructure synthesis and controlled growth methods. Nano 12(18):3226

Harshitha V, Kumar KS, Nethravathi PC, Nagaraju DH, Suresh D (2024) Multifunctional Pr-V_2O_5/rGO hybrid nanomaterial for sensitive heavy metal ion detection and efficient photocatalytic dye degradation. Ionics 30:1–18

Hossain SI, Kukushkina EA, Izzi M, Sportelli MC, Picca RA, Ditaranto N, Cioffi N (2023) A review on montmorillonite-based nanoantimicrobials: state of the art. Nano 13(5):848

Huang S, Yang E, Yao J, Chu X, Liu Y, Xiao Q (2019) Nitrogen, phosphorus and sulfur tri-doped carbon dots are specific and sensitive fluorescent probes for determination of chromium (VI) in water samples and in living cells. Microchim Acta 186:1–10

Iman Amir MN, Julkapli NM, Bagheri S, Yousefi AT (2015) TiO_2 hybrid photocatalytic systems: impact of adsorption and photocatalytic performance. Rev Inorg Chem 35(3):151–178

Jin Y, Li A, Hazelton SG, Liang S, John CL, Selid PD, Pierce DT, Zhao JX (2009) Amorphous silica nanohybrids: synthesis, properties and applications. Coord Chem Rev 253(23–24): 2998–3014

Jin M, Mou ZL, Zhang RL, Liang SS, Zhang ZQ (2017) An efficient ratiometric fluorescence sensor based on metal-organic frameworks and quantum dots for highly selective detection of 6-mercaptopurine. Biosens Bioelectron 91:162–168

Kamran MA, Abbas A, Ullah S, Alomar M, Malik S, Aziz MH, Bukhari SF (2024) Boosting electrochemical and photocatalytic performance of cadmium sulfide/zinc telluride nanocomposites via nickel doping. Mater Sci Eng B 299:116932

Kang DJ, Park GU, Lee HH, Park HY, Park J-U (2013) Photopatternable and refractive-index-tunable sol–gel-derived silica–titania nanohybrid materials. Curr Appl Phys 13(8):1732–1737

Khan DM, Kausar A, Salman SM (2016) Exploitation of nanobifiller in polymer/graphene oxide–carbon nanotube, polymer/graphene oxide–nanodiamond, and polymer/graphene oxide–montmorillonite composite: a review. Polym-Plast Technol Eng 55(7):744–768

Khaorapapong N, Khumchoo N, Ogawa M (2015) Preparation of copper oxide in smectites. Appl Clay Sci 104:238–244

Khoj MA, Awwad NS, Ibrahium HA, Awad AM, Hassan AF (2024) Enhanced adsorption and photo-Fenton degradation of diazinon pesticide utilizing nanomagnetite/graphene oxide composite: kinetic and thermodynamic studies. J Inorg Organomet Polym Mater 34:1–18

Köhler JM (2021) Challenges for nanotechnology. Encyclopedia 1(3):618–631

Li W, Hu X, Li Q, Shi Y, Zhai X, Xu Y, Li Z, Huang X, Wang X, Shi J, Kang S (2020) Copper nanoclusters@ nitrogen-doped carbon quantum dots-based ratiometric fluorescence probe for lead (II) ions detection in porphyra. Food Chem 320:126623

Liaw W-C, Cheng Y-L, Liao Y-S, Chen C-S, Lai S-M (2011) Complementary functionality of SiO_2 and TiO_2 in polyimide/silica–titania ternary hybrid nanocomposites. Polym J 43(3):249–257

Liu X, Zhang N, Bing T, Shangguan D (2014) Carbon dots based dual-emission silica nanoparticles as a ratiometric nanosensor for Cu^{2+}. Anal Chem 86(5):2289–2296

Liu W, Cui Y, Li T, Diao H, Wei S, Li L, Chang H, Zhang B, Wei W (2018a) Green and facile synthesis of highly photoluminescent nitrogen-doped carbon dots for sensors and cell imaging. Chem Lett 47(4):421–424

Liu Q, Zhu J, Zhang L, Qiu Y (2018b) Recent advances in energy materials by electrospinning. Renew Sust Energ Rev 81:1825–1858

Liu H, Jia L, Wang Y, Wang M, Gao Z, Ren X (2019) Ratiometric fluorescent sensor for visual determination of copper ions and alkaline phosphatase based on carbon quantum dots and gold nanoclusters. Anal Bioanal Chem 411:2531–2543

Ma R, Fu L, Long N, Guo H, Hou Y, Li Y, Li P, Wang J, Zhou L, Kong W (2023) Gold nanoclusters and silica-coated carbon dots-assisted ratiometric fluorescent nanosensors for ultrasensitive detection of glyphosate. ACS Sustain Chem Eng 11(13):5093–5104

Madhav S, Ahamad A, Singh AK, Kushawaha J, Chauhan JS, Sharma S, Singh P (2020) Water pollutants: sources and impact on the environment and human health. In: Sensors in water pollutants monitoring: role of material. Springer, pp 43–62

Manisalidis I, Stavropoulou E, Stavropoulos A, Bezirtzoglou E (2020) Environmental and health impacts of air pollution: a review. Front Public Health 8:505570

Martín-Yerga D, González-García MB, Costa-García A (2012) Use of nanohybrid materials as electrochemical transducers for mercury sensors. Sensors Actuators B Chem 165(1):143–150

McCrink-Goode M (2014) Pollution: a global threat. Environ Int 68:162–170

Mirzaei A, Leonardi SG, Neri G (2016) Detection of hazardous volatile organic compounds (VOCs) by metal oxide nanostructures-based gas sensors: a review. Ceram Int 42(14): 15119–15141

Mishra S, Bharagava RN, More N, Yadav A, Zainith S, Mani S, Chowdhary P (2019) Heavy metal contamination: an alarming threat to environment and human health. In: Environmental biotechnology: for sustainable future. Springer, pp 103–125

Mohandoss S, Ganesan S, Palanisamy S, You S, Velsankar K, Sudhahar S, Lo HM, Lee YR (2023) Nitrogen, sulfur, and phosphorus Co-doped carbon dots-based ratiometric chemosensor for highly selective sequential detection of Al3+ and Fe3+ ions in logic gate, cell imaging, and real sample analysis. Chemosphere 313:137444

Mourdikoudis S, Pallares RM, Thanh NT (2018) Characterization techniques for nanoparticles: comparison and complementarity upon studying nanoparticle properties. Nanoscale 10(27): 12871–12934

Nethravathi PC, Manjula MV, Devaraja S, Suresh D (2022) Ag and BiVO4 decorated reduced graphene oxide: a potential nano hybrid material for photocatalytic, sensing and biomedical applications. Inorg Chem Commun 139:109327

Nguyen QK, Nguyen DT, Pham TMA, Pham B, Nguyen TAH, Pham TD, Sharma S, Pham DT, Gangavarapu RR, Pham TNM (2024) A highly sensitive fluorescence nanosensor for determination of amikacin antibiotics using composites of carbon quantum dots and gold nanoparticles. Spectrochim Acta A Mol Biomol Spectrosc 305:123466

Nunes D, Pimentel A, Gonçalves A, Pereira S, Branquinho R, Barquinha P, Martins R (2019) Metal oxide nanostructures for sensor applications. Semicond Sci Technol 34(4):043001

Oladejo M, Paulishak W, Wood L (2022) Synergistic potential of immune checkpoint inhibitors and therapeutic cancer vaccines. Semin Cancer Biol 88:81–95

Panda S, Jadav A, Panda N, Mohapatra S (2018) A novel carbon quantum dot-based fluorescent nanosensor for selective detection of flumioxazin in real samples. New J Chem 42(3): 2074–2080

Pang LF, Wu H, Fu MJ, Guo XF, Wang H (2019) Red emissive boron and nitrogen co-doped "on-off-on" carbon dots for detecting and imaging of mercury (II) and biothiols. Microchim Acta 186:1–9

Parasuraman B, Kandasamy B, Vasudevan V, Thangavelu P (2024) Enhanced dye degradation performance enabled by swift electron mediator decorated WO_3/g-C_3N_4/V_2O_5 hybrid nanomaterials. Environ Sci Pollut Res 31(34):46591–46601

Perumal V, Hashim U, Gopinath SCB, Haarindraprasad R, Liu WW, Poopalan P, Balakrishnan SR, Thivina V, Ruslinda AR (2015) Thickness dependent nanostructural, morphological, optical and impedometric analyses of zinc oxide-gold hybrids: nanoparticle to thin film. PLoS One 10(12):e0144964

Perumal V, Hashim U, Gopinath SCB, Haarindraprasad R, Poopalan P, Liu WW, Ravichandran M, Balakrishnan SR, Ruslinda AR (2016) A new nano-worm structure from gold-nanoparticle mediated random curving of zinc oxide nanorods. Biosens Bioelectron 78:14–22

Rahman DS, Ghosh SK (2016) Manipulating electron transfer in hybrid ZnO–Au nanostructures: size of gold matters. J Phys Chem C 120(27):14906–14917

Ramesh T, Foo KL, Sam AJ, Solayappan M (2019) Gold-hybridized zinc oxide nanorods as real-time low-cost nano biosensors for detection of virulent DNA signature of HPV-16 in cervical carcinoma. Sci Rep 9(1):17039

Rathee G, Bartwal G, Rathee J, Mishra YK, Kaushik A, Solanki PR (2021) Emerging multimodel zirconia nanosystems for high-performance biomedical applications. Adv Nano Biomed Res 1(9):2100039

Saleh NB, Afrooz AN, Bisesi JH Jr, Aich N, Plazas-Tuttle J, Sabo-Attwood T (2014) Emergent properties and toxicological considerations for nanohybrid materials in aquatic systems. Nanomaterials 4(2):372–407

Shafiei H, Hassaninejad-Darzi SK (2023) Simultaneous electrocatalytic measurement of dopamine and acetaminophen by nanosensor based on Ag@ polyoxometalate@ reduced graphene oxide and ionic liquid. Electrocatalysis 14(6):811–828

Sun X, Liu P, Wu L, Liu B (2015) Graphene-quantum-dots-based ratiometric fluorescent probe for visual detection of copper ion. Analyst 140(19):6742–6747

Sun J, Ma Q, Xue D, Shan W, Liu R, Dong B, Shao B (2021) Polymer/inorganic nanohybrids: an attractive materials for analysis and sensing. TrAC Trends Anal Chem 140:116273

Tang Y, Jiang Z, Tay Q, Deng J, Lai Y, Gong D, Dong Z, Chen Z (2012) Visible-light plasmonic photocatalyst anchored on titanate nanotubes: a novel nanohybrid with synergistic effects of adsorption and degradation. RSC Adv 2(25):9406–9414

Tang S, Wang Y, Guo G, Li T, Xing H, Hu H, Leng X, Gu C, Chen D (2023) Activated cascade effect for dual-mode ratiometric and smartphone-assisted visual detection of curcumin and F– based on nitrogen-doped carbon dots. Sci Total Environ 872:162277

Tümay SO, Şenocak A, Sarı E, Şanko V, Durmuş M, Demirbas E (2021) A new perspective for electrochemical determination of parathion and chlorantraniliprole pesticides via carbon nanotube-based thiophene-ferrocene appended hybrid nanosensor. Sensors Actuators B Chem 345:130344

Verma DK, Shukla N, Kumar B, Singh AK, Shahu K, Yadav M, Rhee KY, Rastogi RB (2020) Synergistic tribo-activity of nanohybrids of zirconia/cerium-doped zirconia nanoparticles with nano lamellar reduced graphene oxide and molybdenum disulfide. Nanomaterials 10(4):707

Wang H, Zhong W, Xu P, Du Q (2005) Polyimide/silica/titania nanohybrids via a novel non-hydrolytic sol–gel route. Compos A: Appl Sci Manuf 36(7):909–914

Wang Y, Zhang C, Chen X, Yang B, Yang L, Jiang C, Zhang Z (2016) Ratiometric fluorescent paper sensor utilizing hybrid carbon dots–quantum dots for the visual determination of copper ions. Nanoscale 8(11):5977–5984

Xiang G, Ren Y, Zhang H, Fan H, Jiang X, He L, Zhao W (2018) Carbon dots based dual-emission silica nanoparticles as ratiometric fluorescent probe for chromium speciation analysis in water samples. Can J Chem 96(1):72–77

Yin C, Liu T, Wu M, Liu H, Sun Q, Sun X, Niu N, Chen L (2022) Smartphone-integrated dual-emission fluorescence sensing platform based on carbon dots and aluminum ions-triggered aggregation-induced emission of copper nanoclusters for on-site visual detecting sulfur ions. Anal Chim Acta 1232:340460

Yuan C, Liu B, Liu F, Han MY, Zhang Z (2014) Fluorescence "turn on" detection of mercuric ion based on bis (dithiocarbamato) copper (II) complex functionalized carbon nanodots. Anal Chem 86(2):1123–1130

Zhang Z, Pei K, Yang Q, Dong J, Yan Z, Chen J (2018) A nanosensor made of sulfur–nitrogen co-doped carbon dots for "off–on" sensing of hypochlorous acid and Zn (II) and its bioimaging properties. New J Chem 42(19):15895–15904

Zhang X, Tan X, Hu Y (2021a) Blue/yellow emissive carbon dots coupled with curcumin: a hybrid sensor toward fluorescence turn-on detection of fluoride ion. J Hazard Mater 411:125184

Zhang Y, Zhu Y, Zeng Z, Zeng G, Xiao R, Wang Y, Feng C (2021b) Sensors for the environmental pollutant detection: are we already there? Coord Chem Rev 431:213681

Zhu L, Shi L, Tan Y, Zhang H, Yin J, Xu C, Wu D, Ma Y (2022) Dual-emissive ratiometric fluorescent nanosensor based on multi-nanomaterials for Ag^+ determination in lake water. RSC Adv 12(46):30113–30119

Chapter 6
Green Nanotechnology to Remedy Oil Spill

Saurabh Mishra, Shweta Shukla, and Aviral Chaurasia

Abstract Oil spills pose a grave threat to marine environments, triggering a cascade of devastating consequences that ripple through ecosystems. Water and air pollution, soil contamination, and ecological devastation are just a few of the dire consequences that follow. While various cleanup methods have been employed to mitigate these disasters, restoring the delicate balance of affected ecosystems remains a formidable challenge. Traditional approaches, often relying on chemical dispersants and mechanical containment, can inadvertently exacerbate the problem, inflicting further harm on marine life and habitats. In recent years, green nanotechnology has emerged as a beacon of hope, offering sustainable and environmentally friendly alternatives for oil spill cleanup. This chapter delves into the immense potential of green nanotechnology to address this global challenge. By harnessing eco-friendly synthesis techniques and drawing inspiration from nature, scientists have developed green nanomaterials with remarkable oil-binding capabilities. Utilizing natural resources, such as plant extracts, these nanomaterials effectively remove oil from water surfaces, minimizing the ecological footprint. Furthermore, green nanotechnology holds significant promise for enhancing the biodegradation of oil pollutants, accelerating the natural recovery process of fragile marine ecosystems. This chapter will explore the diverse array of green nanomaterials, their synthesis methods, and their specific applications in oil spill cleanup. By integrating green nanotechnology into oil spill response strategies, we can pave the way for more effective, eco-conscious cleanup efforts, safeguarding marine ecosystems and preserving the invaluable biodiversity they harbor.

Keywords Cleanups · Ecofriendly · Green-nanotechnology · Oil spills · Nanomaterial

---

S. Mishra (✉) · A. Chaurasia
Centre for Advanced Studies, Dr. APJ Abdul Kalam Technical University, Lucknow, India
e-mail: saurabhm@cas.res.in

S. Shukla
Department of Chemistry, National Institute of Technology, Tiruchirappalli, India

© The Author(s), under exclusive license to Springer Nature Switzerland AG 2025

J. M. Al-Khayri et al. (eds.), *Green Nanotechnology Applications for Ecosystem Sustainability*, Nanotechnology in Plant Sciences 4,
https://doi.org/10.1007/978-3-031-90766-1_6

6.1 Introduction

Pipeline breaches, tanker accidents, mishandling during offshore oil exploration, drilling, transportation, and illegal dumping often result in crude oil spills, leading to environmental and biological risks such as water and air pollution, soil contamination, biological toxicity, and economic consequences for many decades (Sun et al. 2019; Wang et al. 2019). The world has witnessed several oil spills which had devastated the marine ecosystem and had impacted the health of the people living around or people involved in remediation (Mishra et al. 2022).

Marine pollution from oil spills has exhibited a concerning upward trend over time. Oil spills over marine surfaces can cover several square kilometers of area and destroy the ecosystem as well as the shores. One notable incident occurred in 1967 when the Torrey Canyon super-tanker struck rocks off the Cornwall coast in the English Channel. This event led to the release of 119,000 mt of crude oil from Kuwait, resulting in the deaths of over 25,000 seabirds and numerous marine organisms. Similarly, in 1989, the Exxon Valdez tanker ran on the Bligh Reef in northern Prince William Sound, causing the spillage of 42 million liters of Alaskan North Slope crude oil. This incident contaminated approximately 1990 Km of shoreline and resulted in significant wildlife mortality. Another significant oil spill took place in 2002 involving the tanker Prestige, which suffered structural failure and sank near the Cies Islands. This event led to the release of over 60,000 mt of oil, contaminating more than 1300 Km of coastline. In 2017, the collision between the BW Maple, a liquid petroleum gas carrier, and the MT Dawn Kanchipuram, a lubricant oil carrier, off the coast of Southern India, resulted in the spillage of 75 mt of oil, contaminating approximately 25 miles of coastline. In the year 2020, MV Wakashio traveling from China to Brazil ran aground here in Mauritius's environmentally sensitive area, off the shore of Pointe d'Esny resulting in release of an estimated 3894 tons of low-Sulphur fuel oil, 207 tons of diesel and 90 tons of lubricant oil that affected marine ecosystem, the corals, and other species detrimentally (https://www.imo.org/en/MediaCentre/HotTopics/Pages/WakashioFAQ.aspx).

In the same year, 2020, Norlisk diesel oil spill was an industrial disaster near Krai, Russia, a fuel storage tank at Norlisk-Taimyr Energy's thermal power plant failed, and flooding local rivers with up to 17,500 tons of diesel oil affecting an immediate area of 18 ha (44 acres), in the nearby Daldykanriver and contaminating are of 350 Km^2. It was estimated that immediate cost of emergency relief activities was USD 47.5 million including clean-ups (https://tass.com/emergencies/1164423; Seddon 2020). The irreversible damages from the oil spill must be prevented by putting more efforts to improve oil spill response (Singh et al. 2013). Efforts to mitigate the oil spills involve strict regulations on shipping and drilling operations, improved safety standards, and better emergency response planning. It is crucial to find effective and environmentally friendly solutions to mitigate these spills and minimize the long-term damage from the disastrous oil spills. Researchers around the world are actively exploring strategies and showing keen interest in addressing oil spill disasters and their subsequent effects.

The Traditional Strategies for addressing oil spills can be broadly categorized into three main groups: physical, chemical, and biological methods (Lee et al. 2023). Physical methods primarily involve the use of equipment such as oil booms and skimmers to contain and remove the spilled oil from the affected area (Hou et al. 2019). Chemical methods utilize dispersants or surfactants sprayed onto the spilled oil to reduce its interfacial tension, causing it to break up into smaller droplets that can be more easily recovered through physical means (Yap and Tan 2021). The clean-up processes usually involve the combination of these methods.

Biological methods encompass two distinct approaches: bio-stimulation and bioaugmentation. Bio-stimulation involves the introduction of nutrients to stimulate the growth and activity of naturally occurring microorganisms capable of degrading the oil (Prendergast and Gschwend 2014). On the other hand, bioaugmentation involves the addition of specific microorganisms to enhance the existing native population's ability to degrade the oil. Both approaches leverage the natural capabilities of microorganisms to aid in the remediation of oil-contaminated environments (Liu et al. 2000). Nevertheless, these approaches exhibit restricted effectiveness and are not extensively implemented in real-world scenarios. Their limited utilization largely stems from the potential for secondary pollution caused by the remediation techniques, which in turn pose challenges for subsequent mitigation or oil extraction endeavors (Liu et al. 2000).

Traditional methods employed for oil spill cleanup often prove inefficient and environmentally detrimental (Singh et al. 2021). However, the emergence of nanotechnology presents a potential remedy to this issue, particularly in the context of marine water oil spill cleanup. By harnessing nanostructured materials like nanofibers or nanoparticles, significant enhancements in the efficiency and effectiveness of oil remediation can be achieved. These nanostructured materials provide high surface areas and the capability to selectively adsorb or absorb oil molecules, facilitating a more thorough removal of oil from water bodies (Singh et al. 2021). Moreover, nanomaterials can be tailored to possess specific properties that enhance their efficiency in breaking down hydrocarbons found in oil, thereby converting them into less harmful substances (Bottero et al. 2009). This innovative application of nanotechnology in oil spill cleanup offers a more efficient and environmentally sound solution and mitigates the adverse impacts on marine ecosystems. Consequently, nanotechnology is a promising solution for addressing oil spills in marine water, utilizing nanostructured materials to efficiently remove and degrade oil molecules (Qu et al. 2012). Furthermore, nanotechnology holds the potential to improve the monitoring and detection of oil spills (Bottero et al. 2009). Overall, the application of nanotechnology in oil spill cleanup over marine water presents a promising avenue that is more efficient, environmentally friendly, and reduces the negative impacts on marine ecosystems, ultimately contributing to the restoration of healthy oceans (Bottero et al. 2009).

This chapter delves into an exploration of the manifold opportunities afforded by green nanotechnology in the realm of oil spill remediation within marine ecosystems. It encompasses a comprehensive examination of the diverse applications of green nanotechnology in mitigating oil spills across marine water bodies.

Additionally, it prognosticates the future trajectory of green nanotechnology, particularly its role in safeguarding the marine ecosystem against the deleterious effects of oil spills.

6.2 Green Nanotechnology and Its Environmental Significance

Nanotechnology strives to achieve deliberate design, fabrication, characterization, manufacturing, and utilization of structures, devices, and systems through precise control of their dimensions at the nanometer level. The nanotechnology focuses on materials with dimensions smaller than 100 nm, encompassing nanostructures, nanocomposites, and nanoparticle assemblies (Bhushan 2017). The defining characteristics of nanomaterials lie in their size-dependent behaviors and properties, stemming from their unique surface reactivity, as well as their electrical, magnetic, and optical attributes inherent to their nanoscale dimensions (Bhainsa and D'Souza 2006; Shahverdi et al. 2007). Veeresh et al. (2024) highlighted the use of engineered nanomaterials, specifically nanoparticles, to remove persistent pollutants from the environment. The authors emphasized the versatility of these nanomaterials, noting their effectiveness in addressing a range of pollutants across different environmental mediums like water, soil, and air. Furthermore, they pointed out the potential of modifying these nanomaterials to enhance their reactivity and efficiency, suggesting a dynamic field with room for further innovation. The nanotechnology can be positioned as a promising and sustainable alternative to traditional remediation methods, emphasizing its potential to revolutionize environmental cleanup efforts.

In recent decades, there has been significant progress in developing nanomaterials for various applications. These include their utilization as intelligent and versatile materials in the fields of medicine and pharmaceuticals, particularly in the diagnosis and treatment of cancer. These nanomaterials can be categorized into the following, as shown in Fig. 6.1. Additionally, nanostructured electrodes have been employed in batteries, single-walled carbon nanotubes have found applications in communication technology devices, and nanomaterials have been utilized as antimicrobial agents in industries such as cosmetics, food, and clothing (Chithrani et al. 2006; Sawle et al. 2008; De Gusseme et al. 2010). In the past, the production and utilization of synthetic nanomaterials incurred high costs and often resulted in the generation of environmentally harmful by-products. Moreover, their usage in medicine has been restricted due to associated risks and adverse effects. As a result, contemporary efforts by scientists are focused on employing greener synthesis routes for nanomaterials to mitigate these drawbacks and minimize potential side effects. Green nanotechnology, a specialized branch within the field of nanotechnology, is dedicated to devising environmentally conscious solutions for diverse applications. The features of the green-nanotechnology are shown in the Fig. 6.2. The researchers

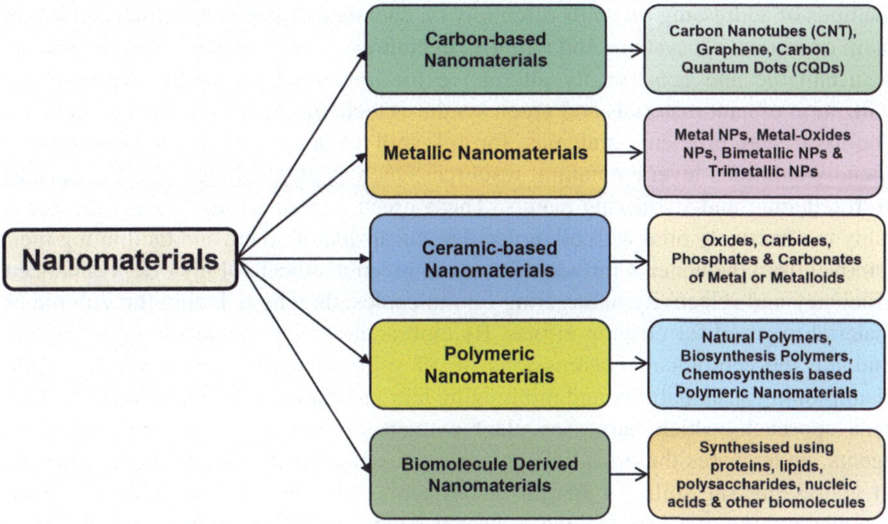

Fig. 6.1 Nanomaterials classification depending on their variety. (Mishra et al. 2022)

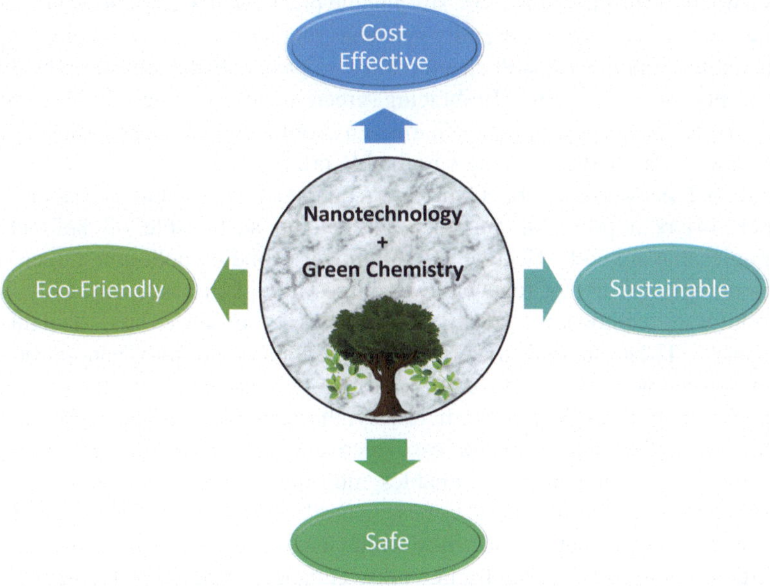

Fig. 6.2 The features of green nanotechnology

and scientists are focused to adopt green routes to synthesize the nanomaterials to prevents its adverse effects stated above.

One area ripe for the impactful intervention of green nanotechnology is the mitigation of oil spills in marine environments (Alqarni et al. 2022). Conventional methods of addressing oil spills often rely on chemical dispersants, which can inflict harm on marine ecosystems and wildlife. In contrast, green nanotechnology presents a sustainable and eco-friendly alternative for remedying oil spills. Through the utilization of nanomaterials and green synthesis techniques, researchers can develop non-toxic and efficient strategies for oil spill cleanup. These environmentally friendly solutions leverage natural resources, such as plant extracts, which serve as both reducing and stabilizing agents. These green nanomaterials possess the capability to effectively bind with oil molecules, encapsulating them and facilitating their removal from the water's surface. Moreover, green nanotechnology offers enhanced efficiency and selectivity in targeting oil molecules, thereby reducing the volume of material required for cleanup efforts. By embracing green nanotechnology, we can mitigate the ecological repercussions of oil spills in marine environments while championing sustainability and minimizing harm to marine ecosystems and wildlife. This approach, which harnesses plant extracts as both reducing and stabilizing agents, underscores the potential for a more sustainable and eco-friendly approach to addressing oil spills in marine ecosystems. Through the adoption of green nanotechnology, we can achieve a cleaner, safer, and more environmentally conscious method compared to conventional chemical approaches.

The integration of green nanotechnology into efforts to mitigate oil spills represents a paradigm shift towards more sustainable practices. By capitalizing on natural resources and employing eco-friendly synthesis methods, green nanotechnology not only facilitates effective oil spill cleanup but also promotes the long-term health and sustainability of our oceans. Embracing green nanotechnology holds immense promise for addressing oil spills in marine environments while upholding environmental stewardship and advancing sustainable principles.

Figure 6.2 demonstrate the Combining nanotechnology and green chemistry principles offers a powerful approach to develop sustainable oil spill cleanup technologies that are both effective and minimize secondary pollution. Green chemistry encourages the use of renewable resources like plant-based oils, polysaccharides (chitosan, cellulose), and microorganisms to produce biosurfactants and biopolymers. These natural materials can be fabricated into nanoparticles or nanocomposites with high surface area and tailored surface chemistry for efficient oil absorption. Iron oxide nanoparticles, synthesized using green methods, offer magnetic properties that allow for easy recovery and reuse after oil absorption. This reduces material waste and enables multiple cycles of oil removal. Green chemistry promotes the use of enzymes, nature's catalysts, for biodegradation of oil pollutants. Encapsulating enzymes within nanoreactors, such as porous silica nanoparticles or metal-organic frameworks, enhances their stability, activity, and reusability in harsh marine environments. Semiconductor nanomaterials like titanium dioxide and zinc oxide can be synthesized using green methods and employed as photocatalysts. Upon sunlight irradiation, these nanomaterials generate reactive oxygen species that degrade oil into less harmful substances, offering a sustainable approach to oil spill cleanup. Green synthesis of metallic nanoparticles (gold, silver) enables the development of highly sensitive SERS nano-sensors. These sensors can detect and monitor oil spills in real-time, even at trace levels, facilitating rapid response and targeted cleanup efforts. Green chemistry emphasizes the design of nanomaterials that degrade naturally in the environment, minimizing long-term ecological impact. Thorough eco-toxicity assessments are crucial to ensure the safety of these materials for marine life. By integrating the principles of green chemistry into nanomaterial design and synthesis, we can develop eco-friendly, sustainable, and cost-effective solutions for oil spill cleanup. This synergistic approach holds immense promise for mitigating the environmental impact of oil spills and protecting our valuable marine ecosystems.

6.3 Innovative Green Nanotechnologies for Mitigating Oil Spills

Innovative green nanotechnologies are pivotal in safeguarding marine ecosystems against the detrimental impacts of oil spills, offering sustainable and environmentally conscious methods for remediation and biodiversity preservation. Researchers and scientists have determined that effective cleanup of oil spills requires materials possessing specific properties such as super-hydrophobicity and super-oleophilicity, enabling the permeation of oil while repelling water from the material's porous structure. Significant advancements have been made in this field to enhance process efficiency while minimizing environmental impact.

Abdullah et al. (2018) introduced a rapid, eco-friendly methodology for synthesizing magnetite nanoparticles (MNPs). Surface modification using hydrophobic biocomponents derived from *Anthemis pseudocotula* (AP) facilitated the formation of APH-MNPs or APC-MNPs, depending on the presence of n-hexane extract (APH) or chloroform extract (APC). Analysis demonstrated APH-MNPs' superior efficiency in collecting heavy crude oil compared to APC-MNPs. In a subsequent study (Abdullah et al. 2017), green recyclable magnetic iron oxide nanomaterials, coated with hydrophobic plant extracts, were synthesized for efficient oil spill cleanups.

Addressing crucial marine climate parameters, Kalliola et al. (2016) developed green chemicals for oil spill treatment in Arctic regions. They synthesized carboxymethylated chitosan nanoparticles crosslinked with Ca-ions, reporting their stability in increased salinity and pH and promising potential for oil spill cleanups. Atta et al. (2020) explored the potential of using water hyacinth (*Eichhornia crassipes*) alongside magnetic nanoparticles capped with natural Myrrh extracts for treating freshwater contaminated with crude petroleum oil. They synthesized magnetic nanoparticles composed of magnetite and coated them with Myrrh extracts, characterizing their chemical structures, morphologies, particle sizes, and magnetic properties. These nanoparticles were then employed to adsorb heavy components of crude oil. To assess their effectiveness, the researchers cultivated water hyacinth plants in Nile River freshwater, exposing them to varying levels of crude oil contamination over a month. Following harvest, the analysis was done to determine the plant's sulfur content, total aromatic and saturated hydrocarbons, and individual aromatic constituents. The findings suggested that water hyacinth exhibits the potential to remediate slightly polluted water contaminated with crude petroleum oil. Moreover, the presence of magnetite nanoparticles coated with Myrrh extracts was found to enhance the remediation process, particularly in water bodies highly polluted by petroleum crude oil. This research sheds light on the promising synergistic effects of natural plant materials and advanced nanotechnology in combating freshwater pollution, offering insights into sustainable and effective methods for environmental remediation.

Wu et al. (2017) introduced a simple method for synthesizing crystalline particles of calcite, the most stable polymorph of $CaCO_3$, using oleic acid as an additive in a one-pot synthesis approach. They confirmed the well-crystallized nature of the synthesized particles as calcite with X-Ray Diffraction (XRD) analysis, and with Fourier-Transform Infrared Spectroscopy (FTIR) results, they reported the presence of alkyl groups from oleic acid, leading to the formation of 'rice-shaped' morphology with hydrophobic properties. Poddar et al. (2022) attempted to synthesize green algae-mediated titanium oxide (TiO_2) nanoparticles and polyacrylonitrile (PAN) nanofiber mats for effective oil spill cleaning. Hethnawi et al. (2023) proposed novel protocol for green synthesis of nanoparticles of titanomagnetite (NTM) under various conditions. The authors utilized, the characterization techniques such as Brunauer-Emmett-Teller (BET), XRD, X-ray photoelectron spectroscopy (XPS), High-Resolution Transmission Electron Microscopy (HRTEM), Scanning Electron Microscopy (SEM), and contact angle measurements to optimize the synthesis conditions and assess nanoparticles properties. The optimized nanoparticles were then utilized as sorbents to remove crude oil spills from the water, following a modified ASTM protocol. Additionally, the dominant mechanisms involved in sorption were investigated using a cationic dye (methylene blue) as a model pollutant in batch adsorption mode. The optimal nanoparticles demonstrated outstanding performance in crude oil spill removal (uptake of 38 ± 2 g crude oil/g NTM) and methylene blue adsorption (uptake of 36 mg MB/g). Furthermore, the spent NTM can be fully regenerated, maintaining over 95% removal efficiency after three successive cycles without altering surface chemistry.

Soares et al. (2017) fabricated chitosan-silica hybrid nano-sorbent for oil removal from the water. The synthesized spherical chitosan/silica hybrid nanoparticles utilizing a sol-gel non-emulsion method. Fe_3O_4 nanoparticles were then encapsulated with chitosan hybrid siliceous thin shells. This green synthesized nano-sorbent resulted in potential oil uptake from water. Gu et al. (2020) attempted to combine magnetic nano-cellulose and nano magnetic aerogel to improve their adsorption capacity and functionality. This method was also an attempt of delving green nanotechnology in oil-water separation. Before it, Lavoine and Bergström (2017) has confirmed that nanocellulose is a renewable and biocompatible nanomaterial with a high strength, low density and tunable surface chemistry. It is a porous structure and can be utilized potentially in oil spill cleanup job. Phanthong et al. (2018) also have attempted to fabricate a super hydrophobic nanocellulose sponge, utilizing a facile method: freeze drying of nanocellulose aqueous suspension to sponge state. Chao et al. (2020) Designed and fabricated a compressible wood-sponge, coated with thermally reduced graphene-oxide. They utilized in-situ photothermal thermogenesis using environment friendly material derived from natural wood. So, this was an attempt to use natural material in oil spill cleanup job.

Leveraging green nanotechnology in addressing oil spills represents a sustainable and eco-friendly approach, allowing us to combat the adverse impacts of such incidents while championing environmental stewardship. By harnessing the power of green nanotechnology, we can effectively mitigate the environmental repercussions of oil spills in marine ecosystems, thereby promoting ecological sustainability. Green nanotechnology presents a promising avenue for tackling the environmental challenges posed by oil spills in marine environments. It offers an eco-friendly and sustainable means of oil spill cleanup, minimizing harm to marine ecosystems and wildlife. Indeed, the potential of green nanotechnology to revolutionize oil spill cleanup in marine ecosystems cannot be overstated, offering sustainable, eco-friendly, and efficient solutions. Moreover, the utilization of green nanotechnology can substantially reduce reliance on harmful chemicals and mitigate the release of nanoparticles into aquatic environments. Furthermore, the development of nanocomposites holds promise in minimizing the release or mobilization of nanomaterials, thereby mitigating their potential adverse effects on the environment and human health. Consequently, the application of green nanotechnology in addressing oil spills holds significant potential for curbing environmental damage and fostering the long-term resilience of marine ecosystems. In summary, green nanotechnology offers innovative and sustainable strategies for mitigating the impacts of oil spills on marine ecosystems. Through the utilization of environmentally friendly synthesis methods and benign reaction precursors, green nanotechnology enables the production of metal and metal oxide nanoparticles, thereby contributing to effective oil spill remediation while prioritizing environmental sustainability (Table 6.1).

6.4 Assessing the Effectiveness of Green Nanotechnology

Assessing the effectiveness of green nanotechnology in oil spill response is crucial to determining its feasibility and potential for mitigating environmental damage (Rathod et al. 2024; Sharma et al. 2024). By conducting comprehensive studies and experiments, we can evaluate the efficiency of green nanotechnology in cleaning up oil spills and its overall impact on the marine ecosystem. Green nanotechnology offers a sustainable and environmentally friendly approach to mitigating oil spills in marine ecosystem. The use of green nanotechnology in oil spill cleanup is a promising approach that can minimize environmental impacts and provide effective solutions compared to traditional methods (Yao et al. 2024).

The advantages of green nanotechnology in oil spill mitigation include:

Table 6.1 The nanomaterials and nanostructures utilized for oil recovery over the water surface (Mishra et al. 2022)

	Nanomaterial/nanostructure	Product (oil)	Maximum sorption capacity (g/g)	Recovery rate	Reusability	References
Aerogel	Graphene aerogel	Crude oil	–	68.8 g/s	Reusable	Wadhwani (2017)
	Carbon nanofiber (CNF) aerogels	Crude oil	–	2000% (sorption kinetics) per 100 s. Approximately	–	Wu (2014)
	Octanoyl cellulose	Crude oil	57 g/g	5.7 g/s	Reusable	Bidgoli et al. (2019)
	NFC/PVA aerogels		95 g/g	–	Reusable	Chhajed (2019)
CNT/ graphene	3D graphene-CNT hybrid foam	Compressor oil	85 g/g	–	Significant reusability reported	Dong et al. (2012)
	CNF/carbon foam	Wash oil	28.4 g/g	–	Reusable until fully consumed	Xiao et al. (2013)
	CNT/PDMS-coated polyurethane (PU) sponge	Organic oil	2.4 g/g	–	Excellent reusability	Wang and Lin (2013)
	CNT sponge	Crude oil	92.3 g/g	–	Reusable	Zhu et al. (2013)
	CNTs-SiO_2 coated polyurethane sponge	Pump oil	18 g/g	–	Recyclable	Ge et al. (2014)
	MWCNT@Teflon	Crude oil	8.4 g oil/0.5 gm sorbent	8.4 g/0.5 g/12 cycles	Recyclable	Sasikala et al. (2018)
	PVDF/magnetic fibrous composite nanostructure & PVDF	Motor oil (HC)	18.07 & 23.6		Not reported	Dorneanu et al. (2017)
	Joule-heated graphene-wrapped sponge (GWS)	Viscous crude oil	–	2320 kg h^{-1} per kg of MS@RGO	Recyclable until fully consumed	Ge (2017)

Magnetic nanoparticle	Magnetic shell cross-linked needlelike nanoparticles	Crude oil	16.8 g/g	–	12 working cycle	Pavia et al. (2013)
	Polyvinylpyrrolidone (PVP)-coated iron oxide					Palchoudhary and Lead (2014)
	MNP-Fe_3O_4	Crude oil				Atta (2015)
	Polyvinylpyrrolidone (PVP)-coated iron oxide	Crude oil	150 v/v			Mirshahghassemi and Lead (2015)
	Polyurethane with polydopamine (PDA)					Ma and Wang (2019)
Membrane	'Tea-bag' nanocomposite system	Motor oil & vacuum pump oil	2.5 g/g–40 g/g	0.32 g/g/min–0.80 g/g/min	Reported recyclable	Avila et al. (2014)
	ABSTRACT bioinspired polymer–nanoparticle–fluorosurfactant complex composite	Oil (hexadecane)			50 times	Ritchie (2019)
Nanomachine	Mobile nano-, micro-, and millimeter-scale autonomous systems	Oil	–	–	–	Zhao (2011)
	The hydrophobic micro submarine,	Oil	–	–	–	Guix (2012)
	Trimetallic (Pt/Ni/Au) microcaps	Oil	–	–	–	Safdar et al. (2015)
	Self-propelled Janus foam motor	Oil	–	–	–	Li (2018)

1. The use of eco-friendly and sustainable materials in the form of nanocomposites, which can help minimize the release or mobilization of harmful nanoparticles while maintaining their reactivity and effectiveness in cleaning up oil spills.
2. It reduces the need for hazardous chemicals, minimizing harm to marine life and ecosystems during oil spill response.
3. It offers a more efficient and effective approach to oil spill cleanup in comparison to traditional methods.
4. It has the potential to enhance oil spill cleanup efforts by increasing the efficiency and effectiveness of remediation techniques.
5. The synthesis of nanoparticles using green chemistry approaches reduces the use of hazardous compounds and harsh reaction conditions, minimizing potential environmental damage.
6. It can improve the dispersibility and solubility of oil, making it easier to remove from water surfaces. Additionally, green nanotechnology can be used to modify the properties of sorbents and adsorbents, enhancing their capacity to absorb and capture oil.
7. The use of green nanotechnology can accelerate the biodegradation process of oil, ensuring a faster and more complete breakdown of hydrocarbons.
8. It can be cost-effective in the long run, as it reduces the dependence on expensive and finite resources used in traditional oil spill response methods such as dispersants and mechanical containment.

However, it is important to acknowledge that there are still challenges and uncertainties associated with the use of nanomaterials in oil spill mitigation. Further research and testing are needed to fully understand the behavior and fate of nanomaterials in marine environments, as well as their potential impacts on aquatic organisms and ecosystems. It is crucial to prioritize the development and implementation of safety protocols and regulations to ensure the responsible use of green nanotechnology in oil spill mitigation. Nanocomposites synthesized through green nanotechnology techniques hold great potential in mitigating oil spills in marine ecosystems. They minimize the release or mobilization of harmful nanoparticles while maintaining their reactivity and effectiveness in cleaning up oil spills. Green nanotechnology can also contribute to the restoration and rehabilitation of affected ecosystems after an oil spill, by promoting the growth of beneficial microorganisms and facilitating the breakdown of pollutants.

6.5 Compatibility of Green Nanotechnology with Marine Life

Environmental safety is a critical aspect to consider when applying green nanotechnology in oil spill mitigation. It is important to assess the potential toxic effects of nanoparticles on marine life and ensure that the materials used in green nanotechnology are environmentally friendly and non-toxic. One approach for promoting

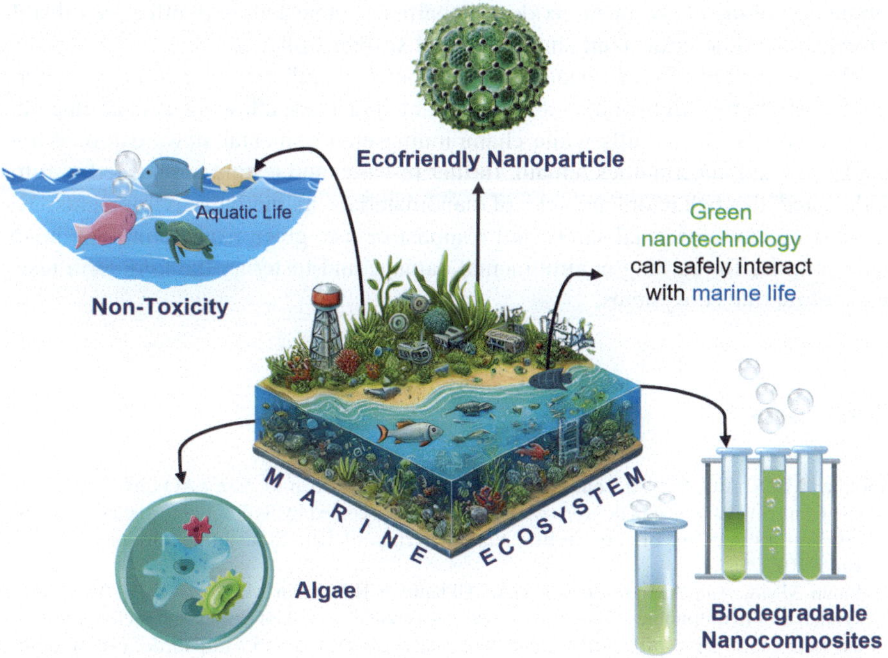

Fig. 6.3 Green nanotechnology is compatible with marine life, showing no adverse effects on their ecosystem

environmental safety is to conduct thorough toxicity assessments of the nanoparticles used, taking into consideration their potential impact on marine organisms and ecosystems. Additionally, the development of biodegradable nanocomposites can further enhance environmental safety by ensuring that the materials used in oil spill cleanup will eventually degrade and minimize any long-term environmental impacts (Fig. 6.3).

Furthermore, it is crucial to understand the behavior and fate of nanomaterials in water and wastewater treatment processes to ensure their safe application. Minimize any long-term impacts on the marine ecosystem. Minimize any long-term impacts on the marine ecosystem. Not persist in the marine environment, minimizing potential long-term effects on marine life.

6.6 Conclusion and Prospects

Green nanotechnology offers innovative and sustainable solutions for mitigating the impacts of oil spills on marine ecosystems. By utilizing eco-friendly synthesis methods and benign reaction precursors, green nanotechnology enables the production of metal and metal oxide nanoparticles, contributing to effective oil spill remediation while prioritizing environmental sustainability.

The integration of green nanotechnology into oil spill response strategies represents a paradigm shift towards more sustainable practices, allowing us to combat the adverse effects of oil spills while championing environmental stewardship. While challenges and uncertainties remain, further research and testing are needed to fully understand the behavior and fate of nanomaterials in marine environments. By prioritizing environmental safety and responsible use, green nanotechnology holds great promise for curbing environmental damage and fostering the long-term resilience of marine ecosystems.

References

Abdullah MMS, Atta AM, Al-Lohedan HA, Alkhathlan HZ, Khan M, Ezzat AO (2017) Synthesis of green recyclable magnetic iron oxide nanomaterials coated by hydrophobic plant extracts for efficient collection of oil spills. Nanomaterials 9(10):1505. https://doi.org/10.3390/nano9101505

Abdullah MMS, Atta AM, Allohedan HA, Alkhathlan HZ, Khan M, Ezzat AO (2018) Green synthesis of hydrophobic magnetite nanoparticles coated with plant extract and their application as petroleum oil spill collectors. Nanomaterials 8(10):855. https://doi.org/10.3390/nano8100855

Alqarni LS, Alghamdi MD, Alshahrani AA, Nassar AM (2022) Green nanotechnology: recent research on bioresource-based nanoparticle synthesis and applications. J Chem:1–31. https://doi.org/10.1155/2022/4030999

Atta A, Al-Lohedan H, Al-Hussain S (2015) Functionalization of magnetite nanoparticles as oil spill collector. Int J Mol Sci 16(12):6911–6931. https://doi.org/10.3390/ijms16046911

Atta AM, Mohamed NH, Hegazy AK, Moustafa YM, Mohamed RR, Safwat G, Diab AA (2020) Green technology for remediation of water polluted with petroleum crude oil: using of Eichhornia crassipes (Mart.) solms combined with magnetic nanoparticles capped with myrrh resources of Saudi Arabia. Nano 10(2):262. https://doi.org/10.3390/nano10020262

Avila AF, Munhoz VC, De Oliveira AM, Santos MC, Lacerda GR, Gonçalves CP (2014) Nano-based systems for oil spills control and cleanup. J Hazard Mater 272:20–27. https://doi.org/10.1016/j.jhazmat.2014.02.038

Bhainsa KC, D'Souza SF (2006) Extracellular biosynthesis of silver nanoparticles using the fungus Aspergillus fumigatus. Colloids Surf B: Biointerfaces 47(2):160–164. https://doi.org/10.1016/j.colsurfb.2005.11.026

Bhushan B (2017) Introduction to nanotechnology. Springer Handbooks. https://link.springer.com/book/10.1007/978-3-662-54357-3

Bidgoli H, Khodadadi AA, Mortazavi Y (2019) A hydrophobic/oleophilic chitosan-based sorbent: toward an effective oil spill remediation technology. J Environ Chem Eng 7(5):103340. https://doi.org/10.1016/j.jece.2019.103340

Bottero J, Rose J, Wiesner MR (2009) Nanotech: tools for sustainability in a new wave of water treatment processes. Integr Environ Assess Manag 2(4):391–395. https://doi.org/10.1002/ieam.5630020411

Cao KLA, Rahmatika AM, Kitamoto Y, Nguyen MTT, Ogi T (2020) Controllable synthesis of spherical carbon particles transition from dense to hollow structure derived from Kraft lignin. J Colloid Interface Sci 589:252–263. https://doi.org/10.1016/j.jcis.2020.12.077

Chao W, Wang, Li S, Cao Y, Zhao G, Sun Y, Wang X, Ho SH (2020) Natural sponge-like wood-derived aerogel for solar-assisted adsorption and recovery of high-viscous crude oil. Chem Eng J 400:125865. https://doi.org/10.1016/j.cej.2020.125865
Chhajed M, Yadav C, Agrawal AK, Maji PK (2019) Esterified superhydrophobic nanofibrillated cellulose based aerogel for oil spill treatment. Carbohydr Polym 226:115286. https://doi.org/10.1016/j.carbpol.2019.115286
Chithrani BD, Ghazani AA, Chan WC (2006) Determining the size and shape dependence of gold nanoparticle uptake into mammalian cells. Nano Lett 6(4):662–666. https://doi.org/10.1021/nl052396o
De Gusseme B, Sintubin L, Baert L, Thibo E, Hennebel T, Vermeulen G, Uyttendaele M, Verstraete W, Boon N (2010) Biogenic silver for disinfection of water contaminated with viruses. Appl Environ Microbiol 76(4):1082–1087. https://doi.org/10.1128/AEM.02433-09
Dong X, Ma Y, Zhu G, Huang Y, Wang J, Chan-Park MB, Wang L, Huang W, Chen P (2012) Synthesis of graphene–carbon nanotube hybrid foam and its use as a novel three-dimensional electrode for electrochemical sensing. J Mater Chem 22(33):17044. https://doi.org/10.1039/c2jm33286h
Dorneanu PP, Cojocaru C, Olaru N, Samoila P, Airinei A, Sacarescu L (2017) Electrospun PVDF fibers and a novel PVDF/Co Fe_2O_4 fibrous composite as nanostructured sorbent materials for oil spill cleanup. Appl Surf Sci 424:389–396. https://doi.org/10.1016/j.apsusc.2017.01.177
Ge B, Zhang Z, Zhu X, Men X, Zhou X (2014) A superhydrophobic/superoleophilic sponge for the selective absorption oil pollutants from water. Colloids Surf A Physicochem Eng Asp 457:397–401. https://doi.org/10.1016/j.colsurfa.2014.06.020
Ge J, Shi L, Wang Y, Zhao H, Yao H, Zhu Y, Zhang Y, Zhu H, Wu H, Yu S (2017) Joule-heated graphene-wrapped sponge enables fast clean-up of viscous crude-oil spill. Nat Nanotechnol 2(5):434–440. https://doi.org/10.1038/nnano.2017.33
Gu H, Zhou X, Lyu S, Pan D, Dong M, Wu S, Ding T, Wei X, Seok I, Wei S, Guo Z (2020) Magnetic nanocellulose-magnetite aerogel for easy oil adsorption. J Colloid Interface Sci 560: 849–856. https://doi.org/10.1016/j.jcis.2019.10.084
Guix M, Orozco J, García M, Gao W, Sattayasamitsathit S, Merkoçi A, Escarpa A, Wang J (2012) Superhydrophobic alkanethiol-coated microsubmarines for effective removal of oil. ACS Nano 6(5):4445–4451. https://doi.org/10.1021/nn301175b
Hethnawi A, Kashif O, Jeong R, Sagala F, Hashlamoun K, Manasrah AD, Nassar NN (2023) Green synthesis of novel titanomagnetite nanoparticles for oil spill cleanup. Colloids Surf A Physicochem Eng Asp 664:131191. https://doi.org/10.1016/j.colsurfa.2023.131191
Hou J, Zhao G, Zhang L, Wang G, Li B (2019) High-expansion polypropylene foam prepared in non-crystalline state and oil adsorption performance of open-cell foam. J Colloid Interface Sci 542:233–242. https://doi.org/10.1016/j.jcis.2019.02.028
International Maritime Organization. Responding to MV Wakashio oil Spill. https://www.imo.org/en/MediaCentre/HotTopics/Pages/Wakashio-FAQ.aspx
Kalliola S, Repo E, Sillanpää M, Arora JS, He J, John VT (2016) The stability of green nanoparticles in increased pH and salinity for applications in oil spill-treatment. Colloids Surf A Physicochem Eng Asp 493:99–107. https://doi.org/10.1016/j.colsurfa.2016.01.011
Lavoine N, Bergström L (2017) Nanocellulose-based foams and aerogels: processing, properties, and applications. J Mater Chem A 5(31):16105–16117. https://doi.org/10.1039/C7TA02807E
Lee H, Trinh BM, Crawford EA, Mekonnen TH (2023) Thermoplastic elastomer melt-blown fiber mats for oil spill remediation: fabrication, oil uptake, and gel formation studies. Sep Purif Technol 326:124761. https://doi.org/10.1016/j.seppur.2023.124761
Li X, Mou F, Guo J, Deng Z, Chen C, Xu L, Luo M, Guan J (2018) Hydrophobic Janus foam motors: self-propulsion and on-the-fly oil absorption. Micromachines 9(1):23. https://doi.org/10.3390/mi9010023
Liu Y, Qian J, Wang J (2000) Pyrolysis of polystyrene waste in a fluidized-bed reactor to obtain styrene monomer and gasoline fraction. Fuel Process Technol 63(1):45–55. https://doi.org/10.1016/S0378-3820(99)00066-1

Ma W, Wang H (2019) Magnetically driven motile superhydrophobic sponges for efficient oil removal. Appl Mater Today 15:263–266. https://doi.org/10.1016/j.apmt.2019.02.004

Ma L, Niu S, Zhao F, Tang R, Zhang Y, Su W, Wei L, Tang G, Wang Y, Pang A, Li W (2022) A high-performance polyurethane–polydopamine polymeric binder for silicon microparticle anodes in lithium-ion batteries. ACS Appl Energy Mater 5(6):7571–7581. https://doi.org/10.1021/acsaem.2c01033

Mirshahghassemi S, Lead JR (2015) Oil recovery from water under environmentally relevant conditions using magnetic nanoparticles. Environ Sci Technol 49(19):11729–11736. https://doi.org/10.1021/acs.est.5b02687

Mishra S, Chauhan G, Verma S, Singh U (2022) The emergence of nanotechnology in mitigating petroleum oil spills. Mar Pollut Bull 78:113609. https://doi.org/10.1016/j.marpolbul.2022.113609

Palchoudhury S, Lead JR (2014) A facile and cost-effective method for separation of oil–water mixtures using polymer-coated iron oxide nanoparticles. Environ Sci Technol 48(24): 14558–14563. https://doi.org/10.1021/es5037755

Pavía-Sanders A, Zhang S, Flores JA, Sanders JE, Raymond JE, Wooley KL (2013) Robust magnetic/polymer hybrid nanoparticles designed for crude oil entrapment and recovery in aqueous environments. ACS Nano 7(9):7552–7561. https://doi.org/10.1021/nn401541e

Phanthong P, Reubroycharoen P, Kongparakul S, Samart C, Wang Z, Hao X, Abudula A, Guan G (2018) Fabrication and evaluation of nanocellulose sponge for oil/water separation. Carbohydr Polym 190:184–189. https://doi.org/10.1016/j.carbpol.2018.02.066

Poddar M, Lakshmi GBVS, Sharma M, Chaudhary N, Nigam S, Joshi M, Solanki PR (2022) Environmental friendly Polyacrylonitrile nanofiber mats encapsulated and coated with green algae mediated Titanium oxide nanoparticles for efficient oil spill adsorption. Mar Pollut Bull 182:113971. https://doi.org/10.1016/j.marpolbul.2022.113971

Prendergast DP, Gschwend PM (2014) Assessing the performance and cost of oil spill remediation technologies. J Clean Prod 78:233–242. https://doi.org/10.1016/j.jclepro.2014.04.054

Qu X, Brame J, Li Q, Alvarez PJJ (2012) Nanotechnology for a safe and sustainable water supply: enabling integrated water treatment and reuse. Acc Chem Res 46(3):834–843. https://doi.org/10.1021/ar300029v

Rathod S, Preetam S, Pandey C, Bera SP (2024) Exploring synthesis and applications of green nanoparticles and the role of nanotechnology in wastewater treatment. Biotechnol Rep 41: e00830. https://doi.org/10.1016/j.btre.2024.e00830

Ritchie A, Cox H, Barrientos-Palomo S, Sharples G, Badyal J (2019) Bioinspired multifunctional polymer–nanoparticle–surfactant complex nanocomposite surfaces for antibacterial oil–water separation. Colloids Surf A Physicochem Eng Asp 560:352–359. https://doi.org/10.1016/j.colsurfa.2018.10.030

Safdar M, Itkonen T, Jänis J (2015) Bubble-propelled trimetallic microcaps as functional catalytic micromotors. RSC Adv 5(17):13171–13174. https://doi.org/10.1039/c4ra16589f

Sasikala SP, Mahesh K, Prabhakaran P, Mohammed AP, Ananthakumar S (2018) Multiwall carbon nanotube reinforced teflon fibrils for oil spill clean up and its effective recycling as textile dye sorbent. J Environ Manag 211:198–205. https://doi.org/10.1016/j.jenvman.2018.01.011

Sawle BD, Salimath B, Deshpande R, Bedre MD, Prabhakar BK, Venkataraman A (2008) Biosynthesis and stabilization of Au and Au–Ag alloy nanoparticles by fungus, Fusarium semitectum. Sci Technol Adv Mater 9(3):035012. https://doi.org/10.1088/1468-6996/9/3/035012

Seddon M (2020) Siberia fuel spill threatens Moscow's Arctic ambitions. Financial Times. Accessed 5 June 2020

Shahverdi AR, Fakhimi A, Shahverdi HR, Minaian S (2007) Synthesis and effect of silver nanoparticles on the antibacterial activity of different antibiotics against Staphylococcus aureus and Escherichia coli. Nanomedicine 3(2):168–171. https://doi.org/10.1016/j.nano.2007.02.001

Sharma K, Shah G, Singh H, Bhatt U, Singhal K, Soni V (2024) Advancements in natural remediation management techniques for oil spills: challenges, innovations, and future directions. Environ Pollut Manag. https://doi.org/10.1016/j.epm.2024.08.003

Singh V, Kendall RJ, Hake K, Ramkumar S (2013) Crude oil sorption by raw cotton. Ind Eng Chem Res 52(18):6277–6281. https://doi.org/10.1021/ie4005942

Singh N, Malik MA, Hashmi AA (2021) Nanotechnology and its application in wastewater treatment. In: Sustainable practices in the textile industry. https://doi.org/10.1002/9781119818915.ch13

Soares SF, Rodrigues MI, Trindade T, Daniel-da-Silva AL (2017) Chitosan-silica hybrid nanosorbents for oil removal from water. Colloids Surf A Physicochem Eng Asp 532:305–313. https://doi.org/10.1016/j.colsurfa.2017.04.076

Sun Y, Cao X, Liang F (2019) Investigation on underwater spreading characteristics and migration law of oil leakage from damaged submarine pipelines. Process Saf Environ Prot 127:329–347. https://doi.org/10.1016/j.psep.2019.05.030

TASS Moscow (2020) Diesel fuel spill in Norilsk in Russia's Arctic. Assessed 6 June 2020

Veeresh DA, Bhutia DD, Pokhrel I, Chakraborty S, Kumar A, Mishra A, Sarkar A (2024) Nanoremediation and role in environmental clean-up. In: Microbiome research in plants and soil. Microbiome-assisted bioremediation. Academic, pp 381–400. https://doi.org/10.1016/B978-0-443-21911-5.00007-6

Wadhwani R (2017) Oil spill cleanup using aerographene. Paper presented at the SPE oil and gas India conference and exhibition, Mumbai, India

Wang C, Lin S (2013) Robust superhydrophobic/superoleophilic sponge for effective continuous absorption and expulsion of oil pollutants from water. ACS Appl Mater Interfaces 5(18): 8861–8864. https://doi.org/10.1021/am403266v

Wang Y, Zhou L, Luo X, Zhang Y, Sun J, Ning X, Yuan Y (2019) Solar-heated graphene sponge for high-efficiency clean-up of viscous crude oil spill. J Clean Prod 230:995–1002. https://doi.org/10.1016/j.jclepro.2019.05.178

Wu Z, Li C, Liang H, Zhang Y, Wang X, Chen J, Yu S (2014) Carbon nanofiber aerogels for emergent cleanup of oil spillage and chemical leakage under harsh conditions. Sci Rep 4(1). https://doi.org/10.1038/srep04079

Wu MN, Maity JP, Bundschuh J, Li CF, Lee CR, Hsu CM, Lee WC, Huang CH, Chen CY (2017) Green technological approach to synthesis hydrophobic stable crystalline calcite particles with one-pot synthesis for oil–water separation during oil spill cleanup. Water Res 123:332–344. https://doi.org/10.1016/j.watres.2017.06.040

Xiao N, Zhou Y, Ling Z, Qiu J (2013) Synthesis of a carbon nanofiber/carbon foam composite from coal liquefaction residue for the separation of oil and water. Carbon 59:530–536. https://doi.org/10.1016/j.carbon.2013.03.051

Yao S, Ouyang S, Zhou Q, Tao Z, Chen Y, Zheng T (2024) Environmental remediation and sustainable design of iron oxide nanoparticles for the removal of petroleum-derived pollutants from water: a critical review. Environ Res:120009. https://doi.org/10.1016/j.envres.2024.120009

Yap KY, Tan MC (2021) Oil adsorption onto different types of microplastic in synthetic seawater. Environ Technol Innov 24:101–994. https://doi.org/10.1016/j.eti.2021.101994

Zhao G, Seah TH, Pumera M (2011) External-energy-independent polymer capsule motors and their cooperative behaviors. Chem Eur J 17(43):12020–12026. https://doi.org/10.1002/chem.201101450

Zhu K, Shang Y, Sun P, Li Z, Li X, Wei J, Wang K, Wu D, Cao A, Zhu H (2013) Oil spill cleanup from sea water by carbon nanotube sponges. Front Mater Sci 7(2):170–176. https://doi.org/10.1007/s11706-013-0200-1

Chapter 7
Green Nanotechnology for Water Purification

B. Deepa, Merin Sara Thomas, Cintil Jose Chirayil, Rini Thresia Varghese, Kavya Manoj, L. A. Pothan, and Sabu Thomas

Abstract This chapter delves into the pressing issues surrounding water purification, emphasizing the importance of addressing common contaminants and developing sustainable solutions. It critically examines the shortcomings of conventional purification methods and introduces green nanotechnology as a promising alternative to overcome these limitations. The focus is on biopolymer-based nanomaterials, which are explored in detail, including their various types and applications in water treatment. The chapter highlights the eco-friendly nature of these materials and their efficacy in removing contaminants from water sources. A comprehensive analysis of the environmental impact and safety considerations associated with green nanotechnology is presented, providing a balanced perspective on both the advantages and potential risks of this innovative approach. In addition, the chapter describes the synergistic potential of combining green nanotechnology with other emerging water treatment technologies, such as advanced membrane nanofiltration, photocatalysis and biologically mediated processes. It explores how these integrated approaches can lead to more comprehensive and efficient water purification solutions, capable of addressing a broader spectrum of water quality issues. The chapter concludes by outlining future research directions and anticipated challenges in the field, emphasizing the significant role that green nanotechnology is expected to play in advancing global water purification efforts. By addressing these crucial aspects, the chapter underscores the potential of this technology to contribute to more sustainable and

B. Deepa · K. Manoj · S. Thomas (✉)
School of Nano Science and Nano Technology, Mahatma Gandhi University, Kottayam, Kerala, India

M. S. Thomas
Department of Chemistry, Mar Thoma College, Tiruvalla, Kerala, India

C. J. Chirayil · R. T. Varghese
Department of Chemistry, Newman College, Thodupuzha, Kerala, India

L. A. Pothan
Department of Chemistry, Bishop Moore College, Mavelikara, Kerala, India

© The Author(s), under exclusive license to Springer Nature Switzerland AG 2025
J. M. Al-Khayri et al. (eds.), *Green Nanotechnology Applications for Ecosystem Sustainability*, Nanotechnology in Plant Sciences 4,
https://doi.org/10.1007/978-3-031-90766-1_7

effective water treatment solutions, ultimately supporting global efforts to ensure access to clean and safe water resources.

Keywords Biopolymer-based nanomaterials · Green nanomaterials · Green nanotechnology · Membrane nanofiltration · Photocatalysis · Water purification

7.1 Introduction

Green nanotechnology (GN) is a revolutionary method that puts environmental sustainability first in the creation and use of nanotechnology and nanotechnological processes. This emerging discipline attempts to reduce the harmful effects of nanotechnology by utilizing the ideas associated with sustainable chemistry and green engineering. It entails the production of environmentally friendly nanomaterials, the application of sustainable synthesis techniques, and the use of nano in applications that support resource efficiency, renewable energy, and cleaner air and water. Consequently, green nanotechnology offers solutions that increase product durability, decrease waste, and increase energy efficiency, all of which help to contribute to a future that is more environmentally friendly.

The main objective of this chapter is to explore the integration of green nanotechnology in water purification as a sustainable solution to current global water challenges. The chapter aims to highlight the challenges of water contamination and the limitations of traditional purification methods, introduce green nanotechnology as an eco-friendly, efficient alternative for addressing these challenges, discuss various green nanotechnology-based water purification techniques, focusing on their potential to offer sustainable, scalable solutions and evaluate the environmental impacts and safety concerns associated with the use of nanomaterials in water treatment. The chapter is divided into different sections. The Sects. 7.1.1, 7.1.2 and 7.1.2.1 provides an overview of the global water crisis, discussing the types of contaminants commonly found in water, the importance of ensuring access to clean water, and the shortcomings of conventional water purification techniques. It concludes by highlighting the pressing need for more sustainable water purification technologies.

The second Sect. 7.2 introduces green nanotechnology as a sustainable alternative to traditional purification methods. The following Sect. 7.2.1 details how green nanotechnology can tackle various water purification challenges and the Sects. 7.3, 7.3.1 and 7.3.2 emphasizing the role of green nanomaterials and various types of biopolymer-based nanomaterials for water purification process. In the Sect. 7.4, different water treatment techniques that utilize green nanotechnology are explored. These include membrane nanofiltration (Sect. 7.4.1), adsorption (Sect. 7.4.2), photocatalysis (Sect. 7.4.3) and biologically mediated processes (Sect. 7.4.4). Each method is assessed for its potential to efficiently remove contaminants while minimizing environmental impact. The Sect. 7.5 focuses on the environmental implications of employing nanotechnology in water treatment. It examines the life cycle of nanomaterials, addressing both the potential benefits and the risks associated with their production, use, and disposal.

The final section conclusion (Sect. 7.6) summarizes the key insights gained from the chapter, highlighting the potential of green nanotechnology as a transformative solution for sustainable water purification. This section also looks ahead, discussing the future perspectives challenges that must be addressed for green nanotechnology to achieve widespread adoption.

7.1.1 Water Purification and Challenges

The major globally concerned issue is the increasing need for clean water for various uses in domestic, agricultural, industrial and drinking purposes. The contamination of water by several forms of wastes causes great concern over the health of various biological systems and other components of the environment. Water is the lifeline of all living forms on the globe. The major goal of water purification is the process whereby undesirable chemicals, biological contaminants, suspended solids, and gases present in water are removed to make it safe and useful for various applications, including drinking, cooking, and irrigation and other industrial purposes. Water purification is critical for maintaining public health and ensuring access to clean water, especially in areas where natural water sources are contaminated. There is a huge scarcity of clean and portable water supply that leads to a number of challenges faced globally (Chaturvedi and Dave 2019; Gupta et al. 2022).

The challenges faced are the enormous cost of water purification technologies and discarding of contaminants. There are many cases of organo chloro chlorine pesticides deposited into the water resulting in major deterioration of water bodies like rivers, lakes (Ahuja 2025). This can cause the pollution of both ground and surface water. About 1.2 billion people around the world have no access to fresh potable water (Journal and Reuse 2019). Also, the increasing water demand combined with climate change prospects, severely threatens the regional landscape, cultural heritage, development and food security of the human beings.

7.1.2 Common Water Contaminants

There are several synthetic or naturally occurring contaminants and microorganisms in the environment that have the potential to enter the aquatic system and to possess adverse ecological or human health effects. There are different chemicals contaminants in the surface and groundwater and are mostly present in traces in the unit range of parts per trillion (ppt or ng/L) to parts per billion (ppb or μg/L) and these are coming from various sources as shown in Fig. 7.1. These are different emerging contaminants like pharmaceuticals, drugs, pesticides, personal care products, per fluorinated compounds, industrial chemicals that are present to a few nanograms per

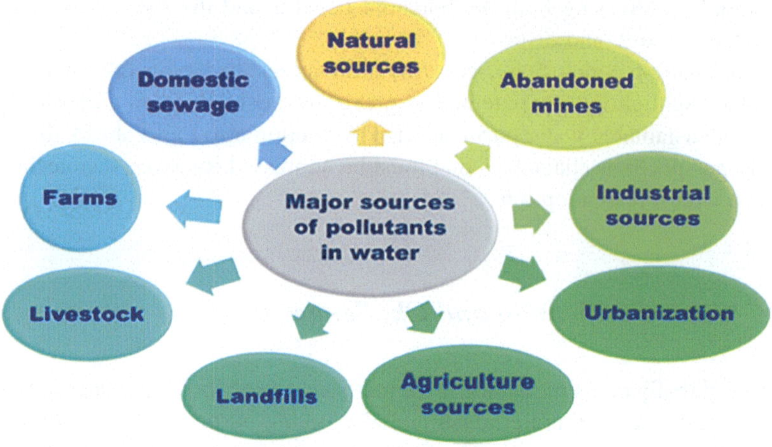

Fig. 7.1 Common sources causing contamination in water. (Chelu et al. 2023)

liter of solution to grams per liter of solution. All of the different types of water resources, wastewater, groundwater, and surface waters have been polluted with various contaminants. The contaminants present and redistributed in various water resources have toxic effects on the living organisms present in water and it disrupts their natural endocrine activity. When their residual concentrations reach the surface and groundwater resources, it is tough to remove these harmful compounds efficiently. Hence, their residual concentrations reach the surface and groundwater resources, thus, posing a risk to the environment. The different contaminants get accumulated into the aquatic systems via various paths, that can be point sources (municipal sewage, industrial wastewaters, landfill) and non-point sources (agricultural runoff underground contamination) (Kumar et al. 2022).

There are different types of contaminants that occur which can be either geological or anthropogenic (man-made). Geological contaminants enter the water system when the groundwater flows through sedimentary rocks and picks up certain ions like chloride, arsenate, nitrate and when their concentration reaches an unacceptable level in water, it serves as a contaminant.

Other type of contaminants include those that are obtained via agricultural and industrial processes and their by-products like improper disposal of wastes, heavy metals, chemicals, pesticides. Also, microbial contaminants that include different pathogens like bacteria, viruses, and parasites like microscopic protozoa and worms. These microbial living organisms can be spread by careless disposal of human and animal wastes. Also, the presence of clay, slits, sands, or organic clay, algae and leaf particles increases turbidity of the water solution which can also cause some health effects (Sharma and Bhattacharya 2017).

Basically there are four different types of contaminants that cause water pollution,

(i) Inorganic contaminants: dissolved amounts of carbonate or magnesium leads to hardness of the solution. About 300–400 mg/L hardness in water is suitable for drinking purposes. The presence of several other salts or ions like lead, fluoride, chromium, mercury, arsenic, copper, antimony, cyanide and their prolonged exposure at levels above 500 mg/L can cause kidney problems. Also the oxidation state of the ions also pave way for toxicity. Arsenic present in oxidation III is ten times more toxic than arsenic (V). Arsenic contamination can cause various problems to the skin, diarrhea, and also a disease called arsenicosis. These ions may enter the drinking water through agricultural means or industrial particles (Pontius 2000).

(ii) Organic contaminants: these mainly include the different kinds of pesticides released from the agricultural field. Different types of organic solvents and dyes that are discarded from industries. All these contaminations mainly occur due to their improper handling and application methods, and disposal.

(iii) Biological contaminants: The living organisms present in almost all aquatic water system, algae, bacteria, protozoa, viruses are the biological contaminants. The presence of these in undesired levels can cause certain infections and diseases.

(iv) Radiological contaminants: These are present mostly in ground water than surface water due to erosion and explosion of natural deposits of minerals emitting α and β rays. Radioactive elements cause these contamination and cancer is one of the most risk that can occur due to radiological contamination. (Sharma and Bhattacharya 2017)

An important pervasive problem afflicting the people throughout the world is inadequate access to clean water and sanitation. Water-related issues are predicted to worsen over the next few decades, with water scarcity occurring everywhere, including in areas that are currently thought to be water-rich. It is common knowledge that the shortage of pure, fresh water causes numerous issues around the world. 3900 children die every day from diseases spread by contaminated water or human excrement, and 2.6 billion people without access to adequate sanitation. Additionally, 1.2 billion people lack access to safe drinking water. The need to develop robust water treatment methods using a minimum amount of chemicals became even more pressing after the occurrence of several severe outbreaks of waterborne pathogens through water treatment systems. We need more reliable, affordable, and strong ways to clean and disinfect water from the source to the point of consumption without adding to environmental stress or posing a health risk to people (Shannon et al. 2008). Water shortages may have important consequences in energy and climate policies.

7.1.2.1 Challenges in Traditional Water Purification Methods

Due to growing issues including extensive urbanization, the use of chemical products, and the existence of developing toxins, there are several obstacles associated with the supply of drinking water. The lack of access to clean water services has been caused by factors like climate change, population growth, and excessive water use.

This has frequently resulted in poor sanitation and water-borne illnesses in the impacted areas. From time immemorial, there have been different methods employed for the removal of pollutants and disinfecting water resources. Today these traditional water purification methods have been replaced by new technologies which are much more efficient and faster. However in rural areas, the traditional methods of water purification are still prevalent. Through these methods, plant debris, soil fractions, microorganisms are removed. Conventional filtration techniques encounter many obstacles. The sluggish filtration rate, continuous attention needed to avoid overflow, and requirement to remove the filtering material to cease filtration are the drawbacks of these techniques, which include funnels, cotton, and filter paper. Most of the traditional methods rely on the use of chemicals such as chlorine or alum. Hence the traditional methods of water purification faced many challenges that paved way for researchers to search for novel methods with less environmental consequences (Adesina et al. 2024). Some of the major challenges of the traditional methods of water purification from the environmental point of view include,

(i) **Chemical usage**: Involvement of chemicals such as chlorine, which is a common disinfectant, and its improper disposal into water resources can produce harmful byproducts like trihalomethanes (THMs) that are linked to health issues. For almost a century, chlorine disinfection has been recognized as an economical and reliable method of disinfection, and this is expected to continue. Trihalomethane production is the most significant unfavorable consequence of adding chlorine to water. These substances are produced when free chlorine reacts with the organic materials in the feed water. The sludge generated during water treatment procedures need to be properly disposed to prevent water and soil contamination. Also, most of the coagulants and flocculants often leave behind metal-rich sludge, necessitating appropriate disposal (Gitis and Hankins 2018).

(ii) **Energy Consumption:** The traditional methods require a lot of energy affecting the overall energy chain. Conventional techniques may be energy-intensive, which raises the need for fossil fuels and increases greenhouse gas emissions, both of which may have an effect on climate change.

(iii) **Water loss:** Traditional methods may also require a lot of water. In areas where water is scarce, water loss from evaporation and backwashing during filtration procedures is especially concerning (Adesina et al. 2024).

7.2 Green Nanotechnology as a Sustainable Approach

Despite the fact that millions of people globally lack consistent access to safe drinking water, access to clean water is a fundamental human right. Particularly in underprivileged areas, sustainable purifying technologies can help close this gap by providing scalable and reasonably priced solutions. In order to lessen the harmful effects that conventional membrane science and technologies have on the environment, it is now essential to increase their efficiency. It is imperative to find long-term solutions to stop freshwater reserves from being depleted as a result of population growth and irresponsible usage of existing stocks. Different types of sustainable solutions have been discussed nowadays to improve the availability of clean and potable water without causing any distress to the environment. Sustainable purification technologies can help optimize water usage and preserve precious resources. Reusing membranes, repurposing waste brine or sludge, capturing energy from wastes, and reducing waste through membrane antifouling techniques are a few techniques for sustainable water treatment (Domb 2017).

Circular economy solutions, zero-waste and zero-liquid discharge (ZLD), emerging membrane synthesis materials like quantum dots (QDs) and non-toxic solvents, emerging membrane synthesis technologies like additive manufacturing (AM), and emerging modeling approaches like molecular dynamic simulations (MDS) and artificial intelligence (AI) approaches are some of the strategies for sustainable desalination and wastewater treatment. Compared to the reuse option, the recycle option is a less popular circular economy idea since, in most circumstances, recycled membranes would still go through the manufacture phase in which additional materials and energy are used (Labhasetwar and Yadav 2023). Reuse of Reverse Osmosis (RO) driven membranes has gained popularity in the field of desalination. As per the different studies conducted, in order to retain high separation efficiency, the repurposed RO membranes need to undergo regular cleaning and backwashing. Reusing old RO membranes (those at their end-of-life stages) as UF and NF filters, after they have been treated with $KMnO_4$ or NaOCl, have been drastically able to reduce material wastes in desalination. UF and NF nano-filter membranes are normally created from the majority of spent RO membranes that have reached the end of their useful life, either to be employed as a pretreatment step in the desalination of water or in different applications. Also, wastes such as reject brine, waste heat, and cooling water from one membrane process can be utilized in another process to reduce the overall energy consumption. Brine reuse has also found a lot of recent applications in zero liquid discharge (ZLD) processes. ZLD methodologies have also been used recently for almost full retrieval of fresh water from saline groundwater and surface water. About 97% water recovery was achieved from surface water by using ultra nano filters (Yusuf et al. 2020). Nanotechnology has emerged as a novel technology to address the problem of water scarcity and wastewater treatment (Domb 2017). There are various innovative, ecologically friendly and sustainable nanotechnology based water purification methods that include biosorption, adsorption, nanofiltration, sensing technologies and photocatalytic disinfection (Thanigaivel et al. 2022).

In order to effectively address environmental issues, green nanotechnology should be introduced as a replacement for synthetic nanotechnology. This should be done in a manner that adheres to the green science tenets, which include creating nanomaterials and products not endangering human health or the environment, using the least amount of energy possible, and generating the least amount of waste or pollution possible. In order to achieve this, it is crucial to use nanotechnological techniques and methods, such as solid-state reactions, energy conservation, the use of water in place of hazardous organic solvents, structural modifications that lessen the toxicity of nanomaterials, and aspects related to recycling and reusing them. Green technology, in general, refers to all inventions that are beneficial to the environment and include energy efficiency, recycling, health and safety concerns, resources that are renewable, etc. Environmental technology is used for monitoring and evaluations, pollution control and prevention, remediation, and restoration. Green technology encompasses the generation as well as the consumption of resources. It is imperative to monitor detrimental pollution, be it man-made or natural (Nasrollahzadeh et al. 2019). Using green chemistry and green engineering concepts, sustainable nanotechnology constitutes one of the areas of green technology, in addition to green energy, greener information technology, and green construction (Schulte et al. 2013).

The primary goal of GN is to develop technologies in contrast to present manufacturing procedures, to manufacture nano-pigments in an ecologically friendly manner. Similarly, their applications are significantly eco-friendly materials. In order to achieve the following aims, GN must work to achieve the following objectives (Gurtu and Sharma 2021):

- methods for enhancing colored nano clays' dispersibility, fatigue process, temperature resistance, and light-fastness;
- Improved colorant compound fixation using surface treatment procedures for nano clays;
- Use of environmentally friendly and sustainable nano-pigments in all photovoltaic equipment
- Use of environmentally friendly and sustainable nanoparticles in adhesives, paints, and colorants.
- Environmental impact of more environmentally friendly nano powders is being evaluated
- Greener nanoparticles are being commercialized and dispersed.
- Preparation of nanoparticles based on the use of nano clays and desired natural and synthetic colorants

We must weigh the benefits of nano-based products against their drawbacks in order to investigate the trade-offs, as the production of nanomaterials may have negative effects on the environment in addition to potential health and safety risks. Manufacturers are exempt from having to define nanoparticles in their products, which adds to the worry that the policies in place solely cover conventional chemical substances (Som et al. 2010). Two potential alternatives include environmentally friendly and sustainable nano synthesis methods, sometimes known as "Green Alternatives," and assessment paradigms that incorporate risk assessment (RA) and life cycle assessment (LC), such as comprehensive environmental analysis (CEA) and life cycle risk assessment (LCRA).

The US Environmental Protection Agency (EPA) on their GC webpage has emphasized the following GC suggestions, which can be included into nanomanufacturing processes.

- Make chemical synthesis waste-free.
- Both chemicals and ecologically friendly items should be produced.
- Make fewer hazardous chemical mixtures.
- Use feedstocks and renewable raw materials together.
- Make use of catalytic reactions to cut down on waste.
- Avoiding chemical compounds is advised.
- Boost the atom economy to the greatest extent you can.
- Reaction conditions and safe solvents should be applied.
- Boost the efficiency of your energy use.
- Make substances and goods that, when used, decompose naturally.
- Pollution issues need to be analyzed in real time.
- Decrease the probability of accidents

The seemingly endless potential of nanotechnology spurs innovation across a wide range of diverse applications. Even while nanotechnology has great promise, there is a noticeable sense of unease about nanoparticles because of their unclear effects on the environment and human health (Karn and Bergeson 2009). In order to minimize chemical dangers to human health and the environment, reduce waste, and prevent pollution, a wide range of chemical products and processes are designed using the 12 principles associated with green chemistry (Anastas and Eghbali 2010). Application of these concepts results in decreased usage of dangerous solvents and reagents, increased material and energy efficiency in chemical processes, and improved end-of-life product design.

It is essential to assess nanoparticles' whole lifecycle, from manufacture to disposal, to make sure they don't harm the environment. This all-inclusive method, called lifecycle assessment (LCA), includes the analysis of environmental effects related to every phase of a nanomaterial's life. This covers the extraction, synthesis, manufacture, distribution, use, and, at the end, disposal of raw materials. It is feasible to identify and prevent potential environmental threats by reviewing each of these steps, which in turn promotes sustainability and reduces ecological footprints. Doing a comprehensive effect evaluation is one of the main goals of a nanomaterial's lifecycle assessment. Various phases in the lifecycle of a nanomaterial might have unique and noteworthy effects on the environment. For instance, pollution, resource depletion, and habitat loss can arise from the extraction and processing of raw materials. Manufacturing operations have the potential to produce hazardous waste, emit pollutants, and use a lot of energy. It is feasible to identify the locations of the biggest environmental consequences and create effective mitigation plans by looking at the full lifespan. This holistic approach makes sure that advancements made in one area don't unintentionally worsen conditions in another.

Furthermore, unlike bulk materials, nanomaterials can exhibit distinct behaviors and toxicities due to their special qualities, which include their shape, size, area and chemical makeup. Throughout their lifetime, nanomaterials may interact with many environments, changing these qualities. For example, during use, nanomaterials may break down and release nanoparticles into the surrounding area. Both people and wildlife may be at health risk from such emissions.

Researchers and producers can gain insight into how these qualities change over time and impact possible health and environmental issues by performing a lifecycle evaluation. Producing safer nanomaterials and creating suitable handling and disposal protocols require this information. Lifecycle evaluations not only address health and environmental issues, but also encourage resource efficiency. Through an analysis of the material and energy flow during the nanomaterials' lifecycle, possibilities to minimize waste, conserve resources, and enhance energy efficiency can be found. One way to preserve raw materials and lessen the requirement for energy-intensive manufacturing processes is to recycle and reuse nanomaterials when their useful lives are coming to an end. Similarly, the environmental impact can be greatly decreased by streamlining production processes to use less energy and waste (Karn and Bergeson 2009). Thus, lifecycle assessments promote the circular economy and the creation of more environmentally friendly industrial techniques. Regulatory compliance is yet another crucial lifecycle evaluation component. Globally, governments and regulatory agencies are paying more attention to how nanoparticles affect human health and the environment. Thorough lifecycle analyses offer the information required to confirm that nanoparticles satisfy sustainability and safety standards established by regulatory bodies. This helps producers meet current regulatory requirements and gets them ready for new ones down the road. Businesses can lower liability, evade fines from regulators, and win over customers by proactively managing possible risks.

Lifecycle evaluations help with informed decision-making. Stakeholders in the creation, application, and disposal of nanomaterials, such as producers, regulators, consumers, and legislators, depend on precise and thorough data to guide their decisions. Lifecycle assessments offer a comprehensive comprehension of the advantages and disadvantages linked to various materials and procedures. Making decisions that balance social, environmental, and economic factors requires knowledge of this kind. Finally, reducing the environmental impact of nanomaterials requires assessing their lifespan. In particular, end-of-life disposal is a crucial phase where poor management can seriously contaminate the ecosystem. Techniques like recycling, secure disposal, or creating nanoparticles that break down into non-toxic parts can all help to considerably lessen their influence on the environment. Manufacturers may create nanostructures that are not only affordable and efficient, but also environmentally safe by taking into account the complete lifespan. In conclusion, assessing a nanomaterial's whole lifecycle is crucial because it can guarantee thorough impact assessments, detect and reduce risks, encourage resource efficiency, guarantee regulatory compliance, support well-informed decision-making, and reduce environmental impact. The sustainable development and application of nanoparticles depend on this all-encompassing strategy to guarantee that they advance sustainability objectives without inadvertently endangering human health or the environment (Anastas and Eghbali 2010).

The use of green nanotechnology in water treatment is quickly developing, with new approaches that prioritize efficiency and sustainability. The utilization of plant-based materials, ecological nanoparticles, and energy-efficient production techniques are some of the most promising developments. These methods not only improve the efficiency of water filtration but also reduce their negative effects on the environment, which is in line with the objectives of global sustainability.

Making nanoparticles using materials derived from plants is one novel approach. These substances provide a greener substitute for conventional chemical procedures in the environmentally friendly production of nanoparticles by acting as both stabilizing and reducing agents. For instance, extracts from a variety of plants, including neem, green tea, and Moringa oleifera, have been effectively utilized to create gold and silver nanoparticles. These biogenic nanoparticles may successfully remove pathogens from the water and have outstanding antibacterial qualities. By using plant-based resources, the synthesis of nanoparticles has a less overall environmental impact and requires fewer toxic chemicals.

Another noteworthy development in green nanotechnology is biodegradable nanoparticles. The purpose of these nanoparticles is to remove contaminants from the environment and then decompose into non-toxic byproducts, so avoiding secondary pollution. For example, biopolymers that are naturally found in crab shells, such as chitosan, can be used to create nanoparticles that have high potential for adsorbing organic contaminants and heavy metals in water. Because of its biodegradability and strong attraction for contaminants, chitosan is a great option for environmentally friendly water treatment applications (Mittal et al. 2013). Furthermore, biodegradable nanocomposites for the removal of dyes and heavy metals from contaminated water have been created using cellulose nanocrystals obtained from plant biomass.

Innovative approaches in green nanotechnology also include energy-efficient production techniques for nanomaterials. Conventional nanoparticle synthesis is expensive and harmful to the environment since it frequently calls for high pressures, temperatures, and energy inputs. Green synthesis techniques, on the other hand, use renewable energy sources and ambient conditions to create nanomaterials. Utilizing microwave-assisted synthesis is one such technique that drastically cuts down on energy usage and response times. High purity and homogeneity nanoparticles of metals and metal oxides have been produced using this method (Iravani 2011).

7.2.1 Role of Green Nanotechnology in Addressing Water Purification Challenges

A worldwide issue is the availability of safe and clean drinking water. Treatment techniques for water and wastewater include a lot of chemicals, are expensive, and can have negative environmental effects (Naushad et al. 2016). Instead of totally converting the pollutants into an environmentally benign final product, conventional treatment systems only move the contaminants from one phase into another, requiring a significant initial investment, a sizable area, and a high maintenance expense. Currently, wastewater is purified using a wide range of physicochemical techniques, including carbon adsorption, distillation, ion-exchange, reversible osmosis, and nanofiltration. These techniques have a number of significant disadvantages, such as membrane distortion, expensive handling, sludge development, and disposal issues. However, since most of these pollutants need to be treated immediately, non-biodegradable pollutants represent a severe concern. This calls for the use of an advanced, more economical, and environmentally friendly way of treating wastewater (Khan 2020).

Water purification is a global concern as a result of pollution and scarcity, which pose serious hazards to this essential resource for life. Hazardous chemicals, heavy metals, diseases, and plastic debris have all been found contaminating water bodies as a result of industrial activity, agricultural practices, and urbanization. Because of the major health hazards that these contaminants represent to both humans and ecosystems, effective purification technologies are required to guarantee the safety and quality of the water (Khatri and Tyagi 2015). The presence of pollutants including pesticides, pharmacological residues, and chemicals from industry that are difficult for typical treatment procedures to eliminate is one of the main problems with purifying water. For example, materials such as endocrine-disrupting chemicals and microplastics may linger in treated water, with long-term effects on the environment and human health (Shannon et al. 2008). Microplastics are a major issue because of their tiny size and widespread presence in aquatic habitats. They come from a variety of sources, such as cosmetic goods and deteriorated plastic debris. Toxic compounds can be absorbed and concentrated by these particles, which can endanger human health and wildlife by entering the food chain. Protozoa, viruses, and bacteria are examples of pathogens that pose serious obstacles to the purification of water. Waterborne infections can have catastrophic effects on public health; hence it is imperative to prevent them using effective disinfection procedures. While efficient against many diseases, conventional techniques like chlorination may not be enough to inactivate all germs and can result in hazardous byproducts. The situation is made more complex by newly emerging contaminants such bacteria that are resistant to antibiotics, which calls for the creation of more sophisticated and reliable treatment methods.

Apart from pollution, another significant issue is the paucity of water. Due to over-extraction of groundwater, shifting climatic patterns, and inadequate infrastructure, many areas, particularly in developing nations, experience chronic shortages of clean water (Khatri and Tyagi 2015). People are forced to depend on contaminated water sources due to this scarcity, which increases financial hardships and health hazards. Sustainable and reasonably priced purification methods are desperately needed in places where water is limited so that people have consistent access to pure water.

By changing precipitation patterns, climate change makes water scarcity worse by causing extended droughts in some areas and catastrophic flooding in others. It can be extremely difficult to guarantee a consistent supply of pure water because of the potential for these severe weather conditions to interfere with water distribution systems and taint water sources. Moreover, increasing global temperatures cause surface water bodies to evaporate more quickly, which further depletes the supply of water. To overcome these obstacles, creative methods of water filtration are needed.

Reverse osmosis, sophisticated oxidation techniques, and nanofiltration are among the technologies that have demonstrated promise in eliminating a variety of pollutants (Shannon et al. 2008). For example, reverse osmosis works incredibly well at eliminating organic molecules, dissolved salts, and even some kinds of bacteria. Nevertheless, these techniques frequently have significant energy and cost needs, which makes them less practical for general use, especially in low-income areas. Since these technologies increase greenhouse gas emissions, their energy-intensive nature may potentially limit their sustainability. Demand for low-cost, energy-efficient, and simple-to-maintain purifying systems that may be installed in a variety of locations, including isolated villages and urban centers, is rising. The development of new filtering media that may effectively remove impurities at less expense and with reduced energy usage is being driven by advances in material science and nanotechnology. For instance, the large surface area & mechanical strength of graphene-based filters have demonstrated potential, enabling the efficient removal of contaminants while preserving durability and a long service life.

The worldwide scope of issues related to water purification necessitates cooperative efforts in policy-making, technical development, and research. Collaboratively, engineers, scientists, and policymakers must devise and execute strategies that tackle pollution and scarcity. This entails funding studies to comprehend the characteristics and effects of newly developing pollutants as well as encouraging global collaboration to exchange information and resources. Long-term water security requires policies that support sustainable water management techniques and provide incentives for the use of cutting-edge purification technologies.

Nanoscale particles with a strong affinity for adsorbing chemicals, derived from either organic or inorganic materials, are known as nano adsorbents. Nano adsorbents not only have the ability to sequester pollutants with varying size of molecules, hydrophobicity, and speciation actions due to their active surface, small size, large porosity, but they also facilitate the efficient consumption of raw materials by the manufacturing process without releasing their toxic payload (Pacheco et al. 2006). In addition to their quick rate of operation, nano adsorbents feature significant pollutant-binding capabilities. When they run out, they may also be chemically renewed (Yang and Xing 2007). These factors have contributed to the sharp rise in global interest in nanotechnology among academics. Materials exhibit distinct properties at the nanoscale, and due to their small size, they have a high surface area and "surface area to volume ratio." These features increase the nanoparticles' ability to adsorb substances. These particles exhibit special qualities, such as strong reactivity and catalytic potential, in addition to their enormous surface area, which makes them superior adsorbing materials than ordinary materials. Nanoparticles have more active locations for interacting with various chemical species due to their large surface area. Nanoparticles are emerging as new wastewater treatment solutions in order to get better outcomes on the removal of contaminants from wastewater (Santhosh et al. 2016; Yusuf et al. 2020; Mittal et al. 2013).

Nano photocatalysts, nano reductives, nanomembranes, and nano sorbents are the four main types of nanomaterial that can play a crucial part in the wastewater treatment process (Fig. 7.2). The distinctive attributes of nanomaterials, such as

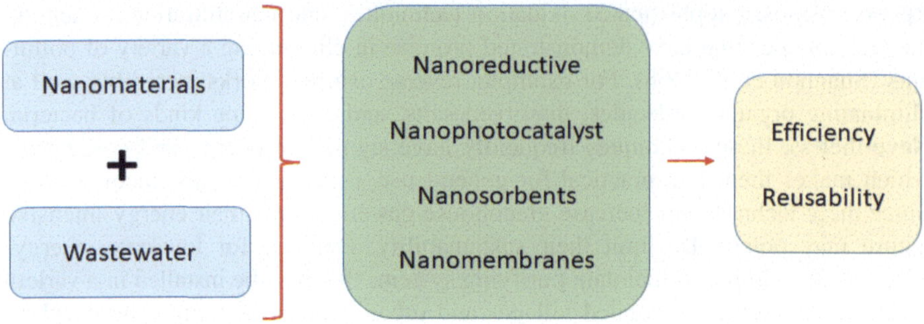

Fig. 7.2 Nanomaterials and processes in the waste water treatment. (Mbarek et al. 2022)

their high surface area, porosity, high reactivity, strong mechanical and thermal properties, ease of dispersion, and hydrophobic/hydrophilic characteristics, have been demonstrated by researchers to be suitable candidates for wastewater technology (Mbarek et al. 2022). There are different types of nanomaterials used in water purification some of them are discussed in the following sections.

7.3 Green Nanomaterials for Water Purification

Green nanomaterials offer an innovative and sustainable approach to water purification. Green nanomaterials are increasingly being explored for their potential in water purification, as they provide an environmentally friendly alternative to traditional methods. Furthermore, green nanomaterials enhance water purification efficiency by selectively capturing contaminants, reducing chemical usage, and minimizing waste, thus offering a more cost-effective and sustainable approach for addressing the challenges of water scarcity (De et al. 2022; Nasrollahzadeh et al. 2021a, b; Neeti et al. 2023; Romanovski 2021a, b; Tewari et al. 2022). The potential of green nanomaterials to address the issue of water scarcity has led to increased research in the field and the development of innovative applications. As a result, the use of green nanomaterials in water purification and desalination processes has gained significant attention, with promising results in improving water quality and accessibility, particularly in regions facing water scarcity (Ali and Hassaan 2017; Alsaiari et al. 2023; Neeti et al. 2023; Romanovski 2021a, b; Shahid-ul-Islam et al. 2023). Various types of green nanomaterials are used for water purification, such as metal nanoparticles (silver, Iron), metal oxide nanoparticles, carbon nanotubes and biopolymeric nanoparticles (cellulose, lignin, chitosan, starch and alginate) have demonstrated their effectiveness in removing contaminants and improving the taste and odor of water (Bhatt et al. 2023; Nasrollahzadeh et al. 2021a, b; Qamar et al. 2022; Sajjadi et al. 2021). Among them biopolymer-based nanomaterials are gaining attention in water purification due to their renewable nature, low toxicity, and environmentally friendly characteristics. These nanomaterials, derived from various natural and renewable sources, offer unique properties that can be leveraged to address different water contaminants. In addition to their ability to remove heavy metals, these plant-derived nanomaterials also possess antimicrobial properties, making them effective in the removal of bacterial contaminants in water sources.

7.3.1 Biopolymer Based Nanomaterials for Water Purification

Biopolymers are naturally occurring polymers derived from biological sources, including plants, animals, and microorganisms (Bhatt et al. 2023; Nasrollahzadeh et al. 2021a, b; Qamar et al. 2022; Sajjadi et al. 2021). Nanomaterial based on polysaccharide entities have been investigated as novel and more efficient (nano) catalysts or sorbents in eliminating an array of aqueous pollutants and contaminants, including ionic metals and organic/inorganic pollutants from wastewater. In this respect, cellulose, lignin, starch, chitin and alginate are important examples of natural, renewable, sustainable and environmental-friendly organic biopolymers (Fig. 7.3) preferred for water purification. All these biopolymeric materials possess unique properties due to their nanoscale dimensions, including high surface area-to-volume ratio, enhanced reactivity, and tunable physical and chemical characteristics.

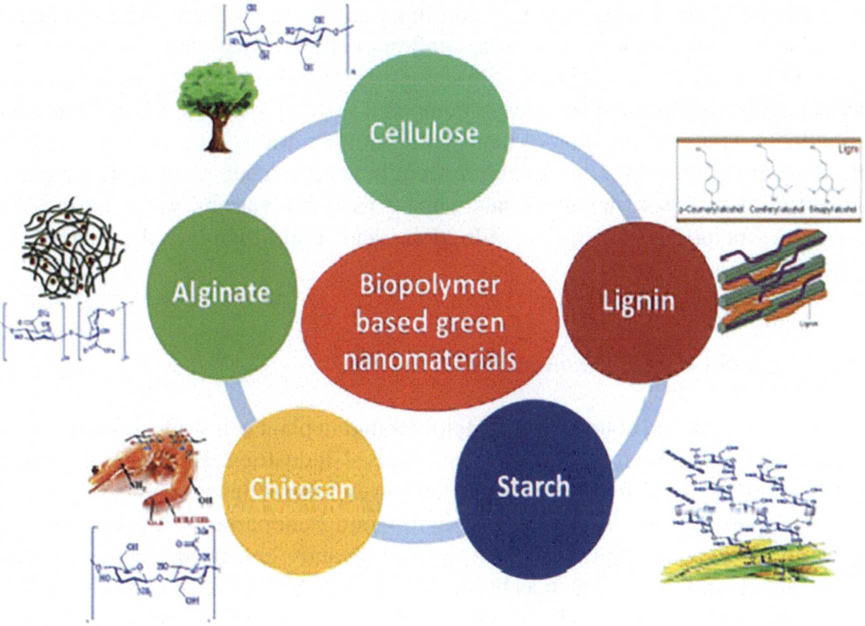

Fig. 7.3 Sustainable and environmental-friendly biopolymer-based nanomaterials for water purification. (Nasrollahzadeh et al. 2021a, b)

When combined with biopolymers, these nanomaterials can significantly enhance water purification processes by improving adsorption capacity, catalytic activity, and mechanical strength. These materials are biodegradable, renewable, and often exhibit functional properties such as adsorption, flocculation, and antimicrobial activity, making them suitable for water purification applications (Udayakumar et al. 2021). Herein, the following chapter section summarizes the remediation and elimination of aqueous pollutants and toxic contaminants from water using some of the most abundant natural biopolymer resources, namely cellulose, starch, lignin, chitin, starch and alginate-based nanomaterials.

7.3.2 Types of Biopolymer-Based Nanomaterials

7.3.2.1 Cellulose Based Nanomaterials

Cellulose, a natural polymer found in the cell walls of plants, is the most abundant biopolymer on Earth. It consists of linear chains of glucose molecules linked by $\beta(1\rightarrow4)$ glycosidic bonds, forming a crystalline. It can be transformed into nanocellulose with exceptional mechanical properties and high surface area. Nanocellulose includes cellulose nanocrystals (CNCs), cellulose nanofibrils (CNFs) and bacterial nanocellulose (BNC), which are obtained by breaking down cellulose fibers into nanoscale particles. Their natural abundance, renewability, environmental friendliness, and versatile functionality make them ideal candidates for addressing the challenges of water pollution (Fig. 7.4) It also possesses some interesting features such as high tensile strength, high stiffness, high flexibility, good thermal and visco-elastic properties as compared to other synthetic fibers (Chakkour et al. 2023). The incorporation of nanocellulose into various biopolymeric matrices offer great possibilities for the development of novel green nanocomposite materials. These Cellulose-based nanomaterials can be used for various water purification processes, including adsorption, filtration, and photocatalytic degradation of pollutants.

7.3.2.2 Lignin Based Nanomaterials

Lignin, a complex and abundant biopolymer found in plant cell walls, is traditionally considered a byproduct of the paper and biofuel industries. However, its unique chemical structure and functional properties have garnered interest in its application for water purification. By converting lignin into nanoparticles, its efficacy and versatility as a water treatment agent are significantly enhanced. Lignin contains various functional groups, such as hydroxyl, methoxyl, and carboxyl groups, which can interact with pollutants. These functional groups can chelate with heavy metal ions, making lignin nanoparticles effective adsorbents for removing metals like lead,

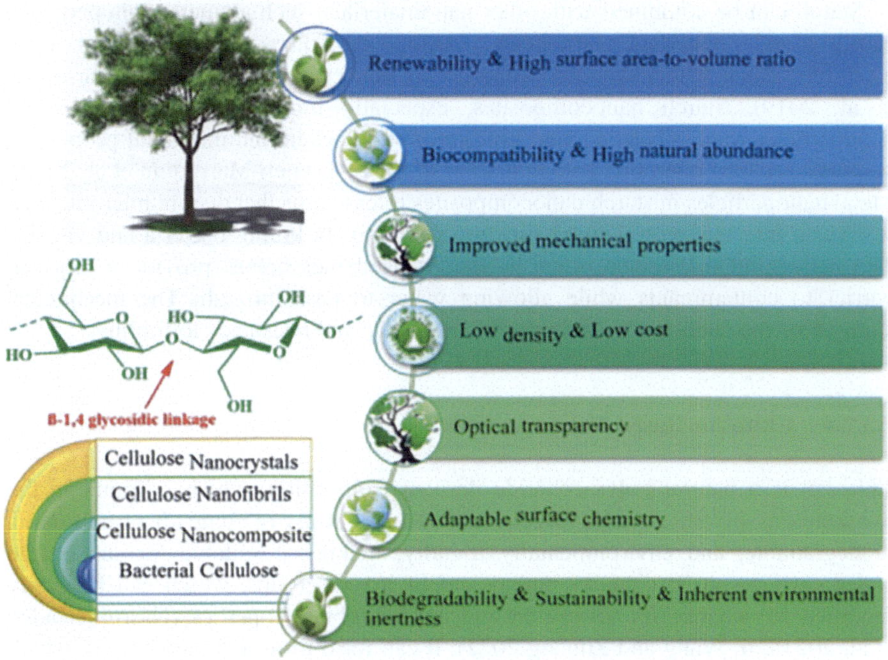

Fig. 7.4 Outstanding properties of cellulosic nanomaterials. (Nasrollahzadeh et al. 2021a, b)

cadmium, and mercury and various dyes from industrial effluents, including methylene blue from water (Rivière et al. 2020; Santander et al. 2021; Zhang et al. 2019).

7.3.2.3 Starch-Based Nanomaterials

Starch, a polysaccharide composed of glucose units, is derived from various plants, including corn, potatoes, and rice. It consists of two main components: amylose (linear chains) and amylopectin (branched chains). Starch nanoparticles (SNPs) can be prepared by breaking down starch into nanoscale particles using methods like nanoprecipitation, ultrasonication, and high-pressure homogenization. SNPs have high surface area and can be functionalized for specific applications. The high surface area and tunable surface chemistry of SNPs allow for efficient adsorption of a wide range of pollutants (Nasrollahzadeh et al. 2021a, b; Wang et al. 2021). Chemical modifications can enhance specific interactions with target contaminants. Chemically modified SNPs can adsorb heavy metals such as lead, cadmium, and arsenic (Nasrollahzadeh et al. 2021a, b). Functional groups like carboxyl, amino, and thiol can be introduced to enhance binding affinity's can also remove dyes and organic pollutants from industrial effluents through physical adsorption and chemical interactions (Azmana et al. 2021; Chen et al. 2019; Wang et al. 2021).

Starch can be combined with other nanomaterials, such as metal nanoparticles, carbon-based materials (graphene oxide), or other polymers, to form nanocomposites with enhanced properties (Chen et al. 2019; Lomelí-Marroquín et al. 2019). Starch nanocomposites, especially those combined with metal nanoparticles (e.g., silver, copper, gold) exhibit excellent antimicrobial properties, helping to reduce microbial contamination in water (Lomelí-Marroquín et al. 2019). Metal nanoparticles in starch nanocomposites release ions that disrupt microbial cell walls and interfere with cellular functions, effectively killing bacteria and viruses. The nano- and microscale pores in starch-based membranes provide a physical barrier to contaminants while allowing water to pass through. The mechanical strength of the nanocomposites ensures durability and resistance to fouling.

7.3.2.4 Chitosan-Based Nanomaterials

Chitosan is a natural polysaccharide derived from chitin, which is found in the exoskeletons of crustaceans, insects, and the cell walls of fungi. It is non-toxic, biodegradable, and environmentally friendly, making it suitable for sustainable water purification. It contains amino and hydroxyl groups, which can be easily modified to enhance its adsorption and antimicrobial properties (Nasrollahzadeh et al. 2021a, b; Wang and Zhuang 2022). It can form films and membranes, which are useful in filtration applications. Chitosan-based nanomaterials can be synthesized using various methods such as nanoprecipitation, ionic gelation, emulsion techniques and electrospinning. These nanomaterials are utilized in various water purification processes, including adsorption, filtration, and antimicrobial activity. Chitosan nanoparticles and nanocomposites can adsorb heavy metals such as lead, cadmium, and mercury due to their amino and hydroxyl groups, which bind with metal ions (Nasrollahzadeh et al. 2021a, b; Thakur and Voicu 2016). These nanomaterials can also remove dyes and organic pollutants from wastewater through electrostatic interactions and chemical bonding. Chitosan-based nanofiber membranes offer high porosity, mechanical strength, and chemical stability. They effectively filter out bacteria, viruses, and other microorganisms, as well as particulate matter. Chitosan exhibits inherent antimicrobial properties, which can be enhanced by incorporating metal nanoparticles (e.g., silver, copper) or other antimicrobial agents (Morsi et al. 2017). These nanocomposites effectively reduce microbial contamination in water (Babaee et al. 2022; Nasrollahzadeh et al. 2021a, b).

7.3.2.5 Alginate-Based Nanomaterials

Alginate is a natural polysaccharide and extracted mainly from brown seaweed by alkali treatment. It is a linear copolymer composed of (1–4)-linked β-D-mannuronate (M) and α-L-guluronate (G) residues, covalently linked together in different sequences, which determine the physico-chemical properties of alginate (Deepa et al. 2016; Nasrollahzadeh et al. 2021a, b). Alginate can be crosslinked with the divalent cations such as Ca^{2+} for the preparation of hydrogels because of the presence of abundant carboxyl groups in the molecular chains (Naseri et al. 2016). The gelling property of alginate in an aqueous medium is due to the ionic interaction between the carboxylic group and divalent cations (Deepa et al. 2016; Nasrollahzadeh et al. 2021a, b). Alginate can be used in various water purification processes, including adsorption, filtration, and flocculation. Alginate beads and gels can adsorb heavy metals such as lead, cadmium, and copper due to their carboxyl groups, which bind with metal ions (Wang et al. 2016). Alginate-based adsorbents can be further functionalized with other materials, like activated carbon or metal oxides, to enhance adsorption capacity (Amiri et al. 2020). Alginate-based nanocomposites, including alginate-clay and alginate-metal oxide nanoparticles, are employed in water purification for their high adsorption capacity and ability to remove a wide range of pollutants (Amiri et al. 2020; Asadi Haris et al. 2023; Kausar et al. 2020).

7.4 Green Nanotechnology Based Water Treatment Techniques

Green nanotechnology offers several solutions for water treatment, which offers efficient, sustainable, and eco-friendly paths. Some of them are membrane nanofiltration, adsorption, photocatalysis, biologically mediated processes etc. (Rabiee et al. 2023; Rethinam et al. 2020).

7.4.1 Membrane Nanofiltration

In this technique the pore size of membranes plays an important role, and the pore size should be in the nanometer range. These membranes filter the contaminants at the molecular level. High removal efficiency and lower energy consumption are the advantages of this method. Fabrication of green nanomembranes control the usage of petroleum-based membranes (Rethinam et al. 2020). The green nanomembranes can be prepared from plants, biopolymers and microorganisms (Rabiee et al. 2023). Figure 7.5 shows the schematic representation of some methods for preparation of green nanomembranes.

Biopolymeric materials have widely used to remove contaminants from water. Chitosan, alginate, cellulose etc. based membranes were widely used for the removal of heavy metal ions (Mohammed et al. 2023), organic compounds, bacteria and viruses from water. Reinforcement by nanofibers based on biopolymers offer significant benefits for enhancing the stability and durability of membranes used in water treatment. These materials not only improve the membranes' contaminant removal capabilities but also contribute to their mechanical strength.

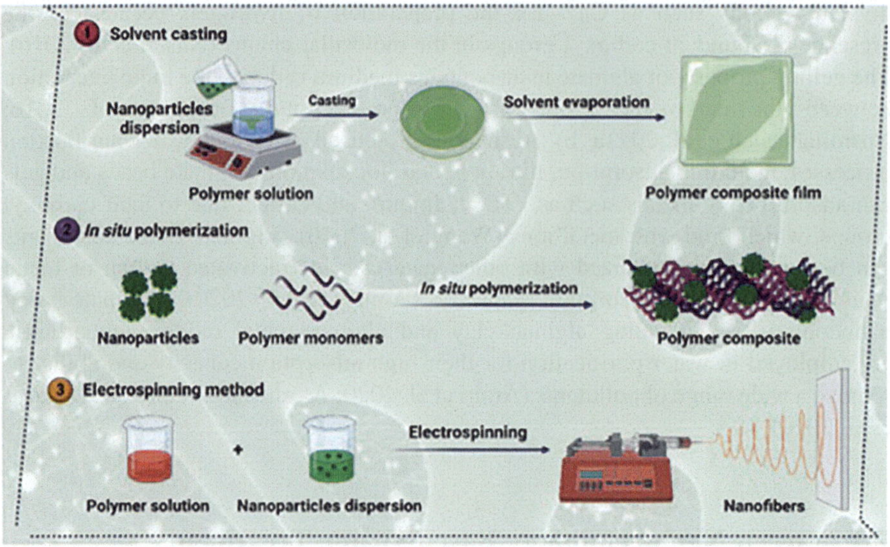

Fig. 7.5 Schematic representation of some methods for preparation of green nanomembranes. (Rabiee et al. 2023)

Koriem et al. (2022) fabricated eco-friendly cellulose acetate membranes by electrospinning method for dye removal. They enhanced the performance of the membrane by incorporating nano activated carbon as a nanofiller into the membrane. Fu et al. (2019) established the synthesis of chitosan-functionalized activated coke for incorporating gold (Au) nanoparticles. Chitosan act as a reductant, linker, and stabilizer for gold nanoparticles.

Metal-Organic Framework (MOF) membranes have significant impact in the field of wastewater treatment due to their unique structural properties like porous nature, selective permeation abilities, and high surface area make them particularly suitable for applications in water filtration and purification. MOF membranes are effectively removing heavy metals and organic contaminants from wastewater. Their ability to selectively permeate water molecules while rejecting salt ions makes them a promising alternative to traditional desalination technologies. The porous structure and functional groups on MOF membranes can be designed to capture and remove pathogens, including bacteria and viruses, ensuring safer drinking water.

Recent advancements involve the development of hybrid membranes that combine MOFs with other materials like polymers or nanomaterials. Modifying MOF membranes with specific functional groups can improve their selectivity and adsorption capacities for targeted contaminants. Functionalized MOFs can be tailored to address specific water treatment challenges (Rego et al. 2021; Yu et al. 2021).

Kebria et al. (2019) a thin film composite (TFC) membrane by coating polyvinylidene fluoride (PVDF) with an ultrathin layer of ZIF-8/chitosan for use in membrane distillation (MD) for water desalination. The liquid entry pressure of water for the TFC membrane increased considerably. The increase in LEP can be attributed to the enhanced surface porosity and the decreased mean surface pore size due to the ZIF-8/chitosan coating.

7.4.2 Adsorption

Adsorption is a surface phenomenon that involves the transfer of a molecule or ion (adsorbate) from a liquid or gaseous bulk phase onto a solid surface (adsorbent). As a result, formation of a monomolecular layer on the surface of the adsorbent occurs via physicochemical or chemical interactions under specific conditions (Jain et al. 2021). Biosorption is a special case of adsorption where biological materials like bacteria, algae, or fungi act as adsorbents. These biological materials have an intrinsic ability to bind and collect heavy metals from even very dilute aqueous solutions (Jain et al. 2021).

Adsorption is often preferred over conventional metal removal techniques due to several significant advantages like cost-effective, selective removal of specific metals, chemical efficiency, regenerative etc. Nanoparticles are widely used for adsorption due to non-toxicity, high adsorption capacity and recyclability.

Metal oxide nanoparticles (MONPs) are potential adsorbents due to their high adsorption capacity, low cost, simplicity of separation, and improved stability.

Alumina nanoparticles (Al_2O_3NPs), exhibits numerous beneficial properties such as a large surface area, low cost, high surface reactivity, and significant adsorption ability. Its nanoparticles are intrinsically hydrophilic, making surface modification essential for enhancing their functionality (Koopi and Buazar 2018). Koopi and Buazar (2018) employed the brown marine alga *Sargassum ilicifolium* for the green synthesis of Al_2O_3NPs. The hydroxyl and carbonyl groups present in the *S. ilicifolium* extract added to the functionalization of the Al_2O_3NPs, enhancing their functional properties. The high adsorption ability of Al_2O_3NPs makes them ideal for removing contaminants such as heavy metals, dyes, and organic pollutants from wastewater.

Khilji et al. (2022) reported ZnO nanoparticles (ZnONPs) derived from *Oedogonium sp.* algae for their ability to eliminate heavy metals from wastewater used in leather production. The significant reduction in heavy metal concentrations and other contaminants demonstrates the high efficiency of ZnONPs. The use of algae for nanoparticle production supports environmentally friendly practices, making this method sustainable and beneficial for large-scale industrial applications.

7.4.3 Photocatalysis

Photocatalysis is a process that utilizes UV or visible light, to activate a photocatalyst, which then facilitates the breakdown of pollutants in water. In photocatalysis the catalyst absorbs light energy, generating charge carriers. These charge carriers breaking down contaminants into less harmful substances such as water and carbon dioxide. Titanium dioxide (TiO_2), zinc oxide (ZnO), copper oxide (CuO), iron oxide (Fe_2O_3) nanoparticles are mainly used for photocatalytic degradation (Ghamarpoor et al. 2023; Goudjil et al. 2024; Khasawneh and Palaniandy 2021; Mrabet et al. 2023; Umar and Aziz 2013; Yuan et al. 2013). Photocatalysis offers sustainable, efficient, and effective solutions for tackling water pollution.

Bouafia et al. (2023) utilized porous nanoparticles for the degradation of hydrocarbons in oily water and the removal of Total Suspended Solids (TSS) and heavy metals such as Bismuth (Bi), Chromium (Cr), Molybdenum (Mo), Antimony (Sb), and Selenium (Se). The photocatalytic degradation of oily water saturated with hydrocarbons and total suspended solids (TSS) associated with wastewater from crude oil production was studied using several nanoparticles, including α-Fe_2O_3, CuO, and ZnO, prepared by green synthesis. The nanoparticles α-Fe_2O_3, CuO, and ZnO were highly efficient in degrading hydrocarbons and removing TSS from the wastewater. The total photo-mineralization of these organics to carbon dioxide (CO_2) and water was achieved within 1 h in an air-equilibrated solution. In addition to hydrocarbon degradation and TSS removal, these nanoparticles effectively removed heavy metals from the wastewater.

ZnONPs demonstrated significant photocatalytic activity in degrading methylene blue (MB) and removal of Cr (VI) (Elhousseini et al. 2020). The use of *Ulva fasciata* Delile extract as a bio reducing and stabilizing agent for the synthesis of ZnONPs (Fouda et al. 2022) provides excellent photocatalytic and antibacterial properties, along with a high capacity for removing heavy metals from wastewater.

7.4.4 Biologically Mediated Processes

Biofilm-based systems are biologically active filters which utilize biofilms of microorganisms that degrade organic contaminants and remove pathogens. Biofilms produced from biopolymers have great potential in various fields due to their biodegradability and high biocompatibility. The extracellular polymeric substances (EPS) in biofilms can vary significantly depending on the microorganism and environmental conditions.

The presence of polysaccharides in the EPS matrix, have the capacity to chelate heavy metals from waste water (Huang et al. 2022). Many bacterial biofilms have been used to eliminate harmful heavy metal ions and xenobiotic chemicals (Saini et al. 2023). Industrial wastewater often contains heavy metals like manganese (Mn) and zinc (Zn), which are unsafe to both the environment and public health. Certain bacterial isolates have demonstrated the ability to tolerate high concentrations of such heavy metal ions (Pani et al. 2017).

Phytoremediation is an environmentally friendly and cost-effective biologically mediated process in which aquatic plants absorb and accumulate heavy metals and nutrients from water. Aquatic plants absorb heavy metals and nutrients through their roots and store it into their shoots and leaves. Water Hyacinth (*Eichhornia crassipes*), duckweed (*Lemna minor*), cattail (*Typha spp.*) and water lettuce (*Pistia stratiotes*) are some examples of aquatic plants which are capable of effective removal of a wide range of contaminants including heavy metals and organic pollutants. Phyto-extraction, phytovolatilization, phyto-stabilization, phyto-reduction, phyto-filtration are some important mechanisms of phytoremediation.

Enzymes often immobilized on nanomaterials, can degrade organic pollutants like dyes and pharmaceuticals. This technique enhances the stability, reusability, and catalytic efficiency of enzymes, making them highly effective for wastewater treatment. Laccase immobilized on TiO_2 nanoparticles used for the degradation of synthetic dyes in textile wastewater (Maryskova et al. 2021; Mohajershojaei et al. 2015). Horseradish peroxidase (HRP) was immobilized on multi-walled carbon nanotubes (MWCNTs) and used to treat water samples containing phenolic compounds and pharmaceutical residues. The immobilized HRP on MWCNTs demonstrated high degradation efficiency (Maryskova et al. 2021).

7.5 Environmental Impact and Safety

The availability of clean and safe drinking water is indeed a global challenge. Wastewater treatment methods, often involve the use of chemicals that can be both costly and harmful to the environment. This has led to an increasing interest in alternative methods, such as green nanotechnology. Green nanotechnology for water purification has gained attention due to its potential to address water pollution while minimizing adverse environmental impacts. Resource efficiency reduced chemical use and waste minimization are some of the environmental impact of green nanotechnology.

Conventional methods usually involve the use of chemicals like chlorine, chloramines, or ozone for disinfection. These chemicals can generate harmful by products such as trihalomethanes (THMs) and haloacetic acids (HAAs) that pose risks to human health and the environment. Chemicals like aluminium sulphate and ferric chloride are often used in coagulation and flocculation procedures to remove suspended particles. Nanomaterials have gained widespread attention in the field of environmental clean-up and remediation due their unique properties, including high surface reactivity and a large surface-to-volume ratio, which enhance their effectiveness in removing contaminants. Residual chemicals can stay in treated water and may create environmental pollution when discharged while green synthesis of nanomaterials minimize the use of harmful chemicals. For example, silver nanoparticles can be biosynthesized using plant extract, reducing the need for toxic reducing agents (Keat et al. 2015; Naseem et al. 2020). Nanomaterials like nanocellulose (Das et al. 2022), nano chitosan (Isibor et al. 2024; Olajire and Bamigbade 2021) and graphene oxide (Coros et al. 2020; Mahmoud 2020), when produced through green methods, are usually more biocompatible and biodegradable, creating less hazard to the environment and human health.

Renewable materials like plant-based polymers can be used to reduce raw material costs and reliance on non-renewable resources. Many green nanomaterials can be recovered and reused and thus, reducing the overall material costs. Magnetic nanoparticles, for example, can be easily separated from treated water using a magnet and reused. Green synthetic approaches synthesizing iron nanoparticles using green methods generates fewer toxic byproducts. Iron nanoparticles with diverse morphology and size have been synthesized using biological extracts by Kumar et al. (2023). Microbial (bacteria, fungi, algae etc.) and plant extracts have been used in green synthesis of iron nanoparticles due to the presence of various metabolites and biomolecules.

Very close approaches are needed to evaluate the long-term environmental and health impacts of nanomaterials. Research is ongoing to develop guidelines for the safe use and disposal of nanomaterials in wastewater treatment. Understanding the lifecycle of nanomaterials, from production to disposal, is crucial for assessing their sustainability and minimizing potential environmental risks.

7.6 Conclusion and Future Perspectives

Worldwide population growth and climate change provides the need for novel, innovative water purification technologies to satisfy the requirement for clean and safe water while safe guarding the eco-friendly requirements. The upcoming issues regarding the use of nanotechnology methods for the remediation of waste water opens a new window towards sustainably tackling the water challenges faced globally. The forthcomings of traditional purification methods are mitigated through this environmentally benign nanotechnology derived methods for efficient removal of contaminants. Combining and utilizing the unique properties of nanomaterials can result in novel engineered materials that can compete with the conventional water treatment technologies. Different nano engineered materials, such as nano adsorbents, nano membranes, nano metals, and photocatalysts, provide the potential for novel water treatment technologies utilizing the properties of nanostructured materials like small size, high surface area, high specificity, more surface modifications etc. Furthermore, various cost-effective and robust technologies need to be developed, focusing on improving the present drawbacks, especially the economic reasons in the area of nanotechnology based water purification. Additionally, various studies have to be conducted to incorporate nanomaterials into the existing water purification treatment systems. Also, the stability of nanomaterials in both natural and engineered systems have to be examined before implementation. However, proper research and development into this field emphasizing the green principles hold a lot of promise for cleaner water and healthier environment.

The greener and bio-synthesized versions of nanoparticles, in the form of nanocatalysts and nanoadsorbents, nanometals, nanomembranes exhibit significant promise for wastewater treatment for the removal of contaminants while also reducing toxicity. Water and wastewater treatment alternatives are offered by the rapidly evolving fields of nanotechnology and nanoscience research. The degradation and remediation of hazardous organic and inorganic pollutants have recently focused on nanostructured materials due to their special physicochemical qualities, which include high catalytic activity, high physical/chemical and thermal stability, large specific surface area, significant chemical reactivity, and strong electron transfer ability, among others. Chemically modified nanoadsorbents with high surface area offer excellent water treatment ability. Although, the use of nanomaterials for water treatment purposes have increased and its low production cost serves as an advantage, further research should focus on improving these nanomaterials economic feasibility and assessing their interactions in water systems. Issues pertaining to the toxicity and biosafety standards of the nanomaterial adsorbents need to be comprehensively evaluated and elaborative studies on their mechanistic approaches need to be done. More thorough research is still needed to determine the green fabricated nanocatalysts and nanomaterials recyclability as well as their low cost, high adsorption capacity, and high selectivity of the manufacturing process (Nasrollahzadeh et al. 2021b).

The major challenges associated to nanotechnology-based water treatment methods include high cost, technical constraints and certain potential risk to humans and also the environment. The nanoparticles involved in water treatment can lead to accumulation in the environment for longer periods of time. Also, due to their high specific surface area, nanoparticles have a tendency to agglomerate and can be prevented by adding proper surfactants and after adsorption, the nanoparticles can be removed by sedimentation, filtration or flocculation (Gehrke et al. 2015). Nevertheless, nanotechnology based water purification systems need to be more innovatively developed enough to challenge the conventional methods by providing more robust and sustainable solutions.

References

Adesina OB, William C, Oke EI (2024) Evolution in water treatment: exploring traditional self-purification methods and emerging technologies for drinking water and wastewater treatment: a review. World News Nat Sci 53(February):169–185

Ahuja S (2025) Chapter 2—global water challenges and solutions. In: Advances in water purification techniques. Elsevier Inc. https://doi.org/10.1016/B978-0-12-814790-0.00002-8

Ali HR, Hassaan MA (2017) Applications of bio-waste materials as green synthesis of nanoparticles and water purification. Adv Mater 10(1):85–101

Amiri S, Asghari A, Vatanpour V, Rajabi M (2020) Fabrication and characterization of a novel polyvinyl alcohol-graphene oxide-sodium alginate nanocomposite hydrogel blended PES nanofiltration membrane for improved water purification. Sep Purif Technol 250:117216

Anastas P, Eghbali N (2010) Green chemistry: principles and practice. Chem Soc Rev 39(1): 301–312

Asadi Haris S, Dabagh S, Mollasalehi H, Ertas YN (2023) Alginate coated superparamagnetic iron oxide nanoparticles as nanocomposite adsorbents for arsenic removal from aqueous solutions. Sep Purif Technol 310:123193. https://doi.org/10.1016/J.SEPPUR.2023.123193

Azmana M, Mahmood S, Hilles AR, Rahman A, Arifin MAB, Ahmed S (2021) A review on chitosan and chitosan-based bionanocomposites: promising material for combatting global issues and its applications. Int J Biol Macromol 185:832–848

Babaee M, Garavand F, Rehman A, Jafarazadeh S, Amini E, Cacciotti I (2022) Biodegradability, physical, mechanical and antimicrobial attributes of starch nanocomposites containing chitosan nanoparticles. Int J Biol Macromol 195:49–58

Bhatt P, Joshi S, Bayram GMU, Khati P, Simsek H (2023) Developments and application of chitosan-based adsorbents for wastewater treatments. Environ Res 226:115530

Bouafia A, Meneceur S, Chami S, Laouini SE, Daoudi H, Legmairi S, Mohammed HA, Aoun N, Menaa F (2023) Removal of hydrocarbons and heavy metals from petroleum water by modern green nanotechnology methods. Sci Rep 13(1):5637

Chakkour M, Ould Moussa M, Khay I, Balli M, Ben Zineb T (2023) Towards widespread properties of cellulosic fibers composites: a comprehensive review. J Reinf Plast Compos 42(5–6): 222–263

Chaturvedi S, Dave PN (2019) Water purification using nanotechnology an emerging opportunities. Chem Methodol 3:115–144. https://doi.org/10.22034/CHEMM.2018.143461.1069

Chelu M, Musuc AM, Popa M, Moreno JMC (2023) Chitosan hydrogels for water purification applications. Gels 9(8):664

Chen Y, Dai G, Gao Q (2019) Starch nanoparticles-graphene aerogels with high supercapacitor performance and efficient adsorption. ACS Sustain Chem Eng 7(16):14064–14073. https://doi.org/10.1021/ACSSUSCHEMENG.9B02594/SUPPL_FILE/SC9B02594_SI_001.PDF

Coros M, Pogacean F, Turza A, Dan M, Berghian-Grosan C, Pana I-O, Pruneanu S (2020) Green synthesis, characterization and potential application of reduced graphene oxide. Phys E 119: 113971

Das R, Lindstrom T, Sharma PR, Chi K, Hsiao BS (2022) Nanocellulose for sustainable water purification. Chem Rev 122(9):8936–9031

De A, Singh NB, Guin M, Barthwal S (2022) Water purification by green synthesized nanomaterials. Curr Pharm Biotechnol 24(1):101–117. https://doi.org/10.2174/1389201023666220507030548

Deepa B, Abraham E, Pothan LA, Cordeiro N, Faria M, Thomas S (2016) Biodegradable nanocomposite films based on sodium alginate and cellulose nanofibrils. Materials 9(1):50. https://doi.org/10.3390/MA9010050

Domb AJ (2017) Nanotechnology for water. In: Water purification. Elsevier Inc. https://doi.org/10.1016/B978-0-12-804300-4/00002-2

Elhousseini MH, Isık T, Kap Ö, Verpoort F, Horzum N (2020) Dual remediation of waste waters from methylene blue and chromium (VI) using thermally induced ZnO nanofibers. Appl Surf Sci 514:145939

Fouda A, Eid AM, Abdelkareem A, Said HA, El-Belely EF, Alkhalifah DHM, Alshallash KS, Hassan SE-D (2022) Phyco-synthesized zinc oxide nanoparticles using marine macroalgae, Ulva fasciata Delile, characterization, antibacterial activity, photocatalysis, and tanning wastewater treatment. Catalysts 12(7):756

Fu Y, Qin L, Huang D, Zeng G, Lai C, Li B, He J, Yi H, Zhang M, Cheng M (2019) Chitosan functionalized activated coke for Au nanoparticles anchoring: green synthesis and catalytic activities in hydrogenation of nitrophenols and azo dyes. Appl Catal B Environ 255:117740

Gehrke I, Geiser A, Somborn-schulz A, Gehrke I, Somborn-schulz A (2015) Innovations in nanotechnology for water treatment innovations in nanotechnology for water treatment. Nanotechnol Sci Appl 8:1–17. https://doi.org/10.2147/NSA.S43773

Ghamarpoor R, Jamshidi M, Fallah A, Eftekharipour F (2023) Preparation of dual-use GPTES@ZnO photocatalyst from waste warm filter cake and evaluation of its synergic photocatalytic degradation for air-water purification. J Environ Manag 342:118352

Gitis V, Hankins N (2018) Water treatment chemicals: trends and challenges. J Water Process Eng 25(May):34–38. https://doi.org/10.1016/j.jwpe.2018.06.003

Goudjil MB, Dali H, Zighmi S, Mahcene Z, Bencheikh SE (2024) Photocatalytic degradation of methylene blue dye with biosynthesized Hematite α-Fe2O3 nanoparticles under UV-irradiation. Desalin Water Treat 317:100079

Gupta M, Savla N, Pandit C, Pandit S, Kumar P, Pant M (2022) Use of biomass-derived biochar in wastewater treatment and power production: a promising solution for a sustainable environment. Sci Total Environ 825:153892. https://doi.org/10.1016/j.scitotenv.2022.153892

Gurtu MG, Sharma M (2021) Major applications of green nanotechnology in sustainable development and its impact on global environment: a review. Nat Vol Essent Oil 8(4):12823–12851

Huang L, Jin Y, Zhou D, Liu L, Huang S, Zhao Y, Chen Y (2022) A review of the role of extracellular polymeric substances (EPS) in wastewater treatment systems. Int J Environ Res Public Health 19(19):12191

Iravani S (2011) Green synthesis of metal nanoparticles using plants. Green Chem 13(10): 2638–2650. https://doi.org/10.1039/C1GC15386B

Isibor PO, Agbontaen DO, Adebayo OO (2024) Nanochitosan-based water-quality enhancement. In: Nanochitosan-based enhancement of fisheries and aquaculture: aligning with sustainable development goal 14–life below water. Springer, pp 159–179

Jain K, Patel AS, Pardhi VP, Flora SJS (2021) Nanotechnology in wastewater management: a new paradigm towards wastewater treatment. Molecules 26(6):1797

Journal TA, Reuse W (2019) Nanotechnology for water purification: electrospun nanofibrous membrane in water and wastewater treatment I. Tlili and Tawfeeq Abdullah Alkanhal. J Water Reuse Desalin 9(13):232–248. https://doi.org/10.2166/wrd.2019.057

Karn BP, Bergeson LL (2009) Green nanotechnology: straddling promise and uncertainty. Nat Resour Environ 24:9

Kausar A, Sher F, Hazafa A, Javed A, Sillanpää M, Iqbal M (2020) Biocomposite of sodium-alginate with acidified clay for wastewater treatment: kinetic, equilibrium and thermodynamic studies. Int J Biol Macromol 161:1272–1285

Keat CL, Aziz A, Eid AM, Elmarzugi NA (2015) Biosynthesis of nanoparticles and silver nanoparticles. Bioresour Bioprocess 2:1–11

Kebria MRS, Rahimpour A, Bakeri G, Abedini R (2019) Experimental and theoretical investigation of thin ZIF-8/chitosan coated layer on air gap membrane distillation performance of PVDF membrane. Desalination 450:21–32

Khan SH (2020) Green nanotechnology for the environment and sustainable development. In: Green materials for wastewater treatment. Springer, pp 13–46

Khasawneh OFS, Palaniandy P (2021) Removal of organic pollutants from water by Fe2O3/TiO2 based photocatalytic degradation: a review. Environ Technol Innov 21:101230

Khatri N, Tyagi S (2015) Influences of natural and anthropogenic factors on surface and groundwater quality in rural and urban areas. Front Life Sci 8(1):23–39

Khilji SA, Munir N, Aziz I, Anwar B, Hasnain M, Jakhar AM, Sajid ZA, Abideen Z, Hussain MI, El-Habeeb AA (2022) Application of algal nanotechnology for leather wastewater treatment and heavy metal removal efficiency. Sustain For 14(21):13940

Koopi H, Buazar F (2018) A novel one-pot biosynthesis of pure alpha aluminum oxide nanoparticles using the macroalgae Sargassum ilicifolium: a green marine approach. Ceram Int 44(8):8940–8945

Koriem OA, Kamel AM, Shaaban W, Elkady MF (2022) Enhancement of dye separation performance of eco-friendly cellulose acetate-based membranes. Sustain For 14(22):14665

Kumar R, Qureshi M, Kumar D, Al-ansari N (2022) A review on emerging water contaminants and the application of sustainable removal technologies. Case Stud Chem Environ Eng 6 (May):100219. https://doi.org/10.1016/j.cscee.2022.100219

Kumar V, Kaushik NK, Tiwari SK, Singh D, Singh B (2023) Green synthesis of iron nanoparticles: sources and multifarious biotechnological applications. Int J Biol Macromol 253(Pt 4):127017

Labhasetwar PK, Yadav A (2023) Water supply systems and the need for point-of-use treatment systems. IWA Publishing, pp 29–58. https://doi.org/10.2166/9781789062724

Lomelí-Marroquín D, Medina Cruz D, Nieto-Argüello A, Vernet Crua A, Chen J, Torres-Castro A et al (2019) Starch-mediated synthesis of mono-and bimetallic silver/gold nanoparticles as antimicrobial and anticancer agents. Int J Nanomedicine 14:2171–2190

Mahmoud AED (2020) Nanomaterials: green synthesis for water applications. In: Handbook of nanomaterials and nanocomposites for energy and environmental applications. Springer, pp 1–21

Maryskova M, Linhartova L, Novotny V, Rysova M, Cajthaml T, Sevcu A (2021) Laccase and horseradish peroxidase for green treatment of phenolic micropollutants in real drinking water and wastewater. Environ Sci Pollut Res 28(24):31566–31574

Mbarek WB, Escoda L, Saurina J, Pineda E, Alminderej FM, Khitouni M, Suñol JJ (2022) Nanomaterials as a sustainable choice for treating wastewater: a review. In: Materials. (Vol. 15, Issue 23). MDPI. https://doi.org/10.3390/ma15238576

Mittal AK, Chisti Y, Banerjee UC (2013) Synthesis of metallic nanoparticles using plant extracts. Biotechnol Adv 31(2):346–356

Mohajershojaei K, Mahmoodi NM, Khosravi A (2015) Immobilization of laccase enzyme onto titania nanoparticle and decolorization of dyes from single and binary systems. Biotechnol Bioprocess Eng 20:109–116

Mohammed AM, Thalji MR, Yasin SA, Shim J-J, Chong KF, Guda AA, Ali GAM (2023) Recent advances in electrospun fibrous membranes for effective chromium (VI) removal from water. J Mol Liq 383:122110

Morsi RE, Alsabagh AM, Nasr SA, Zaki MM (2017) Multifunctional nanocomposites of chitosan, silver nanoparticles, copper nanoparticles and carbon nanotubes for water treatment: antimicrobial characteristics. Int J Biol Macromol 97:264–269

Mrabet C, Jaballah R, Mahdhi N, Boukhachem A, Amlouk M (2023) CuO-ZnO nanocomposites-based thin films: characterization, physical properties and sunlight photocatalytic degradation of organic pollutants. J Alloys Compd 968:172252

Naseem K, Zia Ur Rehman M, Ahmad A, Dubal D, AlGarni TS (2020) Plant extract induced biogenic preparation of silver nanoparticles and their potential as catalyst for degradation of toxic dyes. Coatings 10(12):1235

Naseri N, Deepa B, Mathew AP, Oksman K, Girandon L (2016) Nanocellulose-based interpenetrating polymer network (IPN) hydrogels for cartilage applications. Biomacromolecules 17(11):3714–3723

Nasrollahzadeh M, Sajadi M, Atarod M, Sajjadi M (2019) An introduction to green nanotechnology. Interface Sci Technol 28:1–337. Academic Press, Elsevier, ISBN:978-0-12-813586-0

Nasrollahzadeh M, Sajjadi M, Iravani S, Varma RS (2021a) Green-synthesized nanocatalysts and nanomaterials for water treatment: current challenges and future perspectives. J Hazard Mater 401:123401. https://doi.org/10.1016/J.JHAZMAT.2020.123401

Nasrollahzadeh M, Sajjadi M, Iravani S, Varma RS (2021b) Starch, cellulose, pectin, gum, alginate, chitin and chitosan derived (nano)materials for sustainable water treatment: a review. Carbohydr Polym 251:116986. https://doi.org/10.1016/J.CARBPOL.2020.116986

Naushad M, Ansari AA, Alothman ZA, Mittal J (2016) Synthesis and characterization of YVO4: Eu3+ nanoparticles: kinetics and isotherm studies for the removal of Cd2+ metal ion. Desalin Water Treat 57(5):2081–2088. https://doi.org/10.1080/19443994.2014.986205

Neeti K, Singh R, Ahmad S (2023) The role of green nanomaterials as effective adsorbents and applications in wastewater treatment. Mater Today Proceedings 77:269–276. https://doi.org/10.1016/J.MATPR.2022.11.300

Olajire AA, Bamigbade LA (2021) Green synthesis of chitosan-based iron@ silver nanocomposite as adsorbent for wastewater treatment. Water Resour Ind 26:100158

Pacheco S, Medina M, Valencia F, Tapia J (2006) Removal of inorganic mercury from polluted water using structured nanoparticles. J Environ Eng 132(3):342–349. https://doi.org/10.1061/(ASCE)0733-9372(2006)132:3(342)

Pani T, Das A, Osborne JW (2017) Bioremoval of zinc and manganese by bacterial biofilm: a bioreactor-based approach. J Photochem Photobiol B Biol 175:211–218

Pontius FW (2000) Guest perspectives. 7(10):41–43

Qamar SA, Qamar M, Basharat A, Bilal M, Cheng H, Iqbal HMN (2022) Alginate-based nano-adsorbent materials—bioinspired solution to mitigate hazardous environmental pollutants. Chemosphere 288:132618. https://doi.org/10.1016/J.CHEMOSPHERE.2021.132618

Rabiee N, Sharma R, Foorginezhad S, Jouyandeh M, Asadnia M, Rabiee M, Akhavan O, Lima EC, Formela K, Ashrafizadeh M (2023) Green and sustainable membranes: a review. Environ Res 231:116133

Rego RM, Kuriya G, Kurkuri MD, Kigga M (2021) MOF based engineered materials in water remediation: recent trends. J Hazard Mater 403:123605

Rethinam S, Basaran B, Vijayan S, Mert A, Bayraktar O, Aruni AW (2020) Electrospunnano-bio membrane for bone tissue engineering application—a new approach. Mater Chem Phys 249: 123010

Rivière GN, Korpi A, Sipponen MH, Zou T, Kostiainen MA, Österberg M (2020) Agglomeration of viruses by cationic lignin particles for facilitated water purification. ACS Sustain Chem Eng 8(10):4167–4177. https://doi.org/10.1021/ACSSUSCHEMENG.9B06915/SUPPL_FILE/SC9B06915_LIVESLIDES.MP4

Romanovski V (2021a) Agricultural waste based-nanomaterials: green technology for water purification. In: Aquananotechnology: applications of nanomaterials for water purification, pp 577–595. https://doi.org/10.1016/B978-0-12-821141-0.00013-6

Romanovski V (2021b) Agricultural waste based-nanomaterials: green technology for water purification. In: Aquananotechnology. Elsevier, pp 577–595

Saini S, Tewari S, Dwivedi J, Sharma V (2023) Biofilm-mediated wastewater treatment: a comprehensive review. Mater Adv 4(6):1415–1443

Sajjadi M, Ahmadpoor F, Nasrollahzadeh M, Ghafuri H (2021) Lignin-derived (nano)materials for environmental pollution remediation: current challenges and future perspectives. Int J Biol Macromol 178:394–423. https://doi.org/10.1016/J.IJBIOMAC.2021.02.165

Santander P, Butter B, Oyarce E, Yáñez M, Xiao LP, Sánchez J (2021) Lignin-based adsorbent materials for metal ion removal from wastewater: a review. Ind Crop Prod 167:113510. https://doi.org/10.1016/J.INDCROP.2021.113510

Santhosh C, Velmurugan V, Jacob G, Jeong SK, Grace AN, Bhatnagar A (2016) Role of nanomaterials in water treatment applications: a review. Chem Eng J 306:1116–1137. https://doi.org/10.1016/J.CEJ.2016.08.053

Schulte PA, Mckernan LT, Heidel DS, Okun AH, Dotson GS, Lentz TJ, Geraci CL, Heckel PE, Branche CM (2013) Occupational safety and health, green chemistry, and sustainability: a review of areas of convergence. Environ Health 12:31. http://www.ehjournal.net/content/12/1/31

Shahid-ul-Islam, Bairagi S, Kamali MR (2023) Review on green biomass-synthesized metallic nanoparticles and composites and their photocatalytic water purification applications: progress and perspectives. Chem Eng J Adv 14:100460. https://doi.org/10.1016/J.CEJA.2023.100460

Shannon MA, Bohn PW, Elimelech M, Georgiadis JG, Mariñas BJ, Mayes AM (2008) Science and technology for water purification in the coming decades. Nature 452(7185):301–310. https://doi.org/10.1038/nature06599

Sharma S, Bhattacharya A (2017) Drinking water contamination and treatment techniques. Appl Water Sci 7(3):1043–1067. https://doi.org/10.1007/s13201-016-0455-7

Som C, Berges M, Chaudhry Q, Dusinska M, Fernandes TF, Olsen SI, Nowack B (2010) The importance of life cycle concepts for the development of safe nanoproducts. Toxicology 269: 160–169. https://doi.org/10.1016/j.tox.2009.12.012

Tewari C, Tatrari G, Kumar S, Pandey S, Rana A, Pal M, Sahoo NG (2022) Green and cost-effective synthesis of 2D and 3D graphene-based nanomaterials from Drepanostachyum falcatum for bio-imaging and water purification applications. Chem Eng J Adv 10:100265. https://doi.org/10.1016/J.CEJA.2022.100265

Thakur VK, Voicu SI (2016) Recent advances in cellulose and chitosan based membranes for water purification: a concise review. Carbohydr Polym 146:148–165. https://doi.org/10.1016/J.CARBPOL.2016.03.030

Thanigaivel S, Priya AK, Gnanasekaran L, Hoang TKA, Rajendran S, Soto-Moscoso M (2022) Sustainable applicability and environmental impact of wastewater treatment by emerging nanobiotechnological approach: future strategy for efficient removal of contaminants and water purification. Sustain Energy Technol Assess 53:102484. https://doi.org/10.1016/j.seta.2022.102484

Udayakumar GP, Muthusamy S, Selvaganesh B, Sivarajasekar N, Rambabu K, Sivamani S, Sivakumar N, Maran JP, Hosseini-Bandegharaei A (2021) Ecofriendly biopolymers and composites: preparation and their applications in water-treatment. Biotechnol Adv 52:107815. https://doi.org/10.1016/J.BIOTECHADV.2021.107815

Umar M, Aziz HA (2013) Photocatalytic degradation of organic pollutants in water. In: Organic pollutants-monitoring, risk and treatment, vol 8. IntechOpen, pp 196–197

Wang J, Zhuang S (2022) Chitosan-based materials: preparation, modification and application. J Clean Prod 355:131825. https://doi.org/10.1016/J.JCLEPRO.2022.131825

Wang F, Lu X, Li XY (2016) Selective removals of heavy metals (Pb2+, Cu2+, and Cd2+) from wastewater by gelation with alginate for effective metal recovery. J Hazard Mater 308:75–83. https://doi.org/10.1016/J.JHAZMAT.2016.01.021

Wang F, Chang R, Ma R, Tian Y (2021) Eco-friendly and superhydrophobic nano-starch based coatings for self-cleaning application and oil-water separation. Carbohydr Polym 271:118410. https://doi.org/10.1016/J.CARBPOL.2021.118410

Yang K, Xing B (2007) Desorption of polycyclic aromatic hydrocarbons from carbon nanomaterials in water. Environ Pollut 145(2):529–537. https://doi.org/10.1016/J.ENVPOL.2006.04.020

Yu S, Pang H, Huang S, Tang H, Wang S, Qiu M, Chen Z, Yang H, Song G, Fu D (2021) Recent advances in metal-organic framework membranes for water treatment: a review. Sci Total Environ 800:149662

Yuan R, Zhou B, Hua D, Shi C (2013) Enhanced photocatalytic degradation of humic acids using Al and Fe co-doped TiO2 nanotubes under UV/ozonation for drinking water purification. J Hazard Mater 262:527–538

Yusuf A, Sodiq A, Giwa A, Eke J, Pikuda O, De Luca G, Di Salvo JL, Chakraborty S (2020) A review of emerging trends in membrane science and technology for sustainable water treatment. J Clean Prod 266:121867. https://doi.org/10.1016/J.JCLEPRO.2020.121867

Zhang Y, Ni S, Wang X, Zhang W, Lagerquist L, Qin M, Willför S, Xu C, Fatehi P (2019) Ultrafast adsorption of heavy metal ions onto functionalized lignin-based hybrid magnetic nanoparticles. Chem Eng J 372:82–91. https://doi.org/10.1016/J.CEJ.2019.04.111

Chapter 8
Water and Wastewater Treatment by Membrane Separation Technology and Nanofiltration Membrane Applications

Mona A. Abdel-Fatah, Randa M. Osman, and Salwa I. Hawash

Abstract This chapter illustrates water and wastewater treatment (WWT) technologies that support membrane separation. It covers a range of membrane filtration methods, from microfiltration to reverse osmosis, as well as membrane distillation and membrane-integrated hybrid process technologies. Following discussions on membrane materials, flow regimes, membrane modules, and transport mechanisms, various membrane-based processes are demonstrated through examples. Nanofiltration (NF) membranes are a critical technology in wastewater treatment. As a relatively recent advancement in membrane technology, NF membranes can be used with both aqueous and non-aqueous solutions. Their characteristics lie between Ultrafiltration (UF) and Reverse Osmosis (RO), utilizing both pore-size flow (convective) and solution-diffusion mechanisms. Membrane surface charges, often negative in NF membranes, play a crucial role in their function. NF technology is applied in various wastewater treatment processes across various industries, primarily for the selective removal of ions and organic substances. It is also employed in specific seawater desalination applications. This chapter highlights the key applications of NF membranes in water reuse, tertiary wastewater treatment, water softening, and desalination.

Keywords Membrane technology · Water treatment · Microfiltration · Ultrafiltration · Nanofiltration · Reverse osmosis · Forward osmosis

8.1 Introduction

Prior to the membrane filtration process, wastewater undergoes pretreatment using appropriate methods to remove the majority of suspended or undissolved materials, such as inorganic and organic compounds. This step is crucial for protecting the membrane from potential damage, especially given its high manufacturing cost. The

M. A. Abdel-Fatah (✉) · R. M. Osman · S. I. Hawash
Chemical Engineering & Pilot Plant Department, Institute of Engineering Research and Renewable Energy, National Research Centre (NRC), Giza, Egypt

© The Author(s), under exclusive license to Springer Nature Switzerland AG 2025

J. M. Al-Khayri et al. (eds.), *Green Nanotechnology Applications for Ecosystem Sustainability*, Nanotechnology in Plant Sciences 4,
https://doi.org/10.1007/978-3-031-90766-1_8

residual contaminants are primarily heavy metal salts, so in the treatment technique, it is important to increase the pollutant's molecular size and select the suitable membrane filtration procedure for pollutant treatment (Abdel-Fatah 2018).

The concept that a membrane may effectively purify components as a semipermeable barrier can be found all across nature. Membranes provide a screening service to prevent the survival of living cells. This concept allows a slew of membrane-based separation and purification applications. Membranes, for instance, are active in artificial dialysis of the kidneys to help cleanse the blood by transferring Urea with low molecular weight to a stripping solution. A method to reduce the energy demands of NF is by utilizing more permeable NF membranes, which lowers both pressure and energy consumption, thereby influencing membrane performance. However, it is essential to balance minimizing energy requirements and maintaining optimal operation conditions (Sombekke et al. 1997).

NF is the most favorable dye-rejection technique, associated with low energy demand and high antifouling properties (Subramanian 2022). UF membranes are utilized in the pharmaceutical industries to generate injection-grade water on a comparable scale. The RO was initially used and developed for brackish water and seawater desalination in rural areas (Abdel-Fatah 2018). For seawater desalination, hollow fiber modules of RO have been broadly used. They are the common membrane modules consisting of hollow fiber bundles in a pressure vessel (Shah and Rodriguez-Couto 2022).

UF was later developed and commercialized to bridge the gap between RO, which focuses on salt rejection, and MF, which retains particles while allowing salts to pass through. The crossflow mode should be utilized for both RO and UF processes to achieve cost-effective operation. While the crossflow mode may pose certain operational challenges under specific conditions, the advancements in RO and UF technologies mark a significant leap forward. The RO membrane has been used in large-scale seawater desalination since the 1970s. Regardless of substantial proof of purification potential, membrane-based technology has been hampered by the high synthesis cost and the membranes' short life. Membranes with improved selectivity, durability, and mechanical strength at lower costs are settled over time, laying the groundwork for the large-scale application of membrane-based technologies in industrial wastewater treatment (Belleville and Vaillant 2016).

Membranes may be customized for various uses, and membrane-based technologies have several advantages over traditional separation purifying systems. Because no phase shift is involved in the purifying process, this technique is naturally energy efficient. Membrane technology with a modular design also allows for operational flexibility regarding capacity utilization. Compared to conventional technology, the High selectivity of customized membranes can guarantee a high efficiency of system purification in a compact plant configuration.

Furthermore, a membrane-integrated separation and purification production system is environmentally favorable because no strong chemicals or chemical reagents are required, and multistep operations can typically be reduced to a single step. To put it another way, membrane technology leads to considerable intensification of the process, which gives a product that is energy efficient, low-cost, and ecologically friendly while maintaining a high purity level.

However, a Membrane synthesis employs a wide range of materials, from polymers to ceramics; current challenges include polymeric membrane incompatibility in numerous solvent systems, low-temperature resistance of most polymeric membranes, manufacture restrictions of ceramic membranes outside the MF common range, membrane fouling in the majority of the existing modules, a lack of scale-up confidence, and a reluctance to approve a new technology. Apart from these drawbacks, the technologies of the membranes are advancing quickly in process magnification and water-treatment patterns (Belleville and Vaillant 2016).

Fouling remains a significant challenge in membrane technology for water treatment. Researchers have explored the combined effects of coagulation and rotating UF membranes to control fouling and remove organic matter, including both macro- and micropollutants, in the treatment of drinking water (Giacobbo and Bernardes 2022).

Furthermore, regarding the operating conditions optimization importance for ensuring better performance of membrane technology, the effects of membrane molecular weight cut-off, trans-membrane temperature, pressure, permeate quality, crossflow velocity on permeate fluidity, and fouling physical appearance during UF were evaluated (Giacobbo and Bernardes 2022).

While these technologies offer many benefits, several challenges hinder their widespread, large-scale application. These challenges include high initial investment, significant energy requirements, and membrane fouling. Combining two or more membrane processes is advantageous for addressing the limitations of each method, reducing energy consumption, and improving overall efficiency. The substantial impact of membrane-based technologies on the global economy largely relies on the reuse, recycling, and minimization of membranes (Shehata et al. 2023).

8.2 Classification of Membrane-Based Processes

The separated materials or molecule size is employed to classify membrane-based separation processes, though size alone does not define the separation degree. Also, membranes have been classified according to the availability of pores and their sizes, and they can filter soluble macromolecules and microorganisms. Unlike UF membranes, MF membranes have the largest pore size, allowing them to filter or reject bigger particles and numerous organisms. RO membranes are nonporous and adapted to exclude low molar mass particles/ions such as salts, organics, etc. NF membranes, also known as “loose RO” membranes, are transformed formulas of RO membranes with a pore size of ten angstroms or less (Shah and Rodriguez-Couto 2022).

The degree of purification is frequently determined by the membrane physicochemical features and surface nature, the solute-membrane interaction, and the

Table 8.1 Membrane process classification (Pal et al. 2009)

Membrane process	Driving force	Mass transfer mechanism	Applications
Microfiltration	Pressure gradient	Convection	Clarification and cold-sterilization
Ultrafiltration	Pressure gradient	Convection	Concentration, fractionation of macromolecular solutions
Nanofiltration	Pressure gradient	Diffusion/ convection	Concentration, purification of small organic compounds, separation of selected salts
Reverse osmosis	Pressure gradient	Solubilization/ diffusion	Concentration/desalination
Electrodialysis	Electrical potential gradient	Ion exchange (Donnan exclusion)	Separation of ions from water and non-ionic solutes
Pervaporation	Concentration gradient	Absorption/diffusion/ desorption	Separation of mixtures of volatile liquids
Membrane distillation/ osmotic membrane distillation	Partial pressure gradient	Evaporation/ diffusion/ condensation	Desalination, concentration

materials' respective sizes and membrane holes. The classification, separating mechanism, and driving force are shown in Table 8.1.

8.3 The Terminology of Membrane Separation

Volume Flux

The solvent volume collected per unit membrane area for each unit time on the permeate side is identified as the volumetric flux and is expressed as $L/(m^2 \cdot h)$ or LMH.

Solute Mass Flux

$$Mass\ flux = mass\ of\ permeated\ solute/membrane\ surface\ area\ experimental\ time$$

Flux of Solvent

Only solvent permeation is used to compute flux.

$$Total\ flux = solute\ flux + solvent\ flux$$

8.3.1 Retention of Solutes

The two most important membrane evaluation criteria are solvent flux and solute retention. The solute holding or the refusal R can be calculated by measuring the difference in concentration across the membrane divided by the bulk concentration on the feed or concentrate side. (the solute remaining fraction in the feed stream). The feed, permeate, and retentate streams of a typical membrane module are shown in Fig. 8.1. The retention of a solute can be calculated as follows:

$$R = (C_1 - C_2)/C_1 = 1 - \frac{C_2}{C_1} \quad (8.1)$$

Where C_1 = in the bulk liquid phase, the concentration of the diffusing solute (F); C_2 = the solute concentration in the permeate (P).

8.3.2 Polarization of Concentration (CP)

CP occurs while solutes on the membrane's surface accumulate, lowering flow. It is a key impediment to the long-term usage of membranes.

8.4 Modes of Flow

Fluid flow may be in a dead-end mode in any membrane module, as shown in Fig. 8.2, or in crossflow mode, as shown in Fig. 8.3. Fouling is more likely in dead-end mode due to fast CP, but fouling is less likely in crossflow mode due to the cleaning mode of fluid on the surface of the membrane.

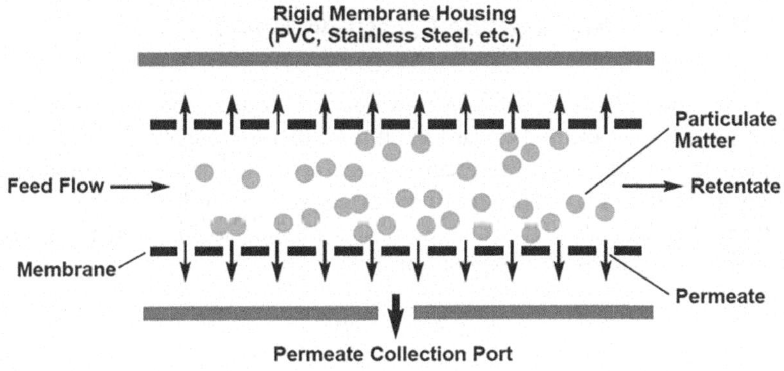

Fig. 8.1 The streams in a conventional membrane module

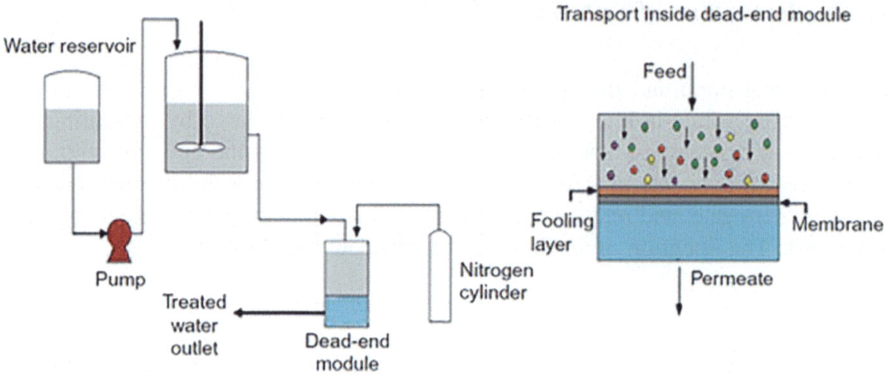

Fig. 8.2 A membrane module that was in a dead-end state and had a significant fouling potential was used

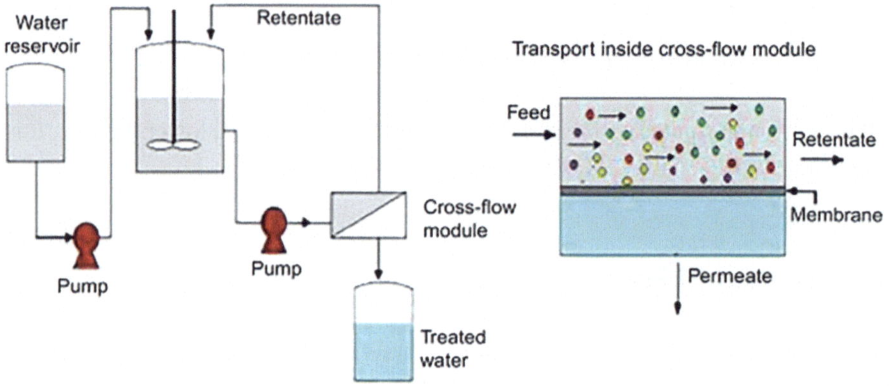

Fig. 8.3 Membrane modules that were operated in crossflow mode and had a minimal chance of fouling

8.4.1 Pressure Drop and the Crossflow Model

In crossflow mode, the bulk solution flows parallel to the membrane surface. The deposited particles are cleared toward the filter outlet by a relatively high flow rate tangential to the surface, leaving a relatively thin deposited cake layer comparable to the gel layer created in UF. This crossflow effectively controls the CP.

From the flow pattern and pressure drop in relation to Fig. 8.4, the following trans-membrane pressure drop expressions are:

$$\Delta P = P_i - P_o \tag{8.2}$$

Trans-membrane pressure:

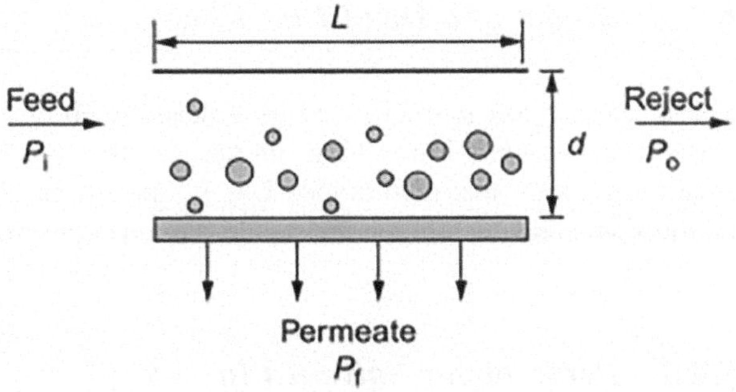

Fig. 8.4 Pressure drop in a crossflow module

$$\Delta P_{TM} = \left(\frac{P_i - P_o}{2}\right) - P_f \tag{8.3}$$

ΔP_{TM} = Average driving force.

Because the permeate side is exposed to the environment, $\mathbf{P_f}$ is usually considered to be zero.

For the purpose laminar flow (*Poiseuille flow*)

$$\Delta P = \left(\frac{C_1 \mu \, L v}{d_2}\right)$$

$$\Delta P = \left(\frac{C_2 \mu \, L Q}{d_4}\right) \tag{8.4}$$

For turbulent flow (*Fanning equation*)

$$\Delta P = \left(\frac{C_3 f \, L v^2}{d}\right)$$

$$\Delta P = \left(\frac{C_4 f \, L Q^2}{d_5}\right) \tag{8.5}$$

Where

C_1, C_2, C_3, and C_4 = Constants based on channel geometry;
f = factor linked to Reynolds's number;
v = velocity;
Q = volumetric flow rate;
L = length of filter;
d_5 = fluid channel height overhead the membrane;
μ = viscosity.

8.4.2 Model of a Dead-End Flow

The batch method is often used as a dead-end flow in various applications. The particles accumulate in the long run and create a block, forcing the illuminated to penetrate across the membrane. Block opposition increases gradually as cake buildup, whereas membrane resistance remains constant.

8.4.3 Permeate or Solvent Flux (N_w)

$$N_W = \frac{\Delta P}{\mu(R_m + R_C)} \tag{8.6}$$

Where:

N_w = solvent flux (kg/(m^2 . s));
ΔP = pressure difference, Pa;
R_m = membrane opposition, m^2/kg;
R_c = block opposition (m^2/kg);
μ = solvent viscosity (Pa . s).

8.5 The Materials of Membrane

Membranes are formed of glass, polymer, metal, ceramic, liquid, zeolite, or complex materials and are porous or nonporous. Nonporous membranes can transport electrical charges, catalysts, or functional groups, and can be dense homogeneous or dense heterogeneous. Cellulose acetate, polyamide, poly-sulfones, Poly-vinylidene di-fluoride (PVDF), poly-acrylo-nitrile (PAN), polypropylene (PP), and poly-ether-sulfone (PES) are the most commonly utilized polymers in membrane synthesis. Zeolites are a relatively new technology. The three-dimensional crystalline material, which has a lattice structure of aluminum, silicon, and oxygen, features well-defined pores and passages that can be used for both molecular sieving and catalysis. Recently, zeolite-based membranes have been used in gas departure and high temperature high temperature applications. Liquid membranes are vigorously produced membranes that are primarily created from surface stabilizers, organic liquid, and embedded carrier molecules (Starzak 1984).

The complex membranes are made up of two coats: a thin top coat that screens or selects supplies that pass across the membrane and a deep bottom coat (100–300 μm) that gives mechanical assistance to the thin top coat (0.1–5.0 μm). Thin film composite (TFC) membranes with a top coat of polyamide or polyimide as well Asan assistance coat of poly-sulfones are used in RO and NF membranes. A NF membrane can be made via placing a thin coat of polyamide on a UF polysulfone membrane. The structure of membranes can be symmetric (isotropic) or else asymmetric (anisotropic).

In symmetric membranes (isotropic), the membrane assembly is consistent during the cross-section; however, in asymmetric membranes (anisotropic), it changes. The flow properties of symmetric membranes, commonly employed in dialysis, electrodialysis, or MF, are alike across the cross-section, and the flux is nearly proportionate to the membrane width (30–500 μm). This class's mechanical stability is often low. Pressure-driven high mechanical stability membranes and high flux are asymmetric membranes with excellent mechanical stability (Starzak 1984).

Symmetric membranes are manufactured from material identical to asymmetric membranes. When asymmetric membranes could be made using a single step phase inversion method, the skin coat and the support coat are formed differently by the same material. Alternatively, a thin skin coat is placed on an assistance coat of dissimilar material. Hollow-fiber Flat-sheet, tubular, and spiral-wound membrane modules are examples of membrane modules. Flat sheet membranes are symmetric or asymmetric, but capillary and hollow fiber membranes are almost always asymmetric.

8.6 Membrane Modules

To conduct the separation required for membrane plants for industry, tens of thousands to millions of square meters of surface of membrane are frequently required. Membranes must be packaged affordably and efficiently for large-scale use. Membrane modules are divided into four categories: plate and frame, tubular, spiral-wrapped, hollow fiber, and flat-sheet crossflow (Starzak 1984).

8.6.1 Plate-and-Frame Membrane Module

In a plate-and-frame module, the feed spacers, membrane, and product spacers are placed together between two end plates, as depicted in Fig. 8.5. The feed mix is driven through the membrane surface. The fluid travels across the membrane and into the permeate channel to a central collection of the permeate. With heavily fouling feeds, this module is employed in electrodialysis, pervaporation, and rough water purification before NF, MF, UF, and RO uses (Starzak 1984).

Cleaning and replacing the membrane surface are simple procedures on procedures. Fine flow control on the permeated and feed sides of the membrane could be possible. However, this module is only appropriate for small-scale applications because it is costly and needs a significant number of spacer sheets and seals. This module also has maintenance issues due to the vast number of gaskets used between the plates, which can lead to leaks.

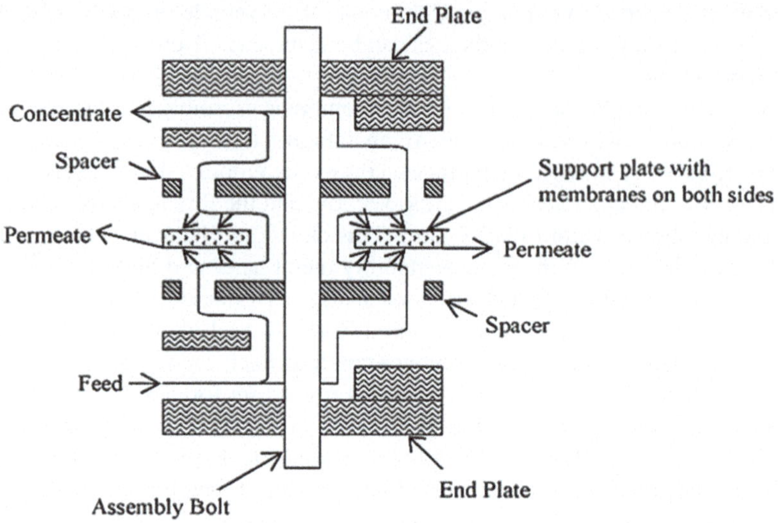

Fig. 8.5 Schematic of a plate-and-frame module (Starzak 1984)

Fig. 8.6 Modules of shell tube and tube membrane

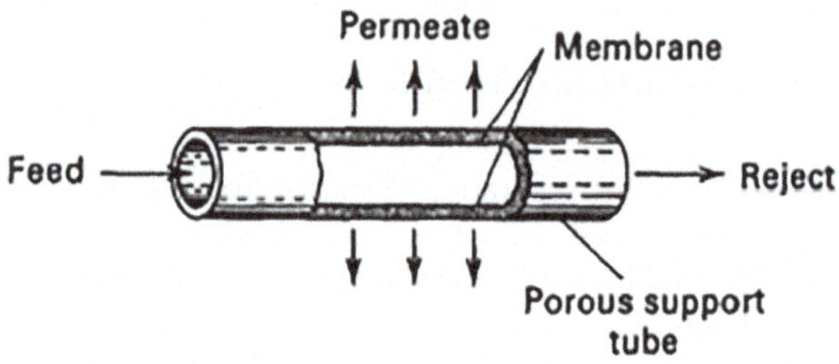

Fig. 8.7 Module of shell and membrane of tube

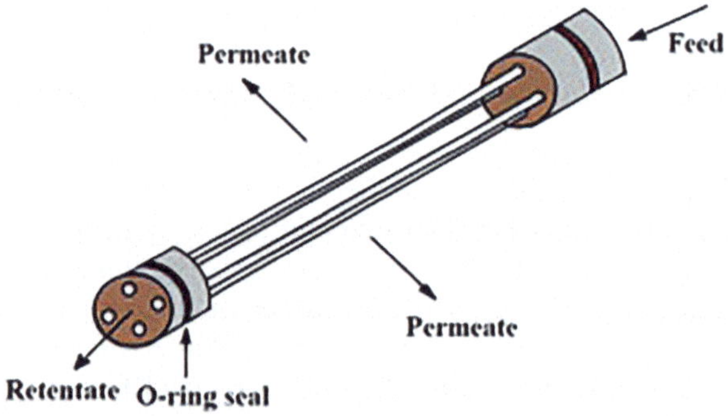

8.6.2 Shell and Tube or Tubular Membrane Modules

Tubes of the membrane, which are 5–15 nm (diameter), are grouped in a shell like a shell and tube heat exchanger in a shell and tube membrane module. Figures 8.6 and

8.7 show how large-number-of-tubes can exist joined in a chain. The feed solution (FS) is pushed across the tubes in sequence, then permeate could be collected in every tube then transferred to a common header for permeate group. The membrane is generated on the inside of the tubes, which are made of porous paper or fiber-glass.

The style of shell & tube disqualified for a huge membrane area per unit volume, but membrane creation may be difficult since the assistant and alternative coats could be turned together as units of integral cylindrical. Fouling and volumetric holdup are both exacerbated by a large surface area.

Tubular modules are presently mostly used in UF applications, where the membrane fouling resistance benefit because of strong fluid hydrodynamics outweighs their high price. This suggests that once a turbulent flow regime is desired because of a tremendous solid concentration in the fluid (where Re $>$ 10,000), this type of module should be employed (Strathmann 2011; Donnan 1924).

8.6.3 Spiral-Wound Membrane Modules

A wrapper of spacers and a membrane looped all over the perforated center collection tube make up this design, as seen in Fig. 8.8. Inside a tubular pressure vessel, the module is installed. The feed travels axially along the module, passing through the membrane wrapper. Portion of the membrane feed the wrapper, spiraling in the direction to the center then out the collection tube. The module can be built quickly and cheaply, and then the hydrodynamics (flow rate, trans membrane pressure) are modified through varying the spacer wideness to compensate for CP and fouling. In this module, when the surface of membrane area per unit volume is small. Because a high-pressure vessel is required for housing, the system may be costly for high-pressure applications. Feed bypass can also occur as a result of non-uniform spacer wrapping. In the natural gas industry, spiral wound modules may be the best prevalent module design for UF and RO besides high pressure gas separation applications.

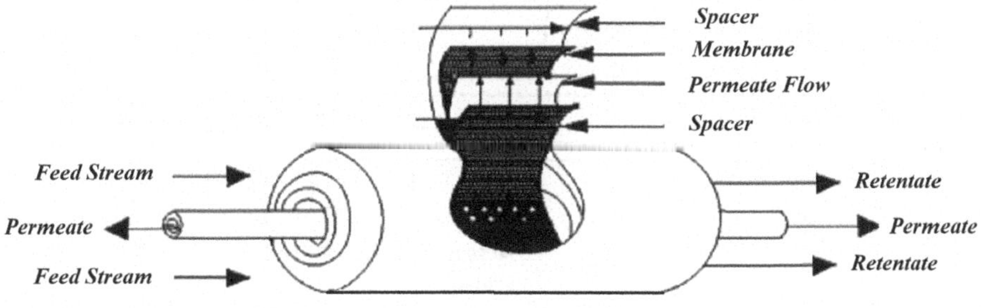

Fig. 8.8 Module of spiral wound membrane

8.6.4 Hollow-Fiber Membrane Modules

The shell and bore side feed styles are the two main module types of hollow-fiber membranes. Figure 8.9 depicts the shell-side feed style. Shell side feed modules could become commonly utilized for high-pressure units, as much as 70 bars in the shell and tube style. A loop or a shut package of fibers is confined in a pressure vessel in such a module. When the permeate travels across the fiber wall and then leaves from the open fiber ends, the system is pressured from the shell side. Because the fiber wall must withstand significant hydrostatic pressure, the fibers typically possess small diameters and thick walls, an internal diameter of 50 μm, and around 100–200 μm of exterior diameter (Strathmann 2011; Donnan 1924).

Bore Side Feed Style
The two ends of the fibers are open in a bore-side design, and the feed fluid can be cycled across the bore of the fibers. The sizes are normally bigger than those of the thin fibers utilized in a shell side feed system for decreasing pressure difference inside the fibers, and they are usually manufactured by solution spinning. The pressure of the feed is usually kept under 10 bars. In UF, pervaporation, and other low to medium-pressure usages, bore side feed forms are recommended.

By cramming a huge membrane area into a single module, a hollow fiber module using a bore side feed style, as shown in Fig. 8.10, provides an extremely an extremely dense style. Fouling on the feed side of the membrane in the shell side feed module, on the other hand, is frequently an issue, necessitating the feed stream is pretreated before introducing the stream into this module. Bore side feed modules may be often utilized for medium-pressure feed streams as much as 10 bars, where thin flow control is required to reduce fouling and CP on the membrane feed side.

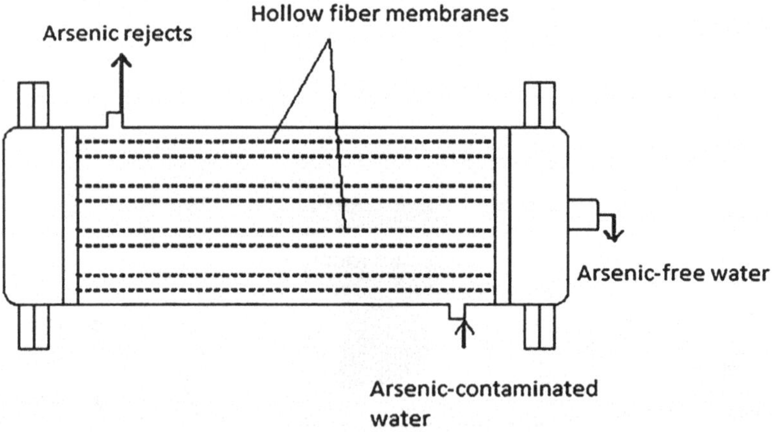

Fig. 8.9 Hollow-fiber modules shell side feed style

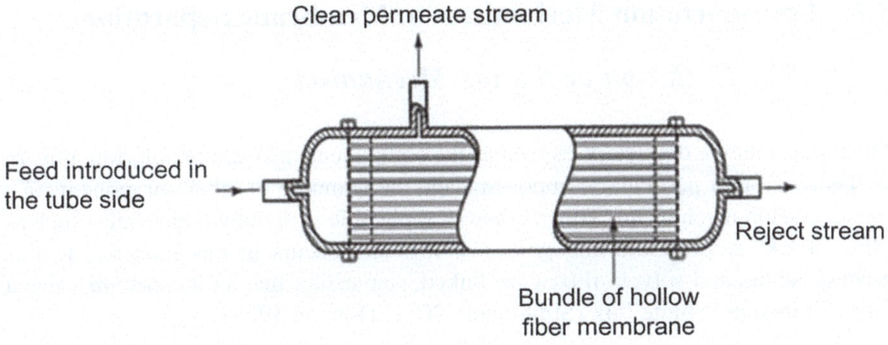

Fig. 8.10 Hollow fiber modules bore side feed style

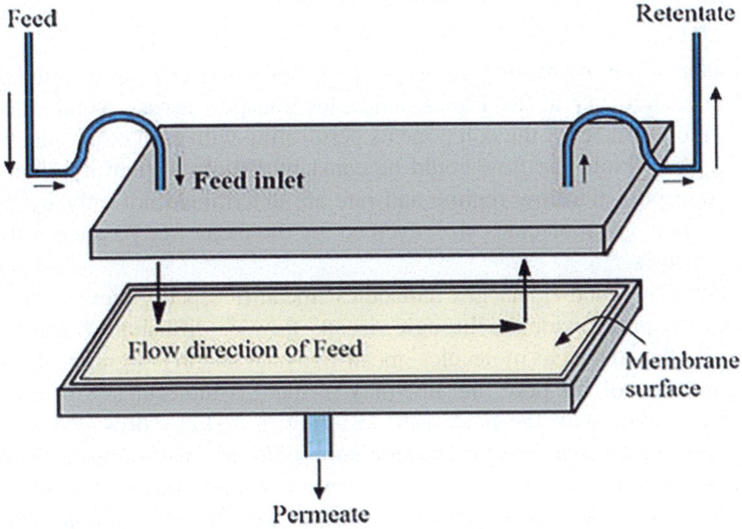

Fig. 8.11 Flat-sheet, crossflow membrane module

8.6.5 *Flat Sheet, Crossflow Membrane Module*

The membrane modules made of flat sheet crossflow in Fig. 8.11 are identical to plate and frame membrane modules. Nevertheless, the fluid dynamics are substantially different. The fluid enters the module from the top surface at one end of the membrane box, travels tangentially across the surface of the membrane horizontally, and then exits the module from the bottom at the other end when crossing the entire surface of the membrane. The tangentially running fluid's sweeping motion has the surface of the membrane generally has no fouling.

8.7 Transportation Mechanisms in Membrane Separation

8.7.1 *Size-Exclusion or Sieving-Mechanism*

The pores' relative diameters, as well as the components mixture or solution, govern the penetration of particular components and the retaining of other components in a size-exclusion mechanism. This method is applicable to spongy membranes such as MF and UF. Appropriate transportation of solute occurs in this instance. In this process, solute and solvent fluxes are linked, suggesting that an increase in solvent flux will increase solute flux (Strathmann 2011; Donnan 1924).

8.7.2 *Mechanism of Knudsen-Diffusion*

The diameter of the permeating components or molecules can be substantially less than the pore diameter in the liquid molecules transport across membrane pores, leading to more contact of the components permeating with each other than with the pore wall. As a result, the flow could be constantly sticky. When it comes to gas molecule transport, the flow regime and rate are determined not only by the pore diameter and the gas molecules size, even so, by the mean free passage distance of the gas molecules.

Depending on whether the gas molecules' mean-free path length is smaller or larger than the pore diameter, through viscous flow or diffusion. When the pore diameter is high and the gas molecules' mean-free path length is significantly smaller than the diameter of the pore, the interplay of the gas molecules with each other exceeds the contact with the pore wall, resulting in a sticky flow. When the gas molecule's mean free passage exceeds the pore width in small pores, however, the interplay between the permeable wall and permeating components takes precedence, and then Knudsen diffusion assumes control (Strathmann 2011; Donnan 1924).

The distance a penetrating gas molecule travels between two consecutive collisions with other gas molecules is defined as the deciding factor mean free path interval (λ) of the gas molecules. The mean free path interval (λ) is determined by the absolute temperature (T); pressure (p);as well as gas molecule diameter (d_{gas}); and is stated mathematically through this relation:

$$\lambda = \frac{kT}{\pi d_{gas}^2 p\sqrt{2}} \tag{8.7}$$

Where: k is the Boltzmann-constant.

The flux (J) of the permeating component (i) in Knudsen diffusion through a cylindrical membrane pore is influenced by several factors, including the pressure drop (Δp) across the membrane, the pore radius (r), the number of pores (n), the thickness (Δz) of the membrane, and the tortuosity factor (τ). These factors can be represented as follows:

$$J_i = \frac{n\pi r^2 D_i^k \Delta p}{RT_\tau \Delta z} \tag{8.8}$$

where D_i^k is the component (i) Knudsen-diffusion coefficient is in relation with the permeating component molecular weight (M_i) as:

$$D_i^k = 0.66r\sqrt{\frac{8RT}{\pi M_i}} \tag{8.9}$$

Equation (8.9) establishes that the component diffusion (i) is inversely proportional to the square root of the molecular weight. So, the separation ($\alpha_{i,j}$) of the two components (i & j) that exist in a mixture depends on the molecular weights ratio (M_i & M_j) and can be stated as:

$$\alpha_{i,j}\alpha = \sqrt{\frac{M_j}{M_i}} \tag{8.10}$$

In Knudsen-diffusion, the component's selective retardation happens by the membrane pores when the pore diameters are close to molecular size.

8.7.3 Solution-Diffusion Model

The solutes in the membrane phase material dissolve and diffuse across the membrane to the opponent. That movement is clearly a result of compressed membranes in the RO form, pervaporation, otherwise gas separation processes. The Donnan principle is partially followed by NF membranes, and partly by charge repulsion. The membrane is believed to be in equilibrium with its neighboring forms in solution diffusion, and then the concentration and flexibility of penetrating the membrane phase components dictate transportation. The component i permeability can be calculated as follows:

$$P_i = S_i \times D_i \tag{8.11}$$

Where: P_i = the membrane permeability; S_i = solubility; D_i = diffusivity. The component flux (Ji) depends on its concentration C_i, mobility mi, and chemical potential μ_i as:

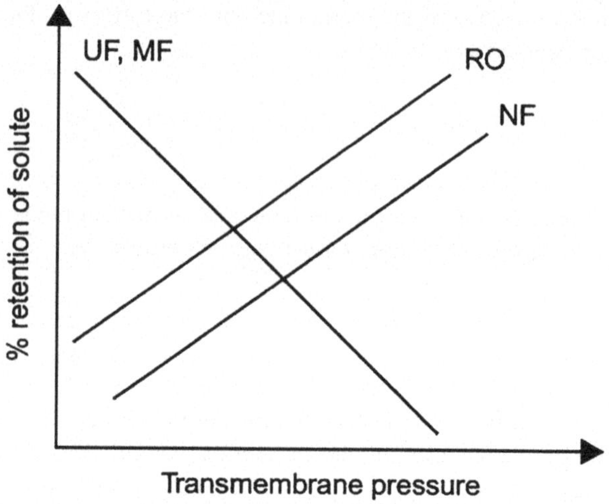

Fig. 8.12 Solute retention performances of vs membranes under different pressure

$$J_i = -C_i^m m_i^m \frac{d\mu_i^m}{dz} \tag{8.12}$$

Where: $\frac{d\mu_i}{dz}$ is the chemical potential gradient in the direction "z".

The membrane phase concentration C_i^m depends on bulk phase concentration C_i^b by the distribution coefficient k_i^{mb} as:

$$C_i^m = k_i^{mb} \times C_i^m \tag{8.13}$$

The component mobility mi, which relies on diffusivity coefficient D, viscosity (μ), temperature (T), the solute radius (r), can be shown as:

$$D_i^m = \frac{RT}{6\pi\eta_b r_i} \tag{8.14}$$

$$m_i = \frac{D_i^m}{RT} \tag{8.15}$$

The solvent and solute fluxes are separated in the solution-diffusion (SD) transportation mechanism. As a result, arise in solvent flux doesn't always imply a rise in solute flux. Figure 8.12 depicts that flow dynamics as a function of transmembrane pressure. Solute retention rises with increased transmembrane pressure for dense RO and NF membranes, and once solvent flow rises, solute flux reduces, lead to in enhanced solute retention in SD transport. However, because the size-exclusion mechanism, solvent, and solute fluxes are all linked, UF and MF membranes indicate the opposite trend in solute retention. The solvent flux increases as the pressure rises, resulting in an equal increase in solute flux (Strathmann 2011).

8.7.4 Donnan-Exclusion Mechanism

A membrane procedure is an open system exchanging mass and heat with the nearby environment. Therefore, for an open system containing composition change, the Gibbs free energy that governs the driving force in transit can be represented as:

$$dG = Vdp - SdT + \sum \left(\frac{\partial G}{\partial n_i}\right)_{p.T.n_j} dn_i (n_i.n_1.n_2.\cdots\cdots\cdots.i \neq j) \tag{8.16}$$

Where: $\left(\frac{\partial G}{\partial n_i}\right)_{p.T.n_j}$ is the component i partial-Gibbs-free-energy of constant pressure, temperature, and composition. This partial-Gibbs-free-energy is described as the component chemical potential. Accordingly, the chemical potential (μ_i) of component "i" at constant pressure, temperature, and composition can be stated as:

$$\mu_i = \left(\frac{\partial G}{\partial n_i}\right)_{p.T.n_j} \tag{8.17}$$

This chemical potential μ_i is a partial molar property shows the component "i" energy in a system has over one component. Likewise, we can define the partial molar volume (V_i), partial molar pressure (p_i), and partial molar entropy (S_i) as:

$$V_i = \left(\frac{\partial V}{\partial n_i}\right)_{p.T.n_j} ; \quad P_i = \left(\frac{\partial p}{\partial n_i}\right)_{p.T.n_j} ; \quad S_i = \left(\frac{\partial S}{\partial n_i}\right)_{p.T.n_j} \tag{8.18}$$

To put it another way, the entropy, total Gibbs free energy, pressure, and volume are state functions which can be deduced from these partial state functions:

$$G = \sum_i n\mu_i ; S = \sum_i n_i S_i ; p = \sum_i n_i p_i ; V = \sum_i n_i V_i \tag{8.19}$$

Because Gibbs free energy varies with pressure, temperature, and composition, the chemical potential also varies with these three variables. It can thus be expressed as a function of partial differentials:

$$d\mu_i = \left(\frac{\partial \mu_i}{\partial p}\right)_{T.n_i} dp + \left(\frac{\partial \mu_i}{\partial T}\right)_{P.n_i} dT + \sum \left(\frac{\partial \mu_i}{\partial n_j}\right)_{p.T} dn_j \tag{8.20}$$

Where: the term $\left(\frac{\partial \mu_i}{\partial n_j}\right)_{p.T}$ dn_j is an abbreviation of the component i chemical potential expressed as a function of the remaining components in the mixture at constant temperature and pressure.

In a perfect mix, the component i chemical potential at constant pressure and temperature is associated to molar fraction X_i as:

$$(\mu_i)_{p.T} = \mu_i^0 + RT \ln X_i \tag{8.21}$$

Where: μ_i^0 is the component chemical potential at normaltemperature and pressure.

For a non-ideal mixture, the ionic action, i.e., the action of a component "i" in a solution, is identified as $a_i = a_{i=\gamma_i} X_i$, where γ_i is the action coefficient and X_i is the molar fraction. Consequently, the adjustment in the component i chemical potential in a solution can be showed as:

$$d\mu_i = V_i dp + RTd \ln a_i \tag{8.22}$$

When there are charged species in a mixture, the Gibbs free energy have to be identified as a summation of chemical and electrical potential. As a result, electrochemical potential can be written as:

$$d\overline{\mu_i} = d\mu_i + zFd\varphi \tag{8.23}$$

Where: zi is an abbreviation of the charge number, ϕ is the electrical potential, and F is the Faraday constant. From Eqs. (8.20 and 8.21), the electrochemical potential gradient, which performs the role of the driving force in mass transportation through the membrane in the direction "z" perpendicular to the membrane surface, can be stated as (Strathmann 2011; Donnan 1924):

$$\frac{\partial \mu_i}{\partial z} = V_i dp + RT \frac{d \ln a_i}{dz} \tag{8.24}$$

$$\frac{\partial \overline{\mu_i}}{\partial z} = V_i dp + RT \frac{d \ln a_i}{dz} + zF \frac{d\varphi}{dz} \tag{8.25}$$

If a mixture has charged components or ions, and the membrane is permeable to one ion as a minimum, the membrane will reach equilibrium with the surrounding solution when the electrochemical potential of all ions in the solution equals the electrochemical potential of the ions within the membrane phase, where the membrane phase electrochemical potential is described by $\overline{\mu}_i^m$ and the electrochemical potential in the solution is designated by $\overline{\mu}_i^s$. The chemical potential in the membrane phase is denoted by the letter μ_i^m.

$$\overline{\mu}_i^m = \overline{\mu}_i^s = \mu_i^m + z_i F \varphi^m \tag{8.26}$$

The Donnan potential (φ Donnan) is the electrical potential change between the membrane and the solution phases stated as (Wang et al. 2020):

$$\varphi_{Donnan} = \varphi^m - \varphi^s = \frac{1}{z_i F}\left(RT\ln\frac{a_i^s}{a_i^m} + V_i(p^s - p^m)\right)$$

$$\varphi_{Donnan} = \frac{1}{z_i F}\left(RT\ln\frac{a_i^s}{a_i^m} + V_i\Delta\pi\right) \tag{8.27}$$

Where: V_i is the partial molar volume $\left(\frac{\partial V}{\partial n_i}\right)_{p.T.n_j}$ and $\Delta\pi$ is the osmotic pressure drop between the bulk solution (s) and membrane phase (m).

Counter ions are partitioned favorably, whereas co-ions are partitioned un-favorably between the adjacent solution and a charged membrane at Donnan equilibrium. This causes cation and anion distribution among the membrane and solution phases, with ions with similar electrical charge where the membrane's fixed ions are rejected from the membrane and those with the opposite charge being adsorbed on it.

Functional groups of ions in the solution separation or adsorption can cause a charge on a membrane surface. Steric and Donnan affect dynamic roles in separating the components in any feasible membrane separation system using NF membranes, and that Donna steric partitioning is identified:

$$\frac{C_i}{C_i^0} = \varphi_i e^{-\frac{z_i F}{RT}\Delta\psi_d} \tag{8.28}$$

Where: c_i is the solute concentration i in membrane pore, C_i^0 is the same solute concentration at the surface of the membrane, zi is the valence or stoichiometric coefficient, F is Faraday's constant, $\Delta\psi_d$ is the dimensionless Donnan potential, and φ_i is the steric separating coefficient. A steric separating influencers from the sieve influence because of intrinsic membrane porosity (Wang et al. 2020).

When electro-neutrality is maintained both within the membrane and in the neighboring solutions, ion separation occurs by the Donnan-exclusion mechanism charge repulsion. Within the membrane, electro-neutrality is expressed as:

$$\sum_{i=1}^{n} z_i c_i = -X_d \tag{8.29}$$

where X_d is the volumetric membrane charge density (mol/m^3), z_i is the valence, stoichiometric coefficient of ion I, or charge number, and c_i is the of the ion concentration i. The electro-neutrality in the bulk solution or permeate solution can be stated as:

$$\sum_{i=1}^{n} z_i c_i = 0 \tag{8.30}$$

8.7.5 Dielectric Exclusion

Separation can occur via sieving, solution diffusion, Donnan exclusion, and dielectric elimination in a partially porous, incompletely compressed membrane like the NF membrane, while transport can occur via diffusion, convection, and electromigration. Dielectric exclusion occurs when the solvent in the bulk and inside the membrane pore dielectric constants disagree, resulting in solute refusal. Almost all of the mechanisms outlined so far, with the exception of Knudsen diffusion (mostly used for gas separation), play a role in transportation through a partially porous, partially compressed membrane like the NF membrane. In the following part, we will analyze how various mechanisms provide to the overall mass movement by describing the NF membrane separation transport models (Ezugbe and Rathilal 2020).

8.8 Choice of Membrane Technology in Water Treatment

Membrane technology has developed as a preferred option for water reclaiming from diverse wastewater streams to potable water (Ezugbe and Rathilal 2020). The water/wastewater watercourse physicochemical properties, needed purification, a specific application of the treated water, and treatment price all factor into the decision to agree with membrane technology in treating polluted water or wastewater. The MF membrane technique should be used for extremely rough preliminary treatment of water containing predominantly suspended particles. UF can be used to separate particles that are significantly smaller than those found in MF feed water.

The comparative membrane pores sizes and the impurities would dictate the membrane type that has to be used in both circumstances because separation-purification is related to the sieving mechanism in the two cases. In general, UF and MF membranes can be employed for particle concentration, pretreatment for NF, RO feed water, sterilization of many watercourses in the pharmaceutical industry, ultrapure water production, and injection grade water production.

RO is utilized to recover freshwater from seawater or salty water, to remove ions from wastewater, to pre-treat boiler feed water, also to remove color from wastewater. Insignificant quantities of water can be removed from biological solutions and insignificant quantities of organics can be removed from water using pervaporation membranes. Membrane distillation is applied to remove trace impurities from water while also purifying it to a great degree. Electrodialysis membranes can be applied to desalinate brackish water or deionize boiler feed water by applying the electrical potential gradient as the driving force in ionic transportation. Large-scale ion-exchange membranes are used to remove acid or alkali from water (Ezugbe and Rathilal 2020).

8.9 MF Technology in Water Treatment

MF is a membrane separation procedure that uses the static pressure change as a driving force and the filtering parting of the membrane as a parting method. Microfiltration has a similar separation mechanism to ordinary filtration, but it has better filtration accuracy and can steal 0.13–15 um particles or organic macromolecules (Sun et al. 2015). MF separation mechanisms vary due to the various structures of microporous membranes, and they are loosely categorized into adsorption interference, mechanical interference, and linking, with physical interference being the most common (Torki et al. 2017).

In general terms, the microfiltration membrane's pore size ranges from 0.1 to 1 μm, allowing macromolecular organic chemicals and liquefied solids to flow over, however it cannot straight capture heavy metal ions or eliminates liquefied solid contaminants smaller than 1 μm. Additionally, MF is not an impenetrable viral barrier. MF, when employed in conjunction with disinfection, can manage these microorganisms in water (Abdel-Fatah et al. 2020). To treat polluted water, microfiltration technology can be used with other techniques. It is vital for the treatment of radioactive and heavy metal wastes.

8.9.1 Application in Radioactive Wastewater Treatment

In the 1980s, the microfiltration method was used to purify radioactive wastewater in an industrial plant (Zhang et al. 2009). Microfiltration is commonly employed as a pretreatment in treating wastewater containing radioactive compounds, and it can be combined with adsorption and flocculation. Cesium concentrations may be decreased from (106.87 μg/l–0.59 μg/l) using a combination of adsorption complication and immersion microfiltration membrane reactors, with a removal rate of 99.44% (Gao et al. 2004). The microfiltration technique, when used in conjunction with ferric chloride as a flocculants, is capable of successfully removing 241 Am from wastewater. When the original wastewater has a radioactivity of 809.2 Bq/l and the effluent has a radioactivity of less than 1 Bq/l, the elimination rate can approach above 99.9%, demonstrating that the procedure is effective in the radioactive wastewater treatment (Zhao et al. 2007). In the method made of flocculent containing ferrous sulfate and mutual MF, after the amount of Fe^{2+} is 35–60 mg/L, the elimination rate of plutonium can approach 99.9% (Li and Yu 2009). In radioactive wastewater treatment, microfiltration technology will advance.

8.9.2 Heavy Metal Wastewater Treatment

Huge quantities of heavy metal wastewater are discharged as a result of the fast growth of industry and additional industries in China, which would be indirectly dumped into watercourses, causing a serious danger to people's health. Reduce precipitation, adsorption, and other pretreatment processes for heavy metal wastewater treatment by microfiltration transform heavy metals into unsolvable particles with a particle size above 0.1 μm. Metal ions inside wastewater create co-precipitation in the co-precipitation microfiltration system, which is subsequently eliminated by microfiltration technology, which has a significant elimination influence. When there is heavy metal wastewater containing extremely organic matter, ferric hydroxide is commonly utilized as a coagulant in adsorption microfiltration systems since it does not only cause heavy metal ions to create co-precipitation but adsorb some organic molecules or chelates. To obtain a large removal impact, precipitation and adsorption are frequently coupled (Wan and Wang 2008).

A combination of neutralization and microfiltration to treat wastewater having Zn^{2+} with Pb^{2+} was used (Song et al. 2007). When adding flocculants, the elimination level of Zn^{2+}, Pb^{2+}, and Zn^{2+} may reach 99.9% and 99.8%, respectively, indicating an efficient removal impact. When the modification of the surface of the microfiltration membrane made of poly-vinyl-idene fluoride was completed via a pH of 5 and a heat-induced polymerization (Wang et al. 2020), Song et al. observed that the elimination level of 0.5 mmol/l Cu^{2+} can approach 75.18%. Due to the heavy metal ions small particle size, the microfiltration membrane cannot effectively remove them under normal conditions. Furthermore, the corresponding alteration of the microfiltration membrane has yet to yield satisfactory results. Heavy metal ions can be eliminated more effectively using a combined microfiltration membrane and microfiltration membrane technique. The use of microfiltration technology in wastewater treatment is summarized in Table 8.2.

Table 8.2 The use of MF technology in wastewater treatment (Huang et al. 2015)

Microfiltration technology	Wastewater type	Pollutant	Removal rate
Adsorption + microfiltration	Radioactive wastewater	Cs	99.4%
Flocculation + microfiltration	Radioactive wastewater	Am	99.9%
Flocculation + microfiltration	Radioactive wastewater	Pu	99.9%
Microfiltration	Artificial wastewater	Cu	75.18%
Microfiltration + reverse osmosis	Landfill leachate	Fe, Cr	>99.5%
Precipitation + microfiltration	Ground water	Zn	99.9%
Microfiltration + reverse osmosis	Artificial wastewater	Ca, Mn	76.12%, 91.61%
Complexation + microfiltration	Artificial wastewater	Zn	99%

8.10 UF Technology in Water Treatment

In comparison to MF, UF is extensively utilized in the water treatment industry. UF can separate almost all water pollutants from water, albeit to various degrees, as long as the contaminants' diameters are between 10 and 50 nanometers. This unequal membrane could be defined as MWCO, the smallest molecular weight (in Dalton) of the molecules the membrane retains 90%. CP is a serious issue when employing UF membranes CP. In order to run at a set flux, more pressure must be supplied in the UF to resolve the CP influence on flux. When pressure (Δp) rises, the flux of the solvent rises due to increased convective force. In contrast, back diffusion rises as well, and at a constant state, the entire solute convective flux from bulk to surface of the membranes equivalent to the summation of the solute flux at the permeate side plus the flux as a result of back diffusion (Huang et al. 2015).

The solute concentration on the surface (C_s) grows as p rises to a limiting estimate known as gel concentration (C_g). As C_s approaches C_g, solute retaining turns out to be finished, and solute concentration in the permeate side (C_p) tends to zero. Manufacturers advocate acid-alkali cleaning cycles and reverse flushing to overcome the membrane fouling concern. The fouling of polymeric UF membranes is lower than that of ceramic membranes. Ceramic types, on the other hand, are simple to clean and disinfect (Huang et al. 2015).

The size exclusion concept is used in MF and UF membranes to isolate objects. While the filtrate is regularly, they wanted output in MF, the retentate is the wanted result in UF. UF membranes, for instance, maintain the required components in food manufacturing, protein separation, and biological macromolecules. As a result, UF membranes could typically be used in crossflow mode, where a fluid flow with a high rate parallel to the membrane surface aids in the removal and collection of retentate. This also helps to prevent CP. During UF, the usual crossflow velocity in frame, plate, and tubular modules is 2–5 ms^{-1}. Substantial crossflow speed causes a great pressure difference, which usually ranges from 0.1 to 0.5 bar/m of the membrane length surface. High crossflow velocity is frequently blamed for insufficient UF recovery.

Separating metal ions by UF using a typical polymer membrane is ineffective, but specific UF membrane designs may efficiently remove metals from treated water. For instance, in a crossflow operation mode, an improved ultrafiltration polymer membrane is extremely successful in binding mercury (Hg-II) from wastewater, as illustrated in Fig. 8.13. The polyvinyl amine (PVAm) enriched PES UF membrane by 0.3 wt.% of PVAm was discovered operative in eliminating mercury at a flux of 200 LMH below only 2 bar adequate pressure at a removal efficiency of 99% (Garcia-Ivarsa et al. 2015).

The related improved UF polymer membranes even though with a little less effectiveness than the PVAm enhanced sort is metal binding polyethyleneimine and polyacrylic acid enhanced UF membranes. PES UF membrane adapted with poly-vinyl-pyrrolidone as pore generating agent was discovered to eliminate iron from water with 15 mg/L initial concentration with a removal efficiency of 98%.

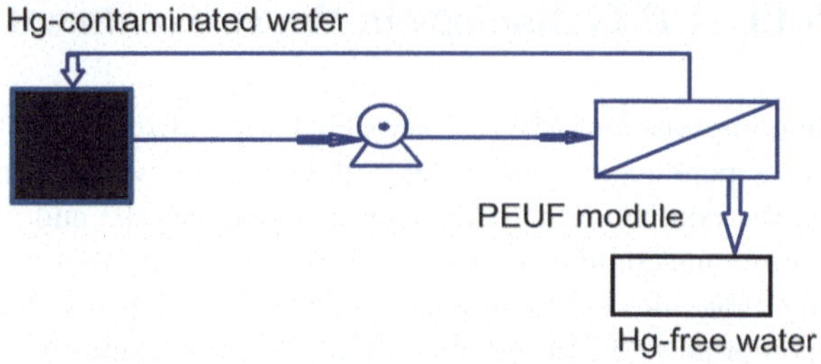

Fig. 8.13 PEUF, Polymer-enhanced UF

Likewise, UF membranes (photo-modified) are utilized to recover phenolic chemicals from olive oil effluent (Liu and Liu 2016). In the treatment of wastewater contaminated with fluoride from the aluminum, fertilizer, and semiconductor industries, a UF membrane (PVDF, 30 kD MWCO) coupled with chemical precipitation (Urgun-Demirtas et al. 2013) can effectively treat fluoride-contaminated industrial wastewater by adding calcium chloride in a molar ratio (Ca^{+2}/F^{-}) of 0.7 to fluoride polluted wastewater. Although used in conjunction with other approaches, UF has been proven to be highly effective in eliminating different metallic and metalloid contaminants from water. Cadmium, copper, chromium, Cobalt, Nickel, and mercury have all been recovered from wastewater using UF membranes. However, in most cases, metal recovery is achieved by UF-coupled hybrid technology. In the leather business, UF membrane is widely used. In the leather business, UF can be used to recover hexavalent chromium from tanning baths, sulfite from the unhearing liming process, enzyme from the enzymatic unhearing liming step, and surfactants from the degreasing step.

UF technology is also well-established in the pulp and paper industry for wastewater treatment. UF is utilized in paper machine circulation water, which is reused after treatment on a huge scale. UF is especially good for removing lignin from black liquor. Dye, latex, and lignosulfonates are all recovered by UF. UF technology is used in the pharmaceutical industry to recover antibiotics from fermentation broth, separate proteins, and purify streams of viruses and bacteria.

UF membranes have been demonstrated to be effective in oily wastewater treatment. Normally, a PES membrane having a molecular weight of 50 kDa (MWCO 50 kDa) be able to remove oil from water effectively. Mercury (22.7–1.3 ppt) was effectively eliminated using an aeration-coupled UF method from refinery effluents, this method is known as Zee Weed low-energy membrane technology. A resonating fiber membrane module is constantly ventilated from the bottommost in this membrane technology to establish turbulence and maintain a clean membrane module. There has been reported elimination of above 97% turbidity and mercury to a concentration lower than 0.5 ppt (Barakat and Schmidt 2010).

In the presence of aluminum oxide and hydrophilic glycol, UF membranes with modified surfaces using UV radiation may effectively remove phenolic chemicals from wastewater. Such surface changes greatly improve the antifouling properties of UF membranes. The UF membranes cannot retain metal ions directly because of their high pore size compared to hydrated metal ions. However, those metals can be eliminated through UF by micelle formation or complexation when micelle or complexation reagents are added to water. MEUF (micelle-enhanced UF) or CEUF (continuous improved UF) are the terms used to describe these processes CEUF. The appliance of CEUF is usually determined by a complexation process or chelation, in which wastewater is treated with chelating polymer, and the metal needs to be improved. Through electrostatic interactions, the polymer ligands' vigorous functional groups assist in fastening metal ions to two or three donor atoms of the polymer ligands (Langelier 1936).

The water-soluble polymer carboxy methyl cellulose facilities with the heavy metals cationic types presented in water in the polymer enhanced UF process for the elimination of heavy metals like Ni (II), Cu (II), and Cr (II) from wastewater. The water with the compounds is then passed over a PES UF membrane (MWCO 10,000 Da). UF was first used to purify water. UF has since been widely used in the pharmaceutical and food industries, but its function in producing clean drinking water remains critical. In water disinfection, the UF membrane (10 nm) adopts an important function where it nearly splits all pathogens like bacteria (0.5–10 μm), viruses (20–80 nm), cysts, and oocysts (3–15 μm) from treated water, while many of these pathogens are unaffected to disinfection using chlorine-based.

In 2017–2018, the worldwide UF membrane marketplace was worth more than US$900 million; by 2023, it is estimated to be worth over US$2140 million. Almost every year, main creators like Hy-flux Ltd. (Singapore), DowDuPont Inc. (US), Ing GmbH (Germany), GE Water DuPont Process Technologies (US), Toray Industries Inc. (US), Hydranautics (US), and Pentair plc (US) announce new projects involving the development of new centers, the expansion of manufacturing facilities, and the diversification of product ranges. These findings point to a thriving membrane market around the world, as well as the sector's potential for expansion.

8.11 NF Technology in Water Treatment

NF is increasingly gaining importance in treating water, coinciding with unfamiliar contaminants in water and more strict water quality requirements. NF offers certain distinct benefits over RO and UF, although it is the result of research and development work in the fields of UF and RO. NF exhibits characteristics that lie between UF and RO, functioning through both pore-size flow and solution-diffusion mechanisms. The membrane's charge is a key factor in its performance, and NF membranes frequently have negatively charged surfaces. NF technique is employed for treatment of different water and wastewater treatment in various industrial applications (Abdel-Fatah 2018).

The primary function of NF is the selective removal of ions and organic compounds, and it is also utilized in certain specialized sea. NF membranes comprise groundwater treatment, surface water treatment, brackish water treatment, water reuse, and many other applications. We not only reviewed the removal of key pollutants, such as hardness, pathogens, and natural organic matter, but also addressed the elimination of high-priority micropollutants, including disinfection byproducts, arsenic, per-and-polyfluoroalkyl substances (PFAS) (Guo et al. 2022). The fundamental benefit of NF is that it generates water that encounter nearly all of the standards for harmless and hygienic drinking water (with the exception of certain monovalent ions such as Na^+ and Cl^-) at a minor operational pressure compared to the RO.

Since the development of NF in the mid-1980s, membrane-based water treatment has become a more appealing choice than traditional multistage water treatment procedures. Today, a single-step water treatment unit with an NF membrane can remove practically all pollutants from water, including viruses, bacteria, and protozoa. Rigidity components, sulfates, nitrates such as Nickel, Cr, Iron, and Mn, inorganic and organic pollutants, developing contaminants, pesticides, and disinfection by-products have been removed using NF technology. Monovalent ions such as sodium or chlorine seep through NF significantly, making it unsuitable for brackish water. While RO keeps even beneficial elements like magnesium and calcium, the NF allows for partial passage of these minerals in healthful drinking water. Because of NF's susceptibility to water, which contains elevated solid content in water, pretreatment is frequently a concern. Water pretreatment is frequently required to protect sensitive NF from excessive fouling (Alkaisi et al. 2017).

Water's ability to create scale is another delicate matter in water treatment and transportation, where many gauges could harm transportation systems. Also, a lack of gauging causes the water to be destructive since there is no defensive scale to shield pipes and transfer lines from water-damaging effects. NF-treated water has a higher potential for forming scale than RO-treated water, preventing pipework from significant corrosion and maintaining drinking water free of liquefied heavy metals like Pb and Cd. The recovery rate for once-through NF is roughly 85%.

In a plant, modules must be connected in series or parallel. Modules are essential in the production of scale and the prevention of significant on-membrane surface fouling. Regarding fouling, tubular membrane modules outperform spiral-wound versions, and cleaning is a breeze. Tubular membranes, on the other hand, have a lower packing density with just 0.05 m^2 surfaces of membrane to each meter of module length, and the fabrication is considerably more costly. The spiral-wound module is easier to build and has a higher density of packing than the tubular module membrane, which is more fouling-prone. The water quality when it comes to dissolved solids, pH, ionic strength, and temperature could be crucial in scale formation because the water potential scale formation is determined by these characteristics, which has a parameter characterized by Langelier as the Langelier Saturation Index (LSI) (Alkaisi et al. 2017).

8.11.1 Langelier Saturation Index

LSI is characterized (Alkaisi et al. 2017) by the water pH minus saturation pH, which is the pH at which water turns out to be saturated with calcium carbonate.

Therefore,

$$LSI = pH_{actual} - pH_s$$

pH_s: is the temperature (°C) dependent recorded by a temperature-dependent parameter "A," ionic strength or TDS recorded by the parameter "B," calcium hardness (mg/L $CaCO_3$) recorded by the parameter "C," and total alkalinity (mg/L $CaCO_3$) recorded by the parameter "D" as stated in the following rule:

$$pH_s = A + B - C - D \tag{8.31}$$

"A" is reduced at temperatures 2.60 at 0 °C and 1.15 at 80 °C, and intermediate values at 4 °C, 12 °C, 20 °C, 30 °C, 50 °C, and 70 °C are 2.5, 2.3, 2.0, 1.9, and 1.6, respectively. "B" depends on the ionic strength and rises with TDS, which is 9.70 when TDS is 0.0 mg/L and 9.90 when TDS is 1000 mg/L. Intermediate parameter "B" values at 100, 200, 400, and 600 mg/L when TDS values are 9.77, 9.83, 9.86, and 9.89, respectively. "C" or "D" are dependent on total alkalinity or calcium stiffness (expressed in terms of $CaCO_3$). The values increase with calcium stiffness or total alkalinity, with 1.0 at 10 mg/L calcium stiffness to 2.78 at 600 mg/L. Intermediate values at total alkalinity or calcium stiffness (mg/L $CaCO_3$) of 20, 30, 40, 60, 80, 100, 200, 300, 400, and 500 mg/L are 1.3, 1.5, 1.6, 1.8, 1.9, 2.0, 2.3, 2.5, 2.6, and 2.7, respectively. If LSI = 0, water is shown to be stable since $CaCO_3$ precipitation or dissolution is possible.

When $LSI > 0$, precipitation or scale formation with $CaCO_3$ may occur, which acts as a protecting covering on pipelines, boilers, and waterlines, protecting them from rust? This says that water isn't destructive, and that hazardous heavy metals like Pb and Cd won't persist in dissolved states in water, posing serious health concerns when the treated water is utilized for drinking purposes (Alkaisi et al. 2017).

When: $LSI < 0$, suggesting $pH_{actual} < pH$, water would be corrosive or destructive, when containingharmful heavy metals dissolved in water without any defensive $CaCO_3$ coat forming on infrastructure and metal pipes to shield them from corrosion. Other factors that interfere with $CaCO_3$ precipitation, like the existence of ions like sulphate and chloride, also alter corrosive water. As a result, LSI May not be able to predict the water corrosive nature in all solid content levels. This LSI value is correct for water with over 40 mg/L alkalinity, pH 6.5–9.6, and enough $CaCO_3$. For other scenarios, an aggressive index (AI) is utilized, which is a more straightforward but less accurate indication. AI is defined as: $AI = pH_{actual} + C + D$. If the value >12, water is nondestructive, if the value ranging 10–12, water is temperately destructive and if the value <10, it will be greatly destructive.

8.11.2 Nanofiltration Needs Pretreatment

Due to their sensitivity to fouling and scale-forming supplies, NF membranes have to be run with caution for long-term flux. Pre-filtration of the inlet stream before NF is therefore highly recommended, especially when treating inlet streams including high loads of solids. NF membranes may come into touch with bases, free chlorine, acids, and free oxygen during filtration service, all of which potentially harm the membrane. Metal oxides, organic and inorganic colloidal particles, besides organic materials are instances for fouling supplies, while sulphate, fluorides, barium sulphate, carbonates of calcium, and silica are scale forming materials common examples. Acidic cleansers are recommended for eliminating biofouling materials, whereas cleansing agents and chelating chemicals are beneficial in eliminating foulant by reducing adsorptive pressures and surface tensions, as well as preventing calcium scaling. Enzymatic cleansing of lipids and proteins is recommended. Once flux drops by over 10% or the permeate salt concentration watercourse surpasses 10%, the membrane should be cleaned regularly.

To minimize quick NF performance decline, most NF membrane manufacturers recommend lowering the NF inlet-water turbidity lower than 1 NTU, ideally to 0.2 NTU. Prior to NF, inlet water with great fouling potential has to be prepared with predictable oxidation, MF/UF, or coagulation precipitation. When common salts like $CaCO_3$, SiO_2, $CaSO_4$, $BaSO_4$, and numerous metal hydroxides are rejected by the membrane, they commonly exceed solubility limits on the surface of the membrane, and their precipitation can cause substantial membrane fouling or scaling. This frequently demands the use of a scale inhibitor.

Mery-sur-Oise, in a Paris suburb, was an early and effective water treatment plant that used NF technology, with a capacity of 140 km^3/day and a population of 8×10^5. This plant, which was developed using the NF70 NF membrane manufactured by Dow Filmtec, was able to remove about 90% of pesticides, liquefied organic carbon, and disinfection by-products. The issue with that membrane is its significant refusal of calcium and bicarbonates (83%), which made scaling control more difficult. Dow Filmtec created NF200B to replace NF70, which could only remove 50–70% of calcium hardness instead of the high 83% of NF70.

In 1999, the Mery Sur Oise was built as a present conventional physicochemical water treatment plant postponement for the Oise River's surface water, to meet severe water quality standards while also permitting water softening. This NF plant had several subunits, including an ozonation unit, coagulation-flocculation unit, dual media filtration and in-line coagulation outline, NF unit, back-washable MF unit, carbon dioxide stripping unit (when CO_2 permeates across an NF membrane), and UV disinfection unit, where it was only a postponement of a physicochemical treatment unit. The goal of the ozonation unit operated using a low-dose regime is to improve coagulation-precipitation downstream. Sand and activated carbon make up the dual-media filter. This plant's water is combined with water treated in the

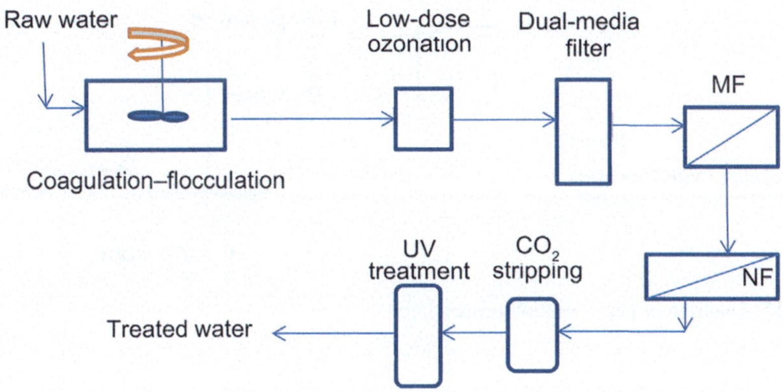

Fig. 8.14 Action structure of Mery Sur Oise NF plant in Paris

conventional treatment plants, chlorinated, and changed for the appropriate LSI value by the addition of sodium hydroxide, as shown in Fig. 8.14.

The worldwide NF market was around US$644 million in 2017–2018, and it is predictable to increase to US$955 million by 2025. This equates to an annual growth rate of more than 5%. Industrialization and urbanization are contributing to the establishment of even stringent water quality criteria in developing and great population countries like Brazil, China, and India, as a result of rising call for harmless and hygienic water for domestic and industrial usage. The growth of general medications, medical infrastructure, and a flow in call for low price medications, also the spread of pharmaceutics making services throughout developing nations are all expected to enhance demand for NF membrane technology.

8.12 Pervaporation Technology in Water Treatment

Pervaporation has been employed to remove volatile pollutants from water, such as hydrocarbon vapor (VOC). When the usual distillation 'procedure can't split the components at a definite degree due to azeotropism, this approach comes in handy. This method, for example, is capable of separating water from ethanol at a rate of more than 96%. For example, using the treatment system indicated in Fig. 8.15, a combination of benzene and cyclohexane can be separated.

8.13 Opposite Osmosis Technology in Water Treatment

The RO market is growing at an average annual rate of around 8.6%, despite a massive mismatch between mandate and availability of fresh and harmless water around the creation, mainly in water stressed countries. By 2025, the global market,

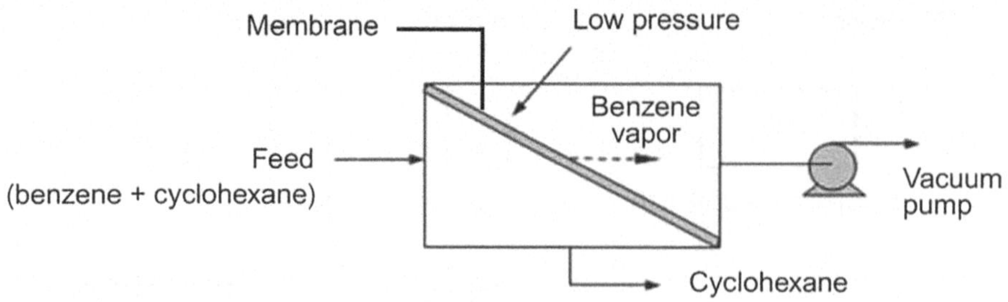

Fig. 8.15 Diagram of pervaporation process

worth roughly US$6201 million in 2017–2018, is predicted to have doubled. The Asia-Pacific region is the fastest-growing market. Concerns about the safety of drinking water, population growth, industrialization, and the rapid reduction of sweet water supplies have affected unprogressively reliant on seawater or brackish water desalination to encounter overall sweet water request. Many water stressed areas, like the Caribbean islands, the Mediterranean, and the Spain Canary Islands, are now heavily reliant on freshwater supply through desalination. Gulf nations like Kuwait, Qatar, and Saudi Arabia are seriously reliant on desalination to secure freshwater demand.

Qatar and Kuwait completely rely on the desalination of seawater to secure their freshwater requirements. Due to the lack of groundwater resources, many metropolitan cities worldwide are experiencing severe freshwater shortages. Desalination of salt water is usually done by thermal technologies such as multiple effect distillation, multistage flash evaporation, and thermal vapor compression in water-stressed locations like Middle Eastern countries. On the other hand, thermal desalination requires large energy inputs, which come from burning fossil fuels, which contribute to increased emissions of GHG (greenhouse gas) and global warming. To produce 1000 m^3 of freshwater from seawater, a whopping 10,000 tons of fossil fuel must be burnt through thermal desalination (Zaro and Prats 2018). As a result, there is a water-energy nexus because water production requires a lot of energy, and energy generation requires a lot of water.

The situation of oil-rich countries like Saudi Arabia, where approximately one-quarter of the produced gas and oil created is used to provide drinkable water and electricity, exemplifies this point. Though the RO process energy consumption is reduced compared to thermal processes, it is still relatively significant at roughly 4–5 kWh/m^3. In the 1970s, energy ingesting was around 20 kWh/m^3, which has now dropped to around 4 kWh/m^3 thanks to continued improvement in membrane characteristics, increased pumping productivity, energy retrieval via pressure exchangers, hydraulic turbines, and the usage of adjustable frequency drives. The membrane desalination process accounts for over 70% of total energy consumption (https://www.jersywater.je/about-us/learn-more/desalination-plant/).

The urgent demand for energy has shifted the balance in favor of membrane-based desalination. RO technology is being used in several large-scale desalination

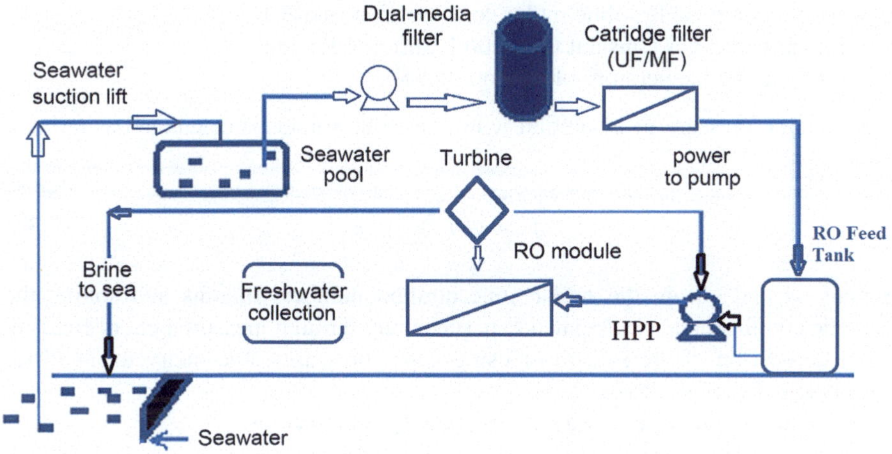

Fig. 8.16 The reverse osmosis treatment process for seawater at the La Rosiere Desalination Plant

plants around the world. The world's largest RO plants are built in water-strained countries like Saudi Arabia. The RO-based desalination plant, which cost $400 million and was completed in 2013, generates more than 620,000 m^3/day of freshwater through desalination of seawater. The plant uses a cathodic protection system to protect the infrastructure from severe seawater corrosion. Two offshore intake heads provide saline water to the RO module after pretreatment. To have entrainment properties to the lowest and hold off from marine species, the suction heads are only delivered with a low speed of 0.15 m/s. The facility gets its seawater from two open intake heads from sea about 1.15 kilometers offshore.

For pretreatment of seawater before RO, the La Rosiere Desalination facility (Long and Yu 1999) uses a dual-media filter and MF or UF membrane cartridges, as shown schematically in Fig. 8.16. In a turbine, the high-pressure brine's energy is retrieved. Desalination plants are divided into brackish water RO (BWRO) and seawater RO (SWRO). Although BWRO plants have a salinity range of 500–10,000 mg/L, SWRO plants have a salty water salinity level of 30,000 mg/L. Because each 100 mg/L TDS provides around 1 psi osmotic pressure, a TDS of 30,000 mg/L causes about 300 psi osmotic pressure, or about 20 bars. For 3000 mg/L salinity, the applied pressure must be significantly higher than 20 bars to oppose the water flow in the RO system.

The osmotic pressure in ideal dilute solutions with a single solute is calculated by the van't Hoff equation as follows:

$$\Pi = CRT \tag{8.32}$$

Where:

C is the non-permeable solute molar concentration (mol/L),
R is the universal gas constant (0.08206 L atm/mol K), and
T is the absolute temperature of the solution(K).

The osmotic pressure of a solution with numerous solutes is calculated as (Abdel-Fatah et al. 2016):

$$\Pi = \varphi RT \sum v_i c_i \tag{8.33}$$

Where: vi and Ci are the molar concentration numbers of ions and solute, the osmotic coefficient ϕ is calculated experimentally through measurement of freezing point depression of the system or using vapor pressure osmo-metry using Pitzer correlation (Pal et al. 2016).

The osmotic pressure in seawater is roughly computed as:

$$\Pi = 1.12T \sum m_i \tag{8.34}$$

Where: $\sum m_i$ signifies the molalities summation for all ionic and non-ionic species dissolved in the solution.

The major companies in RO technology are Hydranautics, Toyobo Co. Ltd., Toray Industries Inc., Axeon Water Technologies, The Dow Chemical Co., Koch Membrane Systems, MECO Incorporated, MICRODYN-NADIR GmbH, RO Chem Marine, MEMBRANIUM, Pure Aqua Inc., Bionics Advanced Filtration Systems, IMPAC, United States, Aspen Water Inc., Wananchi UK, Lifestream Life stream Water Systems, and B&P Water Technologies.

8.14 Forward Osmosis Technology in Water Treatment

Forward osmosis (FO) is a membrane separation process that does not rely on pressure or temperature. Under osmotic pressure, FO-based technology imitates the natural process of osmotic transport. When water contaminants are not easy to remove from water due to their complicated characteristics, it is preferable to separate the water from the pollutants, as indicated graphically in Fig. 8.17, this is precisely what is done in FO Fig. 8.17 (Shawky et al. 2019).

If a semipermeable membrane separates a chamber of pure water from a chamber of concentrated saline water, accordingly, pure water will move from the treated water side to the concentrated saline side under osmotic pressure until osmotic equilibrium is reached. This method is used to treat exceedingly complicated and unclean wastewater, using a semipermeable FO membrane placed in the middle of the wastewater compartment and the concentrated saline water compartment by the separation device. The draw solution (DS) is concentrated salty water that, under osmotic pressure, draws clean water from the wastewater compartment via the FO

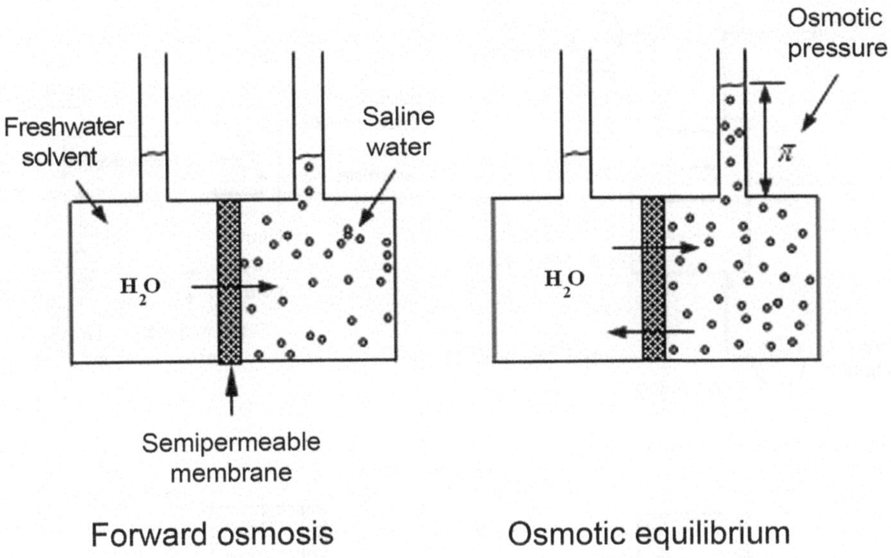

Fig. 8.17 Forward osmosis

membrane. NaCl, $CaCl_2$, $MgCl_2$, $MgSO_4$, KCl, sugar, and ammonium bicarbonate are the most commonly utilized draw solutes.

Depending on the separation requirements, there may be numerous different DS compositions. The saline water concentrated with high salt concentrations, or DS, converts to a diluted phase with time and loses its capability to extract fresh water. As a result, to draw pure water from the wastewater side to the DS side continually, using a downstream separation device, the diluted DS must be constantly concentrated. This downstream separation device separates pure water from diluted DS, yielding two products: clean and salt-free water and concentrated DS. Both products are now usable, with one being recycled into FO and the other collected as pure water.

Typically, hydrophilic and mechanically robust cellulose acetate has been chosen as a membrane material for FO modules. An ultrathin skin layer can also be applied to a polyether UF membrane to create a FO membrane. Because hydraulic pressure is no longer used as a driving force in the FO process, membrane fouling is far more unlikely than in a pressure-driven process. However, in the membrane module vertical orientation, as shown in Fig. 8.18, the CP of the membrane is rather considerable. External CP and internal CP (ICP) are the two forms of CP (ECP).

The ECP occurs when a shallow, stationary fluid layer forms at the interface of the membrane fluid, and transport occurs exclusively perpendicular to the membrane surface via advection and diffusion. The ICP is caused by the salt solution axial flow through the porous layer, which occurs when the diffused salt reaches the thick active layer but cannot penetrate it. The solute builds up within the porous layer due

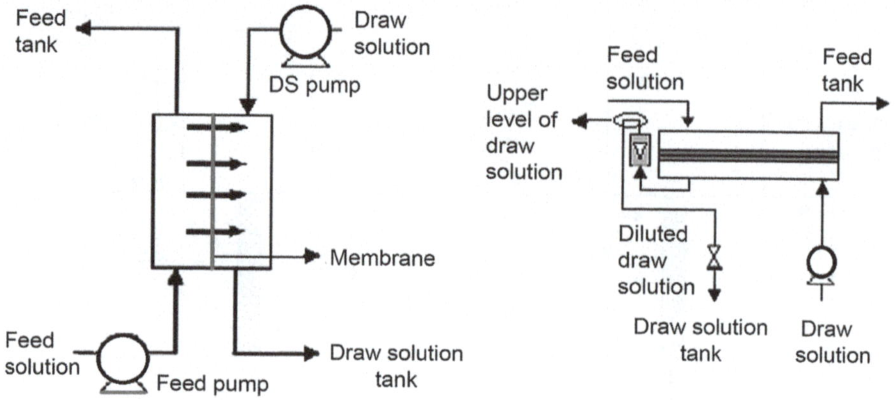

Fig. 8.18 Forward osmosis

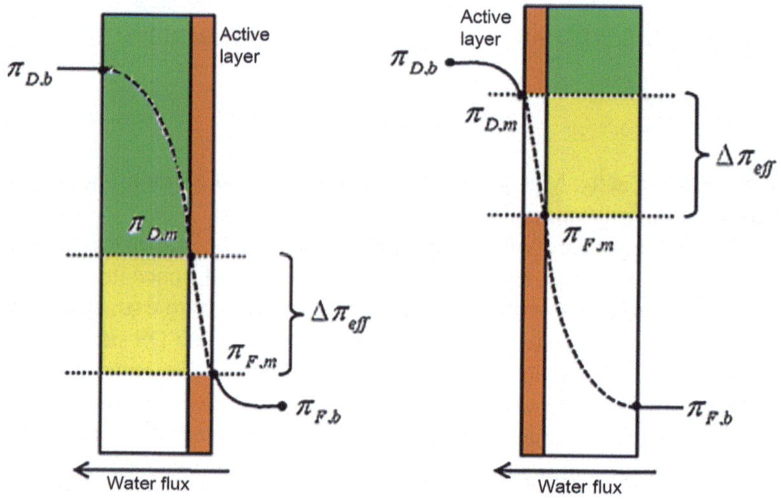

Fig. 8.19 ICP accumulation as a function of osmotic pressure in the active layer facing the feed solution and the active layer facing the draw solution

to salt solution diffusion and reverse diffusion. As a result, the driving force of water flux from the wastewater side to the DS side is reduced, as shown in Fig. 8.19.

In the FO process, ECP and ICP cause a reduction in pure water flux. The ECP and ICP are relatively substantial in a traditional vertical orientation, significantly restricting membrane performance regarding water flux (Pal et al. 2016) constructed a FO-NF incorporated module with horizontal membrane module orientation, as shown in Figure, to avoid this CP constraint Fig. 8.19; the effects of CP were significantly decreased, resulting in a significant rise in the flux of pure water. The unique design ensures nearly CP-free mass transfer with sweeping tangential fluid

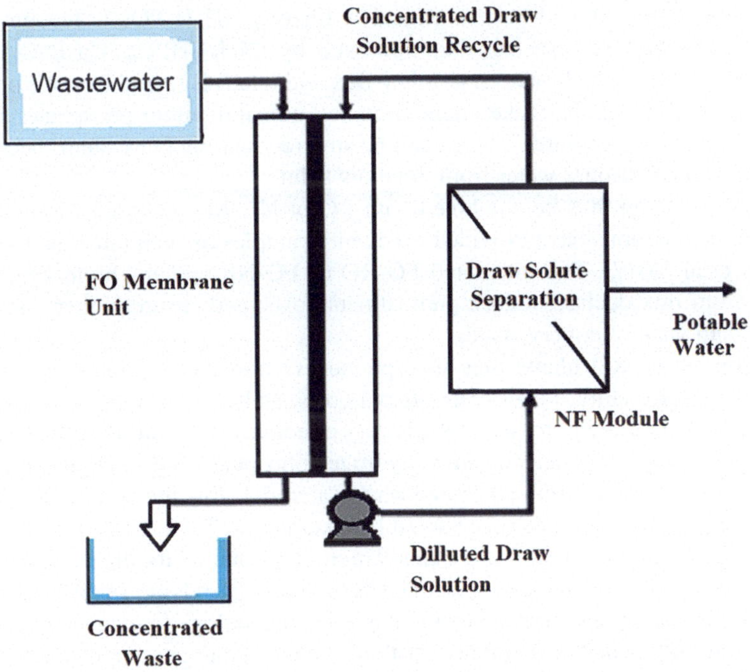

Fig. 8.20 A typical FO-NF treatment scheme is used to recover pure water from wastewater

flow across the membrane surface counter-currently on both sides, offering clear advantages of FO over RO in terms of energy usage and economics.

A regular strategy for industrial wastewater purification using FO is shown in Fig. 8.20. Seawater, brackish water, or any other sort of wastewater can be used as the feed-water stream, and the DS can be of many types. The DS in a system like FO might be thermos-lytic substances like CO_2 or SO_2 that, after osmotic dilution, can be transformed into volatile dioxide gases by applying heat. As a result, these volatile molecules can be reused in the system (Shawky et al. 2019).

When the solutes are recovered by UF and recycled, hydrophilic nanoparticles are also used in DS preparation. Divalent Na_2SO_4 has been used as a draw solute in brackish water desalination, and the draw solute is recovered downstream by NF modules. Ammonium bicarbonate is an excellent DS for extracting water from saline water because it is extremely soluble in water and has the potential to produce osmotic solid pressure. To achieve high flux in a desalination system, ammonium bicarbonate can be used as the draw solute. The diluted draw solute is degraded into ammonia and carbon dioxide in a packed column, and then the ammonia and carbon dioxide are absorbed, and the ammonium bicarbonate solution is reused (Zhao et al. 2012).

In this situation, a high osmotic pressure causes a high flow. Chemical fertilizers, such as $(NH_4)_2HPO_4$, $NH_4H_2PO_4$, $(NH_4)_2SO_4$, and $Ca(NO_3)_2$, are another appropriate class of draw solutes for the FO system in the framework of water extraction from saline water. Out of such chemical fertilizers, $NH_4H_2PO_4$ is recognized as having the lowest reverse salt flux, followed by $(NH_4)_2HPO_4$, Ca $(NO_3)_2$, and $(NH_4)_2SO_4$. Due to the low reverse solute flux and relatively rapid osmotic pressure increase, divalent anions make ammonium sulfate, ammonium phosphate, and Ca $(NO_3)_2$ ideal as draw solutes. There can't be any freestanding FO systems because of the need to recover pure water from the draw solute.

FO systems can thus be combined with RO or NF. Many desalination or water purification situations have provided excellent examples of such integrated systems (Alkaisi et al. 2017). The integrated FO-RO or FO-NF systems protect the entire system from flux decline, reduce pretreatment costs, and ensure higher flux in the downstream water recovery stage.

Several active RO plants may incorporate FO into their systems, resulting in enhanced water quality, lower pretreatment costs, and a more stable and improved flow. An FO-RO system, for example, is very effective in separating sulfate, sodium hexa-meta-phosphate, silica, disodium hydrogen phosphate, sodium lignosulfonates, and other contaminants from coal mine wastewater, reducing the wastewater volume by 80% and allowing it to be discharged (Thakura et al. 2015; Kim et al. 2014).

In 2002, Hydration Technology Innovations was one of the first companies to commercialize FO technology for the US military based on sugar-based draw solutions. This company was the first to apply FO technology to recover and recycle water from wastewater in hydraulic drilling. Other manufacturers entered the FO industry afterward, including Oasis Water, Trevi Systems, Aquaporin A/S, Fluid Technology Solutions, Toyobo, Modern Water, and Porifera.

8.15 Integrated Hybrid Processes in Water Treatment

Applications of hybrid and integrated desalination processes were applied in the treatment of a great mixture of water and wastewater sources (Awad et al. 2020). Hybrid technologies are developed to provide water treatment solutions. As the demand for closed-loop water treatment plants increases, protecting surface water bodies from toxic industrial or municipal wastewater is essential while also reducing utilization of freshwater. The water treatment method developed by the University Cape Town (UCT) is based on membrane bioreactors (MBR), which combine membrane separation and biological treatment to specify high-quality water for reuse. Figure 8.21 shows a hybrid treatment strategy that effectively removes 96% carbon, 89% nitrogen, and 86% phosphorus from municipal wastewater (Thiruvenkatachari et al. 2016). The collected water at the RO membrane module permeate side after treatment is reusable.

Anaerobic, anoxic, and aerobic bioreactors are combined with a hollow fiber MF module of 0.1 m in the UCT configuration of the MBR. The aerobic and the membrane reactors are blown with air. TSS (total suspended solids) sensors are fitted in the anaerobic reactor and the membrane section, while an oxidation-reduction potential sensor is fixed in the anoxic reactor. The matching probes

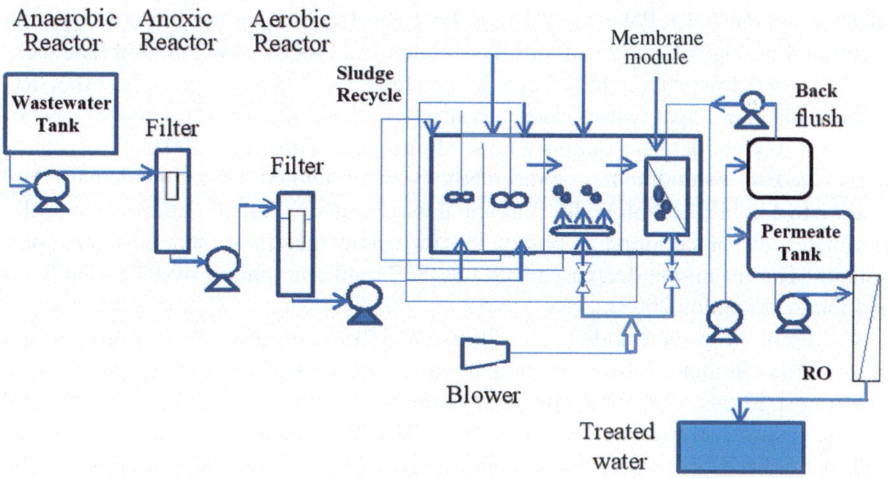

Fig. 8.21 Hybrid water treatment developed by the University of Cape Town MBR Configuration (Thiruvenkatachari et al. 2016)

added in the bioreactor to measure the aerobic reactor's temperature, pH, and DO. Throughout air diffusers and the related proportional-integral-derivative (PID) controller, D_o is kept at 1.5 mg/L in the reactor.

Aeration, backwashing, hydraulic holding time, sludge retention time, mixed liquor suspended particles, and recycle streams are controlled using programmed control techniques. Before charging the municipal wastewater to the MBR system, hydraulic holding time runs through coarse screening (1 cm) and two stages of particle filtering (1 mm and 0.6 mm). MBR is well-known for efficiently removing both N and P. De-nitrification requires the anoxic zone. The successful removal of organic phosphorus in MBR systems is attributed to the growth of phosphate-accumulating bacteria in the anaerobic reactor. This is where the assimilation of organic compounds, such as volatile fatty acids, occurs, leading to the release of accumulated phosphorus.

Phosphorus is ultimately consumed by the aerobic reactor's phosphate-accumulating microbes. Aeration in the MBR reduces membrane fouling. Sludge produced through the membrane tank is recovered into the aerobic and anoxic reactors. The sludge from the aerobic reactor is directed into the anoxic reactor, while the sludge from the anoxic reactor is directed to the anaerobic reactor. Similar hybrid water treatment systems have been developed, combining biological and chemical treatment technologies with membrane separation technologies, with nutrients recaptured as valuable byproducts from wastewater (Monclus et al. 2010).

Combining membrane technologies with activated sludge yields several advantages, including high quality of the discharge stream and a low carbon footprint. Because MBR uses membrane-based separation rather than gravity sedimentation, the effluent has low turbidity and is microorganism-free. Furthermore, when secondary clarifiers become obsolete, the treatment plant becomes compact and smaller

(Kumar and Pal 2015; Pal et al. 2015). Other hybrid techniques involving integrating traditional biological, chemical, or physicochemical processes with membrane technologies used downstream have been documented. One such system is the MBR-RO system for municipal wastewater advanced treatment, which removes 97% of chemical oxygen demand but only 66% of nitrogen (Xing et al. 2001).

The MBR technique has additionally been combined with downstream NF, leading to a hybrid technique for municipal wastewater reclamation that successfully eliminates dissolved organic pollutants, inorganic materials, and nitrogen compounds. The technique decreases COD by 86% and ammonium by 84% (Dialynas and Diamadopoulos 2008).

Carrier biofilms were added to traditional MBRs to improve carbon, nitrogen, and phosphorus elimination from municipal wastewater, resulting in increases in COD, total phosphorous, and total nitrogen, elimination efficiency of 3.8%, 1.7%, and 13.7%, respectively (Abdel-Fatah et al. 2021). Membrane separation was linked with a fluidized-bed bioreactor containing granular activated carbon sponge. The submerged MBR successfully eliminated 95% of NH_4-N, 90% of dissolved organic matter, and 70% of PO_4-P (Liu et al. 2010).

The staged anaerobic fluidized-bed reactor (SAF-MBR) was designed to reduce the expense of membrane cleaning by combining two reactors, including an anaerobic fluidized-bed and an aerobic fluidized-bed membrane. This SAF-MBR system eliminated 84% of COD and 92% of biochemical oxygen demand (Nguyen et al. 2013).

When dealing with complex wastewater streams, hybrid technology is frequently required. Coke-oven wastewater is an example of a harmful wastewater stream that is challenging to adequately treat with a single treatment method. The treatment of a large volume of such wastewater using biological methods can be cost-effective. The presence of toxic impurities in high concentrations, like phenol, prevents direct wastewater stream biological treatment because microbes cannot survive in the presence of a toxic environment. Before chemical treatment is required to achieve significant reductions in such harmful pollutants for downstream biological treatment to be successful.

As a result, Fenton's chemical treatment of wastewater before using biological methods is an excellent choice. Because Fenton's treatment cannot handle ammonium nitrogen, separating the total waste stream into a stream rich with cyanide- and phenol compounds and another stream containing ammonium nitrogen-enriched flow by filtering the wastewater using membrane modules results in improved wastewater treatment efficiency. With the addition of magnesium compounds, ammonium nitrogen can be retrieved as a helpful byproduct called struvite. Through the successful development of hybrid treatment technology, a combination of chemical and biological treatments with membrane-based separation has been achieved (Yoo et al. 2012; Kumar and Pal 2013).

8.16 MD Technology in Water Treatment

As a favorable water treatment procedure, Membrane distillation (MD) has been applied to many wastewater treatments, including oily wastewater, desalination brine, textile wastewater, and radioactive wastewater. The effectiveness of MD in these applications has been completely evaluated. Additionally, the practical study discussed obstacles to industrialization, such as membrane fouling, membrane wetting, and high energy consumption. To address these challenges, several strategies have been implemented to enhance MD performance. These strategies involve integrating membrane processes and creating membranes with specific surface characteristics. Furthermore, nutrient recovery and waste heat utilization have been emphasized (Yan et al. 2021).

The only membrane-based water purification method that involves phase shift is membrane distillation (MD), or a slightly unique style of pervaporation. As a result, the energy required for an MD process is projected to be more than that needed for a membrane separation method driven by pressure difference. On the other hand, the heat energy requirement for a thermally driven membrane separation process is minimal, and the true driving force is the difference in vapor pressure caused by the temperature difference across the membrane (Yan et al. 2021).

MD can be used to purify water from 30 to 80 °C. As in traditional distillation, the equilibrium between vapor and liquid phases influences the components' separation in MD. In MD, however, low-grade heat energy can cause the needed separation, unlike in classical distillation. MD, unlike traditional distillation, is ineffective when the solution contains many volatile components. Water can be purified using MD technology to remove trace impurities such as arsenic, lead, boron, fluoride, and other toxins. Such a technique is more likely to be viable if the requisite low-grade thermal energy can be collected from sunshine. To put it another way, a solar-powered MD that purifies water of trace yet poisonous or hazardous chemicals might be deemed a long-term solution. Later in this chapter, we'll look at a technology that was recently invented (Abdel-Fatah et al. 2019).

8.16.1 Membrane Distillation Operating Conditions

The membrane must have micro-pores (0.1–1 μm) to allow water vapors to pass through for effective separation. The temperature difference across the membrane creates a difference in water vapor pressure, which drives water vapor molecules to move through the dry pores. When the vapor molecules reach the membrane pore's exit point, they should descend into a comparatively chilly environment for quick condensation. The liquid entry pressure (LEP) is the minimum hydrostatic pressure required for liquid water molecules to enter the membrane pores, and it is a crucial membrane design parameter.

The manufactured membrane must have a superior LEP to prevent liquid infiltration. Where reasonable thermal stability is important within the temperature range of 30–90 °C, MD membranes can be made from a variety of polymers such as poly-tetra-fluoro-ethylene (PTFE), PP, and poly-vinylidene-di-fluoride (PVDF) and other chemically resistant polymers. The feedstock material must have minimal thermal conductivity to avoid heat loss. The MD process operates at low temperatures and pressures, resulting in relatively modest equipment and operating costs. Membrane fouling is essentially non-existent in this technology compared to the pressure-driven membrane process. Pore size (rp), tortuosity (τ), membrane thickness (Δx), porosity (ε), and flow mechanism (Knudsen diffusion) or viscous flow through (Lawson and Lloyd 1997), all play a role in the molar flux (N) through the membrane pore.

$$N\alpha\frac{r^{a}\in}{\tau\delta} \tag{8.35}$$

Where: the factor "a" is equal to 1 for Knudsen diffusion and 2 for viscous flow.

8.16.2 Membrane Distillation Configurations

There are four main designs for MD: Direct contact membrane distillation (DCMD), air-gap membrane distillation (AGMD), sweeping gas membrane distillation (SGMD), and vacuum membrane distillation (VMD). These designs differ based on how the penetrating vapor is collected on the opposite side of the membrane. The feed and permeate liquids are in explicit contact with the membrane within the respective compartments of the feed and permeate in the DCMD design. In a pure liquid pool, the vapor phase is condensed and stored on the other side of the membrane. We investigate such a configuration in the context of a flash vaporization DCMD module. Despite the heat loss due to membrane conduction, this module type has more flux than other modules.

In the design of an AGMD system, the condensing surface is separated through the membrane by a stagnant air gap on the permeate side. Permeated vapor molecules should cross the air gap to condense on a cold surface. As this air gap widens, the heat loss by conduction drops quickly at first, then gradually becomes constant. As the air gap widens, the flux reduces as well. In SGMD, vapor condensation is accomplished in the permeate compartment by a cool sweeping gas. Condensation on the permeate side occurs under vacuum in VMD.

8.16.3 *MD Technology for Purification of Water from Trace-Metalloid*

MD's intrinsically low flux is a key limitation for industrial or large-scale applications. As a result, we will present the creation of an MD technology in which low flux has been effectively disabled (Pal et al. 2013) by increasing the flux within the module design. Furthermore, because it is a solar-driven module, the system is autonomous in terms of supplying the energy requirements for the phase shift. The module type is predicted to be particularly effective in sun-drenched Southeast Asian countries, where substantial arsenic-affected areas exist, and solar-powered pumps are employed in agriculture. A carefully designed feature in this modern module dubbed a flash-vaporization module, causes flash vaporization of the heated polluted feed-water, virtually doubling the pure water flux, almost double the commonly reported values.

The flux-enhancing module (FVMD) consists of a direct contact FVMD module and a solar-heating loop formed of an evacuated glass panel, as shown in Fig. 8.22. The flat-sheet crossflow membrane module, situated between the feed and permeate cells, is constructed of a polycarbonate-sandwiched flat membrane. As a result of the poor thermal conductivity, the polycarbonate material aids in preventing thermal loss. The module is set up in a DCMD configuration, which allows for high flux due to flash vaporization. The feed is heated using a solar energy collector with evacuated glass tubes. The membrane surfaces are in direct contact with the hot-liquid feed and cold-liquid permeate in this DCMD.

8.17 Conclusion and Prospects

This chapter has provided a comprehensive overview of membrane separation technologies in water and wastewater treatment, emphasizing the versatility and effectiveness of various membrane filtration methods. The detailed exploration of membrane materials, flow regimes, modules, and transport mechanisms has laid a solid foundation for understanding the operational principles and applications of these technologies.

NF membranes, in particular, have emerged as a pivotal technology due to their unique characteristics that bridge the gap between UF and RO. Their ability to selectively remove ions and organic substances makes them indispensable in diverse wastewater treatment processes, including water reuse, tertiary treatment, water softening, and desalination.

Looking ahead, the continued development and optimization of NF membranes hold significant promise for advancing water and wastewater treatment technologies. Key areas of focus include:

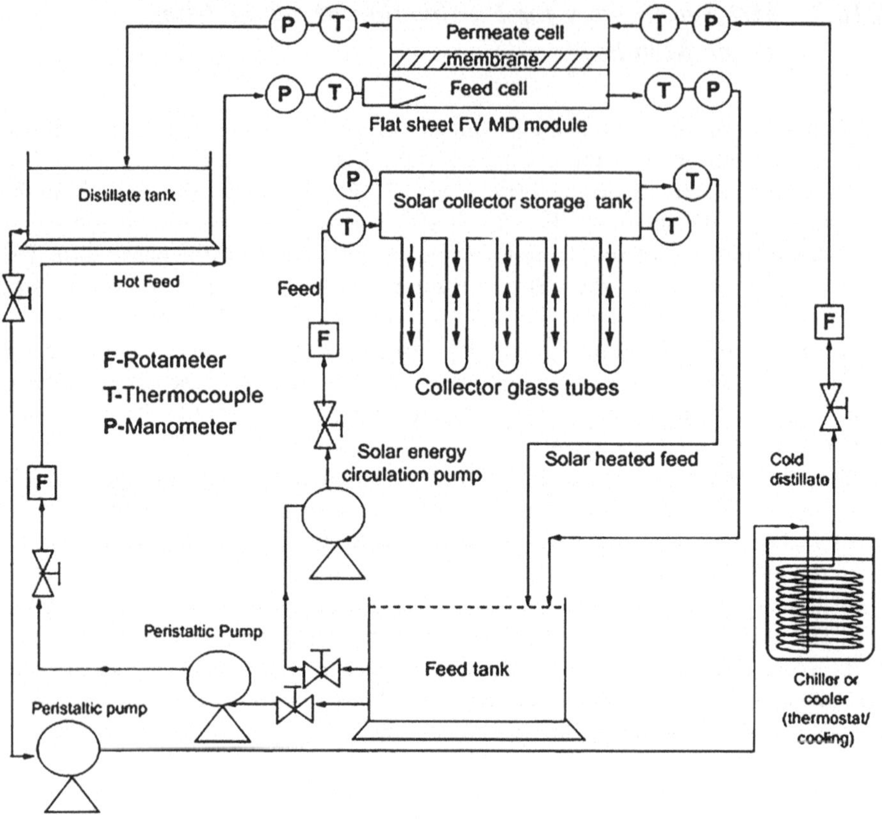

Fig. 8.22 A flash-vaporization direct-contact (MD) module (Pal et al. 2013).

1. Material Innovation: Developing new membrane materials with enhanced selectivity, permeability, and fouling resistance to improve efficiency and longevity.
2. Energy Efficiency: Reducing the energy consumption of membrane processes through innovative design and operational strategies.
3. Hybrid Systems: Integrating NF membranes with other treatment technologies to create hybrid systems that maximize treatment efficacy and resource recovery.
4. Sustainability: Enhancing the sustainability of membrane processes by minimizing waste generation and promoting the use of environmentally friendly materials.
5. Broader Applications: Expanding the use of NF membranes in emerging fields such as pharmaceutical wastewater treatment, food and beverage processing, and advanced industrial applications.

By addressing these areas, NF membrane technology can continue to evolve, offering more efficient, sustainable, and versatile solutions for water and wastewater treatment challenges.

References

Abdel-Fatah MA (2018) Nanofiltration systems and applications in wastewater treatment: review article. Ain Shams Eng J 9(4):3077–3092

Abdel-Fatah MA, Shaarawy HH, Hawash SI (2019) Integrated treatment of municipal wastewater using advanced electro-membrane filtration system. SN Appl Sci 1(10):1153

Abdel-Fatah MA, Khater EMH, Hafez AI, Shaaban AF (2020) Performance of fouled NF membrane as used for textile dyeing wastewater. Membr Water Treat 11(2):111–121

Abdel-Fatah MA, Amin A, Elkady H (2021) Industrial wastewater treatment by membrane process. In: Membrane-based hybrid processes for wastewater treatment. IntechOpen, pp 341–365

Alkaisi A, Mossad R, Barforoush S (2017) A review of water desalination systems integrated with renewable energy. Energy Procedia 110:268–274

Awad Abouelata AM, Abdallah SMA, Sorour MH, Shawky NA, Abdel-Fatah MA (2020) Modification and ionic stimulation of hollow fiber membrane by electric field for water treatment. J Appl Polym Sci 137(39):49190

Barakat MA, Schmidt E (2010) Polymer-enhanced ultrafiltration process for heavy metal removal from industrial wastewater. Desalination 256:90_3

Belleville M-P, Vaillant F (2016) Membrane technology for production of nutraceuticals. Taylor & Francis Group, LLC

Dialynas E, Diamadopoulos E (2008) Integration of immersed membrane ultrafiltration with coagulation and activated carbon adsorption for advanced treatment of municipal wastewater. Desalination 230:113–127

Donnan FG (1924) The theory of membrane equilibria. Chem Rev 1:73–90

Ezugbe EO, Rathilal S (2020) Membrane technologies in wastewater treatment: a review. Membranes 10:89

Gao Y, Zhao J, Zhang GH, Zhang D, Cheng WW, Yuan GQ, Liu XJ, Ma BZ, Zeng JH, Gu P (2004) Treatment of the wastewater containing low-level 241 Am using flocculation-microfiltration process. Sep Purif Technol 40(2):183–189

Garcia-Ivarsa J, Iborra-Clara M-I, Alcaina-Miranda M-I, Mendoza-Roca J-A, Pastor-Alcaniz L (2015) Treatment of table olive processing wastewater using novel photo-modified ultrafiltration membranes as first step for recovering phenolic compounds. J Hazard Matter 290:51–59

Giacobbo A, Bernardes AM (2022) Membrane separation process in wastewater and water purification. Membranes 12:259

Guo H, Li X, Yang W, Yao Z, Ying Mei L, Peng E, Yang Z, Shao S, Tang CY (2022) Nanofiltration for drinking water treatment: a review. Front Chem Sci Eng 16(5):681–698. https://www.jersywater.je/about-us/learn-more/desalination-plant/

Huang Y, Du J, Zhang Y, Lawless D, Feng X (2015) Removal of mercury(II) from wastewater by polyvinyl amine-enhanced ultrafiltration. Sep Purif Technol 154:1–10

Kim Y, Lee JH, Kim YC, Lee KH, Park IS, Park S-J (2014) Operation and simulation of pilot scale forward osmosis desalination with ammonium bicarbonate. Chem Eng Res Design. Available from: https://doi.org/10.1016/j.cherd.2014.08.015

Kumar R, Pal P (2013) Membrane-integrated hybrid bioremediation of industrial wastewater: a continuous treatment and recycling approach. Water Reuse Desalin 3(1):26–38

Kumar R, Pal P (2015) Assessing the feasibility of N and P recovery by struvite precipitation from nutrient-rich wastewater: a review. Environ Sci Pollut Res 22:17453–17464

Langelier WF (1936) The analytical control of anticorrosion water treatment. J Am Water Works Assoc 28:1500

Lawson KW, Lloyd DR (1997) Membrane distillation. J Membr Sci 124:1–25

Li CM, Yu N (2009) Analysis about the application of membrane separation technique in heavy metal wastewater treatment. Environ Ecol Three Gorges 2(1):46–49

Liu C-C, Liu JC (2016) Coupled precipitation-ultrafiltration for treatment of high fluoride content wastewater. J Taiwan Inst Chem Eng 58:259–263

Liu Q, Wang XC, Liu Y, Yuan H, Du Y (2010) Performance of a hybrid membrane bioreactor in municipal wastewater treatment. Desalination 258:143–147
Long S, Yu S (1999) Concentration polarization in crossflow reverse osmosis. AICHE J 45:921–928
Monclus H, Simpa J, Ferrero G, Rodriguez-Roda I, Comas J (2010) Biological nutrient removal in an MBR treating municipal wastewater with special focus on biological phosphorus removal. Bioresour Technol 101:3984–3991
Muthu SS, Khadir A (eds) (2022) Membrane based methods for dye containing wastewater, Technical aspects of nanofiltration for dyes wastewater treatment, recent advances. Springer Nature Singapore, pp 23–35
Nguyen TT, Ngo HH, Guo W (2013) Pilot scale study on a new membrane bioreactor hybrid system in municipal wastewater treatment. Bioresour Technol 141:8–12
Pal P, Sikder J, Roy S, Giorno L (2009) Process intensification in lactic acid production: a review of membrane-based processes. Chem Eng Process Process Intensif 48:1549–1559
Pal P, Manna AK, Linnanen L (2013) Arsenic removal by solar-driven membrane distillation: modeling and experimental investigation with a new flash vaporization module. Water Environ Res 85:63–76
Pal P, Abrar I, Kumar R (2015) Managing hazardous municipal wastewater: a membraneintegrated hybrid approach for fast and effective treatment in low temperature environment. J Membr Sep Technol 4(2):1–13
Pal P, Das P, Chakrabortty S, Thakura R (2016) Dynamic modeling of a forward osmosisnanofiltration integrated process for treating hazardous wastewater. Environ Sci Pollut Res 23:21604–21618
Shaaban AMF, Hafez AI, Abdel-Fatah MA, Abdel-Monem NM, Mahmoud MH (2016) Process engineering optimization of nanofiltration unit for the treatment of textile plant effluent in view of solution diffusion model. Egypt J Pet 25(1):79–90
Shah MP, Rodriguez-Couto S (2022) Membrane-based hybrid processes for wastewater treatment. Elsevier
Shawky NA, Abdallah SMA, Sorour MH, Abouelata AMA, Abdel-Fatah MA (2019) Influence of the electric potential applied on hollow fiber (HF) membrane and surface morphology. Int J Recent Technol Eng 8(3):1608–1611
Shehata N, Egirani D, Olabic AG, Inayat A, Abdelkareem MA, Chaef K-J, Sayed ET (2023) Membrane-based water and wastewater treatment technologies: issues, current trends, challenges, and role in achieving sustainable development goals, and circulareconomy. Chemosphere 320:137993
Sombekke HDM, Voorhoeve DK, Hiemstra P (1997) Environmental impact assessment of groundwater treatment with nano filtration. Desalination 113:293–296
Song LZ, Zhang ZJ, Chen MZ, Zhao J (2007) Surface modification of polyvinylidene fluoride microfiltration membrane. Membr Sci Technol 27(5):46–50
Starzak ME (1984) The physical chemistry of membranes. Academic, New York
Strathmann H (2011) Introduction to membrane science and technology. Wiley-VCH
Sun QJ, Sun CS, X JL, Jui HJ (2015) The application of microfiltration-reverse osmosis process in printing and dyeing wastewater reuse. Shandong Chem Ind 44(2):141–142, 145
Thakura R, Chakrabortty S, Pal P (2015) Treating complex industrial wastewater in a new membrane-integrated closed loop system for recovery and reuse. Clean Techn Environ Policy 17(8):2299–2310
Thiruvenkatachari R, Francis M, Cunnington M, Su S (2016) Application of integrated forward osmosis and reverse osmosis for coal mine wastewater desalination. Sep Purif Technol 163: 181–188
Torki M, Nazari N, Mohammadi T (2017) Evaluation of biological fouling of RO/MF membrane and methods to prevent it. Eur J Adv Eng Technol 4(9):707–710

Urgun-Demirtas M, Negri MC, Gillenwater PS, AgwuNnanna AG, Yu J (2013) Meeting world's most stringent Hg criterion: a pilot-study for treatment of oil refinery wastewater using un ultrafiltration membrane process. J Environ Manag 117:65–75

Wan JB, Wang JY (2008) Experimental study on neutralization/microfiltration process for treatment of acid wastewater containing heavy metals. China Water Wastewater 24(1):62–64

Wang C, Wang Y, Qin H, Lin H, Kong C (2020) Application of microfiltration membrane technology in water treatment. IOP Conf Ser Earth Environ Sci:571. https://doi.org/10.1088/1755-1315/571/1/012158

Xing CH, Qian Y, Wen XH, Wu WZ, Sun D (2001) Physical and biological characteristics of a tangential-flow MBR for municipal wastewater treatment. J Membr Sci 191:31–42

Yan Z, Jiang Y, Liu L, Li Z, Chen X, Xia M, Fan G, Ding A (2021) Membrane distillation for wastewater treatment: a mini review. Water 13:3480

Yoo R, Kim J, McCarty PL, Bae J (2012) Anaerobic treatment of municipal wastewater with a staged anaerobic fluidized membrane bioreactor (SAF-MBR) system. Bioresour Technol 120: 133–139

Zaro D, Prats D (2018) Desalination and energy consumption. Can we expect future? Desalination 427:1–9

Zhang CP, Gu P, Zhao J, Zhang D, Deng Y (2009) Research on the treatment of liquid waste containing cesium by an adsorption-microfiltration process with potassium zinc hexacyanoferrate. J Hazard Mater 167:1057–1062

Zhao J, Wang T, Zhang D, Liu XJ, Fu YB (2007) Treatment of low-level wastewater of Pu using coagulation and microfiltration technological process. J Nucl Radiochem 29(2):113–117

Zhao S, Zou L, Mulcahy D (2012) Brackish water desalination by a hybrid forward osmosisnanofiltration system using divalent draw solute. Desalination 284:175–181

Chapter 9
Green Nanotechnology in Food Packaging and Preservation

Dhriti P. Shah and Amee Krishnakumar

Abstract Food packaging and preservation remains central to the core aspects of the food industry, including safety, preservation, processing, and supply chain management. As the demand and consumption of packaged and processed foods continue to grow, the need for safe, healthy, unadulterated food, complete information regarding its nutritional value, shelf-life, constituents and manufacturing process becomes an absolute necessity. The conventional ways of food packaging poses potential threats to the consumer's health and often leads to environmental pollution, thus necessitating greener options for food packaging and preservation. The advanced eco-friendly options to food preservation and packaging, aims not only at enhancing the quality of food, shelf-life and sustainability but also at mitigating environmental hazards. The current chapter sheds light on the significance of the newer advents of green nanotechnology in food packaging and preservation. While the wide variety of conventional techniques are available for food preservation, modern day technologies including smart packaging comprising of active and intelligent packaging, as well as the inclusion of green nanomaterials such as nanoparticles, nanoencapsulation, nanoemulsion, nanocomposites, nanofillers and nanoclays provide noteworthy enhancements in the food packaging arena. These alternatives ensure improved quality of packaging and food, thereby addressing the prevalent limitations of conventional methods. Furthermore, these advanced automated technologies enable the understanding and monitoring of food quality and signals the quality deterioration over the period in real-time, tackling the issues of conventional packaging and focusing on consumer beneficial packaging.

Keywords Biopolymers · Food packaging · Food preservation · Green nanotechnology · Nanomaterials · Smart packaging

---

D. P. Shah · A. Krishnakumar (✉)
Institute of Science, Nirma University, Ahmedabad, Gujarat, India
e-mail: 16ftphds39@nirmauni.ac.in; ameenair@nirmauni.ac.in

© The Author(s), under exclusive license to Springer Nature Switzerland AG 2025

J. M. Al-Khayri et al. (eds.), *Green Nanotechnology Applications for Ecosystem Sustainability*, Nanotechnology in Plant Sciences 4,
https://doi.org/10.1007/978-3-031-90766-1_9

9.1 Introduction

Modernization deals with improving the quality of life with evolution in the traditional methods, more specifically revolving around the basic necessities of survival. Food constitutes the most basic necessity for survival of living organisms. With newer advents and modernization, food related challenges have also evolved necessitating the need for new ecofriendly technology, to resolve them.

Food technology involves formulation, preparation, manufacture, preservation (Yan et al. 2022), quality assessment, packaging as well as continuous research (Tumuluru 2023) to improve the problems underlying food supply, enhancing the food quality and reliability (Maurizzi et al. 2022), which is central to the smooth functioning of the food industry. Food industry focuses on understanding the needs, while providing constructive solutions to the underlying problems and thereby aiding in meeting the demands of the consumers. With the increase in demands for packaged and processed foods, there arises the need for efficient packaging, that ensures clean, edible and safe foods with appropriate treatment for its longevity (Yan et al. 2022). Food packaging and preservation makes up the core part of food industry (Garba 2023; Muller and Schmid 2019), that entails a coordinated system aiming at maintaining and ensuring the highest safety and quality for packaged foods (Maurizzi et al. 2022). This also includes maintaining the shelf-life and product related communications between the manufacturer and the consumer (Yan et al. 2022). Furthermore, when food adulterations and wastage are at peak, a safe, unhampered packaging of the food with appropriate preservation to avoid depletion, becomes an absolute necessity (Maurizzi et al. 2022). Figure 9.1 emphasizes the importance of food preservation and safety in food industry.

The earlier techniques of food packaging are efficient, however is deemed risky as well as harmful to the human health and environment at large, besides being costly. These concerns are chiefly addressed by the use of green nanotechnology while being eco-friendly and renewable. The current chapter focuses on the green nanotechnological advancements in the field of food packaging and preservation, discussing the alternatives that are preferred while being nonhazardous and maintaining the efficiency of preservation.

9.2 Food Preservation

Food safety is reliant on to its edibility, and integrity, and food spoilage is one of the major issues while considering packaging and preservation. Environmental alterations, moisture content of food, temperature, microbial invasion and chemical actions can cause food spoilage (Sridhar et al. 2021). Appropriate food is needed for cautious food preservation to prevent occurrence of food borne illnesses. Food preservation entails an action or method aimed at maintaining the quality of food at a desired level (Prokopov and Tanchev 2007). Food preservation includes processing

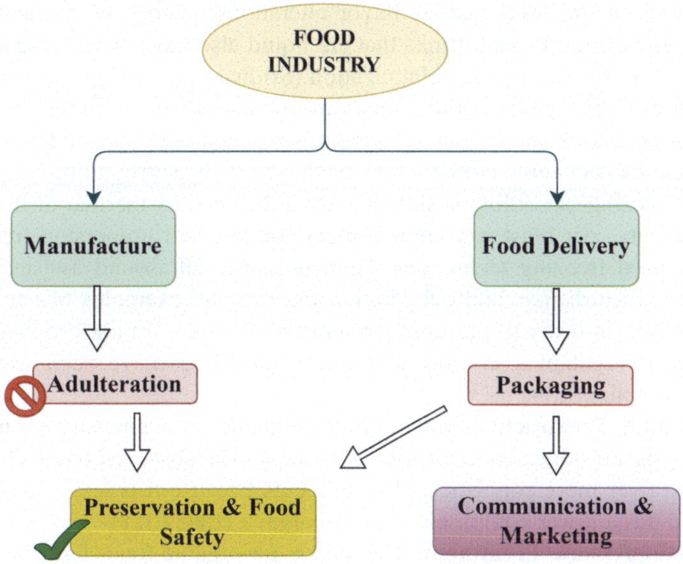

Fig. 9.1 Rationale of food preservation in food industry. (Figure constructed by Shah and Krishnakumar)

of food, to enhance its shelf life, and maintain the quality of food in a desired manner until a stipulated period of time, while retaining its original color, shape, size and texture. Food preservation intends to overcome challenges of food spoilage caused by a variety of chemical or biochemical reactions and microbial spoilage of foods, which can be preserved by application of a variety of conventional and primitive techniques like, freezing, drying, chilling, pasteurization, along with chemical preservation (Amit et al. 2017).

9.2.1 Food Preservation Techniques

The major techniques employed for food preservation include Thermal treatment, Freezing, Ultrasound treatment, Ozonation/Ozone treatment, Dense phase Carbon Dioxide (DPCD), High Hydrostatic Pressure (HHP), Pulse Electric Field, Ionic liquids, Osmotic dehydration, ultra-sonication, microwave vacuum drying and thermoelectric refrigeration (Sridhar et al. 2021).

The following are the major traditional techniques of food preservation:

1. **Thermal Treatment:** One of the cutting-edge methods for preserving food is heat, or thermal treatment. Foods are typically heated during the procedure to a temperature of 75–90 °C or higher and held for 25–30 s. Foods that have been heated lessen the germs. However, a wealth of study has also found that there are

reductions in the food matrix, flavor alterations, energy waste, and nutrient losses. Juices, foods, and drinks that are liquid also have an adverse effect that can result in browning and gelatinization (Sridhar et al. 2021).

2. **Freezing:** Heat transfer is the primary mechanism involved in the freezing and cooling processes. A cooling agreement is reached when heat energy from the meal and its packaging container is transferred to the surrounding air. Thus, the rate of freezing or cooling is significantly influenced by thermal diffusivity and conductivity. Air blast, cryogenic, direct contact, and immersion freezing are widely used freezing techniques. High pressure, ultrasound assisted, electromagnetic disturbance, and dehydration freezing are examples of sophisticated approaches. In order to preserve the nutritional value of milk products, spices, and green vegetables, cooling and freezing methods have been widely used (Sridhar et al. 2021).
3. **Ultrasound Treatment:** Utilizing high frequency and intensity sound waves, food items are subjected to ultrasound therapy. The cost-effective technology is selected over more sophisticated instruments because it is less complicated to use and has a lower equipment need (Bhargava et al. 2021).
4. **Ozonation/Ozone Treatment:** The market for organic foods has grown significantly as consumers gradually shift their needs towards sustainable lifestyles and healthy eating. Customers require a useful product with a respectable shelf life that is devoid of additives and preservatives. As a result, ozone therapy technology has gained popularity in recent years. Ozone was selected because of its many characteristics and rapid breakdown. To put it simply, ozone is an oxygen allotrope. When oxygen splits in the presence of light or ultraviolet radiation into a single oxygen or nascent oxygen, the molecule is created. Compared to chemical preservatives like chlorine (1.35 V), hydrogen peroxide (1.78 V), and hypochlorous acid (1.79 V), the compound has a high oxidation potential (2.07 V) and breaks down quickly into oxygen molecules, making it an effective antibacterial and antiviral agent. In addition, because ozone can be produced instantaneously, it eliminates the need to stockpile dangerous chemicals. In addition, less energy is needed than with thermal treatment, which priorities shelf life (Sridhar et al. 2021).
5. **Pulse Electric Fields:** A more sophisticated pre-drying method that uses a shorter residence period to treat food is called pulse electric field technology. Because of its continuous functioning and low electric field demand (1–5 kV/cm), the approach gained widespread recognition. Given that the process needs a very low temperature of 40 °C to operate, it may be used in place of thermal drying and perhaps improve food drying. The food (fruit, vegetable, milk or any juice) is placed between two electrodes as part of the pulse electric field technique. A high voltage (50 kV/cm) pulse is then applied for brief periods of time. Combining electroporation with electro permeabilization is the basic idea. The food's cell membrane matrix is broken by the electric field, improving its nutritional value, safety, and shelf life. The temperature, polarity, treatment duration, pulse width, frequency, and field strength all affect the pulse electric field (Sridhar et al. 2021).

6. **High Hydrostatic Pressure (HHP):** Enhancing food safety and shelf life can be achieved by inactivating enzymes and destroying microorganisms by the application of high hydrostatic pressure (HHP) treatment. The following procedures can be used to apply classical HHP or single-pulsed HHP (spHHP) treatment: compression to target pressure, holding at the target pressure for a predetermined amount of time, and decompression to air pressure (Aganovic et al. 2021; Buzrul 2015).
7. **Dense phase Carbon Dioxide (DPCD):** Fruit juices, drinks, and other liquid meals are examples of highly heat-sensitive perishable food items that can be preserved non-thermally using dense phase carbon dioxide (DPCD). In food processing, pressurized CO_2 (pressure 1/10th of the HHP) is used in a continuous process at a moderate temperature of 30–50 °C. In addition to other non-thermal methods, this technology can be employed in conjunction with irradiation, high hydrostatic pressure (HHP), ultrasound, and pulsed electric field (PEF) (Balaban and Duong 2014; Devraj et al. 2021).
8. **Ionic Liquids:** Salts that have melting temperatures lower than 100 °C are known as ionic liquids. They are composed of an anion that can be either organic or inorganic and a bulky organic cation. These salts are liquid at room temperature owing to the strong degree of asymmetry between the ions, which prevents the formation of a crystalline structure. The association between their structure and particular functionality is directly responsible for their undetectable volatility, high chemical and strong thermal stability, good electrical conductivity, and vast tuning capacity, thereby proving to be a good preservant than traditional solvents (Fiorentini et al. 2022).
9. **Osmotic Dehydration:** Osmotic dehydration is a phenomenon that occurs when water is transferred from a solution with a lower solute concentration to one with a greater concentration through a semi-permeable membrane, creating an equilibrium on both sides of the membrane. Because osmotic dehydration reduces the water activity of fruits and vegetables, it is widely used in the preservation of food materials. The advantages of osmotic dehydration over alternative techniques are their color, fragrance, nutritional value, and ability to retain flavor compounds. The solutes used in osmotic dehydration are often sugar syrup combined with cubes or slices of fruit and brine or salt (sodium chloride) combined with vegetables (Yadav and Singh 2014).
10. **Ultra sonication:** As a non-thermal food processing technology, ultrasound is used to improve food processing in ways like food preservation, enhanced mass transfer, support for thermal treatments, texture modification, and food analysis. Ultrasonic waves, which are also referred to as supersonic waves, are sound waves that fall between 20 and 100 kHz in frequency. The medium's particle composition experiences rarefaction (decompressions) and compressions as a result of the acoustic waves' propagation. Because of the increased mass transfer and turbulence, this in turn generates a large quantity of energy. The fundamental idea is that sound waves scatter and reflect similarly to light waves. In liquid media, ultrasound causes "cavitation," in gas media, pressure changes, and in

solid media, liquid movement. It can be thought of as a type of high-frequency vibration that produces microscale shear forces and fluid mixing (Bhargava et al. 2021; Chavan et al. 2022).

11. **Microwave Vacuum Drying:** With its fast drying times and low operating costs, microwave vacuum drying (MWVD) has been seen as a competitive alternative for producing crisp fruits and vegetables that are high in nutrients and sensory appeal. Water and other vapours are produced inside the product by microwave heating under vacuum, which encourages pressure gradients between the food's interior and surface. The vapor and liquid water in the pores flow as a result of these gradients. Because thermos-sensitive foods require water evaporation at moderate temperatures, the hoover produces high-quality dried products with less shrinking (González-Cavieres et al. 2021; Monteiro et al. 2020).
12. **Thermoelectric Refrigeration:** Due to its wide temperature range, compact design, and silent operation (as it doesn't require a compressor), thermoelectric refrigeration is a possible substitute. Thermoelectric devices have a robust structure, excellent stability, great dependability, and no greenhouse gas evolution. They transform electricity into heat and vice versa. The creation of thermoelectric refrigeration systems, which remove heat by applying direct current (DC) across two dissimilar materials connected together and creating a temperature difference, is possible with the use of thermoelectric devices. These systems are based on the Peltier effect. Therefore, a thermoelectric refrigeration system was conceived and built in order to address the growing worry of food waste caused by the lack of scientific cold chain infrastructure for rural fruit and vegetable sellers and the concurrent increase in demand for fresh fruits and vegetables (Chavan et al. 2022; Sridhar et al. 2021).

Beyond these techniques, food processing involves the use of nanomaterials which includes certain viscosifying agents, gelation agents, nutraceuticals, as well as agents targeting nutrient delivery, minerals and vitamins fortification and nanoencapsulation of innate and added flavors, all of which aids in food preservation (Huang et al. 2010; Pradhan et al. 2015; Weiss et al. 2006).

9.3 Food Packaging as a Part of Food Preservation

Food packaging can be defined in simple terms as mainly a means of transiting food from one place to another by providing it a safe, non-damageable, physical support in the form of a protective covering (Lloyd et al. 2019; Ranken et al. 1997) that aims at easy, efficient, clean, food supply while maintaining its integrity and ensuring sterile, non-pathogenic infestation while its transportation from the manufacturers to the consumers (Pascall et al. 2022; Tumuluru 2023). Packaging functions to be a barrier to the prevalent atmospheric changes while being protective against gases like oxygen, nitrogen, carbon dioxide along with aromas and moisture while protecting the food from direct light thereby maintaining the nutritive values and color of the food (Ranken et al. 1997). Packaging also provides consumers to be more informed regarding the food content, its nutritive value, place and date of manufacturing, date of expiry, and thus enriching the customers with necessary information regarding the edible product (Yan et al. 2022). Moreover, the food packaging also enables the food items to endure shocks during transit, and withstand the changes in gases, temperature and pressure (Marsh and Bugusu 2007). The food package should also render a creative visual effect amongst the consumers as it is critical from the sales aspect (Khandeparkar et al. 2024).

9.4 Significance of Food Packaging and Preservation

With increasing food adulteration and microbial infestation, the need for efficient, impenetrable food packaging becomes vital, and hence based on the food type, respective packaging changes to ensure protection. Food borne illnesses are common health hazards globally, making the entire concept of food packaging vital. Food packaging is important, not just as a protective sealant for preventing food spoilage, but also remains essential for food sustainability, food waste reduction, transportation, traceability, storage, convenience, integrity, marketing as well as product information (Pascall et al. 2022) and a means to attract the consumer towards its purchase (Sulochana 2024). Lockhart has described the central idea behind food packaging to be the protection, utility and communication pertaining to the food product in concern (Lockhart and Paine 1996).

Packaging of food holds importance owing to the following key reasons (Marsh and Bugusu 2007; Yan et al. 2022):

1. Long-term storage
2. Transit from manufacturers to the consumers
3. Protection from microbial infestation, chemical degradation as well as physical deterioration, environmental effects including extreme temperatures, dirt and dust
4. Increased shelf-life of food items
5. Branding of product
6. Communication pertaining to the food product, its content and nutritional information along with providing the instructions for use
7. Reduction of food-waste owing to poor food packaging
8. Traceability of the packages

If the food packages are not efficient enough to demonstrate the aforementioned characteristics, it leads to food wastage (Marsh and Bugusu 2007; Risch 2009; Sulochana 2024; Yan et al. 2022).

9.5 History of Food Packaging and Preservation

Food packaging has evolved leaps and bounds from the primitive time where leaves and animal skins were used for transit of food materials, to today where the packaging are interactive, with sensors that can provide information in real-time, about the quality of food. Though no records of the initial packaging materials are available, it is believed during the Nomadic era, the food was being transmitted using leaves, animal skins, shells of nuts or gourds (Berger 2003; Risch 2009), while grass, wood and bamboo was used to make baskets out of them for food storage (Risch 2009). As early as 7000 B.C., pottery and glass was used to prepare storage vessels to preserve food. Later paper also sought its application thereby for preparing storage containers (Berger 2003; Risch 2009).

Ancient Egyptians used to sun-dry their foods to protect from spoilage (Pradhan et al. 2015). Later, around 1500 B.C. they discovered the techniques for blowing up glasses which weren't transparent, that were later shaped in glass vessels to hold food and water (Risch 2009). Romans introduced the concept of pickling in order to prevent microbial infestation and protect the food, while the Greeks suggested jellying of food by adding honey or sugar in the preservation technique (Pradhan et al. 2015).

In the first or the second centuries, 105 A.D., Chinese used treated mulberry bark to wrap food. Progressively, they developed flexible paper, that was used for packaging food as well as medicines and tea parcels (Berger 2003). During the medieval ages, fifth—fifteenth century, wood gained popularity. Wooden barrels, boxes as storage and transport carriers for food products like rum, dried food and water.

During the Industrial Revolution (1760–1840), several packaging materials were developed and used. In 1784, Cullen devised a crude method for preparing artificial refrigeration. Rubber printing, introduced printing of real pictures of the food product on the package. In early 1800s, a method to treat the food with salt to keep it fresh and preserve it was introduced. Later the idea of canning was introduced (Pradhan et al. 2015). Nicolas Appert, Father of Canning, has been known to develop the method of boiling food and sealing them in airtight glass container with corks secured with wires to provide closure while heating (Risch 2009) for preservation (Pradhan et al. 2015), which was later replaced with metal cans, that were introduced with an idea of providing a barrier while maintaining moisture levels thereby aiding in preserving food flavors (Risch 2009).

While in 1810, Peter Durand, patented the tin-coated iron cans, instead of glass for preservation of food like cookies and tobacco (Pradhan et al. 2015). In 1817, the first non-corrugated cardboard box was used in the Silk industry to transport worms, eggs and moths across the continents (Risch 2009). The first commercial paper bags were reportedly manufactured in 1844, which resembled large envelopes from mills (Berger 2003).

Louis Pasteur in 1862, familiarized the concept of pasteurization to protect milk, wine and beer from spoilage (Pradhan et al. 2015). In 1867, cellulose fiber from wood pulp was derived and started to gain commercial importance (Berger 2003). In

1871, the corrugated cardboard boxes were developed (Risch 2009). Later in 1879, first cartons were manufactured by Robert Gair (Berger 2003). In 1889 Owens patented the automatic rotary bottle making machines and led to the manufacture of fancy bottles. William Painter in 1892, patented crown corks, a metal cap with a layer of cork that can seal the glass bottle ensuring tight closure and protection (Risch 2009).

In 1906, Kellog brothers, formulated the cereal boxes, that have a layer of heat-sealed plastic bag surrounding the cardboard box, which eventually reversed, that would contain the cereals to maintain its crispiness and freshness (Berger 2003). Introducing the idea of cellophane, was what brought revolution for the packaging materials. Jacques E. Brandenberger in 1908, while trying to prepare a cloth that would not absorb liquid, formulated using wood cellulose by serendipity.

In 1950s, Aluminum and plastic found a great deal of innovations as packaging material during this phase and were employed for canning (Risch 2009). Following it, in 1950s and 60 s, cellophane became the choice of packaging owing to its transparency. Yet another serendipitous discovery was that of Saran wraps, which was remade as dark-green film, called Saran resin. It was earlier used to protect military planes from external elements, later upon removal of the green color of Saran, was used as food packaging material, i.e. after World War II. Post world war II, plastics became more frequent as packaging material for commercialization of food. During the phase of 1950s–1980s, aluminum trays, Polyethylene Terephthalate (PET) and Polyethylene (PE) bags and films were created as covers and aluminum cans with cost-effective and simpler designs were made.

In 1954, Prof. Giulio Natta introduced isotactic polypropylene, that is stretched in crosswise direction giving rise to oriented polypropylene (OPP) that showcases better vapor barrier, more clarity and stiffness and used as an overwrap for foods (Hanlon 1992; Risch 2009). In 1960, Bubble wrap was invented, which was earlier used by IBM to wrap its computers, and now is used for wrapping all fragile materials. One of the first plastics that was used for food packaging was PE. By 1966, about 25% of the bread was sold in Low Density Polyethylene (LDPE) plastic bags. The pop-tabs on cans also came into design in 1963. However, with the concern of ingesting it, later in 1975, the stay tabs were introduced that held the pop-tabs/ring tabs to the can (Risch 2009). PET was the first plastic bottle, patented by Nathaniel Wyeth in 1973, which could carry carbonated drinks and was a cheaper option against glass bottles (Risch 2009; Wyeth and Roseveare 1973).

Currently, with the advents on internet and modernization, the picture of food packaging has changed completely. Efficient vacuum packaging is in practice. Since, 1990s active packaging became popular. Crown-caps now are aligned with special liners that enable removal of residual oxygen from headspace in the bottle to eliminate or slower the pace of content oxidation.

In 2008, packaging contained detailed information as well as with Quick Response (QR) codes that could provide quick information by scanning using smart phones. Now recyclable packaging is commonly being used. Furthermore, in the recent years nanoenabled antioxidant packaging materials are made available to inhibit food deterioration. Incorporation of metal nanoparticles like AgNPs, ZnO-NPs, CuO-NPs along with nano clays, nano tubes have been incorporated to generate a renewable packaging in order to combat the hazards imposed to the environment. Bio-based nanomaterials are being employed as packaging materials. Smart packaging techniques employing nanomaterials, detects early food spoilage, contamination, alteration in temperature levels, moisture, gases by the use of sensors are also in practice in certain countries. These bio-based nano-materials forms the present and future of food packaging with an idea of preventing food spoilage and providing enhanced quality of food to the consumers. Figure 9.2 illustrates the key events in the history of food packaging and preservation.

Fig. 9.2 Timeline for history of food packaging. (Figure constructed by Shah and Krishnakumar)

9.6 Material Used for Food Packaging

Packaging of foods depends on the type of food, its purpose and duration of maintenance and the mode of transit. It also considers the shape and size of the material, the cost involved and keeping in mind to minimize environmental hazards.

Food packaging is usually classified as primary and secondary packaging. Primary packaging forms the most important form as it remains in direct contact with the food product thereby playing a significant role in maintaining the quality and integrity of food while preserving it from pathogenic infestation (Garba 2023). The secondary packaging is more of a branding suite, but it also proves as a supplementary measure of protection, a containment-like facility to carry the bulk of foods packaged in primary packaging (Garba 2023; Yan et al. 2022). At instances with more than two layers packaging, secondary packaging is made from multiple layers of primary packaging material. These form the tertiary and quaternary layers of packaging. Tertiary packages involve multiple secondary packaging or even insulating materials, while the quaternary packaging serves to sell off the compact packaging, and would also serve for the branding purpose (Lalpuria et al. 2012). Figure 9.3 shows the different layers of packaging involved in providing a tight and efficient package for any food product.

Conventionally, food packaging materials are composed of a variety of types of starting materials including plastics, wood, cloth fiber, glass, metals and complex polymers (Marsh and Bugusu 2007).

Glass containers used for food packaging are often surface coated, not just to provide lubrication, but also to increase resistance. Food containers, bottles of milk

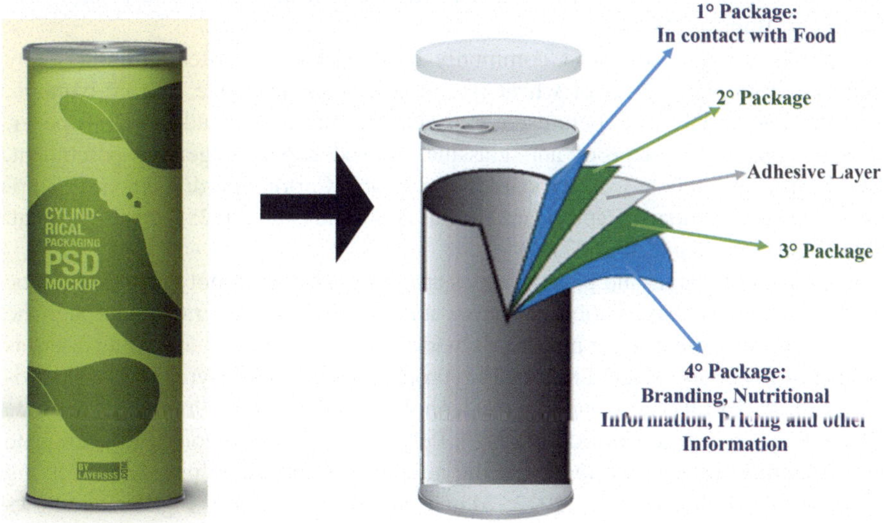

Fig. 9.3 Layers of packaging material for food preservation. (Figure constructed by Shah and Krishnakumar)

are usually made of glass, which in turn is prepared of silica. Glass is prepared with a combination of silica, limestone, soda-ash and alumina (Marasinghe et al. 2024). Furthermore, glass being inert and odorless, possesses the advantage of being impermeable to gases, vapor, moisture and odors, thereby aiding in maintaining freshness and flavors for a longer duration of time (Lalpuria et al. 2012; Marsh and Bugusu 2007).

Metal based packaging are used more for materials to protect it from environmental changes while demonstrating better physical protection as well as enhanced barrier properties. Most commonly used metals include—Steel, Aluminum, Aluminum foil, Tin and Chromium to prepare cans from different material combinations. Aluminum, derived from the bauxite ore, often in combination with oxygen as Alumina, or with magnesium and manganese, presents the property of resistance to corrosion, providing it efficient barrier properties to air, moisture, temperature, odors and even microbes (Kim et al. 2020). Different thickness aluminum foils have also been derived and used for food packaging. Metallized films are plastic containing films with a thin layer of metal, more commonly aluminum, that have suggested to have good barrier properties against moisture, gases and odor and used for food packaging (Marsh and Bugusu 2007). Tin-steel, derived from low-carbon steel and tin-free steel, known as electrolytic chromium are used for preparing drums for bulk-storage containers. Steel cans are used for packaging canned fruits, vegetables, soups, and meats, aluminum cans packaging carbonated beverages like soda and beer, aluminum wrapping food, metallized films with a thin layer of metal for packaging snacks, chocolates, and other products requiring high barrier against gases and moisture, collapsible aluminum for packaging condiments like mustard, mayonnaise, and ketchup, while metal lids or caps are used for sealing glass jars and bottles containing foods such as jams, pickles, and sauces (Lalpuria et al. 2012; Marsh and Bugusu 2007).

Paper is amongst the most commonly used packaging materials and usually produced from plant fibers of wheat rye, barley, sugarcane, rice, cotton, bamboo, corn etc. The variety of papers used as packaging material includes—Kraft paper, Bleached paper, Greaseproof and glassine, Waxed paper, Vegetable parchment, Paperboards, Whiteboard, Linerboard, Food board, Carton board/Boxboard, Chipboard, Corrugated board, Paperboard cartons (Lalpuria et al. 2012; Marasinghe et al. 2024; Marsh and Bugusu 2007).

In contrast to metals and glass, plastics are a unique class of polymers that can be molded into an extensive variety of shapes at relatively low temperatures through the use of controlled heat and pressure. A wide variety of plastics are being used as packaging materials—including PE, Polypropylene (PP), Polystyrene (PS), Polyvinyl alcohol (PVOH)/ethylene vinyl alcohol (EVOH), PET, Polyvinyl chloride (PVC), Polyvinylidene chloride (PVDC), Polyamides (PA or nylon), Polycarbonate (PC), Ionomers along with their specific auxiliary additives, helping in modifying the materials (Lalpuria et al. 2012; Marsh and Bugusu 2007). Figure 9.4 shows different types of food and the respective packaging materials used for these foods.

Fig. 9.4 Types of food and packaging materials used. (Figure constructed by Shah and Krishnakumar)

These are the materials used more traditionally, while the modern day packaging focuses on the interactions of food and the packaging, as well as the external environment. Hence, the newer packaging is more focused at maintaining the quality and shelf-life of foods.

9.7 Pros and Cons Associated with Conventional Food Packaging and Preservation

Plastics and paper form the primary form of packaging material. However, plastics being non-degradable, persist in the environment for long durations, accumulate causing environmental pollution (Maurizzi et al. 2022). Harmful synthetic plastic waste generated from the industries, become difficult to manage and leads to litter and unmanaged waste accumulation. Also, over the years the partial degeneration of plastics, forms micro-plastics which enter the food chain and lead to damaging effect on the terrestrial and marine lives, adding up to the air, water and soil pollution further depleting to the normal flora and fauna of the ecosystem. Paper packaging also harms environment, since it involves cutting down of trees. Furthermore, the derivatives of these materials, are eco-toxic and most of them are not reusable or recyclable (Ncube et al. 2020).

While considering efficient and environment friendly options, Cellulose, Chitosan, N-Acetly-D-Glucosamine, Soy protein, Whey protein, Gelatin, Poly-Lactic acids, are used as favored raw materials of choice. These materials, because of their properties of renewability, easy availability as well cost-effectiveness, are chosen while aiming at generation of green nanoparticles. Many technologies have conventionally and evolutionarily used for food preservation to improve its shelf-life, quality and edibility. However, these techniques and technologies also have inherent drawbacks which are of utmost concern to the environment and health hazards thereby. The following are the major drawbacks associated with food packaging and preservation industries (Marsh and Bugusu 2007):

1. **Waste Generation and its Management:** Improper food packaging increases the risks of food spoilage and generation of food and packaging waste causes pollution as well as unhealthy gases generation that paves way for severe health hazards (Maurizzi et al. 2022).
2. **Limited recyclability and Generation of Landfills and Micro-plastics:** Majority of the plastics and its derived polymers along with certain types of glasses, paper, and aluminum sheets show limited or negligible recyclability. The supplementary materials used in food packaging are also not completely recyclable and continue to persist in the environment for a long time, causing a built up of non-degradable waste packaging materials and pollution. Thus, poor recyclability climaxes into hazardous consequences resulting into severe environmental pollution, littering and improper waste disposal along with poor waste management. The accumulated wastes when discarded at a dumping yard, generates landfills, that undergoes years of decaying and plastics being non-renewable, breakdowns into smaller non-degradable plastic waste called the micro-plastic.
3. **Pollution, Global Warming and Generation of Greenhouse gases:** Decay of the non-degradable materials causes generation of greenhouse gases besides releasing a lot of smoke during production and hence causing air pollution. Harvesting of trees ultimately lead to deforestation and soil erosion are the starting effects, followed by waste generation, disposal, which might eventually flow into water bodies causing water pollution. These together can be one of the major contributors of global warming.
4. **Hazards to health and safety:** Certain food packaging release chemicals and toxins that may seep into its foods packaged causing health hazards. Renewable paper, plastic and metal packaging, that are recycled and reused may pose the risk of severe disorders if not recycled hygienically results in environmental pollution and generation of toxic chemicals raising health issues specifically concerning, neuronal, endocrine and reproductive health disruptions.
5. **Escalated costing and its Implications:** Manufacturing of packaging materials and the preservation techniques contributes to the overall cost of food. As raw materials of non-renewable sources cannot be reused, it adds up to escalation. Even the process of recycling adds to the cost and affordability of the products.
6. **Limited shelf-life and product usability:** Certain conventional packaging do not focus on proper packaging resulting in spoilage of food material and affecting the freshness and quality of food. This questions the product usability further leading to wastage. Example includes limited shelf life of vegetables and fruits like tomatoes, that are spoiled post-harvest in a matter of five-seven days with conventional plastic film wrapping (Lalpuria et al. 2012).
7. **Risk of Contamination and Failure to Trace:** Inappropriately recycled packages or improperly packed or preserved foods are always prone to contamination and microbial infestation which may pose health risks. Many of the packaging do

not provide complete details of manufacturing dates and address and lack the details of expiry date, which not only questions the integrity of food, but lacks traceability and fails to prevent any known contamination risks, as in cases of contamination the whole batch can be withdrawn from market by the tracing details.

8. **Unregulated Energy Expenditures:** All the packaging materials, their manufacture, use, recycling and reuse after recycle is energy intensive, exhaustive and which most of the times goes unaccounted for.

9.8 Green Nanotechnology

An amalgamation of biotechnology, chemistry, material science (Martins and Kaczerewska 2021), while formulating the matter, more typically at the nanoscale (1–100 nm) (Dutta and Das 2021; Huston et al. 2021), leads to the field of nanotechnology (Martins and Kaczerewska 2021). With these solutions and fabrication of matter, more focused on health, societal and environmental benefits gave rise to the field of green nanotechnology (Ahmad et al. 2024). Nanomaterial synthesis is attributed to either traditional or green methods of synthesis. Nanomaterials with high scalability and controlled morphology are accounted for their extensive applicability subordinated to the traditional nanomaterials synthesis. However, traditional methods are amenable to numerous hazards to the nature including use of organic solvents, high pressure and heat conditions, while giving rise to volatile vapor and carbon dioxide adding up to the greenhouse effect (Huston et al. 2021).

The phenomenon termed as green synthesis lies central to the concept of green nanotechnology, which incorporates the application of biological systems to produce a variety of nanomaterials, managing its morphology as well as regulating the characteristics of these materials, while retaining a cost effective and environmental friendly synthesis also guaranteeing cleanliness as well as safety. Bacteria, Fungi, Yeast, Algae and plants serve as substrates for green nanotechnological synthesis of nanomaterials (Huston et al. 2021).

9.8.1 Transition from Conventional to Green Nano-technological Applications

Green nanotechnology involves blending nanotechnology with environmental sciences with a perspective of enhancing sustainability, incorporating the use of renewability, and cost effectiveness (Dutta and Das 2021; Martins and Kaczerewska 2021). It facilitates enhanced energy efficiency, reduction in greenhouse gases, wastes as well as minimizing the consumption of non-renewable raw materials (Dutta and Das 2021). The concept of green nanotechnology relies on core 12 principles (Anastas et al. 2000; Huston et al. 2021; Madani et al. 2022), as shown in Fig. 9.5:

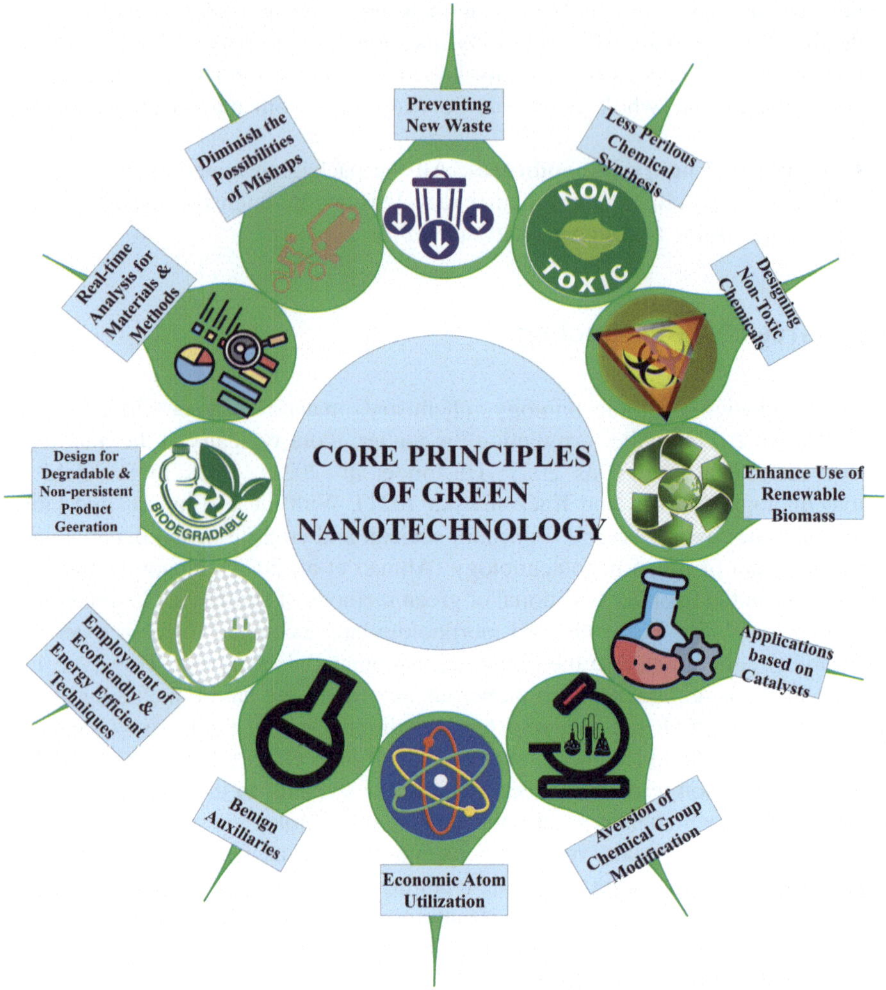

Fig. 9.5 Core principles of green nanotechnology. (Figure constructed by Shah and Krishnakumar)

1. **Preventing New Waste:** This aims at developing nanomaterial that tends to decrease or prevent production of waste while considering pollution aversion.
2. **Less Perilous Chemical Synthesis:** Production focused on to synthesis of nanoparticles with desired properties, with little to no generation of toxic waste.
3. **Designing nontoxic chemicals:** Practice designing and development of reagents and chemicals, using environmental friendly techniques that reduces or eliminates noxious effects to human health and environment.
4. **Enhancing the use of Renewable Biomass:** Practicing the use of sustainable, environmental friendly, renewable raw materials or feedstock viz. biomass substrate or the starting raw material is emphasized rather than the use of perishable goods.

5. **Applications based on catalysts:** Rather than the use of conventional stoichiometric reactions for the generation of green chemicals, developing catalytic techniques or chemicals based on the use of nano-catalysts or green nano-catalysts is encouraged that can not only improve the reaction selectivity but also enhance the efficacy of the process.
6. **Aversion of chemical group modification to reduce waste generation:** It is preferred to reduce or avoid unnecessary modifications by either blocking groups, protection or deprotection or even transient alteration in physical/chemical processes, as these processes may involve additional steps and lead to generation of waste. Instead, use of selective nano-synthesis should be emphasized to exclude solvent-based synthesis.
7. **Ensure economic atom utilization:** Method of synthesis should be planned and established for economic and environment friendly final product generation that would enable incorporation of all or as many possible raw ingredients or by-products generated in the entire process for the generation of the final product.
8. **Use Benign Auxiliaries:** Use of solvent systems, separating agents, auxiliary chemicals that are toxic should be avoided while employing a bottom-up technique ensuring minimal waste generation.
9. **Employment of Ecofriendly and Energy Efficient Techniques:** Techniques employed for chemical synthesis should be ecofriendly, and function at ambient temperatures and pressures, preferably at room temperature and pressure. The processes should favor energy conservation.
10. **Design for degradable and non-persistent product generation:** The materials being designed or synthesized, should be studied for their interaction with other chemicals and materials in the environment and must be synthesized in a way that their end products are degraded or recycled naturally. Thus, it ensures that the end product undergoes degradation which leaves no remnants in the environment.
11. **Practice real-time analysis for materials and methods:** Methodologies, governance bodies or assessment methodologies must be developed while focusing on real-time, in process monitoring while practicing control before generation of toxic waste and assess nanomaterials to manage risks associated with nanoscale development, production and commercialization.
12. **Diminish the possibilities of mishaps:** The potential elements, used for generation of nano materials should be selected with a thought of minimizing the possibility of chemical accidents like chemical explosions, fires and unusual and harmful releases into the environment.

9.8.2 Common Biopolymers Used in Green Synthesis

Polymers enable convenient production possibilities, strong molecular cross-linkages and efficient properties like barrier to oxygen and moisture, while preventing any spoilage. Commonly used non-biodegradable polymers include PP, PE, Polyethylene co-vinyl acetate (EVA), Poly vinyl chloride (PVC) and PET (Diez-Pascual 2020; Yan et al. 2022).

Biopolymers are biodegradable and extracted either from renewable biomasses which includes polysaccharides, proteins, certain biodegradable polymers, or chemical polymerization or those that are generated when they undergo fermentation by means of microbes. These form the raw material for a wide range of food packaging materials. Most of these are natural, renewable, accessible and non-hazardous and hence advantageous over the non-biodegradable polymers (Murugan et al. 2024; Singh et al. 2023). Figure 9.6 shows the classification of these biopolymers based on their source of origin.

1. **Polysaccharides:** These are complexes resulting by formation of linkages between the monosaccharides or disaccharide unit repeats by means of glycosidic bond (Ferreira et al. 2016; Marasinghe et al. 2024). These are naturally available, has easy accessibility, renewable, non-toxic, demonstrate mechanical resistance and barrier properties against oxygen and carbon dioxide gases.

   (a) **Chitosan:** β-(l-4)-2-Amino-2-Deoxy-D-Glucopyranose, commonly known as Chitosan, is a polysaccharide formed of N-acetyl-glucosamine and N-glucosamine units (Marasinghe et al. 2024; Nesic et al. 2020). It is biodegradable, biocompatible, displays a wide range of availability, and in non-toxic. Chitin can be obtained from certain fungi, insect cocoons and waste of shellfish. Chitosan is widely used in preparation of biofilms with strong mechanical strength (Maurizzi et al. 2022) that can limit oxygen and carbon dioxide passage (Machado et al. 2020) by means of cross-linked films

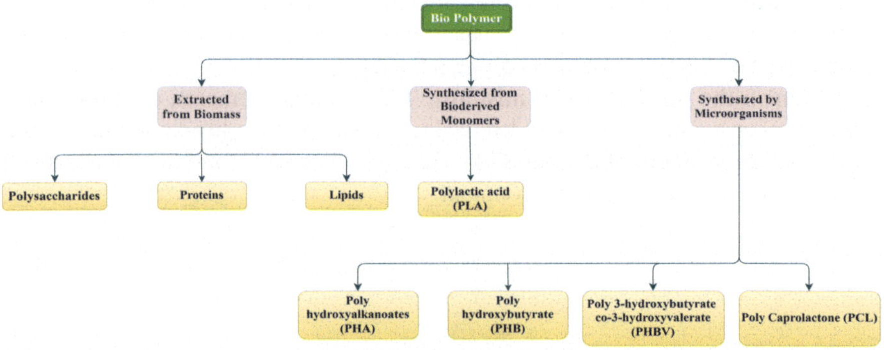

Fig. 9.6 Classification of biopolymers based on their source of origin. (Figure constructed by Shah and Krishnakumar)

and for its antimicrobial properties against gram-positive and negative bacteria, filamentous fungi and yeasts, have gel forming properties recyclability. However, they also demonstrate high sensitivity towards moisture, that restricts its applicability for fresh or fatty products (Maurizzi et al. 2022).

(b) **Cellulose and their derivatives:** (1 → 4)-β-D-Glucopyranosyl or Cellulose is a polysaccharide formed through β-(1 → 4) glycosidic bond between anhydrous glucose rings, with varying number or repeats based on the source of cellulose (Marasinghe et al. 2024; Wang et al. 2016). This polymer in water-insoluble with high crystallinity. Sources include wood, cotton, hemp and other plant components with high abundance, non-toxic (Xu et al. 2016), and stable and develop biodegradable films that are transparent, odorless, resistant to oxidation and that shows mechanical and gas barrier (Maurizzi et al. 2022). Derivatives of cellulose that are employed for films include carboxymethylcellulose (CMC), methylcellulose (MC), hydroxypropyl methylcellulose (HPMC), hydroxypropyl cellulose (HPC) (Han et al. 2018). Nonetheless, these polymers demonstrate high sensitivity to vapor transmission regarding their hydrophilic nature, preventing their applications to dried and low-fat foods (Maurizzi et al. 2022). To overcome this, Shahbazi and colleagues used CMC films in combination with sodium benzoate and glutaraldehyde vapor, followed by gelatin cross-linking, which showed improved hydrophobicity (Shahbazi et al. 2016). Cellulose acetate is amongst the most thermoplastic polymer, is widely employed in coverings and wraps for fresh foods and bakery products as well as accountable for its antimicrobial activity, gel forming properties, recyclability and sturdy film formation (Paunonen 2013), which cannot be punctured or penetrated easily.

(c) **Starch:** Starch is a polysaccharide, found naturally in plants and formed of two different chains formed of glucose moieties—Amylose, which is a linear chain composed of (1 → 4)-α-D-Glucopyranosyl, while the other is Amylopectin, which is a branched chain composed of (1 → 6)-α-D-Glucopyranosyl side chains (Marasinghe et al. 2024; Yun et al. 2008). Starch seeks its application as a biodegradable plastic and starch based films with low glucose permeability as packaging material for foods sensitive to oxidation or humidity. It showcases the properties of renewability, biodegradability, excellent film formation, mechanical strength, elasticity, transparency and lesser permeability towards oxygen, while being brittle and hydrophilic reasoning for poor vapor barrier property (Zhong et al. 2020). Commercial starch includes Ecofram, Solanyl, Biocool, Bioplast and Pantic.

(d) **Pectin:** Pectin, that is a core component of the plant cell-wall, is yet another heteropolysaccharide, with (1 → 4)-α-D-Galacturonic acids or homogalacturonan as the building unit of the linear chain which gives it the smooth texture, which in turn is cross-linked with another chain containing galactose, xylose and arabinose by means of the rhamnose residues, (1 → 2)-α-L-Rhamnopyranose, forming a domain called rhamnogalacturonan I which is connected to the third domain rhamnogalacturonan II, both of which gives rise to the hairy regions of pectin

(Espitia et al. 2014; Marasinghe et al. 2024). Orange pulp and apple pomace are rich sources of pectin, which is used as packaging material in the form of edible films (Cazón et al. 2017). Pectin seeks its other applications as a gelling, thickening and stabilizing agents for the preparation of jams, drinks and ice-creams in the food industry (Satapathy et al. 2020). Complex of pectin with certain hydrophobic lipid derivatives are employed as a layer for moisture and vapor resistance and functioning as a barrier.

2. **Proteins:** Complex protein molecules, composed of amino acid residues, serve as an excellent source as biopolymers, with its properties of being biodegradable along with having excellent physicochemical, optical, mechanical, and barrier performance (Murrieta-Martínez et al. 2018).

   (a) **Gelatin:** Gelatin, a water-soluble protein, synthesized by means of a partial hydrolysis of collagen, consisting of triple-helix residues of glycine-proline-hydroxyproline residues, by a mixture of α, β and γ chains (Chen et al. 2019). Gelatin exhibits the formation of thermo-reversible gels, implying its abundant use as stabilizing agent in food and pharmaceutical industry (Łupina et al. 2019). These gelatin residues are also employed as films with low permeability for oxygen (Ramos et al. 2016), along with its role as a carrier for natural antioxidant and antimicrobial agent (Maurizzi et al. 2022). Lin and coworkers have demonstrated crosslinking network between gelatin and chitosan (Lin et al. 2019) while Xia and colleagues worked on similar blending between gelatin and zein protein (Xia et al. 2019), leading to synthesis of hybrid films enhancing their performances and bioactivity.

   (b) **Zein Protein:** Zein protein, sourced from corn, is alcohol soluble and biodegradable and has its application as edible films (Moeini et al. 2021). These films are usually smooth and thermally stable and deliver function as selective barriers to oxygen, carbon dioxide and oils, also used as medium for targeted drug delivery. Zhang and team demonstrated a unidirectional water barrier film composed of chitin and zein (Zhang et al. 2022). Wang and colleagues also showed preparation of antimicrobial nanofiber by using perillaldehyde within gelatin and zein (Song et al. 2022; Wang et al. 2022a).

   (c) **Soy Protein:** Soy protein showcases a wide range of functional properties for it to be an excellent protein-based films of intermolecular binding, while being biocompatible and biodegradable (Song et al. 2022). Tao et al. have demonstrated carvacrol and glycerol based films with properties of opacity, vapor permeability and antibacterial property with enhanced ductility (Song et al. 2022; Tao et al. 2022).

3. **Lipids:** Lipids are biomaterials that serve as source of energy. Their bioavailability as their presence in living systems, makes them easily accessible. Lipids are bio-renewable and favors food packaging industry owing to their exceptional mechanical, barrier, antibacterial, antioxidant and other functional properties (Zubair et al. 2021). Lipids impart the property of hydrophobicity, cohesion and flexibility to the polymers while the polymers impart strength and toughness

to the lipids (Souto et al. 2021). Lipids are classified into simple lipids—fats, oils and waxes; compound lipids—phospholipids and glycolipids; and derived lipids—steroids, terpenes and carotenoids. Lipids are employed as food packaging films, membranes, surface coatings provided their efficient oxygen and moisture barrier properties (Zubair et al. 2021).

(a) **Fats and Oils:** These are generally composed of triglycerides, which upon hydrolysis gets converted to fatty acids and glycerol. Majority of this is sourced from vegetable oils and a small proportion from the animal origin. Fatty acids are composed of a varying length of carbon chains and are sub categorized into saturated or unsaturated fatty acids (Zubair et al. 2021).
(b) **Wax:** Waxes are made from long and medium-length carbon chains, usually solid and non-polar lipids at room temperature. These are esters of alcohols and long chains of fatty acids, and derived from natural plant or animal or even artificial sources (Zubair et al. 2021). Carnauba wax, Paraffin wax, Bees wax are the examples. Beeswax is purified from the honeycomb of the bees, which gives rise to yellow and white wax. Pure beeswax, with its low emulsifying power, increases the viscosity of formulations, to stabilize the emulsions (Souto et al. 2021).
(c) **Squalene:** Squalene is structured of 30 carbons, unsaturated chain, and is usually sourced from shark liver oil, some vegetable oils as well as in lower proportions from peanut and soyabean oils (Souto et al. 2021).

4. **Biodegradable Polymers:** These are a class of biocompatible polymers that could be degraded following a slow reaction, that causes the breakdown of residues.

(a) **Polylactic Acid:** Polylactic acid (PLA) is a biocompatible, thermoplastic polyester, that can be degraded by breaking of its ester bonds, while its synthesis is by condensation of Lactic acid monomers (Zhong et al. 2020) and synthesized by microbial fermentation of cane sugar pulp and corn (Marra et al. 2016). PLA showcases properties like high tensile strength, thermal stability and gas permeability. Its best applications are demonstrated in the food-packaging for manufacture of disposable cutlery, plates and lids (Moeini et al. 2021). Environment friendly applications with antimicrobial agents to form PLA based film used for packaging of dry and perishable foods with short shelf life like fruits, vegetables and meats is also popular (Ncube et al. 2020).
(b) **Poly-hydroxyalkanoates:** Polyhydroxyalkanoate (PHA) is synthesized by a variety of gram positive and gram negative bacteria. It is a thermoplastic polymer formed by the monomers of (R)-3-hydroxyalkanoic acids by fermentation of lipids and sugars. PHAs are known for their properties of being biocompatible, biodegradable, eco-friendly and stable to UV radiation (Nanda et al. 2022; Zhong et al. 2020).

(c) **Polyhydroxybutyrate:** Polyhydroxybutyrate (PHB) is yet another biodegradable, crystalline, thermoplastic polyester synthesized by gram positive bacteria *Bacillus megaterium* by means of microbial fermentation. It is characterized with excellent properties like superior water vapor barrier and low carbon footprint (Zhong et al. 2020). However, it is also highly brittle, has low thermal stability and lesser process-ability (Panaitescu et al. 2018).
(d) **Poly-3-hydroxybutyrate-co-3-hydroxyvalerate:** Poly-3-hydroxybutyrate-co-3-hydroxyvalerate (PHBV) is a linear aliphatic thermoplastic polyester, synthesized by copolymerization of 3-hydroxybutanoic acid and 3-hydroxypentanoic acid. Catalyst aluminoxane also converts Butyrolactone and valerolactone to PHBV. *Escherichia coli* strain of bacteria enables the synthesis of this polyester as an intracellular carbon and energy storage compound, generated using glucose and propionate. PHBV exhibits the characteristics of being biodegradable, non-toxic and biocompatible plastic (Mohan et al. 2016), while having an excellent absorption capacity, low cytotoxicity, high crystallinity and resistance to UV radiation. It also showcases the properties of excellent oxygen barrier, chemical inactivity, high viscosity in fluid state. It also has better flexibility and increased surface tension in comparison to PHB, and seeks its applications with a variety of nanomaterials, metal oxides and nano particles (Rivera-Briso and Serrano-Aroca 2018).
(e) **Polycaprolactone:** Polycaprolactone (PCL) is a biodegradable yet non-renewable biosynthetic polymer that is synthesized either from ε-Caprolactone or polycondensation of hydroxycarboxylic acid, leading to production of PCL of varying molecular weights, with low molecular weights being crystalline, brittle and forming hard film, while the higher molecular weights being elastic, tough and poorly crystalline (Thakur et al. 2021; Zhong et al. 2020). It also demonstrates characteristics of partial or complete solubility in organic solvents while being insoluble in water, barrier properties against oxygen and water vapor as well as non-toxicity. It finds its applications as combined polymer for gas barrier applications and for drug-delivery systems (Ilyas et al. 2022).

9.9 Interaction Between Food, Food Package and Environment

When food products are being packaged, they come in contact with the food, and thereby react with the food (Garba 2023). Interactions between the food, its packaging and environment plays a key role in regulation of the food quality, maintenance of its shelf-life and its freshness. Hence, the foods with packaging that can prevent the alterations to flavors and color, vitamin loss, microbial activities are favored (Floros 1993; Robertson 2009). Besides the food quality the interactions between food and its packaging should not be injurious and rather aiding in preventing undesirable interactions while favoring interactions that absorb oxygen, prevent microbial growth, remove unintended flavors of sorption. The packaging should be able to have a preventive or absorptive actions towards the free radicals being generated in foods like dairy and meat products, maintain moisture in and around the food in desired quantity and reabsorption of the additional moisture, and enhancing barrier properties of food (Garba 2023; Lalpuria et al. 2012).

These types of food package interactions with food involve—permeation, migration and sorption and these occur amongst the external environment, food and the food package (Garba 2023).

(i) **Permeation:** Permeation is defined as the molecular movement of gases like oxygen, carbon dioxide, nitrogen, water vapor, as well as volatile substances, aroma and other low molecular weight from either the external environment to the food through the packaging material or the other way (Caner 2018). Porosity of the packaging material or its poor intactness is mainly responsible for this. This process incorporates three key steps—Solution or Absorption, Diffusion and Emergence or Desorption.

   (a) **Absorption:** Encompasses solubilization of the food on the surface of the packaging material following the Henry's law of adsorption that suggests adsorption to be dependent on the partial pressure of the gasses present within the package and the condition of the food prior to packaging.
   (b) **Diffusion**: Entails the material that gets exchanged through the packaging film where the material moves across the difference of concentrations, which is usually from higher concentration to the lesser concentration following the Fick's law, depending on its concentration, time and inversely on the thickness of the packaging material, either from the environment to the internal environment or vice versa.
   (c) **Emergence/Desorption:** The material that gets exchanged, when leaves the surface and moves to the opposite side of the packaging film, is termed as Desorption or emergence and is based on Graham's Law of desorption.

   Thus, on the whole, the factors governing permeation include, concentration gradient, density, solubility, internal and external pressure and temperature, permeability of the packaging film, time, thickness of packaging film and relative humidity, thereby affecting the shelf-life, color, odor and flavor profile of the product (Garba 2023).

(ii) **Migration:** Here chemicals or toxins from the packaging material enters the food by means of diffusion, based on Fick's law. This migration is usually in the direction of higher concentration to lower concentration, which results in altered food quality or food toxicity. This process encompasses of Diffusion, 1st Desorption, Sorption and 2nd Desorption phases (Ferrara et al. 2001) and modulated by factors like temperature, packaging size, physico-chemical properties of food, storage time, type and quality of packaging material and type of the contact (Garba 2023). During ideal conditions, an inert packaging material like steel, ceramics or glass is preferred, where in the migration does not occur from packaging material (Abbès et al. 2015) to the food rather from food to the packaging material, that can cause wear-tear of the package (Bhunia et al. 2013; Garba 2023).

(iii) **Sorption:** Sorption or Scalping involves either mass adsorption or absorption of food flavors, aromas, moisture and lipids on to or through the food packaging material (Caner 2011). This has the ability to hamper and alter the quality of food. Thus, governed by the Fick's law, the mechanism of sorption is regulated by temperature mainly, apart from some certain chemical components. For solid foods, sorption depends on the solubility and vapor component, while in case of liquid foods it depends on the partition coefficient. Silvestre and coworkers, have demonstrated the effect of sorption based on temperature and pressure, showcasing that increased pressure would reduce the quality of the food along with causing a disruption in the packaging, thereby increasing the rate of sorption onto the packaging film (Silvestre et al. 2011). Pieper and colleagues, have demonstrated no significant effect on the food quality (Pieper et al. 2006). While studying the effect on orange juice packaged in a glass bottle and packaging made of polyethylene-laminated cardboard packages, it noticed the later showed adsorption of d-limonene and small aldehydes as well as alcohols absorbed on the cardboard packaging while keeping the sensory properties of the juice intact (Garba 2023; Pieper et al. 2006).

Adverse effects posed by food-package interaction involves changes in the physical appearance—color, water content or damage to the packaging; chemical changes—oxidation, alteration in flavor, decarbonation; microbial growth—contamination, toxins generated by the microbes, pathogenicity; and sensory alterations—loss of crunchiness, loss of flavor, loss of aroma, taste alteration. Hence, it is crucial to understand the effect of interaction between food and its packaging (Garba 2023). Figure 9.7 emphasizes the interaction between food, its packaging and the external environment, while considering the factors affecting the interaction between them, as adapted from (Garba 2023).

Properties like pH of the food, water activity, nutrient content, antimicrobial properties, redox potential, biochemical characteristics along with extrinsic properties like temperature, humidity, atmospheric gases content influence the quality and shelf-life of food (Garba 2023; Lalpuria et al. 2012). While considering food interaction, the factors pertaining to food like acidity of the food, its viscosity, extent till which the food has aroma, molecular weight of the food component, presence of the functional group and the polarity of the molecules plays a crucial role (Dury-Brun et al. 2007; Guth and Rusu 2008). Also besides, the composition of food, the nature and quality of packaging material contributes equally. As suggested, ideally the packaging material should be inert—non-reactive with the food material, non-corrosive, impermeable, sealable. Semipermeable or porous material like paper enables rapid migration (Castle 2007). Thus factors like, molecular weight, density, temperature, crystal lability regulate the nature of the interaction with food (Silvestre et al. 2011). Preferred packaging material involves flexibility, low density,

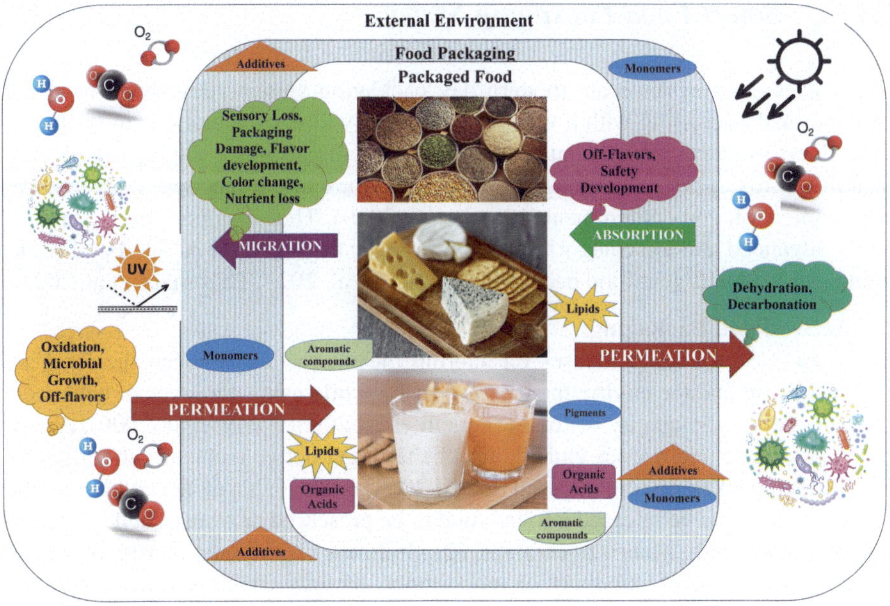

Fig. 9.7 Food, package and environment interaction. (Figure constructed by Shah and Krishnakumar)

strength, cost, controlled hydrophobicity and malleability into different sizes and shape. Assessment of the health risk assessment, degree of stability as well as health impact of the food-packaging interactions (Garba 2023; Lalpuria et al. 2012).

9.10 Application of Green Nanotechnology in Food Packaging and Preservation

The growing concerns associated with health hazards, problems associated with food spoilage, wastage and environmental pollution, more specifically with traditionally synthesized materials for food packaging that use fossils, plastics, paper and glass posing serious challenges like exhaustion and non-availability of resources as well as pollution. During the times of impending global warming, it necessitates the use for healthy, ecofriendly and sustainable means of food packaging and preservation. Bioactive films and smart techniques incorporating the use of green synthesized nanoparticles and nanocomposites serve as a promising outlook towards such effective and affordable solutions (de Sousa et al. 2023).

9.10.1 Smart Food Packaging System

Smart packaging systems are modern day packaging systems developed from the conventional packaging with a view to monitor any kind of alterations in packaged food due to either environmental changes or microbial attack aiding in reducing health hazards, while continuing the basic functions like protection, convenience, communication and containment (Lloyd et al. 2019). These systems are technologically advanced and are categorized into two sub-categories: (a) Active Food packaging and (b) Intelligent food packaging (Naskar et al. 2023; Thirumalai et al. 2023).

(a) **Active Food Packaging:**

Active packaging focuses on altering the interaction between the food, its packaging and the environment consciously, while using additives or enhancers to maintain the quality of food (Dainelli et al. 2008; Kerry et al. 2006). Active packaging encompasses embedding of the target active element either directly through packaging film/material (Garba 2023) or through a sachet containing the active agent to be released into the food to be preserved or whose shelf life has to be maintained (Maurizzi et al. 2022). It forms a polymer matrix based on polymerization of polymers or biopolymer, thereby enabling releasing of the bioactive agents naturally to the food or surrounding environment (Wicochea-Rodriguez et al. 2019; Yan et al. 2022). It is one of the most efficient ways to protect the food from environmental deterioration, increase shelf life and preserve the food more specifically for the oxidation-prone foods. The mechanism is based on the polymerization of the active agent and its targeted release within the food. Active packaging agents are classified into categories like—absorbers/scavengers for gases like oxygen and carbon dioxide, moisture, ethylene, UV light, and flavors or odors; emitters for ethanol, antioxidants, preservatives, sulfur dioxide and flavors; elimination of lactose and cholesterol like components; temperature controlled packaging, insulating materials in temperature regulated packaging, self-heating and cooling packages, temperature-sensitive packaging, microwave susceptors; and microbials and UV- and surface treated packaging materials (Lee et al. 2015; Lloyd et al. 2019). Figure 9.8 shows the active agents being used widely for active packaging of food, as adapted from (Maurizzi et al. 2022).

(i) **Antimicrobials:** Antimicrobials are agents subdue the growth of microbes in packaged food (Malhotra et al. 2015; Yan et al. 2022). These are usually enveloped in a biofilm or a polymer film for its action (Sung et al. 2013) and are usually considered alternatives to direct addition of bioactive agents into the food. Antimicrobials are classified as migrating and non-migrating antimicrobial agents (Lloyd et al. 2019; Thirumalai et al. 2023). These involve controlled and slow release of these antimicrobials over a stipulated period of time combatting any microbial attack (Lee et al. 2015), and are usually used in preservative free systems. This movement is based on molecular structures of the polymer and antimicrobial compounds, the

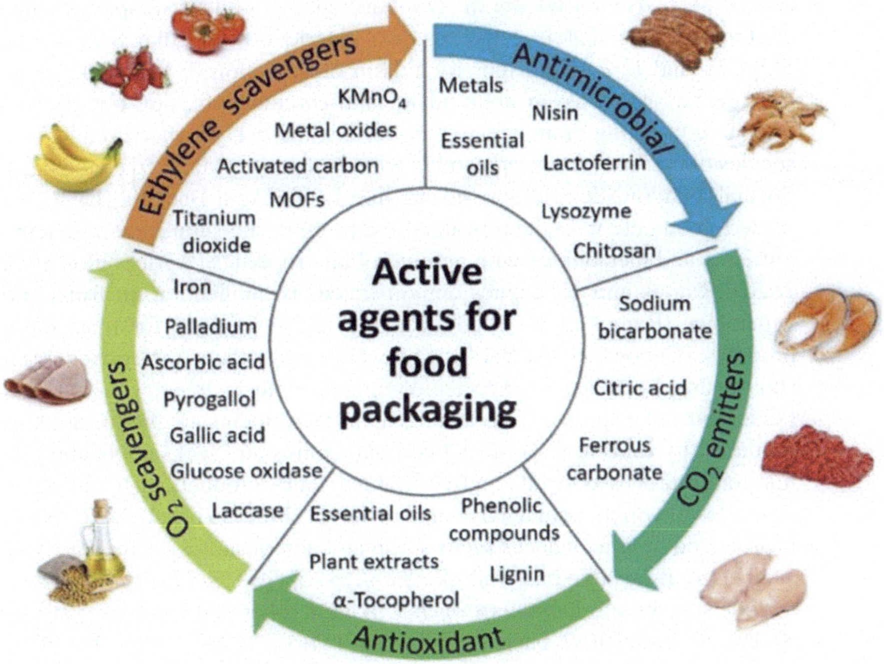

Fig. 9.8 Active agents used for smart packaging. (Adapted from Maurizzi et al. 2022)

physicochemical characteristics of the packaging item, and the internal and external environmental conditions of the packaging (Gonçalves 2017). The food product in consideration, the internal environment of the package, targeted microorganism, the packaging material and the active agent in consideration together decide the design of the packaging. The methods involved in designing the packaging includes antimicrobial sachets that are placed into the packaging and it releases the active ingredient gradually. For another approach the active molecules are directly mixed into the polymer matrix. Yet another method includes adsorption of a specific matrix that functions as the carrier of the antimicrobial agent. One of the methods involves the active agent to be immobilized on the polymeric matrix through ionic/covalent bonds. Lastly the antimicrobial agent is directly added into the food. The mechanism involves destruction of the cell walls of the microorganisms being targeted (Sofi et al. 2018). Meat, bakery products like cake and pastry and dairy based products like cheese are maximally prone to microbial attacks. There are a number of commonly used antimicrobials available. These include chlorine dioxide—against pathogens like *E-coli* and *Listeria monocytogenes*, chelating agents, organic acids like benzoic acid, sorbic acid, acetic acid, citric acid and propionic acid, enzymes like lysozymes for undesired butyric fermentation caused by *C. tycobutyricum* in semi-hard cheese. Other examples include Lactoperoxidase that is active against enteric bacteria like Salmonella, Shigella, and *E. coli* and thereby aids in preservation of raw milk, while Lactoferrin serves as an antioxidant, anti-carcinogenic, anti-obesity and antibiotic targeting gram negative bacteria while being bactericidal against species like *E. coli*, *Klebsiella* and *L. monocytogenes* (Maurizzi et al. 2022). Special antagonistic microorganisms like Lactic Acid Bacteria (LAB) are made to compete with the unwanted microorganisms alongside producing nutrients and metabolites with antimicrobial properties (Agriopoulou et al. 2020). Ramos and colleagues demonstrated its application on fruits and vegetables (Ramos et al. 2020), Cheong and coworkers for fresh dairy products (Cheong et al. 2014), while Gao and others for cooked meat showcasing Lactobacillus species competing with *L. monocytogenes* and Penicillium like species (Gao et al. 2015). Bacteriocins are the metabolites produced by LAB against other microbial strains and seeks application on minimally processed fruits and vegetables, dairy products, fish and meat, examples of which include Nisin and pediocin (Maurizzi et al. 2022). Nisin is produced by *Lactococcus lactis* strain and demonstrates a strong activity against gram-positive bacterial like *Staphylococcus*, *Bacillus cereus*, *Clostridium* species and *L. monocytogenes* like microbes and has a low activity against gram negative bacteria and fungi (Tiwari et al. 2009). One of its most remarkable applications have been demonstrated in limiting the growth of *L. monocytogenes* and *Clostridium* species in dairy products (Ibarra-Sanchez et al. 2020). Nisin has been demonstrated by Blanco and team as an active agent for an antimicrobial film employed to cover raw and processed meat against the action of Listeria species (Blanco Massani et al. 2014). Pediocin is produced by pediococcus bacteria of the Lactobacillaceae family that has its action in preservation of vegetables, dairy products (Verma et al. 2017) and processed meat (Castellano et al. 2017). Certain fungicides, oxygen absorbers like butylated hydroxytoluene, Silver based materials, natamycin-based antifungal coating, allyl isothiocyanate, a natural antibacterial and antifungal substance are the most commonly used antimicrobials targeting meat and dairy sectors as well as fruits and vegetables. Chemical additives in these are often replaced by natural antimicrobial agents like essential oils extracted from clove, cinnamon, rosemary (Choulitoudi et al. 2017), ginger (Bonilla et al. 2018), basil, lemongrass, oregano (Lee et al. 2019), tea tree (Souza et al. 2017), thyme (Sharma et al. 2022) and fennel.

(ii) **Antioxidants:** Antioxidants are more focused on inhibiting the oxidative reactions along with the microbial changes and thereby delay the oxidation rate of packaged foods (Sanches-Silva et al. 2014; Trajkovska Petkoska et al. 2021; Yan et al. 2022). These agents act to delay the lipid peroxidation. These majorly deploy the use on natural antioxidants to enhance the shelf life of foods (Vilela et al. 2018), thereby enabling prevention of food spoilage and deterioration without coming in direct contact with the foods.

Antioxidants act either directly or as preventative agents and are selected based on the target's aptness and compatibility with the polymer matrix to ensure an equal homogeneous distribution all through the food product (Salgado et al. 2021). Antioxidant components in food includes plant extracts from stems, roots, leaves, fruit seeds that are abundant in polyphenols, alkaloids, terpene substances, flavonoids and quinones (Yan et al. 2022). Essential oils from cumin, black tea, sage, fennel, thyme, cinnamon, rosemary, mint, garlic and onion (Jamroz and Kopel 2020; Sharma et al. 2021; Sirocchi et al. 2017) are believed to be rich in antioxidants and are incorporated within biopolymer matrix composed of gelatin and gelatin-montmorillonite films (Jamroz and Kopel 2020; Yan et al. 2022). Natural antioxidants comprise of vitamins like ascorbic acid, α-tocopherol, carotenoids like carotenes, xanthophylls, polyphenols which form the most abundant resource are one of the strongest free radical scavengers includes the flavonoids viz. flavanols, flavones, isoflavones, anthocyanins, and non-flavonoids (Nguyen et al. 2016; Salgado et al. 2021; Sanches-Silva et al. 2014). Polyphenols importantly constitute the part of the packaging in the form of complexes with plant extracts and essential oils (Maurizzi et al. 2022; Yan et al. 2022; Zhang et al. 2021).

(iii) **Scavengers:** One of the major problems associated with food packaging is the presence of oxygen in the packaged food, which might lead to oxidation of food, eventually causing its spoilage as well as change in flavors, color, loss of nutrients and generation of toxins (Lee et al. 2015; Lloyd et al. 2019; Suppakul et al. 2006; Yan et al. 2022). Ethylene scavengers are mainly used for preservation of fruits and vegetables (Lloyd et al. 2019; Yan et al. 2022), thereby averting the color or flavor changes. Oxygen, ethylene, moisture, odors and volatile compounds present in the food, that in turn react with the targeted elements and prevent deterioration of food while maintaining freshness and quality of food. Scavengers added to the packaged food scavenge oxygen, ethylene, moisture, odor and volatile compounds present in the food, that in turn reacts with targeted elements which acts by reducing the levels of oxygen thereby inhibiting the formation of aerobic microbes and thus, prevent deterioration of food while maintaining freshness and quality of food. These scavengers are added in the form of sachets, or through packaging films or other packaging materials (Suppakul et al. 2006). Figure 9.9 shows the image of oxygen and ethylene scavenger sachets available in food packages. One such instance, as demonstrated by Kartal and coworkers for preservation of strawberries using oxygen scavengers. These substances include iron powder for oxygen scavenging with strawberries is observed to elongate the shelf life of the fruit beyond 4 weeks (Kartal et al. 2012). Packaged orange juice revealed better quality and shelf life with added oxygen scavengers rather than the presence of barrier packaging as showcased by Zerdin and colleagues (Zerdin et al. 2003). Furthermore, it is also beneficial to packaged meat, wherein it

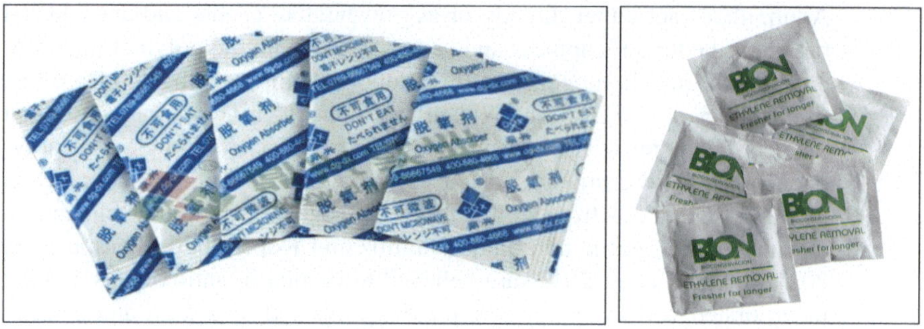

Oxygen Scavenger **Ethylene Scavenger**

Fig. 9.9 Example of scavengers used in packaged food. (Figure constructed by Shah and Krishnakumar)

functions as carbon dioxide generator and thereby aids by preventing the growth of microbes and prevents color deterioration (Lee et al. 2015).

$$Fe \rightarrow Fe^{2+} + 2e^{-}$$

$$\frac{1}{2}O_2 + H_2O + 2e^{-} \rightarrow 2OH^{-}$$

$$Fe^{2+} + 2OH^{-} \rightarrow Fe(OH)_2$$

$$Fe(OH)_2 + \frac{1}{4}O_2 + \frac{1}{4}H_2O \rightarrow Fe(OH)_3$$

The above set of equations describe the chemical reactions involved in creation of reactive hydrated metallic reducing agent that in turn scavenges oxygen converting it to a stable oxide form.

Ethylene, as a plant hormone is associated with the function of initiating ripening, fruit and leaf abscission, leaf senescence and seed germination. Ethylene scavengers seek its prompt application in delaying the maturation rate of climacteric fruits that help in maintaining the time gap and regulates time for ripening during the transit for import/export of fresh fruits and vegetables thereby assisting in delayed degradation and maintaining freshness of fruits and vegetables during this stage (Lee et al. 2015).

Other examples of scavengers include silica embedded with potassium permanganate (Lloyd et al. 2019), zeolite and activated charcoal for targeting ethylene, plastics with anti-fog additives added between packaging films and even starch polymers and copolymers localized between two superabsorbent micro-porous polymer sheets to tackle high water activity or high moisture bearing foods, Sodium chloride, silica gel, molecular sieves, calcium oxide, natural clays like montmorillonite are used as desiccant (Rooney 1995), filled in pouches to address humidity, combination of calcium oxide and silica gel is used in coffees to scavenge the formed carbon-dioxide, ascorbic acid, catechol, enzymes like glucose oxidase, ethanol oxidase (Lee et al. 2015), enables oxidation of alcohols to their respective aldehyde forms, which ultimately reduces to hydrogen peroxide, Unsaturated fatty acids like oleic acid, linoleic acid, linolenic acid, rice extract, immobilized yeast. Carbon dioxide absorbers are also used in flexible coffee packets to combat the carbon dioxide being generated there (Lloyd et al. 2019; Vermeiren et al. 1999).

(iv) **Emitters:** Emitters are used as a means of increasing shelf life of food by releasing certain gases. Carbon dioxide is one of the major gases used due to its ability to protect food from oxidation and its antimicrobial activity, along with being soluble in foods. Carbon dioxide generators reduce the rate of metabolism of foods, while carbon dioxide also reduce the oxygen concentration (Lee et al. 2015) that has the ability to spoil the food, and thus often used complimentary to the scavenging system. Carbon dioxide is used as emitter in fish, meat, cheese and poultry products seldom in combination with an oxygen scavenger, ethanol sprays or as embedded in a carrier material for breads and other bakery products, or even in combination with refrigerated fresh and fermented foods are widely used.

$$C_6H_8O_7 + 3NaHCO_3 \rightarrow 3H_2O + Na_3C_6H_5O_7 + 3CO_2$$

The above chemical equation shows the reaction of citric acid with sodium bicarbonate, that results into the formation of water, sodium citrate and carbon dioxide. This generated carbon dioxide can further function as carbon dioxide emitter, in complementarity to oxygen scavenging system. Jin and contemporaries have also suggested for post-harvest treatment of sweet melons with ethanol vapor that benefits in delayed senescence, while maintaining efficient storage (Jin et al. 2013).

Modified Atmosphere Packaging (MAP) is the technique that utilizes active or passive modulation of the gaseous system depending on the product to prevent spoilage (Yan et al. 2022). MAP entraps the air surrounding the food products in a vapor barrier material, by removing or replacing it (McMillin 2008). It regulates the respiration and metabolism of food microbes while managing the permeability of the packaging to maintain the quality of food. Oxygen (Lee et al. 2015), Carbon dioxide (McMillin 2008) and Nitrogen, singly or in combination are the majorly used gases (Alasalvar 2005; Fernández-León et al. 2013; Martínez-Ferrer et al. 2006) with MAP technology that have demonstrated successful extension of shelf-life, deterioration in microbial growth and survival, delayed ripening, which in turn facilitated fruits, vegetables and meat products an extended life. Other gases like helium, neon, nitrous oxide, argon, ethylene oxide, Sulphur dioxide have shown application for the technique (Church 1994; McMillin 2008).

(b) **Intelligent Food Packaging:** Intelligent packaging systems are aimed at providing timely information about the product in question, including the details regarding the physicochemical conditions like freshness of the product, environment inside like the temperature, pH levels, and surrounding the packaging during the storage and transit (Muller and Schmid 2019; Rodrigues et al. 2021; Yan et al. 2022) as well as involves recording, detecting, sensing, tracing and communicating information to the consumer (Rodrigues et al. 2021; Yan et al. 2022). These essentially do not interact with the foods (Biji et al. 2015; Muller and Schmid 2019), but express the conditioning of the food while being located on the primary, secondary or tertiary packaging's surface and provide visual or sound information indicating the condition of the product. These mainly includes Sensors—gas sensors, biosensors and Indicators—Time Temperature Indicators (TTIs), ripeness indicators, toxin indicators and Radio Frequency Identification (RFID) (Kerry et al. 2006). However, using these entails a lot of legal considerations and are currently not used widely except for the USA, Japan and Australia (Muller and Schmid 2019).

A general classification for intelligent packaging systems can be provided on the basis of environmental conditions, quality characteristics and lastly data carriers (Muller and Schmid 2019)

(i) **Sensors:** Sensors are composed of a receptor, that converts chemical or physical information into a form of energy and a transducer, that converts that energy into signals, thereby aids in providing information pertaining to the internal and external environment of the product (Sohail et al. 2018; Yan et al. 2022). The size of these nano-sensors ranges between 1 and 100 nm. These enable for quality assurance to the consumers as well as timely monitoring and spoilage prevention to the manufacturers by locating, detecting and quantifying the energy. There are several kinds of sensors that are employed. These include—Chemical sensors, Biosensor, Gas sensor, Fluorescence-based oxygen sensors, Printed sensors and Electronic nose.

Chemical sensors have the ability to detect presence, activity, composition as well as concentrations of specific chemical or gases as per the design of the sensor, and are used for detection of pathogen, chemical contaminants, spoilage, product tampering, presence of trace elements, over and above product traceability (Liu et al. 2007; Lloyd et al. 2019).

Biosensors are designed to assess, transmit and record biological activities in the product detected by bio-receptors targeted to certain enzymes, microbes, nucleic acids, hormones and antigens that are actively being altered owing to possible microbial or pathogenic infestation or environmental alterations (Lloyd et al. 2019; Yam et al. 2005), which are converted into signals by the transducers to generate electrochemical or optical response. These are often employed to assess fertility amongst animals and determine livestock, as well infected crops. Food Sentinel system and Toxin Guard are available in the market for its application as biosensor (Muller and Schmid 2019).

Gas sensors, as the name suggests are used to detect the presence of certain gases like oxygen, carbon dioxide, water vapor, ethanol, pollutants in the system (Kerry et al. 2006), which might change with the degree of microbial growth and respond by means of signals on an external device. One of its major applications has been suggested to be in combination with MAPS technology to detect for any possible gas leakage (Garba 2023). Ageless Eye, Shelf Life Guard, Tell-Tab, O_2Sense, Tufflex GS are examples of available gas sensors (Lloyd et al. 2019).

Fluorescence based oxygen sensors incorporate the property of luminescence quenching of oxygen and is mainly employed in meat packaging, wherein the oxygen is quantified by the sensors while altering the luminescence parameters that is in contact with the liquid or gas being monitored, by means of oxygen sensing elements (Biji et al. 2015; Lloyd et al. 2019).

While printed sensors are a unique printed receptor over a printed transducer and are mainly used for lightweight packaging and portability (Lloyd et al. 2019; Vanderroost et al. 2014).

Electronic nose is a mimetic system that mimics mammalian olfactory system and identifies volatile compounds generated during ripening and is applied to classify fragrances and assess the freshness and quality of food materials (Biji et al. 2015; Lloyd et al. 2019).

(ii) **Indicators:** Indicators, as labeled indicate regarding the quality of the product which includes temperature, leakage, presence or absence (Biji et al. 2015), microbiological concentration, freshness, color, odor (Rodrigues et al. 2021), reaction or concentration between substances through a color change. They assist the consumers in verifying the quality of the products (Yan et al. 2022). They are of a variety of types including TTI that can detect changes with time and can accurately notify about the remaining shelf-life of the product (Lloyd et al. 2019; Muller and Schmid 2019; Pennanen et al. 2014) with its complete timeline as well as indicate about the history of temperature variations (Muller and Schmid 2019). Commercially available TTIs include Checkpoint, Fresh-Check, Monitor Mark, Keep-it, Colour-Therm; Freshness indicators that indicate regarding microbial growth and product quality, by reacting to the presence of certain chemical or metabolites generated during ripening or spoilage (Kerry et al. 2006; Smolander 2003); while Thermochromic Inks are dynamic color changing inks, that change color with a variation in temperature and help consumers in knowing about the previous storage temperatures (Vanderroost et al. 2014); Oxygen indicators bring to notice for any leakage or quality integrity; Color indicators, indicate spoilage by means of tags in the form of adhesive labels; while pathogen indicators and breakage indicators are also employed (Lloyd et al. 2019). An example of freshness indicator uses a sensor ticket that indicates the growth of bacterial change in color of the sticker. Commercially available indicators include Fresh Tag, SensorQ (Muller and Schmid 2019; O'Grady and Kerry 2008), Ripe Sense and Food fresh. Gas indicators are also employed to detect the presence of oxygen, carbon dioxide, as well as water vapor, ethanol, and hydrogen sulphide (Fang et al. 2017; Muller and Schmid 2019). The spoilage is usually linearly proportional to the levels of these gases. For redox reactions, redox indicator dyes are used as a part of indicator. However, these systems are not full-proof, as any external leakage of these gases might also alter the results (Lloyd et al. 2019; Muller and Schmid 2019). Figure 9.10 shows the use and application of indicators in Intelligent packaging system, as adapted from (Yan et al. 2022).

(iii) **Traceability technology:** As a part of food safety, RFID involves a wireless communication, based on tags on food products and containers, tracks, records and provides real-time product information (Yan et al. 2022) through its entirety aiding with transparency and quality assessment. It provides information like temperature, shelf-life, relative humidity as well as nutritional information. RFID consists of a microchip, that is put up as a tag, connected to a small antenna, its respective reader that gauges its signals upon triggering radio waves to receive the information (Kumar et al. 2009). RFIDs store information like temperature, relative humidity and nutritional information as well as maintenance and cooking instructions

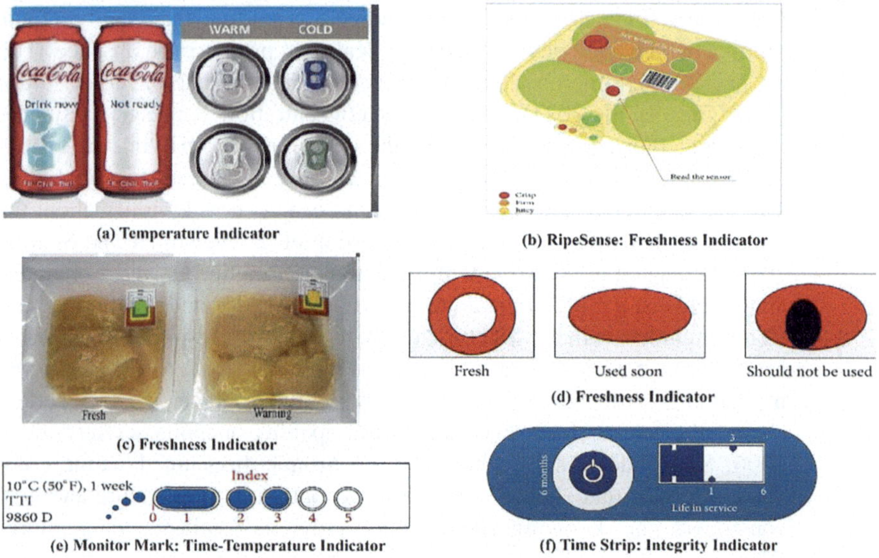

Fig. 9.10 Applications of indicators as a part of intelligent packaging system. ((**a**, **c**) Adapted and modified from Yan et al. (2022) (under CC @ 4.0). (**b**, **d**, **e**, **f**) Adapted and modified from Fuertes et al. (2016) (under CC @ 4.0)). (**a**) Temperature Indicator indicates the changes in temperature that the package has gone through (**b**) ripeSense Indicator indicates the stages of ripeness of the food with a color indication (**c**) and (**d**) Freshness Indicator indicates towards the food being fresh or stale with a change in color of the indicator (**e**) Time-Temperature Indicator suggests the periodical changes in the temperature the package has undergone (**f**) Time strip Integrity indicator indicates the remaining shelf life of the food product

(Lee et al. 2015). Easy2log, Intelligent Box, Temptrip are some of the examples for RFID tags available (Muller and Schmid 2019). An illustration emphasizing the mechanism of RFID tags has been shown in Fig. 9.11, as adapted from (Fuertes et al. 2016).

Barcodes are also used for application as tracing the products. One-dimensional as well as two-dimensional barcodes are now printed on the products. 1D barcodes are made up of parallel lines and gaps, while 2D barcodes are a combination of dots and spaces forming an abstract shape giving rise to an array or a matrix (Fang et al. 2017), known as Quick Response (QR) code. Also 2D barcodes offer more memory, and upon scanning provides details on packaging date, batch number, expiry date, weight, batch number, product composition as well nutritional details (Muller and Schmid 2019). Figure 9.12 shows an illustration of barcode and QR code, commonly seen on products.

Most recent TTI integrates the concept of both barcodes and RFID, wherein upon scanning the QR code, the information about the product temperature is sent (Muller and Schmid 2019). A company Bioett AB, has developed tags based on these principles. The TTI-RFID indicates changes in the product temperature as it moves across the market (Fuertes et al. 2016). Also a smart system, that informs about pathogenic infestation by the use of pathogen specific antibody on the barcode, that forms patch on the barcode making it unreadable (Muller and Schmid 2019; Yam et al. 2005).

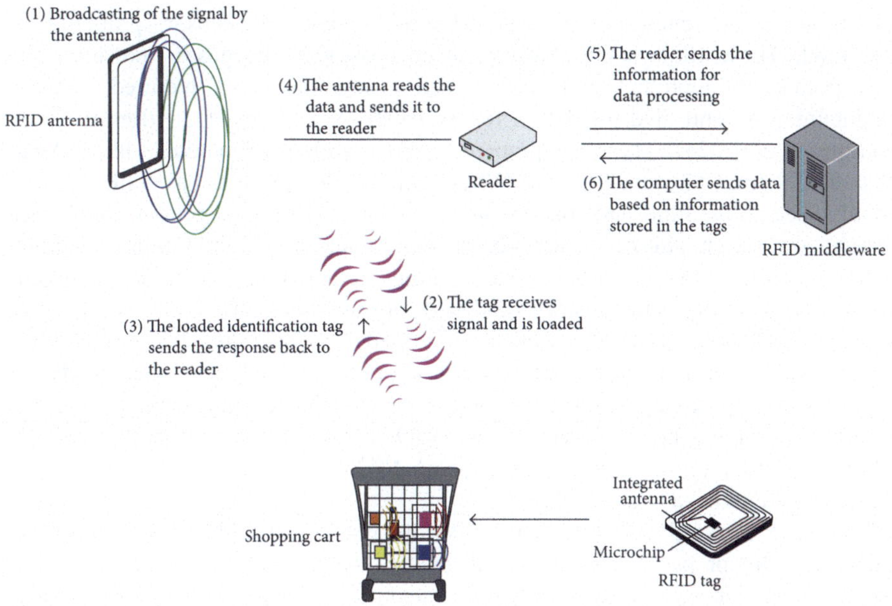

Fig. 9.11 Schematic representation of RFID Tag. (Adapted from Fuertes et al. (2016) (under CC @ 4.0))

Fig. 9.12 Examples of Barcode and QR code used actively for traceability technology. (Figure constructed by Shah and Krishnakumar)

Chen and associates have demonstrated reuse of food barcode as a colorimetric sensor integrated with camera and vapor sensors entrapped in resin microbeads that monitors and indicates the end of shelf life of food and indicates deterioration of food when exposed to different temperature (Chen et al. 2017; Salgado et al. 2021).

9.11 Edible Films and Coatings

Edible films or coatings are prepared from natural materials that are edible and non-toxic, aiming at preservation of the food material that is being coated with it, and the food is eaten along with the film or coating (Trajkovska Petkoska et al. 2021). These are one of the most sustainable, environment friendly, safe food packaging and also contributes to the nutritive value of the food. They have natural antimicrobial property, are renewable, biodegradable and enhance the sensory properties of the food. These films act as barriers against oxygen and gases, moisture, can incorporate antimicrobial and release active ingredients as required, thereby maintaining a controlled regulation for exchange of materials (Lloyd et al. 2019; Martins et al. 2024). These are prepared from a variety of materials like—Agar, Starch, Gelatin, Chitosan, Alginate, Whey proteins, Soy proteins based films. Commonly used edible films and coatings include nisin, soy protein, gelatin, starch, rice starch, corn starch, chitosan, carrageenan, agar. Agar based edible films containing active ingredients like green-tea extract and cinnamon essential oil target pathogens from food enabling better quality food with boosted shelf-life. Other edible films include starch-based film with essential oils added, Chitosan films, Soy protein films coated with nisin and grape-seed or green tea extracts used more specifically for meat packaging. Biodegradable films have shown to have less elasticity, less thermal stability and high water sensitivity. These edible films aid in increasing the shelf life of the food product, while maintaining the quality of the product (Lloyd et al. 2019; Trajkovska Petkoska et al. 2021).

Wu and workfellows showcased development of an edible food packaging nanofiber film prepared from zein and gelatin, loaded with cinnamaldehyde and thymol as the bioactive agent and demonstrated properties like vapor barrier, tensile strength, UV barrier, antioxidant capacity, antibacterial property and preserved strawberries packaged thereby fresh for 7 days at room temperature (Wu et al. 2015).

9.12 Nano-composites

Nanoparticles of the dimension <100 nm, called as nanofillers, are incorporated into the biopolymer material with an intent of a technological advancement is termed as nanocomposite or bio-nanocomposite (Sharma et al. 2020). Thus, a nanocomposite involves two phases, a continuous one that comprises of the biopolymer, while a discontinuous one, that is the nano dimensional agent. Primary function associated with the nanofillers is as the structural reinforcing agent owing to its small size, high aspect ratio and large interfacial areas. They interact with the biopolymers and create a network of hydrogen bonds (Chawla et al. 2021). These also serve with efficient barrier properties as gas barrier to prevent leakage of carbon dioxide in carbonated beverages. Nanocomposites composed of cellular nano reinforcements generate inexpensive, light-weight materials and are flexible while providing low permeable polymer matrix (Flanagan and Singh 2006; Pradhan et al. 2015). Apart from the structural aspect, nanofillers are also attributed with the function of bioactive agent and can confer the properties of antimicrobial, antioxidant as well as scavenging properties beyond its structural role (Nile et al. 2020). A wide variety of nanocomposites are employed for food packaging and preservation. These includes nano clays, nano starch, carbon nanotubes to name a few.

9.13 Nanofillers

Bio-nanofillers are fine structures that are biodegradable, renewable, and contribute as structural strengthening agents that aid in recruitment and delivery of the bioactive components (Johansson et al. 2012). Most common example includes that of cellulose and its derivatives. Wang and colleagues (Wang et al. 2022b), Yu and coworkers (Yu et al. 2021) and Menon and associates (Menon et al. 2017) have demonstrated the isolation of cellulose from natural sources and later its applications in the form of nano crystals and nanofibers. Oyeko and coworkers demonstrated biofilms with enhanced water absorption and biodegradation of films prepared using cellulose nanocrystals (Oyeoka et al. 2021).

9.14 Nano-encapsulations

Nanoencapsulation is intended at providing a solid, protective covering to the bioactive agent, and its controlled release while preserving it from the interactions of the external environment (Maurizzi et al. 2022; Pabast et al. 2018). Thus, nanoencapsulation, in simple terms means, putting the active agent which can be in either of the solid, liquid or gaseous states in a capsule. Nanoencapsulation encompasses benefits like stability, solubility, prevention of deterioration, masking tastes and flavors, and most important is controlled release (Chaudhry et al. 2008; Marsh and Bugusu 2007). The encapsulation is mediated by an encapsulating material which includes lipid systems like liposomes, micelle formation, polymer systems like alginate, carrageenan, chitosan, cyclodextrin, pectin, polyethylene glycol, poly lactic acid, as well as protein systems and metallic nanoparticles (Prieto and Lagaron 2024; Taouzinet et al. 2023).

Nanoencapsulation has been employed for milk and milk products, for improving its nutritional values, in fruit juices for antioxidant activity, in meat and meat products for nutritional supplementation, enhance lipid profile and oxidative protection. Apart from this carbohydrate, protein vitamins, minerals, essential oils, flavoring agents, food colorants, supplementation have all been encapsulated for enrichment of food nutritive value (Pateiro et al. 2021).

Dutra and coworkers have demonstrated encapsulation of curcumin in orange juice provides better pH regulation, color stability as well as maintains freshness for a period of 3 days at temperature of 8 °C (Dutra et al. 2021). Liu and associates demonstrated the formulation and application of gelatin based biofilms containing polyphenols from tea and chitosan nanoparticle as bioactive agent and showed a sustained release of the polyphenols, slowly (Liu et al. 2015). Yet another application was demonstrated by Cui and colleagues who used zein protein for the bio filler and pomegranate polyphenol in chitosan nanoparticles as the bioactive agent (Cui et al. 2020). These applications emphasized on the slow and timely release of the bioactive agent. Encapsulation of α-tocopherol in fat droplets has been suggested as a way of preventing milk degradation by Chaudhary and coauthors (Chaudhry et al. 2008). Parvej, Waani and Wasoodi showed nanoencapsulation of green tea extract within maltodextrin causes lowering of its crystallinity, controlled release within intestinal conditions, improved physical properties, better hygroscopicity and better bioactivity control (Parvez et al. 2022).

9.15 Nano-emulsions

Nano emulsion aims at having an improved stability and bioavailability of the active agent while ensuring its efficient delivery in the environment (Robledo et al. 2018), where the system comprises of two immiscible liquids, homogenously mixed giving rise to nano droplets and have high delivery efficacy (Hasan et al. 2019). These liquids are dispersed either oil in water or water in oil equally and the droplets that are formed there, each of them are surrounded by surfactants to ensure its stability as well as bioavailability. It is most aptly used for sustained release of bioactive agents into solid foods, more specifically with delivery of essential oil directly into food via the polymer matrix with minimal concentration of the active agent, while maintaining flavors and preventing their oxidation (Maurizzi et al. 2022; Pradhan et al. 2015).

Neem oil acts as a natural bactericide, fungicide, antioxidant, insecticide and pesticide and thereby it is widely used as a nano emulsion for improved antioxidant activity, reduced cytotoxicity and increased antibacterial activity against foodborne pathogens and thereby plays a role on food preservation and storage. Other essential oils have also sought their application as nano emulsions (Kumar et al. 2022).

Louis and colleagues prepared nano emulsion with different concentrations of cinnamaldehyde, with Tween 80 as the surfactant in a glycerol containing alginate based coating and used for preservation of mushrooms. The addition of the nanocomposite comprising of the nano emulsion increased the quality of the button mushrooms along with a decrease in the respiration rate and enhanced firmness, color, and antioxidant property, thereby providing an elongated shelf life to the mushrooms (Louis et al. 2021). Yet another study by Youssef and workfellows, incorporates a nano emulsion made up of cinnamon essential oils and TiO_2 nanoparticles with polyvinyl alcohol (PVA) and chitosan. The resulting bio nanocomposite were observed to be with improved barrier properties, antioxidant and antimicrobial activities and when applied to chicken breast fillets, its shelf life extended the preservation by about another 14 days and proved to be efficient (Youssef et al. 2022).

9.16 Nanoparticles

Nanoparticles made up of metal ions like copper, gold, silver and their alloys are employed to prepare nanocomposites (Bajpai et al. 2018) owing to the antimicrobial and antioxidant property of these metal ions. The nano dimension modifies the physical, chemical and biological properties of these particles. Furthermore, incorporation of metal nanoparticle also enhances the barrier property of the material (de Moura et al. 2012).

Iron (II) nanoparticles had been designed by El-Saadony and coworkers and used for enhancing the shelf-life of yoghurt. These Bio—Fe (II) nanoparticles were characterized with being crystalline, UV absorbing, antimicrobial and scavengers. Yoghurt supplemented with Fe(II)NPs, demonstrated inhibition of lipid oxidation by 72% post 3 week of cold preservation, and these nanoparticles did not affect the count of lactobacillus in the yoghurt. Moreover, the yoghurt demonstrated enhanced color, taste and flavor as well as better quality, shelf-life with an increased iron (II) content without being toxic (El-Saadony et al. 2021). Rodino and associates demonstrated silver based nanoparticles for the preservation and enhancing the shelf-life of berries. Silver nanoparticles are the most widely used nanoparticles owing to its bactericidal as well as bacteriostatic properties and demonstrated improved quality and enhanced shelf-life, which was maintained during the 7 days of its preservation (Rodino et al. 2019). Furthermore, Chowdhary and coauthors showed the application of gold nanoparticles and graphene oxide amalgamated into polyvinyalcohol crosslinked with glyoxal and glutaraldehyde improved mechanical properties, tensile strength, vapor transmission rate, water solubility and when applied to banana showed an enhanced shelf life with improved antimicrobial properties (Chowdhury et al. 2020).

9.17 Metal Oxides

Metal oxides prove to be efficient agents to be used for replacing organic agents owing to their excellent antioxidant properties (Jagadish et al. 2018). Examples of metal oxides included for food packaging include titania (TiO_2), silica (SiO_2), magnesium oxide (MgO) and zinc oxide (ZnO). TiO_2 and ZnO demonstrate maximum application in the field of food preservation subjected to their biocompatibility, and chemical stability (Oun et al. 2020). It seeks its application as a reinforcing agent improving the quality and strength of edible films (Rajakumar et al. 2018). Siripatrawan and Kaewklin developed a chitosan film enriched with TiO_2 that showcased enhanced ethylene photo-degradation, besides showing enhanced antimicrobial property against both gram positive and negative bacteria (Siripatrawan and Kaewklin 2018). Marra and associates made use of ZnO, by adding it to PLA matrix, giving rise to PLA/ZnO film nanocomposite that is amorphous in nature with enhanced barrier property and better tensile strength (Jagadish et al. 2018; Marra et al. 2016; Maurizzi et al. 2022).

9.18 Nano-clay

Conventional clays have been used in combination with suitable biopolymers to produce a nanocomposite with excellent barrier properties. The incorporation of the nanofiller takes place either by means of intercalation or exfoliation. The nature of the polymer matrix, clay, interaction between the polymer and the clay and the conditions of reaction all contribute to the firmness and characteristics of the clay reinforced biofilms. Some of the best examples of nano clay include montmorillonite, bentonite, palygorskite and sepiolite. Montmorillonite, which is composed of hydrated aluminum silicate, is one of the largely used nano clays with properties of abundance, lesser costing, and easy compatibility with a wide variety of biopolymers and confers high surface ratio and interfacial area that ensures it's even distribution (Maurizzi et al. 2022).

9.19 Carbon Dots

Carbon dots, previously known as carbon quantum dots, are one of the most sought advances of nanotechnology, that includes zero dimensional carbon based nanomaterial with sizes below 10 nm, spherical shaped nanomaterials (Gupta et al. 2023; Wang and Qiu 2016). Though carbon dots on non-biomass origin are also synthesized, but not used for food preservation applications owing to their toxicity and non-biocompatibility. Biomass including meats, egg, dairy products, seafoods, fruits, vegetables, fruit juice, tea waste, paper waste, are the preferred carbon sources for synthesis of carbon dots used for food grade applications. These are having a wide range of properties including low cost, enhanced tensile strength, UV barrier, photo-stability, low toxicity, biocompatibility, antioxidant, antimicrobial, abundance as well as photoluminescence and hence seeks applications as food packaging material (Gupta et al. 2023). Its applications ranging from biosensing with combination of metal ions incorporating the phenomenon of photo sensing and photo quenching, bioimaging, targeted drug delivery, photocatalysis, as biofilms with antioxidant and antimicrobial properties (Cui et al. 2021).

Figure 9.13 shows different active agents and use of nano materials for food packaging and preservation. Figure 9.13a demonstrates the method of bio coating fresh fruits and vegetables, as a means of its preservation, as adapted from (Ungureanu et al. 2023), while Fig. 9.13b showcases the use of biopolymers in food preservation and packaging, as adapted from (Perera et al. 2023).

9.20 Barrier Properties

Oxidation based food deterioration is a major concern associated with food packaging and preservation (Garba 2023; Yan et al. 2022). Green nanoparticles contribute to the development of barrier films with enhanced gas and moisture barrier properties. Nanocomposite films incorporating nanoparticles derived from renewable resources, such as cellulose nanocrystals or chitosan, exhibit superior barrier performance compared to traditional packaging materials. These barrier films effectively prevent oxygen permeation (Yan et al. 2022), moisture ingress, and aroma loss, thereby preserving the freshness, flavor, and nutritional quality of packaged foods for an extended period. Figure 9.14 demonstrates the packaging barrier as created with the use of different barrier guards, where the one with no barrier for a PET packaging shows maximum oxygen penetration within the package, one with PET barrier, shows restricted penetration, while biopolymer line PET shows the most efficient barrier properties.

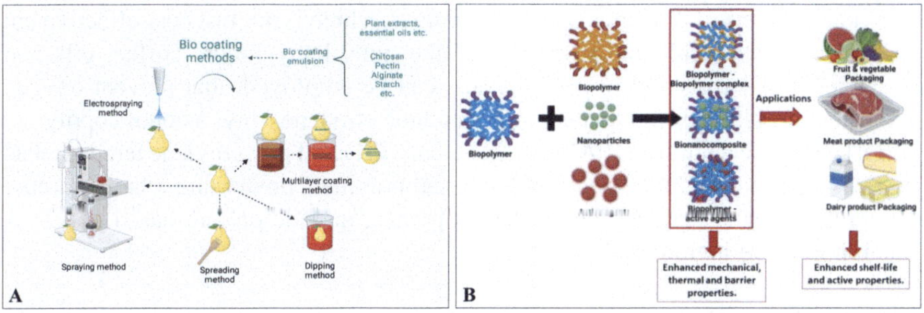

Fig. 9.13 Application of active agents and nanomaterials for food preservation (**a**) Methods of bio coating fresh fruits and vegetables as adapted from (Ungureanu et al. 2023) (under CC @ 4.0). (**b**) Use of common biopolymers application in food packaging as adapted from (Perera et al. 2023) (under CC @ 4.0)

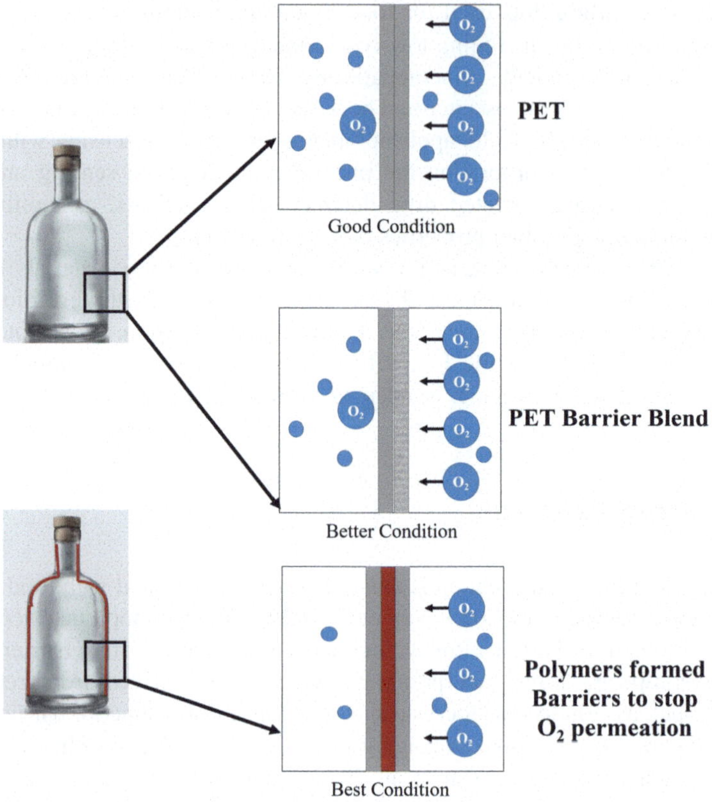

Fig. 9.14 Barrier properties demonstrated to prevent oxygen permeation. (Adapted and modified from Yan et al. (2022) (under CC @ 4.0))

Furthermore, the use of green nanoparticles in barrier coatings helps minimize food spoilage, waste, and the need for synthetic additives or preservatives, aligning with the growing consumer demand for clean label and natural products. Adding antioxidants is one of the quick and easy solution, which is not just cost-effective but also effective for oxidation prone foods like nuts, baby foods, coffee, oils and vegetables. Polymer based oxygen barriers can be employed, that prevent oxygen permeation. Widely used oxygen barriers include ethylene-vinyl alcohol copolymer (EVOH), poly-acrylonitrile (PAN), poly-ketone (PK) and polyethylene terephthalate (PET) (Yan et al. 2022). Polysaccharide based polymers possessing a large number of hydrogen bonds regulate the barrier properties, include protein-based polymers containing active ingredients.

9.21 Significance of Green Nanotechnology in Food Packaging and Preservation

The conventional techniques and technologies, though efficient, demonstrated an enormous range of shortcomings concerning the safety, purity and quality of food that has the potential to not just harm the environment but also the health and lives of living organisms, more specifically humans including adulterations, pollutants, and pathogenic infestations. Green nanotechnology paves a methodological way to overcome these shortcomings by the use of biological sources as means of manufacturing products at the nanoscale. Such an approach cuts down the extent of food industrialization involved, thereby overpowering the drawbacks presented by the conventional techniques and technologies, by improving the quality and shelf-life of the products than that showcased by traditional food packaging and preservation (Dutta and Das 2021).

With advancements in nanotechnology, enriching the foods with the use of antimicrobials, antioxidants and edible films, has the potential to inhibit the growth of pathogens in the packaged food thereby providing enhanced food safety, extended shelf-life to perishable foods and ultimately reducing the health hazards of food-borne diseases. Food packaging criteria focuses on characteristics that were observed to be lacking in the conventional methods of packaging. Nanocomposites are selected with enhanced barrier properties, shielding against oxygen, moisture, and other gases, built-up of which has the potential to degrade food and its quality, thereby ensuring its quality and freshness (Maurizzi et al. 2022). Also the use of active components like antioxidants or oxygen scavengers in packaging material, enables interaction of these components with the food that not just increases the shelf-life of foods but also aids in preserving flavor, odors and protect from oxidative damage. Intelligent packaging systems provide information, extend shelf-life of products, enhance the quality of food, and ensure safety and real-time monitoring of the food product. Intelligent packaging system also aids in reducing the cost of frequent material analysis and also reducing the time utilized thereby. Use of barcodes and QR codes have become easy, user friendly and cost effective. Real-time monitoring eases the quality assessment and assurance (Yan et al. 2022). Eco-sustainable packaging with natural sources and recyclable material can be employed. Starch and protein-based edible films and coatings are preferred which incorporates natural antimicrobial agents. Innovative food packaging reduces the amount of waste generated by maintaining the quality of food as well as managing their shelf-life, furthermore using renewable packaging reduces the harm to the environment as well. Active packaging prevents oxidation, growth of pathogens, loss of flavor and odor, and chiefly slows down the metabolism of food. Intelligent packaging ensures the modulating by the naked eyes, and observing external conditions continuously enabling real-time monitoring of food and issuing warning as and when required (Lee et al. 2015; Lloyd et al. 2019; Yan et al. 2022).

Thus, the core significance of green nanotechnology for food packaging and preserving relies on the ability to improve the food safety, increase its shelf-life, reduce waste, increase environmental sustainability while taking into account, constant innovations encouraging safer, healthier and ecofriendly food products, meeting every consumer needs (Table 9.1).

9.22 Limitations of Green Nanotechnology in Food Packaging and Preservation

Green nanotechnology, as it is transitioning, seeks its applications commercially while currently being practiced at lab scale. A wide variety of literature has been made available citing different methods for synthesis of a various nanomaterials including nanoparticles, nano emulsions, nano clays, however each of these literatures are indicative of different conditions, suitable for synthesis, including different starting materials, different reaction conditions, and thus the synthesis remains inconclusive. Illustratively in literature indicating the reduction of nano particles, the active agent triggering the reduction has not been mentioned distinctly. Also, different scientist groups have reported pH ranging from neutral to acidic to alkaline to be most suitable for synthesis of zinc oxide nano particles, which only causes confusion regarding the optimal condition for synthesizing these nano particles (Dutta and Das 2021). Intelligent packaging systems, despite of their benefits, adds up to the cost of the food products, more specifically RFID (Radio Frequency Identification). TTIs (Time-Temperature Indicators) can be used at lower temperatures and protected from radiations to maintain its effectiveness and are thereby expensive as well as need thorough check all through (Lee et al. 2015). It also presents the risk of particle migration from the packaging material into the food. Active and intelligent packaging system majorly have limitations to be used for liquid packaging, as well as lack of recyclability. Antimicrobials and anti-oxidant containing packaging needs to be evaluated to ensure no health hazards or for any possible side effects. Packaging costs add up to nearly 50–95% of the product cost. Intelligent packaging is not suitable for all types of food products. Moreover, the reliability of automated packaging in times of system failures or errors cannot be taken. Essential oils serve as the most efficient antimicrobial agents however their strong aromatic characteristic would also alter the flavors of food thereby obstructing their potential use. The compatibility of biopolymers and the respective bioactive agent is a matter of concern, and needs to be tested with pilot studies (Maurizzi et al. 2022). While employing nanotechnology for food packaging and preservation, the risk of health deterioration due to chemicals and their additives, their interaction with food and environment, their migration and resulting effects being ailing or carcinogenic must be considered. Thus, metal contaminants, toxicity of the nanoparticles, risk of neurological, cardiovascular, endocrinological complications along with death should be foreseen and discussed and thus, becomes the most important aspect while considering the applications in the field of food packaging and preservation.

Table 9.1 Unique properties demonstrated by different nano-materials used in food packaging

Nano materials used in food packaging	Unique properties
Ag-nano particles	High antimicrobial activity against drug resistant pathogens
ZnO-nano particles	Photocatalytic activity; High stability; Bactericidal effects; Antibacterial activity against spores that are resistant to high temperature and high pressure; Antioxidant, anti-diabetic, anti-inflammatory potential
Au-nano particles	Nontoxicity; Non-induction of any ROS-related processes; Polyvalent effects; Ease of detection; Photo-thermal activity
CuO-nano particles	Bactericidal effects; High stability; Antifungal activity
MgO with chitosan	Enhanced antimicrobial activity with tensile strength, film thickness, transparency, and barrier like mechanical properties
Nano emulsion with cinnamaldehyde and Tween 80	Enhanced shelf-life, quality to mushrooms; antimicrobial activity; improved antioxidant capacity
Montmorillonite clay with potato starch and micro-crystalline cellulose	Improved thermal stability; increased opacity
TiO_2nanopowder with chitosan or zein	Enhanced mechanical and water barrier properties; Ethylene photo-degradation; Antibacterial effect; Thermal stability
Cellulose nanofibrils with Soy protein isolate	Improved barrier to UV-light; Prevents photo-oxidation
PLA/ZnO: Cu/Ag Bio-nano composite	Good mechanical, thermal and barrier properties to UV, water vapor, oxygen and carbon dioxide; Antibacterial activity
Cellulose absorber/Cu Bio-nano composite	Excellent antifungal activity; reducing spoilage related yeasts and molds
Nanozinc oxide–neem oil–chitosan Bio-nano composite	Efficient inhibitive antibacterial activity against *E. coli*
Alginate/nano-Ag Bio-nano composite	Spoilage reduction; improvement of sensory attributes
Nanoemulsion coating of chitosan and mandarin essential oil	Reduction in the population of *L. monocytogenes*
Chitosan-Gelatin based edible coating containing nano-encapsulated tarragon essential oil	Inhibits quality deterioration of Pork slices; Improved antioxidant effect, antibacterial potential and sensory properties
AgNPs-organoclay with Cellulose acetate	Enhanced UV andO_2 blockade, tensile, antioxidant, and antimicrobial activity with reduced optical clarity and thermal stability
Nanoclay	Improved tensile strength and water vapor barrier like mechanical properties
Zein active film—Pomegranate peel extract—Chitosan nanoparticles	Enhanced thermal stability; Antimicrobial property inhibitive for *L. monocytogenes*
Nano-encapsulated Curcumin	

(continued)

Table 9.1 (continued)

Nano materials used in food packaging	Unique properties
	Enhanced antimicrobial and antibacterial activity effective against *Staphylococcus aureus* and *Alicyclobacillus acidoterrestris*; Improved anti-oxidant capacity; Cytotoxicity against carcinoma cell lines
Gelatin films with Chitosan nanoparticles and encapsulated/free tea polyphenols	Ferric reducing capacity; Radical scavenging power; Efficient oxidation inhibitory effect
Thymol nanoemulsion in quinoa protein/chitosan edible films	Antifungal effect for *B. cinerea* in cherry tomatoes
Chitosan—TiO_2 nanocomposite film	Ethylene scavenging; Antimicrobial effect; enhanced tensile strength, water barrier, ethylene photolytic degradation
Chitosan—TiO_2 nanoparticles with anthocyanin rich black plum peel extract	Strong free radical scavenging; High ethylene scavenging; Antimicrobial activity

9.23 Conclusion and Prospects

Green nanotechnology promises a futuristic option in the field of food packaging and preservation. Industrial production of biopolymer for replacing plastics need to be worked on and addressed, as the cost of these biopolymers is much higher than that of plastic. Preparation of nano emulsions, nanoparticles cost an ample amount of money and is not a practical approach for its application. However, environmental pollution, hazards to nature and human health is a grave concern and hence, replacement for plastics and similar polymers is a must, for which green nanomaterial is a better option, which needs to be planned properly before executing. Necessary financial planning to apply these approaches practically and which can be commonly employed without the discrimination is to be thought for. Current research largely focuses on development of nanomaterials in vitro, however more of its in vivo applications and approach needs to be tested. A more extensive use of smart packaging system, by overcoming its disadvantages can be practiced in future considering its advantages. The worth of the products should be checked, its cost, shelf life and focus should be made on wider application and consumer satisfaction. The substantial gap between the lab scale research and its real time application and commercialization needs to be addressed.

Smart food packaging and nanomaterials based application in food industries should be under appropriate jurisdiction issued by bodies like FDA (U.S. Food and Drug Administration). Compulsion on describing the packaging methods and product must be made mandatory to make the consumer aware of the hazards.

Thus, green nano-technology has the potential to balance the benefits and harmful effects presented by conventional food packaging and preservation, towards achieving sustainable and secure waste management practices by means of taste nano-encapsulation, nutrition delivery, mineral and vitamin fortification, and nutraceutical integration. Understanding the importance of these techniques and their purpose, aids in the timely transition from conventional packaging materials to green nanoparticles. A better understanding of these applications and their properly weighed pros and cons might enable to build a greener and more safe future for the field of food packaging and preservation.

References

Abbès B, Abbes F, Guo Y-Q (2015) Interaction phenomena between packaging and product. In: Polymers for packaging applications, 1st edn. Apple Academic Press, pp 39–70

Aganovic K, Hertel C, Vogel RF, Johne R, Schluter O, Schwarzenbolz U, Jager H, Holzhauser T, Bergmair J, Roth A, Sevenich R, Bandick N, Kulling SE, Knorr D, Engel KH, Heinz V (2021) Aspects of high hydrostatic pressure food processing: perspectives on technology and food safety. Compr Rev Food Sci Food Saf 20(4):3225–3266. https://doi.org/10.1111/1541-4337.12763

Agriopoulou S, Stamatelopoulou E, Sachadyn-Krol M, Varzakas T (2020) Lactic acid bacteria as antibacterial agents to extend the shelf life of fresh and minimally processed fruits and vegetables: quality and safety aspects. Microorganisms 8(6). https://doi.org/10.3390/microorganisms8060952

Ahmad R, Naikoo NB, Ahanger SA (2024) Green nanotechnology for a sustainable future. In: Green nanomaterials in energy conversion and storage applications. Apple Academic Press, pp 189–213

Alasalvar C (2005) Effect of chill storage and modified atmosphere packaging (MAP) on antioxidant activity, anthocyanins, carotenoids, phenolics and sensory quality of ready-to-eat shredded orange and purple carrots. Food Chem 89(1):69–76. https://doi.org/10.1016/j.foodchem.2004.02.013

Amit SK, Uddin MM, Rahman R, Islam SMR, Khan MS (2017) A review on mechanisms and commercial aspects of food preservation and processing. Agric Food Secur 6(1). https://doi.org/10.1186/s40066-017-0130-8

Anastas PT, Warner JC, Anastas PT, Warner JC (2000) Principles of green chemistry. In: Green chemistry: theory and practice. Oxford University Press, p 0. https://doi.org/10.1093/oso/9780198506980.003.0004

Bajpai VK, Kamle M, Shukla S, Mahato DK, Chandra P, Hwang SK, Kumar P, Huh YS, Han YK (2018) Prospects of using nanotechnology for food preservation, safety, and security. J Food Drug Anal 26(4):1201–1214. https://doi.org/10.1016/j.jfda.2018.06.011

Balaban MO, Duong T (2014) Dense phase carbon dioxide research: current focus and directions. Agric Agric Sci Proc 2:2–9. https://doi.org/10.1016/j.aaspro.2014.11.002

Berger KR (2003) A brief history of packaging: ABE321/AE206, 12/2002. EDIS 2003(17). https://doi.org/10.32473/edis-ae206-2002

Bhargava N, Mor RS, Kumar K, Sharanagat VS (2021) Advances in application of ultrasound in food processing: a review. Ultrason Sonochem 70:105293. https://doi.org/10.1016/j.ultsonch.2020.105293

Bhunia K, Sablani SS, Tang J, Rasco B (2013) Migration of chemical compounds from packaging polymers during microwave, conventional heat treatment, and storage. Compr Rev Food Sci Food Saf 12(5):523–545. https://doi.org/10.1111/1541-4337.12028

Biji KB, Ravishankar CN, Mohan CO, Srinivasa Gopal TK (2015) Smart packaging systems for food applications: a review. J Food Sci Technol 52(10):6125–6135. https://doi.org/10.1007/s13197-015-1766-7

Blanco Massani M, Molina V, Sanchez M, Renaud V, Eisenberg P, Vignolo G (2014) Active polymers containing Lactobacillus curvatus CRL705 bacteriocins: effectiveness assessment in Wieners. Int J Food Microbiol 178:7–12. https://doi.org/10.1016/j.ijfoodmicro.2014.02.013

Bonilla J, Poloni T, Lourenço RV, Sobral PJA (2018) Antioxidant potential of eugenol and ginger essential oils with gelatin/chitosan films. Food Biosci 23:107–114. https://doi.org/10.1016/j.fbio.2018.03.007

Buzrul S (2015) Multi-pulsed high hydrostatic pressure treatment of foods. Foods 4(2):173–183. https://doi.org/10.3390/foods4020173

Caner C (2011) Sorption phenomena in packaged foods: factors affecting sorption processes in package–product systems. Packag Technol Sci 24(5):259–270. https://doi.org/10.1002/pts.935

Caner C (2018) Fundamentals of the sorption (scalping) phenomena in packaged foods. In: Reference module in food science. Elsevier, pp 1–12

Castellano P, Perez Ibarreche M, Blanco Massani M, Fontana C, Vignolo GM (2017) Strategies for pathogen biocontrol using lactic acid bacteria and their metabolites: a focus on meat ecosystems and industrial environments. Microorganisms 5(3). https://doi.org/10.3390/microorganisms5030038

Castle L (2007) Chemical migration into food: an overview. In: Chemical migration and food contact materials. Woodhead Publishing, pp 1–13

Cazón P, Velazquez G, Ramírez JA, Vázquez M (2017) Polysaccharide-based films and coatings for food packaging: a review. Food Hydrocoll 68:136–148. https://doi.org/10.1016/j.foodhyd.2016.09.009

Chaudhry Q, Scotter M, Blackburn J, Ross B, Boxall A, Castle L, Aitken R, Watkins R (2008) Applications and implications of nanotechnologies for the food sector. Food Addit Contam Part A Chem Anal Control Expo Risk Assess 25(3):241–258. https://doi.org/10.1080/02652030701744538

Chavan P, Sharma P, Sharma SR, Mittal TC, Jaiswal AK (2022) Application of high-intensity ultrasound to improve food processing efficiency: a review. Foods 11(1). https://doi.org/10.3390/foods11010122

Chawla R, Sivakumar S, Kaur H (2021) Antimicrobial edible films in food packaging: current scenario and recent nanotechnological advancements—a review. Carbohydr Polym Technol Appl 2. https://doi.org/10.1016/j.carpta.2020.100024

Chen Y, Fu G, Zilberman Y, Ruan W, Ameri SK, Zhang YS, Miller E, Sonkusale SR (2017) Low cost smart phone diagnostics for food using paper-based colorimetric sensor arrays. Food Control 82:227–232. https://doi.org/10.1016/j.foodcont.2017.07.003

Chen H, Wang J, Cheng Y, Wang C, Liu H, Bian H, Pan Y, Sun J, Han W (2019) Application of protein-based films and coatings for food packaging: a review. Polymers (Basel) 11(12). https://doi.org/10.3390/polym11122039

Cheong EYL, Sandhu A, Jayabalan J, Kieu Le TT, Nhiep NT, My Ho HT, Zwielehner J, Bansal N, Turner MS (2014) Isolation of lactic acid bacteria with antifungal activity against the common cheese spoilage mould Penicillium commune and their potential as biopreservatives in cheese. Food Control 46:91–97. https://doi.org/10.1016/j.foodcont.2014.05.011

Choulitoudi E, Ganiari S, Tsironi T, Ntzimani A, Tsimogiannis D, Taoukis P, Oreopoulou V (2017) Edible coating enriched with rosemary extracts to enhance oxidative and microbial stability of smoked eel fillets. Food Packag Shelf Life 12:107–113. https://doi.org/10.1016/j.fpsl.2017.04.009

Chowdhury S, Teoh YL, Ong KM, Rafflisman Zaidi NS, Mah S-K (2020) Poly(vinyl) alcohol crosslinked composite packaging film containing gold nanoparticles on shelf life extension of banana. Food Packag Shelf Life 24. https://doi.org/10.1016/j.fpsl.2020.100463

Church N (1994) Developments in modified-atmosphere packaging and related technologies. Trends Food Sci Technol 5(11):345–352. https://doi.org/10.1016/0924-2244(94)90211-9

Cui H, Surendhiran D, Li C, Lin L (2020) Biodegradable zein active film containing chitosan nanoparticle encapsulated with pomegranate peel extract for food packaging. Food Packag Shelf Life 24. https://doi.org/10.1016/j.fpsl.2020.100511

Cui L, Ren X, Sun M, Liu H, Xia L (2021) Carbon dots: synthesis, properties and applications. Nanomaterials (Basel) 11(12). https://doi.org/10.3390/nano11123419

Dainelli D, Gontard N, Spyropoulos D, Zondervan-van den Beuken E, Tobback P (2008) Active and intelligent food packaging: legal aspects and safety concerns. Trends Food Sci Technol 19: S103–S112. https://doi.org/10.1016/j.tifs.2008.09.011

de Moura MR, Mattoso LHC, Zucolotto V (2012) Development of cellulose-based bactericidal nanocomposites containing silver nanoparticles and their use as active food packaging. J Food Eng 109(3):520–524. https://doi.org/10.1016/j.jfoodeng.2011.10.030

de Sousa MS, Schlogl AE, Estanislau FR, Souza VGL, dos Reis Coimbra JS, Santos IJB (2023) Nanotechnology in packaging for food industry: past, present, and future. Coatings 13(8). https://doi.org/10.3390/coatings13081411

Devraj L, Naik MG, Modupalli N, Thangaraju S, Natarajan V (2021) Dense phase carbon dioxide (DPCD)-aided preservation of foods. In: Handbook of research on food processing and preservation technologies. Apple Academic Press, pp 169–194

Diez-Pascual AM (2020) Antimicrobial polymer-based materials for food packaging applications. Polymers (Basel) 12(4). https://doi.org/10.3390/polym12040731

Dury-Brun C, Chalier P, Desobry S, Voilley A (2007) Multiple mass transfers of small volatile molecules through flexible food packaging. Food Rev Int 23(3):199–255

Dutra TV, de Menezes JL, Mizuta AG, de Oliveira A, Moreira TFM, Barros L, Mandim F, Pereira C, Goncalves OH, Leimann FV, Mikcha JMG, Machinski Junior M, Abreu Filho BA (2021) Use of nanoencapsulated curcumin against vegetative cells and spores of Alicyclobacillus spp. in industrialized orange juice. Int J Food Microbiol 360:109442. https://doi.org/10.1016/j.ijfoodmicro.2021.109442

Dutta D, Das BM (2021) Scope of green nanotechnology towards amalgamation of green chemistry for cleaner environment: a review on synthesis and applications of green nanoparticles. Environ Nanotechnol Monit Manag 15. https://doi.org/10.1016/j.enmm.2020.100418

El-Saadony MT, Sitohy MZ, Ramadan MF, Saad AM (2021) Green nanotechnology for preserving and enriching yogurt with biologically available iron (II). Innov Food Sci Emerg Technol 69. https://doi.org/10.1016/j.ifset.2021.102645

Espitia PJP, Du W-X, Avena-Bustillos RDJ, Soares NDFF, McHugh TH (2014) Edible films from pectin: physical-mechanical and antimicrobial properties—a review. Food Hydrocoll 35:287–296. https://doi.org/10.1016/j.foodhyd.2013.06.005

Fang Z, Zhao Y, Warner RD, Johnson SK (2017) Active and intelligent packaging in meat industry. Trends Food Sci Technol 61:60–71. https://doi.org/10.1016/j.tifs.2017.01.002

Fernández-León MF, Fernández-León AM, Lozano M, Ayuso MC, Amodio ML, Colelli G, González-Gómez D (2013) Retention of quality and functional values of broccoli 'Parthenon' stored in modified atmosphere packaging. Food Control 31(2):302–313. https://doi.org/10.1016/j.foodcont.2012.10.012

Ferrara G, Bertoldo M, Scoponi M, Ciardelli F (2001) Diffusion coefficient and activation energy of Irganox 1010 in poly(propylene-co-ethylene) copolymers. Polym Degrad Stab 73(3):411–416. https://doi.org/10.1016/s0141-3910(01)00121-5

Ferreira AR, Alves VD, Coelhoso IM (2016) Polysaccharide-based membranes in food packaging applications. Membranes (Basel) 6(2). https://doi.org/10.3390/membranes6020022

Fiorentini EF, Llaver M, Oviedo MN, Quintas PY, Wuilloud RG (2022) State-of-the-art analytical methods based on ionic liquids for food and beverage analysis. Green Anal Chem 1. https://doi.org/10.1016/j.greeac.2022.100002

Flanagan J, Singh H (2006) Microemulsions: a potential delivery system for bioactives in food. Crit Rev Food Sci Nutr 46(3):221–237. https://doi.org/10.1080/10408690590956710

Floros JD (1993) Aseptic packaging technology. In: Handbook of food safety engineering. Wiley

Fuertes G, Soto I, Carrasco R, Vargas M, Sabattin J, Lagos C (2016) Intelligent packaging systems: sensors and nanosensors to monitor food quality and safety. J Sensors 2016:1–8. https://doi.org/10.1155/2016/4046061

Gao Y, Li D, Liu X (2015) Effects of Lactobacillus sakei C2 and sakacin C2 individually or in combination on the growth of Listeria monocytogenes, chemical and odor changes of vacuum-packed sliced cooked ham. Food Control 47:27–31. https://doi.org/10.1016/j.foodcont.2014.06.031

Garba AI (2023) Food preservation packaging. In: Food processing and packaging technologies-recent advances. IntechOpen

Gonçalves AA (2017) Safety and quality of antimicrobial packaging applied to seafood. MOJ Food Process Technol 4(1). https://doi.org/10.15406/mojfpt.2017.04.00079

González-Cavieres L, Pérez-Won M, Tabilo-Munizaga G, Jara-Quijada E, Díaz-Álvarez R, Lemus-Mondaca R (2021) Advances in vacuum microwave drying (VMD) systems for food products. Trends Food Sci Technol 116:626–638. https://doi.org/10.1016/j.tifs.2021.08.005

Gupta D, Kumar L, Gaikwad KK (2023) Carbon dots for food packaging applications. Sustain Food Technol 1(2):185–199. https://doi.org/10.1039/d2fb00020b

Guth H, Rusu M (2008) Food matrices–determination of odorant partition coefficients and application of models for their prediction. Food Chem 108(4):1208–1216

Han Y, Yu M, Wang L (2018) Physical and antimicrobial properties of sodium alginate/carboxymethyl cellulose films incorporated with cinnamon essential oil. Food Packag Shelf Life 15:35–42. https://doi.org/10.1016/j.fpsl.2017.11.001

Hanlon J (1992) Moisture-barrier properties of flexible materials. In: Handbook of package engineering, 2nd edn. Technomic Publishing, Lancaster, pp 4–12

Hasan SMK, Ferrentino G, Scampicchio M (2019) Nanoemulsion as advanced edible coatings to preserve the quality of fresh-cut fruits and vegetables: a review. Int J Food Sci Technol 55(1): 1–10. https://doi.org/10.1111/ijfs.14273

Huang Q, Yu H, Ru Q (2010) Bioavailability and delivery of nutraceuticals using nanotechnology. J Food Sci 75(1):R50–R57. https://doi.org/10.1111/j.1750-3841.2009.01457.x

Huston M, DeBella M, DiBella M, Gupta A (2021) Green synthesis of nanomaterials. Nanomaterials (Basel) 11(8). https://doi.org/10.3390/nano11082130

Ibarra-Sanchez LA, El-Haddad N, Mahmoud D, Miller MJ, Karam L (2020) Invited review: advances in nisin use for preservation of dairy products. J Dairy Sci 103(3):2041–2052. https://doi.org/10.3168/jds.2019-17498

Ilyas RA, Zuhri MYM, Norrrahim MNF, Misenan MSM, Jenol MA, Samsudin SA, Nurazzi NM, Asyraf MRM, Supian ABM, Bangar SP, Nadlene R, Sharma S, Omran AAB (2022) Natural fiber-reinforced polycaprolactone green and hybrid biocomposites for various advanced applications. Polymers (Basel) 14(1). https://doi.org/10.3390/polym14010182

Jagadish K, Shiralgi Y, Chandrashekar BN, Dhananjaya BL, Srikantaswamy S (2018) Ecofriendly synthesis of metal/metal oxide nanoparticles and their application in food packaging and food preservation. In: Impact of nanoscience in the food industry, pp 197–216. https://doi.org/10.1016/b978-0-12-811441-4.00008-x

Jamroz E, Kopel P (2020) Polysaccharide and protein films with antimicrobial/antioxidant activity in the food industry: a review. Polymers (Basel) 12(6). https://doi.org/10.3390/polym12061289

Jin YZ, Lv DQ, Liu WW, Qi HY, Bai XH (2013) Ethanol vapor treatment maintains postharvest storage quality and inhibits internal ethylene biosynthesis during storage of oriental sweet melons. Postharvest Biol Technol 86:372–380. https://doi.org/10.1016/j.postharvbio.2013.07.019

Johansson C, Bras J, Mondragon I, Nechita P, Plackett D, Šimon P, Gregor Svetec D, Virtanen S, Giacinti Baschetti M, Breen C, Aucejo S (2012) Renewable fibers and bio-based materials for packaging applications—a review of recent developments. Bioresources 7(2):2506–2552. https://doi.org/10.15376/biores.7.2.Johansson

Kartal S, Aday MS, Caner C (2012) Use of microperforated films and oxygen scavengers to maintain storage stability of fresh strawberries. Postharvest Biol Technol 71:32–40. https://doi.org/10.1016/j.postharvbio.2012.04.009

Kerry JP, O'Grady MN, Hogan SA (2006) Past, current and potential utilisation of active and intelligent packaging systems for meat and muscle-based products: a review. Meat Sci 74(1): 113–130. https://doi.org/10.1016/j.meatsci.2006.04.024

Khandeparkar AS, Paul R, Sridhar A, Lakshmaiah VV, Nagella P (2024) Eco-friendly innovations in food packaging: a sustainable revolution. Sustain Chem Pharm 39:101579

Kim J, Oh S, Cho SM, Jun J, Kwak S (2020) Oxygen barrier properties of polyketone/EVOH blend films and their resistance to moisture. J Appl Polym Sci 137(47). https://doi.org/10.1002/app.49537

Kumar P, Reinitz HW, Simunovic J, Sandeep KP, Franzon PD (2009) Overview of RFID technology and its applications in the food industry. J Food Sci 74(8):R101–R106. https://doi.org/10.1111/j.1750-3841.2009.01323.x

Kumar S, Singh N, Devi LS, Kumar S, Kamle M, Kumar P, Mukherjee A (2022) Neem oil and its nanoemulsion in sustainable food preservation and packaging: current status and future prospects. J Agric Food Res 7. https://doi.org/10.1016/j.jafr.2021.100254

Lalpuria M, Anantheswaran R, Floros J (2012) Packaging technologies and their role in food safety. In: Microbial decontamination in the food industry, pp 701–745. https://doi.org/10.1533/9780857095756.4.701

Lee SY, Lee SJ, Choi DS, Hur SJ (2015) Current topics in active and intelligent food packaging for preservation of fresh foods. J Sci Food Agric 95(14):2799–2810. https://doi.org/10.1002/jsfa.7218

Lee JY, Garcia CV, Shin GH, Kim JT (2019) Antibacterial and antioxidant properties of hydroxypropyl methylcellulose-based active composite films incorporating oregano essential oil nanoemulsions. Lwt 106:164–171. https://doi.org/10.1016/j.lwt.2019.02.061

Lin J, Pan D, Sun Y, Ou C, Wang Y, Cao J (2019) The modification of gelatin films: based on various cross-linking mechanism of glutaraldehyde at acidic and alkaline conditions. Food Sci Nutr 7(12):4140–4146. https://doi.org/10.1002/fsn3.1282

Liu Y, Chakrabartty S, Alocilja EC (2007) Fundamental building blocks for molecular biowire based forward error-correcting biosensors. Nanotechnology 18(42):424017. https://doi.org/10.1088/0957-4484/18/42/424017

Liu F, Antoniou J, Li Y, Yi J, Yokoyama W, Ma J, Zhong F (2015) Preparation of gelatin films incorporated with tea polyphenol nanoparticles for enhancing controlled-release antioxidant properties. J Agric Food Chem 63(15):3987–3995. https://doi.org/10.1021/acs.jafc.5b00003

Lloyd K, Mirosa M, Birch J (2019) Active and intelligent packaging. In: Encyclopedia of food chemistry, pp 177–182. https://doi.org/10.1016/b978-0-08-100596-5.22421-9

Lockhart H, Paine FA (1996) Introduction to the packaging of pharmaceuticals and healthcare products. In: Packaging of pharmaceuticals and healthcare products, pp 1–12. https://doi.org/10.1007/978-1-4615-2125-9_1

Louis E, Villalobos-Carvajal R, Reyes-Parra J, Jara-Quijada E, Ruiz C, Andrades P, Gacitúa J, Beldarraín-Iznaga T (2021) Preservation of mushrooms (Agaricus bisporus) by an alginate-based-coating containing a cinnamaldehyde essential oil nanoemulsion. Food Packag Shelf Life 28. https://doi.org/10.1016/j.fpsl.2021.100662

Łupina K, Kowalczyk D, Zięba E, Kazimierczak W, Mężyńska M, Basiura-Cembala M, Wiącek AE (2019) Edible films made from blends of gelatin and polysaccharide-based emulsifiers—a comparative study. Food Hydrocoll 96:555–567. https://doi.org/10.1016/j.foodhyd.2019.05.053

Machado BR, Facchi SP, de Oliveira AC, Nunes CS, Souza PR, Vilsinski BH, Popat KC, Kipper MJ, Muniz EC, Martins AF (2020) Bactericidal pectin/chitosan/glycerol films for food pack coatings: a critical viewpoint. Int J Mol Sci 21(22). https://doi.org/10.3390/ijms21228663

Madani M, Hosny S, Alshangiti DM, Nady N, Alkhursani SA, Alkhaldi H, Al-Gahtany SA, Ghobashy MM, Gaber GA (2022) Green synthesis of nanoparticles for varied applications: green renewable resources and energy-efficient synthetic routes. Nanotechnol Rev 11(1): 731–759. https://doi.org/10.1515/ntrev-2022-0034

Malhotra B, Keshwani A, Kharkwal H (2015) Antimicrobial food packaging: potential and pitfalls. Front Microbiol 6:611. https://doi.org/10.3389/fmicb.2015.00611

Marasinghe WN, Jayathunge KGLR, Dassanayake RS, Liyanage R, Bandara PC, Rajapaksha SM, Gunathilake C (2024) Structure, properties, and recent developments in polysaccharide- and aliphatic polyester-based packaging—a review. J Compos Sci 8(3). https://doi.org/10.3390/jcs8030114

Marra A, Silvestre C, Duraccio D, Cimmino S (2016) Polylactic acid/zinc oxide biocomposite films for food packaging application. Int J Biol Macromol 88:254–262. https://doi.org/10.1016/j.ijbiomac.2016.03.039

Marsh K, Bugusu B (2007) Food packaging—roles, materials, and environmental issues. J Food Sci 72(3):R39–R55. https://doi.org/10.1111/j.1750-3841.2007.00301.x

Martínez-Ferrer M, Harper C, Pérez-Muntoz F, Chaparro M (2006) Modified atmosphere packaging of minimally processed mango and pineapple fruits. J Food Sci 67(9):3365–3371. https://doi.org/10.1111/j.1365-2621.2002.tb09592.x

Martins R, Kaczerewska OB (2021) Green nanotechnology: the latest innovations, knowledge gaps, and future perspectives. Appl Sci 11(10). https://doi.org/10.3390/app11104513

Martins VFR, Pintado ME, Morais R, Morais A (2024) Recent highlights in sustainable bio-based edible films and coatings for fruit and vegetable applications. Foods 13(2). https://doi.org/10.3390/foods13020318

Maurizzi E, Bigi F, Quartieri A, De Leo R, Volpelli LA, Pulvirenti A (2022) The green era of food packaging: general considerations and new trends. Polymers (Basel) 14(20). https://doi.org/10.3390/polym14204257

McMillin KW (2008) Where is MAP going? A review and future potential of modified atmosphere packaging for meat. Meat Sci 80(1):43–65. https://doi.org/10.1016/j.meatsci.2008.05.028

Menon MP, Selvakumar R, Ramakrishna S (2017) Extraction and modification of cellulose nanofibers derived from biomass for environmental application. RSC Adv 7(68):42750–42773

Moeini A, Germann N, Malinconico M, Santagata G (2021) Formulation of secondary compounds as additives of biopolymer-based food packaging: a review. Trends Food Sci Technol 114:342–354. https://doi.org/10.1016/j.tifs.2021.05.040

Mohan S, Oluwafemi OS, Kalarikkal N, Thomas S, Songca SP (2016) Biopolymers–application in nanoscience and nanotechnology. Recent Adv Biopolym 1(1):47–66

Monteiro RL, Gomide AI, Link JV, Carciofi BAM, Laurindo JB (2020) Microwave vacuum drying of foods with temperature control by power modulation. Innov Food Sci Emerg Technol 65. https://doi.org/10.1016/j.ifset.2020.102473

Muller P, Schmid M (2019) Intelligent packaging in the food sector: a brief overview. Foods 8(1). https://doi.org/10.3390/foods8010016

Murrieta-Martínez CL, Soto-Valdez H, Pacheco-Aguilar R, Torres-Arreola W, Rodríguez-Felix F, Márquez Ríos E (2018) Edible protein films: sources and behavior. Packag Technol Sci 31(3):113–122. https://doi.org/10.1002/pts.2360

Murugan B, Fatimah I, Hossain MM, Sagadevan S (2024) Biopolymer-based nanomaterials and their applications. In: Green synthesis of nanomaterials: biological and environmental applications. Wiley, pp 71–88

Nanda S, Patra BR, Patel R, Bakos J, Dalai AK (2022) Innovations in applications and prospects of bioplastics and biopolymers: a review. Environ Chem Lett 20(1):379–395. https://doi.org/10.1007/s10311-021-01334-4

Naskar A, Sanyal I, Nahar N, Ghosh DD, Chakraborty S (2023) Bionanocomposites films applied as active and smart food packaging: a review. Polym Eng Sci 63(9):2675–2699. https://doi.org/10.1002/pen.26415

Ncube LK, Ude AU, Ogunmuyiwa EN, Zulkifli R, Beas IN (2020) Environmental impact of food packaging materials: a review of contemporary development from conventional plastics to polylactic acid based materials. Materials (Basel) 13(21). https://doi.org/10.3390/ma13214994

Nesic A, Moeini A, Santagata G (2020) Chapter 4: marine biopolymers: alginate and chitosan. In: Sustainability of polymeric materials, pp 73–92. https://doi.org/10.1515/9783110590586-004

Nguyen VT, Pham HNT, Bowyer MC, Altena IAV, Scarlett CJ (2016) Influence of solvents and novel extraction methods on bioactive compounds and antioxidant capacity of Phyllanthus amarus. Chem Pap. https://doi.org/10.1515/chempap-2015-0240

Nile SH, Baskar V, Selvaraj D, Nile A, Xiao J, Kai G (2020) Nanotechnologies in food science: applications, recent trends, and future perspectives. Nanomicro Lett 12(1):45. https://doi.org/10.1007/s40820-020-0383-9

O'Grady MN, Kerry JP (2008) Smart packaging technologies and their application in conventional meat packaging systems. In: Meat biotechnology. Springer, pp 425–451

Oun AA, Shankar S, Rhim JW (2020) Multifunctional nanocellulose/metal and metal oxide nanoparticle hybrid nanomaterials. Crit Rev Food Sci Nutr 60(3):435–460. https://doi.org/10.1080/10408398.2018.1536966

Oyeoka HC, Ewulonu CM, Nwuzor IC, Obele CM, Nwabanne JT (2021) Packaging and degradability properties of polyvinyl alcohol/gelatin nanocomposite films filled water hyacinth cellulose nanocrystals. J Bioresour Bioprod 6(2):168–185. https://doi.org/10.1016/j.jobab.2021.02.009

Pabast M, Shariatifar N, Beikzadeh S, Jahed G (2018) Effects of chitosan coatings incorporating with free or nano-encapsulated Satureja plant essential oil on quality characteristics of lamb meat. Food Control 91:185–192. https://doi.org/10.1016/j.foodcont.2018.03.047

Panaitescu DM, Frone AN, Chiulan I, Nicolae CA, Trusca R, Ghiurea M, Gabor AR, Mihailescu M, Casarica A, Lupescu I (2018) Role of bacterial cellulose and poly (3-hydroxyhexanoate-co-3-hydroxyoctanoate) in poly (3-hydroxybutyrate) blends and composites. Cellulose 25(10):5569–5591. https://doi.org/10.1007/s10570-018-1980-3

Parvez S, Wani IA, Masoodi F (2022) Nanoencapsulation of green tea extract using maltodextrin and its characterisation. Food Chem 384:132579

Pascall MA, DeAngelo K, Richards J, Arensberg MB (2022) Role and importance of functional food packaging in specialized products for vulnerable populations: implications for innovation and policy development for sustainability. Foods 11(19). https://doi.org/10.3390/foods11193043

Pateiro M, Gomez B, Munekata PES, Barba FJ, Putnik P, Kovacevic DB, Lorenzo JM (2021) Nanoencapsulation of promising bioactive compounds to improve their absorption, stability, functionality and the appearance of the final food products. Molecules 26(6). https://doi.org/10.3390/molecules26061547

Paunonen S (2013) Strength and barrier enhancements of cellophane and cellulose derivative films: a review. Bioresources 8(2). https://doi.org/10.15376/biores.8.2.3098-3121

Pennanen K, Focas C, Kumpusalo-Sanna V, Keskitalo-Vuokko K, Matullat I, Ellouze M, Pentikäinen S, Smolander M, Korhonen V, Ollila M (2014) European consumers' perceptions of time–temperature indicators in food packaging. Packag Technol Sci 28(4):303–323. https://doi.org/10.1002/pts.2105

Perera KY, Jaiswal AK, Jaiswal S (2023) Biopolymer-based sustainable food packaging materials: challenges, solutions, and applications. Foods 12(12). https://doi.org/10.3390/foods12122422

Pieper G, Borgudd L, Ackermann P, Fellers P (2006) Absorption of aroma volatiles of orange juice into laminated carton packages did not affect sensory quality. J Food Sci 57(6):1408–1411. https://doi.org/10.1111/j.1365-2621.1992.tb06870.x

Pradhan N, Singh S, Ojha N, Shrivastava A, Barla A, Rai V, Bose S (2015) Facets of nanotechnology as seen in food processing, packaging, and preservation industry. Biomed Res Int 2015:365672. https://doi.org/10.1155/2015/365672

Prieto C, Lagaron JM (2024) Nanoencapsulation and nanocoating of bioactives of application interest in food, nutraceuticals and pharma. Nanomaterials (Basel) 14(3). https://doi.org/10.3390/nano14030313

Prokopov T, Tanchev S (2007) Methods of food preservation. In: Food safety. Springer US, Boston, pp 3–25

Rajakumar G, Thiruvengadam M, Mydhili G, Gomathi T, Chung IM (2018) Green approach for synthesis of zinc oxide nanoparticles from Andrographis paniculata leaf extract and evaluation of their antioxidant, anti-diabetic, and anti-inflammatory activities. Bioprocess Biosyst Eng 41(1):21–30. https://doi.org/10.1007/s00449-017-1840-9

Ramos M, Valdés A, Beltrán A, Garrigós M (2016) Gelatin-based films and coatings for food packaging applications. Coatings 6(4). https://doi.org/10.3390/coatings6040041
Ramos B, Brandao TRS, Teixeira P, Silva CLM (2020) Biopreservation approaches to reduce Listeria monocytogenes in fresh vegetables. Food Microbiol 85:103282. https://doi.org/10.1016/j.fm.2019.103282
Ranken MD, Kill RC, Baker C (1997) Food packaging. In: Food industries manual, pp 544–556. https://doi.org/10.1007/978-1-4613-1129-4_16
Risch SJ (2009) Food packaging history and innovations. J Agric Food Chem 57(18):8089–8092. https://doi.org/10.1021/jf900040r
Rivera-Briso AL, Serrano-Aroca A (2018) Poly(3-hydroxybutyrate-co-3-hydroxyvalerate): enhancement strategies for advanced applications. Polymers (Basel) 10(7). https://doi.org/10.3390/polym10070732
Robertson GL (2009) Food packaging and shelf life: a practical guide. CRC Press
Robledo N, Vera P, Lopez L, Yazdani-Pedram M, Tapia C, Abugoch L (2018) Thymol nanoemulsions incorporated in quinoa protein/chitosan edible films; antifungal effect in cherry tomatoes. Food Chem 246:211–219. https://doi.org/10.1016/j.foodchem.2017.11.032
Rodino S, Butu M, Butu A (2019) Application of biogenic silver nanoparticles for berries preservation. Digest J Nanomater Biostruct 14(3):601–606
Rodrigues C, Souza VGL, Coelhoso I, Fernando AL (2021) Bio-based sensors for smart food packaging-current applications and future trends. Sensors (Basel) 21(6). https://doi.org/10.3390/s21062148
Rooney ML (1995) Active packaging in polymer films. In: Active food packaging, pp 74–110. https://doi.org/10.1007/978-1-4615-2175-4_4
Salgado PR, Di Giorgio L, Musso YS, Mauri AN (2021) Recent developments in smart food packaging focused on biobased and biodegradable polymers. Front Sustain Food Syst 5. https://doi.org/10.3389/fsufs.2021.630393
Sanches-Silva A, Costa D, Albuquerque TG, Buonocore GG, Ramos F, Castilho MC, Machado AV, Costa HS (2014) Trends in the use of natural antioxidants in active food packaging: a review. Food Addit Contam Part A Chem Anal Control Expo Risk Assess 31(3):374–395. https://doi.org/10.1080/19440049.2013.879215
Satapathy S, Rout JR, Kerry RG, Thatoi H, Sahoo SL (2020) Biochemical prospects of various microbial pectinase and pectin: an approachable concept in pharmaceutical bioprocessing. Front Nutr 7:117. https://doi.org/10.3389/fnut.2020.00117
Shahbazi M, Ahmadi SJ, Seif A, Rajabzadeh G (2016) Carboxymethyl cellulose film modification through surface photo-crosslinking and chemical crosslinking for food packaging applications. Food Hydrocoll 61:378–389. https://doi.org/10.1016/j.foodhyd.2016.04.021
Sharma R, Jafari SM, Sharma S (2020) Antimicrobial bio-nanocomposites and their potential applications in food packaging. Food Control 112. https://doi.org/10.1016/j.foodcont.2020.107086
Sharma S, Barkauskaite S, Jaiswal AK, Jaiswal S (2021) Essential oils as additives in active food packaging. Food Chem 343:128403. https://doi.org/10.1016/j.foodchem.2020.128403
Sharma P, Ahuja A, Izrayeel AMD, Samyn P, Rastogi VK (2022) Physicochemical and thermal characterization of poly (3-hydroxybutyrate-co-4-hydroxybutyrate) films incorporating thyme essential oil for active packaging of white bread. Food Control 133:108688
Silvestre C, Duraccio D, Cimmino S (2011) Food packaging based on polymer nanomaterials. Prog Polym Sci 36(12):1766–1782. https://doi.org/10.1016/j.progpolymsci.2011.02.003
Singh NK, Baranwal J, Pati S, Barse B, Khan RH, Kumar A (2023) Application of plant products in the synthesis and functionalisation of biopolymers. Int J Biol Macromol 237:124174
Siripatrawan U, Kaewklin P (2018) Fabrication and characterization of chitosan-titanium dioxide nanocomposite film as ethylene scavenging and antimicrobial active food packaging. Food Hydrocoll 84:125–134. https://doi.org/10.1016/j.foodhyd.2018.04.049
Sirocchi V, Devlieghere F, Peelman N, Sagratini G, Maggi F, Vittori S, Ragaert P (2017) Effect of Rosmarinus officinalis L. essential oil combined with different packaging conditions to extend the shelf life of refrigerated beef meat. Food Chem 221:1069–1076. https://doi.org/10.1016/j.foodchem.2016.11.054

Smolander M (2003) The use of freshness indicators in packaging. In: Novel food packaging techniques, pp 127–143. https://doi.org/10.1533/9781855737020.1.127

Sofi SA, Singh J, Rafiq S, Ashraf U, Dar BN, Nayik GA (2018) A comprehensive review on antimicrobial packaging and its use in food packaging. Curr Nutr Food Sci 14(4):305–312. https://doi.org/10.2174/1573401313666170609095732

Sohail M, Sun DW, Zhu Z (2018) Recent developments in intelligent packaging for enhancing food quality and safety. Crit Rev Food Sci Nutr 58(15):2650–2662. https://doi.org/10.1080/10408398.2018.1449731

Song T, Qian S, Lan T, Wu Y, Liu J, Zhang H (2022) Recent advances in bio-based smart active packaging materials. Foods 11(15). https://doi.org/10.3390/foods11152228

Souto EB, Yoshida CMP, Leonardi GR, Cano A, Sanchez-Lopez E, Zielinska A, Viseras C, Severino P, Silva CFD, Barbosa RM (2021) Lipid-polymeric films: composition, production and applications in wound healing and skin repair. Pharmaceutics 13(8). https://doi.org/10.3390/pharmaceutics13081199

Souza VGL, Fernando AL, Pires JRA, Rodrigues PF, Lopes AAS, Fernandes FMB (2017) Physical properties of chitosan films incorporated with natural antioxidants. Ind Crop Prod 107:565–572. https://doi.org/10.1016/j.indcrop.2017.04.056

Sridhar A, Ponnuchamy M, Kumar PS, Kapoor A (2021) Food preservation techniques and nanotechnology for increased shelf life of fruits, vegetables, beverages and spices: a review. Environ Chem Lett 19(2):1715–1735. https://doi.org/10.1007/s10311-020-01126-2

Sulochana MB (2024) Food processing and preservation. In: Yaradoddi JS, Meti BS, Mudgulkar SB, Agsar D (eds) Frontiers in food biotechnology. Springer Nature Singapore, Singapore, pp 81–90. https://doi.org/10.1007/978-981-97-3261-6_6

Sung S-Y, Sin LT, Tee T-T, Bee S-T, Rahmat AR, Rahman WAWA, Tan A-C, Vikhraman M (2013) Antimicrobial agents for food packaging applications. Trends Food Sci Technol 33(2): 110–123. https://doi.org/10.1016/j.tifs.2013.08.001

Suppakul P, Miltz J, Sonneveld K, Bigger SW (2006) Active packaging technologies with an emphasis on antimicrobial packaging and its applications. J Food Sci 68(2):408–420. https://doi.org/10.1111/j.1365-2621.2003.tb05687.x

Tao R, Sedman J, Ismail A (2022) Characterization and in vitro antimicrobial study of soy protein isolate films incorporating carvacrol. Food Hydrocoll 122:107091

Taouzinet L, Djaoudene O, Fatmi S, Bouiche C, Amrane-Abider M, Bougherra H, Rezgui F, Madani K (2023) Trends of nanoencapsulation strategy for natural compounds in the food industry. Processes 11(5). https://doi.org/10.3390/pr11051459

Thakur M, Majid I, Hussain S, Nanda V (2021) Poly(ε-caprolactone): a potential polymer for biodegradable food packaging applications. Packag Technol Sci 34(8):449–461. https://doi.org/10.1002/pts.2572

Thirumalai A, Harini K, Pallavi P, Gowtham P, Girigoswami K, Girigoswami A (2023) Nanotechnology driven improvement of smart food packaging. Mater Res Innov 27(4):223–232

Tiwari BK, Valdramidis VP, O'Donnell CP, Muthukumarappan K, Bourke P, Cullen PJ (2009) Application of natural antimicrobials for food preservation. J Agric Food Chem 57(14): 5987–6000. https://doi.org/10.1021/jf900668n

Trajkovska Petkoska A, Daniloski D, D'Cunha NM, Naumovski N, Broach AT (2021) Edible packaging: sustainable solutions and novel trends in food packaging. Food Res Int 140:109981. https://doi.org/10.1016/j.foodres.2020.109981

Tumuluru JS (2023) Introductory chapter: food processing, preservation, and packaging—a brief overview. In: Food processing and packaging technologies-recent advances. IntechOpen

Ungureanu C, Tihan G, Zgârian R, Pandelea G (2023) Bio-coatings for preservation of fresh fruits and vegetables. Coatings 13(8). https://doi.org/10.3390/coatings13081420

Vanderroost M, Ragaert P, Devlieghere F, De Meulenaer B (2014) Intelligent food packaging: the next generation. Trends Food Sci Technol 39(1):47–62. https://doi.org/10.1016/j.tifs.2014.06.009

Verma SK, Sood SK, Saini RK, Saini N (2017) Pediocin PA-1 containing fermented cheese whey reduces total viable count of raw buffalo (Bubalis bubalus) milk. LWT Food Sci Technol 83: 193–200. https://doi.org/10.1016/j.lwt.2017.02.031

Vermeiren L, Devlieghere F, van Beest M, de Kruijf N, Debevere J (1999) Developments in the active packaging of foods. Trends Food Sci Technol 10(3):77–86. https://doi.org/10.1016/s0924-2244(99)00032-1

Vilela C, Kurek M, Hayouka Z, Röcker B, Yildirim S, Antunes MDC, Nilsen-Nygaard J, Pettersen MK, Freire CSR (2018) A concise guide to active agents for active food packaging. Trends Food Sci Technol 80:212–222. https://doi.org/10.1016/j.tifs.2018.08.006

Wang J, Qiu J (2016) A review of carbon dots in biological applications. J Mater Sci 51(10): 4728–4738. https://doi.org/10.1007/s10853-016-9797-7

Wang S, Lu A, Zhang L (2016) Recent advances in regenerated cellulose materials. Prog Polym Sci 53:169–206. https://doi.org/10.1016/j.progpolymsci.2015.07.003

Wang D, Sun J, Li J, Sun Z, Liu F, Du L, Wang D (2022a) Preparation and characterization of gelatin/zein nanofiber films loaded with perillaldehyde, thymol, or ε-polylysine and evaluation of their effects on the preservation of chilled chicken breast. Food Chem 373:131439

Wang J, Han X, Zhang C, Liu K, Duan G (2022b) Source of nanocellulose and its application in nanocomposite packaging material: a review. Nanomaterials (Basel) 12(18). https://doi.org/10.3390/nano12183158

Weiss J, Takhistov P, McClements DJ (2006) Functional materials in food nanotechnology. J Food Sci 71(9). https://doi.org/10.1111/j.1750-3841.2006.00195.x

Wicochea-Rodriguez JD, Chalier P, Ruiz T, Gastaldi E (2019) Active food packaging based on biopolymers and aroma compounds: how to design and control the release. Front Chem 7:398. https://doi.org/10.3389/fchem.2019.00398

Wu J, Liu H, Ge S, Wang S, Qin Z, Chen L, Zheng Q, Liu Q, Zhang Q (2015) The preparation, characterization, antimicrobial stability and in vitro release evaluation of fish gelatin films incorporated with cinnamon essential oil nanoliposomes. Food Hydrocoll 43:427–435. https://doi.org/10.1016/j.foodhyd.2014.06.017

Wyeth N, Roseveare R (1973) Biaxially oriented poly (ethylene terephthalate) bottle. Google Patents

Xia C, Wang W, Wang L, Liu H, Xiao J (2019) Multilayer zein/gelatin films with tunable water barrier property and prolonged antioxidant activity. Food Packag Shelf Life 19:76–85. https://doi.org/10.1016/j.fpsl.2018.12.004

Xu Q, Chen C, Rosswurm K, Yao T, Janaswamy S (2016) A facile route to prepare cellulose-based films. Carbohydr Polym 149:274–281. https://doi.org/10.1016/j.carbpol.2016.04.114

Yadav AK, Singh SV (2014) Osmotic dehydration of fruits and vegetables: a review. J Food Sci Technol 51(9):1654–1673. https://doi.org/10.1007/s13197-012-0659-2

Yam KL, Takhistov PT, Miltz J (2005) Intelligent packaging: concepts and applications. J Food Sci 70(1):R1–R10. https://doi.org/10.1111/j.1365-2621.2005.tb09052.x

Yan MR, Hsieh S, Ricacho N (2022) Innovative food packaging, food quality and safety, and consumer perspectives. Processes 10(4). https://doi.org/10.3390/pr10040747

Youssef AM, El-Sayed HS, El-Sayed SM, Fouly M, El-Aziz MEA (2022) Novel bionanocomposites based on cinnamon nanoemulsion and TiO2-NPs for preserving fresh chicken breast fillets. Food Bioprocess Technol 16(2):356–367. https://doi.org/10.1007/s11947-022-02934-w

Yu S, Sun J, Shi Y, Wang Q, Wu J, Liu J (2021) Nanocellulose from various biomass wastes: its preparation and potential usages towards the high value-added products. Environ Sci Ecotechnol 5:100077. https://doi.org/10.1016/j.ese.2020.100077

Yun Y-H, Wee Y-J, Byun H-S, Yoon S-D (2008) Biodegradability of chemically modified starch (RS4)/PVA blend films: part 2. J Polym Environ 16(1):12–18. https://doi.org/10.1007/s10924-008-0084-9

Zerdin K, Rooney ML, Vermuë J (2003) The vitamin C content of orange juice packed in an oxygen scavenger material. Food Chem 82(3):387–395. https://doi.org/10.1016/s0308-8146(02)00559-9

Zhang Y, Wang B, Lu F, Wang L, Ding Y, Kang X (2021) Plant-derived antioxidants incorporated into active packaging intended for vegetables and fatty animal products: a review. Food Addit Contam Part A Chem Anal Control Expo Risk Assess 38(7):1237–1248. https://doi.org/10.1080/19440049.2021.1885745

Zhang L, Li K, Yu D, Regenstein JM, Dong J, Chen W, Xia W (2022) Chitosan/zein bilayer films with one-way water barrier characteristic: physical, structural and thermal properties. Int J Biol Macromol 200:378–387

Zhong Y, Godwin P, Jin Y, Xiao H (2020) Biodegradable polymers and green-based antimicrobial packaging materials: a mini-review. Adv Ind Eng Polym Res 3(1):27–35. https://doi.org/10.1016/j.aiepr.2019.11.002

Zubair M, Pradhan RA, Arshad M, Ullah A (2021) Recent advances in lipid derived bio-based materials for food packaging applications. Macromol Mater Eng 306(7). https://doi.org/10.1002/mame.202000799

Chapter 10
Green Nanotechnology for Harnessing Energy

Soumyadeep Mitra, Sudipta Ray, and Dilip K. Maiti

Abstract One of the newest and most promising fields in sustainable energy is nanotechnology, which has the potential to make significant advancements in the areas of energy generation, storage, and conversion. In this chapter we have reviewed how nanotechnology can be used to advance renewable energy solutions in a variety of applications, such as fuel cells, solar energy, wind energy, and energy storage. For the production of nanoparticles, green synthesis provides an excellent substitute for conventional techniques. This method works under ambient circumstances and is not only safe and benign but also economical, facile, and simple. It has proven possible to develop notable nanoparticles, including gold, silver, and iron oxides, by employing a variety of bio-reductants made from plant extracts. These biological agents play the twin roles of capping and reducing agents by stabilizing the nanoparticles and accelerating the reduction process. It is well known that today's use of fossil fuels is unsustainable and increases greenhouse gas emissions. In order to harness future energy for a sustainable environment, relevant technologies are examined, such as hydrogen fuel, solar cells based on nanotechnology, and relevant patents. It is also noted that strategies for improving the use of current energy and its byproducts, as well as technologies that ensure the least amount of environmental impact, should be developed. The green synthesis of nanoparticles made from plant extracts and biogenic wastes, highlighting their potential applications in various field of energy production and storage has been presented in this

S. Mitra
Centre for Research in Nanoscience and Nanotechnology, University of Calcutta, Kolkata, India

Department of Physics, Brainware University, Barasat, India
e-mail: dsm.ph@brainwareuniversity.ac.in

S. Ray
A. K. Mitra Institution for Girls, Kolkata, India

D. K. Maiti (✉)
Centre for Research in Nanoscience and Nanotechnology, University of Calcutta, Kolkata, India

Department of Chemistry, University of Calcutta, Kolkata, India
e-mail: dkmchem@caluniv.ac.in

© The Author(s), under exclusive license to Springer Nature Switzerland AG 2025
J. M. Al-Khayri et al. (eds.), *Green Nanotechnology Applications for Ecosystem Sustainability*, Nanotechnology in Plant Sciences 4,
https://doi.org/10.1007/978-3-031-90766-1_10

chapter. This abstract emphasises how important nanotechnology is to speeding up the shift to clean and renewable energy sources by providing a thorough analysis of current research and development initiatives.

Keywords Bio-mass energy · Fuel cell · Energy sources · Green energy · Green synthesis · Nanoparticles · Solar cell

10.1 Introduction

Sustainability is more important than ever, especially in light of the recent extreme weather events, greater storms, global warming, etc. The ozone layer, increasing sea levels, and resource depletion should all be taken seriously since they may endanger the ability of future generations to survive. Global climate change, rapid population development, and an increase in anthropogenic activities have all contributed to a wide range of pollutants in various environmental conditions and the soil quality has also been degraded, which has mostly resulted in a loss in crop yield worldwide. Various eco-friendly methods are being employed to mitigate the harmful effects of abiotic stressors, improve plant stress tolerance, and address the global issue of environmental pollution (Pokrajac et al. 2021; Simpa 2024).

10.1.1 Nano Technology

Nanotechnology is a multidisciplinary field that includes chemistry, physics, microbiology, biotechnology, and, engineering, all of which work together to advance and improve technology (Mohammad Zahra et al. 2022). It is defined as the utilization of research-based knowledge about nanoparticles for practical purposes or applications, whether in industry or in our everyday life. On the other hand, Nanomaterials are those materials which are composed of tiniest particles (like grains, powders, nanospheres, etc.) that have at least one dimension (1D) in the nanoscale of range 1–100 nm. They are highly dense because of their small size. Due to their large surface area to volume ratio, nanomaterials possess enhanced and modified properties of bulk materials (materials made up of many components and all the dimensions lie above 100 nm in the nanoscale). Nanoparticles have unique qualities that set them apart from bulk materials, such as magnetic properties, electrical conductivity, optical characteristics, surface reactivity, physical strength, and stability. NPs are used in a variety of disciplines, including agriculture, energy, optical devices, electronics, sensors, optics, food, textiles, biomedical, health care, cosmetics, electrochemistry, and pharmaceutics. Through the use of nanoparticles' distinct characteristics, like their heightened surface area and catalytic activity, scientists can generate energy solutions that are both ecologically sustainable and highly efficient. The synthesis, characterization, and application of nanodevices and nanomaterials in energy production, utilization, storage, and conversion are all included in the category of clean and sustainable nanotechnologies. The technology based on nanomaterials includes a variety of parts/devices and processes, including integrated energy systems, flexible and self-powered energy systems, fuel cells, hydrogen production system, photovoltaics, batteries, thermoelectric materials, and/or opto-electronic devices. The photovoltaic device, which directly converts sunlight into electric power, is one of the most significant renewable energy technologies among all the energy storage devices (Devasahayam et al. 2020; Serrano-Garcia et al. 2023; Wang et al. 2022). Nanotechnology has been a key component in the creation of solar cell devices as an electrode material. Furthermore, it is important to note that fuel cell devices, which convert chemical energy directly into electrical energy, demonstrate an astonishing impact of nanotechnology.

10.1.2 Green Nanotechnology

In order to attain sustainability, green nanotechnology can be defined as the development of the nanomaterials by various biosynthesis processes using bioactive agents such as several biowastes like fruit peels, vegetable waste, eggshells, etc. and also by using naturally occurring bioactive agents like microbes, and other plant materials. It is an inexpensive, straightforward, risk-free, nontoxic, and ecologically friendly method (Zhang et al. 2020; Moradi et al. 2021). A key component of clean technologies for environmental remediation and the transformation of extra-bioactive products into more profitable and sustainable green nanomaterials is green nanotechnology. The amount of biowaste produced by humans is growing along with the population, which is seriously endangering the purity of the environment. Finding a safe and affordable solution to this issue has a significant environmental cost therefore has become essential. Recently, a lot of research works are going on to find the possible ways to recycle these bioresources into products that are both environmentally friendly and acceptable in order to maximize their use. The amount of non-renewable energy resources used can be reduced thanks to green technologies. In addition, it may result in a healthy environment free of pollution. Green technology's primary goal is to meet societal demands without endangering the earth's renewable natural resources. It is a simple, low cost and biosafe technique for the fabrication of the nanomaterials having potential applications in various fields (Hassanisaadi et al. 2021; Bukhari et al. 2021; Singh et al. 2018).

Therefore, it is essential to synthesize nanomaterials using ecologically benign, sustainable processes. This could potentially mitigate the hazards linked to the application of nanotechnology, boost the output of nanomaterials, and lower carbon emissions. To put it briefly, adopting sustainable and green nanotechnology concepts is critical for applications involving energy conversion and storage. This chapter describes the recent developments in the biosynthesis of specific nanoparticles. We also discuss the benefits of biosynthesis over conventional synthesis techniques, emphasizing its affordability, ease of use, and environmental friendliness. The main objective to present the green synthesis process, go over the many characterization techniques used to investigate nanoparticles that have been green-synthesised, and provide insight into the sustainable uses of these materials in the energy production and storage sector.

10.2 Green Synthesis

While there exist numerous techniques for creating nanostructures, it is imperative to avoid hazardous chemicals and advocate for green synthesis methods that yield nanoparticles with characteristics akin to those achieved through conventional procedures. The primary advantages of green synthesis are enumerated by researchers as follows: it may be used in large-scale manufacturing; it is an environmentally friendly technology; it uses a biological component as a capping and reducing agent; and it saves energy because it doesn't require high pressure or energy. Green nanotechnology helps prevent unfavourable results. Furthermore, bio-nanotechnology significantly influences the evolution of nanostructures and aids in the resolution of current environmental issues by lowering or removing pollutants (Sanchez-Botero et al. 2017; Anastas et al. 1998).

10.2.1 Green Synthesis of NPs from Biogenic Wastes

Effective solid waste management is necessary to reduce potential health concerns and mitigate environmental effects. Both home and industrial waste streams have the potential to significantly raise the risk of infection spread and negatively affect the environment and public health if they are not adequately managed. However, when managed and recycled properly, biowaste can offer long-term solutions and mitigate its adverse effects (Tun et al. 2018). Waste management has a big impact on people, animals, and the environment, both locally and globally. In rising nations, waste generation increases in lockstep with population growth and economic expansion. Non-biodegradable and biodegradable wastes represent hazards and challenges that can negatively impact communities if disregarded or improperly disposed of. It's interesting to note that plant and animal wastes naturally decay (Aswathi et al. 2023). The four "R's" of trash management strategies—reduction, reuse, recycling, and recovery—help to mitigate the risks associated with inappropriate waste management. Moreover, food industry waste and leftover agricultural crops, particularly fruit waste, are abundant and typically don't require pre-treatment (Aswathi et al. 2023). Biodegradable wastes have the potential to serve as environmentally beneficial nanoparticle synthesis raw materials due to their accessibility. Using these wastes for the synthesis of nanoparticles helps minimize waste, reuse, and enhance the economy while also decreasing environmental impact by lowering the accumulation of high-energy waste. Green synthesis techniques harness natural biological processes for developing nanomaterials (Huston et al. 2021).

10.2.1.1 Microbial and Algal-Mediated Synthesis

Microalgae present themselves as a promising feedstock for the production of numerous products with additional value (Joy and Anju 2023). Conventional physical and chemical techniques are acknowledged for their high energy and time requirements as well as the use of chemicals that are bad for the environment. Stabilizing agents, reducing agents, and metal precursors, are needed for chemical synthesis techniques. A number of procedures are recommended, such as reducing silver nitrate in an aqueous solution while both stabilizing and reducing agents are present. Stabilizing agents can be various reducing agents such as hydrogen gas, borohydride, ascorbate, citrate, ligands, and surfactants, or polymers like polyethylene glycol and polyvinylpyrrolidone. The size of the final nanoparticles is greatly influenced by variables such as the initial concentration of the silver nitrate solution, the concentration of the stabilizing agent, and the molar ratio of the reducing agent to the silver nitrate (Gudikandula et al. 2016). As a result, chemical approaches lack ecologically friendly synthesis techniques and offer little advantages.

The focus of research has turned to greener and cleaner synthesis methods. It is generally acknowledged that using bio-based techniques to synthesize nanomaterials is preferable to traditional methods and has great potential for producing nanoparticles. Several research works have demonstrated that microbes are capable of producing nanomaterials. These particles can be accumulated intracellularly by some bacterial species to aid in navigation. Furthermore, the production of silver nanoparticles is largely dependent on the use of particular biomolecules, such as enzymes, for the in situ reduction of metal ions, such as silver ions, by living things including plants, algae, bacteria, fungus, and even mammals (Gupta et al. 2023; Kulkarni et al. 2023). In many cases, biological synthesis outperforms physicochemical techniques in terms of uniformity of size and well-defined nanoparticle production, as well as safety and purity of the final product. Because these aspects of biological synthesis coincide with a number of fundamental ideas in green chemistry, the process of biosynthesis is considered environmentally benign. Furthermore, adding biomolecules to nanoparticles produced by biological processes as stabilizing and capping agents increases their microbial activity.

Numerous beneficial characteristics, including faster rates of multiplication, the synthesis of a variety of secondary metabolites, quick growth in small areas, and the ability to deactivate contaminants in concert, are seen as a result of the synergistic interaction between microorganisms. As a result, microbes are an appropriate substrate to develop nanozerovalent iron particles. (Monga et al. 2020). The ability of various acidophilic microbe strains to reduce iron oxide varies; also, the features listed above can be further altered by the presence of electron donors and the selection of growth conditions (Wang et al. 2016; Xie et al. 2017). For example, Bacillus subtilis enhances the reduction of uranium (VI) by using zerovalent iron nanoparticles (Ding et al. 2015). Additionally, Lee et al. (2015) reported that the breakdown of trichloroethane was increased by a factor of 14 when zerovalent iron nanoparticles carrying di halogen were present. Because of their many uses and beneficial characteristics, microorganisms are a good medium for producing zerovalent iron particles, which improves reduction processes and the breakdown of pollutants.

Furthermore, 20–50 nm iron particles have been produced by synthesizing spherical-shaped iron nanoparticles from iron chloride precursor using *Chlorococcum sp.* algae (Subramaniyam et al. 2015). The produced nanoiron particles were primarily located on the surface of the microalgal cells rather than inside the cells, as shown by an investigation using transmission electron microscopy. This finding implies that biomolecules—more especially, chemicals originating from polysaccharides and glycoproteins—were present in the algal cells and were essential for the synthesis of nanoiron particle. The role of these biomolecules in the creation of nanoiron particle was further supported by the Fourier transform infrared analysis. The use of microorganisms in the synthesis of nanozerovalent nanoparticles offers advantages over traditional techniques.

10.2.1.2 Fungal-Mediated Synthesis

Fungal-mediated green nanoparticle synthesis has many benefits, such as low complexity and ease of amplification, processing simplicity, economic viability, biomass processing, substantial surface distance recovery, and ideal mycelium development. Fungi therefore hold great promise for the creation of nanoparticles. Wang et al. (2021a, b) synthesized silver nanoparticles in vitro using the growth supernatant of *Aspergillus sydowii. Penicillium chrysogenum* was utilized by Mohamed et al. (2021) to develop copper oxide and zinc oxide nanoparticles using in vitro microbial production (Fig. 10.1).

10.2.2 Green Synthesis of NPs from Plant Extracts

The use of natural resources, such as plant extracts, is referred to as “green synthesis” (Milne et al. 2018). The principal source of metals and metal oxide nanostructures is plants (Abuzeid et al. 2018). Plant extracts are better biological agents for the creation of nanoparticles than microbes. With a wide range of metabolic functions, they are plentiful, stable, safe to use, kind to the environment, and affordable (Katata-Seru et al. 2018; Singh et al. 2016; Hussain et al. 2016). Moreover, plant extracts are more scalable and provide quick metal ion reduction. As a result, plants are the main focus of research on nanoparticle synthesis, as they regularly show greater success rates when compared to other “green” approaches (Das et al. 2017).

Using plant extracts to create environmentally friendly NPs has attracted a lot of attention in the last 30 years. Several plant species have been investigated for their potential as sources of nanoparticles, and each has a distinct makeup of organic reducing agents (Samuel et al. 2022). The benefits of using plant extracts in this green synthesis approach are scalability, medical relevance, and biocompatibility

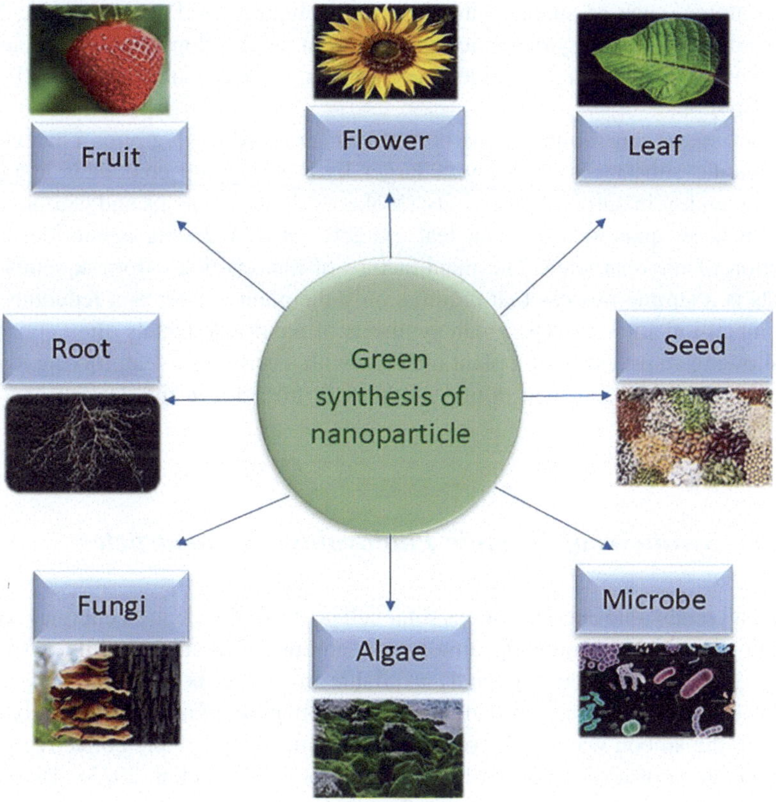

Fig. 10.1 Green synthesis of nanoparticle. (Original Figure made by authors, not a published one & authorised to publish)

(Ahmad et al. 2022). These environmentally friendly manufactured nanoparticles have a wide range of uses in the pharmaceutical industry, including the development of functional nanodevices, targeted drug delivery, and innovative drug formulations (Bar et al. 2009). Thus, this strategy is essential for developing novel therapies to address numerous risks to global health (Cruz et al. 2010). The search for sustainable nanoparticle production techniques has accelerated due to the growing commercial demand for nanomaterials (Iravani et al. 2014). Numerous plant parts, such as cultured tissues, bark, tubers, latex, fruits, flowers, seeds, and leaves have been linked to the synthesis of nanoparticles, according to research.

Rich in polyphenols, commonly consumed fruits, vegetables, and herbs can act as reducing or antioxidant agents during the nanoparticle manufacturing process. Plant extracts have several benefits when it comes to green nanoparticle synthesis: they are readily available, energy-efficient, and can be produced at room temperature in ambient circumstances (Yallappa et al. 2015). The antioxidant qualities present in the plant extracts are what cause the creation of metallic nanoparticles (Souri et al. 2018). These plants are abundant in bioactive substances that aid in the reduction of ions to atoms, such as sugars, vitamins, and polyphenols (Purnomo et al. 2023). Specifically, the main phytochemicals involved in the development of nanoparticles are carboxylic acids, flavones, aldehydes, amides, ketones, and terpenoids (Prathna et al. 2010).

An aromatic ring bound to carbon atoms and hydroxyl groups characterizes polyphenolic substances. Notably, a larger hydroxyl group content is associated with an increase in antioxidant activity (Makarov 2014). Proteins and carbohydrates found in large quantities in plant leaf extracts act as reducing agents during the formation of nanoparticles. The manufacture of nanoparticles from aqueous plant extracts is a simple process that requires only the plant extract as a reducing agent and a metal-ion solution. The green synthesis of nanoscale metals often consists of the following steps: extracting plant material with a certain solvent, mixing it with a metal salt solution, and then doing another step to achieve the desired nanoscale metal.

10.2.3 Synthesis of Organic Photovoltaic Nanoparticles

Among all renewable energy sources solar cell is one of the cleanest energies sources till date as it is not producing almost any amount of residues while generating energy. The solar cell commercially available in market is predominantly silicon based solar cells which has good efficiency and longevity of nearly 20 years but after its tenure the silicon wafer will completely become scrapped and collectively, they will generate enormous amount of electronic waste (Mitra et al. 2023). That's why scientists have come out with some better alternative like organic solar cells, perovskite solar cells those are less hazardous to the environment. In organic photovoltaics the commercially available acceptor materials are mostly fullerene based that have stability and problem at open air due to oxygen and moisture. Moreover, these type of materials possesses some solubility problem as fullerene is a bulky nanomaterial that's why some other non-fullerene acceptors should be recommended. One of the good alternatives is Perylenediimide (PDI) based material. In recent days, researchers have developed a non-fullerene based ANTPABA-PDI nanoparticle and used it as non-fullerene acceptor to fabricate inverted organic solar cells (Mitra et al. 2023). The synthesis scheme is presented in Fig. 10.2.

10.3 Applications

In the era of renewable energies utilizing the green nanotechnologies, there are various available applications of harnessing energies like: hydrogen fuel/fuel cell, solar energy/solar cell, biofuel, wind energy, ocean energy or tidal energy, nuclear energy etc. (Mitra et al. 2021b).

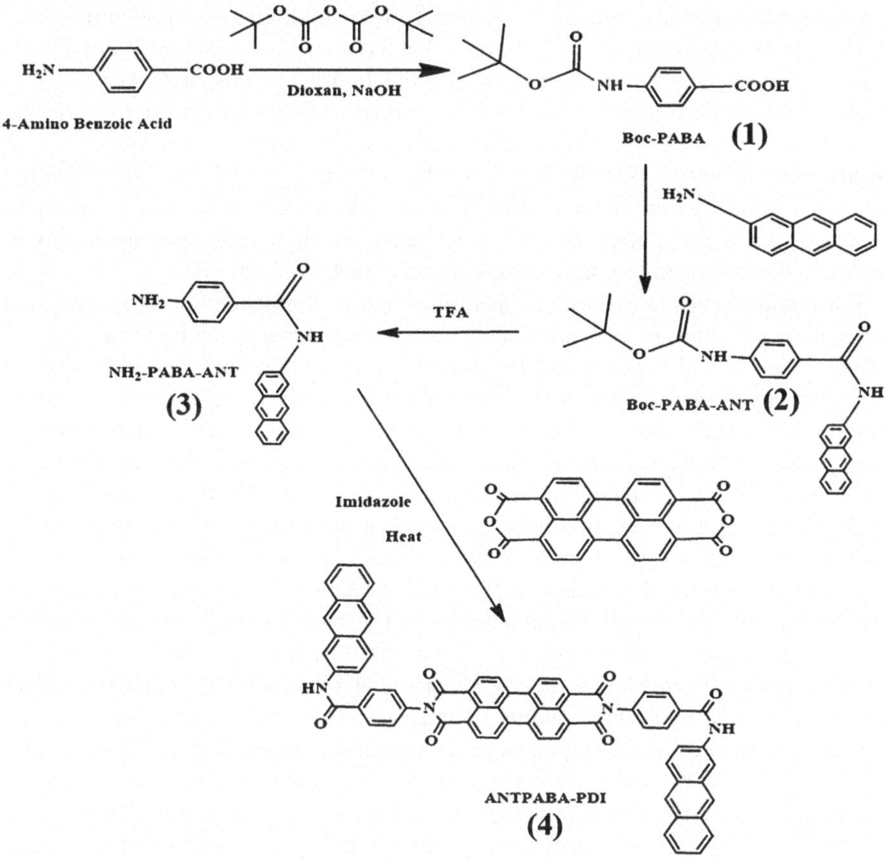

Fig. 10.2 Schematic representation of synthesis of ANTPABA-PDI nanoparticle. (Original Figure made by authors, not a published one & authorised to publish)

10.3.1 Hydrogen Fuel/Fuel Cell

An electrochemical reaction between the oxygen that can be carried in a separate storage vessel after taking from the ambient air and the hydrogen like fuel or natural gas is a very good example for the generation of low-voltage direct current, and other auxiliary equipment that are related to it are the pumps that are used for the removal of the produced water and for the transportation of the fuel over the system, a compressor or blower that can be used for cooling the stack and for transportation of oxygen to the fuel cell, and polymer electrolyte membrane cells like humidifier. The design and construction of practical fuel cell systems are extremely difficult, particularly the compact, tough ones for automobiles, trucks, and buses that have to withstand temperature changes and bumps. The early prototypes weren't put on the road for a long time for this reason. That being said, the fundamental idea of a fuel cell's operation is really simple. Molecular gaseous hydrogen is supplied in on the anode side of a hydrogen–oxygen fuel cell that has an acid electrolyte. There, the H_2 molecules lose their electrons, leaving behind hydrogen ions that are positively charged. When they reach the positively charged cathode through the electrolyte layer, they mix with oxygen to generate water. Concurrently, the freed electrons, or electric current, move from the anode to the cathode via a wire or other metallic substance, carrying out tasks like driving an electric vehicle, turning on a lightbulb, or recharging a cell phone. On the other hand, the fuel cell's performance will decline when the operating temperature drops (Seselj et al. 2023).

Furthermore, because these temperatures imply significant heat transfer and hence need a cooling system additionally, that would increase the fuel cell's overall volume, they are still insufficient for portable devices. The fact that each of these processes involves a chemical reaction or more expensive equipment is a more significant concern. The associated so-called "green" micro- and nanotechnology used in fuel cell manufacturing have to be provided in an industrial setting in accordance with eco-friendly regulations (Varcoe et al. 2009; Peighambardoust et al. 2010). Green synthesis is very important in the making of crystalline porous materials for energy applications. A solvent-free approach has been developed in order to prepare a highly resilient and porous covalent organic framework based on pyrazine and highly crystalline olefin linkages. The resultant COF will have a perfect structure with strong hydrophilicity, plenty of nitrogen sites, and well-defined one-dimensional nanochannels that can host and stabilize an H_3PO_4 network within the pores through hydrogen-bonding interactions. The resultant material shows an ultrahigh proton conductivity in a large temperature window from 25 to 80 °C and RH from 10% to 90% with a low Ea of 0.06 eV. Using olefin-linked COF as the solid electrolyte, a realistic proton exchange membrane fuel cell is able to deliver up to 135 mW cm^{-2} maximum power density and 676 mA cm^{-2} current density (Wang et al. 2021a). Highly effective bio-inspired platinum nanoparticles (Pt NPs) are being developed from green synthesis of bio-waste usage. They are intended to be used as an electrocatalyst for the methanol oxidation reaction (MOR) with superior intrinsic kinetics and high performance. Under optimal circumstances of 1.5 mM bio-Pt NPs, 1.05 M CH_3OH, and 2.14 mg, the optimized current density generated by bio-inspired Pt NPs was 640.11 mA mgPt−1. In the end, the bio-inspired Pt NPs-powered passive DMFC single-cell produces power density with Pmax values of 5.70, 6.67, and 8.28 mW cm^{-2} at 25, 80, and 100 °C. As a result, it has been demonstrated that bio-inspired Pt NPs produced from green synthesis methods and biomass-mediated extract are excellent anode electrocatalysts that may be used to convert renewable energy sources into useful energy (Ishak et al. 2024).

10.3.2 Solar Energy/Solar Cell

For a century, technologists have worked to produce electricity using solar thermal energy. In order to meet the affordability criterion for commercial applications, the cost of high efficiency must be reduced. But in order to achieve cost-effectiveness, high-efficiency solar cells need thinner less than 200 nm semiconductor wafers and the replacement of present expensive photolithography technologies with low-cost, straightforward processing methods.

At the moment, several costly photolithography techniques utilizing thick silicon wafers are used to create silicon solar cells of high-efficiency throughout the fabrication process. Technology is now on the verge of reaching adulthood. Enhancing the surface of the device exposed to solar radiation's capacity for solar absorption will increase the solar cell's conversion efficiency. Using nanotechnology could help reduce expenses and increase efficiency while using solar energy (Mitra et al. (2021a). The physical device and the device architecture have been shown in Fig. 10.3 which has been fabricated in author's laboratory.

Developing an impenetrable vapor barrier to stop water molecules from flowing inside and reducing the efficiency is a significant technical challenge with flexible solar cells, though. Increasing efficiency across all wavelengths is an additional strategy. Current single-junction solar cells are unable to capture the vast majority of the sun's infrared radiation. Solar cells rely on the p-n junction, a narrow semiconductor layer in which solar photons generate electron-hole pairs that flow as current. On either side of the junction, different semiconductor materials are doped, with one n-type to collect electrons and the other p-type to collect holes. The p–n junction has the ability to absorb solar photons at a particular wavelength, contingent on the semiconductors selected. Multijunction thinfilm solar cells efficiently generate energy from a range of wavelengths, including the infrared, by stacking many p–n

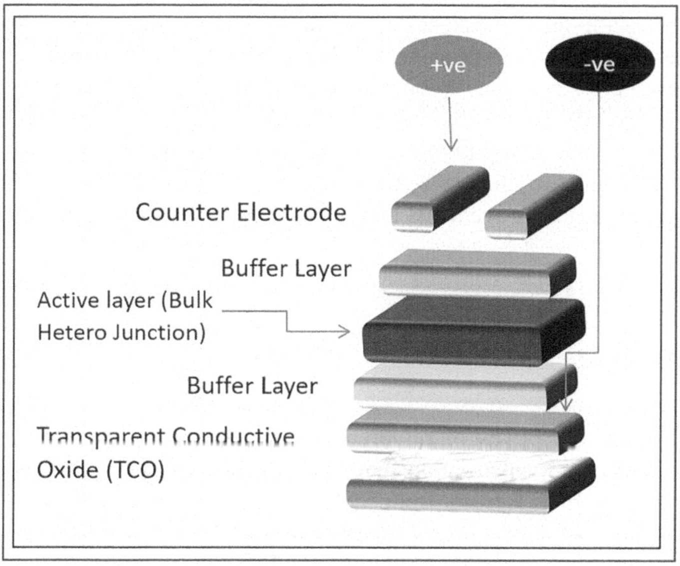

Fig. 10.3 Diagram of organic solar cells device and its architecture. (Original Figure made by authors, not a published one & authorised to publish)

junctions in thin layers (Strümpel et al. 2007; Li et al. 2009; Basol et al. 2010). Mitra and his group (in 2023) have developed a stable inverted organic solar cell utilizing non-fullerene acceptor in open atmosphere shown in Fig. 10.3. Quantum dot solar cells, which absorb solar infrared and produce electron-hole pairs from thermal gradients, have the potential to be much more efficient. Theoretically, all wavelengths of solar light, from infrared to ultraviolet, may be harvested by optimizing nanoscale quantum dots of different sizes and compositions (Fafard 2011).

It should be noted that, aside from the developed photovoltaic cells made of nanowires and nanocrystals, quantum dots offer additional clear benefits over photosensitive dyes when it comes to solar cell manufacturing. Metal halide perovskite quantum dots have inherited some of the beneficial properties of perovskite materials, such as long exciton lifetime and defect tolerance, besides sharing the common qualities with regular quantum dots. Now, after a few years of research and development, perovskite quantum dot-based solar cells achieved a Power Conversion Efficiency of 18.1% (Hao et al. 2024). Moreover, because the size of the particle or dot can alter the absorption spectra of quantum dots, they are better able to match the solar spectrum. It has been shown that quantum dots can produce several electron-hole pairs for every photon, which could lead to improved solar cell conversion efficiency. It is anticipated that the use of quantum dot technology to the production of solar cells will satisfy the two essential needs of reduced manufacturing costs and increased conversion efficiency.

To attain higher conversion efficiencies, scientists are exploring alternative topologies and potential materials for quantum dots and nanowires. Furthermore, including zinc oxide nanorods based on nanotechnology will greatly increase the solar cell's conversion efficiency. Moreover, researchers are focusing on photovoltaic (PV) devices with good efficiencies that are simpler and far less expensive to manufacture than conventional solar cells by combining the inorganic semiconductors having superior electronic properties of with the process flexibility of organic polymers. Hybrid PV solar cells based on conducting polymers and semiconductor nanorods also hold special promise. In the lab, early prototypes have shown power conversion efficiency of about 8%; these levels are anticipated to rise sharply as a result of engineering improvements and novel material compositions (Brabec et al. 2011).

10.3.3 Biofuels

One of the main objectives of research and development is to find sustainable feedstock substitutes for petroleum fuel, with biologically produced gasoline substitutes coming in first. Results from the use of alcohol generated from plants in engines already point to benefits for the environment. However, feedstocks based on oil seeds, starch, and sugar are currently used in the commercial production of biofuels. For instance, canola seeds are frequently used as a feedstock for the manufacturing of biodiesel, whereas maize starch, soybeans, palm fruits, and rape are used to generate bioalcohol. The discussion about fuel vs food/feed will intensify as biofuel output continues to grow. Therefore, non-food feedstock should be employed for sustainable biofuel production. In the current approaches, starch is also explored and used as fuel or a fuel component for a combustor such as a furnace, dryer, kiln, or boiler. Starch has several advantages over other fuels in terms of cost and convenience in handling and manufacture (Munasinghe et al. 2010; Anders et al. 2009).

Therefore, more effective biomass conversion methods would aid in making biofuels more economically viable for the usage of fuels that produce biomass. The availability of land and crop choice are important factors in the use of biomass fuels. It's reasonable to anticipate that biomass alternatives will expand dramatically in order to supply fuel.

While biomass offers several benefits over fossil fuels for both electricity and heat, a significant expansion in the production of biomass energy has the potential to have detrimental effects on the environment. Major concerns are land use issues and pollution. It would be necessary to protect areas with endangered species and delicate ecosystems. Food production would also face competition from agricultural lands. Overuse can lead to a loss of soil fertility, which is concerning. Sustainable and diverse biomass production would be required to protect regional ecosystems. Increased fertilizer and bioengineered organism use in energy farms may lead to pollution issues. Increased air and water pollution could come from the agricultural system absorbing dangerous chemicals from MSW (Xu et al. 2024).

10.3.4 Wind Energy

The transformation of wind energy into a useful form of energy is known as wind power. Examples of this include the usage of wind turbines to generate electricity, wind mills to provide mechanical power, wind pumps to pump water or drain debris, and sails to drive ships. In the field of materials science, it will undoubtedly be crucial to the development of lubricants, coatings, and lightweight, durable composite materials needed to build turbine blades and towers that can withstand stresses, particularly those that offshore installations are subjected to (such as corrosion and wind speeds). It is possible to create embedded sensors and sensing materials that can detect damage and stability and provide immediate protection. Offshore turbines exposed to UV light, salt air, and pollution will also experience corrosion. To lower maintenance costs and increase the operating life of wind energy devices, improved, long-lasting protective coatings must be developed further (Elizondo et al. 2009). Electroless nickel-phosphorus (ENP) coatings on glass fiber-reinforced plastic composites, which are often used in wind turbine blades, were investigated by Lee (2008) for their wear-corrosion and corrosion resistance.

10.3.5 Ocean Energy

The energy of the ocean waves is immense. Wave energy cannot be efficiently captured everywhere, because wave power varies greatly around the globe. More advancements in the most advanced ocean technologies, including as tidal barrages, tidal current turbines, and wave turbines, are needed to fully realize the potential of producing electricity from wave and tidal sources. This needs to be combined with advances in less advanced technologies that use salinity gradient energy, also known as blue energy. These technologies operate on the principles of electrodialysis, which transfers salt from a highly concentrated solution to a low concentrated solution, or osmosis, which moves water from a low salinity concentration to a high salinity concentration. This calls for additional advancements that lower costs or boost membrane efficiency in order to greatly enhance the process's economics. Similar to wind energy, corrosion is an issue with wave and tidal energy, necessitating the ongoing development of sophisticated, long-lasting protective coatings to lower maintenance costs and increase operating life (Malayeri et al. 2009).

10.3.6 Nuclear Energy

In the end, there isn't a single right response. Renewables, however, are a crucial component of the answer. Imagine the energy supply system as a pyramid, with coal and nuclear power serving as its foundation. Natural gas is in the centre, and efficiency and renewable energy are at the top. To make the base of the pyramid consume energy significantly more efficiently, it should be reversed. The middle is made up of some combined heat and power and renewable energy sources.

In order to help balance supply and demand, this component is connected to real-time pricing, demand response, and smart discharging and charging of electrified vehicles. The final fossil fuels and nuclear power will be phased out gradually at the top of the pyramid. The world, however, did not wait; it required an increasing amount of power and depended more and more on oil and other fossil fuels.

Nuclear energy is a low-carbon electricity source that significantly adds to the diversity of energy sources, and it is a crucial medium-term solution to the energy problems. The carbon emissions would continue to rise in the absence of the current nuclear power plants. For this reason, the production of nuclear energy is a crucial part of the energy mix (Van Goethem et al. 2007). Exploring both fission and fusion technology, the technical challenge is to harness nuclear energy safely and efficiently. Low- or zero-emission sources must account for a specific portion of a region's power generation, according to Standards for Low-Carbon Energy Portfolios (LC-EPS). The study employed variance decomposition analysis to predict the factors that will have the biggest impact on carbon emissions in the US economy, Granger causality to draw causal inferences, and the ARDL-Bounds testing approach to estimate coefficient parameters between 1995 and Attallah et al. 2020.

An inverted U-shaped environmental Kuznets curve shows how nuclear power considerably affects carbon emissions, but long-term innovation impacts emissions lessening them. Conversely, exporting advanced technology lowers carbon emissions. There is a noticeable correlation between economic growth and nuclear power, innovation, carbon emissions, and investment for ecologically friendly projects (Liu et al. 2024). Advanced nuclear technology has the best chance of achieving carbon neutrality in terms of safety, waste reduction, real renewability for millennia, process heat for manufacturing, and a workable way to wean ourselves off of fossil fuels for chemical production. "There are few opportunities for chemical engineers in nuclear," some of my colleagues tell me. I'm not on board. Opportunities include designing and running 550–750 °C plants that involve liquid metal, helium, and molten salts; using this high temperature capability for heating industrial processes; recycling "waste" from nuclear plants to fuel more advanced reactors; incorporating the hydrogen economy into the design and operation of nuclear plants; advancing pebble-bed advanced reactor technology; improving uranium and thorium mining, including uranium extraction from seawater; and molten salt storage systems (Rehm 2023).

10.3.7 Hydroelectric Nanogenerators (HENGs)

Global warming has been exacerbated by the extensive use of conventional energy sources, such as fossil fuels. To fulfil the ever-increasing demand for energy, it is imperative that the usage of green and sustainable energy resources be increased. Numerous cutting-edge technologies can be used to capture clean energy sources as geothermal, solar, and wind energy. The environmental catastrophe may be lessened by the efficient and sustainable energy production from these clean energy sources. However, the conventional methods of generating electricity are bound to have certain drawbacks, including comparatively low conversion efficiency, the need for huge, costly equipment, and dependence on particular climate and geographic conditions. Another major difficulty is supplying the electricity needed to keep up with the continuously expanding number of portable electronic devices, particularly in rural areas. Thus, it is imperative to create new tactics using widely accessible power sources that have a high energy conversion efficiency and are small in size. To meet this demand, nanogenerators have been developed to convert solar, hydrodynamic, mechanical, and thermal energy into electrical energy using various technologies such as triboelectric (TENGs), pyroelectric/thermoelectric (PyENGs), piezoelectric (PENGs), and hydroelectric nanogenerators. Publications on nanogenerators have grown monotonically fast during the past 10 years, indicating that they have become a prominent topic for research. But in order to produce energy, some technologies—like PyENGs and TENGs, for example—need certain circumstances, including appropriate weather outside and mechanical motion, respectively. This restricts their use and lowers their capacity to produce electricity continually. Conversely, HENGs rely solely on the surrounding moisture/water to produce electricity because of the intense interactions that exist between functional materials and water, including moisture, droplets, waves, and moving water (F. Zhao et al. 2015; Král et al. 2001).

10.3.8 The Triboelectric Nanogenerator (TENG)

Researching portable and renewable energy sources is crucial given the increasing need for electronics. There are many different types of energy in our surroundings that can be captured and used, such as mechanical, chemical, solar, wind, and thermal energy. By scavenging several sorts of energy sources from the environment, researchers have built a variety of self-sufficient power sources (Kumar et al. 2024). Conventional technologies, however, were unable to meet the demands for great efficiency and sustainability. Therefore, in order to address the power supply concerns of future functioning electronics, new energy harvesters must be developed. A triboelectric nanogenerator (TENG) that can efficiently convert different mechanical energy into electricity was created in 2012 by Wang et al. The TENG has numerous benefits, including high conversion efficiency, low cost, huge output power, and simple production. Many TENGs, mostly based on the coupling between the triboelectric effect and electrostatic induction, have been developed by Wang and colleagues. Outstanding TENGs with an area power density of up to 500 W/m^2 and a corresponding conversion total efficiency of roughly 85% have been designed by researchers. The use of flexible TENGs as power sources has been thoroughly researched due to the rapidly increasing need for flexible electronics in applications such as bendable displays, wearable electronics, and electronic skin. Appropriate materials and well-designed structures are two essential components that can influence the flexible TENGs' mechanical stability and output performance. Considerable research has been conducted in the past 5 years to enhance the functionality and use of flexible TENGs.

Many studies conducted in the last few years have concentrated on enhancing TENG's output performances and providing examples of its applications. There are various research aspects and efforts that need to be addressed for the creation of flexible TENGs.

1. Enhancing flexible TENG's capabilities. Many cutting-edge materials, like graphene, carbon nanotubes, and nano-Ag ink, would be excellent options. It is discovered that graphene's nanoscale roughness can raise surface charges, which can boost the TENG's output power. In addition to material optimization, unique structural designs such as grating, sponge, and wavy structures still need to be tuned to increase device durability and output stability.
2. Concentrating on TENG's operational processes. Although some basic mechanisms have been identified, no significant conclusion has been put out. Comprehending the corresponding mechanisms is crucial for the development of novel high-performance TENGs.

3. TENG is integrated with additional power generators to create hybrid energy cells. When dealing with various functional requirements in human life, hybrid energy cells might improve the output performances of the TENG and provide solutions to some important issues (Wang et al. 2017).

10.3.9 Biomass Waste

Waste from agriculture is a tremendous source of energy but has no nutritional value for humans. Many agricultural wastes are created at various phases of the process, such as weeds and plant leaves, hay and cereal stubble, and garden products, which can be used as renewable energy sources (Hosseini et al. 2013). The process of extracting biogas from livestock faces and obtaining energy from these waste materials holds significant economic and environmental value. The growing rate of contaminated municipal solid waste (MSW) creation can increase the occurrence of infectious diseases.

Thus, it is necessary to implement certain rules in order to properly manage the massive volumes of MSW that are produced. On the other hand, MSW can be a great source of biogas, and energy recovery can be carried out by implementing sustainable policies (Hosseini et al. 2013). According to, a novel composite material known as egg shell hydroxyapatite-humic acid nanoparticles made of biomass waste is suggested for the removal of radio-lanthanides and actinides from nuclear liquid waste and contaminated aquatic environments (Attallah et al. 2020). As a catalyst, nano-NiO made from leftover nickel slag can be used to enhance the pyrolysis of rain tree biomass, a horticultural waste (Guo et al. 2020). Because of its large porosity, carbon curvature, and other characteristics that promote hydrogen electroadsorption, the nano biomass made from walnut shells offers a great deal of economic potential for electrochemical hydrogen storage (Hajialigol et al. 2019).

10.3.10 Production of Biofuel and Bioethanol

Forests are excellent suppliers of biofuel products, including biodiesel and bioethanol. The three primary sources of bioenergy that can be used are animal waste, municipal solid waste (MSW), and agricultural residue. Additionally, the primary sources of bioethanol and biodiesel production may be farmed crops and woods. The use of liquid biofuel in diesel engines has opened up new avenues for the production of green energy, and significant sums of money have been invested globally in the development of biofuels because of their favourable qualities, which include lowering greenhouse gas emissions, promoting rural development, securing natural energy supplies, and providing energy for future generations. Not only have industrialized nations such as France, Italy, and Germany, implemented biodiesel production and consumption, but developing nations such as Argentina,

Indonesia, Brazil, and Malaysia, have also made significant investments in biodiesel production. The main sources of bioethanol include a range of crops, including sugar beets, cotton, sugar cane, barley, pistachio nuts, wheat, and rice. Other oil seeds that can be used to make biodiesel include corn, sesame, sunflower, and olive.

10.3.11 Hydropower

Hydropower is more affordable than other renewable energy sources, at a cost per kWh of $0.04 (including transmission costs). Building hydro-dams is mostly done for the provision of water, irrigation of agricultural area, flood control, and electricity generation. The production of electricity from hydropower can eliminate greenhouse gas emissions. Triboelectric nanogenerators (TENGs) are devices that use a combination of the triboelectrification effect and electrostatic induction to generate electricity from mechanical energy sources such as wind, water flow, vibration, and human motion (Bai et al. 2013; Jiang et al. 2019). Stable and effective acrylate resin coated triboelectric nanogenerator as the friction layer material. Its benefits include ease of use, affordability, and preparation of a vast area. (Wang et al. 2020).

10.3.12 Thermoelectric Generators

Because of the enormous amount of solar thermal energy and the high demand for waste heat recovery, thermoelectric generators (TEGs) are particularly appealing as a low-cost energy source. The goal is to create versatile, affordable, environmentally friendly, and possibly energy-efficient organic thermoelectric materials. These kinds of materials are shown themselves to be highly efficient in the conversion of green energy. Thermoelectric generators function as effective heat engines because of their ability to efficiently transform temperature differences into electrical voltages through the Seebeck effect. When given an electrical energy source, TEG devices can also function as Peltier-effect heat pumps (Zhang et al. 2014). A TEG gadget is extremely dependable and environmentally friendly since it doesn't require any moving parts and doesn't use any liquid or gas medium. Additionally, low-quality heat that would normally evaporate can be used by TEG devices. (Zhang et al. 2014). Organic materials are requiring more attention because the majority of inorganic thermoelectric materials are too heavy and brittle for daily use. Furthermore, organic materials are selected because to their low cost, light weight, flexibility, and ease of processing solutions. As a result of the amazing advancements in organic electronics, hundreds of organic semiconductors with nearly equivalent field effect mobilities to polycrystalline silicon have been developed. (Zhang et al. 2014).

In applications where temperatures are low, organic thermoelectric materials are also incredibly effective. Pentacene (C60) and fullerene (C60), two small conjugated molecules, have been doped and employed as p- and n-type TEG legs, respectively.

When applied, nanofluids outperform base fluid water in photovoltaic and thermoelectric generators. (Sami et al. 2019).

10.3.13 Organic Thermoelectric Materials-Based on Conductive Polymers

Approximately three different categories of study objects may be identified in conductive polymers, such as conjugated polymers and specific coordination polymers, which form the basis of the majority of organic thermoelectric materials that have been described so far. The first category consists of highly conductive polymers that have been thoroughly investigated in the past for various uses. In this instance, the conditions of synthesis and/or processing have a significant impact on the thermoelectric properties of the polymers. Although there are comparatively few monomer species available, this effective technique is the most widely used. The second kind of object is chosen from the tens of solution-processable polymeric semiconductors for organic electronics that have been created during the last 20 years. These semiconductors' thermoelectric properties have been refined by blending, doping, and other techniques. The third strategy involves creating new chemical structures by inventive design and derivation in an effort to find prospective building blocks suitable for thermoelectric applications. Further investigation is necessary to gain further understanding of the working mechanisms behind the significant and successful way of adjusting the thermoelectric property of conjugated polymers through tuning carrier concentration. (Zhang et al. 2014). Nanotechnology has been used to develop coating for solar thermal applications, efficiently converting sunlight into heat.

10.3.14 Other Applications (Environmental Applications of Green Synthesized Nanoparticles)

10.3.14.1 Economically Attractive Battery Recycling

Batteries are used in industries, public transportation, hospitals, power tools, cars, boats, toys, cameras, remote controls, flashlights, computers, alarm clocks, mobile phones, watches, and other devices. They are an essential component of modern life. The majority of batteries include heavy metals like lead, cadmium, nickel, and mercury, which can be hazardous to human health if improperly disposed of in the environment (Chen et al. 2013).

10.3.14.2 Energy Storage

Direct solar energy conversion is among the most innovative and adaptable forms of renewable energy technology. In quantum dot and polymer-based solar cells, a variety of carbon nanomaterials, including carbon nanotubes, C60 fullerenes, and graphene, are thought to be extremely effective electron acceptors. Dye-sensitized solar cells have been employed recently; these cells are made of nanocrystalline mesoporous TiO_2 film with a monolayer of dye attached to its surface and mounted on a transparent conductive substrate. The chemisorption of dye and brief charge movement of nanoparticles are triggered by their enormous surface area, which also contributes to their power conversion efficiency. In contrast, fuel cell devices—which can control the conversion of chemical energy into electricity—are greatly impacted by nanotechnology in the context of solar cells. Highly effective electrocatalysts for oxidation-reduction reactions at the electrode of fuel cells are metals with low specific densities, rich surface chemistry, high and nanoporous surface area (Karn et al. 2013).

10.3.14.3 Radioactive Waste Removal

Radioactive material, which is harmful to both living things and the environment, is present in radioactive waste. Zero-valent iron nanoparticles are a rapidly developing technology used to clean up contaminated groundwater. Numerous researchers identified various nanotechnology-based methods for radioactive waste cleaning. Utilizing titanate nanofibers as an adsorbent, radioactive ions that pose a risk are removed from wastewater. Currently, scientists have developed an environmentally friendly process for creating highly selective nanomaterials that have distinctive structural characteristics and can remove radioactive cesium and iodine ions. Metal ions, heavy metals, and organic molecules exhibit notable alterations in their adsorption ability when exposed to metal-based nano-adsorbents, carbon nanotubes, and fullerene. Green-produced nanomaterials shown remarkable characteristics like optical, magnetic, and electrochemical properties (Shabalin et al. 2019).

10.3.14.4 Construction Industry

Despite the usage of nanotechnology in the biomedical and electrical industries, advanced conventional building materials have been used in the construction sector. In the building sector, nanomaterials have been applied to improve material functions, qualities, and energy efficiency. Using nanoparticles, common building materials like steel, concrete, and cement will also be upgraded to create concrete that is quicker to pack, easier to clean, self-healing, fire resistant, and harder. Carbon nanotubes, carbon nanofibers, silica fume, and nanostructured metals are a few examples of these nanoparticles.

Due to the increasing effects of global warming and energy scarcity, experts are focusing increasingly on clean energy and green technologies in the sector. It has been researched how to use nanoparticles for more green applications. The enhanced characteristics of nanoparticles have made it possible to create new packaging films with excellent mechanical, exfoliating, barrier, and fireproofing qualities. These characteristics increase food safety and shelf life, especially in areas without easy access to refrigeration. Additionally, due of their ability to hold water molecules firmly, nanoporous zeolites have been used as permanent water basins or as a slow-releasing carrier of fertilizers. Because of their unique qualities, nanoparticles are very appealing for a wide range of creative, sustainable, and environmentally friendly applications. Advanced physicochemical features of nanomaterials, including as size, chemical composition, surface area, and protein corona on ecosystems, in vitro models, and in vivo models, can be used to determine their biological uses (Papadaki et al. 2018).

10.4 Advantages and Disadvantage of Green Nanotechnology

Due to the fact that the bio-green route of NP synthesis does not use any hazardous chemicals, it has garnered a lot of attention. Therefore, bio-green produced nanoparticles (NPs) may be a promising material that opens up new avenues for environmental, energy, and clinical study.

Main advantages of green chemistry are as follows:

- Reducing waste of clean water, fuels, and energy consumption.
- Recycling paper, plastic, cans, and batteries.
- Refusing to use plastic bags and materials made of plastic.
- Repurposing energy sources like wastewater, wind, water and solar energy etc.
- Limiting the use of fuel, food, water, and electricity.

To put it briefly, the objective is to create cutting-edge techniques and their uses without depleting natural resources, raise production while lowering waste and pollution, and improve environmental quality. Utilizing green chemistry and green engineering concepts, green nanotechnology is a subfield of green technology that is grouped alongside green energy, green information technology, and green building. A practical application of green chemistry, green processes, and the production and consumption of green goods are necessary to fully appreciate the advantages of sustainability. Green nanotechnology has a number of disadvantages, such as a lack of knowledge, high cost, the lack of substitute chemicals and process technologies, health hazards, performance implications caused by errors, and a shortage of personnel and expertise.

10.5 Conclusion and Prospect

In recent years, there has been a significant increase in the demand for safe and affordable materials that can be used for a wide range of applications, including dye degradation, cosmetics, textiles, biosensors, the medical field, cancer therapies, drug delivery, biomedicine, water treatment, and energy storage devices. Researchers are looking more closely at these problems as a result of the growing interest in them. The race is on to develop cheap and environmentally friendly processes, knowing that new, stable, non-toxic, and affordable materials will be needed in the future. Further investigation for the biosynthesis of nanoparticles may be very beneficial in this regard.

This chapter discusses the green synthesis of nanoparticles and compares it to traditional procedures. This chapter focuses on the possibilities of various natural extracts as substitutes for traditional components or solvents, avoiding the need for additional capping agents or common industrial surfactants that are difficult to remove after nanoparticle synthesis. Green synthesis appears as a scalable, energy-efficient, cost-effective, environmentally friendly, and approach simple for producing nanomaterials. Special emphasis has been given to the application of green synthesized nanoparticles in different fields of the development of energy production and storage devices.

References

Abuzeid HM et al (2018) Electrochemical performance of nanosized MnO2 synthesized by redox route using biological reducing agents. J Alloys Compd 746:227–237

Ahmad W et al (2022) A review on current trends in the green synthesis of nickel oxide nanoparticles, characterizations and their applications. Environ Nanotechnol Monitor Manag 18:100674

Anastas PT et al (1998) Green chemistry: theory and practice. Oxford University Press, New York

Anders VN et al (2009) Processes for production of a starch hydrolysate. EP2032713 (A2)

Aswathi VP et al (2023) Green synthesis of nanoparticles from biodegradable waste extracts and their applications: a critical review. Nanotechnol Environ Eng 8:377–397

Attallah MF et al (2020) Preparation of novel nano composite materials from biomass waste and their sorptive characteristics for certain radionuclides. Radiochim Acta 108:137–149

Bai P, Zhu G, Liu Y, Chen J, Jing Q, Yang W, Ma J, Zhang G, Wang ZL (2013) Cylindrical rotating triboelectric nanogenerator. ACS Nano 7(7):6361–6366. https://doi.org/10.1021/nn402491y

Bar H et al (2009) Green synthesis of silver nanoparticles using latex of Jatropha curcas. Colloids Surf A Physicochem Eng Asp 339:134–139

Basol BM et al (2010) Technique and apparatus for depositing layers of semiconductors for solar cell and module fabrication US7736940

Brabec C et al (2011) Inexpensive organic solar cell and method of producing same. US7868405

Bukhari SI et al (2021) Biosynthesis of copper oxide nanoparticles using Streptomyces MHM38 and its biological applications. J Nanomater 6693302:16

Chen JS et al (2013) SnO2-based nanomaterials: synthesis and application in lithium-ion batteries. Small 9:1877–1893

Cruz D et al (2010) Preparation and physicochemical characterization of Ag nanoparticles biosynthesised by Lippiacitriodora (Lemon Verbena). Colloids Surf B Biointerfaces 81:67–73

Das RK et al (2017) Biological synthesis of metallic nanoparticles: plants, animals and microbial aspects. Nanotechnol Environ Eng 2:18

Devasahayam S et al (2020) Thin-film nanocomposite devices for renewable energy current status and challenges. Sustain Mater Technol 26:e00233

Ding C et al (2015) RETRACTED: effects of *Bacillus subtilis* on the reduction of U(VI) by nano-Fe^0. Elsevier:86–107

Elizondo J et al (2009) Experimental study of a small wind turbine for low- and medium-wind regimes. Int J Energy Res 30:33

Fafard S (2011) Solar cell with epitaxially grown quantum dot material. US7863516

Gudikandula K et al (2016) Synthesis of silver nanoparticles by chemical and biological methods and their antimicrobial properties. J Exp Nanosci 11:714–721

Guo D et al (2020) Catalytic pyrolysis of rain tree biomass with nano nickel oxide synthetized from nickel plating slag: a green path for treating waste by waste. Bioresour Technol 315:123831

Gupta D et al (2023) Green and sustainable synthesis of nanomaterials: recent advancements and limitations. Environ Res 231:116316

Hajialigol S et al (2019) Promising electrochemical hydrogen storage properties of nano biomass derived from walnut shell. Int J Hydrog Energy 44:10713–10721

Hao M, Ding S, Gaznaghi S, Cheng H, Wang L (2024) Perovskite quantum dot solar cells: current status and future outlook: focus review. ACS Energy Lett 9(1):308–322

Hassanisaadi M et al (2021) Environmentally safe biosynthesis of gold nanoparticles using plant water extracts. Nano 11:2033

Hosseini SE et al (2013) A review on green energy potentials in Iran. Renew Sust Energ Rev 27: 533–545

Hussain I et al (2016) Green synthesis of nanoparticles and its potential application. Biotechnol Lett 38:545–560

Huston M et al (2021) Green synthesis of nanomaterials. Nanomaterials 8:2130

Iravani S et al (2014) Synthesis of silver nanoparticles: chemical, physical and biological methods. Res Pharm Sci 9:385–406

Ishak NAIM, Kamarudin SK, Mansor M, Yahya N, Bahru R, Rahman S (2024) Effect of tunable composition-shape of bio-inspired Pt NPs electrocatalyst in direct methanol fuel cell: process optimization and kinetic studies. J Clean Prod 440:140637

Jiang C, Li X, Yao Y, Lan L, Shao Y, Zhao F, Ying Y, Ping J (2019) A multifunctional and highly flexible triboelectric nanogenerator based on MXene-enabled porous film integrated with laser-induced graphene electrode. Nano Energy 66:104121. https://doi.org/10.1016/j.nanoen.2019.104121

Joy SR, Anju TR (2023) Microalgal biomass: introduction and production methods. In: Handbook of biomass. Springer

Karn B et al (2013) Ten years of green nanotechnology. In: Sustainable nanotechnology and the environment: advances and achievements. American Chemical Society, pp 1–10

Katata-Seru L et al (2018) Green synthesis of iron nanoparticles using Moringa oleifera extracts and their applications: removal of nitrate from water and antibacterial activity against Escherichia coli. J Mol Liq 256:296

Král P et al (2001) Nanotube electron drag in flowing liquids. Phys Rev Lett 86:131

Kulkarni D et al (2023) Biofabrication of nanoparticles: sources, synthesis, and biomedical applications. Front Bioeng Biotechnol 11:1159193

Kumar R, Lee D, Ağbulut Ü, Kumar S, Thapa S, Thakur A et al (2024) Different energy storage techniques: recent advancements, applications, limitations, and efficient utilization of sustainable energy. J Therm Anal Calorim 149(5):1895–1933

Lee CK (2008) Corrosion and wear-corrosion resistance properties of electroless Ni-P coating on GFRP composite in wind turbine blade. Surf Coat Technol 202:4868

Lee M et al (2015) Relative contributions of *Dehalobacter* and zerovalent iron in the degradation of chlorinated methanes. Environ Sci Technol 49:4481–4489

Li HB et al (2009) Structural defects caused by a rough substrate and their influence on the performance of hydrogenated nanocrystalline silicon n-i-p solar cells. Sol Energy Mater Sol Cells 93:338

Liu A, Imran M, Nassani AA, Binsaeed RH, Zaman K (2024) Reducing carbon emissions with geoscience solutions: a look at the contributions of nuclear energy, technology, and green finance. Geosci Front 15(4):101698

Makarov VV (2014) "Green" nanotechnologies: Synthesis of metal nanoparticles using plants. Acta Naturae 6:35–44

Malayeri MR et al (2009) Application of nano-modified surfaces for fouling mitigation. Int J Energy Res 33:1101–1113

Milne J et al (2018) Application of octanohydroxamic acid for liquid-liquid extraction of manganese oxides and fabrication of supercapacitor electrodes. J Colloid Interface Sci 515:50–57

Mitra S et al (2021a) Environmental problems and management aspects of waste electrical and electronic equipment and use of clean energy for sustainable development. In: Environmental management of waste electrical and electronic equipment. Springer, pp 3–21

Mitra S et al (2021b) Nanotechnology for green energy and sustainable future, in micro and nano technologies. In: Nano tools and devices for enhanced renewable energy, pp 521–533. ISBN 9780128217092

Mitra S et al (2023) Designed and synthesized de novo ANTPABA-PDI nanomaterial as an acceptor in inverted solar cell at ambient atmosphere. Nanotechnology 34:31

Mohamed AA et al (2021) Eco-friendly mycogenic synthesis of ZnO and CuO nanoparticles for in vitro antibacterial, antibiofilm, and antifungal applications. Biol Trace Elem Res 199:2788–2799

Mohammad Zahra H et al (2022) Application of nanotechnology in different aspects of the food industry. Discov Food 2:12

Monga Y et al (2020) Sustainable synthesis of nanoscale zerovalent iron particles for environmental remediation. ChemSusChem 13:3288–3305

Moradi F et al (2021) Review on green nano-biosynthesis of silver nanoparticles and their biological activities: with an emphasis on medicinal plants. Inorg Nano Met Chem 51:133–142

Munasinghe PC et al (2010) Biomass-derived syngas fermentation into biofuels: opportunities and challenges. Bioresour Technol 101:5013

Papadaki D et al (2018) Applications of nanotechnology in construction industry. In: Fundamentals of nanoparticles. Elsevier, pp 343–370

Peighambardoust SJ et al (2010) Review of the proton exchange membranes for fuel cell applications. Int J Hydrogen Energy 35:9349

Pokrajac L et al (2021) Nanotechnology for a sustainable future: addressing global challenges with the international Network4Sustainable nanotechnology. ACS Nano 15:18608–18623

Prathna TC et al (2010) Biomimetic synthesis of nanoparticles: science, technology and applicability. In: Mukherjee A (ed) Biomimetics learning from nature. Intech Open Science, Rijeka, pp 1–20

Purnomo SKR et al (2023) Micro-titer plate assay for measurement of total phenolic and total flavonoid contents in medicinal plant extracts. Arab J Chem 16:105003

Rehm TE (2023) Advanced nuclear energy: the safest and most renewable clean energy. Curr Opin Chem Eng 39:100878

Sami S et al (2019) Modelling and simulation of PV solar-thermoelectric generators using nano fluids. Int J Sustain Energy Environ Res 8:70–99

Samuel MS et al (2022) A review on green synthesis of nanoparticles and their diverse biomedical and environmental applications. Catalysts 12:459

Sanchez-Botero L et al (2017) Oriented growth of α-MnO2 nanorods using natural extracts from grape stems and apple peels. Nano 7:117

Serrano-Garcia W et al (2023) New insights to design electrospun fibers with tunable electrical conductive–semiconductive properties. Sensors 23:1606

Seselj N, Aili D, Celenk S, Cleemann LN, Hjuler HA, Jensen JO et al (2023) Performance degradation and mitigation of high temperature polybenzimidazole-based polymer electrolyte membrane fuel cells. Chem Soc Rev 52(12):4046–4070

Shabalin BH et al (2019) Development of the nano-mineral phases at the steel-bentonite interface in time of the evolution of geological repository for radioactive waste. In: Nanocomposites, nanostructures, and their applications: selected proceedings of the 6th international conference nanotechnology and nanomaterials (NANO2018), August 27–30, 2018, Kyiv, Ukraine. Springer, pp 29–44

Simpa P (2024) Nanotechnology's potential in advancing renewable energy solutions. Eng Sci Technol J 5:1695–1710

Singh P et al (2016) Biological synthesis of nanoparticles from plants and microorganisms. Trends Biotechnol 34:588–599

Singh J et al (2018) "Green" synthesis of metals and their oxide nanoparticles: applications for environmental remediation. J Nanobiotechnol 16:84–24

Souri M et al (2018) Comparison of microwave and ultrasound assisted extraction methods on total phenol and tannin content and biological activity of Dittrichiagraveolens (L.) Greuter and its optimization by response surface methodology. Curr Bioact Compd 14:23

Strümpel C et al (2007) Modifying the solar spectrum to enhance silicon solar cell efficiency-an overview of available materials. Sol Energy Mater Sol Cells 91:238

Subramaniyam V et al (2015) Chlorococcum sp. MM11—a novel phyco-nanofactory for the synthesis of iron nanoparticles. J Appl Phycol 27:1861–1869

Tun M et al (2018) Utilization of biodegradable wastes as a clean energy source in the developing countries: a case study in Myanmar. Energies 11:3183

Van Goethem G et al (2007) Euratom innovation in nuclear fission: community research in reactor systems and fuel cycles. Nucl Eng Des 237:1486

Varcoe JR et al (2009) Fuel cells-alkaline fuel cells Anion-exchange membranes. In: Encyclopedia electrochem power source, p 329

Wang S et al (2016) Integration of organohalide-respiring bacteria and nanoscale zero-valent iron (Bio-nZVI-RD): a perfect marriage for the remediation of organohalide pollutants? Biotechnol Adv 34:1384–1395

Wang S, Liu S, Zhou J, Li F, Li J, Cao X, Li Z, Zhang J, Li B, Wang Y, Gong X (2020) Advanced triboelectric nanogenerator with multi-mode energy harvesting and anti-impact properties for smart glove and wearable e-textile. Nano Energy 78:105291. https://doi.org/10.1016/j.nanoen.2020.105291

Wang Y et al (2017) Triboelectric nanogenerators as flexible power sources. npj flexible. Electronics 1

Wang D et al (2021a) Fungus-mediated green synthesis of nano-silver using Aspergillus sydowii and its antifungal/antiproliferative activities. Sci Rep 11:10356

Wang Z, Yang Y, Zhao Z, Zhang P, Zhang Y, Liu J et al (2021b) Green synthesis of olefin-linked covalent organic frameworks for hydrogen fuel cell applications. Nat Commun 12(1):1982

Wang Q et al (2022) Two-dimensional metal-organic frameworks and covalent organic frameworks. Chin J Chem 40:1359–1385

Xie Y et al (2017) The interactions between nanoscale zero-valent iron and microbes in the subsurface environment: a review. J Hazard Mater 321:390–407

Xu Q, Zhang T, Niu Y, Mukherjee S et al (2024) A comprehensive review on agricultural waste utilization through sustainable conversion techniques, with a focus on the additives effect on the fate of phosphorus and toxic elements during composting process. Sci Total Environ 173567

Yallappa S et al (2015) Phytosynthesis of stable Au, Ag and Au-Ag alloy nanoparticles using J. sambac leaves extract, and their enhanced antimicrobial activity in presence of organic antimicrobials. Spectrochim Acta A 137:236–243

Zhang Q et al (2014) Organic thermoelectric materials: emerging green energy materials converting heat to electricity directly and efficiently. Adv Mater 26:6829–6851

Zhang D et al (2020) Green synthesis of metallic nanoparticles and their potential applications to treat cancer. Front Chem 8:799

Zhao F et al (2015) Direct power generation from a graphene oxide film under moisture. Adv Mater 27:4351–4357

Chapter 11
Green Nanotechnology for the Treatment of Biological Pollutants

K. Aishwarya, S. Aleena Irene, P. R. Devikrishna, Neenu Mary Thomas, and Naijil George

Abstract Green nanotechnology represents a pioneering and eco-friendly strategy for addressing biological pollutants. This approach involves the synthesis of nanomaterials through environmentally benign processes, minimizing the use of hazardous chemicals and producing materials that are both effective and sustainable. Green nanomaterials, with their high surface area and reactivity, are particularly well-suited for targeting and neutralizing a diverse array of biological pollutants. By integrating principles of green chemistry, this technology not only fosters the development of biocompatible and biodegradable nanomaterials but also enhances their functionality and safety. This paradigm shift is crucial for developing sustainable pollution control strategies that safeguard public health and maintain ecological equilibrium. The scope of green nanotechnology spans multiple environmental applications, including water purification, air quality enhancement, and soil decontamination. Studies illustrate the successful deployment of green nanomaterials in various contexts, highlighting their superior performance in terms of efficiency, cost-effectiveness, and sustainability compared to traditional methods. It underscores the transformative potential of green nanotechnology in environmental remediation. Despite its promising benefits, the broader adoption of green nanotechnology encounters several challenges, such as technical limitations, regulatory hurdles, and socio-economic concerns. Addressing these challenges necessitates a collaborative, interdisciplinary approach to advance research and innovation in this field. Green nanotechnology stands as a significant advancement in the management of biological pollutants, offering a sustainable and effective route to environmental remediation. Leveraging the unique properties of green nanomaterials paves the way for cleaner water, air, and soil, thereby contributing to a healthier and more sustainable future.

K. Aishwarya · S. Aleena Irene · P. R. Devikrishna · N. George (✉)
Department of Biotechnology, St. Josephs College (Autonomous), Irinjalakuda, Thrissur, Kerala, India
e-mail: naijil@stjosephs.edu.in

N. M. Thomas
Department of Physics, Morning Star Home Science College, Angamaly, Kerala, India

© The Author(s), under exclusive license to Springer Nature Switzerland AG 2025
J. M. Al-Khayri et al. (eds.), *Green Nanotechnology Applications for Ecosystem Sustainability*, Nanotechnology in Plant Sciences 4,
https://doi.org/10.1007/978-3-031-90766-1_11

Keywords Biocompatibility · Biodegradable nanomaterials · Biological pollutants · Green nanotechnology · Nanoparticle applications · Sustainability

11.1 Introduction

Biological pollutants, encompassing microorganisms such as bacteria, viruses, protozoa, and helminths, present significant threats to both human health and the environment. These pollutants are pervasive, stemming from various sources and affecting air, water, and soil quality. The detrimental impacts of biological pollutants are especially pronounced in drinking water and public health. Contaminated water sources often harbor pathogens responsible for infectious diseases like typhoid, cholera, dysentery, and hepatitis A, which are prevalent in regions with inadequate sanitation infrastructure (Elliott 2003). These diseases not only cause widespread morbidity and mortality but also strain public health systems and undermine social and economic development. The pathways through which biological pollutants enter the environment are diverse. Microorganisms can infiltrate water bodies through untreated or poorly treated sewage, agricultural runoff laden with animal waste, and urban stormwater. Water quality deterioration affects aquatic ecosystems, leading to fish kills, biodiversity loss, and disruption of food webs. Airborne biological pollutants, including bacteria, viruses, fungal spores, and pollen, disperse through natural processes and human activities, contributing to respiratory illnesses, allergies, and pandemics (Seltzer 1994; Hussain et al. 2022). Soil contamination occurs when pathogens enter through polluted water, animal waste, and organic fertilizers, posing risks to agriculture and food safety. Soil pathogens reduce crop yields and compromise food security. Wildlife populations are also at risk, with disease outbreaks caused by environmental pathogens leading to significant declines in certain species. Addressing the challenge of biological pollutants requires an integrated approach that combines public health initiatives, environmental science, and technological innovation. Emerging technologies, such as green nanotechnology, offer promising solutions for detecting and neutralizing biological pollutants, providing efficient and sustainable options for environmental remediation.

Green nanotechnology combines the principles of green chemistry and engineering with nanotechnology to develop eco-friendly solutions for environmental and health challenges. Its main goal is to create nanomaterials and nanoproducts that have minimal environmental impact, use fewer hazardous substances, and improve resource efficiency throughout their lifecycle. One of the key approaches in green nanotechnology is the biological synthesis of nanoparticles. This method utilizes natural processes where biomolecules, such as enzymes and proteins, interact with metal ions to form nanoparticles. These biomolecules are produced by various organisms, including bacteria, fungi, algae, and plant extracts (Salem et al. 2023). By leveraging these natural capabilities, the biological synthesis process can generate nanoparticles with diverse characteristics—such as size and shape—based on factors like the organism's metabolism, pH, temperature, and other environmental conditions. Biological synthesis of nanoparticles involves using different biological sources, including yeast, actinomycetes, and other microorganisms. This method is gaining popularity because it is more environmentally friendly compared to traditional chemical synthesis techniques (Alsharif et al. 2020). The nanoparticles can form either inside the cells (intracellular) or outside them (extracellular), depending on the type of microorganism used. For instance, some microbes produce nanoparticles within their cells, while others release them into their surroundings. Recent research has also focused on using cell extracts for nanoparticle production, which provides a sustainable alternative to chemical methods (Saied et al. 2021; Suba et al. 2021).

Green nanotechnology integrates principles from green chemistry and green engineering with nanotechnology to create sustainable solutions for environmental and health issues. It emphasizes minimizing the environmental impact of nanomaterials, reducing hazardous substances, and improving resource efficiency throughout their lifecycle. A significant aspect of green nanotechnology is its focus on life cycle assessment, which evaluates the environmental and health impacts from production to disposal. This approach ensures that nanomaterials are developed and used in a way that protects human health and the environment (Dhingra et al. 2010). In addition, green nanotechnology supports the circular economy by promoting the recycling and reusability of nanomaterials. This approach encourages processes that minimize waste and facilitate the recovery of valuable resources from used products. As traditional economic models face challenges due to finite resources and environmental constraints, green nanotechnology offers a sustainable path forward (Lateef 2023). It applies nanotechnological innovations to areas such as energy production, environmental cleanup, and infrastructure, aiming to reduce raw material dependence, enhance energy efficiency, and improve ecosystem health.

One of the key areas where green nanotechnology shows promise is in pollution treatment. Nanomaterials have unique properties that enable them to address environmental pollutants more effectively than conventional methods. They can be engineered for high sensitivity and specificity towards contaminants like heavy metals, organic pollutants, and pathogens (Yunus et al. 2012). For instance, nanomaterial-based sensors can detect low concentrations of pollutants in various media, including water, air, and soil, providing early warnings to prevent widespread environmental damage. These sensors can be part of real-time monitoring systems, enabling continuous surveillance and rapid response to pollution events (Baruah and Dutta 2009).

Nanomaterials such as metal oxides, carbon nanotubes, and nanocomposites are known for their large surface areas and distinctive chemical properties, which enhance their ability to adsorb and degrade pollutants. Silver nanoparticles, for example, have antimicrobial properties that can be utilized in water filtration systems to ensure clean water (Durán et al. 2016). Nanofibrous membranes, made from materials like cellulose or chitosan, offer precise filtration, removing contaminants like bacteria, viruses, and suspended solids, thereby improving water treatment efficiency while reducing the need for harmful chemicals. Moreover, nanomaterials can be employed to capture and neutralize air pollutants, including particulate matter, volatile organic compounds (VOCs), and greenhouse gases (Mammadova et al. 2022). They also contribute to soil remediation by immobilizing or degrading pollutants, such as enhancing the biodegradation of organic contaminants through improved conditions for microbial activity. These methods provide sustainable solutions for land restoration and pollution control. This chapter aims to provide a thorough examination of the role of green nanotechnology in the treatment of biological pollutants, highlighting its importance and potential in addressing environmental and public health challenges. The primary objectives are to understand the nature and impact of biological pollutants, explore the principles and applications of green nanotechnology, and assess its advantages and limitations in pollution treatment.

11.2 Types of Biological Pollutants

Biological pollutants encompass a diverse range of organisms that can significantly disrupt the health and stability of ecosystems. These pollutants include microorganisms, invertebrates, algae, and other biological entities, each of which can affect their environments in unique ways. Each category of biological pollutant interacts with its environment in distinct ways, making it essential to monitor and control their introduction and proliferation.

Effective management strategies must consider these ecological roles and interactions to mitigate adverse effects on both environmental and human health.

Microorganisms, such as bacteria, viruses, and protozoa, are often primary concerns when discussing biological pollutants. Bacteria, including species like Escherichia coli and Salmonella, are commonly found in contaminated water and can lead to severe diseases such as typhoid fever and cholera. These bacteria are typically introduced into water systems through wastewater discharges and can accumulate in sediments and filter-feeding organisms, potentially altering microbial communities and impacting ecosystem health (Olaolu et al. 2014). Viruses, including those that target bacteria (bacteriophages), also contribute to biological pollution. Although they are sometimes used intentionally to control bacterial populations, their presence in the environment can influence microbial community dynamics and sediment quality (Armon and Kott 1996; Rogovski et al. 2021).

Protozoa, such as Giardia and Cryptosporidium, are resistant to many conventional water treatment methods and can cause gastrointestinal illnesses, representing a major concern for water safety (Mons et al. 2009). Macrophytes and algae also contribute to biological pollution. The green alga *Caulerpa taxifolia* is notorious for its ability to invade and dominate new environments. Invasive algae can outcompete native plant species, leading to reduced biodiversity and altered habitat conditions. Not all introduced species become invasive, but those that do can have significant impacts on ecosystem dynamics (Wilk-Woźniak and Najberek 2013; Lv et al. 2024).

Invertebrates also play a role in biological pollution. Certain species, such as barnacles (*Elminiusmodestus*), mollusks (*Crepidulafornicata*), and crustaceans (*Eriocheir sinensis*), can become invasive when introduced into new environments, often through ballast water from ships. These invertebrates can outcompete native species and disrupt local ecosystems. For example, the Chinese mitten crab has caused damage by burrowing into flood defenses, illustrating its impact on both human infrastructure and ecological systems (Danilenko and Lukyanova 2014). Polychaetas, like *Marenzelleriaviridis*, can alter sediment structure and nutrient cycling, which can affect benthic communities and overall ecosystem health (Kristensen et al. 2011). Non-native and genetically modified species are another concern. Aquatic species, including genetically modified organisms such as modified salmonids and non-native oysters, can escape from controlled environments like fish farms and establish themselves in the wild. These species can compete with native organisms and potentially cause genetic contamination, leading to further ecological disruptions (Saxena et al. 2020).

11.3 Sources of Biological Pollutants

Biological pollutants arise from diverse sources, each contributing uniquely to their prevalence across various environments. The magnitude of biological pollution is significant and widespread, influencing both natural ecosystems and human-made environments. The impact of these pollutants ranges from health hazards to ecological disruptions, reflecting the pervasive nature of biological contaminants in our surroundings.

Bacteria, including pathogenic strains such as Salmonella, *E. coli*, and Listeria, are prevalent in numerous environments. These microorganisms are transported by human activities, animals, soil, and plant debris. Effective sanitation practices are crucial in controlling bacterial spread. Inadequate sanitation in public spaces, especially in areas with high foot traffic or close proximity to waste, can exacerbate bacterial contamination (Navab-Daneshmand et al. 2018). Moreover, industrial-scale operations, such as large poultry and beef farms, often contribute significantly to biological pollution. These farms can release substantial quantities of waste, which, when improperly managed, can lead to the proliferation of pathogenic bacteria and subsequent contamination of soil and water sources. The runoff from such farms can carry bacterial contaminants into nearby water bodies, further amplifying the pollution (Park et al. 2013).

Mold is commonly associated with damp environments and is a significant indoor pollutant. Mold thrives on materials damaged by water, including wood, paper, and insulation. It can also grow on surfaces exposed to high humidity. Mold releases spores into the air, contributing to indoor air pollution and posing health risks such as allergic reactions, asthma, and other respiratory conditions. Mildew, a variant of Mold, similarly proliferates in moisture-rich environments and contributes to similar health issues (Calonnec et al. 2006). Viruses are a major source of biological pollutants, transmitted through various means including human and animal interactions. They can spread via direct contact, airborne droplets, or contaminated surfaces. Respiratory viruses like influenza and coronaviruses can be dispersed through activities such as coughing, sneezing, or talking, leading to the contamination of indoor environments. This transmission mode heightens the risk of widespread infections, especially in enclosed spaces with poor ventilation (Lv et al. 2021).

Industries can contribute to biological contamination both directly and indirectly. Directly, industrial processes involving organic materials can result in the release of biological pollutants. For example, wastewater from food processing plants may contain high levels of microorganisms that can contaminate surrounding environments if not properly treated. Indirectly, industrial activities can contribute to pollution through the release of pollutants that create conditions favorable for microbial growth. Poorly managed waste disposal and industrial runoff can exacerbate the spread of biological pollutants (Pandey et al. 2014). Plants release pollen, a natural allergen that can trigger respiratory issues and allergic reactions in sensitive individuals. Pollen levels are particularly high during specific seasons when plants are in bloom. Additionally, agricultural activities contribute to biological pollution through the use of fertilizers and pesticides. Fertilizers can increase nutrient levels in soil and water bodies, promoting the growth of harmful microorganisms and invasive species. The introduction of exotic plants and animals into new environments can also lead to ecological disruptions, as these species may outcompete native organisms and become invasive biological pollutants (Cox 2013). Exotic pets and plants can become sources of biological pollution when they escape or are released into the wild. Non-native species may carry diseases or parasites that can disrupt local ecosystems and pose risks to native flora and fauna. Proper management and regulation are essential to prevent the introduction and spread of such biological pollutants (Foster et al. 2021).

11.4 Pathways of Biological Pollutants

Biological pollutants traverse various pathways, each contributing to their dispersion and impact on the environment and human health. Airborne transmission is a significant pathway for biological pollutants, as tiny particles such as pollen, fungal spores, and microorganisms can become suspended in the air and be inhaled by individuals. In environments where ventilation systems are poorly maintained or absent, airborne biological contaminants can accumulate and spread rapidly. For instance, Mold spores released into the air from water-damaged materials can lead to widespread indoor air pollution, exacerbating respiratory conditions and allergies (Binnie 2021). Aerosolized droplets containing viruses or bacteria, expelled through actions such as coughing or sneezing, can also disperse pathogens across considerable distances, increasing the likelihood of infection in densely populated areas (Fig. 11.1).

Contaminated water sources, such as rivers, lakes, and reservoirs, can harbor a variety of pathogens, including bacteria, viruses, and protozoa. These microorganisms can enter the water through sources such as untreated sewage, agricultural runoff, or animal waste. When water from these contaminated sources is used for

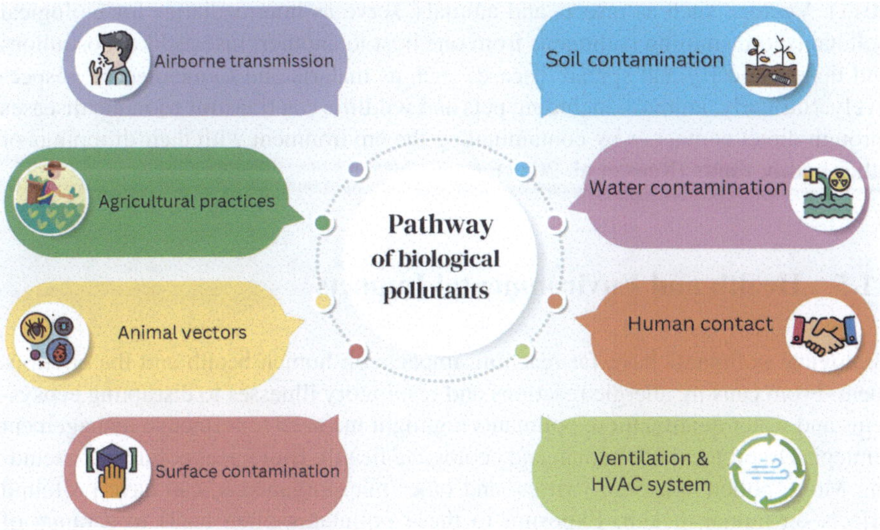

Fig. 11.1 Pathways of Biological Pollutants. Diagram illustrates the diverse pathways through which biological pollutants can spread and impact the environment

drinking, recreation, or irrigation, it can introduce pathogens into human populations and ecosystems. Additionally, improperly treated wastewater can contribute to the spread of harmful microorganisms, impacting both aquatic life and human health (Marutescu et al. 2023). Foodborne transmission is a pathway where biological pollutants manifest through contaminated food products. Pathogenic bacteria, viruses, and parasites can infiltrate the food supply chain through improper handling, inadequate cooking, or cross-contamination. Contaminated food, if consumed, can lead to outbreaks of foodborne illnesses, with symptoms ranging from mild gastroenteritis to severe, life-threatening conditions. The handling and storage practices in food processing and preparation environments play a crucial role in controlling the risk of biological contamination (Heredia and García 2018).

Soil contamination is another pathway through which biological pollutants can spread. Decomposing organic matter in the soil can release microorganisms, including bacteria and fungi, into the surrounding environment. Agricultural practices, such as the use of animal waste as fertilizer, can also contribute to soil contamination by introducing pathogens into the soil. These microorganisms can then spread to water sources or plants, further affecting the environment and human health (Santamaría and Toranzos 2003). Surface contamination is a pathway where pathogens can persist on various surfaces and be transmitted through direct contact. Microorganisms can survive on surfaces such as countertops, door handles, and equipment, potentially leading to infections if individuals touch these contaminated surfaces and then come into contact with their mouths, noses, or eyes. Poor hygiene and sanitation practices, including inadequate cleaning and disinfection, can exacerbate the problem by allowing pathogens to accumulate and spread (Otter et al.

2011). Vectors, such as insects and animals, serve as intermediaries for biological pollutants, transmitting pathogens from one host to another. Insects like mosquitoes and ticks can carry and spread diseases such as malaria and Lyme disease, respectively. Similarly, animals, including pets and wildlife, can transmit zoonotic diseases through direct contact or by contaminating the environment with their droppings or other bodily fluids (Rees et al. 2021).

11.5 Health and Environmental Impacts

Biological pollutants have far-reaching impacts on human health and the environment. From causing allergic reactions and respiratory illnesses to disrupting ecosystems and water quality, these pollutants highlight the need for effective management strategies to protect both human and ecological health. Biological pollutants, including Mold, pollen, bacteria, viruses, and other microorganisms, can have profound effects on human health. Exposure to these pollutants often leads to a range of allergic reactions and respiratory issues. Allergic responses such as hypersensitivity, allergic rhinitis, pneumonitis, and certain types of asthma are commonly triggered by biological contaminants. For instance, Mold and mildew produce mycotoxins that can cause chronic respiratory conditions, including coughing, wheezing, and shortness of breath (Medeiros et al. 2012). Pollen released by plants can exacerbate allergic rhinitis and asthma, particularly during peak pollen seasons.

Airborne biological contaminants, including viruses and bacteria, can spread infectious diseases such as influenza, measles, chickenpox, and tuberculosis (Kumar et al. 2021a, b). These pathogens can be transmitted through aerosols or direct contact with contaminated surfaces, leading to outbreaks and public health concerns. Individuals with preexisting health conditions, such as asthma or compromised immune systems, are particularly vulnerable to severe reactions. In addition to respiratory issues, exposure to biological contaminants can lead to systemic symptoms such as fever, dizziness, lethargy, and digestive problems. Chronic exposure to allergens can heighten sensitivity and result in more severe allergic reactions over time. Indoor environments with inadequate ventilation or moisture control, such as those affected by Mold in HVAC systems, can be breeding grounds for these pollutants, exacerbating health risks (Möritz et al. 2001).

Biological pollutants also have significant implications for the environment and ecosystems. Mold and mildew growth on building materials can deteriorate structures and contribute to the degradation of indoor air quality. In outdoor environments, pollen dispersal can affect plant-pollinator interactions and alter local biodiversity. High pollen concentrations can disrupt ecosystems by impacting plant reproduction and contributing to imbalances in species populations. In aquatic ecosystems, pathogens from sewage and agricultural runoff can degrade water quality and harm aquatic life. Contaminated water sources can lead to the proliferation of harmful microorganisms, affecting fish and other aquatic organisms. This can result in disrupted food chains, decreased biodiversity, and contamination of water resources used by humans for drinking and recreation (Murphy 2017). The introduction of non-native and invasive species can further exacerbate ecological imbalances. Exotic plants and animals, often introduced through human activities, can outcompete native species for resources, alter habitat structures, and disrupt existing ecological relationships. For example, invasive species like the zebra mussel can rapidly colonize and alter freshwater ecosystems, leading to significant ecological and economic impacts (Andersen et al. 2004). Farms, particularly largescale poultry and beef operations, contribute to biological pollution through the release of animal waste, which can contain pathogens and nutrients that lead to environmental contamination. Runoff from these farms can carry pathogens into water bodies, promoting algal blooms and disrupting aquatic ecosystems.

11.6 Green Nanotechnology

Nanotechnology represents an emerging field focused on the manipulation of matter at the atomic, molecular, and supramolecular levels. The term "nano" denotes a scale of one billionth of a meter (10^{-9} meters), signifying the small size at which these materials operate. Nanomaterials are defined as substances with at least one dimension less than 100 nanometers (Roduner 2006). Due to their minuscule size and elevated surface area, these materials exhibit distinctive properties that vary significantly from their bulk forms. The high surface-to-volume ratio inherent to nanomaterials enhances their reactivity, strength, and electrical properties, resulting in advanced mechanical, thermal, and chemical characteristics (Asha and Narain 2020).

Nanomaterials are categorized based on their composition and structural arrangement. Key categories include Carbon-Based Nanoparticles, Dendrimers, and Metal Nanoparticles (Pandit and Gayatri 2020). Carbon-based nanoparticles, such as fullerenes and carbon nanotubes, are notable for their exceptional electrical and mechanical properties. Fullerenes, comprising purely carbon atoms, manifest as hollow spheres, ellipsoids, or tubes and are recognized for their unique strength and conductivity. Carbon nanotubes, with their cylindrical structure, possess remarkable mechanical strength and electrical conductivity, making them invaluable in electronics and nanocomposites. Dendrimers are macromolecules characterized by their highly branched, tree-like architecture centered around a core. Their precise, uniform size and shape enable their use in targeted drug delivery and nanomedicine, as they can encapsulate and transport molecules within their branches. Metal nanoparticles, including those composed of gold, silver, and platinum, display unique optical, electronic, and catalytic properties. These characteristics render them essential for applications spanning medical diagnostics to environmental cleanup (Huang et al. 2020).

Green nanotechnology integrates the principles of green chemistry and engineering with nanotechnology to develop environmentally sustainable and health-conscious solutions. This approach aims to minimize the environmental impact of nanomaterial production and use by employing eco-friendly methods and reducing the reliance on hazardous substances. By fostering the development of safer, more sustainable nanomaterials, green nanotechnology aligns with broader sustainability goals, offering innovative solutions across a range of applications, from energy production to environmental remediation.

11.7 Methods of Synthesizing Green Nanomaterials

The synthesis of nanoparticles is central to nanotechnology, and methods for producing these materials can be categorized into physical, chemical, and green approaches. Each technique offers distinct advantages and is chosen based on the specific requirements of the nanoparticles in terms of their properties and applications. Physical synthesis methods involve physically altering bulk materials to produce nanoparticles. A prominent technique is evaporation–condensation, where metals are evaporated in a vacuum or inert gas environment and then condensed into nanoparticles. Mechanical milling is another approach where bulk materials are ground into nanoparticles using high-energy ball mills. Pulsed laser ablation involves directing high-power laser pulses at a metal target submerged in a liquid, causing the metal to vaporize and form nanoparticles upon cooling (Ijaz et al. 2020). Chemical synthesis methods involve chemical reactions to create nanoparticles. One commonly used technique is the reduction of metal ions in solution with various reducing agents. This method is favored for its simplicity and the ability to control particle size and shape by adjusting reaction conditions. The wet chemical process, where metal salts are reduced in the presence of stabilizing agents in a liquid medium, is another versatile technique capable of producing nanoparticles of different shapes and sizes. The sol–gel process involves the hydrolysis and condensation of metal alkoxides or salts to form a colloidal suspension, which then gels to produce a network of nanoparticles. This method is particularly effective for creating uniform nanoparticles and coatings. Solvothermal synthesis, which employs high temperature and pressure to dissolve and recrystallize materials, allows for the precise control of nanoparticle size and crystallinity (Hasan 2015).

Green synthesis is a more environmentally friendly approach, utilizing natural resources and reducing the reliance on toxic chemicals. This method employs biological sources such as plant extracts, bacteria, fungi, and polysaccharides as reducing and stabilizing agents. Plant extracts, rich in biomolecules like flavonoids, terpenoids, and polyphenols, can reduce metal ions to nanoparticles in a simple, cost-effective, and eco-friendly manner (Ijaz et al. 2020). Microorganisms, including bacteria and fungi, contribute to nanoparticle formation through the secretion of enzymes and reducing agents. Polysaccharides such as starch or cellulose act both as reducing and stabilizing agents, providing a sustainable alternative to traditional methods (Gour and Jain 2019). Green synthesis offers several advantages over conventional methods. It is generally less expensive and simpler, relying on readily available natural resources and avoiding hazardous chemicals. This method adheres to the principles of green chemistry, promoting sustainability and reducing environmental impact. Additionally, green synthesis processes are often safer and require fewer steps and less complex equipment, making them practical for widespread application. Despite their simplicity, these methods can produce high-quality nanoparticles with good yields, suitable for diverse applications (Gour and Jain 2019). While physical and chemical synthesis methods are well-established and provide precise control over particle characteristics, green synthesis represents a promising alternative that aligns with global sustainability goals and supports the development of environmentally conscious nanotechnology (Hussain et al. 2016).

11.8 Significance of Green Nanoparticles

Green nanoparticles represent a significant advancement in nanotechnology due to their unique characteristics and the benefits they offer over conventionally synthesized nanoparticles. One of the foremost advantages of green nanoparticles is their biocompatibility. This characteristic is crucial, especially in biomedical applications where direct interaction with biological systems is necessary. The high surface area-to-volume ratio of green nanoparticles enhances their reactivity and functionality. This property is particularly beneficial in catalytic processes, where increased surface interactions lead to more efficient and effective reactions. Additionally, the stability of green nanoparticles is significantly improved due to the use of natural stabilizing agents during their synthesis. The mild synthesis conditions employed in green nanoparticle production help preserve the functional properties of natural reducing agents and minimize the formation of unwanted byproducts (Kumar et al. 2021a, b).

Green nanoparticles also play a vital role in environmental sustainability. They are employed in water purification processes to remove contaminants such as heavy metals, organic pollutants, and pathogens, thereby improving water quality and safety. In soil remediation, green nanoparticles aid in detoxifying contaminated soils, which promotes healthier ecosystems and supports agricultural productivity (Ahmad et al. 2019). In the energy sector, they are used to enhance the efficiency of solar cells, develop high performance batteries, and create advanced supercapacitors. Their application in environmental contexts provides a sustainable solution to pollution control and resource management challenges. The non-toxic nature of green nanoparticles ensures that they can be used extensively in environmental treatment processes without introducing additional contaminants, aligning with global sustainability goals. Furthermore, the use of natural materials in the synthesis of green nanomaterials supports the principles of the circular economy. By utilizing renewable resources and minimizing waste, green nanotechnology reduces dependence on finite resources and mitigates environmental impact. This approach not only advances cleaner technologies but also fosters innovation in material science, driving the development of new, eco-friendly technologies and contributing to a more sustainable future.

11.9 Mechanisms of Action

Green nanomaterials exhibit complex interaction mechanisms with biological pollutants, driven largely by their unique physicochemical properties. The synthesis of these nanoparticles using environmentally friendly methods often results in materials with high surface area-to-volume ratios and functionalized surfaces that are highly reactive and versatile. The ability of green nanomaterials to adsorb, catalyze, and degrade pollutants through various mechanisms makes them highly effective tools for addressing contamination and promoting cleaner environments. One primary mechanism is adsorption, where pollutants adhere to the surfaces of nanomaterials. This process is governed by electrostatic interactions, van der Waals forces, and hydrophobic interactions between the nanoparticles and the contaminants. For example, the negative charges on certain nanomaterials can attract positively charged pollutants, facilitating their removal from the environment (Rana et al. 2020). Additionally, functional groups on the surface of green nanomaterials—such as hydroxyl, carboxyl, or amino groups—can form strong bonds with pollutants, enhancing their removal efficiency. These surface modifications are often a result of the natural stabilizers used in the green synthesis process, which can include plant extracts, polysaccharides, or microbial metabolites (Virkutyte and Varma 2011).

Green nanomaterials are known to induce oxidative stress in biological pollutants. This occurs through the generation of reactive oxygen species (ROS) such as hydroxyl radicals, superoxide anions, and hydrogen peroxide. Nanoparticles like silver and zinc oxide, known for their antimicrobial properties, can catalyze reactions that produce ROS. These ROS can damage the cellular components of microbes or degrade organic pollutants, leading to their inactivation or decomposition. For instance, silver nanoparticles can disrupt microbial cell membranes, leading to leakage of cellular contents and eventual cell death, while zinc oxide nanoparticles can produce ROS that degrade organic contaminants in water (Balkrishna et al. 2021; Bin-Jumah et al. 2020). Another important interaction mechanism is the catalytic degradation of pollutants. Green nanomaterials can act as catalysts in various chemical reactions that break down pollutants into less harmful substances. For example, certain green synthesized nanoparticles can facilitate the photocatalytic degradation of organic pollutants under UV light, transforming them into simpler molecules such as carbon dioxide and water (Ayodhya et al. 2016). This catalytic action is enhanced by the high surface area and active sites on the nanoparticles, which provide numerous opportunities for chemical reactions to occur. Green nanomaterials can also interact with biological pollutants through the formation of nanocomposites. In these systems, nanoparticles are combined with other materials to create composite structures that have enhanced properties for pollutant removal. For example, composite materials incorporating green nanomaterials and polymers can be used in filtration systems to capture and remove pollutants from air or water (Rani et al. 2021).

11.10 Antimicrobial Properties

Green nanomaterials are renowned for their antimicrobial properties, which are harnessed in various applications, from healthcare to environmental remediation. The antimicrobial properties of green nanomaterials arise from a combination of mechanisms including membrane disruption, ROS generation, and interference with microbial metabolism. Their effectiveness is influenced by their size, shape, and surface chemistry, as well as the incorporation of natural bioactive agents.

One of the principal antimicrobial mechanisms of green nanomaterials is their ability to disrupt microbial cell membranes. For example, silver nanoparticles (AgNPs), synthesized through green methods, exhibit potent antimicrobial activity due to their ability to bind to microbial cell walls and membranes. The interaction of AgNPs with the cell membrane leads to the formation of pits and ruptures, causing leakage of essential intracellular components and ultimately resulting in cell death (Vazquez-Muñoz et al. 2019). The high surface area of these nanoparticles allows for extensive contact with microbial cells, enhancing their effectiveness. Additionally, the release of silver ions (Ag) from the nanoparticles contributes to this antimicrobial effect by generating reactive oxygen species (ROS) that induce oxidative stress in microbes, further exacerbating cell damage (Durán et al. 2016). Another green nanomaterial with notable antimicrobial properties is zinc oxide nanoparticles (ZnO NPs). These nanoparticles exhibit a dual mechanism of action. First, ZnO Nanoparticles can generate ROS, including hydroxyl radicals and hydrogen peroxide, upon exposure to UV light. These ROS are highly reactive and can damage microbial DNA, proteins, and lipids, leading to cell death. Second, ZnO Nanoparticles can directly interact with microbial cell membranes, causing structural damage and compromising cell integrity (Venkataraju et al. 2014). The combination of these mechanisms makes ZnO Nanoparticles effective against a wide range of bacterial and fungal pathogens. Gold nanoparticles (AuNPs), another example of green nanomaterials, exhibit antimicrobial properties through a different mechanism. AuNPs can induce cell death by interacting with microbial proteins and nucleic acids. The surface of AuNPs can absorb proteins and enzymes essential for microbial metabolism, disrupting cellular functions. Additionally, AuNPs can interfere with the replication of microbial DNA, inhibiting growth and proliferation (Zhang et al. 2015). Their stability and ease of functionalization allow for targeted antimicrobial applications, such as in medical devices and coatings.

The antimicrobial effectiveness of green nanomaterials is also influenced by their size and shape. For instance, nanoscale materials with high surface-to-volume ratios exhibit enhanced antimicrobial activity compared to their bulk counterparts. Smaller nanoparticles can penetrate microbial cells more easily, leading to more effective disruption of cellular processes. Shapes such as rods, spheres, and plates can also influence antimicrobial activity, with specific shapes being more effective against particular types of microorganisms (Osonga et al. 2020). Additionally, green nanomaterials often incorporate natural stabilizers or functionalizing agents derived from plants, bacteria, or fungi. These bio-derived molecules not only stabilize the nanoparticles but also enhance their antimicrobial properties. For example, plant extracts used in the synthesis of nanoparticles can impart additional antimicrobial effects through the presence of phytochemicals such as flavonoids, alkaloids, and tannins (Hernández Díaz et al. 2021).

11.11 Role of Surface Chemistry and Particle Size

The effectiveness of green nanomaterials in various applications, including antimicrobial activity, catalysis, and environmental remediation, is significantly influenced by their surface chemistry and particle size. Surface chemistry refers to the properties and interactions of the surface of nanomaterials with their environment. It encompasses factors such as surface charge, hydrophobicity, and the presence of functional groups, all of which impact how nanomaterials interact with biological and chemical entities. One key aspect of surface chemistry is the presence of functional groups or coatings on the nanoparticles. Functionalization with specific chemical groups can enhance the affinity of nanoparticles for target molecules or cells. For example, the attachment of hydroxyl, carboxyl, or amine groups can improve the stability and dispersion of nanoparticles in various media, increasing their interaction with microorganisms or pollutants (Khan et al. 2019). In antimicrobial applications, surface functionalization can increase the binding affinity of nanoparticles to microbial cells, enhancing their bactericidal or fungicidal effects. In addition, the surface charge of nanoparticles plays a critical role in their biological interactions. The presence of natural stabilizing agents or bio-derived molecules on the surface of green nanomaterials can further influence their effectiveness. These stabilizers can act as protective layers that prevent aggregation, enhance biocompatibility, and provide additional functionality(Gour and Jain 2019).

Nanoparticles exhibit unique properties that change with size, primarily due to the high surface-to-volume ratio. Smaller nanoparticles have a larger surface area relative to their volume, which allows for more extensive interaction with target substances. In antimicrobial applications, smaller nanoparticles are often more effective because their increased surface area allows for greater contact with microbial cells, leading to enhanced disruption of cell membranes and increased generation of reactive oxygen species (ROS). Silver nanoparticles with diameters in the range of 10–50 nm have been shown to exhibit higher antimicrobial activity compared to larger particles due to their superior surface area and ability to penetrate microbial cells more efficiently (Kailasa et al. 2019). The size of nanoparticles also affects their ability to penetrate biological barriers. Smaller nanoparticles can more easily traverse cell membranes and tissues, making them useful for targeted drug delivery and imaging applications. This capability is particularly important in medical and environmental applications where precise delivery and interaction are required (Khan et al. 2022).

11.12 Green Nanotechnology in Water Treatment

Green nanotechnology has emerged as a transformative approach in water treatment, offering sustainable and efficient methods for addressing water pollution. Green nanotechnology is used to tackle various water treatment challenges. The use of green nanomaterials for water disinfection represents a significant advancement in water treatment technology. Green nanomaterials, such as silver nanoparticles (AgNPs), zinc oxide nanoparticles (ZnO NPs), and copper nanoparticles (Cu Nanoparticles), are synthesized using environmentally benign methods involving plant extracts, microorganisms, or polysaccharides. These nanoparticles exhibit potent antimicrobial properties that are harnessed for disinfecting water, effectively neutralizing pathogens and ensuring safe drinking water (Kamyab et al. 2023). Silver nanoparticles are among the most widely studied green nanomaterials for water disinfection due to their broad-spectrum antimicrobial activity. Silver nanoparticles synthesized using plant extracts from species such as neem (*Azadirachta indica*) or garlic (*Allium sativum*) have demonstrated high efficacy in eliminating bacteria and viruses from contaminated water (Gour and Jain 2019). These biogenic AgNPs disrupt microbial cell membranes and inhibit essential cellular processes, leading to effective disinfection. AgNPs synthesized using coriander (*Coriandrum sativum*) were used to treat contaminated water samples, achieving significant reduction in bacterial counts, including E. coli and Salmonella (Zahoor et al. 2021). ZnO Nanoparticles exhibit photocatalytic activity under UV light, which generates reactive oxygen species (ROS) that degrade organic pollutants and inactivate microbial contaminants. Green synthesis of ZnO Nanoparticles using plant extracts has been shown to enhance their photocatalytic efficiency while minimizing the use of toxic chemicals (Khan et al. 2019). ZnO Nanoparticles synthesized from *Moringa oleifera* leaves have been effectively employed to treat wastewater, demonstrating high efficiency in disinfection and pollutant degradation.

Beyond disinfection, green nanomaterials are also effective in removing microbial contaminants from water through various mechanisms. Nanoparticles can adsorb, capture, or destroy microorganisms, contributing to improved water quality and safety. Copper nanoparticles, produced using plant-based methods, exhibit strong antimicrobial activity that can effectively remove bacteria and fungi from water. In a study conducted by Khan et al. (2019), CuNPs synthesized from *Catharanthus roseus* were used to treat water contaminated with *Klebsiella pneumoniae* and *Aspergillus niger*. The CuNPs demonstrated excellent removal efficiency, significantly reducing microbial load and improving water quality. Green-synthesized titanium dioxide nanoparticles (TiO2 NPs) have been utilized for their photocatalytic properties to inactivate protozoan cysts in water. TiO2 NPs, synthesized using natural sources such as *Hibiscus rosa-sinensis* (hibiscus) extracts, have shown high efficacy in degrading Cryptosporidium and Giardia cysts under UV light (Hassan et al. 2019). The photocatalytic activity of TiO2 NPs results in the generation of ROS that damage the cyst walls and lead to their inactivation.

When comparing green nanotechnology with traditional water treatment methods, several advantages become evident. Traditional methods, such as chlorination, filtration, and coagulation, have been effective in treating water but often involve the use of hazardous chemicals, generate secondary pollutants, and require extensive infrastructure. Green nanomaterials offer a more sustainable alternative by leveraging natural processes and materials. The use of green-synthesized nanoparticles for water treatment can reduce the reliance on chemical disinfectants and minimize the production of harmful byproducts. Additionally, the high surface area and reactivity of green nanomaterials enhance their efficiency in contaminant removal and disinfection (Tom 2021).

11.13 Green Nanotechnology in Air Purification

Green nanotechnology has emerged as a transformative approach to combatting air pollution, a critical global challenge with far-reaching implications for human health and the environment. Air pollution, stemming from both natural sources and human activities, is characterized by the presence of harmful substances such as particulate matter (PM), surface ozone (O3), nitrogen dioxide (NO2), Sulphur dioxide (SO2), and carbon monoxide (CO). The application of green nanomaterials in air pollution control is multifaceted, addressing several crucial areas: detection of pollutants, source reduction or pollution prevention, and remediation or degradation of pollutants (Saleem et al. 2022). One of the key advantages of green nanomaterials is their exceptionally high surface area-to-volume ratio. This property allows nanoparticles to adsorb pollutants or catalyze chemical reactions with remarkable efficiency. For instance, their large surface area enables them to capture substantial amounts of pollutants or facilitate chemical reactions at accelerated rates, enhancing the overall effectiveness of air purification strategies. This characteristic not only improves pollutant removal efficiency but also reduces the energy required for degradation processes, thus potentially mitigating the release of contaminants into the atmosphere (Sharma et al. 2022). In terms of pollutant detection, green nanotechnology has significantly advanced sensor technology. Nanomaterial-based sensors offer numerous benefits over traditional methods, including low power consumption, high sensitivity, and the ability to detect pollutants at very low concentrations. Sensors incorporating gold nanoparticles have been shown to detect NO2 at ppb levels, providing real-time monitoring capabilities that are crucial for assessing air quality (Rizwan et al. 2022). These sensors also require minimal interaction distances between the sensor and the target pollutant, making them highly effective for continuous environmental monitoring.

Nanomaterials are utilized in advanced nano-filters to separate and remove pollutants from the air. These filters exploit the unique properties of nanomaterials to trap pollutants, enhancing air quality. A notable example is the use of carbon nanotube-based filters, which effectively capture small particles from the air, offering an efficient solution for improving urban air quality (Kong et al. 2016).

Nanofibers represent a promising approach in capturing airborne pathogens and pollutants. Electro spun nanofibers, with their high surface area and small diameter, create a dense network that efficiently traps particles as air passes through. For example, electro spun nanofiber filters have been demonstrated to effectively capture bacterial and viral particles, improving air quality in both residential and industrial settings (Zhang et al. 2020). The high porosity and large surface area of these fibers enhance their ability to entrap a wide range of biological contaminants.

Green nanomaterials are highly effective as adsorbents for various air pollutants. Their high surface area allows them to capture and retain pollutants such as volatile organic compounds (VOCs) and particulate matter. Zinc oxide nanoparticles have demonstrated exceptional performance in adsorbing VOCs, thereby reducing their concentration in the air (Shojaei et al. 2021). Green nanomaterials with photocatalytic properties can degrade pollutants through light-induced chemical reactions. Titanium dioxide (TiO2) nanoparticles are a prominent example, as they can break down organic pollutants under UV light. Studies have shown that TiO2 nanoparticles can effectively decompose hazardous gases like formaldehyde, making them a valuable tool for air purification (Salama and Zafar 2022). Green nanomaterials also facilitate chemical reactions that convert harmful pollutants into less toxic or benign substances. Silver nanoparticles have been used as catalysts in the transformation of NO2 into less harmful nitrogen dioxide (NO2) and nitrogen gas (N2), thereby reducing the environmental impact of nitrogen oxides (Zheng et al. 2017).

Nano-sensors offer a sophisticated method for detecting airborne biological pollutants. These sensors utilize nanomaterials to detect low concentrations of pollutants with high sensitivity. Sensors based on gold nanoparticles have been developed to detect specific airborne viruses, providing early warning systems for potential outbreaks or contamination (Khan and Rao 2020). These sensors enable real-time monitoring and help in managing air quality more effectively. Bio-inspired nano coatings are developed to mimic nature's self-cleaning surfaces, which help reduce the accumulation of pollutants. Inspired by natural processes such as lotus leaf's water-repellent properties, these nanocoating prevent the buildup of dust and microbes on surfaces, thereby maintaining cleaner air and reducing the need for frequent cleaning (Barthwal et al. 2024).

11.14 Green Nanomaterials for Soil Decontamination

Soil contamination is a significant environmental issue that arises when pollutants exceed regulatory standards, adversely affecting food safety, ecosystem health, and human well-being. Biological pollutants, including pathogenic bacteria, fungi, viruses, and organic matter such as animal danger and plant residues, further complicate soil contamination. These biological pollutants can disrupt soil ecosystems, degrade soil structure, and affect plant health, leading to reduced agricultural productivity and ecosystem services. Traditional methods for soil remediation, such as excavation, disposal, soil washing, and thermal desorption, are often costly, time-consuming, and can create secondary pollution. These approaches also tend to be environmentally disruptive, necessitating the development of more sustainable and less invasive techniques. Consequently, there is a growing interest in *in situ* treatment technologies, which offer a promising alternative for soil decontamination. Green nanomaterials represent a valuable tool for addressing soil contamination, offering benefits such as high reactivity, eco-friendly synthesis, and reduced secondary pollution (Mukhopadhyay et al. 2022).

One notable example of green nanomaterials in soil remediation is the use of green tea extracts to synthesize zero-valent iron nanoparticles (nZVI). Chrysochoou et al. (2012) demonstrated that nZVI derived from green tea effectively reduced the concentration of hexavalent chromium (Cr(VI)) in granular media such as silica sand and sand coated with aluminum hydroxide. The reduction efficiency ranged from 30% to 66%, showcasing the potential of green Nanoparticles for heavy metal remediation. However, the study also highlighted a limitation: soil acidification resulting from the application of nZVI, which led to the leaching of metals like Cr, Fe, and Pb. Green silver nanoparticles (Ag NPs) synthesized from plant extracts of *Brassica alba* and *Plantago major* were used to remediate soil contaminated with Fipronil, a pesticide known for its endocrine-disrupting, carcinogenic, and neurotoxic properties (Romeh 2018). The research found that Ag Nanoparticles synthesized from Brassica alba exhibited a removal efficiency of 68.8%, whereas those produced with *Plantago major* leaf extract achieved only 10.1% removal efficiency. This variation in effectiveness highlights the influence of different plant extracts on the performance of green Nanoparticles in soil decontamination.

The application of green nanoparticles extends significantly into the remediation of petroleum hydrocarbons, which are persistent organic pollutants frequently encountered in contaminated soils. Green nanoparticles enhance bioremediation by aiding in the immobilization of hydrocarbon-degrading microorganisms, substrates, and enzymes. These nanoparticles support the stability and activity of biological agents, optimizing their effectiveness in breaking down complex hydrocarbons. Additionally, green nanoparticles improve the solubilization of hydrophobic compounds, making them more accessible for microbial degradation by modifying soil properties and facilitating the dispersion of contaminants. They also stimulate the degradative properties of microbes, boosting their metabolic activity and accelerating the breakdown of pollutants. This multifaceted approach leverages green nanoparticles to significantly advance the efficiency and effectiveness of soil remediation efforts against challenging petroleum hydrocarbon contamination (Kumari and Singh 2016).

Green nanoparticles offer promising solutions for improving soil conditions by enhancing various soil properties and promoting sustainable agricultural practices. Their application spans several areas, from microbial activity enhancement to nutrient management, leading to improved soil health and productivity. One of the significant benefits of nanoparticles in soil management is their effect on soil microbial communities. For instance, nano silica synthesized from rice husk, with particle sizes ranging from 20 to 40 nm, has been shown to significantly enhance microbial diversity and total biomass in the soil (Rangaraj et al. 2013). This improvement is attributed to the high surface area and reactivity of silica nanoparticles, which interact positively with soil microorganisms. The presence of these nanoparticles leads to increased microbial biomass and higher carbon and nitrogen contents in the soil. Compared to traditional silica sources like micro silica or sodium silicate, nano silica exhibits superior performance in promoting plant growth-promoting microorganisms such as those involved in phosphate solubilization, nitrogen fixation, and silicate solubilization. Additionally, nanoparticles can enhance the release of essential nutrients from soil. Rick and Arai (2011) reported that silica nanoparticles coated with iron, carbon, and trace amounts of phosphorus, calcium, sodium, magnesium, and aluminum significantly improved phosphorus release in Ulti sols, a soil type known for its low phosphorus availability. This enhanced nutrient release is crucial for supporting plant growth and improving crop yields. The development of nano-fertilizers represents a breakthrough in sustainable agriculture. Dr. J.C. Tarafdar and his team at the Central Arid Zone Research Institute, Indian Agricultural Research Institute (IARI), New Delhi, pioneered the creation of the world's first nano-fertilizer through biosynthesis (ANI 2014). These nano-fertilizers are eco-friendly and offer several advantages over conventional fertilizers. They enhance soil aggregation, moisture retention, and carbon buildup, thereby improving soil structure and health. Nano-fertilizers also exhibit significantly higher nutrient use efficiency compared to traditional fertilizers, often requiring 80–100 times less quantity to achieve the same effect. Encapsulation techniques, such as coating nutrients with thin polymer films or embedding them in nanoporous materials, regulate the slow release of nutrients. This controlled release helps prevent nutrient loss and reduces direct interaction with soil microorganisms, mitigating potential environmental impacts (Rai et al. 2012). Moreover, engineered nanoparticles such as nano-clays and nanoporous zeolites are utilized to enhance moisture retention in arid soils. These materials possess honeycomb-like structures that can hold and release essential nutrients 'on demand,' thereby reducing nutrient leaching and improving soil fertility (Chinnamuthu and Boopathi 2009). The combination of these nanoparticles with nanofertilizers ensures that nutrients are available to plants for extended periods, supporting sustainable soil management and reducing the need for frequent fertilizer applications (Table 11.1).

11.15 Challenges and Limitations

The green synthesis of nanomaterials, while offering a more sustainable alternative to traditional methods, faces several technical and operational challenges. Key issues include the selection and preparation of suitable raw materials, the optimization of reaction conditions, and the control of product quality. For example, sourcing plant materials such as cotton leaves or *Coffea arabica* for nanoparticle production is often constrained by factors like availability, growth cycles, and processing requirements. Additionally, many green synthesis methods require high temperatures and long

Table 11.1 Green nanoparticles and their applications in the remediation of biological pollutant

Green nanoparticles	Significance in biological pollution management	References
Silver Green nanoparticles	Antibacterial activity Catalytic property Waste water treatment	Palani et al. (2023) and Al-Haddad et al. (2020)
Gold Green nanoparticles	Antibacterial activity Catalytic property	López-Miranda et al. (2019)
	Waste water treatment	
Zinc Green nanoparticles	Antimicrobial activity Photocatalytic activity Agriculture	Manojkumar et al. (2023) and Sabir et al. (2014)
Copper Green nanoparticles	Absorption of pollutants Biosensor Photocatalytic activity	Show et al. (2021)
Titanium Green nanoparticles	Air purification Antimicrobial activity Water treatment	Jassal et al. (2022) and Han et al. (2016)

reaction times, leading to significant energy consumption that may counteract the environmental benefits of using renewable resources. The complexity of the synthesis pathways and the lack of standardized protocols further complicate the reproducibility and scalability of green nanomaterials. Addressing these challenges is essential for making green nanotechnology more practical and commercially viable.

While green nanomaterials offer potential environmental benefits, they also raise several environmental and health concerns. The stability and long-term behavior of nanoparticles in various ecosystems are not fully understood, leading to uncertainties about their potential for bioaccumulation and adverse ecological impacts. Health risks are another significant concern; nanoparticles may pose occupational hazards during production and handling due to their unknown toxicological profiles. Moreover, there are potential risks associated with exposure to nanoparticles, including neurological, carcinogenic, and reproductive effects. Environmental impacts, such as toxicity to soil microorganisms, plants, and aquatic life, further highlight the need for rigorous risk assessments and safety evaluations. Comprehensive studies are required to fully understand the implications of nanoparticle use and to develop strategies to mitigate potential risks.

Economic and scalability challenges are major barriers to the widespread adoption of green nanotechnology. The high costs of research, development, and large-scale production of nanomaterials can limit their commercial feasibility. Achieving cost-effective production methods while maintaining the quality and performance of nanoparticles at scale is a significant hurdle. Additionally, there are logistical challenges related to the supply of raw materials and the establishment of appropriate infrastructure for large-scale manufacturing. Standardization of production processes and quality control measures is necessary to ensure consistency and reliability. Addressing these economic and scalability issues is crucial for the successful commercialization and integration of green nanotechnology into various industries.

The regulatory and policy landscape for nanotechnology is still evolving, presenting several challenges. One major issue is the lack of global standardization for the nomenclature, characterization, and testing of nanomaterials. This lack of uniformity can lead to inconsistencies in regulatory practices and complicate compliance efforts. Regulatory agencies, such as the FDA and EMA, are working to develop guidelines for the safe use of nanomaterials, particularly in pharmaceuticals and consumer products. However, the rapid pace of technological advancements often outstrips regulatory frameworks, creating gaps in oversight and enforcement. Ethical concerns, such as the use of nanotechnology in military applications and surveillance, also require careful consideration and policy development. Ensuring that regulatory frameworks are robust, transparent, and internationally harmonized is essential for managing the risks associated with nanotechnology and supporting its responsible development.

11.16 Future Perspectives and Conclusion

Green nanotechnology is at the forefront of innovative strategies for treating biological pollutants, with emerging trends poised to significantly enhance its effectiveness. One notable development is the creation of multifunctional nanoparticles that can tackle various environmental challenges simultaneously, such as pollutants in water, air, and soil. Additionally, the rise of bio-inspired nanomaterials, which mimic natural biological processes, offers new avenues for sustainable synthesis and application. Future advancements in green nanotechnology are likely to involve the integration of artificial intelligence and machine learning to optimize the design, synthesis, and application of nanomaterials for specific pollutants. Scaling up green nanotechnology while maintaining its environmental benefits is another key area of focus, including innovations in low-energy synthesis methods and improved recycling processes.

The potential of green nanotechnology in addressing biological pollutants is vast and promising. Future research may explore novel applications of green nanoparticles for emerging contaminants like pharmaceuticals and personal care products that persist in the environment. Enhanced strategies for bioremediation could involve using engineered nanoparticles to boost microbial degradation or deliver specific enzymes directly to contaminated sites. The development of nanomaterials with high affinity for specific pollutants, such as antibiotics and hormone disruptors, could significantly improve the remediation of complex environments. Moreover, combining green nanotechnology with traditional bioremediation methods could lead to more sustainable and effective solutions. Interdisciplinary collaboration across fields such as chemistry, biology, environmental science, and engineering will be essential to advance green nanotechnology for biological pollutant treatment. Innovations may include hybrid systems that integrate nanotechnology with biotechnology, leading to novel microorganisms designed to interact with nanoparticles for enhanced pollutant degradation. The integration of green nanotechnology with renewable energy sources and sustainable agricultural practices could also provide comprehensive solutions to global environmental challenges.

In conclusion, green nanotechnology offers significant improvements over traditional remediation methods, providing efficient, cost-effective, and environmentally friendly solutions. The application of green nanoparticles has shown substantial potential in improving environmental conditions, enhancing microbial diversity, and accelerating the degradation of pollutants. By utilizing renewable resources and minimizing harmful byproducts, this technology supports more sustainable environmental practices. The integration of green nanoparticles with existing bioremediation strategies amplifies their effectiveness and broadens their applicability. The unique properties of nanoparticles, such as high surface area and reactivity, enable innovative solutions for managing complex biological pollutants. Tailoring these nanoparticles for specific pollutants and environmental conditions further emphasizes their critical role in biological pollution management. As green nanotechnology advances, it holds the potential to revolutionize the treatment of biological pollutants. Continued interdisciplinary collaboration, innovation, and a commitment to sustainability are essential to overcoming challenges and fully realizing the potential of this transformative approach to environmental remediation.

References

Al-Haddad J, Alzaabi F, Pal P, Rambabu K, Banat F (2020) Green synthesis of bimetallic copper–silver nanoparticles and their application in catalytic and antibacterial activities. Clean Techn Environ Policy 22:269–277

Alsharif SM, Salem SS, Abdel-Rahman MA, Fouda A, Eid AM, Hassan SE, Awad MA, Mohamed AA (2020) Multifunctional properties of spherical silver nanoparticles fabricated by different microbial taxa. Heliyon 6(5)

Andersen MC, Adams H, Hope B, Powell M (2004) Risk assessment for invasive species. Risk Anal Int J 24(4):787–793

ANI (2014) Indian agro-scientist develops world's first nano-fertiliser through biosynthesis. http://www.dnaindia.com/scitech/report-indian-agro-scientist-develops-world-s-first-nanofertiliser-through-biosynthesis-1968243. Accessed 29 July 2024

Armon R, Kott Y (1996) Bacteriophages as indicators of pollution. Crit Rev Environ Sci Technol 26(4):299–335

Asha AB, Narain R (2020) Nanomaterials properties. In: Polymer science and nanotechnology. Elsevier, pp 343–359

Ayodhya D, Venkatesham M, Santoshi Kumari A, Reddy GB, Ramakrishna D, Veerabhadram G (2016) Photocatalytic degradation of dye pollutants under solar, visible and UV lights using green synthesised CuS nanoparticles. J Exp Nanosci 11(6):418432

Balkrishna A, Kumar A, Arya V, Rohela A, Verma R, Nepovimova E et al (2021) Phytoantioxidant functionalized nanoparticles: a green approach to combat nanoparticle induced oxidative stress. Oxidative Med Cell Longev 2021(1):3155962

Barthwal S, Uniyal S, Barthwal S (2024) Nature-inspired superhydrophobic coating materials: drawing inspiration from nature for enhanced functionality. Micromachines 15(3):391

Baruah S, Dutta J (2009) Nanotechnology applications in pollution sensing and degradation in agriculture: a review. Environ Chem Lett 7:191–204

Bin-Jumah M, Al-Abdan M, Albasher G, Alarifi S (2020) Effects of green silver nanoparticles on apoptosis and oxidative stress in normal and cancerous human hepatic cells in vitro. Int J Nanomedicine 2020:1537–1548

Binnie PW (2021) Biological pollutants in the indoor environment. In: Indoor air pollution. CRC Press, pp 13–24

Calonnec A, Cartolaro P, Deliere L, Chadoeuf J (2006) Powdery mildew on grapevine: the date of primary contamination affects disease development on leaves and damage on grape. IOBC WPRS Bull 29(11):67–72

Chinnamuthu CR, Boopathi PM (2009) Nanotechnology and agroecosystem. Madras Agric J 96: 17–31

Chrysochoou M, Johnston CP, Dahal G (2012) A comparative evaluation of hexavalent chromium treatment in contaminated soil by calcium polysulfide and green-tea nanoscale zero-valent iron. J Hazard Mater 201:33–42

Cox GW (2013) Alien species and evolution: the evolutionary ecology of exotic plants, animals, microbes, and interacting native species. Island Press

Danilenko SA, Lukyanova ON (2014) Molecular biomarkers of the physiological state of mitten crab Eriocheir japonica (de Haan, 1835) in the estuarine zones of Peter the Great Bay (Sea of Japan). Inland Water Biol 7:264–272

Dhingra R, Naidu S, Upreti G, Sawhney R (2010) Sustainable nanotechnology: through green methods and life-cycle thinking. Sustain For 2(10):3323–3338

Durán N, Durán M, De Jesus MB, Seabra AB, Fávaro WJ, Nakazato G (2016) Silver nanoparticles: a new view on mechanistic aspects on antimicrobial activity. Nanomedicine 12(3):789–799

Elliott M (2003) Biological pollutants and biological pollution—an increasing cause for concern. Mar Pollut Bull 46(3):275–280

Foster R, Peeler E, Bojko J, Clark PF, Morritt D, Roy HE, Bass D (2021) Pathogens cotransported with invasive non-native aquatic species: implications for risk analysis and legislation. NeoBiota 69:79–102

Gour A, Jain NK (2019) Advances in green synthesis of nanoparticles. Artif Cell Nanomed Biotechnol 47(1):844–851

Han C, Lalley J, Namboodiri D, Cromer K, Nadagouda MN (2016) Titanium dioxide-based antibacterial surfaces for water treatment. Curr Opin Chem Eng 11:4651

Hasan S (2015) A review on nanoparticles: their synthesis and types. Res J Recent Sci 2277:2502

Hassan D, Farghali M, Eldeek H, Gaber M, Elossily N, Ismail T (2019) Antiprotozoal activity of silver nanoparticles against Cryptosporidium parvum oocysts: new insights on their feasibility as a water disinfectant. J Microbiol Methods 165:105698

Heredia N, García S (2018) Animals as sources of food-borne pathogens: a review. Anim Nutr 4(3): 250–255

Hernández Díaz JA, Garza García JJ, Zamudio Ojeda A, León Morales JM, López Velázquez JC, García Morales S (2021) Plant mediated synthesis of nanoparticles and their antimicrobial activity against phytopathogens. J Sci Food Agric 101(4):1270–1287

Huang J, Liu J, Wang J (2020) Optical properties of biomass-derived nanomaterials for sensing, catalytic, biomedical and environmental applications. TrAC Trends Anal Chem 124:115800

Hussain I, Singh NB, Singh A, Singh H, Singh SC (2016) Green synthesis of nanoparticles and its potential application. Biotechnol Lett 38:545–560

Hussain HH, Ibraheem NT, Al-Rubaey NKF, Radhi MM, Hindi NKK, Al-Jubori RHK (2022) A review of airborne contaminated microorganisms associated with human diseases. Med J Babylon 19(2):115–122

Ijaz I, Gilani E, Nazir A, Bukhari A (2020) Detail review on chemical, physical and green synthesis, classification, characterizations and applications of nanoparticles. Green Chem Lett Rev 13(3): 223–245

Jassal PS, Kaur D, Prasad R, Singh J (2022) Green synthesis of titanium dioxide nanoparticles: development and applications. J Agric Food Res 10:100361

Kailasa SK, Park TJ, Rohit JV, Koduru JR (2019) Antimicrobial activity of silver nanoparticles. In: Nanoparticles in pharmacotherapy. William Andrew Publishing, p 461484

Kamyab H, Chelliapan S, Hayder G, Yusuf M, Taheri MM, Rezania S, Nouri J (2023) Exploring the potential of metal and metal oxide nanomaterials for sustainable water and wastewater treatment: a review of their antimicrobial properties. Chemosphere 335:139103

Khan A, Rao TS (2020) Nanobiosensors for virus detection in the environment. In: Nanomaterials for air remediation. Elsevier, pp 61–87

Khan SB, Ali F, Akhtar K (2019) Chitosan nanocomposite fibers supported copper nanoparticles based perceptive sensor and active catalyst for nitrophenol in real water. Carbohydr Polym 207: 650–662

Khan S, Mansoor S, Rafi Z, Kumari B, Shoaib A, Saeed M, Shakeel F (2022) A review on nanotechnology: properties, applications, and mechanistic insights of cellular uptake mechanisms. J Mol Liq 348:118008

Kong LB, Yan W, Huang Y, Que W, Zhang T, Li S (2016) Carbon nanomaterials based on carbon nanotubes (CNTs). Adv Nanomater:25–101

Kristensen E, Hansen T, Delefosse M, Banta GT, Quintana CO (2011) Contrasting effects of the polychaetes Marenzelleria viridis and Nereis diversicolor on benthic metabolism and solute transport in sandy coastal sediment. Mar Ecol Prog Ser 425:125–139

Kumar JA, Krithiga T, Manigandan S, Sathish S, Renita AA, Prakash P, Crispin S (2021a) A focus to green synthesis of metal/metal based oxide nanoparticles: various mechanisms and applications towards ecological approach. J Clean Prod 324:129198

Kumar P, Kausar MA, Singh AB, Singh R (2021b) Biological contaminants in the indoor air environment and their impacts on human health. Air Qual Atmos Health 14(11):1723–1736

Kumari B, Singh DP (2016) A review on multifaceted application of nanoparticles in the field of bioremediation of petroleum hydrocarbons. Ecol Eng 97:98–105

Lateef A (2023) Cola nitida: milestones in catalysis, biotechnology and nanotechnology for circular economy and sustainable development. In: Biocatalysis and agricultural biotechnology. Elsevier, p 102856

López-Miranda JL, Esparza R, Rosas G, Pérez R, Estévez-González M (2019) Catalytic and antibacterial properties of gold nanoparticles synthesized by a green approach for bioremediation applications. 3 Biotech 9:1–9

Lv J, Gao J, Wu B, Yao M, Yang Y, Chai T, Li N (2021) Aerosol transmission of coronavirus and influenza virus of animal origin. Front Vet Sci 8:572012

Lv T, Wang H, Wang Q, Li D, Gao C, Zhang T, Liu C (2024) Invasive submerged plant has a stronger inhibitory effect on epiphytic algae than native plant. Biol Invasions 26(4):1001–1014

Mammadova S, Nasibova A, Khalilov R, Mehraliyeva S, Valiyeva M, Gojayev A, Eftekhari A (2022) Nanomaterials application in air pollution remediation. Eurasian Chem Commun 4(2): 160–166

Manojkumar U, Kaliannan D, Srinivasan V, Balasubramanian B, Kamyab H, Mussa ZH, Palaninaicker S (2023) Green synthesis of zinc oxide nanoparticles using Brassica oleracea var. botrytis leaf extract: photocatalytic, antimicrobial and larvicidal activity. Chemosphere 323: 138263

Marutescu LG, Popa M, Gheorghe-Barbu I, Barbu IC, Rodríguez-Molina D, Berglund F, Chifiriuc MC (2023) Wastewater treatment plants, an "escape gate" for ESCAPE pathogens. Front Microbiol 14:1193907

Medeiros FHV, Martins SJ, Zucchi TD, Melo IS, Batista LR, Machado JC (2012) Biological control of mycotoxin-producing molds. Ciência e Agrotecnologia 36:483–497

Mons C, Dumètre A, Gosselin S, Galliot C, Moulin L (2009) Monitoring of cryptosporidium and Giardia river contamination in Paris area. Water Res 43(1):211–217

Möritz M, Peters H, Nipko B, Rüden H (2001) Capability of air filters to retain airborne bacteria and molds in heating, ventilating and air-conditioning (HVAC) systems. Int J Hyg Environ Health 203(5–6):401–409

Mukhopadhyay R, Sarkar B, Khan E, Alessi DS, Biswas JK, Manjaiah KM, Ok YS (2022) Nanomaterials for sustainable remediation of chemical contaminants in water and soil. Crit Rev Environ Sci Technol 52(15):2611–2660

Murphy H (2017) Persistence of pathogens in sewage and other water types. In: Global water pathogen project, vol 4

Navab-Daneshmand T, Friedrich MN, Gächter M, Montealegre MC, Mlambo LS, Nhiwatiwa T, Julian TR (2018) Escherichia coli contamination across multiple environmental compartments (soil, hands, drinking water, and handwashing water) in urban Harare: correlations and risk factors. Am J Trop Med Hyg 98(3):803

Olaolu DT, Akpor OB, Akor CO (2014) Pollution indicators and pathogenic microorganisms in wastewater treatment: implication on receiving water bodies. Int J Environ Protect Pol 2(6): 205–212

Osonga FJ, Akgul A, Yazgan I, Akgul A, Eshun GB, Sakhaee L, Sadik OA (2020) Size and shape-dependent antimicrobial activities of silver and gold nanoparticles: a model study as potential fungicides. Molecules 25(11):2682

Otter JA, Yezli S, French GL (2011) The role played by contaminated surfaces in the transmission of nosocomial pathogens. Infect Control Hosp Epidemiol 32(7):687–699

Palani G, Trilaksana H, Sujatha RM, Kannan K, Rajendran S, Korniejenko K, Uthayakumar M (2023) Silver nanoparticles for waste water management. Molecules 28(8):3520

Pandey PK, Kass PH, Soupir ML, Biswas S, Singh VP (2014) Contamination of water resources by pathogenic bacteria. AMB Express 4:1–16

Pandit P, Gayatri TN (2020) Introduction to green nanomaterials. In: Green nanomaterials: processing, properties, and applications. Springer, Singapore, pp 1–21

Park S, Navratil S, Gregory A, Bauer A, Srinath I, Jun M, Ivanek R (2013) Generic Escherichia coli contamination of spinach at the preharvest stage: effects of farm management and environmental factors. Appl Environ Microbiol 79(14):4347–4358

Rai V, Acharya S, Dey N (2012) Implications of nanobiosensors in agriculture. J Biomater Nanobiotechnol 3:315–324

Rana A, Yadav K, Jagadevan S (2020) A comprehensive review on green synthesis of natureinspired metal nanoparticles: mechanism, application and toxicity. J Clean Prod 272: 122880

Rani M, Yadav J, Chaudhary S, Shanker U (2021) An updated review on synthetic approaches of green nanomaterials and their application for removal of water pollutants: current challenges, assessment and future perspectives. J Environ Chem Eng 9(6):106763

Rees EM, Minter A, Edmunds WJ, Lau CL, Kucharski AJ, Lowe R (2021) Transmission modelling of environmentally persistent zoonotic diseases: a systematic review. Lancet Planet Health 5(7): e466–e478

Rick AR, Arai Y (2011) Role of natural nanoparticles in phosphorus transport processes in ultisols. Soil Sci Soc Am J 75(2):335–347

Rizwan K, Bilal M, Iqbal HM (2022) Role of nanomaterials in sensing air pollutants. In: Hybrid and combined processes for air pollution control. Elsevier, pp 1–17

Roduner E (2006) Size matters: why nanomaterials are different. Chem Soc Rev 35(7):583–592

Rogovski P, Cadamuro RD, da Silva R, de Souza EB, Bonatto C, Viancelli A, Fongaro G (2021) Uses of bacteriophages as bacterial control tools and environmental safety indicators. Front Microbiol 12:793135

Romeh AAA (2018) Green silver nanoparticles for enhancing the phytoremediation of soil and water contaminated by fipronil and degradation products. Water Air Soil Pollut 229:1–13

Sabir S, Arshad M, Chaudhari SK (2014) Zinc oxide nanoparticles for revolutionizing agriculture: synthesis and applications. Sci World J 2014(1):925494

Saied E et al (2021) The catalytic activity of biosynthesized magnesium oxide nanoparticles (MgO-NPs) for inhibiting the growth of pathogenic microbes, tanning effluent treatment, and chromium ion removal. Catalysts 11(7):821

Salama KF, Zafar M (2022) Purification of ambient air by novel green plant with titanium dioxide nanoparticles. Int J Prev Med 13(1):67

Saleem H, Zaidi SJ, Ismail AF, Goh PS (2022) Advances of nanomaterials for air pollution remediation and their impacts on the environment. Chemosphere 287:132083

Salem SS et al (2023) A comprehensive review of nanomaterials: types, synthesis, characterization, and applications. Biointerf Res Appl Chem 13(1):41

Santamaría J, Toranzos GA (2003) Enteric pathogens and soil: a short review. Int Microbiol 6:5–9

Saxena G, Kishor R, Saratale GD, Bharagava RN (2020) Genetically modified organisms (GMOs) and their potential in environmental management: constraints, prospects, and challenges. In: Bioremediation of industrial waste for environmental safety: volume II: biological agents and methods for industrial waste management. Springer, Singapore, pp 1–19

Seltzer JM (1994) Biological contaminants. J Allergy Clin Immunol 94(2):318–326

Sharma K, Tyagi S, Vikal S, Devi A, Gautam YK, Singh BP (2022) Green nanomaterials for remediation of environmental air pollution. In: Handbook of green and sustainable nanotechnology: fundamentals, developments and applications. Springer, Cham, pp 1–26

Shojaei A, Ghafourian H, Yadegarian L, Lari K, Sadatipour MT (2021) Removal of volatile organic compounds (VOCs) from waste air stream using ozone assisted zinc oxide (ZnO) nanoparticles coated on zeolite. J Environ Health Sci Eng 19:771–780

Suba S et al (2021) Microbial mediated synthesis of ZnO nanoparticles derived from lactobacillus spp: characterizations, antimicrobial and biocompatibility efficiencies. Sens Int 2:100104

Tom AP (2021) Nanotechnology for sustainable water treatment—a review. Mater Today Proc 10: 16–78

Vazquez-Muñoz R, Meza-Villezcas A, Fournier PGJ, Soria-Castro E, Juarez-Moreno K, Gallego-Hernández AL, Huerta-Saquero A (2019) Enhancement of antibiotics antimicrobial activity due to the silver nanoparticles impact on the cell membrane. PLoS One 14(11):e0224904

Venkataraju JL, Sharath R, Chandraprabha MN, Neelufar E, Hazra A, Patra M (2014) Synthesis, characterization and evaluation of antimicrobial activity of zinc oxide nanoparticles. J Biochem Technol 3(5):151–154

Virkutyte J, Varma RS (2011) Green synthesis of metal nanoparticles: biodegradable polymers and enzymes in stabilization and surface functionalization. Chem Sci 2(5):837–846

Wilk-Woźniak E, Najberek K (2013) Towards clarifying the presence of alien algae in inland waters—can we predict places of their occurrence? Biologia 68:838–844

Yunus IS, Harwin KA, Adityawarman D, Indarto A (2012) Nanotechnologies in water and air pollution treatment. Environ Technol Rev 1(1):136–148

Zahoor M, Nazir N, Iftikhar M, Naz S, Zekker I, Burlakovs J, Ali Khan F (2021) A review on silver nanoparticles: classification, various methods of synthesis, and their potential roles in biomedical applications and water treatment. Water 13(16):2216

Zhang Y, Shareena Dasari TP, Deng H, Yu H (2015) Antimicrobial activity of gold nanoparticles and ionic gold. J Environ Sci Health C 33(3):286327

Zhang Q, Li Q, Zhang L, Wang S, Harper DP, Wu Q, Young TM (2020) Preparation of electrospun nanofibrous poly(vinyl alcohol)/cellulose nanocrystals air filter for efficient particulate matter removal with repetitive usage capability via facile heat treatment. Chem Eng J 399:125768

Zheng Y, Hou L, Liu M, Newell SE, Yin G, Yu C, Liu C (2017) Effects of silver nanoparticles on nitrification and associated nitrous oxide production in aquatic environments. Sci Adv 3(8): e1603229

Chapter 12
Green Nanotechnology for Air Pollution Control

Surya Venkatesh, P. L. Praveena, Nageshwari Raja, Vaishali Kesavan, Anitha Kannan, and Shilpa Joy

Abstract Nanotechnology is an inevitable technology nowadays, involved in environmental monitoring and pollution control. Air pollution is the root cause of global warming and other major environmental impacts, which need to be controlled in an environmentally friendly manner. Green nanotechnology gives a better solution for air pollution control by concentrating several advanced technologies and nanoparticles. Various sources and detection methods of air pollution are described like using single walled nanotubes, cantilever sensors, etc. The details of application of nano catalysts, nano filters and nanoparticle based gas adsorption system and photocatalysts are highlighted. The health and environmental impact of green nanotechnology in air pollution control are also projected, which will have wide impact in pollution control. Nanoparticles are used for air pollution control under proper regulatory framework. The creation of ecologically friendly, sustainable nanomaterials that lessen the secondary environmental effects of traditional nanotechnology is also covered in this chapter. Applications of photocatalysis and adsorptive technology are emphasized, demonstrating its efficacy in air purification. Furthermore, chances for creative, energy-efficient pollution control solutions are presented by the combination of green nanotechnology with renewable energy sources like solar power. Promising avenues for the future of green nanotechnology for air pollution management include scalability for urban and industrial applications, regulatory frameworks for safety and efficacy, and developments in nanosensing technologies for real-time monitoring. In order to enhance air quality and contribute to a cleaner, healthier environment, this chapter highlights the revolutionary potential of green nanotechnology in developing long-lasting and practical solutions.

S. Venkatesh · P. L. Praveena · N. Raja · V. Kesavan · A. Kannan
Department of Biotechnology, Sethu Institute of Technology, Kariapatti, India

S. Joy (✉)
Department of Biomedical Engineering, Sri Ramakrishna Engineering College, Tamil Nadu, Coimbatore, India

© The Author(s), under exclusive license to Springer Nature Switzerland AG 2025
J. M. Al-Khayri et al. (eds.), *Green Nanotechnology Applications for Ecosystem Sustainability*, Nanotechnology in Plant Sciences 4,
https://doi.org/10.1007/978-3-031-90766-1_12

Keywords Air pollution · Green nanotechnology · Nanocatalyst · Nanodevices · Nanofilters · Nanofilteration · Nanoparticle sensors · Nanoparticles toxicity · Nanorobotics · Nanosensors · Regulatory framework

12.1 Introduction

In recent years, the issue of air pollution has emerged as one of the most pressing environmental challenges globally, posing significant risks to public health and ecological balance. Rapid industrialization, urbanization, and increased vehicular emissions have contributed to the escalation of air pollution levels, leading to adverse effects on human health, ecosystems and climate. Addressing this complex problem requires innovative approaches that are both effective and environmentally sustainable. Green nanotechnology, a rapidly evolving field at the intersection of nanoscience, environmental science, and sustainable engineering, holds immense promise for mitigating air pollution while minimizing environmental impact. By harnessing the inherent properties of nanomaterials and adopting environmentally benign synthesis routes and sustainable practices, green nanotechnology offers a pathway towards more efficient and eco-friendly air pollution control strategies (Manisalidis et al. 2020). Nanostructured catalysts exhibit enhanced catalytic activity and selectivity, enabling efficient conversion of harmful pollutants into benign products through processes such as oxidation, reduction, and photocatalysis. Nano-engineered membranes offer superior separation performance, enabling the selective removal of pollutants, particulate matter and toxic gases from air streams while minimizing energy consumption and waste generation. Functionalized nanomaterials can be integrated into filtration media for improving capture efficiency, adsorption capacity and antimicrobial properties, thereby enhancing the performance of air filtration systems for indoor and outdoor air quality management. Miniaturized nano sensors capable of detecting and quantifying trace pollutants in air matrices enable continuous monitoring of air quality, facilitating early warning systems and targeted intervention strategies. Nanostructured composites comprising natural or synthetic polymers and nanoparticles exhibit synergistic properties for pollutant sorption, photocatalytic degradation, and soil or water remediation (Das et al. 2022). This chapter illustrates the diverse applications and potential of green nanotechnology in addressing various facets of air pollution, from emission control and pollutant abatement to monitoring and remediation. Additionally, it outlines the challenges and considerations associated with the responsible development and deployment of nanotechnology-based solutions, including nanotoxicity and regulatory frameworks (Mishra et al. 2020). Furthermore, this chapter provides a comprehensive overview to the concept of green nanotechnology for air pollution control, laying the groundwork for subsequent discussions on specific nanomaterial-based approaches, their synthesis, characterization, performance evaluation and environmental implications and emerging standards in ensuring the safe and responsible deployment of nanotechnology in air pollution control. By addressing environmental, health, and safety concerns while promoting sustainability, the regulations and guidelines facilitate the development and adoption of innovative nanotechnology-based solutions for mitigating air pollution and safeguarding human health and the environment.

12.2 Sources and Detection of Air Pollution

Globally, air pollution is a serious issue in both industrialized and developing nations. Due to the quick rise in population and energy consumption, harmful air pollutants are being released into the atmosphere, endangering both human health and the ecosystem. Reductions in the physical, chemical, or biological components of the environment (temperature, climate, air masses, etc.) lead to environmental pollution. Either by introducing hazardous compounds or by raising levels above normal, pollutants damage our ecosystem. Oxides of nitrogen (NOx), Sulphur dioxide (SO_2), carbon monoxide (CO), particulate matter (PM), volatile organic compounds (VOCs), and ozone (O_3) are the main air pollutants that are to blame for the declining quality of the air. The transportation industry is the main cause of air pollution, particularly in metropolitan areas, with vehicle emissions being the main cause of the deterioration of air quality. Additionally, there is the risk of industrial mishaps, in which the local population could perish due to the spread of a toxic cloud. Numerous factors influence the dispersion of contaminants, but atmospheric stability and wind speed are the most important ones (Kelishadi and Poursafa 2010; Figs. 12.1 and 12.2).

Apart from this air pollution is also caused due to rapid industrialization, agricultural activities, burning of fossil fuels, emissions from nuclear power plants etc. Air pollution affects people's health in many ways. Even days with moderate levels of air pollution can have an influence on the health of vulnerable and sensitive people. The following conditions are strongly associated with short-term exposure to air pollution: asthma, respiratory diseases, wheezing, cough, shortness of breath, and high rates of hospitalization (a measure of morbidity). Chronic asthma, pulmonary insufficiency, cardiovascular illnesses, and cardiovascular mortality are the long-term impacts linked to air pollution (Manisalidis 2020).

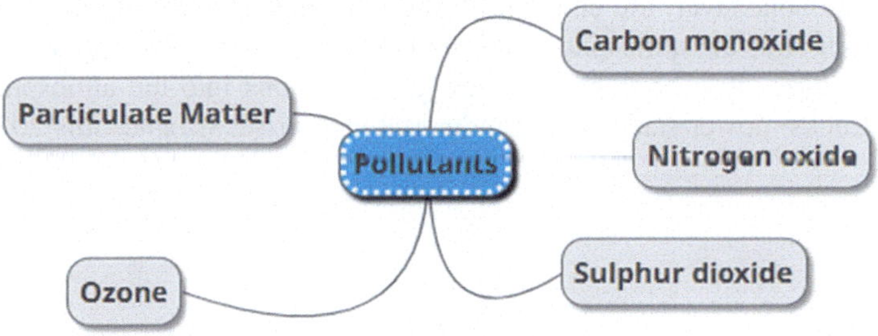

Fig. 12.1 Air pollutants. (Photo constructed by Er. PL Praveena)

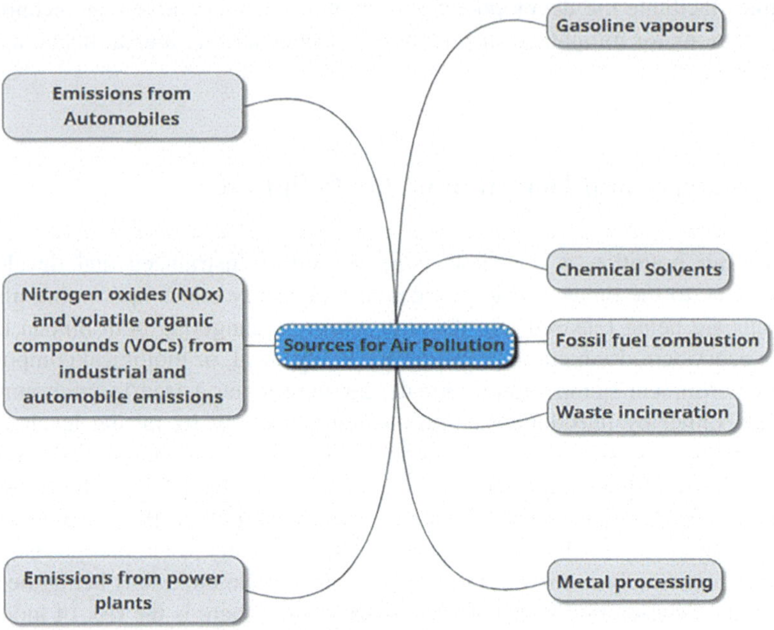

Fig. 12.2 Sources for air pollution. (Photo constructed by PL Praveena)

Air pollution can influence the quality of soil and water bodies by polluting precipitation, falling into water and soil environments (Zuhara and Isaifan 2018). Notably, acid precipitation can modify the chemistry of the soil by impacting water quality, plant life, and cultures (Pathak et al. 2011). Furthermore, soil acidity promotes the transport of heavy metals, which leads to the introduction of metals to wet environments. It is well known that fish and other creatures are poisoned by heavy metals like aluminum. Given that acid rain is more likely to affect soils with low calcium carbonate levels, soil quality appears to be important. In addition to rain, snow, and particle materials fall into water bodies thereby polluting water also.

Climate change and air pollution are intimately associated. The other factor that degrades the quality of our planet is the climate (D'amato et al. 2016). Due to the depletion of ozone layer, life on earth becomes a challenge. When tiny particles are spread throughout the atmosphere, the transparency of the atmosphere decreases, resulting in haze. It is brought about by gas emissions into the atmosphere from vehicles, trucks, power stations, and industrial enterprises (Manisalidis 2020).

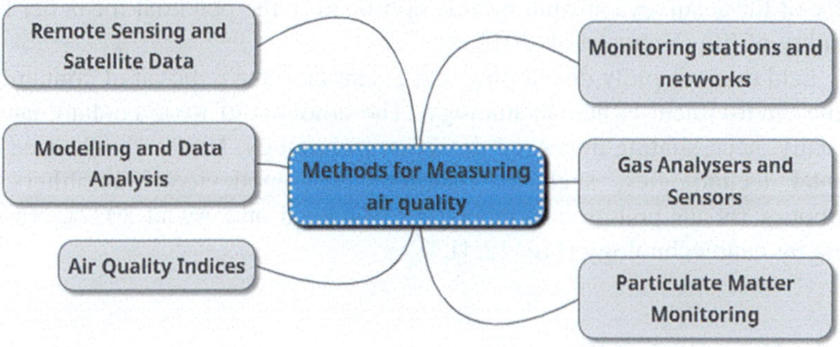

Fig. 12.3 Methods for measuring air quality. (Photo constructed by PL Praveena)

Fig. 12.4 Emerging Nanomaterials for combating air pollution published in 2021–2023. (Data accessed from Web of Science and google scholar) (Photo constructed by PL Praveena)

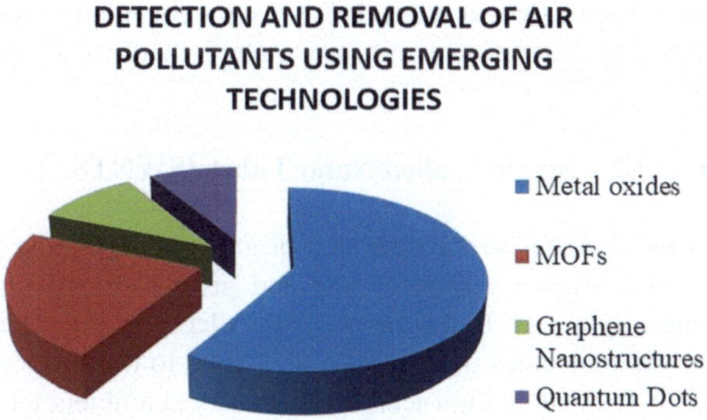

12.2.1 *Detection of Air Pollution*

Environmental and public health sustainability may be better protected by humans with the use of rapid and reliable sensors that can identify contaminants at the molecular level. If pollutant detection technology becomes more affordable and sensitive, significant improvements in process control, ecosystem monitoring, and environmental-based decision-making could ensue. One of the preferred technologies is a continuous monitoring tool that can deliver data, particularly data on pollutants in a relatively short amount of time for analysis (Fig. 12.3).

The air is the main place where contaminants are mixed. Technologies that can monitor, identify, and, if feasible, remove pollutants from the air, water, and soil is therefore necessary. In this regard, a variety of technologies and capacities are available through nanotechnology to enhance the current environment. The power to manipulate matter at the nanoscale and produce materials with certain characteristics and uses is provided by nanotechnology (Zhang and Masciangiolo 2003). Regarding the chances/risk ratio of nanotechnology, surveys from a few European Union (EU) media outlets reveal a comparatively high level of optimism, with the majority of these surveys attributing this optimism to the potential for better health and quality of life (Yunus et al. 2012a).

One field that is rapidly developing and essential to the removal of contaminants from the environment is nanotechnology. The amounts of toxic contaminants are rising daily, necessitating improved detection technology. New and enhanced environmental technologies, such as nanosensors, nanodevices, nanofilters, and nanorobotics for air pollution monitoring (Mohamed and Awad 2022), are made possible by nanotechnology (Fig. 12.4).

12.2.1.1 Nano Contact Sensors

It may be possible to detect some metal ions with a nanocontact sensor without the need for preconcentration. This sensor is especially useful for the on-site detection of radioactive materials and heavy metal ions (Tao 2004).

12.2.1.2 Single Walled Nano Tubes (SWNTS)

When it comes to the detection of gas molecules like NO_2 and NH_3, SWNTs have proven a faster response time and greater sensitivity than the traditional probes currently in use. In this instance, the electrical resistance of the sensor is influenced by gas molecules that are directly bound to the surface of SWNTs. The capacity of SWNTs to attain high sensing sensitivity at ambient temperature is another benefit of using them as sensors. Conventional solid sensors typically function between 200 and 600 °C (Smart et al. 2006).

12.2.1.3 Cantilever Sensors

Cantilever sensors are designed to identify pathogens including salmonella, heavy metals, pesticides, and volatile organic compounds (VOCs). A cantilever sensor is a device made of a silicon cantilever array coated with nano-coating that is sensitive to specific pollutants (Filipponi and Sutherland 2010).

12.2.1.4 Quantum Dots

The size-dependent optical, electrical, and physiochemical features of semiconductor nanocrystals, or QDs, include huge effective surface-to-volume ratios and varying band gaps. QDs release light at specific wavelengths when activated by an electric current or light. Quantum dots (QDs) can be used in optical sensors that give extremely high sensitivity for the detection of air pollutants. A gas sensor based on PbS QDs can detect NO2 gas at room temperature with a low limit of detection (LOD) of 0.15 ppb, excellent selectivity, and high sensitivity. In a similar vein, a

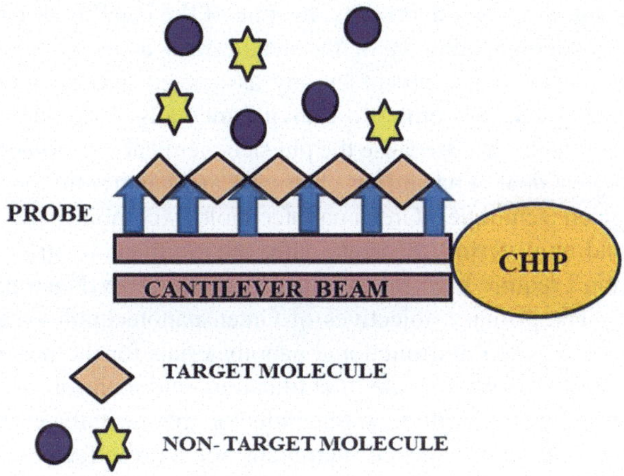

Fig. 12.5 Detection of VOC's by Cantilever sensors. (Photo constructed by PL Praveena)

ZnS QDs-based acetone sensor was created to detect potentially harmful gases in both indoor and outdoor environments (Fig. 12.5).

12.2.2 Advances in Nano Based Sensors

1. Since functionalized tetraphenylsilole nanoparticles can create anionic oxidants bonding, carcinogenic chemicals can be detected at very low quantities. Ar (V) and Cr(VI) are two examples of compounds that can be detected.
2. The idea of a thermocouple has led to the use of peptide nanoelectrodes. A peptide molecule is positioned to create a molecular junction in a "nano-distance" separation gap. The electrical current will result in conductance having a distinct value when a particular metal ion is coupled to the gap. As a result, identifying the metal ion will be simple.
3. To detect chemicals like as organophosphorus insecticides, carbohydrates, and other wood pathogenic compounds in low concentrations, composite electrodes—a combination of nanotubes and copper—have been developed (Yunus et al. 2012b).

12.3 Green Nanotechnology Principles

Advances in nanotechnology are having a significant impact on energy, healthcare, resource management, environmental protection, and energy through the creation of smart materials and connected gadgets. Furthermore, as disciplines, nanoscience and nanotechnology have improved communication to bridge the gaps between the scientific, technical, medical, and other communities, influencing other linked professions in the process. In spite of the benefits of nanoparticles in environmental remediation, toxicity of the synthesized nano particles is a major concern. Green nanotechnology precludes any associated toxicity and offers tools for converting biological systems to environmentally friendly methods of synthesizing nanomaterials. Because the physicochemical creation of these nanoparticles involves a great deal of hazardous chemicals and harsh settings, biological sources are used in green techniques. Green nanotechnology combines the principles of green chemistry and engineering to create safe, environmentally friendly metal nanoparticles that don't require harmful chemicals to synthesize (Nasrollahzadeh et al. 2019).

The primary objectives of Green nanotechnology are processes and goods that need to use multifunctional nanomaterials for the production of sustainable energy, nano sensors for the identification and control of emissions, environmentally friendly packaging, semiconductor photocatalysis for effective environmental cleanup, and a special membrane for more effective wastewater management and purification. In addition to a decrease in greenhouse gas emissions, the use of lighter nanocomposite materials in transportation vehicles can save energy and reduce waste from smaller products. But the nanoparticles themselves—which are made up of a lot of hazardous chemicals—will seriously harm the ecosystem. Thus, nanoparticle purity is crucial in the biogenic oxidation of metal precursors to produce environmentally benign, less expensive, chemical-free nanoparticles for biological and environmental applications. Currently, producing green nanoparticles using algae, bacteria, fungus, and plants shows positive impact on the environment and helps to achieve sustainable development goals (Raja et al. 2022). The majority of green nanotechnologies are exclusively studied in lab settings. To assess if the current technology can be modified for commercial production, initiatives are required. Perhaps the biggest obstacles to most green nanotechnologies are the financial and regulatory restrictions resulting from the uncertain toxicological of nanomaterials. Green nanotechnology frequently necessitates completely novel procedures, which makes going into production expensive at first.

12.4 Green Nanomaterials for Air Pollution Control

The key air pollutants of the environment are particulate matter (PM), surface ozone (O_3), nitrogen dioxide (NO_2), Sulphur dioxide (SO_2), and carbon monoxide (CO). The green nanotechnology approach is an emerging technology that may direct a hasty fixing and effective technology for the air pollution control. The green nanomaterials are used as excellent adsorbents, catalysts, and sensing materials in air pollution control due to specific large surface areas and high reactivity against environmental pollutants (Sharma et al. 2022). The different forms of nanoparticles which act asnano catalysts, nano filters, photocatalyst, etc., involved in the efficient air pollution control are discussed below.

12.4.1 Nanocatalysts for Air Purification

The cerium oxide (CeO_2) nanoparticles were synthesized using the plant extract of *Jatropha curcas* as particle reducing agent. The photocatalytic degradation efficiency of green *Jatropha curcas* extract mediated CeO_2 nanoparticles and chemically synthesized CeO_2 nanoparticles (NH_3 and NaOH) in purifying indoor gaseous pollutant acetaldehyde were compared. The green synthesized CeO_2 nanoparticles are effective than chemically synthesized CeO_2 (Magudieshwaran et al. 2019). Saha et al. (2019) reported the green synthesis of silver nanoparticles from silver nitrate, and Tulsi (*Ocimum tenuiflorum*) leaf extract and impregnation on titanium dioxide (TiO_2). The resulted nanoparticle possesses the catalytic activity in the oxidation reaction of thiophene.

12.4.2 Nanofiltration and Air Cleaning Technologies

Liu et al. (2015) reported a strategy for active air filtration method by constructing the bio-based polyamide-56 nanofiber/nets (PA-56 NFN) membranes with bimodal structures via one-step. The filtration efficiency of PA-56 NFN membranes is about 99.995% with low pressure drop of 111 Pa and large dust holding capacity of 49 g/m^2 which are efficient in filtrating ultrafine airborne particles by safe means. The use of cross-linked protein made of gelatin nanofibers to trap the air pollutants was studied by Souzandeh et al. (2017). This study showed that the biosynthesized protein nanofabrics have potential filtering capacity of air pollutants and also antimicrobial property.

12.4.3 Nanomaterials for Gas Adsorption

The green synthesis of palladium nanoparticles using barberry fruit extract as reducing and stabilizing agent supported on reduced graphene oxide (Pd NPs/RGO) under the minor conditions through reduction of the graphene oxide and Pd^{2+} ions reported as an efficient and heterogeneous catalyst for reduction of nitroarenes using sodium borohydride in an eco-friendly medium (Nasrollahzadeh et al. 2016). The research on carbon-based adsorbents with incorporated metal or metal oxide nanoparticles demonstrated the capacity of nano adsorbents in adsorption of CO_2 and volatile organic compounds. The efficiency of silica-based nano materials for the adsorption of particulate matters in air (Palliyarayil et al. 2021). Saleem et al. (2022) reviewed the importance of nanomaterials and its ability in the adsorption of various contaminants existing in air.

12.4.4 Photocatalysis for Air Quality Improvement

The improvement of air quality by the purification of air by photocatalytic oxidation (PCO) of various pollutants such as volatile organic compounds (VOCs) or inorganic gaseous (NO_x, SO_x, CO, H_2S and ozone, etc.) was reviewed by Boyjoo et al. (2017), and reported the importance of photocatalyst and its reactor design in air purification. Titanium oxide (TiO_2) is a photocatalyst which have the potential to remove toxic NOx but the activity is low. The use of N-doped titanium oxide (TC10) for the removal of noxious nitrogen oxide (NOx) from the environment reported the enhanced photocatalytic activities of 59% and 51% in NO Oxidation (Khan et al. 2021). Nieto-Márquez et al. (2023) tested the photocatalytic effect of TiO_2- based paints which is effective in cleansing indoor air pollutants like $NO_2$and improved air quality. Fifteen bismuth oxyhalides such as BiOX, BiOX/$Bi_xO_yX_z$, heterostructures and $BiOX_xY_{1-x}$ (X, Y = Cl, Br, I) solid solutions holding photocatalytic activity for NOx air purification was also reported (Jimenez-Relinque et al. 2024).

12.5 Health and Environmental Impacts of Green Nanomaterials on Ecosystem

Green nanotechnology has extensive range of potential applications in medicine, environment, biomedical devices, and agriculture. Though there exist wide variety of functional applications of nanomaterials, there exist unforeseen consequences for human fitness and the eco system (Vivek et al. 2021).

12.5.1 Assessment of Potential Risks and Concerns

The source materials that are safe in bulk quantity for synthesis of nanomaterials can develop extreme toxicity at the nano-scale. For example, if they enter and accumulate inside the body, they do not biodegrade and inhaling airborne nano–sized materials affect the lung. So, it is very essential to assess the potential risk associated with the nanomaterial and a special concern have to be made before the application of specific nanomaterials into the ecosystem (Vivek et al. 2021).

12.5.2 Toxicity and Long-Term Effects of Nanomaterials

The toxicity and environmental risk of nanomaterials was reviewed by Ray et al. (2009) and Latha and Kumari (2016). In both the review the authors highlighted the possible risks associated with the different kinds of nanoparticles such as carbon, silver and gold nanomaterials, Quantum dots, titanium oxide nanoparticles. Carbon black and nickel nanoparticles. The possible risks related to the use of nanomaterials such as pulmonary inflammation, granulomas, skin penetration, affecting central nervous system and asthma or chronic obstructive pulmonary diseases were emphasized. White et al. (2003) reported the long-term exposure to silver nanoparticles resulted in Argyria, a blue-gray discoloration of the skin and other organs.

12.5.3 Strategies for Safe and Responsible Use of Nanoparticles

Green synthesis of nanoparticles is the potential and effective mode of minimizing the use of toxic chemicals and formation of their toxic by-products. Green nanomaterials have a maximum level potential activity with minimum toxicity. The strategies of safe applications of these nanomaterials for the safe environment were discussed by Patel (2022) and Khalaj et al. (2020), presented the first scientific study on the green synthesis of nanomaterials which aimed to examine the state of science in this area followed by highlight the knowledge gaps and provide the useful recommendations to transfer the knowledge from lab to real full-scale production and applications. The synthesis of nanomaterials through green technology ruled by proper regulation, systematic control, clean-up activity, bioremediation process will abruptly aid to uplift their nature of environmental stability and suitability.

12.6 Regulatory Framework and Standards

Infested air is one of the most pressing issues confronting the globe today, as it worsens with each passing year, wreaking havoc on the planet and its surroundings. Air pollution is created by both organic and artificial causes, and it appears that human-caused activities contribute more than natural sources. According to the World Health Organization (WHO), the most prevalent airborne contaminants are particulates (PM), surface ozone (O_3), oxides of nitrogen (NO_2), oxides of Sulphur (SO_2), and the gas carbon monoxide (CO). Sustainable nanotechnology is an upcoming technology that can provide a rapid remedy and survive air pollution. We intend to provide complete information regarding the role of developing green nanomaterials in air pollutant management in three stages: detection of contaminants, reduction of sources or preventing pollution, and remediation/degradation of pollutants. Green nanoparticles are effective adsorbents, catalysts, and sensors due to their huge specific surface areas and strong reactivities. Nanomaterials, due to their vast area of surface and high energy density, may absorb significant amounts of airborne pollutants or catalyze processes at a much quicker pace. As a consequence, energy consumption decreased during deterioration, and it may help avoid the release of pollutants. This chapter discussed the use of green nanomaterials to identify and treat air pollution, as well as future prospects in this sector.

Nanomaterials may be engineered to efficiently interact with contaminants and breakdown them into non-toxic byproducts. In addition to identifying and treating areas that are polluted using nanotechnological methods in air pollution management, they can also minimize contamination in the future by substituting hazardous substances with healthier alternatives Lu and Ozcan (2015). Another key use is nanostructured coatings, which are resistant to contaminants and have self-cleaning properties. Nanotechnology is utilized to treat and eliminate air pollutants using three methods: adsorption using nano-absorptive materials, degradation using nano catalysis, and separation using nano filters. Nanoscience and nanotechnology indicate that nano adsorbents, such as carbon nanostructures, can solve and enhance air quality issues. These nano adsorbents offer great specificity, affinity, and capacity due to their physical characteristics. Adsorption relies heavily on highly reactive surface locations and structural connections. The addition of functional groups containing oxygen can provide additional active sites for adsorption. Overall, nano adsorbents have a high prospect of industrial applications (Sharma2022).

Carbon nanostructures such as CNTs and graphene nanosheets are being used to improve the photocatalytic effectiveness of TiO_2, since they allow for simple electron transport across the TiO_2-CNT composite, minimizing electron-hole recombination. The positively charged conduction spectrum of CNTs enables electron transport from TiO_2 to CNTs. Researchers have created novel production techniques for metal oxide nano catalysts, that may assist minimize air pollution. These nanofibers, made of silver, iron, gold, and manganese oxide, can be utilized in environmental management to remove organic compounds that are volatile from industrial smokestacks. Nanogold-based catalysts have demonstrated high efficacy in transforming hazardous air pollutants into safe ones.

Nanostructured membranes with holes tiny enough to segregate distinct contaminants from exhaust are another method for reducing air pollution. The research focuses on improving and optimizing nano-structured membranes for the collection of various gas pollutants. Nanofiber-coated filter media are utilized for dust removal in industrial facilities and input air filtering in gas turbines. (Muralikrishnan et al. 2014).

The capacity of nanomaterials to be coated with several compounds and modify surface area/volume ratio by modifying their shape enables the construction of highly particular, delicate, and customized sensors. Nanoparticles can also be used to clean up air pollution using adsorption, oxidation, and filtering processes. Because of their huge surface area, nanoparticles may adsorb the greatest number of air pollutants in an adsorption removal process. In the oxidative treatment process, nano catalysts catalyze processes at a quicker pace than bulk material, reducing the energy used during deterioration or preventing pollutants from being released. Another strategy is to use nanostructured materials membranes for controlling air pollution and the collection of various contaminants such as VOCs (Mohamed 2017) Nanotechnology is also utilized to avoid the production of pollutants or pollutants by developing an environmentally friendly chemical or material that replaces commonly used harmful materials while reducing environmental effect. The field of nanotechnology additionally has a potential and crucial role in the enhancement of speedy and precise environmental processes to reduce or avoid emissions, or to transform toxins into usable byproducts.

Research on green nanoparticles for pollution control is an evolving field, with efforts to establish standards and guidelines. Regulatory bodies such as the EPA and ECHA may develop guidelines, and organizations like ISO and ASTM could work on standards for production, characterization, and application. Emphasis is placed on risk assessment, toxicity studies, and standardized methods for nanoparticle characterization. Guidelines may cover environmental fate and transport, life cycle assessment, safe use, monitoring, and surveillance. Collaborative research initiatives and public engagement strategies are encouraged to ensure ethical considerations and transparency. Staying updated on the latest research and regulatory developments is crucial.

NMs can permeate cell walls, tissues, and organs, resulting to a rise in bioactivity and bioavailability. Nanomaterials have distinctive characteristics but may also be harmful. Regulatory agencies in several nations have developed laws and regulations for the usage of nanomaterials. Internationally, there is currently a lack of standards and regulations for producing, processing, testing, and evaluating nanoproducts.

The US Environmental Protection Agency (USEPA), Food and Drug Administration (FDA), and Institute for Food and Agricultural Standards (IFAS) established standard protocols for dealing with nanomaterial production and use, such as potential threats to people, animals, and the environment. The toxicity of nanomaterials can be assessed by several parameters, such product concentration, particle size, shape, surface area, crystal structure, dosage and exposure time, and pre-exposure and crystal structure influence. The great potential that nanotechnology and nanomaterials can bring to society, such as the removal of pollutants and the treatment of contaminated water (Maipa et al. 2001), was confirmed by the results obtained. Nevertheless, a discussion between the interested parties is required in order to assess any potential risks. Its consequences, as it is crucial to establish a legal framework to address the demands that advancements in the field of nanotechnology bring.

As a result, it was noted that Brazilian legislation is currently completely unprepared to offer any kind of legal support to anyone who might require it in terms of potential implications that nanotechnology may bring, rendering researchers, industry professionals, legal professionals, and even the final consumer of these products helpless. This indicates that Brazil is moving in the opposite direction from what was revealed in the present work, as European nations are already well-advanced in terms of laws and legal systems that support and seek to take a form of precaution regarding the treatment of products that contain nanotechnology. This serves as evidence of Brazil's current lack of readiness for new technologies in terms of legal support.

Because the State does not support and secure researchers, industry, commerce, or users of these products, it creates uncertainty for all parties involved. This uncertainty can arise from a delay in research and technological advancement in

India, as well as the previously mentioned risks to human health and the environment. In this way, the lack of regulation and legal order ultimately leads to the generation of multiple insecurities and negatively interferes with the development of nanotechnology (Neto et al. 2023).

12.7 Case Studies and Applications

Green nanotechnology offers innovative solutions for mitigating air pollution. Here are some real-world examples:

12.7.1 Nanocatalysts for Air Purification

Nanomaterials, such as gold nanoparticles embedded in porous manganese oxide, act as room temperature catalysts to break down volatile organic compounds (VOCs) in the air. Crystals with nano-sized pores can trap carbon dioxide effectively. Nanocatalysts. (Das et al. 2022). Nanotechnology has revolutionized the field of air pollution by introducing innovative applications that utilize advanced nanomaterials to efficiently remove pollutants and improve overall air quality. These nanomaterials exhibit unique properties that make them highly effective in various air purification techniques, including photocatalysis, adsorption, and catalytic degradation.

12.7.2 Nano Filters for Air Filtration

Similar to water purification methods (Yunus et al. 2012a, b) nano filters find applications in buildings to purify indoor air volumes. These. Nanomaterials can be incorporated into filters to enhance their efficiency in trapping pollutants, such as particulate matter, volatile organic compounds (VOCs), and pathogens. For example, carbon nanotube (CNT)-modified filters have demonstrated superior adsorption efficiency compared to control filters and those modified with graphitic carbon nitride (g-C3N4) or reduced graphene oxide (r-GO)

12.7.3 Green Nanomaterials for Air Pollution Control

Green nanotechnology involves using renewable resources to create nanomaterials. In the context of air pollution, green nanomaterials play a crucial role in three steps:

12.7.3.1 Detection of Air Pollutants

(a) High Sensitivity and Selectivity: Green nanomaterials, such as metal oxides and carbon-based nanomaterials, exhibit enhanced sensitivity and selectivity for specific pollutants. This allows for real-time monitoring of air quality and early detection of harmful substances, such as nitrogen oxides (NOx), sulfur dioxide (SO2), and volatile organic compounds (VOCs).
(b) Miniaturization and Portability: The small size of nanomaterials enables the development of compact and portable sensors. These devices can be deployed in various environments, including urban areas, industrial sites, and remote locations, facilitating widespread air quality monitoring.
(c) Cost-Effectiveness: Many green nanomaterials can be synthesized from abundant and inexpensive natural resources, making them a cost-effective alternative to traditional air quality sensors. This affordability can lead to broader adoption and implementation in various settings.

12.7.3.2 Source Reduction or Pollution Prevention

Nanomaterials can absorb air pollutants or catalyze reactions, reducing energy consumption during degradation (Cao et al. 2021).

(a) Catalytic Activity: Green nanomaterials can serve as catalysts that facilitate the breakdown of harmful gases at lower temperatures, thus reducing energy consumption. For example, catalysts made from sustainable materials can effectively convert NOx into nitrogen and oxygen, mitigating their impact on air quality.
(b) Absorption and Adsorption: Nanomaterials with high surface areas, such as activated carbon nanoparticles and metal-organic frameworks (MOFs), can effectively absorb and adsorb a wide range of air pollutants. This property allows them to capture harmful substances before they disperse into the atmosphere.
(c) Biodegradable Solutions: Many green nanomaterials are derived from biodegradable sources, ensuring that they do not contribute to environmental pollution after their use. This characteristic aligns with the principles of sustainability and waste reduction.

12.7.3.3 Remediation/Degradation of Air Pollutants

Green nanomaterials aid in cleaning up contaminated air by adsorbing pollutants or facilitating their breakdown. (Sharma et al. 2022).

(a) Photocatalytic Degradation: Green nanomaterials, such as titanium dioxide (TiO2) nanoparticles, can be activated by sunlight to catalyze the degradation of organic pollutants in the air. This process transforms harmful compounds into non-toxic byproducts, effectively cleaning the air.

(b) Enhanced Filtration Systems: Incorporating green nanomaterials into air filtration systems can significantly improve their efficiency. For example, filters enhanced with nanofibers or nanoparticles can capture smaller particulate matter and gaseous pollutants, leading to cleaner indoor air quality.
(c) Sustainable Remediation Technologies: Green nanomaterials can be integrated into innovative remediation technologies, such as biofilters and bioreactors that utilize biological processes to degrade pollutants. This approach not only removes contaminants but also promotes the recovery of valuable resources from waste

12.7.4 Overall Benefits of Green Nanotechnology

Eco-friendly and sustainable: Green nanotechnology uses biological materials and renewable resources, making it less expensive and environmentally friendly. Potential (Bhardwaj et al. 2023).

Green nanotechnology represents a significant advancement in the pursuit of sustainable and eco-friendly solutions across various sectors, including agriculture, medicine, and environmental management. By leveraging biological materials and renewable resources, this innovative approach minimizes environmental impact while enhancing efficiency and effectiveness.

Eco-Friendly and Sustainable Practices includes one of the primary benefits of green nanotechnology is its commitment to eco-friendliness. The use of biological materials, such as plant extracts, in the synthesis of nanomaterials reduces reliance on toxic chemicals and harmful processes typically associated with conventional nanotechnology. This approach not only lowers production costs but also mitigates environmental risks, making it a more sustainable option for developing nanomaterials.

The clean Energy Supply includes the integration of green nanotechnology into energy systems holds the potential for cleaner energy production. Nanomaterials can enhance the efficiency of solar cells, fuel cells, and batteries, leading to reduced reliance on fossil fuels and lower greenhouse gas emissions. This transition towards cleaner energy sources is crucial in combating climate change and promoting sustainable development.

The environmental cleanliness and pollution reduction where the green nanotechnology contributes significantly to environmental cleanliness. Nanomaterials can be employed in various applications, such as water purification, air filtration, and the remediation of contaminated sites. Their high surface area and reactivity allow for effective removal of pollutants, including heavy metals and organic compounds, thereby improving overall environmental quality.

In summary, green nanotechnology presents a multifaceted approach to addressing some of the most pressing challenges of our time. By utilizing biological materials and renewable resources, it fosters eco-friendly practices that benefit agriculture, healthcare, and environmental management. The potential for improved biological diagnostics, enhanced agricultural efficiency, and cleaner energy solutions underscores the importance of advancing green nanotechnology as a means to achieve a sustainable future.

12.7.4.1 Ambient Air Purification by Nanotechnology

Researchers have made significant progress in using nanomaterials for air purification. Filtration, adsorption, photocatalysis (Sharma et al. 2022) and ambient-temperature catalysis are key techniques. Nanocatalysts embedded in materials like gold nanoparticles in porous manganese oxide break down volatile organic compounds (VOCs).Nano filters can purify indoor air and be applied to automobile tailpipes and smokestacks to prevent contaminants from entering the atmosphere (Cao et al. 2021).

12.7.4.2 Nanocatalysts for Air Pollution Mitigation

Zeolites, titanium dioxide, and carbon nanotubes exhibit exceptional adsorption and catalytic properties. These materials can be tailored to target specific pollutants, providing a versatile toolkit for pollution abatement (Cao et al. 2021). Air pollution is a significant environmental concern, with severe impacts on human health, climate change, and ecosystems. The development of effective technologies for air pollution mitigation is crucial to address this issue. Nanocatalysts, such as zeolites, titanium dioxide, and carbon nanotubes, have emerged as promising materials for air pollution abatement due to their exceptional adsorption and catalytic properties.

12.7.4.3 Real-World Examples

Gold nanoparticles embedded in porous manganese oxide act as room temperature catalysts for VOC breakdown (Cao et al. 2021) Nano filters are used for air filtration in buildings and smokestacks. Cobalt. Green nanotechnology focuses on using renewable resources for nanomaterial production. Nanosensors play a crucial role in monitoring air quality and using renewable resources reduces environmental impact (Sharma et al. 2022).

12.7.5 Gold Nanoparticles in VOC Breakdown

Gold nanoparticles embedded in porous manganese oxide are utilized as effective catalysts for the breakdown of volatile organic compounds (VOCs) at room temperature. This application highlights the potential of nanocatalysts to operate under mild conditions, making them suitable for various environmental remediation processes.

The high surface area and reactivity of these nanoparticles facilitate the efficient degradation of harmful VOCs, contributing to cleaner air in urban and industrial settings .

12.7.6 Nano Filters for Air Filtration

Nanofiber-based filters are increasingly being used in buildings and smokestacks for air filtration. These filters are engineered to trap pollutants such as particulate matter, bacteria, and viruses effectively. The use of nanofibers enhances the filtration efficiency while maintaining low pressure drops, which is critical for energy savings in ventilation systems. This technology not only improves indoor air quality but also helps in reducing emissions from industrial sources.

12.7.7 Cobalt and Platinum Nanocatalysts

Cobalt and platinum nanocatalysts are employed to remove nitrogen oxides (NOx) from smokestack emissions. These nanomaterials facilitate catalytic reactions that convert harmful NOx gases into less harmful nitrogen and oxygen, thereby mitigating the impact of industrial pollution on air quality. The effectiveness of these nanocatalysts at lower temperatures makes them particularly advantageous for application in real-world scenarios where energy efficiency is a priority.

12.7.8 Nanosensors for Air Quality Monitoring

Nanosensors play a crucial role in monitoring air quality by detecting various pollutants in real time. These sensors, often made from nanomaterials such as metal oxides and carbon nanotubes, offer high sensitivity and selectivity for specific contaminants, enabling timely responses to pollution events. The integration of nanosensors into air quality management systems allows for better data collection and analysis, leading to more informed decision-making regarding air pollution control .

12.7.9 Renewable Resource Utilization

The focus of green nanotechnology on using renewable resources for nanomaterial production significantly reduces the environmental impact associated with traditional manufacturing processes. By sourcing materials from biological or sustainable

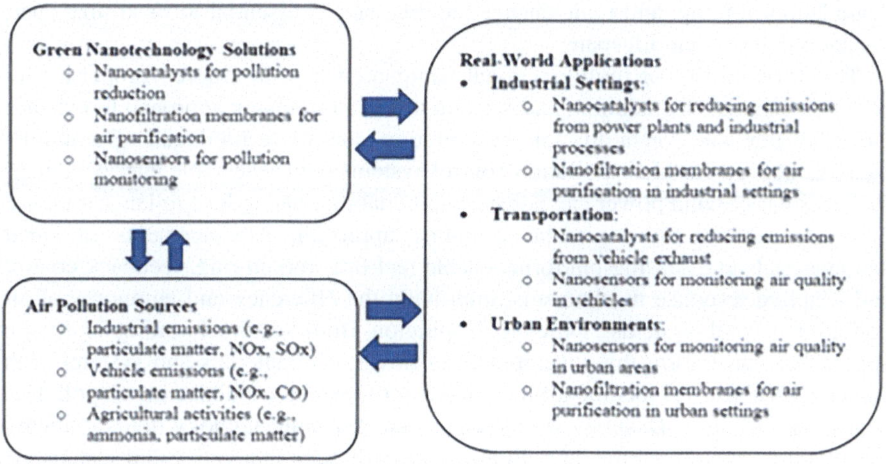

Fig. 12.6 Case studies and applications. (Photo constructed by V. Surya)

origins, the lifecycle of nanomaterials can be made more eco-friendly, aligning with global sustainability goals. This approach not only minimizes waste and pollution but also enhances the overall effectiveness of air pollution control technologies (Fig. 12.6).

12.8 Challenges and Future Directions

Green nanotechnology offers promising solutions for addressing the pressing issue of air pollution while minimizing environmental impact. However, the widespread adoption of nanotechnology-based approaches in air pollution control faces several challenges that must be addressed to realize its full potential (Saleem et al. 2022). Despite their beneficial applications, certain nanomaterials may pose risks to human health and the environment. Understanding the toxicity and fate of nanomaterials in air pollution control applications is essential to mitigate potential adverse effects and ensure environmental safety (Juárez-Maldonado 2021). The scalability of nanotechnology-based air pollution control technologies remains a significant challenge. Transitioning from laboratory-scale experiments to large-scale implementation requires overcoming technical, economic, and regulatory barriers to facilitate widespread adoption and commercialization. The cost of nanomaterials and fabrication processes is often prohibitive for widespread deployment, particularly in resource-constrained regions. Developing cost-effective synthesis methods and affordable air pollution control solutions is essential to ensure accessibility and equity in environmental protection efforts. Ensuring the long-term performance and durability of nanotechnology-based air pollution control systems is crucial for their effectiveness and sustainability. Addressing issues such as material degradation, fouling, and maintenance requirements is essential to maximize operational efficiency and lifespan.

The development of multifunctional nanomaterials with tailored properties for pollutant capture, degradation, and sensing offers a promising approach to enhance the efficiency and versatility of air pollution control technologies. Combining nanotechnology-based air pollution control systems with renewable energy sources such as solar or wind power can enhance sustainability and reduce reliance on fossil fuels, contributing to a circular economy approach. Advancements in smart nanomaterials and sensing platforms enable real-time monitoring, feedback control and adaptive response mechanisms, improving the efficiency and autonomy of air pollution control systems. Drawing inspiration from nature, biomimetic design approaches can inform the development of novel nanomaterials and systems with enhanced efficiency, biocompatibility, and environmental resilience for air pollution control applications. Despite the challenges, green nanotechnology holds immense potential for revolutionizing air pollution control through innovative and sustainable solutions. By addressing key challenges and embracing future directions, researchers and stakeholders can contribute to the development of effective, affordable, and environmentally friendly nanotechnology-based approaches to safeguard air quality and promote human well-being. Research should focus on developing standardized protocols for assessing the ecotoxicity and bioaccumulation potential of nanomaterials to ensure their safe deployment in environmental applications. Addressing regulatory gaps and uncertainties surrounding the safe use and disposal of nanomaterials in air pollution control requires interdisciplinary collaboration between scientists, policymakers, and regulatory agencies.

12.9 Conclusion and Prospects

Nanotechnology is involved in all advance technologies of various field, in which environmental role is emphasized here. The implementation of safe and environmentally friendly nanotechnology principles for air pollution monitoring, assessment and control gains immense global attention. The development of completely biodegradable and non-toxic nanomaterials will be given more attention as green nanotechnology advances. This will guarantee that there are no secondary environmental dangers associated with the solutions themselves. Technological developments in nanosensing may result in inexpensive, extremely sensitive sensors that can identify a variety of contaminants in real time. These sensors can offer data-driven insights for pollution management measures and aid in the early identification of dangerous air quality. Enhancing the catalytic effectiveness of nanomaterials for the degradation of contaminants, especially in the presence of visible light, may be the main focus of future studies. More energy-efficient air purification systems that operate under ambient settings may result from this.

By combining several kinds of nanomaterials (such as metal oxides and carbon nanotubes), hybrid nanocomposites that target many contaminants at once can be created. Because of its multiple uses, air purification systems can be more flexible and able to adjust to different types of pollutants. Growing output while preserving cost-effectiveness and environmental requirements will be a major problem as green nanotechnology develops. Commercializing these technologies for use in urban and industrial settings will need the development of large-scale production and deployment techniques for green nanomaterials. Comprehensive regulatory frameworks that guarantee safety and reduce hazards are also necessary for the development of green nanotechnology for the management of air pollution. To guarantee that these technologies are sustainable over the long run, future regulations will probably cover lifespan evaluation, disposal recommendations, and environmental effect. Green nanotechnology has the potential to be integrated with renewable energy systems. For example, photocatalytic nanoparticles used in solar-powered air filtration systems may offer effective and sustainable air cleaning solutions, especially in isolated or off-grid locations. Various possible green nanoparticles, their different modes of advanced utilization with examples and case studies highlight the potential of green nanotechnology in safe pollution control. These key studies can be utilized for better understanding of monitoring and control systems in other types of pollutions. The limitations of green nanotechnology in pollution control need to be addressed with prior importance. A better solution for alleviating the negative impacts need to be studied and implemented for pollution less good future.

In conclusion, green nanotechnology has the ability to completely transform air pollution management by utilizing creative, environmentally responsible methods. This technology may be essential to the creation of healthier, cleaner air and a more sustainable future if scientists, legislators, and industry stakeholders maintain their study, development, and cooperation.

References

Bhardwaj LK, Singh VV, Dwivedi K, Rai S (2023) A comprehensive review on sources, types, impact and challenges of air pollution. https://doi.org/10.20944/preprints202311.1245.v1

Boyjoo Y, Sun H, Liu J, Pareek VK, Wang S (2017) A review on photocatalysis for air treatment: from catalyst development to reactor design. Chem Eng J 310:537–559

Cao JJ, Huang Y, Zhang Q (2021) Ambient air purification by nanotechnologies: from theory to application. Catalysts 11(11):1276

D'amato G, Pawankar R, Vitale C, Lanza M, Molino A, Stanziola A et al (2016) Climate change and air pollution: effects on respiratory allergy. Allergy, Asthma Immunol Res 8(5):391

Das K, Singh AA, Eram R (2022) Clearing the air: technologies for monitoring and control of air pollution. In: Environmental sustainability and industries. Elsevier, pp 85–116

Filipponi L, Sutherland D (2010) NANOYOU teachers training kit in nanotechnologies. In: Interdisciplinary Nanoscience Centre, iNANO. Aarhus University

Juarez-Maldonado (2021) Impact of nanomaterials on plants: What other implications do they have?. BIOCELL 46:651–654

Jimenez-Relinque E, Dappozze F, Berhault G, Gilbert C, Leonard D, Guillard C (2024) Bismuth oxyhalide as efficient photocatalyst for water, air treatment and bacteria inactivation under UV and visible light. J Photochem Photobiol A Chem 115554

Kelishadi R, Poursafa P (2010) Air pollution and non-respiratory health hazards for children. Arch Med Sci 6(4):483–495

Khalaj M, Kamali M, Costa MEV, Capela I (2020) Green synthesis of nanomaterials-a scientometric assessment. J Clean Prod 267:122036

Khan TT, Bari GAR, Kang HJ, Lee TG, Park JW, Hwang HJ et al (2021) Synthesis of N-doped TiO2 for efficient photocatalytic degradation of atmospheric NOx. Catalysts 11(1):109

Latha PP, Kumari RS (2016) Toxicity and environmental risks of nanomaterials: challenges and future needs. Indian J Res Pharm Biotechnol 4(6):247

Liu B, Zhang S, Wang X, Yu J, Ding B (2015) Efficient and reusable polyamide-56 nanofiber/nets membrane with bimodal structures for air filtration. J Colloid Interface Sci 457:203–211

Lu Y, Ozcan S (2015) Green nanomaterials: on track for a sustainable future. Nano Today 10(4): 417–420

Magudieshwaran R, Ishii J, Raja KCN, Terashima C, Venkatachalam R, Fujishima A, Pitchaimuthu S (2019) Green and chemical synthesized CeO2 nanoparticles for photocatalytic indoor air pollutant degradation. Mater Lett 239:40–44

Maipa V, Alamanos Y, Bezirtzoglou E (2001) Seasonal fluctuation of bacterial indicators in coastal waters. Microb Ecol Health Dis 13(3):143–146

Manisalidis I, Stavropoulou E, Stavropoulos A, Bezirtzoglou E (2020) Environmental and health impacts of air pollution: a review. Front Public Health 20(8):14. https://doi.org/10.3389/fpubh.2020.00014

Mishra AK, Tiwari A, Arya R, Panchal D (2020) Environmental nanotechnology: global framework and integrative strategies of nanowaste management. In: Handbook of environmental materials management. Springer, Cham, pp 1–31

Mohamed EF (2017) Nanotechnology: future of environmental air pollution control. Environ Manage Sustain Dev 6(2):429

Mohamed EF, Awad G (2022) Development of nano-sensor and biosensor as an air pollution detection technique for the foreseeable future. In: Comprehensive analytical chemistry, pp 163–188. https://doi.org/10.1016/bs.coac.2021.11.003

Muralikrishnan R, Swarnalakshmi M, Nakkeeran E (2014) Nanoparticle-membrane filtration of vehicular exhaust to reduce air pollution-a review. Int Res J Environ Sci 3(4):82–86

Nasrollahzadeh M, Sajadi SM, Rostami-Vartooni A, Alizadeh M, Bagherzadeh M (2016) Green synthesis of the Pd nanoparticles supported on reduced graphene oxide using barberry fruit extract and its application as a recyclable and heterogeneous catalyst for the reduction of nitroarenes. J Colloid Interface Sci 466:360–368

Nasrollahzadeh M, Sajjadi M, Sajadi SM, Issaabadi Z (2019) Green nanotechnology. In: Interface science and technology, vol 28. Elsevier, pp 145–198

Neto GR, da Silva WL, Martins MO (2023) Contemporary challenges in face of nanotechnology regulatory gaps. Disciplinarum Sciential Naturais e Tecnológicas 24(2):41–58

Nieto-Márquez A, de Mateo M, Barrios A, de la Fuente MDM, Narros A (2023) Improving indoor air quality by using photocatalytic paints. Real scale study at the technical university of Madrid. Atmos Pollut Res 14(8):101827

Palliyarayil A, Saini H, Vinayakumar K, Selvarajan P, Vinu A, Kumar NS, Sil S (2021) Advances in porous material research towards the management of air pollution. Emerg Mater 4:607–643

Patel M (2022) Green synthesis of nanoparticles: a solution to environmental pollution. In: Handbook of solid waste management: sustainability through circular economy. Springer, Singapore, pp 1965–1993

Pathak RK, Wang T, Ho KF, Lee SC (2011) Characteristics of summertime PM2. 5 organic and elemental carbon in four major Chinese cities: implications of high acidity for water-soluble organic carbon (WSOC). Atmos Environ 45(2):318–325

Raja RK, Hazir S, Balasubramani G, Sivaprakash G, Obeth ESJ, Boobalan T et al (2022) Green nanotechnology for the environment. In: Handbook of microbial nanotechnology. Academic, pp 461–478

Ray PC, Yu H, Fu PP (2009) Toxicity and environmental risks of nanomaterials: challenges and future needs. J Environ Sci Health C 27(1):1–35

Saha B, Kumar S, Sengupta S (2019) Green synthesis of nano silver on TiO2 catalyst for application in oxidation of thiophene. Chem Eng Sci 199:332–341

Saleem H, Zaidi SJ, Ismail AF, Goh PS (2022) Advances of nanomaterials for air pollution remediation and their impacts on the environment. Chemosphere 287:132083

Sharma K, Tyagi S, Vikal S, Devi A, Gautam YK, Singh BP (2022) Green nanomaterials for remediation of environmental air pollution. In: Handbook of green and sustainable nanotechnology: fundamentals, developments and applications. Springer, Cham, pp 1–26

Smart SK, Cassady AI, Lu GQ, Martin DJ (2006) The biocompatibility of carbon nanotubes. Carbon 44(6):1034–1047

Souzandeh H, Molki B, Zheng M, Beyenal H, Scudiero L, Wang Y, Zhong WH (2017) Cross-linked protein nanofilter with antibacterial properties for multifunctional air filtration. ACS Appl Mater Interfaces 9(27):22846–22855

Tao N (2004) A nanocontact sensor for heavy metal ion detection. Proceedings EPA nanotechnology and the environment: applications and implications STAR Progress review. Workshop:28–29

Vivek P, Rekha M, Steephen A, Kowsalya M, Magesh M (2021) Meticulous review on potential nano–sized catalysts for air and water purifiers. Iran J Catal 11(4):331–345

White JML, Powell AM, Brady K, Russell-Jones R (2003) Severe generalized argyria secondary to ingestion of colloidal silver protein. Clin Exp Dermatol 28(3):254–256

Yunus IS, Harwin AK, Adityawarman D, Indarto A (2012a) Nanotechnologies in water and air pollution treatment. Environ Technol Rev 1(1):136–148

Yunus IS, Harwin KA, Adityawarman D, Indarto A (2012b) Nanotechnologies in water and air pollution treatment. Environ Technol Rev 1(1):136–148

Zhang WX, Masciangioli T (2003) Environmental technologies at the nanoscales. Environ Sci Technol 37(5):103–108

Zuhara S, Isaifan R (2018) The impact of criteria air pollutants on soil and water: a review. J Environ Sci Pollut Res 4(2):278–284

Chapter 13
Green Nanotechnology for Ocean Pollution Control

Sherin Antony, Sonia John, and Simi Joseph

Abstract Marine pollution is an increasingly serious threat to the world's oceans and marine ecosystems, with detrimental effects on living resources, human health, marine activities, water quality, and coastal amenities. Ocean pollution is characterized as the introduction of deleterious substances or energy into marine environments. Over 80% of this pollution is attributed to terrestrial sources, including agricultural runoff, plastic refuse, and untreated sewage. Although traditional methodologies have been utilized to mitigate the ramifications of marine pollution, they frequently prove inadequate in tackling the fundamental causes. This limitation necessitates a more comprehensive approach that incorporates governmental action, public awareness campaigns, and investments in sustainable practices. Recently, green nanotechnology has surfaced as a promising instrument for addressing ocean pollution in an ecologically sustainable manner. It provides innovative methodologies for both the prevention and reduction of pollution at its source, employing novel materials and processes that minimize ecological damage. This chapter focuses on the various pollutants affecting marine ecosystems and explores sustainable techniques for ocean pollution control. Special attention is given to the application of green nanotechnology, highlighting its potential to revolutionize pollution mitigation. Despite its potential, green nanotechnology is not devoid of challenges, including potential high financial implications, scalability concerns, and the imperative for additional research regarding its long-term environmental consequences. By integrating advanced technological innovations with regulatory frameworks and public involvement, a more efficacious and sustainable strategy for managing marine pollution can be realized. This chapter aims to provide a comprehensive overview of these strategies to inspire further innovation in the field.

S. Antony (✉) · S. John · S. Joseph
Department of Zoology, Bharata Mata College (Autonomous), Thrikkakara, Ernakulam, Kerala, India
e-mail: sherin@bharatamatacollege.in; Sonia@bharatamatacollege.in; simi@bharatamatacollege.in

© The Author(s), under exclusive license to Springer Nature Switzerland AG 2025
J. M. Al-Khayri et al. (eds.), *Green Nanotechnology Applications for Ecosystem Sustainability*, Nanotechnology in Plant Sciences 4,
https://doi.org/10.1007/978-3-031-90766-1_13

Keywords Green nanotechnology · Nanoparticles · Nanoscale sorbents · Ocean pollutants · Ocean pollution control · Nanosensors

13.1 Introduction

Pollution denotes the infusion of various substances or forms of energy, such as light or thermal energy, into the natural environment at concentrations that may pose significant risks to humans, fauna, and flora. The integrity of ecosystems may be jeopardized by pollution, as it can inflict harm or even result in mortality among the organisms residing within these systems (Freedman 2013). The magnitude of the adverse effects of pollution is contingent upon the specific characteristics of the pollutant, including its nature and spatial context. Pollution is primarily constituted of synthetic or anthropogenic substances; however, naturally occurring substances such as sediments, nutrients, and carbon dioxide may also become pollutants when they exceed defined thresholds. Nevertheless, when natural substances surpass acceptable levels, it is highly probable that anthropogenic activities are the underlying cause (Appannagari 2017). The classification, investigation, and regulation of pollution are contingent upon the Earth system that is influenced. Pollution has the potential to impact various components of Earth's systems, encompassing terrestrial environments, aquatic ecosystems (including groundwater, rivers, lakes, bays, and oceans), atmospheric conditions, and climatic patterns.

Oceans play an indispensable role in sustaining life, fostering diverse ecosystems, and regulating climatic conditions, covering over 70% of the Earth's surface. Pollution represents one of the remarkable threats that anthropogenic actions impose on these vast marine ecosystems. Marine pollution, resulting from the introduction of a variety of pollutants such as plastics, heavy metals, pesticides, and hydrocarbons, poses significant threats to marine organisms, ecosystems, and human health (Todd et al. 2010).

The rapidly evolving field of nanotechnology has shown considerable potential in recent years for addressing environmental challenges, including marine pollution (Mehndiratta et al. 2013; Esposito et al. 2021; Mishra et al. 2022). This field involves the manipulation of materials at the nanoscale, typically ranging from 1 to 100 nanometers, to create novel features and functionalities. This technique is referred to as nanotechnology. Nanotechnology offers innovative strategies for the prevention of ocean pollution that are both effective and environmentally sustainable (Marcharla et al. 2024).

This chapter explores the concept of "green nanotechnology" as a pragmatic approach to mitigating ocean pollution. "Green nanotechnology" refers to the development and application of nanomaterials and nanotechnologies that are environmentally benign, energy-efficient, and commercially viable (Khan 2020). As per Nasrollahzadeh et al. (2019), by harnessing the unique properties of nanoparticles, such as their enhanced surface area-to-volume ratio, reactivity, and customizable features, green nanotechnology provides promising avenues for both the reduction and prevention of pollution in marine environments.

This chapter aspires to contribute to the ongoing dialogue regarding sustainable strategies for the protection of marine ecosystems and the health and welfare of both current and future generations. By embracing the principles of green nanotechnology, we can endeavour to achieve healthier, cleaner oceans that are resilient to the challenges posed in the twenty-first century.

13.2 Ocean Pollution

Ocean pollution, "refers to the introduction of substances or energy into the marine environment, resulting in harmful effects on living resources, human health, marine activities, water quality, and amenities" (United Nations Joint Group of Experts on the Scientific Aspects of Marine Pollution- https://wedocs.unep.org/20.500.11822/29015).

The conventional belief of the ocean's ability to absorb and dilute pollution is now challenged by the widespread and significant impact of the current ocean pollution crisis. We are realizing that the ocean, once considered a disposal site for industrial and domestic waste, has now reached its capacity, leading to the return of pollution in various forms such as contaminated fish and seafood, entangled marine life, extensive plastic pollution, and expanding dead zones. These dead zones, caused by climate change, pollution, and warming waters, have quadrupled in size since 1950 and now cover an area equivalent to the European Union (Breitburg et al. 2018). Additionally, coastal areas with low oxygen due to nutrient, organic matter, and sewage runoff have increased tenfold (Kessouri et al. 2021). Human activities have brought drastic changes to coastal ecosystems within a short period of time, resulting in immense ecological impacts. Toxic chemicals like persistent organic pollutants (POPs) and heavy metals have been transported through the atmosphere and have contaminated both land and the world's oceans.

There exist multiple methods through which pollution infiltrates the ocean. One such method is through sewage, as polluting substances flow directly into the ocean via sewage systems, rivers, or drainages. Industries also contribute to ocean pollution by discharging toxic chemicals directly into the oceans. These hazardous and toxic chemicals have adverse effects on marine life and also raise the ocean's temperature, resulting in thermal pollution. As a consequence, aquatic animals and plants face difficulties in surviving at higher temperatures. Moreover, land runoff plays a significant role in marine pollution, accounting for approximately 80% of the problem. Land-based sources, including agricultural runoff, discharge of nutrients and pesticides, and untreated sewage, introduce harmful contaminants such as plastics, fertilizers, petroleum, pesticides, and other soil contaminants into the ocean. Ships are another significant source of ocean pollution, particularly through large-scale oil spills. The devastating effects of oil spills are long-lasting as crude oil remains toxic to marine life for years, suffocating marine animals and posing challenges for cleanup efforts. Additionally, ocean mining for silver, gold, copper, cobalt, and zinc creates sulfide deposits deep in the ocean, up to three and a half

Table 13.1 Current trends of Ocean Pollution—Sources and quantity

Polutant	Annual amount entering ocean	Main sources
Plastic	8–10 million metric tons	Littering, Inadequate waste disposal and management, industrial activities.
Chemicals	300–400 million tons	Waste water, industrial waste
Oil	Varies (dependent on oil spills	Oil spills
Agriculture waste	80% of global marine pollution	Agriculture runoff, untreated sewage, discharge of nutrients and pesticides

Source: https://www.greenmatch.co.uk/ocean-pollution-facts (2017)

Table 13.2 Marine pollutants of global concern

Pollutant	Description
Petroleum hydrocarbons	Crude oil and some of its refined products
Sewage	Effects of sewage pollution are generally local ones but sewage may contain some of the other toxic chemicals.
Halogenated hydrocarbons	Including compounds such as DDT and the PCBs together with their degradation and combustion products (e.g. polychlorinated di- oxins and dibenzofurans).
Other organic compounds	Endocrine disrupting chemicals (also referred to as oestrogenic chemicals).
Trace metals	Particularly mercury, cadmium and lead
Radionuclides	Particularly caesium-137, strontium-90 and plutonium isotopes
Litter (solid waste)	Particularly persistent plastics and including lost fishing nets

Source: https://www.greenmatch.co.uk/ocean-pollution-facts (2017). Krishnakumar (2017)

thousand meters (Sharma 2015). Table 13.1 shows current trends of ocean pollution—Sources and quantity (source-https://www.greenmatch.co.uk/ocean-pollution-facts) and Table 13.2 depicts Marine pollutants of global concern (source-Krishnakumar 2017).

Plastic pollution is another significant concern, "with every square mile of ocean estimated to contain 46,000 pieces of floating plastic" (Ornell and Finn 2011). They undergo weathering and degradation in the marine environment, breaking down into smaller pieces which enables the contaminants absorption. As a result, the amounts of these chemical pollutants that naturally occur in seawater can be much higher in microplastic fragments or on their surface. These micro-plastics, along with plastic pellets, can be found on most beaches globally. Marine pollution can also lead to the concentration of plastic materials and other litter in specific areas called gyres, due to oceanic currents. For example, "the North Pacific Gyre, also known as the Great Pacific Garbage Patch, accumulates waste material from across the North Pacific Ocean, including coastal waters off North America and Japan" (Lebreton et al. 2018). In addition to this, climate change and carbon dioxide have a major influence on the oceans, particularly on fish communities and ecosystems (Brierley and

Kingsford 2009). The acidity of the ocean is caused by growing carbon dioxide levels. Moreover, a number of activities that threaten coastal and marine environments include mining, fishing, aquaculture, port and harbor developments, river damming, urban growth, and tourism along the coast. Effects of Ocean pollution and the outcome of Toxic Wastes on Marine life is “long term effect like cancer, failure in the reproductive system, behavioral changes, and even death” (Weis 2013). As per reports of Doerffer (2013), “Oil spill floats on the surface of the water and prevents sunlight from reaching marine plants and affects the process of photosynthesis and reduces the oxygen content of water”: The majority of marine debris does not break down and stays in the water for years at a time. As a result, oxygen levels drop, which lowers the likelihood that marine species like whales, turtles, sharks, dolphins, and penguins would survive for an extended period of time. A greater number of low oxygen (hypoxic) areas, or “dead zones,” where the majority of marine species cannot survive, have been caused by excessive nutrients from sewage outfalls and agricultural runoff, which has led to the collapse of some ecosystems. Eutrophication: An excessive algal bloom is caused by an overabundance of nutrients and minerals in a body of water. The water body loses oxygen as a result of this process. Failure of Sea Animal Reproductive Systems: Chemicals from pesticides can build up in animals’ fatty tissues, causing reproductive system failure (Khan and Law 2005). Impact on the Food Chain: The whole food chain is impacted when huge animals consume little creatures that have ingested the released toxins.

13.3 Ocean Pollutants

The health of our seas, their inhabitants, and those who depend on them for food, culture, and even survival are all being negatively impacted by marine contaminants. Every day, a never-ending tidal flood of waste—especially plastic waste—enters our rivers and the marine environment, along with an ever-increasing cocktail of purposeful and unintentional chemical releases. Figure 13.1 shows some of the sources of marine pollution and their ill impact on marine ecosystem.

13.3.1 Metals

In recent times, the contamination of marine ecosystems due to the presence of heavy metals has emerged as a significant concern both nationally and globally. Nonetheless, the heightened awareness regarding metal pollution in the sea can be traced back to the tragic incident of the ‘Minamata’ disease, which resulted from the consumption of mercury-infected shellfish and fin fish from Minamata Bay in Japan (Harada 1995). Subsequently, the occurrence of the ‘Itaiitai’ disease was observed due to the ingestion of food contaminated with cadmium from Niigata in Japan (Kasuya et al. 1992). “The term ‘heavy metal’ is a broad one, encompassing

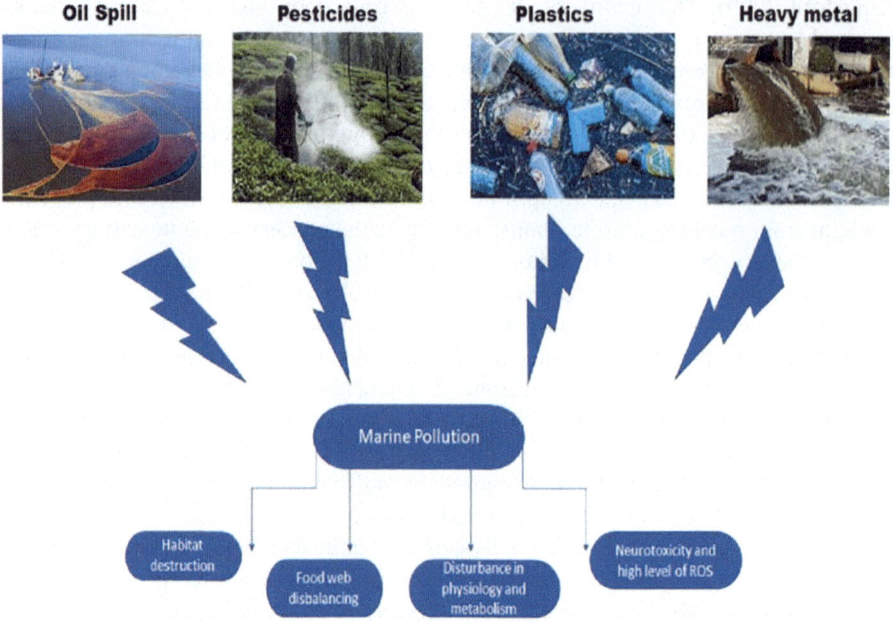

Fig. 13.1 Shows sources of marine pollution and their ill impact on marine ecosystem. (Image source –Belure et al. 2020)

transition metals such as chromium, cobalt, nickel, copper, zinc, cadmium, mercury, lead, arsenic, antimony, and bismuth. Among these, the most pernicious metallic pollutants are mercury, lead, zinc, and copper" (Koller and Saleh 2018). The maritime environment naturally contains heavy metals, which are also introduced into aquatic systems via surface runoff, direct discharge from urban and industrial effluents, and indirect deposition from the air. Certain metals play a crucial role in the normal growth and development of marine organisms, while others can be harmful. The essentiality of an element can be determined based on three criteria: (i) the organism's inability to grow or complete its life cycle without the element, (ii) the irreplaceability of the element by any other, and (iii) the direct impact of the element on the organism and its metabolism (Hewitt 1951). Similarly, an element is considered toxic if it impairs the growth or metabolism of an organism when present in concentrations higher than a certain level. Although all metals are toxic at high concentrations, some can be highly toxic even at lower concentrations. Copper, mercury, lead, cadmium, zinc, and chromium are particularly toxic, with the exception of copper and zinc which are non-essential (Gavhane et al. 2021).

13.3.2 Industrial Wastes

A significant amount of industrial wastewater is discharged into the ocean loaded with s heavy metals like mercury and lead, which are often found in marine life that is consumed by humans. Additionally, the ocean is also contaminated with various chemical and biological contaminants, such as pesticides, pharmaceutical agents and microorganisms. Paper industry releases dioxins, which can cause genetic degradation in marine animals and lead to uncontrolled cell growth in humans (Singh and Chandra 2019). Polychlorinated biphenyls (PCBs)cause reproductive issues in marine fauna (Ngoubeyou et al. 2022). Poly-aromatic hydrocarbons (PAH), which come from oil pollution, as well as burning wood and coal serves as another source (Patel et al. 2020). The ship industry is responsible for significant damage to sensitive marine environments. Passenger ship can produce large amounts of wastewater and sewage that are directly discharged into the ocean along with solid waste which are burned at sea as toxins and ash settle back into the ocean, posing a risk to marine life and seafood contamination (Copeland 2008). Harbors are also polluted with tin, specifically tributyltin (TBT), which is used in antifouling paints on ships. This tin accumulates in shellfish and can be toxic to both the shellfish and consumers like human. (Antizar-Ladislao 2008). When many of these substances move through the web of food to larger longer-lived animals, their concentration increases, a process known as bioaccumulation.

13.3.3 Agricultural Wastes

Nutrient pollution is a significant form of marine contamination resulted by agricultural runoff, containing fertilizers and growth stimulants, as well as nitrogen compounds from automobile exhaust, industrial pollution, and manure ammonia. These factors contribute to nutrient rich conditions in coastal areas. Eutrophication is mainly caused by excessive nitrogen overspill from farm fertilizers, sewage, and industrial pollutants, resulting in reduction of water clarity and dissolved oxygen (Ansari et al. 2011). The diminished water clarity can lead to the starvation of sea grasses and algae living in coral reefs due to insufficient light, ultimately stunting their growth or causing their demise. The algae cannot thrive well due to nutrient depletion resulting in harmful algal blooms. The deceased phytoplankton sinks to the bottom where decomposers consume them. However, the decomposition of algae by these organisms consumes dissolved oxygen, leading to oxygen level decrease. This can be detrimental to the health of fish, and if dissolved oxygen drops below 2 mg/1, it can cause mass fish mortality, known as hypoxia. The areas where hypoxia occurs are referred to as Dead Zones (Islam and Tanaka 2004). As per reports from United Nations Environment Programme reports (2006), "There are currently approximately 150 dead zones worldwide, some of which cover an area as large as 27,000 square miles".

13.3.4 Marine Garbage

The dumping of marine garbage is another important way that ocean pollution manifests itself. The vast oceans serve as a virtual repository for refuse, incorporating items like fishing nets, plastics, and domestic waste. The presence of garbage in marine environments poses a grave concern, as fish become ensnared in fishing nets, and animals occasionally ingest trash items, resulting in their death (Verma et al. 2020). Marine waste often infiltrates animal digestive tracts; animals may unintentionally suffocate on plastic pop tab rings, among other hazards (Verma et al. 2020). The regulation of this type of pollution is vital in order to uphold a salubrious oceanic ecosystem.

13.3.5 Plastics

After being released into the environment for several decades, plastic was once thought to be innocuous, but there are now many related issues. Nowadays, it is generally acknowledged that plastic pollution poses a significant environmental threat especially in the oceans where the degradation of plastics takes longer. Plastics are of varying types based on their chemical composition and environmental impact. The dominant plastics in use are polypropylene (PP), polyethylene (PE), Polyvinyl chloride (PVC), polyurethane (PU), Polyethylene terephthalate (PET), polystyrene (PS), and Polycarbonate (PC) (Schnurr et al. 2018). In addition to their chemical makeup, plastics can also be categorized by size, such as microplastics, mesoplastics, and nano plastics (da Costa et al. 2016; Eriksen et al. 2014). A comprehensive study has compiled information on the presence of micro-, meso-, and macroplastics floating on the ocean's surface. This study combined data from 24 expeditions conducted between 2007 and 2013, which involved a total of 680 net tows for micro- and mesoplastics, as well as survey transects to assess large macroplastic debris (Eriksen et al. 2014).

Recent findings have demonstrated the scope and size of micro(nano)plastic (MnP) pollution, even in the most distant regions of the globe (Allen et al. 2019). The last 10 years have been revolutionary for the study of MnP pollution (Materić et al. 2020; Napper et al. 2020). Originally believed to be primarily a marine problem (Barnes et al. 2009), MnP pollution mostly originates from land-based sources (Jambeck et al. 2015), and its impacts mostly affect marine biota (Gall and Thompson 2015).

The majority of these plastic items were microplastics. The Joint Group of Experts on the Scientific Aspects of Marine Environmental Protection (GESAMP), defines microplastics as 'plastic particles $<$ 5 mm in diameter, which include particles in the nano-size range (1 nm)' (GESAMP 2015, 2016) and it helped further spreading the definition worldwide. Due to their high molecular weight and lack of natural counterparts, conventional plastics do not easily biodegrade in marine or terrestrial environments. Instead, they tend to physically disintegrate when subjected to mechanical forces, like waves or the abrasive action of sediment grains, resulting in smaller plastic fragments. However, this fragmentation does not alter the overall mass of the plastic, only its size distribution. When exposed to UVB radiation from sunlight and oxygen, polymers in plastics can undergo oxidation, leading to polymer chain scission. This process can take several decades to centuries in natural soils (Andrady et al. 2022). In the ocean, these processes occur even more slowly, as the mechanical and photolytic forces are significantly reduced, especially in deeper waters where denser plastics can accumulate (Andrady and Koongolla 2022). Furthermore, when plastic objects enter the ocean, they inevitably become coated with bacteria, algae, other organisms, and sediment, which reduces their surface area exposed to UV radiation and oxygen shortly after entering the marine environment (Zettler et al. 2013). Consequently, most plastic items that reach the marine environment remain intact for centuries and accumulate in a similar manner as persistent organic pollutants (POPs). The most commonly observed impacts of plastic pollution are entanglement and ingestion of macroplastic debris. According to the Secretariat of the Convention on Biological Diversity, a significant proportion of mammal, turtle, and bird species ingest plastic (40%, 100%, and 46% of species in these three taxa, respectively) or become entangled in it (46%, 100%, and 26% respectively) (Harding 2016). Research has shown that plastic components interfere with feeding and reproduction in oysters, resulting in negative effects on their ability to produce offspring and their overall fitness (Sussarellu et al. 2016). Similarly, when exposed to polystyrene (PS) microplastics at concentrations found in coastal environments in the Baltic Sea, fertilized perch eggs experience reduced hatching success, and larvae from these waters exhibit various negative effects, including delayed development, increased mortality, and compromised predator avoidance behaviours (Lönnstedt and Eklöv 2016).

13.3.6 Oil

Oil pollution is the primary form of contamination in the sea, and it is deserving of extensive discussion. Various sources contribute to the introduction of oil into the sea. The sources of oil contamination in the marine environment can be categorized as natural, sea-based, and land-based, with four primary types of pollutant discharges: natural seepage, discharges during oil extraction, discharges during oil transportation, and discharges during oil consumption (including both sea-based and land-based sources). These sources result in an estimated average of 706 million gallons of oil pollution entering the oceans each year (Samwel John 2012). The oil disperses over vast areas and continues to cause harm for extended periods. Animals cannot stay warm when there is enough surface oil to cover their fur and feathers, and they run the risk of ingesting the hazardous oil while trying to clean themselves (Thakur and Koul 2022). As a result, poisoning causes many of these oiled animals to die. After ingesting oil, marine life that survives the initial effects of the spill may develop liver illness and have issues with development and reproduction. Thin layers of oil can form when even tiny amounts of oil spread and cover large regions of water. The population of marine animals is reduced as a result of these thin oil sheets endangering marine larvae.

13.3.7 Pesticides

Pesticides that are still in use today contain various constituents that have the potential to be detrimental to marine organisms. Among these constituents are the active ingredients, formulating chemicals like surfactants, as well as impurities and byproducts. Pesticides can enter the marine environment through a variety of channels, such as rivers and streams, sewage and storm water systems, direct runoff, and vapor and spray drift from forestry, aquaculture, roadside vegetation maintenance, and residential activities. Pesticides can have a range of consequences on aquatic life, including cancer, tumors, lesions, inhibition and failure of reproduction, immune system suppression, endocrine system interference, cellular and DNA damage, and death (Gill and Garg 2014). Furthermore, pesticides can induce behavioral changes in marine organisms, which can have implications for their survival, population dynamics, and overall ecosystem equilibrium (Ashauer et al. 2006). Comparative analysis of the toxicity of pyrethroid insecticides and organophosphate on aquatic arthropods showed that pyrethroid insecticides exhibit higher levels of toxicity compared to organophosphates (Halstead et al. 2015). Figure 13.2 shows key Facts and Trends on Ocean Pollution, Fig. 13.3 shows Countries with highest ocean pollution as per reports of the year 2024).

13.4 Green Nanotechnology for Ocean Pollution Control

Conventional methods for eradicating marine pollutants include a variety of approaches intended at minimizing, removing, or mitigating the impact of pollutants in marine habitats. These techniques can be divided into a few major categories like Source Reduction, Cleanup and remediation, Bioremediation and Physical Barriers. Source reduction entails keeping contaminants from entering marine habitats in the first place. Enforcing stronger industrial discharge restrictions, strengthening waste management procedures, encouraging the adoption of environmentally friendly products, and minimizing single-use plastics (Jay and Meegoda 2023). As Aliko et al. 2022 reports clean-up and remediation includes methods which physically eliminating pollutants from marine ecosystems. They include human clean-up activities with nets, skimmers, and booms to gather floating waste, as well as dredging contaminated sediments and treating polluted water.

Bioremediation is the technique of removing or degrading contaminants from marine environments using natural biological processes. Methods, such as

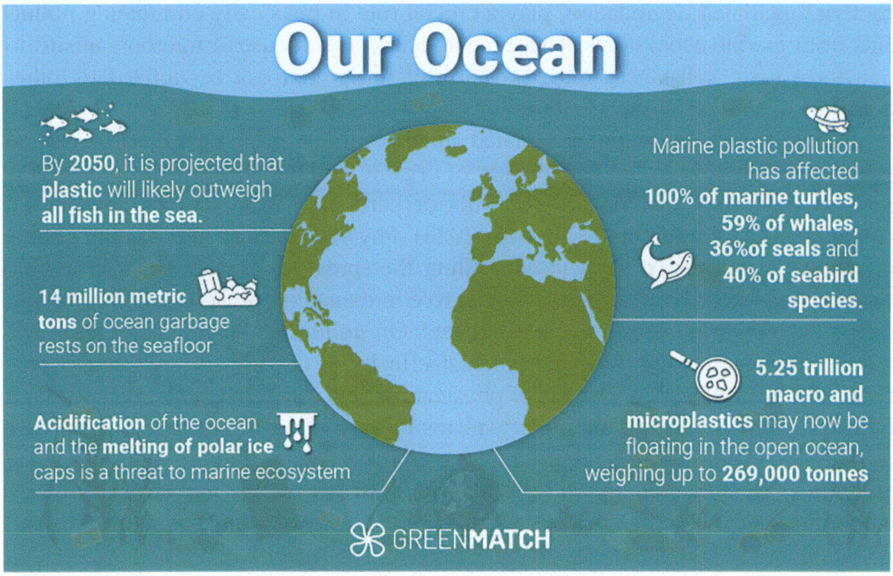

Fig. 13.2 Key Facts and Trends on Ocean Pollution (2024). (Image source-https://www.greenmatch.co.uk/ocean-pollution-facts)

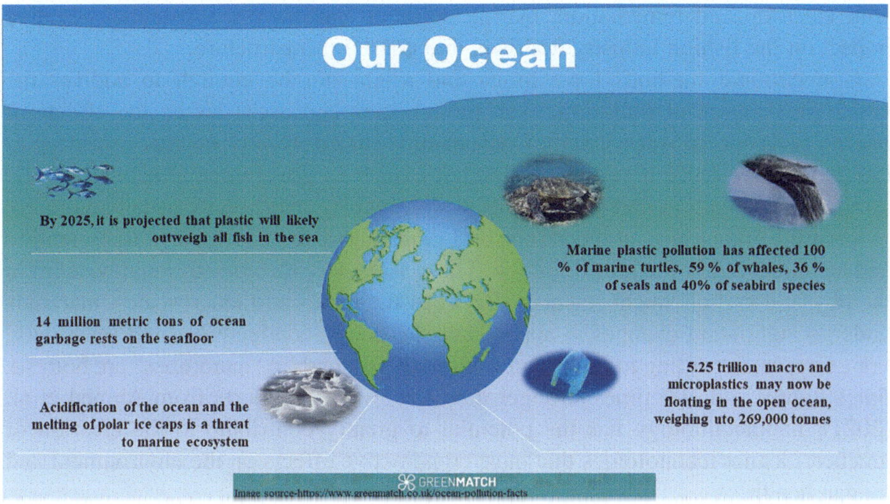

Fig. 13.3 Countries with highest ocean pollution (2024). (Image source-https://www.greenmatch.co.uk/ocean-pollution-facts)

bacterial-mediated degradation, play a crucial role in physically eliminating pollutants from marine ecosystems. These methods involve the use of microorganisms to transform toxic substances into less harmful forms that can be utilized by other organisms, aiding in the clean-up of contaminated areas (Tirkey et al. 2021). Additionally, seaweeds and associated microbes have been identified as natural agents that can help in the remediation of organic pollutants in aquatic ecosystems such as oil-eating bacteria for oil spills or particular algae species for nutrient contamination (Varabih and Nofirman 2023). Physical barriers like floating booms or screens are effective tools in controlling the spread of contaminants in marine ecosystems. These barriers play a crucial role in managing oil spills and preventing debris from entering sensitive environments. Oil pollution, a significant concern in marine ecosystems, can have detrimental effects on aquatic life and ecosystems (Thakur and Koul 2022). The use of physical barriers can help contain oil spills and minimize their impact on marine organisms by restricting the spread of oil (Wang et al. 2022). Additionally, these barriers can aid in keeping debris out of delicate environments, thus reducing the potential harm to marine life and ecosystems. By employing such physical barriers strategically, the risks associated with oil spills and pollution can be mitigated effectively.

The effects of marine pollution and the efficacy of these traditional methods vary based on a number of variables, including the kind and quantity of pollutants involved, the features of the marine environment, and the success of the selected mitigation strategies. Few effects of marine pollution include ecosystem degradation, human health risk and economic impacts. In addition to expenditures associated with clean-up and remediation, marine pollution can have a substantial negative impact on the fishing industry, tourism, and coastal infrastructure.

Conventional methods have limits and might not be enough to address the underlying causes of pollution, even though they can assist lessen the effects of marine pollution to some extent. Consequently, to effectively manage marine pollution and its effects, a variety of strategies, such as governmental actions, public awareness campaigns, and investments in sustainable practices, are usually required.

Creating "materials" at the atomic and molecular size is the goal of the scientific and technical field of nanotechnology. These sizes defy the conventional wisdom in the physical and chemical sciences. Between the nano- and macroscales, materials undergo significant changes in appearance, hardness, conductivity, and sensitivity. For example, according to Janas and Koziol (2016), carbon "nanotubes" are both six times lighter and 100 times stronger than steel. As per reports from Ahmed et al. (2021); nanotechnology has the potential to greatly contribute to the creation of greener, cleaner technologies that have constructive effects on the environment and human health.

Green nanotechnology refers to the wise use of nanotechnology to improve the environmental sustainability of processes (Singh et al. 2024). It involves creating environmentally friendly nano products as well as utilizing them to promote sustainability. In the term Green Nanotechnology, the word "green" has two meanings. It speaks of the environmentally benign methods that are used to create nanoparticles, as well as the production of nanoparticles that is facilitated by organic materials. “The creation of clean technologies to minimize potential environmental and human health risks associated with the manufacture and use of nanotechnology products” is the definition of green nanotechnology. Additionally, it promotes the substitution of new nanoproducts that are more ecologically friendly throughout the course of their lifetimes for old products.

Green nanotechnology emphasizes efficiency, sustainability, and environmental safety in remediation activities, providing a range of tools and solutions to address the problems associated with ocean pollution. It takes lots of years for contamination in the air, water, and soil to be eliminated in the natural world. Nanotechnology aid to clear up the marine habitat in a variety of ways. There are several ways that nanotechnology can contribute to mitigation of water pollution. Marlon et al. (2019) reported that procedures based on nanotechnology are being researched for their potential to help control and mitigate pollution as well as enhance the effectiveness of conventional environmental clean-up procedures. By lowering energy consumption throughout the production and manufacturing processes, enabling items to be recycled after use, and creating and utilizing ecologically friendly materials, nanotechnology can benefit the environment (Kaur and Roy 2021). Several techniques (such as surface assimilation, absorption, synthetic processes, light catalysed reactions, and filtering) are used in the environmental remediation process to remove environmental pollutants (Jiang et al. 2018). Compared to other macroscale materials, nanomaterials often exhibit higher chemical and bio geological reactivity due to their high ratio of surface area to volume. Because of their improved features and efficacy, they are particularly well-suited for carrying out these treatments (Guerra et al. 2018). By lowering energy consumption throughout the production and manufacturing processes, enabling products to be recycled after use, and creating and utilizing ecologically friendly materials, nanotechnology can benefit the environment (Kaur and Roy 2021). According to Khan (2020), nanotechnology now offers a lot of promise for solving sustainability-related problems, but we also need to consider the risks to human health and the environment.

The following are some techniques related to controlling ocean pollution with green nanotechnology:

13.5 Nanomaterials for Water Treatment

13.5.1 Nanoscale Sorbents

Utilizing nanomaterials with high surface area, such as carbon-based nanomaterials (graphene, carbon nanotubes), for adsorption of pollutants in water. Materials with the capacity to adsorb or absorb contaminants at the nanoscale are known as nanoscale sorbents. “Due to high sorption capabilities, nanosorbents are excellent materials for water treatment” as per Montemagno et al. 2007. Nanoscale sorbents can be useful in the context of ocean water pollution in eliminating different types of toxins from the water (Lee et al. 2023).

The following are some applications for nanoscale sorbents: Nanoscale sorbents, like hydrophobic nanoparticles or nano sponges, can be used to clean up oil spills in marine environments (Zhang and Seeger 2011; Wu et al. 2014). These substances can selectively absorb hydrophobic impurities like oil, leaving behind relatively pure water. These materials have the ability to absorb or encase oil selectively, aiding in its recovery and reducing the detrimental impact of oil spills on marine environments either adsorption or absorption is how these sorbents function. Absorbents, which work similarly to sponges, use suction or capillary action to absorb less viscous oils. Adsorbents function best with heavy, sticky oils due to their vast surface area, high porosity, and strong chemical affinity for spilled oil. Certain sorbents have the ability to sequester oil using both methods. Shang et al. (2016), proposed recently wide range of methods based on sorbent materials as potential remedies for pollution caused by oil spills. These absorbing or adsorbing materials are typically readily available, reasonably priced, and environmentally benign, according to Yang et al. (2015). Sorbents are essential for extracting oil from water because they have a high regeneration capacity and allow the oil to be recovered and utilized again. Dispersing agents, gelling agents, and biological agents are most frequently employed to clean up oil spills that are ineffective in retrieving the lost oil in order to effectively remove oil spills from marine ecosystems. Oil and other organic pollutants can be cleaned up while facilitating oil recovery thanks to recent advancements in potassium manganese oxide nanowire technology (Yuan et al. 2008). Because of their water-repellent covering, these nanowires form a mesh that can absorb hydrophobic liquids up to 20 times its weight while rejecting water. It is possible to boil off the oil from the nanowires and reuse both the oil and the nanowires because potassium manganese oxide is extremely stable even at high temperatures (Yuan et al. 2008). Using water-repellent nanowires, Interface Science Corporation successfully introduced a new oil remediation and recovery application to clean up oil spills from damaged oil platforms and refineries (Lamba 2005).

Materials at the nanoscale, like altered nanoparticles or nano clays, can be made that are particularly designed to absorb heavy metals from ocean water. A particular kind of nano-absorbent material that can be utilized to eliminate a variety of contaminants is called nano clays (NC). Made of mineral silicates coated in nanoscale layers, nano clays represent a novel class of adsorbents that are inexpensive, non-toxic, enduring, and possess a great surface reactivity (Abdel-Fadeel et al. 2022). Organic dye elimination and heavy metals from a variety of aquatic habitats has sparked interest in nano clays, such as Halloysite Nano clay (Novikau and Lujaniene 2022; Aigbe et al. 2022). Eliminating heavy metals from the water is essential since they can be dangerous to both human health and marine life.

Current methods for removing nano plastics from potable water include nanotechnology-based strategies like adsorption, photocatalysis, and membrane filtration(Bhore and Kamble 2022). Physicochemical separation processes such as adsorption and magnetic extraction have shown high efficiency in removing microplastics and nano plastics from water sources (Ouda et al. 2023). Specifically, the use of Fe3O4 as nanotricomposites has demonstrated nearly complete removal of diverse microplastics from seawater, regardless of their sizes and shapes, making the process environmentally benign (Urso and Pumera 2022). Additionally, electrocoagulation has been proven effective in removing nano plastics from synthetic wastewater, achieving a removal efficiency of over 95% under specific electrolyte concentrations and pH conditions. These methods offer promising solutions to combat the emerging issue of nano plastic pollution in potable water sources. Promising methods for eliminating microplastics (MPs) from the ocean are provided by green nanotechnology. Green nanotechnology addresses the worldwide problem of nano/microplastics pollution by offering a sustainable solution to remove MPs pollution from the water, safeguarding both human health and marine life. It achieves this by fusing bionanomaterials, nanotechnology, and state-of-the-art carbocatalysis. It is possible to create nanoscale sorbents that specifically target microplastics in the water. By drawing in and absorbing microplastic particles, these substances can contribute to the rising problem of plastic pollution in marine settings. Nanomaterials, such as carbon nanotubes and bionanomaterials, have shown great potential in efficiently degrading and adsorbing MPs (Kamble and Bhore 2023). These nanoparticles are useful in reducing MP pollution in water systems because of their special qualities like adjustable surface charges (Chellasamy et al. 2022). Furthermore, studies on self-propelled nano/microrobots for the purpose of ensnaring and breaking down nano/microplastics have demonstrated sophisticated approaches to water purification (Urso and Pumera 2022).

The removal of excess ocean nutrients, especially phosphorus and nitrogen, which cause eutrophication, has been demonstrated to be possible with nanoscale sorbents (Othman et al. 2018). Engineered nanoparticles have the ability to adsorbed these nutrients, reducing their harmful effects on marine ecosystems. The overabundance of nutrients released by human activities has resulted in environmental imbalances that have an effect on the sustainability and health of ecosystems (Aguilar-Pérez et al. 2022). Low amounts of nitrate and phosphate can be effectively adsorbed from water bodies by nano adsorbents, which include carbon-based materials, metal oxides/hydroxides, and zero-valent metals. This provides an affordable option for nutrient removal (Priya et al. 2022). By collecting, eliminating, and even recycling phosphorus-containing molecules from water, these nanomaterial sorbents can help improve the utilization of phosphorus.

The detrimental effects that pharmaceutical residues and agricultural runoff have on marine environments can be mitigated by using nanoscale sorbents that precisely absorb pesticides. A major environmental problem is the removal of pesticides and pharmaceuticals from saltwater because of their detrimental effects on marine ecosystems. Pharmaceuticals, primarily from brine and wastewater discharges, have been found in seawater at quantifiable amounts. These pharmaceuticals include antidepressants, antibiotics, and anti-inflammatory ones((Durán-Álvarez et al. 2021). Conventional wastewater treatments are ineffective at eliminating these pollutants, which present a risk to non-target organisms (Hejna et al. 2022). Green nanotechnology provides long-term ways to rid water bodies of pollutants including pesticides and medications (Bouafia et al. 2023). Greenly produced nanoparticles have demonstrated excellent efficacy in eliminating chemical compounds and heavy metals from water (Patanjali et al. 2019). Due to their high surface-to-mass ratio and extraordinary reaction rate, nanomaterials are useful for treating water since they offer affordable and environmentally acceptable alternatives (Fukumoto et al. 2022). Green nanotechnology can solve the problems of solubility and long-term toxicological effects related to nanoparticles in medicine by using environmentally friendly nanomaterials. Thus, using green nanotechnology to remove chemicals, medications, and other impurities from seawater effectively shows potential.

13.5.2 Nanoparticles for Catalysis

Green nanoparticles, which are produced from natural resources like plants and algae, present a promising way to accelerate the breakdown of ocean pollutants. These nanoparticles have effective photocatalytic qualities for pollution remediation in water bodies, particularly metal oxide nanoparticles (MONPs) and silver nanoparticles (AgNPs). Algal-mediated MONP synthesis provides an affordable substitute for bioremediation since it is quick, scalable, and ecologically benign (Ahmed et al. 2023a, b). According to Stephen et al. (2023), biosynthesized AgNPs utilizing plant extracts have demonstrated improved catalytic potential for breaking down a variety of environmental contaminants, including those present in industrial effluents.

Due to their eco-friendliness, affordability, and great recyclability, green nanomaterials have become more and more popular as photocatalysts for the degradation of organic contaminants in water (Ahmed et al. 2023b). These environmentally friendly nanoparticles can be extremely helpful in tackling the growing problem of ocean pollution by accelerating the breakdown of dangerous pollutants.

Nanozymes are materials which have the properties of an enzyme (Wei and Wang 2013). They were created in response to the shortcomings of both conventionally produced artificial enzymes and natural enzymes. To clean up oil spills, scientists created nanozymes—nanomaterials having characteristics similar to those of enzymes. These nanozymes convert oil contaminants into non-toxic molecules by acting like natural enzymes (Solomon et al. 2023). These nanozymes are perfect for treating oil-contaminated water and land because of their great bioremediation capabilities, quick destruction of hazardous pollutants, affordability, and specificity. Furthermore, a useful strategy for cleaning up oil spills is the creation of better functional enzyme mimics, or nanozymes, with dynamic catalytic activity in response to outside stimuli (Solra et al. 2022).

Additionally, the use of nanofluids and nano catalysts to increase oil recovery, control oil pollution, and improve the efficiency of oil refining is one way that nanotechnology is being applied in the oil sector (Chauhan et al. 2023). Effective oil spill cleanup can be achieved by utilizing the distinct physicochemical properties of nanoparticles, nanozymes, and nanomaterials. This approach can be both environmentally sustainable and all-encompassing. The effective breakdown of crude oil into biodegradable compounds has been demonstrated by nanozymes. Studies have demonstrated the possibility of using nanomaterials possessing enzyme-like characteristics, including zinc oxide and Fe3O4 nanoparticles, to quicken the biodegradation of crude oil (Rangel-Muñoz et al. 2020; Ehmedan et al. 2021).

These nanozymes provide a versatile platform for environmental remediation because it has been discovered that they imitate the catalytic activity of several other enzymes, such as peroxidase, catalase, and superoxide dismutase. The biodegradation process was greatly accelerated by mixing specific bacterial isolates with Fe2O3 and ZnO nanoparticles, resulting in up to 98% breakdown of crude oil in 3 days (Ehmedan et al. 2021). This novel approach of combining nanoparticles with bacterial species shows how nanozymes may efficiently break down crude oil into biodegradable byproducts, which has potential applications in in situ bioremediation of petroleum contamination. An environmentally friendly substitute for chemical dispersants and mechanical oil spill cleanup techniques is the nanozyme-based strategy.

The efficiency of nanozymes is indeed influenced by various parameters such as the oxidative state of metal ions, pH levels, and other factors, making tuning challenging outside the lab. To address this, researchers are exploring the use of sunshine as a stimulus for nanozymes, aiming for effective and sustainable activation (Waris et al. 2023). Machine learning techniques are being leveraged to guide the rational design of nanozymes, aiding in predicting activity, selectivity, and optimal structures, which could contribute to making nanozymes more cost-effective and easier to use (Li et al. 2023). Efforts in these directions hold potential for creating efficient, accessible, and affordable nanozymes for diverse applications in the future.

13.5.3 Nanostructured Membranes

Green nanotechnology offers innovative solutions for sea water filtration membranes to effectively remove pollutants ((Luque et al. 2023; Khalith et al. 2022; Ahlam et al. 2021). Combining nanomaterials with nanoparticle filter membranes offers a viable way to lessen the harmful impacts of MPs pollution in saltwater. By incorporating nanomaterials like carbonaceous nanomaterials, metal oxide nanoparticles, zeolites, and cellulose nanofibers, these membranes exhibit enhanced performance in terms of pollutant removal, water permeability, and antifouling properties(Kumar et al. 2021; Palacios-Hinestroza et al. 2020. The utilization of nanotechnology in membrane development addresses the challenges of traditional water purification methods by providing cost-effective, energy-efficient, and sustainable solutions. Nanofiltration membranes modified with nanomaterials show promising results in desalination processes, offering high efficiency in removing pollutants from sea water while minimizing production costs. Overall, the integration of green nanotechnology in sea water filtration membranes presents a viable and eco-friendly approach to combat water pollution effectively.

Nanoparticle filtration membranes play a crucial role in removing microplastics (MPs) from seawater (Kamble and Bhore 2023; Malankowska et al. 2021; Jenkins2023). These membranes, such as nanofiltration membranes, offer selective extraction of ions and organic matter, making them effective in water and wastewater treatment processes. Nanomaterials enhance the performance of water treatment processes by adsorption, photocatalysis, and membrane filtration, all of which are vital for combating challenging micro- and nano plastic pollutants (Pizzichetti et al. 2021). Studies have shown that nanomaterials possess different functionalities, making them efficient in removing MPs from aquatic environments. The use of nanoparticle filtration membranes, along with nanomaterials, presents a promising solution to mitigate the adverse effects of MPs pollution in seawater.

13.5.4 Nanosensors

Ocean pollution levels can be tracked in real time by deploying sensors based on nanotechnology. These sensors are capable of detecting a wide range of pollutants, such as hydrocarbons, heavy metals, and poisons, and they can provide useful information for pollution control and response initiatives. By creating eco-friendly nanomaterials, chemical pollutants in marine waters can be reduced. Green nanotechnology, green chemistry, and green engineering provide creative ways to identify and eliminate marine pollutants effectively and safely without endangering the environment or public health (Corsi et al. 2021a, b). According to Ikbal and Choudhury (2023), these nano sensors can be extremely helpful in tracking and mitigating the growing marine pollution that results from industrial processes and human activity. These nano sensors offer ultra-sensitive and selective contamination detection because they are made of nanomaterials like graphene, quantum dots, and gold nanoparticles (Gavela et al. 2019). Furthermore, Çimen et al. (2023) have shown that plasmonicnanosensors based on nanomaterials have demonstrated exceptional sensitivity in detecting pesticides, heavy metal ions, microorganisms, and toxins in water.

Green nano sensors with their high sensitivity, specificity, and environmentally friendly nature, have become highly effective tools for monitoring oceanic harmful algae blooms, or HABs. These nano biosensors which have numerous uses in a variety of industries, including environmental monitoring, make use of nanomaterials produced using green nanotechnology (Antonacci et al. 2023). Effective HAB monitoring is made possible by nano sensors that incorporate biological elements, such as DNA probes, and are made to detect various algal species quickly and sensitively (Noah and Ndangili 2022). Furthermore, in comparison to traditional techniques, biosensors employing nanomaterials have demonstrated improved detection limits and wider linear ranges, indicating their potential for the identification of hazardous algal species in marine environments (Li et al. 2022). Incorporating green nanotechnology into biosensors improves their capacity for detection while also supporting environmentally friendly methods of environmental management and observation. The prompt implementation of response measures, like beach closures and seafood advisories, to safeguard human health and marine ecosystems was made possible by the early discovery of HABs.

Thanks to developments in nanotechnology, inexpensive, small, and environmentally friendly sensors that can identify pollutants at nanomolar to picomolar levels have been developed (Aguilar-Pérez et al. 2020). Nanotechnology makes it possible to create biosensors with fast detection times and wide detection limits, which are essential for real-time monitoring and analysis of contaminated water. This is achieved by employing green synthesis processes.

13.6 Limitations of Green Nanotechnology Solutions

Green nanotechnology is a branch of green technology that utilizes the concepts of green chemistry and green engineering. The ultimate goal of any scientific development is to increase well-being and human health. As we go toward clean and green technology development, society will need new standards for new technologies and novel tactics will be needed to accomplish safe and effective therapeutic treatments beyond the current ones. It is imperative that appropriate technology be used and properly managed to address global societal issues. While there is now much potential in using nanotechnology to address sustainability challenges, we also need to consider the risks to human health and the environment (Khan 2020).

The small size of nanoparticles warrants investigation of the consequences of inhalation and absorption of these particles and their effects inside the body, as they tend to penetrate the skin and diffuse through cell membranes. Xuan et al. (2023) reported that the special properties of nanoparticles inside the body and on environment are unclear and unpredictable. Since nanoparticles are small enough to pass through cell membranes and permeate the skin, it is necessary to investigate the impacts of inhaling and absorbing them as well as their effects inside the body. In order to shed light on the potential harm that the widespread presence of nanoparticles in the environment would cause to the ecosystem, new scientific disciplines like eco-nano toxicology have emerged to investigate the flow of nanomaterials through the biosphere. For preventing nanoparticles becoming new sources of pollution, despite their enormous potential for pollution removal, regulations and policy-making are needed. (Ray et al. 2009).

Risk Assessment and life cycle analysis of nanoparticles used in pollution control need to be assessed before wide spread application in marine ecosystem. Studies have extensively examined the environmental impact and potential toxicity of nanomaterials used in pollution control, particularly in ocean settings. As per Ahamed et al. (2019), Kumah et al. (2023), nanoparticles such as zinc oxide, silicon dioxide, titanium dioxide, silver, and carbon nanotubes, have been found to induce cell death, oxidative stress, DNA damage, apoptosis, and inflammatory responses in various biological models. The release of nanomaterials into the environment due to widespread nanotechnology use poses significant risks to living organisms and their habitats, leading to DNA damage, mitochondrial damage, and cell death in sensitive organs like the kidneys, lungs, and liver (El-Kalliny et al. 2023). Additionally, the fate, impact, toxicity, and risk of nanomaterial contaminants in the environment remain areas of marked knowledge gaps, necessitating the development of detection methods and regulation strategies (Lechuga et al. 2022).

Green synthesis has advantages, but there are drawbacks as well. One such limitation is its possible toxicity. Depending on their size, shape, and chemical makeup, nanoparticles generated using "green" synthesis techniques may nonetheless be hazardous to marine life, with large-scale discharge having the ability to disrupt the entire food chain as per reports of Guo et al. 2019; Solař et al. 2023; Gupta 2023. Furthermore, after synthesis, nanoparticles have a propensity to aggregate, which can lower their catalytic activity (Khadijah and Altammar 2023). Reports regarding the aggregation and stability by Ray et al. (2009) which depicts that nanomaterials can cluster or clump together in the ocean environment, decreasing their efficacy in absorbing pollutants and making it difficult to control their distribution. Another challenges is the evaluation of environmental impact of nanomaterials in the ocean where precise monitoring and detection are essential, but the available methods might not be sensitive enoughor generally accessible. Regarding regulatory obstacles, as reported by Amoabediny et al. (2009); a clear regulatory frameworks for the use of nanomaterials in ocean remediation are still developing, generating ambiguities for future uses.

13.7 Conclusion and Prospects

The potential and promise of nanotechnology have been increasing throughout time. Although there is now a lot of promise in green nanotechnology to address sustainability issues, it is important to remember that nanomaterials have harmful effects on both human and environmental health. While there are still many challenges to be solved before this technology is widely used, science is always improving, evolving, and producing new discoveries. Applying green nanotechnology principles to ocean pollution control not only aims to address immediate environmental challenges but also seeks to do so in a sustainable and ecologically conscious manner. These strategies must be regularly evaluated and improved upon in order to guarantee their long-term efficacy and reduce any possible negative impacts on marine ecosystems.

Although techniques for pollution mitigation and nanoscale bioremediation provide exciting new opportunities, it's important to keep in mind that any potential impacts on the environment and public health should be thoroughly investigated and taken into consideration. Furthermore, in order to avoid unintended consequences, these technologies must be implemented responsibly. Utilizing green nanotechnology in ocean pollution control has the potential to reduce ocean pollution, but before it is widely applied in marine ecosystems, risk assessment and life cycle analysis of the nanoparticles employed in pollution control must be evaluated. As the field of nanotechnology for environmental remediation advances, more research is required to assess the efficacy and safety of these techniques.

Green nanotechnology, with its eco-friendly and sustainable approaches, holds great potential for addressing ocean pollution control. Some key prospects and future research directions in this field are (i) Create biodegradable nanostructures that can selectively capture and neutralize microplastics without harming marine organisms. (ii) Engineer eco-friendly nanomaterials that degrade after use, minimizing environmental impact. (iii) Focus on enhancing the longevity and reusability of these nanomaterials, ensuring that they can effectively handle large-scale applications. (iv) Investigate sustainable, less toxic alternatives to traditional antifouling methods, using naturally derived materials. (v) Develop hybrid materials combining nanotechnology with other eco-friendly approaches, such as biodegradable polymers, to create multifunctional systems that can tackle complex pollutants in ocean waters. (vi) Investigate interactions between nanomaterials and marine organisms at the cellular level to assess bioaccumulation risks and encourage extensive long-term research on the environmental impacts of nanomaterial use in oceans, ensuring that they do not disrupt marine ecosystems or contribute to new forms of pollution over time. By focusing on these prospects and research directions, green nanotechnology can play a pivotal role in reducing ocean pollution and protecting marine ecosystems in a sustainable way.

References

Abdel-Fadeel MA, Aljohani NS, Al-Mhyawi SR, Halawani RF, Aljuhani EH, Salam MA (2022) A simple method for removal of toxic dyes such as Brilliant Green and Acid Red from the aquatic environment using Halloysite nanoclay. Journal of Saudi Chemical Society 26:101475. https://doi.org/10.1016/j.jscs.2022.101475

Aigbe UO, Ukhurebor KE, Onyancha RB, et al. (2022) A facile review on the sorption of heavy metals and dyes using bionanocomposites. Adsorption Science & Technology. https://doi.org/10.1155/2022/8030175

Aguilar-Pérez KM, Heya MS, Parra-Saldívar R, Iqbal HM (2020) Nano-biomaterials in-focus as sensing/detection cues for environmental pollutants. Case Stud Chem Environ Eng 2:100055. https://doi.org/10.1016/j.cscee.2020.100055

Aguilar-Pérez KM, Ruiz-Pulido G, Medina DI, Parra-Saldívar R, Nazish N, Bilal M, Iqbal HM (2022) Nanoadsorbents for environmental remediation of polluting agents. Elsevier eBooks, pp 227–239. https://doi.org/10.1016/b978-0-323-90912-9.00010-1

Ahamed M, Akhtar MJ, Khan MM et al (2019) Oxidative stress mediated cytotoxicity and apoptosis response of bismuth oxide (Bi2O3) nanoparticles in human breast cancer (MCF-7) cells. Chemosphere 216:823–831

Ahlam FM, Alsayed M, Aqeel A (2021) Modified nanofiltration membrane treatment of saline water, pp 25–44. https://doi.org/10.1016/B978-0-12-820644-7.00005-0

Ahmed HM, Roy A, Wahab M, Ahmed M et al (2021) Applications of nanomaterials in agrifood and pharmaceutical industry. J Nanomater 2021(1):1472096

Ahmed E, Alprol A, Mansour T, Abdelwahab A, Mohamed L (2023a) Advances in green synthesis of metal oxide nanoparticles by marine algae for wastewater treatment by adsorption and photocatalysis techniques. Catalysts 13:888–888. https://doi.org/10.3390/catal13050888

Ahmed S, Ara G, Susan MABH (2023b) Green nanomaterials for photocatalytic degradation of toxic organic compounds. Curr Pharm Biotechnol 24(1):118–144

Aliko V, Multisanti CR, Turani B, Faggio C (2022) Get rid of marine pollution: bioremediation an innovative, attractive, and successful cleaning strategy. Sustain For 14(18):11784. https://doi.org/10.3390/su141811784

Allen S, Allen D, Phoenix VR, Duranteza SA, Stéphane B, Galop D (2019) Atmospheric transport and deposition of microplastics in a remote mountain catchment. Nat Geosci 12:339–344

Amoabediny GH, Naderi A, Malakootikhah J, Koohi MK, Mortazavi SA, Naderi M, Rashedi H (2009) Guidelines for safe handling, use and disposal of nanoparticles. In: Journal of physics: conference series 170: 012037. IOP Publishing

Andrady AL, Koongolla B (2022) Degradation and fragmentation of microplastics. In: Plastics and the ocean: origin, characterization, fate, and impacts, pp 227–268. https://doi.org/10.1002/9781119768432.ch8

Andrady AL, Barnes PW, Bornman JF, Gouin T, Madronich S, White CC, Jansen MAK (2022) Oxidation and fragmentation of plastics in a changing environment; from UV-radiation to biological degradation. Sci Total Environ 851:158022. https://doi.org/10.1016/j.scitotenv.2022.158022

Ansari AA, Gill SS, Khan FA (2011) Eutrophication: threat to aquatic ecosystems. In: Eutrophication: causes, consequences and control. Springer, pp 143–170

Antizar-Ladislao B (2008) Environmental levels, toxicity and human exposure to tributyltin (TBT)-contaminated marine environment. A review. Environ Int 34:292–308. https://doi.org/10.1016/j.envint.2007.09.005

Antonacci A, Frisulli V, Carvalho LB, Fraceto LF, Miranda B, De Stefano L, Johanningmeier U, Giardi MT, Scognamiglio V (2023) An all-green photo-electrochemical biosensor using microalgae immobilized on eco-designed lignin-based screen-printed electrodes to detect sustainable nanoherbicides. Int J Mol Sci 24(12):10088. https://doi.org/10.3390/ijms241210088

Appannagari RR (2017) Environmental pollution causes and consequences: a study. North Asian Int Res J Soc Sci Humanit 3(8):151–161

Ashauer R, Boxall A, Brown C (2006) Predicting effects on aquatic organisms from fluctuating or pulsed exposure to pesticides. Environ Toxicol Chem Int J 25(7):1899–1912

Barnes DK, Galgani F, Thompson RC, Barlaz M (2009) Accumulation and fragmentation of plastic debris in global environments. Philos Trans R Soc B Biol Sci 364(1526):1985–1998

Belure A, Rachana A, Teja J, Vishal C, Prabhaka M (2020) Distinctive application of various computational approaches towards remediation of pollutants persisting in marine ecosystem. Special issue of NCCWLS 2019 www.rjlbpcs.com Life Science Informatics Publications, https://doi.org/10.26479/2019.0501.72

Bhore RK, Kamble SB (2022) Nano adsorptive extraction of diverse microplastics from the potable and seawater using organo-polyoxometalate magnetic nanotricomposites. J Environ Chem Eng 10(6):108720. https://doi.org/10.1016/j.jece.2022.108720

Bouafia A, Meneceur S, Chami S, Laouini SE, Daoudi H, Legmairi S, Mohammed HA, Aoun N, Menaa F (2023) Removal of hydrocarbons and heavy metals from petroleum water by modern green nanotechnology methods. Sci Rep 13(1). https://doi.org/10.1038/s41598-023-32938-1

Breitburg D, Levin LA, Oschlies A, Grégoire M, Chavez FP, Conley DJ, Zhang J (2018) Declining oxygen in the global ocean and coastal waters. Science 359(6371):eaam7240

Brierley AS, Kingsford MJ (2009) Impacts of climate change on marine organisms and ecosystems. Curr Biol 19(14):R602–R614

Chauhan P, Imam A, Kanaujia PK, Suman SK (2023) Nano-bioremediation: an eco-friendly and effective step towards petroleum hydrocarbon removal from environment. Environ Res 231: 116224

Chellasamy G, Kiriyanthan RM, Maharajan T, Radha A, Yun K (2022) Remediation of microplastics using bionanomaterials: a review. Environ Res 208:112724. https://doi.org/10.1016/j.envres.2022.112724

Çimen D, Bereli N, DenïZli A (2023) Advanced plasmonic nanosensors for monitoring of environmental pollutants. Curr Anal Chem 19(1):2–17. https://doi.org/10.2174/1573411018666220618155324

Copeland C (2008) Cruise ship pollution: background, laws and regulations, and key issues. In: CRS report for congress, pp 1–30

Corsi I, Bellingeri A, Battocchio C et al (2021a) Ecosafe Nano-based solutions for pollution monitoring and control in the marine environment. In: In 2021 international workshop on metrology for the sea; learning to measure sea health parameters (MetroSea). IEEE, pp 16–20

Corsi A, Bellingeri C, Battocchio A, Fiorati I, Venditti Punta C (2021b) Ecosafe Nano-based solutions for pollution monitoring and control in the marine environment. In: International workshop on metrology for the sea; learning to measure sea health parameters (MetroSea), Reggio Calabria, Italy, vol 2021, pp 16–20. https://doi.org/10.1109/MetroSea52177.2021.9611624

da Costa JP, Santos PS, Duarte AC, Rocha-Santos T (2016) Nanoplastics in the environment–sources, fates and effects. Sci Total Environ 566:15–26

Doerffer JW (2013) Oil spill response in the marine environment. Elsevier

Durán-Álvarez JC, Hernández-Morales VA, Martínez-Avelar C (2021) Strategies to remove pharmaceutical residues from seawater. In: Elsevier eBooks, pp 625–652. https://doi.org/10.1016/b978-0-08-102971-8.00019-6

Ehmedan SS, Ibrahim MK, Azzam AM et al (2021) Acceleration the bacterial biodegradation of crude oil pollution using Fe2O3 and ZnO nanoparticles. Environ Nanotechnol Monit Manage 16:100613

El-Kalliny AS, Abdel-Wahed MS, El-Zahhar AA, Hamza IA, Gad-Allah TA (2023) Nanomaterials: a review of emerging contaminants with potential health or environmental impact. Discov Nano 18(1):68. https://doi.org/10.1186/s11671-023-03787-8. PMID: 37382722; PMCID: PMC10409958

Eriksen M, Lebreton LC, Carson HS, Thiel M, Moore CJ, Borerro JC, Reisser J (2014) Plastic pollution in the world's oceans: more than 5 trillion plastic pieces weighing over 250,000 tons afloat at sea. PLoS One 9(12):e111913

Esposito MC, Corsi I, Russo GL, Punta C, Tosti E, Gallo A (2021) The era of nanomaterials: a safe solution or a risk for marine environmental pollution? Biomol Ther 11(3):441. https://doi.org/10.3390/biom11030441

Freedman B (2013) Environmental ecology: the impacts of pollution and other stresses on ecosystem structure and function. Elsevier

Fukumoto AF, Pimenta JA, Hirooka EY, Kuroda EK (2022) Pesticides removal from water using activated carbons and carbon nanotubes. Environ Technol 1–23. https://doi.org/10.1080/09593330.2022.2112979

Gall SC, Thompson RC (2015) The impact of debris on marine life. Mar Pollut Bull 92(1–2): 170–179

Gavela AF, Herranz S, Chocarro B, Falke F et al (2019) Full integration of photonic nanoimmunosensors in portable platforms for on-line monitoring of ocean pollutants. Sensors Actuators B Chem 297:126758. https://doi.org/10.1016/j.snb.2019.126758

Gavhane SK, Sapkale JB, Susware NK, Sapkale SJ (2021) Impact of heavy metals in riverine and estuarine environment: a review. Res J Chem Environ 25(5):226–233

GESAMP (2015) Sources, fate and effects of microplastics in the marine environment (part 1). Available: http://www.gesamp.org/publications/reports-and-studies-no-90

GESAMP (2016) Sources, fate and effects of microplastics in the marine environment (part 2). Available: http://www.gesamp.org/publications/microplastics-in-the-marineenvironment-part-2

Gill HK, Garg H (2014) Pesticide: environmental impacts and management strategies. Pesticides-Toxic Aspects 8(187):10–5772

Guerra FD, Attia MF, Whitehead DC, Alexis F (2018) Nanotechnology for environmental remediation: materials and applications. Molecules 23(7):1760

Guo Y, Cichocki N, Schattenberg F et al (2019) AgNPs change microbial community structures of wastewater. Front Microbiol 9:3211

Gupta D (2023) Green and sustainable synthesis of nanomaterials: recent advancements and limitations. Environ Res 231:116316–116316. https://doi.org/10.1016/j.envres.2023.116316

Halstead NT, Civitello DJ, Rohr JR (2015) Comparative toxicities of organophosphate and pyrethroid insecticides to aquatic macroarthropods. Chemosphere 135:265–271

Harada M (1995) Minamata disease: methylmercury poisoning in Japan caused by environmental pollution. Crit Rev Toxicol 25(1):1–24

Harding S (2016) Marine debris: understanding, preventing and mitigating the significant adverse impacts on marine and coastal biodiversity. Secretariat of the Convention on Biological Diversity

Hejna M, Kapuścińska D, Aksmann A (2022) Pharmaceuticals in the aquatic environment: a review on eco-toxicology and the remediation potential of algae. Int J Environ Res Public Health 19(13):7717. https://doi.org/10.3390/ijerph19137717

Hewitt EJ (1951) The role of the mineral elements in plant nutrition. Annu Rev Plant Physiol 2(1): 25–52

Ikbal AMA, Choudhury PD (2023) Green nanotechnology- an innovative pathway for sustainability. Paripex Indian J Res:85–86. https://doi.org/10.36106/paripex/9109052

Islam MS, Tanaka M (2004) Impacts of pollution on coastal and marine ecosystems including coastal and marine fisheries and approach for management: a review and synthesis. Mar Pollut Bull 48(7–8):624–649. ISSN 2666-3511,. https://doi.org/10.1016/j.sintl.2022.100166

Jambeck JR, Geyer R, Wilcox C et al (2015) Plastic waste inputs from land into the ocean. Science 347(6223):768–771

Janas D, Koziol KK (2016) Carbon nanotube fibers and films: synthesis, applications and perspectives of the direct spinning method. Nanoscale 8(47):19475–19490

Jay N, Meegoda (2023) A path to a reduction in micro and nanoplastics pollution. Int J Environ Res Public Health. https://doi.org/10.3390/ijerph20085555

Jiang D, Zeng G, Huang D, Chen M, Zhang C, Huang C, Wan J (2018) Remediation of contaminated soils by enhanced nanoscale zero valent iron. Environ Res 163:217–227

Kamble SB, Bhore RK (2023) Versatile nanomaterials for remediation of microplastics from the environment. In: Advances in nano and biochemistry. Academic, pp 107–126

Kasuya M, Teranishi H, Aoshima K et al (1992) Water pollution by cadmium and the onset of Itai-itai disease. Water Sci Technol 25(11):149–156

Kaur S, Roy A (2021) Bioremediation of heavy metals from wastewater using nanomaterials. Environ Dev Sustain 23(7):9617–9640

Kessouri F, McWilliams JC, Bianchi D et al (2021) Coastal eutrophication drives acidification, oxygen loss, and ecosystem change in a major oceanic upwelling system. Proc Natl Acad Sci 118(21):e2018856118

Khadijah A, Altammar (2023) A review on nanoparticles: characteristics, synthesis, applications, and challenges. Front Microbiol 14. https://doi.org/10.3389/fmicb.2023.1155622

Khalith SM, Karuppannan SK, Dowlath MJH et al (2022) Nanostructured catalytic membranes for water filtration. In: Nano-bioremediation: fundamentals and applications. Elsevier, pp 389–412

Khan SH (2020) Green nanotechnology for the environment and sustainable development. In: Green materials for wastewater treatment. Springer, pp 13–46

Khan MZ, Law FC (2005) Adverse effects of pesticides and related chemicals on enzyme and hormone systems of fish, amphibians and reptiles: a review. Proc Pakistan Acad Sci 42(4): 315–323

Koller M, Saleh HM (2018) Introductory chapter: introducing heavy metals. Heavy Metals. https://doi.org/10.5772/intechopen.74783

Krishnakumar PK (2017) Environmental impacts of marine pollution—effects, challenges and approaches; paper in chapter on environmental pollution in book, Mathrubhumi Year Book Plus 2017

Kumah EA, Fopa RD, Harati S et al (2023) Human and environmental impacts of nanoparticles: a scoping review of the current literature. BMC Public Health 23(1). https://doi.org/10.1186/s12889-023-15958-4

Kumar L, Paswan SK, Kumar P et al (2021) Nanotechnology-based filtration membranes for removal of pollutants from drinking water. In: Elsevier eBooks, pp 231–251. https://doi.org/10.1016/b978-0-12-823828-8.00011-6

Lamba B. Nanotechnology for recovery and reuse of spilled oil. (9 September 2005). Available at http://www.physorg.com/news6358.html (10 December 2008)

Lebreton L, Slat B, Ferrari F et al (2018) Evidence that the great Pacific garbage patch is rapidly accumulating plastic. Sci Rep 8(1):1–15

Lechuga M, Fernández-Serrano M, Ríos F et al (2022) Environmental impact assessment of nanofluids containing mixtures of surfactants and silica nanoparticles. Environ Sci Pollut Res 29(56):84125–84136. https://doi.org/10.1007/s11356-022-21598-9

Lee J, Kim C, Liu C et al (2023) Ultra-high capacity, multifunctional nanoscale sorbents for PFOA and PFOS treatment. npj Clean Water 6:62. https://doi.org/10.1038/s41545-023-00263-9

Li P, Qiang L, Han Y et al (2022) A sensitive and portable double-layer microfluidic biochip for harmful algae detection. Micromachines 2022(13):1759

Li Y, Zhang R, Yan X, Fan K (2023) Machine learning facilitating the rational design of nanozymes. J Mater Chem B 11(28):6466–6477

Lönnstedt OM, Eklöv P (2016) Environmentally relevant concentrations of microplastic particles influence larval fish ecology. Science 352(6290):1213–1216

Luque R, Sharma R, Foorginezhad S et al (2023) Green and sustainable membranes: a review. Environ Res 231:116133. https://doi.org/10.1016/j.envres.2023.116133

Malankowska M, Echaide-Górriz C, Coronas J (2021) Microplastics in marine environment: a review on sources, classification, and potential remediation by membrane technology. Environ Sci 7(2):243–258. https://doi.org/10.1039/d0ew00802h

Marcharla E, Vinayagam S, Gnanasekaran L et al (2024) Microplastics in marine ecosystems: a comprehensive review of biological and ecological implications and its mitigation approach using nanotechnology for the sustainable environment. Environ Res 119181

Marlon JR, Bloodhart B, Ballew MT et al (2019) How hope and doubt affect climate change mobilization. Front Commun 4:20

Materić D, Kasper-Giebl A, Kau D, Anten M, Greilinger M, Ludewig E, Holzinger R (2020) Micro- and nanoplastics in alpine snow: a new method for chemical identification and (semi) quantification in the nanogram range. Environ Sci Technol 54(4):2353–2359

Mehndiratta P, Jain A, Srivastava S, Gupta N (2013) Environmental pollution and nanotechnology. Environ Pollut 2(2):49

Mishra S, Chauhan G, Verma S, Singh U (2022) The emergence of nanotechnology in mitigating petroleum oil spills. Mar Pollut Bull 178:113609

Montemagno CD, Schmidt JJ, Tozzi SP (2007) Biomimetic membranes. US7208089

Napper IE, Davies BF, Clifford H et al (2020) Reaching new heights in plastic pollution—preliminary findings of microplastics on Mount Everest. One Earth 3(5):621–630

Nasrollahzadeh M, Sajjadi M, Sajadi SM, Issaabadi Z (2019) Green nanotechnology. In: Interface science and technology, vol 28. Elsevier, pp 145–198

Ngoubeyou PSK, Wolkersdorfer C, Ndibewu PP, Augustyn W (2022) Toxicity of polychlorinated biphenyls in aquatic environments–a review. Aquat Toxicol 251:106284

Noah NM, Ndangili PM (2022) Green synthesis of nanomaterials from sustainable materials for biosensors and drug delivery. Sens Int 3:100166. https://doi.org/10.1016/j.sintl.2022.100166

Ornell C, Finn A (2011) A holistic approach to reducing plastic marine debris in coastal North Carolina. Master's project, Duke University. Retrieved from https://hdl.handle.net/10161/3690

Othman A, Dumitrescu E, Andreescu D, Andreescu S (2018) Nanoporous sorbents for the removal and recovery of phosphorus from eutrophic waters: sustainability challenges and solutions. ACS Sustain Chem Eng 6(10):12542–12561. https://doi.org/10.1021/acssuschemeng.8b01809

Ouda M, Banat F, Hasan SW, Karanikolos GN (2023) Recent advances on nanotechnology-driven strategies for remediation of microplastics and nanoplastics from aqueous environments. J Water Process Eng 52:103543. https://doi.org/10.1016/j.jwpe.2023.103543

Palacios-Hinestroza H, Ureña-Saborío H, Zurita F et al (2020) Nanocellulose and PolycaprolactoneNanospun composite membranes and their potential for the removal of pollutants from water. Molecules 25(3):683. https://doi.org/10.3390/molecules25030683

Patanjali P, Singh R, Kumar A, Chaudhary P (2019) Nanotechnology for water treatment: a green approach. In: Green synthesis, characterization and applications of nanoparticles. Elsevier, pp 485–512.

Patel AB, Shaikh S, Jain KR, Desai C, Madamwar D (2020) Polycyclic aromatic hydrocarbons: sources, toxicity, and remediation approaches. Front Microbiol 11:562813

Pizzichetti ARP, Pablos C, Álvarez-Fernández C et al (2021) Evaluation of membranes performance for microplastic removal in a simple and low-cost filtration system. Case Stud Chem Environ Eng 3:100075. https://doi.org/10.1016/j.cscee.2020.100075

Priya E, Kumar S, Verma C, Sarkar S, Maji PK (2022) A comprehensive review on technological advances of adsorption for removing nitrate and phosphate from waste water. Journal of Water Process Engineering 49:103159. ISSN 2214–7144. https://doi.org/10.1016/j.jwpe.2022.103159

Rangel-Muñoz N, Barrios AFG, Pradilla D et al (2020) Novel bionanocompounds: outer membrane protein a and laccase co-immobilized on magnetite nanoparticles for produced water treatment. Nano 10(11):2278. https://doi.org/10.3390/nano10112278

Ray PC, Yu H, Fu PP (2009) Toxicity and environmental risks of nanomaterials: challenges and future needs. J Environ Sci Health C 27(1):1–35

Samwel John O (2012) Oil and coastal water pollution: a case study of Changamwe division Mombasa County coastal province-Kenya

Schnurr RE, Alboiu V, Chaudhary M et al (2018) Reducing marine pollution from single-use plastics (SUPs): a review. Mar Pollut Bull 137:157–171

Shang B, Wang Y, Peng B, Deng Z (2016) Bioinspired polydopamine particles-assisted construction of superhydrophobic surfaces for oil/water separation. J Colloid Interface Sci 482:240–251. https://doi.org/10.1016/j.jcis.2016.07.081

Sharma R (2015) Environmental issues of deep-sea mining. Proc Earth Planet Sci 11:204–211

Sheng Z, Hu Y (2021) Nanozymes in environmental protection, pp 213–241. https://doi.org/10.1007/978-3-030-68230-9_8

Singh J, Gupta M. & De Matteis, V. Green nanotechnology and nanomaterials for a sustainable environment. Clean Techn Environ Policy (2024). https://doi.org/10.1007/s10098-024-03083-8

Singh AK, Chandra R (2019) Pollutants released from the pulp paper industry: aquatic toxicity and their health hazards. Aquat Toxicol 211:202–216

Solař P, Škorvánková K, Kuzminova A, Kylián O (2023) Challenges in the deposition of plasma polymer nanoparticles using gas aggregation source: rebounding upon impact and how to land them on a substrate. Plasma Process Polym 20(10):e2300070

Solomon E, Edu S, Inam EA et al (2023) Prospects of nanosorption and photocatalysis in remediation of oil spills. J Nigerian Soc Phys Sci:1043–1043. https://doi.org/10.46481/jnsps.2023.1043

Solra M, Das S, Srivastava A (2022) Temporally controlled multienzyme catalysis using a dissipative supramolecular nanozyme. ACS Appl Mater Interfaces 14(40):45096–45109. https://doi.org/10.1021/acsami.2c08888

Sussarellu R, Suquet M, Thomas Y et al (2016) Oyster reproduction is affected by exposure to polystyrene microplastics. Proc Natl Acad Sci 113(9):2430–2435

Thakur AS, Koul B (2022) Impact of oil exploration and spillage on marine environments. In: Elsevier eBooks, pp 115–135. https://doi.org/10.1016/b978-0-323-89978-9.00018-5

Tirkey SR, Ram S, Mishra S (2021) Bacteria-mediated remediation: restoring polycyclic aromatic hydrocarbon (PAHs) contaminated marine ecosystems. In: Advanced oxidation processes for effluent treatment plants. Elsevier, pp 119–136

Todd PA, Ong X, Chou LM (2010) Impacts of pollution on marine life in Southeast Asia. Biodivers Conserv 19:1063–1082

Urso M, Pumera M (2022) Nano/microplastics capture and degradation by autonomous Nano/microrobots: a perspective. Adv Funct Mater 32(20). https://doi.org/10.1002/adfm.202112120

Varabih CA, Nofirman N (2023) The importance of seaweed as bioremediation natural agent. J Oceanogr Aquat Sci 1(2):48–53. https://doi.org/10.56855/joane.v1i2.343

Verma J, Pant H, Sign S, Tiwari A (2020) Marine pollution, sources, effect and management. In: Three major dimensions of life: environment, agriculture and health. Society of Biological Sciences and Rural Development, Prayagraj, pp 270–276

Wang XH, Yu XX, Wang SQ et al (2022) Environmental process, effects and risks of emerging contaminants in the estuary-coastal environment. Huan Jing ke Xue= Huanjing Kexue 43(11): 4810–4821

Waris, Hasnat A, Hasan S, Bano S, Sultana S, Ibhadon AO, Khan MZ (2023) Development of nanozyme based sensors as diagnostic tools in clinic applications: a review. J Mater Chem B 11. https://doi.org/10.1039/d3tb00451a

Wei H, Wang E (2013) Nanomaterials with enzyme-like characteristics (nanozymes): next-generation artificial enzymes. Chem Soc Rev 42(14):6060–6093

Weis JS (2013) Physiological, developmental and behavioral effects of marine pollution, vol 20. Springer, pp 389–412., ISBN 9780128239629. https://doi.org/10.1016/B978-0-12-823962-9.00003-9

Wu L, Zhang J, Li B, Wang A (2014) Magnetically driven super durable superhydrophobic polyester materials for oil/water separation. Polym Chem 5(7):2382. https://doi.org/10.1039/c3py01478a

Xuan L, Ju Z, Skonieczna M (2023) Nanoparticles-induced potential toxicity on human health: applications, toxicity mechanisms, and evaluation models. Med Commun 4(4). https://doi.org/10.1002/mco2.327

Yang Y, Yi H, Wang C (2015) Oil absorbents based on melamine/lignin by a dip adsorbing method. ACS Sustain Chem Eng 3(12):3012–3018. https://doi.org/10.1021/acssuschemeng.5b01187

Yuan J, Liu X, Akbulut O, Hu J, Suib SL, Kong J, Stellacci F (2008) Superwetting nanowire membranes for selective absorption. Nat Nanotechnol 3(6):332–336. https://doi.org/10.1038/nnano.2008.136

Zettler ER, Mincer TJ, Amaral-Zettler LA (2013) Life in the “plastisphere”: microbial communities on plastic marine debris. Environ Sci Technol 47(13):7137–7146

Zhang J, Seeger S (2011) Polyester materials with superwetting silicone nanofilaments for oil/water separation and selective oil absorption. Adv Funct Mater 21(24):4699–4704

Chapter 14
Biomedical Applications of Green Nanotechnology

Joicy George, Emy Tomy, Sharon Varghese, Roshny Roy, Alex Joseph, and Cincy George

Abstract Nanotechnology is one of the promising technologies of this era. It deals with the practical applications of nanoscience theory by observing, manipulating, controlling and manufacturing of matter in the nanometer scale. Nowadays researchers focus on the biological synthesis of nanoparticles. In this process nanoparticles are synthesized using natural capping, reducing and stabilizing agents instead of using high—cost and hazardous materials. Biological processes are environmentally reliable because it uses plants (leaves, flower, fruit, root etc.) and microorganisms (Fungi, bacteria, algae, yeast etc.) as the precursors. Synthesis of nanomaterials using plant origin is termed as green synthesis.

Traditional methods used for the synthesis of nanoparticles are highly expensive and generate toxic materials. It is necessary to find low cost, environmentally benign routes for the synthesis of nanoparticles. Therefore, green nanotechnology grabs the attention of researchers for developing nanomaterials. Less consumption of non—renewable sources, reduction of waste materials, low emission of greenhouse gases, and high energy efficiency are the important advantages of green nanotechnology. Metallic nanoparticles synthesized from medicinal plants show more pronounced therapeutic properties such as antimicrobial activity, antioxidant activity, insecticidal activity, and wound healing properties.

This chapter deals with the biomedical applications of green nanotechnology. Applications of green synthesized iron, selenium, copper, cobalt, cobalt oxide, silver, gold, and platinum nanoparticles are overviewed. Biomedical applications like cancer therapy, targeted drug delivery, antibacterial activity, biomedical Sensors, dentistry, bioimaging and tissue Engineering are discussed below.

J. George · S. Varghese · R. Roy · A. Joseph · C. George (✉)
Mahatma Gandhi University Center for Research in Chemistry, Newman College, Thodupuzha, India

E. Tomy
Mahatma Gandhi University Center for Research in Chemistry, Newman College, Thodupuzha, India

Mahatma Gandhi University Center for Research in Chemistry, Nirmala College, Muvattupuzha, India

© The Author(s), under exclusive license to Springer Nature Switzerland AG 2025
J. M. Al-Khayri et al. (eds.), *Green Nanotechnology Applications for Ecosystem Sustainability*, Nanotechnology in Plant Sciences 4,
https://doi.org/10.1007/978-3-031-90766-1_14

Keywords Bioimaging · Biomedical sensors · Cancer therapy · Dentistry · Green nanotechnology · Targeted drug delivery · Tissue engineering

14.1 Introduction

Nanoscience deals with the molecules and structures in the nanoscale (1–100 nm). Nanoscience includes different branches of science like chemistry, physics, biology, material sciences, electronics, engineering etc. It deals with the synthesis, characterization and exploration of nanomaterials. That is the manipulation of matter in atomic or molecular level. Nanostructured materials should possess at least any one of the dimensions in the nanoscale range.

When a material changes its composition from bulk to nano, some of the chemical and physical properties differ significantly. These nanomaterials are capable of showing outstanding optical, electrical, mechanical and magnetic properties. Due to these properties, nanomaterials make a revolution in various fields like engineering and material sciences (Pokropivny et al. 2007; Bayda et al. 2019).

Nanotechnology is introduced by Richard Feynman (American Physicist) in 1959. It deals with the practical applications of nanoscience theory by observing, manipulating, controlling and manufacturing of the matter in the nanometer scale. Nanotechnology is one of the promising technologies of this era. Nanotechnology is mainly classified in to three types. (i) Computational nanotechnology, (ii) Wet technology and (iii) Dry technology.

The field of nanotechnology is rapidly growing and find applications in various fields. Applications of nanotechnology are,

(a) Nanotechnology in medicine (Biomedicine, nano –oncology, targeted drug delivery, nano biosensors, nano pharmaceuticals, and nano biochips)
(b) Nanotechnology in electronics (Quantum computers, graphene transistors, high density storage devices, Single Electron Transistor, CNT based Nanosensors, and OLED display)
(c) Applications in nanobiotechnology (Nano—mechanical oscillator, nanobots etc.)
(d) Nanotechnology in food (Food processing, smart packaging, nutritional supplements, Food Quality Monitoring, Pathogen Identification in Food, and, Improved Food Storage.)
(e) Nanotechnology in environmental sector (Solar energy, fuel cell, and wind energy)
(f) Nanotechnology in industrial sector (Textile field, construction industry) (Singh et al. 2020) (Fig. 14.1)

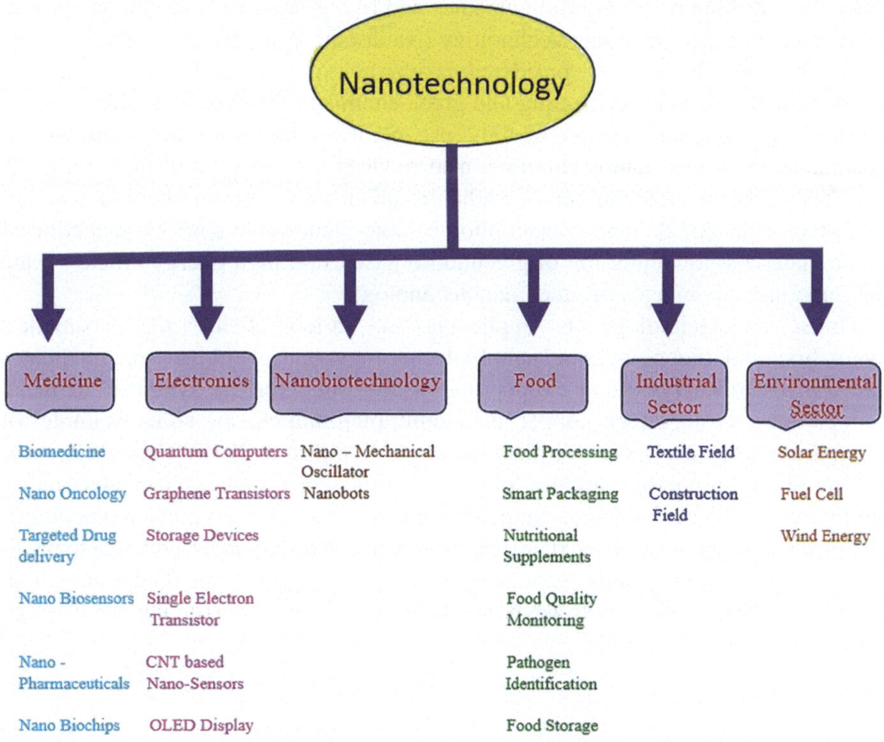

Fig. 14.1 Applications of nanotechnology. (Figure constructed by Joicy George)

14.2 Green Nanotechnology

The synthesis of nanoparticles is mainly divided in to two types. Physical processes involve ultrasonication, laser ablation, irradiation, evaporation, condensation etc. and chemical processes involve mechanical milling, sputtering, etching, thermal decomposition etc.

Nowadays researchers are focuses on the biological synthesis of nanoparticles. In this process nanoparticles are synthesized using natural capping, reducing and stabilizing agents instead of using high—cost and hazardous materials. Biological processes are environmentally reliable because it uses plants (leaves, flower, fruit, root etc.) and microorganisms (Fungi, bacteria, algae, yeast etc.) as the precursors. Synthesis of nanomaterials using plant origin is termed as green synthesis.

Traditional methods used for the synthesis of nanoparticles are highly expensive and generate toxic materials. It is necessary to find low cost, environmentally benign routes for the synthesis of nanoparticles. Therefore, green nanotechnology grabs the attention of researchers for developing nanomaterials (Khan et al. 2022). Green chemistry and green engineering are the basics of green nanotechnology. Green synthesis is more efficient, less expensive, simpler and environmentally friendly than other conventional methods. Biomolecules and living organisms are the precursors of nanomaterials in green nanotechnology (Nallasamy et al. 2023).

Green nanotechnology is considered as a branch of green technology, works on the principles of green engineering and green chemistry (Verma et al. 2019). Green synthesis of nanoparticles are mainly use plant sources for the production of nanomaterials. Green nanotechnology also reduces the energy and fuel consumption. Therefore, green techniques are reliable and ensures environmental sustainability (Khan et al. 2022). Less consumption of non—renewable sources, reduction of waste materials, low emission of greenhouse gases, and high energy efficiency are the important advantages of green nanotechnology.

Green nanotechnology is applicable in various fields like cosmetics, nanofabrication, bioengineering, nanobiotechnology, optical engineering, medicines and drugs. Nowadays, plant extracts are widely used for the synthesis of metal nanoparticles. Gold, silver, copper, palladium, platinum etc. are some examples of nanoparticles synthesized using plant extracts (Verma et al. 2019). Green nanotechnology shows numerous applications in various fields such as nanoelectronics, environmental sector, industries, agricultural sector, biomaterials and nanomedicine. Green technology widely used in health care for the diagnosis and treatment of diseases, drug delivery and preparation of new drug formulations (Nallasamy et al. 2023). This chapter deals with the biomedical applications of green nanotechnology (Fig. 14.2).

14.3 Biomedical Applications

Nanomaterials are not always artificial and their synthesis is not a novel invention. Nature is an efficient nanotechnologist. The nanomaterials that support life on earth include DNA, proteins, and enzymes (Khan et al. 2017a). Nature provides a myriad of nanomaterials ranging from inorganic ash, soot, metal oxides, carbonates etc. present in the air; Sulphur and selenium nanoparticles produced by certain microorganisms (Griffin et al. 2017), iridescent inner surface of the shells of mollusks due to the presence of nanostructured calcium carbonate (Hovden et al. 2015) and birds, butterflies and fishes having highly reflecting and colored surfaces on their body parts due to array of nanostructures (Sun et al. 2013). Nature is skilled at fabricating inorganic and organic structure at the nanoscale. The potential that the nature has, in synthesizing nanoparticles is still to be explored (Gröning 2005).

The field of nanotechnology is expanding continuously and influencing human life in various aspects. The exceptional properties of nanoparticles owing to their greater surface area has made vast development in the field of biotechnology and in biomedical science.

The antibacterial, antimicrobial and the antioxidant properties of nanoparticles mark their importance in the biomedical applications. The green synthesis of nanoparticles is gaining much importance now a days (Shreyash et al. 2021). Synthesis of nanoparticles through green methods is one of the best and ecofriendly

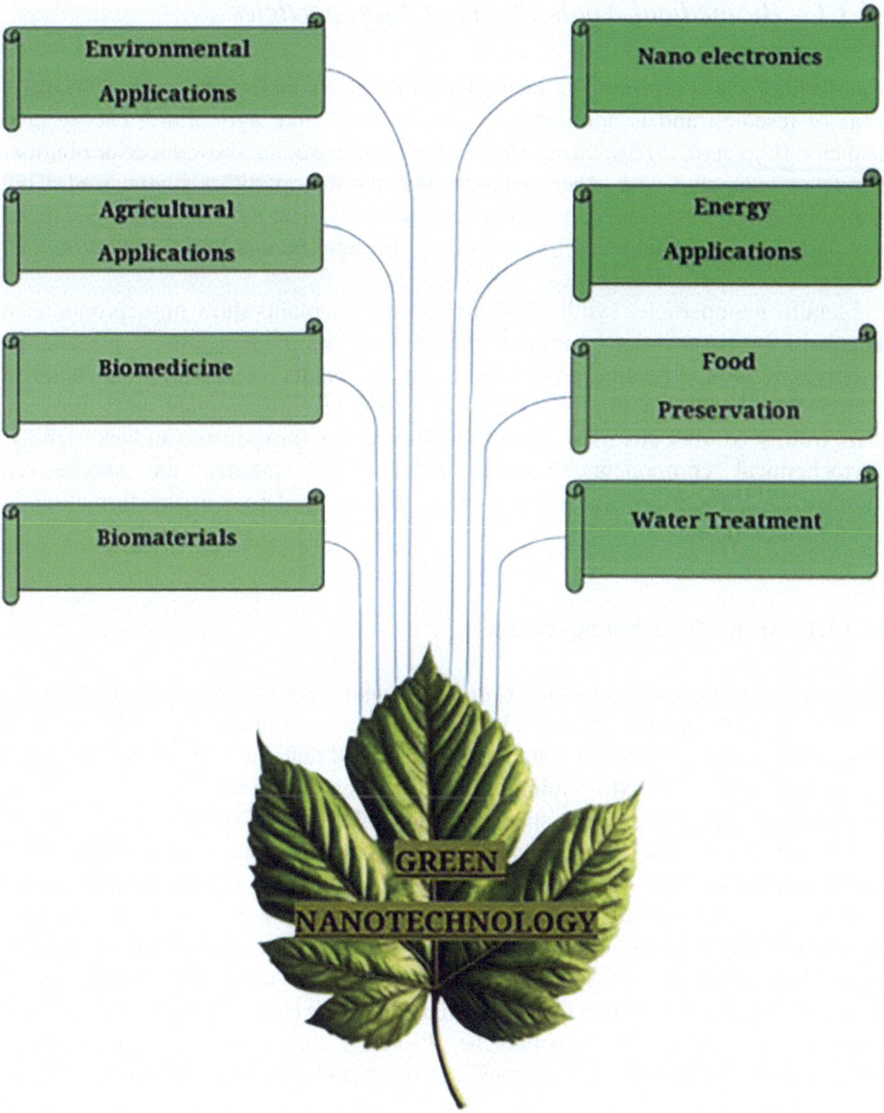

Fig. 14.2 Applications of green nanotechnology. (Figure constructed by Joicy George)

methods. The main advantage of green synthesis is the lack of toxic byproducts during the synthesis. These nanoparticles synthesized via green methods have significant biological activities. They are found to be active against gram positive and gram-negative bacteria. Their anti-fungal and anti-cancer activities are also well studied. The particles of small size and large surface area are of great significance in the medical field (Khan et al. 2017a, b).

14.3.1 *Biomedical Applications of Nanoparticles*

Nanotechnology is a promising technology that offers large opportunities in many areas of research and is applicable in various fields like agriculture, energy, and medicine (Ojo et al. 2020). Green synthesized nanoparticles show superior pharmacokinetics compared with other methods. Metallic nanoparticles synthesized using plant extract are stable and can monodispersed easily. The monodispersing of metal nanoparticles can be achieved by controlling different factors like temperature, pH, mixing ratio and retention time.

Metallic nanoparticles synthesized from medicinal plants show more pronounced therapeutic properties such as antimicrobial activity, antioxidant activity, insecticidal activity, and wound healing properties. Naturally, plants are rich in phytochemical components. For the synthesis of metal nanoparticles various plant parts like leaf, stem, fruit, seed, root etc. used. These components are inexpensive and ecofriendly. Phytochemical components help to reduce and stabilize the synthesized nanoparticles. Therefore, these components play a vital role in the formation of metal nanoparticles.

14.3.1.1 Iron Metal Nanoparticles

Using different plants like lantana (*Lantana Camara*), water-hyacinth (*Eichhornia crassipes*), and sensitive plants (*Mimosa pudica*) are used to synthesize iron nanoparticles in zero—valent state. The synthesized nanoparticles show size range between 20 and 60 nm. Morphological studies revealed that the nanoparticles show irregular and aggregated quasi -spherical shape (Wen et al. 2022).

Iron nanoparticles show significant biomedical applications. Green nanoparticles synthesized from flat—crown leaf (*Albizia adianthifolia*) extract utilized in the treatment of, AMJ-13 (breast cancer cell line) and MCF-7 (breast cancer cell line) cancer cell lines (Sulaiman et al. 2018). Iron nanoparticles synthesized from sugar apple (*Annona squamosa*) leaf extract shows excellent cytotoxic activity against HepG2 (liver cancer cell line) cancer cell line (Banerjee et al. 2021). In addition, iron nanoparticles synthesized from babiche (*Psoralea corylifolia*) leaf extract shows antitumor activity on Caki-2 cells (renal carcinoma cell line) (Rajendran et al. 2021).

14.3.1.2 Selenium Nanoparticles

Selenium nanoparticles synthesized via green methods are non—toxic, cost—effective and produced in large quantities. Selenium nanoparticles can act as good biotherapeutic agents with no side effects. Selenium nanoparticles shown antioxidant action, regulation of immune system and prevention of tumor formation (Menon et al. 2019). Green synthesized garlic (*Allium sativum*) bud extract provides antibacterial response against *Staphylococcus aureus* and *Bacillus subtillis* (Vyas and Rana 2018).

Selenium nanoparticles synthesized from horseshoe geranium (*Pelargonium zonale*) leaf extract revealed antifungal and antibacterial action against fungi like *Colletotrichum coccodes* and *Penicillium digitatum* and bacteria like *Staphylococcus* and *Escherichia coli*. Selenium nanoparticles also show anticancer activity (Vijayaram et al. 2024). Selenium nanoparticles synthesized from garlic (*Allium sativum*) shows cytotoxic response against verocell line (Kidney cell line) (Anu et al. 2017).

14.3.1.3 Copper Nanoparticles

Copper nanoparticles synthesized using green nanotechnology are eco—friendly and non—toxic. Copper nanoparticles are capable of showing biomedical applications such as antifungal, antiviral and antibacterial activity. Copper nanoparticles synthesized from clove bud extract are effective antimicrobial agents.

Copper nanoparticles produced from Kobus magnolia (*Magnolia Kobus*) show significant antibacterial activity against *Escherichia coli* (Lee et al. 2013). Copper nanoparticles synthesized using China—rose (*Hibiscus rosa-sinensis*) leaf extract confirmed the antimicrobial activity against pathogens like *Bacillus subtillis* (Bhuvaneshwari et al. 2022). Copper nanoparticles with size 39.3 nm and spherical shape show cytotoxic effect against skin carcinoma cells (Vijayaram et al. 2024).

14.3.1.4 Cobalt and Cobalt Oxide Nanoparticles

Nanoparticles of cobalt and cobalt oxide show various biomedical applications. Nowadays, people's increased susceptibility to ailments is the main concern over health. Thus, the development of antibiotic drugs with the ability to eradicate harmful bacteria is imperative. Cobalt oxide and cobalt nanoparticles possess high antibacterial activity (Nallasamy et al. 2023).

Cobalt nanoparticles obtained from the leaves of Radish (*Raphanus sativus*) shows potential antibacterial activity against gram—negative bacteria. Nanoparticles of cobalt also show excellent cytotoxicity against Hela cancer cell lines (Cervical cancer cell line) (Koyyati et al. 2016). Nanoparticles of cobalt oxide synthesized from milk bush (*Euphorbia tirucalli*) shows anti proliferative activity against MCF-7 breast cancer cell lines (Kgosiemang et al. 2020). The studies of R. Shanmuganathanet.al revealed that the cobalt oxide nanoparticles obtained from turmeric (*Curcuma longa*) are cytotoxic to MDA-MB -468 cell lines (breast cancer cell line) (Shanmuganathan et al. 2023).

The other biomedical applications of cobalt oxide and cobalt nanoparticles are wound healing, antileishmanial, anticholinergic, larvicidal and anti-diabetic properties (Nallasamy et al. 2023).

14.3.1.5 Silver Nanoparticles

Silver nanoparticles show potential biomedical applications due to its unique properties in the nanoscale. Silver nanoparticles are non—toxic and have high antimicrobial activity. Silver nanoparticles are largely used for antimicrobial applications. These are effective against both gram—positive and gram—negative bacteria. Silver nanoparticles interact with bacterial membrane, then penetrate to the cell leads to structural damage and finally cell death. Other applications of silver nanoparticles are bone healing, wound healing, and drug delivery systems (Burduşel et al. 2018).

Silver nanoparticles synthesized from raspberry jam (*Acacia acuminata*) plant shows antibacterial activity against MDR (Multidrug–resistant) bacterial pathogens (Mishra and Padhy 2018). Silver nanoparticles synthesized from Elephant climber (*Argyreia nervosa)* leaf extract shows good antimicrobial activity. Cytotoxic and antiproliferative studies on KB oral cancer cells show IC_{50} (half—maximal inhibitory concentration) value equal to 58.64 μg/ml. This value indicates the excellent anticancer activity of silver nanoparticles synthesized using *Argyreia nervosa* (Subramanyam et al. 2021).

14.3.1.6 Gold Nanoparticles

Gold nanoparticles prepared through green synthesis can show various biomedical applications. Gold nanoparticles are used in drug delivery systems. The efficiency of gold nanoparticles depends on the size. Small sized nanoparticles are preferred for drug delivery. Gold nanoparticles are also used for cancer treatment. Gold nanoparticles derived Australian cow plant *(Gymnema sylvestre*) shows cytotoxic activity against Hep 2 cells (Liver cell line) (Nakkala et al. 2015). Gold nanoparticles are applicable to other tumors also. Some examples of such cells are breast cancer cells, MCF-7 and HCT- 116 cells.

Biosensing is another application of gold nanoparticles. Gold nanoparticles show excellent sensitivity towards bacteria, pathogens, toxins, biological molecules, pollutants, and cancerous cells. Fluorescent, electrochemical, colorimetric, bio—barcode assay, and surface plasmon resonance sensors are examples for different types of gold nanoparticle sensors. Gold nanoparticles are able to fight against fungi, bacteria, and pathogens. Gold nanoparticles synthesized using Peppermint (*Mentha piperita L*) have antibacterial effect towards *Escherichia Coli* (gram -negative bacteria) and Staphylococcus aureus (gram—positive bacteria) (Teimuri-mofrad et al. 2017). Gold nanoparticles prepared from banana peels exhibit antifungal properties against *Candida albicans* (pathogenic strain) (Lee et al. 2020).

14.3.1.7 Platinum Nanoparticles

Green synthesis of platinum nanoparticles with unique properties finds various biomedical applications. Biomedical applications of platinum nanoparticles are anti—bacterial, anti—oxidant, anti—inflammation, anti—fungal, anti—diabetic, and anti—cancer activities. Antibacterial properties of platinum nanoparticles are related to "Reactive Oxygen Species". ROS reacts directly with platinum nanoparticles cause lipid breakdown and protein degradation leads to down—regulation of DNA and finally cell death.

Size of platinum nanoparticles also influence antibacterial activity. Platinum nanoparticles can inhibit both gram—negative (e.g., *Pseudomonas aeruginosa, Klebsiella oxytoca, S. typhimurium, E. Coli, Aeromonas hydrophila*) and gram—positive (e.g., *Staphylococcus aureus, Streptococcus pneumonia, Listeria innocua, Bacillus subtilis)* bacteria (Riddin et al. 2006).

Platinum nanoparticles synthesized from thorn apple (*Stramonium*) leaves show antifungal activity against *C. tropicalis, Candida albicans, Aspergillus flavus, C. parapsilosis* and *A. niger*. Anticancer activity is another biomedical application of platinum nanoparticles. Plant extracts mediated synthesis of platinum nanoparticles can show anti—cancerous activity. Examples of such plants are Indian aloe (*Aloe vera)*, Hot pepper (*Capsicum annuum*), Tulasi (*Ocimum tenuiflorum*) and Annual vinca (*Catharanthus roseus)*. Other applications of platinum nanoparticles are antioxidant activity, antiinflammation, antidiabetic, photothermal therapy, and radiotherapy (Mikhailova 2022).

14.3.1.8 Tin Oxide Nanoparticles

Green synthesized tin oxide nanoparticles also useful for different biomedical applications. Tin oxide nanoparticles synthesized using sugar apple (*Annona squamosa*) shows moderate cytotoxicity against hepatocellular carcinoma cell lines (HepG2)(Roopan et al. 2015). The other biomedical applications of tin oxide nanoparticles are anti—viral, anti—bacterial, anti—fungal, anti—oxidant and drug delivery (Nallasamy et al. 2023) (Table 14.1).

14.4 Green Nanotechnology in Biomedicine

Green nanotechnology widely used in biomedicine. The main applications of green nanotechnology in medicine includes diagnosis and treatment of diseases (e.g., cancer therapy, dentistry), drug delivery, antibacterial activity, biomedical sensors, bioimaging, and tissue engineering.

Table 14.1 Biomedical applications of various nanoparticles

Nanoparticle	Source	Biomedical application	Reference
Iron	Flat—crown leaf	AMJ-13 (breast cancer cell line) and MCF-7 (breast cancer cell line)	Sulaiman et al. (2018)
	Sugar apple leaf	HepG2(liver cancer cell line)	Banerjee et al. (2021)
	Babiche leaf	Caki-2 cells (renal carcinoma cell line)	Rajendran et al. (2021)
Selenium	Garlic bud	Antibacterial activity (*Staphylococcus aureus* and *Bacillus subtillis)*	Vyas and Rana (2018)
	Horseshoe geranium leaf	Antifungal (*Colletotrichum coccodes, Penicillium digitatum*) and antibacterial (*Staphylococcus, Escherichia coli)* action	Vijayaram et al. (2024)
	Garlic	Cytotoxicity—verocell line (Kidney cell line).	Anu et al. (2017)
Copper	Kobus magnolia	Antibacterial activity (*Escherichia coli*)	Lee et al. (2013)
	China—rose leaf	Antimicrobial activity (*Bacillus subtillis)*	Bhuvaneshwari et al. (2022)
Cobalt	Radish leaves	Cytotoxicity -Hela cancer cell lines (Cervical cancer cell line)	Koyyati et al. (2016)
Cobalt oxide	Milk bush	Antiproliferative activity –MCF-7 (breast cancer cell lines)	Kgosiemang et al. (2020)
	Turmeric	Cytotoxicity—MDA-MB-468 cell lines (breast cancer cell line)	Shanmuganathan et al. (2023)
Silver	Raspberry jam	Antibacterial activity against Multidrug–resistant bacterial pathogens	Mishra and Padhy (2018)
	Elephant climber	Cytotoxicity—KB cell lines (Oral cancer cells)	Subramanyam et al.
Gold	Australian cow plant	Cytotoxicity—Hep 2 cells (Liver cell line)	Nakkala et al. (2015)
	Peppermint	Antibacterial activity (*Escherichia Coli)*	Teimuri-mofrad et al. (2017)
	Banana peels	Antifungal activity (*Candida albicans*)	Lee et al. (2020)
Platinum	Thorn apple	Antifungal activity (*C. tropicalis, Candida albicans, Aspergillus flaves, C. parapsilosis, A. niger*)	Mikhailova (2022)
Tin oxide	Sugar apple	Cytotoxicity—HepG2 cell line (hepatocellular carcinoma cell lines)	Roopan et al. (2015)

14.4.1 Cancer

Cancer is caused by the uncontrolled cell division. It may form malignant tumors and can spread to other parts of the body. Despite the recent advances in medical field, cancer is remaining as a threat to human life. The war to defeat cancer is still in the developing stage. Cancer is primarily a genetic disease (Wishart 2015). It is caused by the mutations in genes that control the function of cells. The increased half-life, delayed release and effective targeting properties made nano-based drugs and drug-delivery systems significant in cancer treatment (Klochkov et al. 2021). Owing to their nano size, which is hundred to ten thousand times smaller than human cells, nano particles have the exceptional ability to interact with biomolecules inside the cells and on the surface of the cells.

14.4.2 Cancer Therapy

Depending on the type of cancer and how advanced it is, there are different types of cancer therapies. Surgery, chemotherapy, and radiation therapy are some among them (Revia et al. 2019). Apart from these traditional methods of cancer treatment, which may affect normal cells and may result in incomplete eradication of cancer cells, nanotechnology can help to target the therapy exactly to the cancer cells. It can guide surgical removal of tumors and the radiation-based treatments can also be made more efficient. Current researches in the nanotechnology have gone beyond targeted drug delivery and is now based on the inherent properties of nanoparticles itself. The major challenge in cancer therapy is to deliver the drugs in the tumor site exclusively, without affecting the healthy tissues (Verma et al. 2018).

Certain physical properties of nano particles like the energy absorption and re-radiation are used to destroy the cancer cells. This method is used in treatments like laser ablation and hyperthermia treatment (Abdullahi et al. 2019). The size of nanoparticles allows them to accumulate at the tumor site. Still, they can encapsulate therapeutic drugs like radionucleotides and other medicines (Baetke et al. 2015). Nanoparticles are having a large surface area that can be functionalized using various ligands such as DNA or RNA strands, peptides, or antibodies. Innovative nanostructured architectures can be utilized to design artificial antigen providing cells which can help for sustained anti-tumor activity (Pucci et al. 2019).

Most of the anti-cancer drugs are water insoluble and are to be dissolved in organic solvents for their administration. But these organic solvents are toxic and have side effects. Nano particles can convert insoluble drugs to soluble suspensions thereby limiting the use of toxic organic solvents (Bharali and Mousa 2010).

14.4.3 Cancer Diagnosis

One of the most effective methods for improving the life quality of cancer patients is the early diagnosis of the disorder. It can help to manage the progress of cancer. Treatment is easier at the early stages. Also, the recovery rate is greater if it is treated at an early stage (Abdolmaleki et al. 2020). Early diagnosis of cancer is aided by nanotechnology by enhanced images and better screening technologies (WINAWER 2003). Quantum dots, owing to their high levels of light emission, photo stability, tunable emission properties and the ability to carry different therapeutic agents, are widely used as probes for optical imaging. MRI scanning is a powerful tool for disease diagnosis. Magnetic nanoparticles having an inorganic nanoparticle core and a surface coating which permits stability in aqueous solutions, is making significant contributions towards improved cancer diagnosis and imaging using MRI (Ni et al. 2017).

14.4.4 Chemotherapy

Chemotherapy is widely used in the treatment of cancer. But it has severe side effects and is not completely satisfactory due to non-specific drug distribution. For minimizing these undesired effects chemotherapy is now used in combination with nanotechnology. This strategy has proved increased therapeutic efficiency with minimal side effects. Nanoparticle based drug delivery systems are made of a core, which acts as a carrier and there are functional groups attached to this core which can target specific sites (Ho et al. 2017).

The use of metal nano particles as a substitute for the drugs used in chemotherapy is now being investigated. Metal nano particles have anti-cancer properties which is inherent in them or these properties can be surface induced on these nanosized particles. Compared to the normal cells specific antigens or receptors are over expressed in tumor cells. Metal nanoparticles functionalized with biological moieties such as antibodies and peptides can specifically target cell surface proteins and receptors on cancer cells. Gold nanoparticles are efficient in gene silencing which are very effective in blocking specific gene expressions in cancer cells (Baranwal et al. 2023).

The success of chemotherapy lies in targeting the drugs to the proper sites without affecting the normal tissues. Targeting the therapeutic drugs to sub cellular organelles is a promising strategy to completely eradicate the tumor cells. Mitochondria—targeted and nucleus—targeted chemotherapeutic methods are being developed with the aid of nanoparticles, which can penetrate these organelles with much ease owing to their small size. Different cancer cells have different microenvironments and therefore the use of size—changeable nanocarriers for drug delivery are much effective (Wei et al. 2021). The controllable self-assembly, effective drug encapsulation and biocompatibility help nanotechnology to overcome the current limitations of chemotherapy (Liang et al. 2010).

14.4.5 Surgical Oncology

Currently surgery is one of the most effective and widely used procedure for treatment of cancer. Recent advances in nanotechnology can help the surgeon to identify the residual tumor cells and to identify whether the tumor cells have been completely removed. Incorporation of nanosized robots in surgical practice is a promising advance in this field. Smart nanodevices are now widely used for diagnostic and therapeutic purposes (Loizidou and Seifalian 2010). Nanoparticles are used as optical contrasting agents in image guided surgery, which can help to decrease the presence of tumor cells left behind after surgery (Wojtynek and Mohs 2020). The structural and functional properties of quantum dots and nano sized colloidal gold particles are conjugated with targeting ligands. They can be used to target tumor cells and its microenvironments (Singhal et al. 2010).

14.4.6 Radiation Therapy

External and internal radiation therapy methods are used for treating cancer. But the low rate of absorption of radiation by the tumor cells demands high doses of radiations which are ionizing. It can lead to severe damages to normal tissues adjacent to the tumor cells. The response of tumors to radiation can be enhanced using nanomedicines. Nanomaterials can act as radio-sensitizers which can help to deposit the radiation energy with in the tumor cells. It will enhance the efficiency of the treatment. Nanocarriers can carry therapeutic radioisotopes to the tumor sites for internal radioisotope therapy (Song et al. 2017a). Radiation sensitizing agents can either be encapsulated by nanomaterials or they can sensitize the cancer cells to the radiation through their physicochemical properties (Bergs et al. 2015). The unusual electronic, magnetic, and optical properties of nanomaterials make them suitable for their use in radiation therapy (Bakht et al. 2012). Specifically designed tumor targeted nanocarriers, which deliver radionucleotide payloads to specific tumor sites enhance the safety of cancer therapy (Ting et al. 2010). Some cancer cells are resistant to ionizing radiations. The Traditional methods used to overcome this problem are increasing the dose of radiation and sensitizing the radioresistant fractions of the tumor cells. Nanotechnology can help to overcome these limitations by combining multiple approaches in one unified therapeutic strategy (Chatterjee et al. 2013).

14.4.7 Immunotherapy

Immunotherapy is using the immune system of the person to fight cancer cells. This treatment method can result in long lasting anticancer responses in the body and reduce the recurrence of the disease. Generally, immunotherapy is the clinical suppression or activation of the immune system for treating a disease. The immune system of the body is activated against the malignant cells in cancer immunotherapy. Nanomaterials with in the size range of 10–200 nm can avoid kidney clearance and it can penetrate tumor cells. This prolongs their retention time in blood stream and thereby enhance their distribution in the tumor sites (Song et al. 2017b). Nanomaterials can easily modify and the preloaded targeting ligands on the surface of the nanomaterials can enhance their capture by specific sites (Gu et al. 2007). One of the major challenges in cancer therapy is the prevention of metastasis. In patients suffering from metastatic cancer nanotechnology has proved improvements through immunotherapeutic treatments (Zhang et al. 2021). Nanomaterials with multiple immunomodulatory functions and exceptional immunogenicity have been developed and proved to be successful (Gong et al. 2021).

14.4.8 Angiogenesis

Angiogenesis is the process by which existing blood vessels grow new ones. In most cases, it is necessary because it promotes wound healing and aids in the delivery of nutrients and oxygen to bodily tissues. But in the case of cancer patients 'angiogenesis' will enhance the growth of tumor by supplying it with oxygen and nutrients. Angiogenesis therapies, deal with angiogenesis inhibition in oncology and its stimulation in tissue engineering and wound healing (Kargozar et al. 2020). The delivery of pro-angiogenic or anti-angiogenic substances with the aid of nanostructured vehicles is a developing field (Clavreul et al. 2019). The green nanoparticles deal with angiogenesis in distinct molecular pathways (Rajasekar et al. 2019).

Tumors are not able to grow beyond 2 mm diameter without angiogenesis (Naumov et al. 2006). The important fact is to target the drug to the proper site. The size of nano particles helps them to reside into the tumor cells by means of enhanced permeability. They also have greater retention effect. They can evade the immune system and will improve the half-life of the drug. These specific properties make nanotherapeutics as a promising field in anti-angiogenesis research (Banerjee et al. 2011).

14.5 Targeted Drug Delivery

Green nanotechnology for targeted drug delivery implicates the use of environmentally sustainable and biocompatible materials and processes to enhance the precision, efficacy, and safety of drug delivery systems. When it comes to the second generation of targeted drug delivery systems, incorporating green nanotechnology principles is aimed at minimizing environmental impact and ensuring biocompatibility (Park 2014). The key aspects of implementing green nanotechnology in targeted drug delivery involves the use of biocompatible nanomaterials, green synthetic methodologies, incorporation of targeted ligands from natural sources, life cycle analysis and biodegradability of nanocarriers, focus on reduced side effects, energy efficient manufacturing, reduced waste generation during manufacturing and throughout life cycle of drugs exploitation of stimuli responsive nanocarriers. In addition to that, safety, efficacy and public awareness should be ensured among the social well beings regarding the guidelines and benefits of green nanotechnology based targeted drug delivery systems.

Due to their nano-size, the drug delivery systems have large surface to volume ratio which provide a lot of benefits such as their enhanced bioavailability; controlled and sustained drug release; high drug stacking capacity; protracted circulation time; superior intracellular penetration and targeted delivery to specific sites or organs; protection of the active ingredient against physiological pH, enzymes, and moisture; procedure of numerous ways of administration, including parenteral, oral, nasal, intraocular; and several more (Kanwar et al. 2019; Dhilip Kumar and Abrahamse 2022).

Biocompatible and biodegradable nanoparticles have been successfully expended as carrier molecules to deliver cancer drugs and photosensitizers at targeted sites due to their material safety in the drug delivery system. They are non-toxic and can ensure high-level biocompatibility within cells, blood, and physiological conditions (Dhilip Kumar and Abrahamse 2022). The use of liposomes, dendrimers, nanoparticles, micelles, magnetic nano particles, gold nano particles and polymeric nanoparticles have shown to be biocompatible materials to construct carriers (Nicolas et al. 2013; Oladipo et al. 2023)). Employment of natural polymers such as chitosan (Andreas Bernkop-Schnürch 2012), alginate (Lee and Mooney 2012), or cellulose (Anirudhan et al. 2021) in the design of nanocarriers would be a good choice as it offers a sustainable methodology (Wang et al. 2016).

The practice of green synthetic methodologies involves eco-friendly synthesis methods to produce nanomaterials, which follows green chemistry principles, using water as a solvent, minimizing the waste and reducing use of hazardous chemicals (Noah and Ndangili 2022). For instance, bio fabrication techniques, such as using plant parts or extracts, bacteria, or fungi are explored to synthesize nanoparticles for targeted drug delivery (Kulkarni et al. 2023). According to Mariadoss et.al, starch encapsulated green synthesized copper oxide nanoparticles shows anticancer properties via targeted drug delivery and are reported to use in breast cancer therapy. The folic acid functionalization on starch potentially retards the cancer growth via apoptosis (Mariadoss et al. 2020). Incorporation of targeting ligands derived from natural sources, such as peptides, antibodies, or aptamers, is need of the hour to enhance the specificity of drug delivery systems and reduce off-target effects (Morshed and Chowdhury 2019). In addition to that, natural receptors that are overexpressed on the surface of target cells can be utilized as recognition elements (Qian 2002).

A targeting strategy which encompasses the development of stimuli-responsive nanocarriers, that are biocompatible and can release drugs in response to specific physiological signals, minimizing unnecessary exposure will be an environmentally benign route (Suner et al. 2018). For instance, Azadpour et.al reported green synthesized silver nanoparticles coated with chitosan nanoparticles that has been decorated with folic acid, which shows smart drug delivery and controlled release behaviour of drug on varying the pH. The nano-complex carrier was potentially useful for the suppression of colon cancer (Azadpour et al. 2022). The specific physiological stimuli can be either pH-responsive, temperature-responsive, hormone levels, and redox potential, or enzyme-triggered release mechanisms etc. Thus, it is crucial to design systems that can be activated only in the presence of disease-specific signals.

It is imperative to ensure that nanomaterials used in drug delivery systems are easily biodegradable, non-toxic and can be metabolized or excreted naturally. And also, the use of materials that break down into non-toxic byproducts, minimizing long-term accumulation is desirable (Su and Kang 2020). Green nanotechnology also lay emphasis on minimizing side effects associated with traditional drug delivery methods by enhancing the precision of drug delivery to the target site with the aid of nanocarriers. They have the ability to increase the concentration of the drug without causing drug toxicity and are capable of reaching remote sites and tissues crossing the blood–brain barrier. Targeted drug delivery employs imaging techniques to monitor the distribution and release of drugs in real-time, allowing for adjustments to optimize therapeutic outcomes (De Jong and Borm 2008).

Green nanotechnology can contribute to resource efficiency by minimizing waste and energy consumption in the manufacturing process. This information can be important for healthcare institutions striving to adopt sustainable practices. The overall energy footprint of the drug delivery system, from production to administration must be cost effective, reliable and the raw materials should be easily accessible (Bosetti and Jones 2019). The overall energy footprint of the drug delivery system, considering the entire life cycle from production to disposal must be evaluated in a conservational perspective (Patra et al. 2018; Madani et al. 2022).

Regarding the waste management strategy, the generation of waste during the production and application of nanocarriers should be reduced throughout their life cycle and recycling or repurposing strategies for nanomaterials after use must be explored (Wu et al. 2017; Ezike et al. 2023) Encouraging awareness among healthcare professionals can stimulate interest and support for further research and innovation in green nanotechnology. This, in turn, can lead to the development of even more advanced and sustainable drug delivery systems. It is vital to adhere to regulatory standards and guidelines to ensure the safety and efficacy of green nanotechnology-based drug delivery systems. Thus, toxicity assessments and biocompatibility studies must be done.

Awareness of the benefits of green nanotechnology in drug delivery can have a positive impact on global health by encouraging the adoption of sustainable practices across healthcare systems worldwide.

14.6 Antibacterial Agents

Green nanotechnology-based antibacterial agents embrace the use of environmentally benign and biocompatible nanomaterials to combat bacterial infections. These materials are designed to be effective against bacteria while curtailing adverse environmental and health impacts. This approach exploits environmentally friendly methods for synthesizing antibacterial nanomaterials, such as plant-mediated synthesis, which involves using plant extracts as reducing and stabilizing agents. Green nanotechnology employs techniques that minimize the use of toxic chemicals and waste generation. It is important to design antibacterial nanomaterials that are biodegradable and can be broken down into non-toxic components after exerting their antimicrobial activity.

Nanomaterials often used as antibacterial agents, fabricated via green synthetic methods are silver nanoparticles (Loo et al. 2018), zinc oxide nanoparticles (Sirelkhatim et al. 2015), chitosan nanoparticles (Qi et al. 2004) and graphene oxide (Ravikumar et al. 2022). Silver nanoparticles have antimicrobial properties and can be synthesized using green methods, such as plant extracts or microbial synthesis (Huq et al. 2022). Chitosan, an abundant carbohydrate polymer derived from the exoskeletons of crustaceans, can be used to create nanoparticles with antibacterial properties (Ikono et al. 2019). Functionalized graphene oxide has shown antibacterial activity and can be produced with minimal environmental impact (Chen et al. 2015; Li et al. 2015).

Antibacterial agents can be incorporated with targeted drug delivery systems to enhance the specificity of, minimizing off-target effects (Kalhapure et al. 2015). The surface of Ligands can be modified with functional groups that specifically recognize bacterial cell components can improve targeting (Lu et al. 2017). Naturally occurring antibacterial compounds, such as essential oils or plant extracts, can be integrated into nanomaterial formulations (Ghosh et al. 2022)). For instance, green tea polyphenols, have been explored for their antibacterial properties when incorporated into nanoparticles (Shetta et al. 2019). The use of biodegradable polymers, such as poly (lactic-co-glycolic acid) (PLGA), can be used in nanoparticle formulations (Essa et al. 2020).

Understanding and optimization of the antibacterial mechanisms of nanomaterials, such as disrupting bacterial cell membranes, interfering with cellular processes, or inducing oxidative stress is needful. Researchers should ensure that the chosen mechanisms do not lead to the development of bacterial resistance. Exploration of combination therapies where green nanotechnology-based antibacterial agents are used synergistically with conventional antibiotics to enhance efficacy and reduce the risk of resistance would be a great notion in an innovative perspective.

14.7 Biomedical Sensors

Biomedical sensors are used in a wide range of applications because of their exceptional capacity to identify and track a vast array of biomarkers, which in turn enables the diagnosis of several illnesses and ailments. Even with the recent significant advancements in this area, there is still a great need to develop the ideal biomedical sensors that can meet the increasing demands and expectations for smart biomedical/diagnostic sensing and accelerate the transition to decentralized/personalized healthcare diagnostics systems. A smart biomedical sensor should have the following qualities: it should be affordable, sensitive, specific, simple to use, able to collect and discard samples, and real-time connected for remote programming, monitoring, reporting, and controlling.

An effective biomedical sensor will be able to accurately and precisely monitor and analyze the concentration of some key bodily physical parameters and (bio) chemicals, alert users about variations in their levels, and share the data it collects with users, doctors, and medical facilities as needed. Some of the main drivers and philosophies behind the shift towards decentralized healthcare through smart biomedical sensing include high costs and limited access to centralized healthcare, a high rate of errors and/or delays in disease diagnosis and, consequently, ineffective and/or delayed therapy, low patient-physician interaction, and not being personalized or tailored to the needs of users (Christodouleas et al. 2018).

For the purpose of future exploitations of broad-impact nanostructured-based innovations and applications, such as novel optical and electrochemical biosensing devices and therapeutic and assessment applications of nanostructured materials (Kiessling et al. 2014), it is imperative to overachieve the challenge of developing dependable environmentally friendly nanotechnologies for nanomaterial synthesis (Petros and DeSimone 2010).

Analytical biochemistry routinely uses biomaterial surfaces' optical and electronic sensing capabilities. Therefore, a wide strategy for the construction of optical or electronic biosensors is provided by the immobilization of biomolecule–nanoparticle conjugates on the surface. So, it is possible to use metal nanoparticles as the visible spectral range contains plasmon absorbance bands that are regulated by the size of the metal nanoparticles, such as Silver and Gold. By attaching to certain molecules, they can change their optical behavior, which makes it possible to detect and quantify analytes using ions. A hybrid system made up of biomolecules and nanoparticles is the basis for the creation of biosensors, which take into account spectrum changes caused by agglomerated metal nanoparticles, such as Au. For instance, when functionalized with two different types of nucleic acid that matched to two different regions of an analyzed DNA, nanoparticles were agglomerated and a red-shifted interparticle plasmon absorbance of the agglomerated nanoparticle was detected (Huo 2007).

14.8 Dentistry

The formulation of nanoparticles such as polymeric nanoparticles, liposomes, ethosomes, proliposomes, solid lipid nanoparticles, and microemulsions offers the possibility of delivering herbal medicines with improved treatment therapy, where the biologically active compounds in herbs could lessen the negative effects of synthetic nanoparticles and, when combined, might gain a better therapeutic potential (Thapa et al. 2013). Because of the vast potential in nanomedicine, the use of herbal remedies has expanded significantly in recent years (Nguyen et al. 2006).

The advantage of using nanoparticle in dentistry has numerous benefits. Nanoparticles are used in herbal medicine to improve selectivity, drug delivery, effectiveness, and safety. Due to their uniform loading and dispersion capacities, herbal drugs can be delivered to target sites even at high concentrations. Herbal nanoparticles increase the stability and solubility of herbal drugs and help localize them to their specific site of action. Herbal nanocarriers carry the optimal amount of the drug to their target site of action, such as tumors and infected cells alone within healthy tissues. Smaller particle size of herbal nanoparticles enhances the permeation to the entire surface area and make it easier. The administration of the medicine by oral and inhalation becomes easier.

Silver nanoparticles (AgNPs) have been extensively studied for their antibacterial properties, and they may be used to a range of dental treatments. When combined with dental materials such as acrylic resins, nanocomposites, adhesives, resin comonomers, intracanal drugs, and implant coatings, AgNPs have a potent antibacterial impact, according to in vitro research. AgNPs are also potential treatments for oral cancers because of their anticancer properties (Noronha et al. 2017). AgNPs was used as a disinfectant in endodontics to get rid of bacteria, toxins, and debris from the root canal system in order to stop microbial growth and avoid infection.

The proven antibacterial effectiveness of AgNP in around 650 bacterial species has contributed significantly to the rapid increase of its utilization in endodontic preparations. To evaluate the complementary benefits of NPs and Aloe vera for root canal disinfection, research was conducted (Haripriya and Ajitha 2017). Nanomaterials encapsulated in A. vera shown long-lasting antibacterial activity against endodontic infections. Dental implant insertion is challenging due to alveolar bone loss, a prevalent disease. The key to alveolar ridge healing is a barrier between the gingiva and the replacement bone that inhibits bacterial infection and fibro-tissue ingrowth while promoting bone formation. AgNP-coated collagen membranes have been shown by Chen et al. to help stop infection following bone graft implantation in alveolar ridge replacement (Chen et al. 2018).

14.9 Bioimaging

Bioimaging is a medical procedure helps to real time analysis of biological functions. This technology does not involve any tool that physically enter the body or attacks the skin. It is an effective tool for visualizing the interior functions of an organism and its disorders. X—ray, MRI (Magnetic Resonance Imaging), CT (Computed Tomography) scan, ultrasound picture etc. are examples of bioimaging in medical industry. Bioimaging gives the cellular level photographs of the body. Therefore, it is very useful for the diagnosis and detection of diseases (Lahoti and Jogdand 2022).

Nanoparticles synthesized using green nanotechnology also have potential for bioimaging. Bioimaging techniques like MRI, PET (Positron Emission Tomography), CT scan etc. use iron oxide nanoparticles as contrast agents. The magnetic iron oxide nanoparticles are also helps to earlier detection of malaria. Gold nanoparticles obtained from Osmundaria obtusiloba shows excellent optical properties used in bioimaging (Nallasamy et al. 2023).

14.10 Tissue Engineering

Tissue engineering helps to produce tissue—like structures using biomaterials, cells, physical and biochemical signals and their combinations. It aims to develop biological substitute which can restore, maintain or improve the functioning of damaged tissues. Applications based on tissue engineering includes the combinations of three pillars, namely, signals (Growth factors, small molecules, mechanical forces), cells (Autologous, Heterologous, differentiated stem cell), and scaffolds (Biomaterials, Hydrogels, Meshes). These three pillars also known as "triad of tissue engineering".

Applications of tissue engineering are implantable organs and tissues, disease studies (using in vitro models), extracorporeal life support units (bioartificial kidney, liver etc.), drug screening, personalized medicine and smart diagnosis. Nanoparticles synthesized using green methods have received more attention in tissue engineering. These types of nanoparticles are used to tissue regeneration and as supporting materials in cell growth (Mhanna and Hasan 2017).

Nanocomposites of magnesium oxide nanoparticles (MgONPs) and bacterial cellulose nanofibers (BC) are applicable in bone regeneration. Green synthesized MgONPs—BC is used as bio scaffold in bone regeneration. MgONPs synthesized from *Camellia sinensis* plant extract. Green synthesized MgONPs loaded with BC scaffold show excellent biological features for cell attachment and proliferation. Based on their studies MgONPs—BC hydrogel is an excellent system for bone regeneration (Ghanbari et al. 2023).

Silver nanoparticles are prepared from gum kondagogu (*Cochlospermum gossypium) via* green synthesis. Next step is the impregnation of silver nanoparticles to gum kondagogu—chitosan (GK–CH–Ag biosheet). Based on their studies, they confirmed antibacterial activity using *S. aureus* and the potential of GK–CH–Ag for wound healing and tissue engineering applications (Rathore et al. 2022).

14.11 Conclusion and Prospects

Nanotechnology is one of the rapidly growing areas of modern research. It deals with the applications of nanoscience theory by observing, manipulating, controlling and manufacturing of the matter in the nanometer scale. Nanoparticles can be synthesized by physical, chemical or biological methods. Biological processes are environmentally benign because it uses plants and microorganisms as the starting material.

Traditional methods used for the synthesis of nanoparticles are highly expensive and generate toxic materials. It is necessary to find low cost, environmentally benign routes for the synthesis of nanoparticles. Therefore, green nanotechnology grabs the attention of researchers for developing nanomaterials. Green synthesis is more efficient, less expensive, simpler and environmentally friendly than other conventional methods. Biomolecules and living organisms are the precursors of nanomaterials in green nanotechnology.

Green nanotechnology shows numerous applications in various fields such as nanoelectronics, environmental sector, industries, agricultural sector, biomaterials and nanomedicine. This chapter mainly focus on the biomedical applications of plant—based metal and metal oxide nanoparticles. Researchers revealed that green synthesized nanoparticles are superior in biomedical applications. Metallic nanoparticles synthesized from medicinal plants show more pronounced therapeutic properties such as antimicrobial activity, antioxidant activity, insecticidal activity, and wound healing properties.

Nanotechnology is widely used in the treatment of cancer and it plays role in detection, therapies, and development of new machineries. Green nanotechnology for targeted drug delivery implicates the use of environmentally sustainable and biocompatible materials. Bio fabrication techniques, such as using plant extracts, bacteria, or fungi are explored to synthesize nanoparticles for targeted drug delivery. Targeted drug delivery employs imaging techniques to monitor the distribution and release of drugs in real-time. Nanomaterials used as antibacterial agents, fabricated via green synthetic methods are silver nanoparticles, zinc oxide nanoparticles chitosan nanoparticles and graphene oxide etc.

Biomedical sensors are one of the advancements of green nanotechnology. It gives the real time analysis of the sample. Green nanotechnology also applicable in the field of dentistry. The advantage of using nanoparticle in dentistry has numerous benefits. Nanoparticles are used in herbal medicine to improve selectivity, drug delivery, effectiveness, and safety. Nanoparticles synthesized using green nanotechnology also have potential for bioimaging. Bioimaging techniques like MRI, PET, CT scan etc. use iron oxide nanoparticles as contrast agents. Nanoparticles synthesized using green methods have received more attention in tissue engineering. These types of nanoparticles are used to tissue regeneration and as supporting materials in cell growth.

It was proven that the synthesis of metal and metal oxide nanoparticles using green methods has various advantages, such as being profitable, efficient, and biocompatible. Furthermore, these nanoparticles have shown remarkable biological, chemical, and physical characteristics that make them useful for a range of pharmacological and therapeutic applications. In addition, the main concern of nowadays is the adverse effects of the medicines that are being used. It is necessary to understand the merits and demerits of green synthesized nanoparticles in a better way. Similarly, green nanotechnology faces a lot of issues and challenges. Hence, additional scientific investigation is required to investigate thoroughly the characteristics of biologically produced nanoparticles.

References

Abdolmaleki A, Asadi A, Gurushankar K et al (2020) Importance of nano medicine and new drug therapies for cancer. Adv Pharm Bull 11:450–457. https://doi.org/10.34172/apb.2021.052

Abdullahi A, Pulipaka S, Sharma R, Pavani S (2019) Recent advances in nanotechnology for cancer therapy-a review. J Emerg Technol Innov Res 6:2026–2048

Andreas Bernkop-Schnürch SD (2012) Chitosan-based drug delivery systems 81:463–469

Anirudhan TS, Manjusha V, Chithra Sekhar V (2021) A new biodegradable nano cellulose-based drug delivery system for pH-controlled delivery of curcumin. Int J Biol Macromol 183:2044–2054. https://doi.org/10.1016/j.ijbiomac.2021.06.010

Anu K, Singaravelu G, Murugan K, Benelli G (2017) Green-synthesis of selenium nanoparticles using garlic cloves (Allium sativum): biophysical characterization and cytotoxicity on vero cells. J Clust Sci 28:551–563. https://doi.org/10.1007/s10876-016-1123-7

Azadpour A, Hajrasouliha S, Khaleghi S (2022) Green synthesized-silver nanoparticles coated with targeted chitosan nanoparticles for smart drug delivery. J Drug Deliv Sci Technol 74:103554. https://doi.org/10.1016/j.jddst.2022.103554

Baetke SC, Lammers T, Kiessling F (2015) Applications of nanoparticles for diagnosis and therapy of cancer. Br J Radiol 88:20150207. https://doi.org/10.1259/bjr.20150207

Bakht MK, Sadeghi M, Pourbaghi-Masouleh M, Tenreiro C (2012) Scope of nanotechnology-based radiation therapy and thermotherapy methods in cancer treatment. Curr Cancer Drug Targets 12: 998–1015. https://doi.org/10.2174/156800912803251216

Banerjee D, Harfouche R, Sengupta S (2011) Nanotechnology-mediated targeting of tumor angiogenesis. Vasc Cell 3(1):3. https://doi.org/10.1186/2045-824X-3-3

Banerjee S, Banerjee I, Dutta M, Pal R (2021) Fabrication of iron nanoparticles using Leptolyngbya valderiana and investigation of its Cr (VI) removal potential in the free and biomass associated forms. Algal Res 58:102373. https://doi.org/10.1016/j.algal.2021.102373

Baranwal J, Barse B, Di Petrillo A, Gatto G, Pilia L, Kumar A (2023) Nanoparticles in cancer diagnosis and treatment. Materials (Basel) 16(15):5354. https://doi.org/10.3390/ma16155354

Bayda S, Adeel M, Tuccinardi T, Cordani M, Rizzolio F (2019) The history of nanoscience and nanotechnology: from chemical–physical applications to nanomedicine. Molecules 25(1):112. https://doi.org/10.3390/molecules25010112

Bergs JWJ, Wacker MG, Hehlgans S, et al (2015) The role of recent nanotechnology in enhancing the efficacy of radiation therapy Biochim Biophys Acta (BBA) Rev Cancer 1856:130–143. https://doi.org/10.1016/j.bbcan.2015.06.008

Bharali DJ, Mousa SA (2010) Emerging nanomedicines for early cancer detection and improved treatment: current perspective and future promise. Pharmacol Ther 128:324–335. https://doi.org/10.1016/j.pharmthera.2010.07.007

Bhuvaneshwari V, Ramasamy NK, Kumar SI et al (2022) Antimicrobial activity of copper nanomaterials: current status and future perspectives. In: Copper nanostructures: next-generation of agrochemicals for sustainable agroecosystems. Elsevier, pp 453–475

Bosetti R, Jones SL (2019) Cost–effectiveness of nanomedicine: estimating the real size of nano-costs. Nanomedicine 14:1367–1370. https://doi.org/10.2217/nnm-2019-0130

Burduşel A-C, Gherasim O, Grumezescu AM, Mogoantă L, Ficai A, Andronescu E (2018) Biomedical applications of silver nanoparticles: an up-to-date overview. Nanomaterials (Basel) 8(9):681. https://doi.org/10.3390/nano8090681

Chatterjee DK, Wolfe T, Lee J, Brown AP, Singh PK, Bhattarai SR, Diagaradjane P, Krishnan S (2013) Convergence of nanotechnology with radiation therapy-insights and implications for clinical translation. Transl Cancer Res 2(4):256–268. https://doi.org/10.3978/j.issn.2218-676X.2013.08.10

Chen X, Huang X, Zheng C et al (2015) Preparation of different sized nano-silver loaded on functionalized graphene oxide with highly effective antibacterial properties. J Mater Chem B 3: 7020–7029. https://doi.org/10.1039/C5TB00280J

Chen P, Wu Z, Leung A et al (2018) Fabrication of a silver nanoparticle-coated collagen membrane with anti-bacterial and anti-inflammatory activities for guided bone regeneration. Biomed Mater 13:065014. https://doi.org/10.1088/1748-605X/aae15b

Christodouleas DC, Kaur B, Chorti P (2018) From point-of-care testing to eHealth diagnostic devices (eDiagnostics). ACS Cent Sci 4:1600–1616. https://doi.org/10.1021/acscentsci.8b00625

Clavreul A, Pourbaghi-Masouleh M, Roger E, Menei P (2019) Nanocarriers and nonviral methods for delivering antiangiogenic factors for glioblastoma therapy: the story so far. Int J Nanomedicine 14:2497–2513. https://doi.org/10.2147/IJN.S194858

De Jong WH, Borm PJ (2008) Drug delivery and nanoparticles: applications and hazards. Int J Nanomed 3(2):133–149

Dhilip Kumar SS, Abrahamse H (2022) Recent advances in the development of biocompatible nanocarriers and their cancer cell targeting efficiency in photodynamic therapy. Front Chem 10. https://doi.org/10.3389/fchem.2022.969809

Essa D, Kondiah PPD, Choonara YE, Pillay V (2020) The design of poly(lactide-co-glycolide) nanocarriers for medical applications. Front Bioeng Biotechnol 8. https://doi.org/10.3389/fbioe.2020.00048

Ezike TC, Okpala US, Onoja UL et al (2023) Advances in drug delivery systems, challenges and future directions. Heliyon 9:e17488. https://doi.org/10.1016/j.heliyon.2023.e17488

Ghanbari E, Khazaei M, Mehdipour A, Khoshfeterat A, Niknafs B (2023) Green synthesized magnesium oxide nanoparticles reinforce osteogenesis properties of bacterial cellulose scaffolds for bone tissue engineering applications: an in vitro assessment. Cell J 25:483–495

Ghosh S, Nandi S, Basu T (2022) Nano-antibacterials using medicinal plant components: an overview. Front Microbiol 12. https://doi.org/10.3389/fmicb.2021.768739

Gong N, Sheppard NC, Billingsley MM et al (2021) Nanomaterials for T-cell cancer immunotherapy. Nat Nanotechnol 16:25–36. https://doi.org/10.1038/s41565-020-00822-y

Griffin S, Masood MI, Nasim MJ, Sarfraz M, Ebokaiwe AP, Schäfer K-H, Keck CM, Jacob C (2017) Natural nanoparticles: a particular matter inspired by nature. Antioxidants (Basel) 7(1):3

Gröning PA (2005) Nanotechnology: an approach to mimic natural architectures and concepts. Adv Eng Mater 7:279–291. https://doi.org/10.1002/adem.200500089

Gu FX, Karnik R, Wang AZ et al (2007) Targeted nanoparticles for cancer therapy. Nano Today 2: 14–21. https://doi.org/10.1016/S1748-0132(07)70083-X

Haripriya S, Ajitha P (2017) Antimicrobial efficacy of silver nanoparticles of Aloe vera. J Adv Pharm Educ Res 7(2):163–167

Ho BN, Pfeffer CM, Singh ATK (2017) Update on nanotechnology-based drug delivery systems in cancer treatment. Anticancer Res 37:10.21873/anticanres.12044

Hovden R, Wolf SE, Holtz ME et al (2015) Nanoscale assembly processes revealed in the nacroprismatic transition zone of Pinna nobilis mollusc shells. Nat Commun 6:10097. https://doi.org/10.1038/ncomms10097

Huo Q (2007) A perspective on bioconjugated nanoparticles and quantum dots. Colloids Surf B Biointerfaces 59:1–10. https://doi.org/10.1016/j.colsurfb.2007.04.019

Huq MA, Ashrafudoulla M, Rahman MM et al (2022) Green synthesis and potential antibacterial applications of bioactive silver nanoparticles: a review. Polymers (Basel) 14:742. https://doi.org/10.3390/polym14040742

Ikono R, Vibriani A, Wibowo I et al (2019) Nanochitosan antimicrobial activity against Streptococcus mutans and Candida albicans dual-species biofilms. BMC Res Notes 12:383. https://doi.org/10.1186/s13104-019-4422-x

Kalhapure RS, Suleman N, Mocktar C et al (2015) Nanoengineered drug delivery systems for enhancing antibiotic therapy. J Pharm Sci 104:872–905. https://doi.org/10.1002/jps.24298

Kanwar R, Rathee J, Salunke DB, Mehta SK (2019) Green nanotechnology-driven drug delivery assemblies. ACS Omega 4:8804–8815. https://doi.org/10.1021/acsomega.9b00304

Kargozar S, Baino F, Hamzehlou S et al (2020) Nanotechnology for angiogenesis: opportunities and challenges. Chem Soc Rev 49:5008–5057. https://doi.org/10.1039/C8CS01021H

Kgosiemang IK, Lefojane R, Direko P et al (2020) Green synthesis of magnesium and cobalt oxide nanoparticles using *Euphorbia tirucalli*: characterization and potential application for breast cancer inhibition. Inorg Nano Met Chem 50:1070–1080. https://doi.org/10.1080/24701556.2020.1735422

Khan ZUH, Khan A, Chen Y et al (2017a) Biomedical applications of green synthesized Nobel metal nanoparticles. J Photochem Photobiol B 173:150–164. https://doi.org/10.1016/j.jphotobiol.2017.05.034

Khan ZUH, Khan A, Chen Y et al (2017b) Biomedical applications of green synthesized nobel metal nanoparticles. J Photochem Photobiol B 173:150–164. https://doi.org/10.1016/j.jphotobiol.2017.05.034

Khan F, Shariq M, Asif M et al (2022) Green nanotechnology: plant-mediated nanoparticle synthesis and application. Nano 12:673. https://doi.org/10.3390/nano12040673

Kiessling F, Mertens ME, Grimm J, Lammers T (2014) Nanoparticles for imaging: top or flop? Radiology 273:10–28. https://doi.org/10.1148/radiol.14131520

Klochkov SG, Neganova ME, Nikolenko VN et al (2021) Implications of nanotechnology for the treatment of cancer: recent advances. Semin Cancer Biol 69:190–199. https://doi.org/10.1016/j.semcancer.2019.08.028

Koyyati R, Kudle KR, Padigya PRM (2016) Evaluation of antibacterial and cytotoxic activity of green synthesized cobalt nanoparticles using Raphanus sativus var. longipinnatus leaf extract. 9(3):466–472

Kulkarni D, Sherkar R, Shirsathe C et al (2023) Biofabrication of nanoparticles: sources, synthesis, and biomedical applications. Front Bioeng Biotechnol 11. https://doi.org/10.3389/fbioe.2023.1159193

Lahoti HS, Jogdand SD (2022) Bioimaging: evolution, significance, and deficit. Cureus. https://doi.org/10.7759/cureus.28923

Lee KY, Mooney DJ (2012) Alginate: properties and biomedical applications. Prog Polym Sci 37:106–126. https://doi.org/10.1016/j.progpolymsci.2011.06.003

Lee H, Song JY, Kim BS (2013) Biological synthesis of copper nanoparticles using *Magnolia kobus* leaf extract and their antibacterial activity. J Chem Technol Biotechnol 88:1971–1977. https://doi.org/10.1002/jctb.4052

Lee KX, Shameli K, Yew YP et al (2020) Recent developments in the facile bio-synthesis of gold nanoparticles (AuNPs) and their biomedical applications. Int J Nanomedicine 15:275–300. https://doi.org/10.2147/IJN.S233789

Li P, Sun S, Dong A et al (2015) Developing of a novel antibacterial agent by functionalization of graphene oxide with guanidine polymer with enhanced antibacterial activity. Appl Surf Sci 355:446–452. https://doi.org/10.1016/j.apsusc.2015.07.148

Liang X-J, Chen C, Zhao Y, Wang PC (2010) Circumventing tumor resistance to chemotherapy by nanotechnology. Methods Mol Biol 596:467–488

Loizidou M, Seifalian AM (2010) Nanotechnology and its applications in surgery. Br J Surg 97:463–465. https://doi.org/10.1002/bjs.7074

Loo YY, Rukayadi Y, Nor-Khaizura M-A-R et al (2018) In vitro antimicrobial activity of green synthesized silver nanoparticles against selected gram-negative foodborne pathogens. Front Microbiol 9. https://doi.org/10.3389/fmicb.2018.01555

Lu HD, Yang SS, Wilson BK et al (2017) Nanoparticle targeting of gram-positive and gram-negative bacteria for magnetic-based separations of bacterial pathogens. Appl Nanosci 7:83–93. https://doi.org/10.1007/s13204-017-0548-0

Madani M, Hosny S, Alshangiti DM et al (2022) Green synthesis of nanoparticles for varied applications: green renewable resources and energy-efficient synthetic routes. Nanotechnol Rev 11:731–759. https://doi.org/10.1515/ntrev-2022-0034

Mariadoss AVA, Saravanakumar K, Sathiyaseelan A et al (2020) Folic acid functionalized starch encapsulated green synthesized copper oxide nanoparticles for targeted drug delivery in breast cancer therapy. Int J Biol Macromol 164:2073–2084. https://doi.org/10.1016/j.ijbiomac.2020.08.036

Menon S, Devi S, Agarwal H, Shanmugam VK (2019) Efficacy of biogenic selenium nanoparticles from an extract of ginger towards evaluation on anti-microbial and anti-oxidant activities. Colloid Interf Sci Commun 29:1–8. https://doi.org/10.1016/j.colcom.2018.12.004

Mhanna R, Hasan A (2017) Introduction to tissue engineering. In: Tissue engineering for artificial organs. Wiley, pp 1–34

Mikhailova EO (2022) Green synthesis of platinum nanoparticles for biomedical applications. J Funct Biomater 13:260. https://doi.org/10.3390/jfb13040260

Mishra MP, Padhy RN (2018) Antibacterial activity of green silver nanoparticles synthesized from Anogeissus acuminata against multidrug resistant urinary tract infecting bacteria in vitro and host-toxicity testing. J Appl Biomed 16:120–125. https://doi.org/10.1016/j.jab.2017.11.003

Morshed M, Chowdhury EH (2019) Gene delivery and clinical applications. In: Encyclopedia of biomedical engineering. Elsevier, pp 345–351

Nakkala JR, Mata R, Bhagat E, Sadras SR (2015) Green synthesis of silver and gold nanoparticles from Gymnema sylvestre leaf extract: study of antioxidant and anticancer activities. J Nanopart Res 17:151. https://doi.org/10.1007/s11051-015-2957-x

Nallasamy L, Murugavelu GS, Ganesh S et al (2023) Green nanotechnology revolution in biomedical application and treatments. In: Nanovaccinology. Springer, Cham, pp 181–191

Naumov GN, Akslen LA, Folkman J (2006) Role of angiogenesis in human tumor dormancy: animal models of the Angiogenic switch. Cell Cycle 5:1779–1787. https://doi.org/10.4161/cc.5.16.3018

Nguyen LA, He H, Pham-Huy C (2006) Chiral drugs: an overview. Int J Biomed Sci 2(2):85–100

Ni D, Bu W, Ehlerding EB et al (2017) Engineering of inorganic nanoparticles as magnetic resonance imaging contrast agents. Chem Soc Rev 46:7438–7468. https://doi.org/10.1039/C7CS00316A

Nicolas J, Mura S, Brambilla D et al (2013) Design, functionalization strategies and biomedical applications of targeted biodegradable/biocompatible polymer-based nanocarriers for drug delivery. Chem Soc Rev 42:1147–1235. https://doi.org/10.1039/C2CS35265F

Noah NM, Ndangili PM (2022) Green synthesis of nanomaterials from sustainable materials for biosensors and drug delivery. Sens Int 3:100166. https://doi.org/10.1016/j.sintl.2022.100166

Noronha VT, Paula AJ, Durán G et al (2017) Silver nanoparticles in dentistry. Dent Mater 33:1110–1126. https://doi.org/10.1016/j.dental.2017.07.002

Ojo OA, Olayide II, Akalabu MC, Ajiboye BO, Ojo AB, Oyinloye BE, Ramalingam M (2020) Nanoparticles and their biomedical applications. Biointerface Res Appl Chem 11:8431–8445. https://doi.org/10.33263/BRIAC111.84318445

Oladipo AO, Lebelo SL, Msagati TAM (2023) Nanocarrier design–function relationship: the prodigious role of properties in regulating biocompatibility for drug delivery applications. Chem Biol Interact 377:110466. https://doi.org/10.1016/j.cbi.2023.110466

Park K (2014) Controlled drug delivery systems: past forward and future back. J Control Release 190:3–8. https://doi.org/10.1016/j.jconrel.2014.03.054

Patra JK, Das G, Fraceto LF et al (2018) Nano based drug delivery systems: recent developments and future prospects. J Nanobiotechnol 16:71. https://doi.org/10.1186/s12951-018-0392-8

Petros RA, DeSimone JM (2010) Strategies in the design of nanoparticles for therapeutic applications. Nat Rev Drug Discov 9:615–627. https://doi.org/10.1038/nrd2591

Pokropivny V, Lohmus R, Hussainova I, Pokropivny A, Vlassov S (2007) Introduction in nanomaterials and nanotechnology. University of Tartu

Pucci C, Martinelli C, Ciofani G (2019) Innovative approaches for cancer treatment: current perspectives and new challenges. Ecancermedicalscience 13. https://doi.org/10.3332/ecancer.2019.961

Qi L, Xu Z, Jiang X et al (2004) Preparation and antibacterial activity of chitosan nanoparticles. Carbohydr Res 339:2693–2700. https://doi.org/10.1016/j.carres.2004.09.007

Qian ZM (2002) Targeted drug delivery via the transferrin receptor-mediated endocytosis pathway. Pharmacol Rev 54:561–587. https://doi.org/10.1124/pr.54.4.561

Rajasekar J, Perumal MK, Vallikannan B (2019) A critical review on anti-angiogenic property of phytochemicals. J Nutr Biochem 71:1–15. https://doi.org/10.1016/j.jnutbio.2019.04.006
Rajendran A, Alsawalha M, Alomayri T (2021) Biogenic synthesis of husked rice-shaped iron oxide nanoparticles using coconut pulp (Cocos nucifera L.) extract for photocatalytic degradation of Rhodamine B dye and their in vitro antibacterial and anticancer activity. J Saudi Chem Soc 25:101307. https://doi.org/10.1016/j.jscs.2021.101307
Rathore HS, Sivagnanam UT, Abraham LS et al (2022) Green synthesized silver nanoparticles-impregnated novel gum kondagogu–chitosan biosheet for tissue engineering and wound healing applications. Polym Bull 79:7215–7227. https://doi.org/10.1007/s00289-021-03832-5
Ravikumar V, Mijakovic I, Pandit S (2022) Antimicrobial activity of graphene oxide contributes to alteration of key stress-related and membrane bound proteins. Int J Nanomedicine 17:6707–6721. https://doi.org/10.2147/IJN.S387590
Revia RA, Stephen ZR, Zhang M (2019) Theranostic nanoparticles for RNA-based cancer treatment. Acc Chem Res 52:1496–1506. https://doi.org/10.1021/acs.accounts.9b00101
Riddin TL, Gericke M, Whiteley CG (2006) Analysis of the inter- and extracellular formation of platinum nanoparticles by *Fusarium oxysporum* f. sp. *lycopersici* using response surface methodology. Nanotechnology 17:3482–3489. https://doi.org/10.1088/0957-4484/17/14/021
Roopan SM, Kumar SHS, Madhumitha G, Suthindhiran K (2015) Biogenic-production of SnO2 nanoparticles and its cytotoxic effect against hepatocellular carcinoma cell line (HepG2). Appl Biochem Biotechnol 175:1567–1575. https://doi.org/10.1007/s12010-014-1381-5
Shanmuganathan R, Sathiyavimal S, Hoang Le Q et al (2023) Green synthesized cobalt oxide nanoparticles using Curcuma longa for anti-oxidant, antimicrobial, dye degradation and anticancer property. Environ Res 236:116747. https://doi.org/10.1016/j.envres.2023.116747
Shetta A, Kegere J, Mamdouh W (2019) Comparative study of encapsulated peppermint and green tea essential oils in chitosan nanoparticles: encapsulation, thermal stability, in-vitro release, antioxidant and antibacterial activities. Int J Biol Macromol 126:731–742. https://doi.org/10.1016/j.ijbiomac.2018.12.161
Shreyash N, Bajpai S, MohdA K et al (2021) Green synthesis of nanoparticles and their biomedical applications: a review. ACS Appl Nano Mater 4:11428–11457. https://doi.org/10.1021/acsanm.1c02946
Singh A, Suki M, Sharma R, Ingle P (2020) Applications of nanotechnology: a review. Int J Adv Res Chem Sci 7. https://doi.org/10.20431/2349-0403.0702004
Singhal S, Nie S, Wang MD (2010) Nanotechnology applications in surgical oncology. Annu Rev Med 61:359–373. https://doi.org/10.1146/annurev.med.60.052907.094936
Sirelkhatim A, Mahmud S, Seeni A et al (2015) Review on zinc oxide nanoparticles: antibacterial activity and toxicity mechanism. Nanomicro Lett 7:219–242. https://doi.org/10.1007/s40820-015-0040-x
Song G, Cheng L, Chao Y et al (2017a) Emerging nanotechnology and advanced materials for cancer radiation therapy. Adv Mater 29. https://doi.org/10.1002/adma.201700996
Song W, Musetti SN, Huang L (2017b) Nanomaterials for cancer immunotherapy. Biomaterials 148:16–30. https://doi.org/10.1016/j.biomaterials.2017.09.017
Su S, Kang PM (2020) Systemic review of biodegradable nanomaterials in nanomedicine. Nano 10: 656. https://doi.org/10.3390/nano10040656
Subramanyam GK, Gaddam SA, Kotakadi VS et al (2021) *Argyreia nervosa (Samudra pala)* leaf extract mediated silver nanoparticles and evaluation of their antioxidant, antibacterial activity, *in vitro* anticancer and apoptotic studies in KB oral cancer cell lines. Artif Cells Nanomed Biotechnol 49:634–649. https://doi.org/10.1080/21691401.2021.1996384
Sulaiman GM, Tawfeeq AT, Naji AS (2018) Biosynthesis, characterization of magnetic iron oxide nanoparticles and evaluations of the cytotoxicity and DNA damage of human breast carcinoma cell lines. Artif Cells Nanomed Biotechnol 46:1215–1229. https://doi.org/10.1080/21691401.2017.1366335
Sun J, Bhushan B, Tong J (2013) Structural coloration in nature. RSC Adv 3:14862. https://doi.org/10.1039/c3ra41096j

Suner SS, Sahiner M, Sengel SB et al (2018) Responsive biopolymer-based microgels/nanogels for drug delivery applications. In: Stimuli responsive polymeric nanocarriers for drug delivery applications, vol 1. Elsevier, pp 453–500

Teimuri-mofrad R, Hadi R, Tahmasebi B, Farhoudian S, Mehravar M, Nasiri R (2017) Green synthesis of gold nanoparticles using plant extract: mini-review. Nanochem Res 2(1):8–19

Thapa RK, Khan GM, Parajuli-Baral K, Thapa P (2013) Herbal medicine incorporated nanoparticles: advancements in herbal treatment. Asian J Biomed Pharm Sci 3(24):7–14

Ting G, Chang C-H, Wang H-E, Lee T-W (2010) Nanotargeted radionuclides for cancer nuclear imaging and internal radiotherapy. J Biomed Biotechnol 2010:1–17. https://doi.org/10.1155/2010/953537

Verma M, Sheoran P, Chaudhury A (2018) Application of nanotechnology for cancer treatment. In: Advances in animal biotechnology and its applications. Springer, Singapore, pp 161–178

Verma A, Gautam SP, Bansal KK et al (2019) Green nanotechnology: advancement in phytoformulation research. Medicines 6:39. https://doi.org/10.3390/medicines6010039

Vijayaram S, Razafindralambo H, Sun Y-Z et al (2024) Applications of green synthesized metal nanoparticles—a review. Biol Trace Elem Res 202:360–386. https://doi.org/10.1007/s12011-023-03645-9

Vyas J, Rana S (2018) Synthesis of selenium nanoparticles using Allium sativum extract and analysis of their antimicrobial property against gram positive bacteria. Pharm Innov 7(9): 262–266

Wang Y, Li P, Truong-Dinh Tran T et al (2016) Manufacturing techniques and surface engineering of polymer based nanoparticles for targeted drug delivery to cancer. Nano 6:26. https://doi.org/10.3390/nano6020026

Wei G, Wang Y, Yang G et al (2021) Recent progress in nanomedicine for enhanced cancer chemotherapy. Theranostics 11:6370–6392. https://doi.org/10.7150/thno.57828

Wen J, Fu W, Ding S et al (2022) Pyrogallic acid modified nanoscale zero-valent iron efficiently removed Cr(VI) by improving adsorption and electron selectivity. Chem Eng J 443:136510. https://doi.org/10.1016/j.cej.2022.136510

Winawer S (2003) Colorectal cancer screening and surveillance: clinical guidelines and rationale? Update based on new evidence. Gastroenterology 124:544–560. https://doi.org/10.1053/gast.2003.50044

Wishart DS (2015) Is cancer a genetic disease or a metabolic disease? EBioMedicine 2(6):478–479

Wojtynek NE, Mohs AM (2020) Image-guided tumor surgery: the emerging role of nanotechnology. WIREs Nanomed Nanobiotechnol 12. https://doi.org/10.1002/wnan.1624

Wu J, Jiang W, Shen Y et al (2017) Synthesis and characterization of mesoporous magnetic nanocomposites wrapped with chitosan gatekeepers for pH-sensitive controlled release of doxorubicin. Mater Sci Eng C 70:132–140. https://doi.org/10.1016/j.msec.2016.08.054

Zhang P, Meng J, Li Y et al (2021) Nanotechnology-enhanced immunotherapy for metastatic cancer. Innovation 2:100174. https://doi.org/10.1016/j.xinn.2021.100174

Chapter 15
Drug Delivery Using Green Nanotechnology

Mohamed A. Selim and Salem S. Salem

Abstract An overview of the application of nanobiotechnology in drug delivery is provided in this chapter. There are numerous varieties of nanoparticles. Due to better absorption, nanoparticulate versions of commonly used medications are more effective than traditional formulations and require lower dosages with fewer side effects. Targeted drug delivery of anticancer medications is made easier by nanobiotechnology. In order to deliver drugs to specific organs, nanoparticles also let drugs pass through biological barriers in the human body. For instance, drugs are able to enter the brain through the blood-brain barrier. Applications of nanobiotechnology in biological therapy delivery are growing, including treatment of cancer, oscular disease, Heart diseases, Respiratory diseases, Dermatology treatment, Neurological disorder, Dentistry treatment and also Gastroenterology diseases. Some nanoparticles have the capacity to contain many therapeutic molecules, allowing for the use of multimodal therapy in conjunction with physical modalities like radiation therapy for cancer patients. The application of nanotechnology to the development of intelligent medication release devices is one of the other predicted advances in this field.

Keywords Nanotechnology · Nanoparticle · Nanomedicine · Drug delivery system

15.1 Introduction

Nanotechnologies have revolutionized the development of medicine, especially in diagnostic procedures, imaging, and drug delivery (Pourmadadi et al. 2023). Researchers at IBM are also working to use nanotechnology to examine DNA in a matter of minutes rather than weeks, to provide cancer patients with a customized treatment plan (Deepa et al. 2023). Nanomedicine is the medical application of nanotechnology, encompassing the use of nanoparticles for illness diagnosis,

M. A. Selim · S. S. Salem (✉)
Botany and Microbiology Department, Faculty of Science, Al-Azhar University, Cairo, Egypt
e-mail: salemsalahsalem@azhar.edu.eg

© The Author(s), under exclusive license to Springer Nature Switzerland AG 2025
J. M. Al-Khayri et al. (eds.), *Green Nanotechnology Applications for Ecosystem Sustainability*, Nanotechnology in Plant Sciences 4,
https://doi.org/10.1007/978-3-031-90766-1_15

monitoring, control, prevention, and therapy. Nanomaterials can be used for therapeutic reasons in three different areas of nanomedicine: regenerative medicine, controlled drug delivery (nano-therapy), and diagnosis (nano-diagnosis). Positive developments in clinical practice are anticipated thanks to nanomedicine. It is possible to design nanomaterials with particular biological goals in mind by altering their surface, size, shape, and compound organization (Kinnear et al. 2017). A successful biological outcome requires the application of a meticulous molecular strategy (Saiding et al. 2023). The treatment of diseases has been greatly impacted by nanotechnology in the medical area, especially with regard to the utilization of sophisticated medication delivery systems made of both synthetic and natural substances (Mohammed et al. 2023). For example, scientists at Harvard University's Wyss Institute created a "nano-robot" that may administer anticancer medications by selectively targeting cancer cells. Additionally, Cardiovascular diseases can be treated with nanorobots, such as those who are able to treat patients with coronary artery blockage or those that can act as artificial platelets to heal blood vessels. Drug delivery systems represent one of the key areas where nanotechnology is being used in medicine. Many traditional/conventional medications are being worked on to increase their efficiency because currently, it is believed the majority of them have poor bioavailability and water solubility, which restricts their uptake and retention in biological systems (Yusuf et al. 2023). Several nanoparticles' reduced size and physical composition are expected to result in enhanced pharmacokinetic features; depending on the type of particle, they may be able to selectively operate on specific cells (Veiga et al. 2023). These particles' easy penetration into target cells and ability to assemble in subcellular composition to change the functions of cells could be useful in the handling of long-term conditions like diabetes, cancer, and kidney problems (Huang et al. 2024). Different techniques can be used to synthesize a range of nanoparticles, such as liposomes, micelles, and gold nano shells. The shapes and sizes of those nanomaterials can be changed depending on the intended functionality during the synthesis process. Larger-sized particles that form from the aggregation of nanoparticles during synthesis may or may not be more harmful, depending on their composition (Baviskar et al. 2024). Reactive groups or molecules, such as antibodies, are added, to the surface of nanoparticles, targeted drug delivery systems can change the surface chemistry of the particles. To address this, more efficient and focused administration of the medicinal drug is achieved through the evolution of nanoparticle drug delivery systems (DSSs) (Finbloom et al. 2023). This prevents the effect of administered drugs from damaging surrounding organs, which would otherwise occur if drug use were free to use. Research on DSSs has made great progress in the previous few decades, and a number of DSSs are already being studied and created to treat diseases including cancer and neurological disorders (Liu et al. 2023). There are numerous varieties of nanoparticles (Salem and Fouda 2021; Salem 2023; Salem et al. 2023). Due to better absorption, nanoparticulate versions of commonly used medications are more effective than traditional formulations and require lower dosages with fewer side effects. Targeted drug delivery of anticancer medications is made easier by nanobiotechnology, which is crucial for the treatment of cancer (Vyas et al. 2023). In order to deliver drugs to specific organs, nanoparticles also let drugs pass through biological barriers in the human body. For instance, drugs can pass reaching the brain across the blood-brain barrier. Applications of nanobiotechnology in biological therapy delivery are growing, including siRNAs, monoclonal antibodies, and cell and gene treatments (Silvestrini et al. 2024).

This chapter highlights the advances in medication delivery that NM-based drug delivery systems have made feasible. The most widely used advanced nano carrier types for drug delivery include polymeric carriers, liposome formulations, protein nanoparticles, micellar formulations, nanocrystals, and many more. These carriers efficiently transport and release drugs at the site of sickness.

15.2 Drug Delivery System (DDS)

Numerous drug delivery and targeting systems are now in development. A feature shared by numerous drug delivery technologies, including liposomes, synthetic polymers, and microcapsules, is the enhancement of drug bioavailability and accumulation at the target location (Pourmadadi et al. 2023). To enhance the efficiency in the administration of active drug components, like medications proteins, peptides, enzymes, antibodies, and vaccinations, etc., drug delivery systems (DDS) are being researched in great detail and are becoming more widely used (Khizar et al. 2023).

The limitations of standard DDS (solutions, suspensions, and emulsions)—such as the effect of the first pass, toxicity, intolerance, changes in drug levels in the plasma, lack of sustained action, low efficacy, poor selectivity, low bioavailability, high dosage, etc. make the development of an advanced type of DDS imperative (Kanwar et al. 2019).

15.3 Different Types of Nanoparticles in Drug Delivery Systems

15.3.1 Inorganic Nanoparticles

Surface plasmon resonance (SPR), a special feature of metal nanoparticles like silver and gold, is absent from liposomes, dendrimers, and micelles (Fig. 15.1) (Daraee et al. 2016). Despite the fact that their uptake may be attributed to either paracellular transport or transcytosis, in vivo transport investigations that focused on the activities associated with their drug delivery but were incapable of determining assuming their toxicity is specifically associated with their ionized or particulate state (Choi et al. 2013). Gold nanoparticles (AuNPs) can attach cancer medicines to their surfaces through physical absorption as well as ionic or covalent interactions. Afterwards, biological or light stimulation can be used to control and deliver these

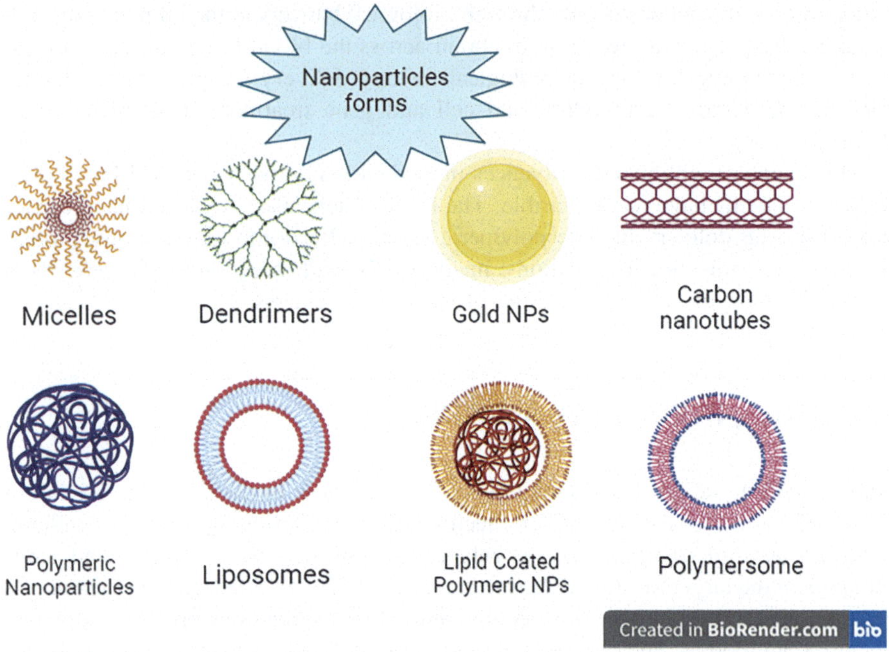

Fig. 15.1 Demonstrates many nanoparticle forms that are employed as anticancer drugs

medications (Kong et al. 2017). Despite the antimicrobial qualities of silver nanoparticles, nothing is known about how they might be used to deliver medication. A multifunctional system of silver nanoparticle-covalently coupled polyacrylamide/dextran nano-hydrogels was developed by Prusty and Swain. Its in vitro release of ornidazole was 98.5%(Prusty and Swain 2018).

15.3.2 Metallic Nanoparticles

Metallic nanoparticles have been used more frequently in medical applications lately, including photoablation therapy, medication transport, and biosensors (Khursheed et al. 2022; Tauseef et al. 2023; Shahrtash et al. 2024). Metallic nanoparticles have a number of benefits, such as being inexpensive, simple to synthesize, and flexible in terms of NP size and shape (Soliman and Salem 2025; Selim et al. 2025; Gad et al. 2025). However, metallic nanoparticles also have certain drawbacks. Among the primary drawbacks is that phagocytic system macrophages can quickly destroy them even before they get to the cancer cell location (Oroojalian et al. 2021). The potential for these materials to aggregate and significantly grow in particle size is another drawback of using oxide nanoparticles in active substance transport systems. Furthermore, by adding particular functional groups to these nanoparticles, they are able to attach medicines, other ligands, and antibodies, which increases the potential of these systems for use in biomedical applications (Yogendra Kumar et al. 2023; Kudr et al. 2017). While the metal nanoparticles that are most studied include copper, iron, silver, and gold, there is growing interest in a number of other metallic nanoparticles, including cerium dioxide, platinum, selenium, palladium, titanium oxide, zinc oxide, and gadolinium (Roshani et al. 2023; Baranwal et al. 2023).

15.3.3 Nanocrystals

Precipitation, homogenization under high pressure, and milling are the methods a majority frequently employed to create nanocrystals (Parveen et al. 2023). As a result, nanocrystals have a number of benefits and drawbacks. Higher solubility, a quicker rate of dissolution than traditional particles, the ability to target medications both passively and actively, and a decreased risk of tissue irritation in the event of subcutaneous or intramuscular delivery are a few benefits (Aldeeb et al. 2024). Lesser drawbacks include the inability to generate a uniform and correct dose, stability issues connected to physiochemistry, and bulking, which necessitates careful handling and transportation (Junyaprasert and Morakul 2015). An additional technique for using nanocrystals to deliver hydrophobic drugs to the lungs is encapsulation of cinaciguat nanocrystals in chitosan microparticle capsules. By taking use of the polymer's swelling and mucoadhesive properties, the nanoparticles were produced for sustained drug delivery (Haider et al. 2024). Further research is necessary to validate the effectiveness of these strategies in medication delivery systems.

15.3.4 Liposomes

Liposomes are one of the frequently researched drug delivery methods and are widely employed for the transportation of different chemicals in the cosmetics and pharmaceutical industries (Dhayalan et al. 2024; Kapre et al. 2024; Farooq et al. 2024). They are a tried-and-true technique for enhancing medication administration. Sized between 50 and 450 nm, these are spherical vesicles made of phospholipids and steroids. Due to the ease of drug integration and their membrane structure's resemblance to cell membranes, they are regarded as better drug delivery transporters (Bozzuto and Molinari 2015). In addition, they are biodegradable and biocompatible, stabilize pharmaceutical molecules, encourage biodistribution, and function with both hydrophilic and hydrophobic medications. Four varieties of liposomes are (1) Conventional liposomes: A lipid bilayer that produces phospholipids, cationic, anionic, and neutral cholesterol envelops an aqueous core. Hydrophobic or hydrophilic substances can be inserted into the lipid bilayer and aqueous space. (2) Types of PEGylated: Steric equilibrium is achieved by adding polyethylene glycol (PEG) to the liposome surface. (3) Ligand-targeting: either at the PEG chain ends, polysaccharides, peptides, and antibodies are attached to the liposome's surface. (4) Ranostic liposomes: this kind of liposome consists of a therapeutic component, targeting, imaging, and nanoparticle (Sercombe et al. 2015). Within the bloodstream, liposomes engage with LDLs, HDLs as well as opsonins. Reticuloendothelial system (RES) is assisted by immunoglobulins and fibronectin in identifying and eliminating liposomes. LDLs and HDLs are destabilized by liposomes. Organs such as the spleen and liver are more likely to accumulate liposomes, which may assist treat pathogenic conditions but also make it more difficult for lipophilic anticancer medications to be cleared from tumors (Jie Li et al. 2020; Gyanani et al. 2021).

15.3.5 Micelles Polymeric

Amphiphilic block copolymer-based nanostructures known as polymeric micelles form a core-shell configuration when they group together in an aqueous solution (Dirany et al. 2024). Hydrophobic medications (such Paclitaxel, Docetaxel, and Camptothecin) able to be combined with hydrophobic core, and the system as a whole is soluble in water due to the hydrophilic shell stabilizing the core. The EPR effect allows polymeric micelles, which are usually fewer than 100 nm in size and have a restricted distribution to impede quick renal clearance, to collect in tumor cells. Additionally, the polymeric shell inhibits biological molecules from reacting in an unintended way. These nanostructures' internal core structures, which enhances medication absorption and increases stability and bioavailability, make them highly promising for hydrophobic drug delivery (Kuna et al. 2024). Drugs for cancer and eye conditions can both be delivered with polymeric micelles (Zheng et al. 2024). Dasatinib was encapsulated by Li et al. in 55 nm nanoparticles produced by PEG-b-PC micellation to cure proliferative vitreoretinopathy (PVR). The nanoparticles had a restricted distribution and exhibited no cytotoxicity towards a spontaneously arising retinal pigment epithelia (ARPE-19) cells (Li et al. 2016).

15.3.6 Dendrimers

Three-dimensional formations known as dendrimers are clearly defined, monodisperse, and strongly bifurcated. They would be ideal drug delivery agents due to their spherical form and ease of regulated functionalization (Kaur et al. 2023). There are two ways to create dendrimers: one is a divergent path that starts at the outside of the dendrimer and goes outward from there, and the other is a convergent path that starts at the outside of the dendrimer (Rastogi et al. 2024). Because of their amine groups, dendrimers are difficult to use in medicine. Dendrimers are frequently changed to reduce these cationic or positively charged groups' toxicity. Drugs can be included into dendrimers by covalent conjugation, electrostatic interaction, or straightforward encapsulation (Martins et al. 2023; Pricl 2023; Pérez-Ferreiro et al. 2023).

15.3.7 Biopolymeric Nanoparticles Used in Drug Delivery System

Drug delivery methods utilize a variety of biopolymeric polymer types (Jacob et al. 2018). The qualities of many of these materials will be emphasized in the following:

15.3.7.1 Chitosan

Nanoparticles based on chitosan are frequently employed in long-term medication delivery systems (Mikušová and Mikuš 2021). Ceftazidime, an antibiotic, is administered to the eye, Silva et al. created an isotonic 0.75% w/w hydroxypropyl methylcellulose (HPMC) solution (Silva et al. 2017). Alginate and chitosan pectin nanoparticles were produced by Pistone et al. as potential means of delivering medication into the buccal cavity. The oral cell line was utilized to compute the cytotoxicity of the nanoparticles and to assess the biocompatibility of the formulations based on their solubility in saliva (Pistone et al. 2017).

15.3.7.2 Xanthan Gum

Menzel et al. examined a novel nasal release excipient. A fundamental polymer called Cys-MNA was connected to Xanthan gum. The quantity of binder, mucoadhesiveness, and degradation resistance of the conjugate were examined. Binder was obtained in 252.52 20.54 mol polymer grams. In comparison to thiolates xanthan, grafted polymer cohesiveness was 1.7 times greater and 2.5 times higher than native. When the polymer was removed from the mucosa, the ciliary beating of nasal epithelial cells remained unaltered and was merely reversible (Menzel et al. 2017).

15.3.7.3 Alginate

Alginate is another type of biopolymer that has been applied to the delivery of drugs (Hariyadi and Islam 2020; Abourehab et al. 2022). Because alginates have carboxyl groups in every residue, it is categorized as a mucoadhesive polymer that is anionic, which has a higher cohesion than neutral and cationic polymers (Kianersi et al. 2021; Lakkakula et al. 2021). Patil and Devarajan employed nicotinamide to penetrate alginate nanoparticles containing insulin in diabetic rats in order to raise insulin levels and decrease glucose. Nicotinamide sublingual nanoparticles (5 IU/kg) exhibited 80% absorption and 100% pharmacology (Patil and Devarajan 2016). Chitosan-coated alginate nanoparticles were developed by Costa et al. to increase daptomycin uptake throughout the ocular epithelium and produce antibacterial activity. To quantify in vitro permeability, models of ocular epithelial cell culture were used (Costa et al. 2015).

15.3.7.4 Cellulose

Drugs delivery systems frequently use cellulose and its derivatives to control the release of drugs by changing the solubility and gelation of the molecule (Ciolacu et al. 2020; Tayeb et al. 2018). Elseoud et al. used chitosan nanoparticles and cellulose nanocrystals to study repaglinide oral release. The combined nanoparticles with a mean size distribution of 198 nanometers were made up of chitosan and cellulose nanocrystals encapsulatingrepaglinide (RPG), whereas chitosan nanoparticles had a mean size distribution of 197 nanometers. The chitosan hybrid nanoparticles and the RPG-containing oxidized cellulose nanocrystals had mean sizes between 251 and 310 nm. Because of the hydrogen interactions between the drug and the oxidized cellulose nanocrystals, lower drug release was seen in nanoparticles created with these nanocrystals (Abo-Elseoud et al. 2018).

15.4 Physiochemical Properties of Nanocarrier

Researchers have known since the early 2000s that a number of widely utilized nanomaterials require in-depth studies on their nano distribution (Zhilong et al. 2024; Zongmin et al. 2019). These studies are crucial not only for creating novel biomedical applications that might be used in a clinical setting, but also for assessing the safety of nanomaterials in the workplace, where there is a lot of mass production of nanomaterials, and it is important to know the safe working doses to prevent adverse side effects in employees (Su et al. 2018; De Crozals et al. 2016). A given material's nanocarriers can be influenced by a wide range of parameters, such as surface chemistry, dissolution, aggregation, size, shape, and chemical composition.

15.4.1 Size

One important aspect influencing how nanomaterials interact with tissues and cells is their size. Size-dependent nano drug carrier systems with therapeutic effects have been reported by numerous researchers in a variety of studies (Zhang et al. 2021; Hickey et al. 2015; Caldorera-Moore et al. 2010). Such size-related medication delivery can be caused by a number of processes, such as cellular uptake, carrier crossing, and binding of the cell membrane surface (Wang et al. 2017). Generally speaking, therapeutic drugs are absorbed by specialized phagocytes when they are in the form of nanoparticles larger than 500 nm, while smaller particles are swallowed through receptor-mediated endocytosis (Youshia and Lamprecht 2016). In human epithelial (HeLa) cells, Chithrani et al. examined the cellular uptake of gold nanoparticles of different sizes (14–100 nm), finding that the highest uptake was seen for 50 nm nanoparticles (Chithrani et al. 2006).

15.4.2 Shape

To increase reliability and improve bioavailability, researchers have looked into the shape, size, and surface characteristics of nanoparticles (Jindal 2017). It is possible to deliver the medicinal chemical to body tissues that could have been challenging to reach in the allotted period by managing these characteristics. If the medicinal chemicals are unable to be released from the nanoparticle lattice, then this training is not critical. Many factors, such as temperature, pH, surface bond, tranquilizer solvency or adsorbed sedate desorption, selective distribution by the structure of the nanoparticles, lattice swelling and breakdown of the nanoparticles, and the matrix of distribution and collapse forms, influence the arrival of compounds from the nanoparticle construct detailing (Xie et al. 2024; Zhoa et al. 2019). Because of the homogenous structure of nanospheres, the polymer chains arrange themselves similarly to surfactants in micelle arrangement. On the other hand, the purpose of nano capsules, which have a heterogeneous structure, is to contain the medicine within a polymer-like vesicle supply (Yu et al. 2018). The spheres are broken up to deliver a nanosphere compound. With bonded active compounds to the significant area of the nanoparticle's surface captured by a sustained release, active chemical release is seen for a brief period of time (Lee and Yeo 2015; Gref et al. 2012). The active chemical is released through the polymeric layer of the nano capsules, which are used. They will construct structures that limit the entry of bioactive chemicals from the case if there is ionic connection among the polymer and the bioactive substances. By adding additional helper operators, like polyethylene oxide-propylene oxide, this can be kept at a strategic distance. In light of the greater notable introduction of therapeutic compounds to the intended tissues, as a result, there will be less communication between the container structure and the therapeutic components (Yue et al. 2023; Upadhyay et al. 2023).

15.4.3 Chemical Composition

The majority of the time, precise results of accessible structure have been used to determine the material plan's work relationship structure in medication delivery, with just limited basic kinds being produced by one study centre. This calls for a deliberate approach. Effective research presents challenges for polymeric frameworks due to the monotonous nature of the combination and the typical polydispersity of most polymers. Rapid progress in combinatorial science enables subatomic level optimization of medicine bearing characteristics. Initially, biocompatible polymers were blended and screened using combinatorial science to coat medical devices and inserts (Yue et al. 2023; Upadhyay et al. 2023).

15.4.4 Surface Functionalization

The ensuing characteristics that release drugs with the mode of delivery, such as targeted delivery, regulated distribution, and augmentation of the drug activity, determine the requirements of the nanocarrier-synthesizing process. Additionally, positively charged nanocarriers—including micelles—have been used, which cancer cells can more easily absorb because to the negative charge of cell membranes (Xue et al. 2024). Positively charged nanocarriers, on the other hand, have a strong interaction with serum components and are easily removed by the reticuloendothelial system (RES), leading to severe aggregation and a short half-life in blood circulation (Woodruff and Hutmacher 2010). Negatively charged nanocarriers, on the other hand, partially resolve this issue and extend blood circulation (Ding et al. 2020). These nanocarriers are challenging to internalize, though. Consequently, enhanced tumor cell selectivity and increased Change in uptake may result from a reversibly charged, negative-to-positive drug carrier surface, i.e., one that remains negatively charged during blood flow and acquires a positive charge when penetrating the tissue of the tumor. Furthermore, the nanocarrier will still come into contact with intracellular compartments (such as late endosomes, early endosomes, and lysosomes) even if it manages to cross blood barriers, enters tumor tissue, and is taken up by tumor cells. Therefore, to achieve effective intracellular drug release, it should be a top priority to find a way for drug nanocarriers to escape from intracellular compartments. Taking into account the pH 4.5 acidic environment of endo/lysosomes (Amarnath Praphakar et al. 2017). A pH-responsive delivery system would be a viable option to accomplish the goal; for instance, nanocarriers that can provide a "effect of proton sponge " to more easily separate from the endo/lysosomes in order to absorb protons in the endo/lysosomes (Guo 2012). When considered together, the pH responsiveness of nanocarriers being delivered from the extracellular medium into the inside of cells should happen at least twice in order to meet the objective of effective intracellular drug release. Unfortunately, either in the inner compartments or in the external environment of the tumor, the majority of these systems are only sensitive to a single pH fluctuation (Xue et al. 2024; Li et al. 2023).

15.5 Using Nanoparticles as a Drug Delivery System

Designing medications at the nanoscale is the newest development in the realm of nanoparticles. This approach offers advantages including the capacity to modify drug release patterns, bioavailability, immunogenicity, solubility, and diffusivity. This can decrease a medicine's toxic effects, reduce any negative effects, boost its bioavailability, expand its shelf life, and simplify and improve drug delivery (Mirza and Siddiqui 2014). The most recent medication delivery systems are either made to release therapeutic ingredients gradually at a designated location or to target a specific area. Self-assembly, in which component pieces join together naturally to form distinct shapes or patterns, is one aspect of their growth. Drugs can be delivered by nanostructures either passively or actively. The hydrophobic effect facilitates medication entry into the interior cavity of the structure. The precise amount of medication is released by nanostructure materials designed for certain areas due to the low pharmacological concentration within a hydrophobic environment (Lu et al. 2016). The latest drug delivery technologies are designed to either target a particular area or release medicinal components gradually at a predetermined spot. One feature of their evolution is self-assembly, in which individual component pieces come together spontaneously to produce unique forms or patterns. Nanostructures have the ability to actively or passively transport drugs. The hydrophobic effect makes it easier for drugs to enter the internal cavity of the structure. Due to hydrophobic environments having low pharmacological concentrations, nanostructure materials tailored for specific applications release precisely the right amount of medication (Patra et al. 2018). Because they are more appropriate for intravenous distribution, nanoparticles have an additional benefit over bigger microparticles. The body's tiniest capillaries have a diameter of 5–6 mm. To prevent the particles from producing an embolism, the size of the particles being injected into the bloodstream must be much smaller than 5 mm, without aggregating. It is evident that a multitude of medications can employ nanoparticulate transporters in various ways. Water-soluble and water-insoluble medications, proteins, vaccinations, biologic macromolecules, etc. can all be delivered using nanoparticles. They can be designed for long-term synthetic circulation or specifically targeted distribution to the brain, artery walls, lungs, liver, and spleen (Hans and Lowman 2002). There are Many Advantages of Using Polymeric Nanoparticles in Drug Delivery (Begines et al. 2020): (1) biocompatible and biodegradable. (2) increase the stability of any volatile drug substances. (3) less toxic. (4) they can be produced in huge quantities using a variety of techniques, quickly, and affordably. (5) possess designed specificity, which enables them to administer a greater concentration of medicinal substances to a targeted area. (6) non-toxicity and lack of immunogenicity.

15.5.1 Controlled Delivery System

The term "Drug action is sustained at a predefined rate" refers to the modern, Controlled delivery methods are delivered accurately (CDS), which minimizes unwanted side effects while preserving a reasonably constant, effective medication level within the body. By strategically arranging a controlled release into a system next to tissues or inside damaged cell, a medication can exert localized effect. Stated differently, the body's physiological or therapeutic requirements dictate the amount and pace of medication release. Typically, a controlled DDS is made to administer the medication at a specific rate for an extended period of time. As long as the drug delivery mechanism is operating, blood levels remain safe and effective for a while. The intended pharmacokinetics of the drug and its therapeutic concentration are the two factors that determine this predefined rate of drug release (Xie et al. 2011). A regulated DDS's development is impacted by a number of variables, including the carrier's morphology, size, shape, and crystalline structure. Similar to this, the surface of the nanoparticle plays a crucial role when concentrating on a certain medication delivery method. Normal negatively charged particles and nanoparticles with no surface modification can be quickly opsonized. And extensively removed by the settling macrophages after they enter the circulatory system. Due of its capacity for recognition DDSs, take them out of the bloodstream, and prevent the successful distribution of the nanospheres to organs other from those that on the reticuloendothelial membrane, the reticuloendothelial membrane, which primarily covers the spleen and liver, is a noteworthy barrier to target dynamically (Kwon and Kataoka 1995). The most well-known method for managing the sonication process and improving the system's surface characteristics is surface alteration of nanocarrier systems using hydrophilic polymers, also known as covering modification with polymers (Khalili et al. 2022). Poly (ethylene glycol) (PEG) chains, for instance, are connected to biodegradable polymers such as poly (lactic-co-glycolic acid) (PLGA) and poly (lactic acid) (PLA). As such, the hydrophilic PEG chain will permit the regulation of cell behavior at the polymer surface as well as the control of biomolecule assimilation (Perinelli et al. 2019; Janrao et al. 2023).

15.5.2 Targeting Delivery System

When employing nanomaterials or nano formulations as drug delivery systems, drug targeting might be either active or passive. Combining drug delivery components, such as peptides and antibodies, with structures of receptors that are seen in target region is known as active targeting (Tiwari et al. 2023; Chen et al. 2023). In passive targeting, the generated drug-carrying complex travels within the bloodstream and is guided to the target site by binding or affinity that is affected by factors including temperature, pH, shape, and molecular site. The body's main targets are cell surface antigens or proteins, lipid membrane components, and cell membrane receptors

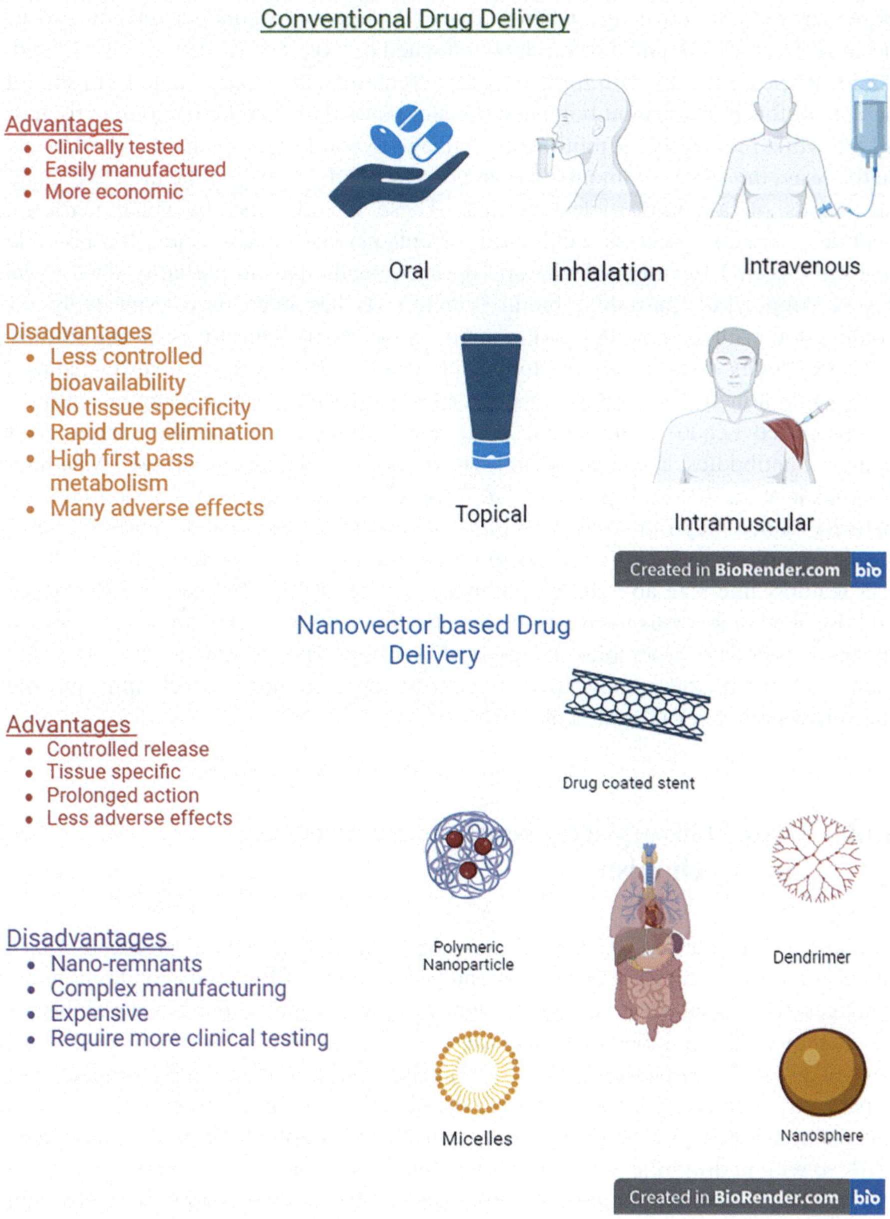

Fig. 15.2 Smart nano vectors for targeted and controlled drug delivery

(Fig. 15.2) (Patra et al. 2018). Because of the larger epithelial crossings, the nanoparticles can shatter stimulated or damaged tissue. This entry can occur passively or actively. Latent targeting occurs when the nanoparticle reaches the target organ due to faulty crossings, whereas dynamic targeting occurs when the therapeutic carrier framework and a tissue are conjugated or, conversely, cell-specific ligand. An ideal release mechanism for therapeutic compounds in nanoparticle form should be able to form an apparent bond and transfer its load to specific pathologic tissues, all the while preventing or minimizing damage to solid tissues caused by sedatives. In this way, the most commonly known process involves coating the targeted ligand or ligands on a nanoparticle's surface. These ligands may be small particles, peptides, complex proteins, antibodies, or aptamers of nucleic acids. It's possible that the targeted ligands lack the appropriate specificity and partiality. Due to its severe streptavidin partiality, biotin (vitamin H) has been used extensively for conjugation with nanoparticles (Pramanik et al. 2016). Due to its strong affinity for endogenous folate receptors, folic acid (vitamin B9) has been investigated as a potential treatment for a variety of illnesses when folate receptor communication is compromised (Zhao et al. 2008). As a result, many more specific sugars, brief peptides, antibodies, and small particles have been created and put to use (Amarnath Praphakar et al. 2016). Liposomes are a useful addition to the discussion on drug delivery. Since they mimic the cell film, one can target the ligand's surface and as previously discussed, outline a specific lipid monomer to customize physicochemical features like size and charge (Guimarães et al. 2021). This structure provides additional advantageous locations; an enhanced lipid-lipid exchange occurs because liposomal structure resembles the targeted cell film. This accelerates the lipophilic medicinal compound's convective movement from the specific cell film into the liposomal lipid layer (Pattni et al. 2015).

15.6 Drug Delivery with Nanocarrier Process and Mechanism

The use of nanocarriers as drug delivery systems (DDS) is very promising. This effect is known as increased retention and permeability (EPR), they can accumulate damaged tissue, lengthen circulation durations, and improve the bioavailability of medications. The number of nanocarriers in clinical studies has risen in the last several years. In particular, polymeric micelles—which are simple, compact, and capable of delivering hydrophobic medications within their cores—have drawn increasing interest (Nelemans and Gurevich 2020). Amphiphilic block copolymers (ABCs) with hydrophilic and hydrophobic blocks make up polymeric micelles (Cho et al. 2015). Aqueous solution containing these ABCs at a concentration higher than their critical aggregation concentration (CAC) will cause the ABCs to aggregate that self-assemble consisting of a hydrophilic corona and a hydrophobic core. These polymeric micelles are comparatively steady in the blood at functional amounts because to their low CAC and further stabilizing brought about by a hydrophobic load (Cho et al. 2015). However, as the actual drug contained and type of cell are important impact in the uptake, in addition to the physicochemical features of the polymeric micelles, the uptake mechanisms of these micelles are difficult to generalize (Ghezzi et al. 2021; Hwang et al. 2020).

15.6.1 The Main Cellular Uptake Mechanism Is Endocytosis

Cells primarily take up nutrients and other materials through a process known as endocytosis. It is also believed that this process absorbs the majority of nanocarriers. Traditionally, there have been two categories for endocytosis: pinocytosis (the absorption of fluids and solutes) and phagocytosis (the absorption of big particles) (Nelemans and Gurevich 2020).

15.6.1.1 Phagocytosis

The largest the size of nanocarriers that are ingested during phagocytosis appears to be contingent upon the volume of cell of the phagocyte (Ngo et al. 2022). On the other hand, the uptake may also be affected by the particle shape. Champion et al. produced polystyrene fragments in a range of sizes and shapes, and they demonstrated that the phagocyte's capacity to absorb the particle was dependent on its curvature at the moment of initial contact (Champion and Mitragotri 2006; Champion et al. 2007).

15.6.1.2 Pinocytosis

Degradative lysosomes are the primary target of nanocarriers that enter cells by clathrin-mediated endocytosis (CME), one of the mechanisms of pinocytosis. Prior to maturing into late endosomes (pH ~ 5), the cargo will first be delivered to early endosomes (pH ~ 6). These late endosomes will combine pre-lysosomal vesicles to become lysosomes, which have an environment rich in enzymes (such as hydrolases) and acidic (pH ~4–5) for destruction. Only when the pharmaceuticals within the nanocarriers are stable in these extreme temperatures can this pathway be used to release the drug through carrier biodegradation. If not, endosome escape mechanisms can be investigated to maximize medication administration (Nelemans and Gurevich 2020).

15.6.1.3 Elucidating Endocytic Pathways of Nanocarriers

A frequent method for examining the mechanisms of nanocarrier uptake is the application of endocytic inhibitors. A route is thought to be in charge of nanocarrier uptake when its blockage significantly reduces the uptake of a nanocarrier (Sousa de Almeida et al. 2021). But the majority of inhibitors can cause other adverse effects and are not limited to a single endocytic pathway. Moreover, a secondary uptake mechanism that may not have been active at first may compensate by blocking a particular pathway. It is advised to utilize numerous inhibitors to confirm the results because these limitations of endocytic inhibitors are frequently disregarded (Jing Liu et al. 2024). The iRNAs usage is another, more accurate method for clarifying a particular mechanism of uptake. The synthesis of some essential endocytosis-related proteins (such as caveolin and clathrin) can be reversibly inhibited with siRNAs, which should lessen off-target consequences (Al Soraj et al. 2012). Additionally, it provides a clearer understanding of how specific proteins function in endocytic pathways (Sousa de Almeida et al. 2021).

15.7 Applications for Disease Treated Using Nanocarriers

15.7.1 Treatment of Cancer

Globally, cancer is the primary cause of mortality. The most enigmatic and complicated illness, cancer is thought to be a multi-phase process of carcinogenesis that includes several physiological cell processes, such as cell signaling and death. Though many advancements have been achieved, conventional cancer treatments like chemotherapy and radiation seem to be far from effective. Because of some drawbacks including chemoresistance and serious side effects. Nanomedicine has the potential to revolutionize cancer therapy and diagnosis through novel biocompatible nanoparticles, which are critical for medication delivery. The use of nanoprobes, nanocrystals, nano shells, nanocantilevers, nano polymers, quantum dots, and dendrimers among other nanotechnologies in cancer treatment is possible (Sahu et al. 2021). The two main characteristics of developing nanoparticles in the treatment of cancer are their ability to identify tumors and their ability to acquire the objective tumor location without harming surrounding cells. Numerous nanoparticles are promising in identifying highly selective and sensitive cancer cells through a variety of processes. These unique properties set these nanoparticles apart from traditional cancer treatments (Misra et al. 2010).

1. The nanoparticles themselves might be designed to carry a wide therapeutic load and exhibit both therapeutic and research qualities.
2. With high specificity and affinity, nanoparticles can attach to polyvalent targeting ligands that specifically connect to target cells.
3. Nanoparticles can accept many medicinal molecules, enabling combination cancer therapy.
4. Typical drug resistance mechanisms can be overcome using nanoparticles.

The most often utilized therapeutic nano formulations for cancer chemotherapy that have been approved by the FDA are Doxil, Daunorubicin (Daunoxome), and Abraxane. Abraxane (Abraxis) is an albumin-bound nanoparticle for paclitaxel and a liposome-mediated drug delivery system for doxorubicin. The treatment of metastatic breast cancer frequently involves the use of both of these nano formulations. Targeted liposomes containing folate receptors have been demonstrated to avoid drug resistance to many drugs in cancer cell lines and to efficiently deliver doxorubicin in vivo (LaRocque et al. 2009). One of the most important factors in tumor therapy is the interchange of antiangiogenic medicines and chemotherapy. Sengupta et al. have discovered that drug resistance factors are expressed more frequently, and that chemotherapy drug delivery may be hampered by tumor dysfunction in the blood vessels. Nanoparticles with two layers were used to create a drug delivery system: first, a core made of polylactic-co-glycolic acid (PLGA), combined with doxorubicin and attached to combretastatin-conjugated liposomes and PEG. Combitastatin is an anti-angiogenic medication, whereas doxorubicin is a cancer treatment. When this nanoparticle was injected intravenously into animals with tumors (carcinomas or cell-derived melanoma), the tumor quickly absorbed the nanoparticles. They significantly slowed the growth of the tumor. This formulation of nanoparticles has demonstrated effectiveness in lung cancer and melanoma in mice models, and similar performance is anticipated in human models (Sengupta et al. 2005). Theranostics, a biomedical technique that integrates treatment and diagnosis in a single step, is emerging. The major stages of healthcare, as therapy and diagnosis, will be integrated with the aid of thermodynamics to expedite, simplify, and improve the effectiveness of treatment (Aslam et al. 2023; Liu et al. 2022). Present research is focused on developing biocompatible nanoparticles as screening of cancer tools that will enable efficient and non-invasive cancer therapy (Augustine et al. 2021). Shim et al. have successfully developed a combined cancer detection and therapy method (theranostics). To explore the potential of combination stimulation, they employed tiny gold nanoparticles that contained Si RNA. -responsive optical imaging and improved gene silencing by stimulation (Shim et al. 2010).

15.7.2 Ocular Applications

One of the primary issues with the topical administration of ophthalmic medications is poor absorption into the eyes, which makes it challenging to keep the drug's concentration in the precorneal area at the proper level. Given that compared to traditional medications ($t1/2 = 1–3$ min), nanoparticles show greater durability and a longer half-life (t1/2) in tear fluids (up to 20 min, they present an alluring substitute for conventional ocular topical medications. Drug delivery systems utilizing nanoparticles have demonstrated the ability to improve therapeutic agent absorption, bioavailability, reduce side effects, and maintain intraocular dose levels (Rizvi and Saleh 2018). Cytomegalovirus, or CMV infection, can permanently harm the choroid, choroid, iris, and surrounding tissues. Merodio et al. have employed bovine serum albumin (BSA) nanoparticles, which contain ganciclovir, a medication used to deal with CMV infection. This medication has been kept alive in an aqueous medium containing BSA, and droplets have been created by adding ethanol using the emulsification process. The resultant particles were roughly 280 nm in diameter.

These nanoparticles were given intravenously after being resuspended in saline. Up to 2 weeks after injection, researchers discovered that the nanoparticles were still present within the retina's thin layer. In comparison to typical eye controls, the histological study revealed no evidence of either improvements in tissue morphology or inflammatory responses (Rijaul et al. 2024).

15.7.3 Application for Heart Diseases

The conditions collectively referred to as cardiovascular diseases (CVDs) include hypertension, stroke, heart failure atherosclerosis, and myocardial infarction. Globally, these diseases are the main reasons why people die. Effective treatment and prevention of cardiovascular disorders depend on early detection. Among the common techniques used to identify CVDs are computed tomography (CT), electrocardiography (ECG), plain X-ray, and magnetic resonance imaging (MRI). However, the limited sensitivity and specificity of these approaches render them ineffective. To get around these challenges, several novel approaches have been developed, including cardiac immunoassays and molecular imaging. The complicated pathophysiology of CVDs makes early-stage identification difficult, despite its many advantages over the aforementioned techniques. Therefore, in order to detect CVDs, quick, precise, sensitive, and targeted approaches are required (Shi et al. 2020; Mohammadi Nasr et al. 2020; Deng 2021). Compared to traditional diagnostic procedures, nanotechnology offers many advantages. Early-stage CVD diagnosis can be achieved by combining cardiac immunoassay with nanotechnology. Certain CVD biomarkers may be found early on by using nanotechnology in a variety of immunoassays, including the Enzyme-Linked Immunosorbent Assay (ELISA), Fluorescence Immunoassay, and Electrochemiluminescence (ECL) Immunoassay. Nanotechnologies have the potential to significantly amplify signals, reduce non-specific binding sites, possess strong cardiac binding targets, and possess tolerable bioavailability, and perform a variety of other tasks. Because of these characteristics, nanoparticles can move freely throughout human bodies with low levels of restriction, form useful imaging vehicles, and significantly raise the standard of diagnosis. Functionalized with visible MOI molecules, injected or swallowed nanoparticles will disperse all across the human body and target various RNAs for diagnosis (Shi et al. 2020). Materials that glow are frequently used to conjugate nanoparticles. For atherosclerotic plaque imaging, for example, luminosity-marked quantum dots (QDs) produce good results. Additionally, when a PET/CT scan is performed, radiolabeled nanoparticles are crucial for atherosclerotic plaques. The MRI can identify macrophage material, intraplaque hemorrhage, fibrous caps, necrotic bases, and plaque neovascularization, among other components of thrombus and plaque composition. Iron oxide nanoparticles and other superparamagnetic nanoparticles can boost the sensitivity of an MRI by improving the signal by producing a dark contrast. Iron oxide fluorescent nanoparticles that are cross-linked were utilized by Jaffer et al. to quickly capture macrophages in order to evaluate the inflammatory reaction of plaques with atherosclerosis (Jaffer et al. 2006). The most common kind of CVD that results in a stroke or heart attack is atherosclerosis. It is distinguished by the artery wall thickening, which is enlarged due to plaque development. Treatment strategies for atherosclerosis based on nanotechnology include controlling lipoprotein levels, lowering inflammatory levels, preventing clotting, and inhibiting neovascularization. Atherosclerotic plaque rupture and atherogenesis are brought on by inflammatory macrophages and monocytes. In order to stop the rupture of the atherosclerotic plaque, Nakashiro et al. employed bioabsorbable nanoparticles to administer the inhibitor of inflammatory response pioglitazone to circulate monocytes (Nakashiro et al. 2016). The most often utilized treatments for hypertension are nano emulsion, liposomes, polymeric nanoparticles, solid lipid nanoparticles (SLNs), and nanostructured lipid carriers. Because of its inability to withstand water and its permeability, the blood pressure medication medoxomil, olmesartan, an angiotensin II-AT1 receptor inhibitor, has inadequate oral accessibility and bioabsorption. Rats' plasma concentration of active olmesartan increased by 2.8 times when combined with nano emulsion and administered orally as opposed to the usual dosage. Furthermore, it has been demonstrated that a threefold reduction in the usual dosage of the medication enhances and increases its antihypertensive effects (Gorain et al. 2014). Patients with heart failure have no other options for treatment except heart transplantation. However, due to the lack of cardiac donors and the significant issue of immunological rejection, relatively few patients have the good fortune to get transplant therapy. Tissue engineering and cell-based therapy have progressively emerged as viable research avenues to address this, but they are not without limitations. Tissue engineering using synthetic biology offers a potential remedy for these issues. According to one study, the incorporation of carbon nanofibers into polylactic acid (PLGA) increased the growth resilience of cardiac muscle. Moreover, the 50:50 ratio of polylactic acid to carbon nanofiber composite enhanced the function of cardiac muscle by imitating the tensile strength and conductivity of cardiac tissue and by improving the adsorption of proteins thought to support function of cardiac muscle (LaRocque et al. 2009).

15.7.4 Application in Respiratory Diseases

In the case of respiratory disorders, the application of medication delivery methods based on nanoparticles has been comparatively low (Zheng Huang et al. 2021). It is known that people with allergic disorders (like asthma) have a decrease in interferon-γ (IFN-γ) production, which leaves them more vulnerable to airway irritation and hyperresponsiveness. The polymer-drug conjugate chitosan/interferon-αpDNA nanoparticles have been shown by Kumar et al. to reduce allergic inflammation of the respiratory tract. The strategy was to supply the polymer-drug combination through the intranasal route in order to compensate for the IFN-γ shortage. The results of the therapy demonstrated that epithelial cells expressed more IFN-γ, which helped reduce inflammation and lung shape in mice that were exposed to allergens in

3–6 h (Mukesh Kumar et al. 2003). Recently, lung cancer treatment with gene therapy has been the focus of research. It has been demonstrated that nanotechnology may specifically target and deliver in situ medications to destroy cancer cells, avoiding exposure to healthy organs and tissues and minimizing adverse effects (Douafer et al. 2020). Researchers are currently treating cancer by combining gene therapy with nanotechnology (Cryer and Thorley 2019). Gopalan et al. employed DOTAP/cholesterol, a non-viral nanoparticle carrier, to deliver tumor suppression genes directly to the tumor site via a controlled release program (Gopalan et al. 2004).

15.7.5 Application in Dermatology

Because of the nature of the epidermis, the skin serves as the primary drug penetrating barrier in dermatology. The medication must first be absorbed by the skin, then pass via the progressively higher layers of the epidermis to obtain dermal contact before it can enter the bloodstream and begin to act (Supe and Takudage 2021). Several drug delivery components, including liposomes, cyclodextrins, nanoparticles, and microparticles like microspheres and microcapsules, are utilized to transfer active ingredients when applied topically. Because they may be put into various formulations, have good physicochemical stability, and are unrestricted by science, Each and every nanoparticle is the most efficient system (Zhao et al. 2008). The current application of nanoparticles is as a transporter of antioxidants to shield the skin from doxorubicin excretion by sweat glands. Likewise, barrier creams containing nanoparticles are superior to moisturizers with high lipid content when it comes to preventing skin moisture loss and reducing the chance of developing contact dermatitis on the hands. They frequently exhibit stronger occlusive effects and antioxidant activity (Cui et al. 2020). In dermatology, formulations that are aseptic are the main use of nanotechnology. One type of nanoparticle formulation that includes CH3 gluconate (Nanochlorex®) exhibits an instant antibacterial impact because of its faster absorbance through the wall of the capsule. Because of its constant release from the particle nucleus, it exerts a persistent effect. After being exposed to UV light, some nanoparticles, such TiO_2, have strong antibacterial effects. After being exposed to UV light, bare TiO_2 functions as a photocatalyst to encourage the polyunsaturated phospholipid's peroxidation section of the bacterial lipid membrane. The most potent and widely accessible antibacterial nanomaterial to date is nano silver, which is used as a room spray, water disinfectant, and treatment for burns and wounds (Mohanta et al. 2023). Its antibacterial action is demonstrated by interacting with thiol groups in internal membrane proteins to generate oxidative stress and mitochondrial damage. Nanoparticle delivery mechanisms have shown to have a crucial role in treating hair disorders by enhancing drug uptake in hair follicles openings and serving as a reservoir for the extended release of drugs within them (Patzelt and Lademann 2020). Additionally, formulations of nanoparticles are thought to be more successful in treating hair disorders like androgenic alopecia and alopecia areata than aqueous solutions and alcohol have been in the past. When included in nanoparticles, hair growth agents outlasted an aqueous control solution in hair follicle areas by 2.0–2.5 times. According to Jain et al., minoxidil encapsulated in liposomes exhibits more pilosebaceous unit penetration than the drug's traditional formulations (Jain et al. 2010). Immunomodulation agents may be integrated into nanoparticles or nanoparticle delivery systems in place of oral medication administration for more dependable and selective topical therapy. As a result, it helps treat autoimmune diseases of hair follicles, like alopecia areata (Cui et al. 2020).

15.7.6 Application in Dentistry

Nanotechnology in dentistry has made some advances. The rising need for cosmetic restorations has resulted in a rise in the manufacturing of tooth-colored items in recent years (Glowacka-Sobotta et al. 2023). Composite nanoparticles are one area of nanotechnology that is frequently applied in dentistry (Fig. 15.3). Studies have indicated that artificial nanocomposite teeth exhibit greater stability and abrasion resistance compared to acrylic and composite micro fill teeth. Following the

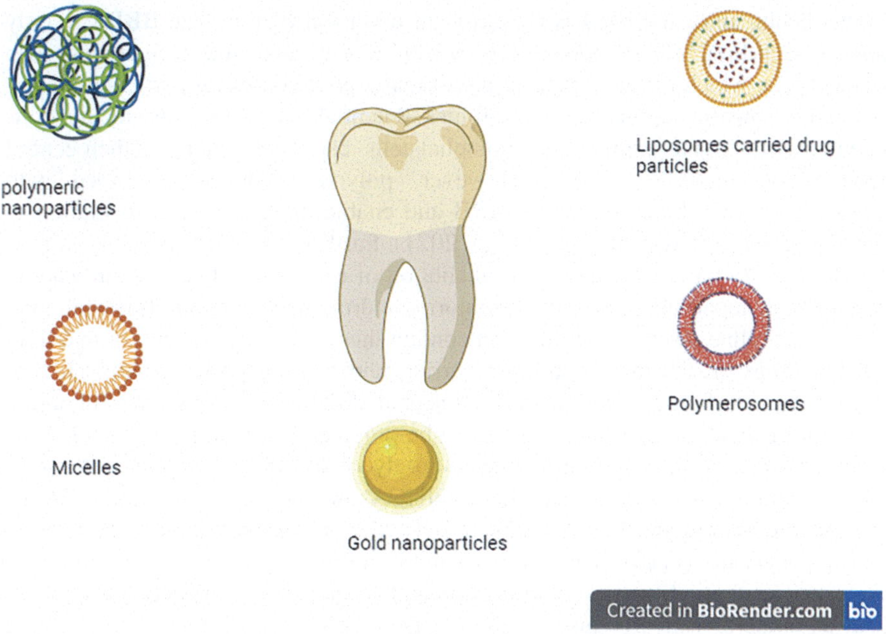

Fig. 15.3 In restorative dentistry, nanoparticles are utilized in composite resin materials to enhance their mechanical properties, resulting in more durable and aesthetic dental fillings. Additionally, nanoparticles are employed in dental adhesives to improve bonding strength between the tooth structure and restorative materials, ensuring longer-lasting restorations

polishing stage, nanoparticles enable the creation of composites with a glossy exterior and give the content better aesthetic qualities. The goal of using nanoparticles is to improve durability rate and surface transparency, which strengthens aesthetics (Jandt and Watts 2020). The quality and effectiveness of dental diagnostics may be enhanced by nanomedicine. There aren't many published studies or treatment plans using nanotechnology for the identification of dental conditions. The genetic and proteomic indicators of oral cancer can be found using AFM (Atomic Force Microscopy)(Zhou et al. 2013). Furthermore, a high sensitivity scanning a live bacterial cell in real time was made possible by AFM cantilevers, which are nanomechanical biosensors. AFM can acquire images of and interact with live cells without changing their appearance or characteristics. This feature makes AFM useful for characterizing germs affixed to dental surfaces or implants. Additionally, trustworthy proof of the biomechanical relationships between antimicrobial medications and bacterial cells is provided by AFM (Wright et al. 2010).

15.7.7 Application in Neurological Disorder

The greatest delicate and intricate organ in the human body protected by the Blood-Brain Barrier (BBB) barrier is the brain. CNS illnesses pose a serious threat to human health since the BBB is a significant treatment barrier. The BBB can only pass through lipophilic compounds or molecules with molecular weights less than 400–600 Da, which means that there aren't many possibilities for particular medicinal and diagnostic approaches in the future. Potential macromolecular medications are not able to pass through brain endothelial cells' capillaries and reach their central nervous system (CNS) target. However, polymer coatings combined with nanoparticles may break down the BBB and enable medication administration to the central nervous system (Zhang et al. 2021). Furthermore, NPs' capacity to pass the blood-brain barrier increases the likelihood of early CNS illness identification. Polymeric nanoparticles make up the majority of drug delivery methods used to treat neurological illnesses because they can contain increased drug content, bridge tight cell junctions, breach the blood-brain barrier, and be coupled to accomplish site-specific release of drugs using ligands (Ding et al. 2020). The most successful option for brain medication distribution seems to be polymeric nanoparticles such human serum albumin (HSA), copolymer (lactide-co-glycolide) (PLGA) or poly(lactic acid) (PLA), andpoly(isohexylcyanoacrylate) (PIHCA) or butyl cyanoacrylate (PBCA). Nonspecific binding needs to be achieved in order to transport drugs to the central nervous system (CNS) with nanoparticles efficiently. Functionalization of nanoparticles through surface modification and conjugation with BBB transporters, such as epidermal growth factor, vascular endothelial growth factor, and growth factor similar to insulin. The existence of 40 or 42 amino acid beta amyloid peptides (Ab40 or Ab42) in the cerebral blood vessels and brain walls indicates the existence of Alzheimer's disease (AD). Wadghiri et al. employed chemically combined magnetic nanoparticles with polyethene glycol (to increase brain permeability) and the Ab1–42 peptide (to identify the presence of amyloid plaque formation). As an advanced contrast agent was administered in vivo to AD transgenic mice, the MRI computation showed a significant imaging capability as compared to wild-type mice. Parkinson's disease (PD) is typified by an accumulation of a-synuclein in the brain stem and the death of dopaminergic neurons in the substantiating. Photoelectrochemical immunosensors utilizing gold and titanium dioxide-incorporated nanotube arrays are used to detect a-synuclein. Moreover, conjugating poly butyl cyanoacrylate nanoparticles with anti-α-synuclein aided in the clearance of a-synuclein in neurons. Gene transfer improves the ability of neuronal cells to spontaneously regenerate and survive, which leads to alterations in gene expression in neurological disorders, mainly in degenerative diseases like Parkinson's and Alzheimer's (Jagaran and Singh 2021). When using the gene transfer approach, viral vectors can be substituted with nanoparticles. They are more efficient, simpler to make and utilize, and they elicit a less immunological reaction. Green fluorescent protein (pGFP) and chitosan- and organically modified silica nanoparticles can be used to generate gene expression vectors and protein expression vectors (Kanwar et al. 2012). The central nervous system (CNS) is difficult to heal, just as the peripheral nervous system. Because ZnO nanoparticles can execute and withstand an electrical charge, which has been demonstrated to control and drive neuronal development, Seil and Webster have employed them to rejuvenate CNS neurons' tissue. This particular form of nanoparticles has proved beneficial (Seil and Webster 2008).

15.7.8 Application in Gastroenterology

An ideal target system for nanotechnology use is the gastrointestinal (GI) tract. It is the location of drug absorption (Dezfuli et al. 2023). The way that nanoparticles behave in the digestive tract can be regulated by a number of factors, including pH, heat, time of transit, and bacterial content. According to research by Bergin et al., gastrointestinal cells absorb nanoparticles at a considerably higher rate than micrometer-scale units when the sizes of the particles are smaller than 100 nm (Bergin and Witzmann 2013). Nano-vehicles need to get beyond disruptive natural barriers including low stomach pH and intestinal wastes' enzymatic processing in order to reach the target location. In addition to the bacterial enzymes found in the local flora, intestinal motility-requiring mechanical forces in the colon also break down nanocarriers. 5-FU is a well-established DNA inhibitor that is frequently used to treat gastrointestinal tumors with chemotherapy. By using ionic gelation, Zhang et al. created chitosan poly aspartic acid nanoparticles that contained 5-FU. They then showed that these nanoparticles had slower release of 5-FU both in vitro and in vivo, longer drug concentration durations, and higher growth rates of tumor inhibition when compared to the control group (Dan-Ying Zhang et al. 2008). The widespread disease colorectal cancer has the potential to be treated more effectively in the future thanks to advancements in nanotechnology (Fig. 15.4). (Al-Joufi et al.

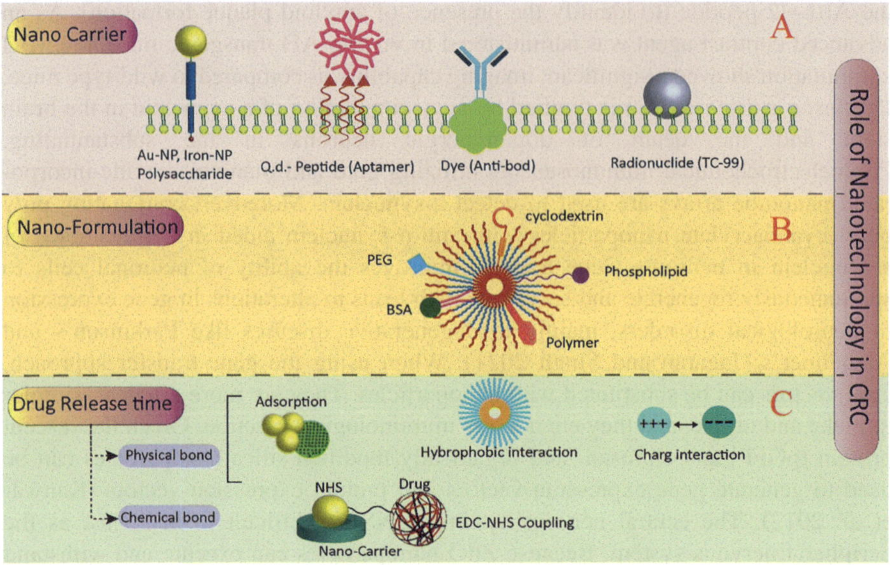

Fig. 15.4 Role of nano-technology in colorectal cancer (Dezfuli et al. 2023)

2022). Several active targeted strategies have been put forth to stop colorectal cancer (Dezfuli et al. 2023). Tumor cells typically overexpress transferrin receptors. Transferring-coated liposomes demonstrated enhanced antitumor effectiveness.

Gold nanoparticles coupled with a single-chain antibody (scFv) and carboxy-terminal phospholipids were employed by Kirui et al. The A33 antigen is overexpressed on the surface of colorectal cancer cells, which causes the nanoparticles to be selectively immobilized on cancer cell surfaces (Kirui et al. 2010). The cancer cells then preferentially occupy the nanoparticles. After 808 nm light absorption, the cancer tissue was selectively destroyed, suggesting that this method is suitable for the identification and therapy of cancer. Liver illness has become a major global contributor to the burden of public health. Liver illness is the 14th most common cause of mortality globally. Hepatic fibrosis, viral infections, and hepatocellular cancer are the main causes of liver disease. New opportunities for drug delivery in the management of liver disease are made possible by nanotechnology. Liver fibrosis can have multiple etiologies, and an individual may have more than one disease. Hepatic fibrosis is caused by a variety of factors, including genetic abnormalities, regular alcohol consumption, viral infections including hepatitis B and C, and chronic liver damage. When fibrotic tissue is compared to normal tissue, different cell types are activated. For example, activated hepatic stellate cells and myofibroblasts stimulate the secretion of different matrix metalloproteinases, which modifies the standard matrix and replaces it with fibrillar collagen structure. Hepatic stellate cells exhibit overexpression of the insulin-like growth factor 2 receptor. Mannose 6-phosphate, when linked to a nanoparticle laden with a medication, directly binds the receptor, releasing the drug for its intended activity. Type VI collagen, a matrix protein involved in cell adhesion, is widely distributed in integrin-expressing hepatic stellate cells.

Targeting the hepatic stellate cells in the fibrotic liver and collagen VI has been done using the RGD (arginine-glycine-aspartate) sequence. When combined with RGD, liposomes loaded with interferon (IFN)a-1b directly target the integrins and exhibit anti-fibrotic effect. The application of silymarin-coated gold particles in the management of liver fibrosis is demonstrated by another investigation. The hepatic stellate cells are downregulated by gold particles. After treatment with silymarin-coated gold nanoparticles, there is a decrease in many inflammatory markers in animal models. These markers are anticipated to be the main indicators for the induction of fibrosis. Lower levels of α-SMA, or alpha-smooth actin muscle, precisely indicate lower fibrogenesis and fibrosis in the liver. To treat liver fibrosis, liposomes loaded with the corticosteroid dexamethasone are manufactured to exhibit anti-inflammatory effects. In order to create the nanoparticles, Berberine was combined with bovine serum albumin. The resulting Berberine BSA nanoparticles are directed towards the liver's hepatic stellate cells, which inhibit the activation of these cells and therefore stop additional fibrogenesis from occurring. Hepatocellular carcinoma is the most prevalent primary liver cancer in adulthood (HCC). It is linked to the prevalence of hepatitis B virus infection, alcohol consumption, and other causes of hepatic cirrhosis, making it the third most common cause of mortality globally. Chemoresistance to numerous antitumor drugs is one of the main reasons why HCC does not respond well to chemotherapy. Nanotherapeutics have the potential to increase the efficacy of treatment, decrease medication availability elsewhere, and improve drug delivery to liver tissue. The doxorubicin-loaded poly (isohexyl cyanoacrylate) nanoparticle formulation known as "doxorubicin trans drug" (DT) has shown a strong anti-tumor effect in vivo against multidrug-resistant protein overexpression in hepatocellular carcinoma. According to one study, folic acid may be targeted by silver/silica nanoparticles in cases of radiation-induced liver cancer. When indocyanine green was combined with silver/silica nanoparticles, it emitted both silver and indocyanine green, which may have anti-folic acid effects and prevent liver cancer. The use of intravenous injections of superparamagnetic iron oxide nanoparticles (SPIONs) in liver imaging has been crucial in the pre-treatment diagnosis of hepatocellular carcinoma (HCC) (Nowak-Jary and Machnicka 2022). Despite this, the liver's Kupffer cells quickly sequester SPIONs with a mean particle width of 150 nm, which improves the comparison between the liver and tumor because the tumor lacks Kupffer cells. Ferucarbotran (Carboxydextran-coated SPION, Resovist) and Ferumoxide (Dextran-coated SPION, Endorem, Ferridex) are the two scientifically authorized SPIONS for liver cancer imaging (Unterweger et al. 2017).

15.8 Conclusions and Prospects

The advancements in medication delivery made possible by NM-based drug delivery systems are highlighted in this chapter. In order to effectively carry and release medications at the illness site, the most popular forms of advanced nano carriers utilized in drug delivery include polymeric carriers, liposome formulations, nanocrystals, micellar formulations, protein nanoparticles, and many more. We concluded from other research studies that Due to better absorption, nonparticulate versions of commonly used medications are more effective than traditional formulations and require lower dosages with fewer side effects. Two highly helpful techniques for intelligent medication delivery by nano carriers are active targeting and stimuli-responsive controlled release of pharmaceuticals. In order to deliver drugs to specific organs, nanoparticles also let drugs pass through biological barriers in the human body as blood-brain barrier. This approach offers advantages including the capacity to modify drug release patterns, bioavailability, immunogenicity, solubility and diffusivity. This can reduce a medicine's toxic effects, reduce any negative effects, boost its bioavailability, expand its shelf life and simply and improve drug delivery. One of the most fascinating areas of research is nanomedicine, and medication delivery methods utilising nanotechnology have been the focus of much recent study. Weaknesses persist despite significant advancements in the production and use of medicinal carriers. From our research we concluded that nanoparticles are used in various application like Cancer treatment, The nanoparticles themselves might be designed to carry a wide therapeutic load and exhibit both therapeutic and research qualities. Also, with high specificity and affinity, nanoparticles can attach to polyvalent targeting ligands that specifically connect to target cell and when this nanoparticle was injected intravenously into animals with tumors (carcinomas or cell-derived melanoma), the tumor quickly absorbed the nanoparticles. They significantly slowed the growth of the tumor. Compared to traditional heart diseases diagnostic procedures, nanotechnology offers many advantages, in early-stage CVD diagnosis can be achieved by combining cardiac immunoassay with nanotechnology. While in Gastroenterology, gastrointestinal cells absorb nanoparticles at a considerably higher rate than micrometer-scale units when the sizes of the particles are smaller than 100 nm. Nano-vehicles need to get beyond disruptive natural barriers including low stomach pH and intestinal wastes' enzymatic processing in order to reach the target location. The same with respiratory diseases, Nanotechnology may specifically target and deliver in situ medications to destroy cancer cells, avoiding exposure to healthy organs and tissues and minimizing adverse effects. By using nanoparticles, we can overcome topical administration of ophthalmic medications as poor absorption into the eyes, which make it challenging to keep the drug's concertation in the precorneal area at the proper level. Also, Drug delivery systems utilizing nanoparticles have demonstrated the ability to improve therapeutic agent absorption, bioavailability, reduce side effects, and maintain intraocular dose levels. While in neurological disorder, polymer coatings combined with nanoparticles may break down the BBB and enable medication administration to the central nervous system. Furthermore, NPs' capacity to pass the blood-brain barrier increases the likelihood of early CNS illness identification. Nanoparticle delivery mechanisms have shown to have a crucial role in Dermatology, as preventing skin moisture loss and reducing the chance of developing contact dermatitis on the hands and in treating hair disorders by enhancing drug uptake in hair follicles openings and serving as a reservoir for the extended release of drugs within them. Targeted medication delivery and controlled drug release are potential therapeutic developments in nanomedicine.

References

Abo-Elseoud WS, Hassan ML, Sabaa MW, Basha M, Hassan EA, Fadel SM (2018) Chitosan nanoparticles/cellulose nanocrystals nanocomposites as a carrier system for the controlled release of repaglinide. Int J Biol Macromol 111:604–613. https://doi.org/10.1016/j.ijbiomac.2018.01.044

Abourehab MAS, Rajendran RR, Singh A, Pramanik S, Shrivastav P, Ansari MJ et al (2022) Alginate as a promising biopolymer in drug delivery and wound healing: a review of the state-of-the-art. Int J Mol Sci 23(16):9035. https://www.mdpi.com/1422-0067/23/16/9035

Al Soraj M, He L, Peynshaert K, Cousaert J, Vercauteren D, Braeckmans K et al (2012) siRNA and pharmacological inhibition of endocytic pathways to characterize the differential role of macropinocytosis and the actin cytoskeleton on cellular uptake of dextran and cationic cell penetrating peptides octaarginine (R8) and HIV-tat. J Control Release 161(1):132–141. https://doi.org/10.1016/j.jconrel.2012.03.015

Aldeeb MME, Wilar G, Suhandi C, Elamin KM, Wathoni N (2024) Nanosuspension-based drug delivery systems for topical applications. Int J Nanomedicine 19(null):825–844. https://doi.org/10.2147/IJN.S447429

Al-Joufi FA, Setia A, Salem-Bekhit MM, Sahu RK, Alqahtani FY, Widyowati R et al (2022) Molecular pathogenesis of colorectal cancer with an emphasis on recent advances in biomarkers, as well as nanotechnology-based diagnostic and therapeutic approaches. Nano 12(1):169. https://www.mdpi.com/2079-4991/12/1/169

Amarnath Praphakar R, Munusamy MA, Sadasivuni KK, Rajan M (2016) Targeted delivery of rifampicin to tuberculosis-infected macrophages: design, in-vitro, and in-vivo performance of rifampicin-loaded poly(ester amide)s nanocarriers. Int J Pharm 513(1):628–635. https://doi.org/10.1016/j.ijpharm.2016.09.080

Amarnath Praphakar R, Alarfaj AA, Munusamy MA, Dusthackeer VNA, Kumar Subbiah S, Rajan M (2017) Phosphorylated κ-carrageenan-facilitated chitosan nanovehicle for sustainable anti-tuberculosis multi drug delivery. Chem Sel 2(24):7100–7107. https://doi.org/10.1002/slct.201701396

Aslam M, Ahmad T, Manzoor MH, Laiba, Verpoort F (2023) MXenes as theranostics: diagnosis and therapy including in vitro and in vivo applications. Appl Mater Today 35:102002. https://doi.org/10.1016/j.apmt.2023.102002

Augustine R, Mamun AA, Hasan A, Salam SA, Chandrasekaran R, Ahmed R et al (2021) Imaging cancer cells with nanostructures: prospects of nanotechnology driven non-invasive cancer diagnosis. Adv Colloid Interf Sci 294:102457. https://doi.org/10.1016/j.cis.2021.102457

Baranwal J, Barse B, Di Petrillo A, Gatto G, Pilia L, Kumar A (2023) Nanoparticles in cancer diagnosis and treatment. Materials 16(15):5354. https://www.mdpi.com/1996-1944/16/15/5354

Baviskar KP, Shah BM, Bedse AP, Raut SS, Dhamane SP, Dave DJ (2024) Nanoparticle properties. In: Nanocarrier vaccines, pp 153–191. https://doi.org/10.1002/9781394175482.ch5

Begines B, Ortiz T, Pérez-Aranda M, Martínez G, Merinero M, Argüelles-Arias F et al (2020) Polymeric nanoparticles for drug delivery: recent developments and future prospects. Nano 10(7):1403. https://www.mdpi.com/2079-4991/10/7/1403
Bergin IL, Witzmann FA (2013) Nanoparticle toxicity by the gastrointestinal route: evidence and knowledge gaps. Int J Biomed Nanosci Nanotechnol 3(1–2):163–210
Bozzuto G, Molinari A (2015) Liposomes as nanomedical devices. Int J Nanomedicine 10 (null):975–999. https://doi.org/10.2147/IJN.S68861
Caldorera-Moore M, Guimard N, Shi L, Roy K (2010) Designer nanoparticles: incorporating size, shape and triggered release into nanoscale drug carriers. Expert Opin Drug Deliv 7(4):479–495. https://doi.org/10.1517/17425240903579971
Champion JA, Mitragotri S (2006) Role of target geometry in phagocytosis. Proc Natl Acad Sci 103(13):4930–4934. https://doi.org/10.1073/pnas.0600997103
Champion JA, Katare YK, Mitragotri S (2007) Making polymeric micro- and nanoparticles of complex shapes. Proc Natl Acad Sci 104(29):11901–11904. https://doi.org/10.1073/pnas.0705326104
Chen Z, Kankala RK, Long L, Xie S, Chen A, Zou L (2023) Current understanding of passive and active targeting nanomedicines to enhance tumor accumulation. Coord Chem Rev 481:215051. https://doi.org/10.1016/j.ccr.2023.215051
Chithrani BD, Ghazani AA, Chan WCW (2006) Determining the size and shape dependence of gold nanoparticle uptake into mammalian cells. Nano Lett 6(4):662–668. https://doi.org/10.1021/nl052396o
Cho H, Lai TC, Tomoda K, Kwon GS (2015) Polymeric micelles for multi-drug delivery in cancer. AAPS PharmSciTech 16(1):10–20. https://doi.org/10.1208/s12249-014-0251-3
Choi S-J, Lee JK, Jeong J, Choy J-H (2013) Toxicity evaluation of inorganic nanoparticles: considerations and challenges. Mol Cell Toxicol 9(3):205–210. https://doi.org/10.1007/s13273-013-0026-z
Ciolacu DE, Nicu R, Ciolacu F (2020) Cellulose-based hydrogels as sustained drug-delivery systems. Materials 13(22):5270. https://www.mdpi.com/1996-1944/13/22/5270
Costa JR, Silva NC, Sarmento B, Pintado M (2015) Potential chitosan-coated alginate nanoparticles for ocular delivery of daptomycin. Eur J Clin Microbiol Infect Dis 34(6):1255–1262. https://doi.org/10.1007/s10096-015-2344-7
Cryer AM, Thorley AJ (2019) Nanotechnology in the diagnosis and treatment of lung cancer. Pharmacol Ther 198:189–205. https://doi.org/10.1016/j.pharmthera.2019.02.010
Cui N, Yang X, Wang B, Geng J, Li J (2020) Secure range query over encrypted data in outsourced environments. World Wide Web 23(1):491–517. https://doi.org/10.1007/s11280-019-00726-5
Daraee H, Eatemadi A, Abbasi E, Fekri Aval S, Kouhi M, Akbarzadeh A (2016) Application of gold nanoparticles in biomedical and drug delivery. Artif Cell Nanomed Biotechnol 44(1):410–422. https://doi.org/10.3109/21691401.2014.955107
De Crozals G, Bonnet R, Farre C, Chaix C (2016) Nanoparticles with multiple properties for biomedical applications: a strategic guide. Nano Today 11(4):435–463. https://doi.org/10.1016/j.nantod.2016.07.002
Deepa K, Sridhar A, Panda T (2023) Biogenic gold nanoparticles: current applications and future prospects. J Clust Sci 34(3):1163–1183. https://doi.org/10.1007/s10876-022-02304-8
Deng Y-H (2021) Common phase and structure misidentifications in high-resolution TEM characterization of perovskite materials. Condens Matter 6(1):1. https://www.mdpi.com/2410-3896/6/1/1
Dezfuli AAZ, Abu-Elghait M, Salem SS (2023) Recent insights into nanotechnology in colorectal cancer. Appl Biochem Biotechnol. https://doi.org/10.1007/s12010-023-04696-3
Dhayalan M, Wang W, Riyaz SUM, Dinesh RA, Shanmugam J, Irudayaraj SS et al (2024) Advances in functional lipid nanoparticles: from drug delivery platforms to clinical applications. *3*. Biotech 14(2):57. https://doi.org/10.1007/s13205-023-03901-8

Ding S, Khan AI, Cai X, Song Y, Lyu Z, Du D et al (2020) Overcoming blood–brain barrier transport: advances in nanoparticle-based drug delivery strategies. Mater Today 37:112–125. https://doi.org/10.1016/j.mattod.2020.02.001

Dirany Z, Smith GN, Aydillo C, Nguewa P, González-Gaitano G (2024) Structure and activity of amphiphilic PEO-PPO-based polymeric micelles and gels incorporating host–guest complexes of miltefosine as novel formulations for the treatment of leishmaniasis. J Mol Liq 400:124455. https://doi.org/10.1016/j.molliq.2024.124455

Douafer H, Andrieu V, Brunel JM (2020) Scope and limitations on aerosol drug delivery for the treatment of infectious respiratory diseases. J Control Release 325:276–292. https://doi.org/10.1016/j.jconrel.2020.07.002

Farooq MU, Lawrie CH, Deng N-N (2024) Engineering nanoparticles for cancer immunotherapy: current achievements, key considerations and future perspectives. Chem Eng J 486:150356. https://doi.org/10.1016/j.cej.2024.150356

Finbloom JA, Huynh C, Huang X, Desai TA (2023) Bioinspired nanotopographical design of drug delivery systems. Nat Rev Bioeng 1(2):139–152. https://doi.org/10.1038/s44222-022-00010-8

Gad ES, Salem SS, Selim S, Almuhayawi MS, Alruhaili MH, Al Jaouni SK, Saddiq AA, Owda ME (2025) A comprehensive study on characterization of biosynthesized copper-oxide nanoparticles, their capabilities as anticancer and antibacterial agents, and predicting optimal docking poses into the cavity of *S. aureus* DHFR. PloS one. 1;20(4):e0319791. https://doi.org/10.1371/journal.pone.0319791

Ghezzi M, Pescina S, Padula C, Santi P, Del Favero E, Cantù L et al (2021) Polymeric micelles in drug delivery: an insight of the techniques for their characterization and assessment in biorelevant conditions. J Control Release 332:312–336. https://doi.org/10.1016/j.jconrel.2021.02.031

Glowacka-Sobotta A, Ziental D, Czarczynska-Goslinska B, Michalak M, Wysocki M, Güzel E et al (2023) Nanotechnology for dentistry: prospects and applications. Nano 13(14):2130. https://www.mdpi.com/2079-4991/13/14/2130

Gopalan B, Ito I, Branch CD, Stephens C, Roth JA, Ramesh R (2004) Nanoparticle based systemic gene therapy for lung cancer: molecular mechanisms and strategies to suppress nanoparticle-mediated inflammatory response. Technol Cancer Res Treat 3(6):647–657. https://doi.org/10.1177/153303460400300615

Gorain B, Choudhury H, Kundu A, Sarkar L, Karmakar S, Jaisankar P et al (2014) Nanoemulsion strategy for olmesartan medoxomil improves oral absorption and extended antihypertensive activity in hypertensive rats. Colloids Surf B: Biointerfaces 115:286–294. https://doi.org/10.1016/j.colsurfb.2013.12.016

Gref R, Domb A, Quellec P, Blunk T, Müller RH, Verbavatz JM et al (2012) The controlled intravenous delivery of drugs using PEG-coated sterically stabilized nanospheres. Adv Drug Deliv Rev 64:316–326. https://doi.org/10.1016/j.addr.2012.09.008

Guimarães D, Cavaco-Paulo A, Nogueira E (2021) Design of liposomes as drug delivery system for therapeutic applications. Int J Pharm 601:120571. https://doi.org/10.1016/j.ijpharm.2021.120571

Guo KW (2012) Green nanotechnology of trends in future energy: a review. Int J Energy Res 36(1):1–17. https://doi.org/10.1002/er.1928

Gyanani V, Haley JC, Goswami R (2021) Challenges of current anticancer treatment approaches with focus on liposomal drug delivery systems. Pharmaceuticals 14(9):835. https://www.mdpi.com/1424-8247/14/9/835

Haider A, Khan S, Iqbal DN, Shrahili M, Haider S, Mohammad K et al (2024) Advances in chitosan-based drug delivery systems: a comprehensive review for therapeutic applications. Eur Polym J 210:112983. https://doi.org/10.1016/j.eurpolymj.2024.112983

Hans ML, Lowman AM (2002) Biodegradable nanoparticles for drug delivery and targeting. Curr Opinion Solid State Mater Sci 6(4):319–327. https://doi.org/10.1016/S1359-0286(02)00117-1

Hariyadi DM, Islam N (2020) Current status of alginate in drug delivery. Adv Pharmacol Pharm Sci 2020:8886095. https://doi.org/10.1155/2020/8886095

Hickey JW, Santos JL, Williford J-M, Mao H-Q (2015) Control of polymeric nanoparticle size to improve therapeutic delivery. J Control Release 219:536–547. https://doi.org/10.1016/j.jconrel.2015.10.006

Huang Z, Kłodzińska SN, Wan F, Nielsen HM (2021) Nanoparticle-mediated pulmonary drug delivery: state of the art towards efficient treatment of recalcitrant respiratory tract bacterial infections. Drug Deliv Transl Res 11(4):1634–1654. https://doi.org/10.1007/s13346-021-00954-1

Huang Y, Ji W, Zhang J, Huang Z, Ding A, Bai H et al (2024) The involvement of the mitochondrial membrane in drug delivery. Acta Biomater 176:28–50. https://doi.org/10.1016/j.actbio.2024.01.027

Hwang D, Ramsey JD, Kabanov AV (2020) Polymeric micelles for the delivery of poorly soluble drugs: from nanoformulation to clinical approval. Adv Drug Deliv Rev 156:80–118. https://doi.org/10.1016/j.addr.2020.09.009

Jacob J, Haponiuk JT, Thomas S, Gopi S (2018) Biopolymer based nanomaterials in drug delivery systems: a review. Mater Today Chem 9:43–55. https://doi.org/10.1016/j.mtchem.2018.05.002

Jaffer FA, Nahrendorf M, Sosnovik D, Kelly KA, Aikawa E, Weissleder R (2006) Cellular imaging of inflammation in atherosclerosis using Magnetofluorescent nanomaterials. Mol Imaging 5(2): 7290.2006.00009. https://doi.org/10.2310/7290.2006.00009

Jagaran K, Singh M (2021) Nanomedicine for neurodegenerative disorders: focus on Alzheimer's and Parkinson's diseases. Int J Mol Sci 22(16):9082. https://www.mdpi.com/1422-0067/22/16/9082

Jain B, Singh B, Katare OP, Vyas SP (2010) Development and characterization of minoxidil-loaded liposomal system for delivery to pilosebaceous units. J Liposome Res 20(2):105–114. https://doi.org/10.3109/08982100903161449

Jandt KD, Watts DC (2020) Nanotechnology in dentistry: present and future perspectives on dental nanomaterials. Dent Mater 36(11):1365–1378. https://doi.org/10.1016/j.dental.2020.08.006

Janrao C, Khopade S, Bavaskar A, Gomte SS, Agnihotri TG, Jain A (2023) Recent advances of polymer based nanosystems in cancer management. J Biomater Sci Polym Ed 34(9):1274–1335. https://doi.org/10.1080/09205063.2022.2161780

Jindal AB (2017) The effect of particle shape on cellular interaction and drug delivery applications of micro- and nanoparticles. Int J Pharm 532(1):450–465. https://doi.org/10.1016/j.ijpharm.2017.09.028

Junyaprasert VB, Morakul B (2015) Nanocrystals for enhancement of oral bioavailability of poorly water-soluble drugs. Asian J Pharm Sci 10(1):13–23. https://doi.org/10.1016/j.ajps.2014.08.005

Kanwar JR, Sun X, Punj V, Sriramoju B, Mohan RR, Zhou S-F et al (2012) Nanoparticles in the treatment and diagnosis of neurological disorders: untamed dragon with fire power to heal. Nanomedicine 8(4):399–414. https://doi.org/10.1016/j.nano.2011.08.006

Kanwar R, Rathee J, Salunke DB, Mehta SK (2019) Green nanotechnology-driven drug delivery assemblies. ACS Omega 4(5):8804–8815. https://doi.org/10.1021/acsomega.9b00304

Kapre S, Palakurthi SS, Jain A, Palakurthi S (2024) DES-igning the future of drug delivery: a journey from fundamentals to drug delivery applications. J Mol Liq 400:124517. https://doi.org/10.1016/j.molliq.2024.124517

Kaur A, Singh N, Kaur H, Kakoty V, Sharma DS, Khursheed R et al (2023) Neurodegenerative diseases and brain delivery of therapeutics: bridging the gap using dendrimers. J Drug Deliv Sci Technol 87:104868. https://doi.org/10.1016/j.jddst.2023.104868

Khalili L, Dehghan G, Sheibani N, Khataee A (2022) Smart active-targeting of lipid-polymer hybrid nanoparticles for therapeutic applications: recent advances and challenges. Int J Biol Macromol 213:166–194. https://doi.org/10.1016/j.ijbiomac.2022.05.156

Khizar S, Alrushaid N, Alam Khan F, Zine N, Jaffrezic-Renault N, Errachid A et al (2023) Nanocarriers based novel and effective drug delivery system. Int J Pharm 632:122570. https://doi.org/10.1016/j.ijpharm.2022.122570

Khursheed R, Dua K, Vishwas S, Gulati M, Jha NK, Aldhafeeri GM et al (2022) Biomedical applications of metallic nanoparticles in cancer: current status and future perspectives. Biomed Pharmacother 150:112951. https://doi.org/10.1016/j.biopha.2022.112951

Kianersi S, Solouk A, Saber-Samandari S, Keshel SH, Pasbakhsh P (2021) Alginate nanoparticles as ocular drug delivery carriers. J Drug Deliv Sci Technol 66:102889. https://doi.org/10.1016/j.jddst.2021.102889

Kinnear C, Moore TL, Rodriguez-Lorenzo L, Rothen-Rutishauser B, Petri-Fink A (2017) Form follows function: nanoparticle shape and its implications for nanomedicine. Chem Rev 117(17):11476–11521

Kirui DK, Rey DA, Batt CA (2010) Gold hybrid nanoparticles for targeted phototherapy and cancer imaging. Nanotechnology 21(10):105105. https://doi.org/10.1088/0957-4484/21/10/105105

Kong F-Y, Zhang J-W, Li R-F, Wang Z-X, Wang W-J, Wang W (2017) Unique roles of gold nanoparticles in drug delivery, targeting and imaging applications. Molecules 22(9):1445. https://www.mdpi.com/1420-3049/22/9/1445

Kudr J, Haddad Y, Richtera L, Heger Z, Cernak M, Adam V et al (2017) Magnetic nanoparticles: from design and synthesis to real world applications. Nano 7(9):243. https://www.mdpi.com/2079-4991/7/9/243

Kumar M, Kong X, Behera AK, Hellermann GR, Lockey RF, Mohapatra SS (2003) Chitosan IFN-γ-pDNA nanoparticle (CIN) therapy for allergic asthma. Genet Vaccin Ther 1(1):3. https://doi.org/10.1186/1479-0556-1-3

Kumar Y, Sinha ASK, Nigam KDP, Dwivedi D, Sangwai JS (2023) Functionalized nanoparticles: tailoring properties through surface energetics and coordination chemistry for advanced biomedical applications. Nanoscale 15(13):6075–6104. https://doi.org/10.1039/D2NR07163K

Kuna K, Baddam SR, Kalagara S, Akkiraju PC, Tade RS, Enaganti S (2024) Emerging natural polymer-based architectured nanotherapeutics for the treatment of cancer. Int J Biol Macromol 262:129434. https://doi.org/10.1016/j.ijbiomac.2024.129434

Kwon GS, Kataoka K (1995) Block copolymer micelles as long-circulating drug vehicles. Adv Drug Deliv Rev 16(2):295–309. https://doi.org/10.1016/0169-409X(95)00031-2

Lakkakula JR, Gujarathi P, Pansare P, Tripathi S (2021) A comprehensive review on alginate-based delivery systems for the delivery of chemotherapeutic agent: doxorubicin. Carbohydr Polym 259:117696. https://doi.org/10.1016/j.carbpol.2021.117696

LaRocque J, Bharali DJ, Mousa SA (2009) Cancer detection and treatment: the role of nanomedicines. Mol Biotechnol 42(3):358–366. https://doi.org/10.1007/s12033-009-9161-0

Lee JH, Yeo Y (2015) Controlled drug release from pharmaceutical nanocarriers. Chem Eng Sci 125:75–84. https://doi.org/10.1016/j.ces.2014.08.046

Li Q, Lai KL, Chan PS, Leung SC, Li HY, Fang Y et al (2016) Micellar delivery of dasatinib for the inhibition of pathologic cellular processes of the retinal pigment epithelium. Colloids Surf B: Biointerfaces 140:278–286. https://doi.org/10.1016/j.colsurfb.2015.12.053

Li J, Tan T, Zhao L, Liu M, You Y, Zeng Y et al (2020) Recent advancements in liposome-targeting strategies for the treatment of gliomas: a systematic review. ACS Appl Bio Mater 3(9):5500–5528. https://doi.org/10.1021/acsabm.0c00705

Li X, Yue R, Guan G, Zhang C, Zhou Y, Song G (2023) Recent development of pH-responsive theranostic nanoplatforms for magnetic resonance imaging-guided cancer therapy. Exploration 3(3):20220002. https://doi.org/10.1002/EXP.20220002

Liu J, Huang M, Zhang X, Hua Z, Feng Z, Dong Y et al (2022) Polyoxometalate nanomaterials for enhanced reactive oxygen species theranostics. Coord Chem Rev 472:214785. https://doi.org/10.1016/j.ccr.2022.214785

Liu R, Luo C, Pang Z, Zhang J, Ruan S, Wu M et al (2023) Advances of nanoparticles as drug delivery systems for disease diagnosis and treatment. Chin Chem Lett 34(2):107518. https://doi.org/10.1016/j.cclet.2022.05.032

Liu J, Cabral H, Mi P (2024) Nanocarriers address intracellular barriers for efficient drug delivery, overcoming drug resistance, subcellular targeting and controlled release. Adv Drug Deliv Rev 207:115239. https://doi.org/10.1016/j.addr.2024.115239

Lu H, Wang J, Wang T, Zhong J, Bao Y, Hao H (2016) Recent progress on nanostructures for drug delivery applications. J Nanomater 2016:5762431. https://doi.org/10.1155/2016/5762431

Martins IJ, Tomás H, Rodrigues J (2023) Carbon dots and dendrimers nanohybrids: from synthesis to applications. Mater Today Chem 32:101662. https://doi.org/10.1016/j.mtchem.2023.101662

Menzel C, Jelkmann M, Laffleur F, Bernkop-Schnürch A (2017) Nasal drug delivery: design of a novel mucoadhesive and in situ gelling polymer. Int J Pharm 517(1):196–202. https://doi.org/10.1016/j.ijpharm.2016.11.055

Mikušová V, Mikuš P (2021) Advances in chitosan-based nanoparticles for drug delivery. Int J Mol Sci 22(17):9652. https://www.mdpi.com/1422-0067/22/17/9652

Mirza AZ, Siddiqui FA (2014) Nanomedicine and drug delivery: a mini review. Int Nano Lett 4(1): 94. https://doi.org/10.1007/s40089-014-0094-7

Misra R, Acharya S, Sahoo SK (2010) Cancer nanotechnology: application of nanotechnology in cancer therapy. Drug Discov Today 15(19):842–850. https://doi.org/10.1016/j.drudis.2010.08.006

Mohammadi Nasr S, Rabiee N, Hajebi S, Ahmadi S, Fatahi Y, Hosseini M et al (2020) Biodegradable nanopolymers in cardiac tissue engineering: from concept towards nanomedicine. Int J Nanomedicine 15(null):4205–4224. https://doi.org/10.2147/IJN.S245936

Mohammed HA, Khan RA, Singh V, Yusuf M, Akhtar N, Sulaiman GM et al (2023) Solid lipid nanoparticles for targeted natural and synthetic drugs delivery in high-incidence cancers, and other diseases: roles of preparation methods, lipid composition, transitional stability, and release profiles in nanocarriers' development. Nanotechnol Rev 12(1). https://doi.org/10.1515/ntrev-2022-0517

Mohanta YK, Mishra AK, Panda J, Chakrabartty I, Sarma B, Panda SK et al (2023) Promising applications of phyto-fabricated silver nanoparticles: recent trends in biomedicine. Biochem Biophys Res Commun 688:149126. https://doi.org/10.1016/j.bbrc.2023.149126

Nakashiro S, Matoba T, Umezu R, Koga J-I, Tokutome M, Katsuki S et al (2016) Pioglitazone-incorporated nanoparticles prevent plaque destabilization and rupture by regulating monocyte/macrophage differentiation in ApoE-/- mice. Arterioscler Thromb Vasc Biol 36(3):491–500. https://doi.org/10.1161/ATVBAHA.115.307057

Nelemans LC, Gurevich L (2020) Drug delivery with polymeric nanocarriers—cellular uptake mechanisms. Materials 13(2):366. https://www.mdpi.com/1996-1944/13/2/366

Ngo W, Ahmed S, Blackadar C, Bussin B, Ji Q, Mladjenovic SM et al (2022) Why nanoparticles prefer liver macrophage cell uptake in vivo. Adv Drug Deliv Rev 185:114238. https://doi.org/10.1016/j.addr.2022.114238

Nowak-Jary J, Machnicka B (2022) Pharmacokinetics of magnetic iron oxide nanoparticles for medical applications. J Nanobiotechnol 20(1):305. https://doi.org/10.1186/s12951-022-01510-w

Oroojalian F, Beygi M, Baradaran B, Mokhtarzadeh A, Shahbazi M-A (2021) Immune cell membrane-coated biomimetic nanoparticles for targeted cancer therapy. Small 17(12): 2006484. https://doi.org/10.1002/smll.202006484

Parveen N, Abourehab MAS, Thanikachalam PV, Khar RK, Kesharwani P (2023) Nanocrystals as an emerging nanocarrier for the management of dermatological diseases. Colloids Surf B: Biointerfaces 225:113231. https://doi.org/10.1016/j.colsurfb.2023.113231

Patil NH, Devarajan PV (2016) Insulin-loaded alginic acid nanoparticles for sublingual delivery. Drug Deliv 23(2):429–436. https://doi.org/10.3109/10717544.2014.916769

Patra JK, Das G, Fraceto LF, Campos EVR, Rodriguez-Torres Md P, Acosta-Torres LS et al (2018) Nano based drug delivery systems: recent developments and future prospects. J Nanobiotechnol 16(1):71. https://doi.org/10.1186/s12951-018-0392-8

Pattni BS, Chupin VV, Torchilin VP (2015) New developments in liposomal drug delivery. Chem Rev 115(19):10938–10966. https://doi.org/10.1021/acs.chemrev.5b00046

Patzelt A, Lademann J (2020) Recent advances in follicular drug delivery of nanoparticles. Expert Opin Drug Deliv 17(1):49–60. https://doi.org/10.1080/17425247.2020.1700226

Pérez-Ferreiro M, Abelairas M, Criado A, Gómez IJ, Mosquera J (2023) Dendrimers: exploring their wide structural variety and applications. Polymers 15(22):4369. https://www.mdpi.com/2073-4360/15/22/4369

Perinelli DR, Cespi M, Bonacucina G, Palmieri GF (2019) PEGylated polylactide (PLA) and poly (lactic-co-glycolic acid) (PLGA) copolymers for the design of drug delivery systems. J Pharm Investig 49(4):443–458. https://doi.org/10.1007/s40005-019-00442-2

Pistone S, Goycoolea FM, Young A, Smistad G, Hiorth M (2017) Formulation of polysaccharide-based nanoparticles for local administration into the oral cavity. Eur J Pharm Sci 96:381–389. https://doi.org/10.1016/j.ejps.2016.10.012

Pourmadadi M, Eshaghi MM, Ostovar S, Mohammadi Z, Sharma RK, Paiva-Santos AC et al (2023) Innovative nanomaterials for cancer diagnosis, imaging, and therapy: drug deliveryapplications. J Drug Deliv Sci Technol 104357

Pramanik AK, Siddikuzzaman, Palanimuthu D, Somasundaram K, Samuelson AG (2016) Biotin decorated gold nanoparticles for targeted delivery of a smart-linked anticancer active copper complex: in vitro and in vivo studies. Bioconjug Chem 27(12):2874–2885. https://doi.org/10.1021/acs.bioconjchem.6b00537

Pricl S (2023) The spicy science of dendrimers in the realm of cancer nanomedicine: a report from the COST action CA17140 Nano2Clinic. Pharmaceutics 15(7):2013. https://www.mdpi.com/1999-4923/15/7/2013

Prusty K, Swain SK (2018) Nano silver decorated polyacrylamide/dextran nanohydrogels hybrid composites for drug delivery applications. Mater Sci Eng C 85:130–141. https://doi.org/10.1016/j.msec.2017.11.028

Rastogi V, Yadav P, Porwal M, Sur S, Verma A (2024) Dendrimer as nanocarrier for drug delivery and drug targeting therapeutics: a fundamental to advanced systematic review. Int J Polym Mater Polym Biomater 73(4):310–332. https://doi.org/10.1080/00914037.2022.2158334

Rijaul SKK, Maity N, Konar A, Hazra S (2024) Topical dexamethasone counters Intravitreal Ivermectin-induced ocular toxicity in a rabbit model. Curr Eye Res:1–9. https://doi.org/10.1080/02713683.2024.2330520

Rizvi SAA, Saleh AM (2018) Applications of nanoparticle systems in drug delivery technology. Saudi Pharm J 26(1):64–70. https://doi.org/10.1016/j.jsps.2017.10.012

Roshani M, Rezaian-Isfahni A, Lotfalizadeh MH, Khassafi N, Abadi MHJN, Nejati M (2023) Metal nanoparticles as a potential technique for the diagnosis and treatment of gastrointestinal cancer: a comprehensive review. Cancer Cell Int 23(1):280. https://doi.org/10.1186/s12935-023-03115-1

Sahu T, Ratre YK, Chauhan S, Bhaskar LVKS, Nair MP, Verma HK (2021) Nanotechnology based drug delivery system: current strategies and emerging therapeutic potential for medical science. J Drug Deliv Sci Technol 63:102487. https://doi.org/10.1016/j.jddst.2021.102487

Saiding Q, Zhang Z, Chen S, Xiao F, Chen Y, Li Y et al (2023) Nano-bio interactions in mRNA nanomedicine: challenges and opportunities for targeted mRNA delivery. Adv Drug Deliv Rev 203:115116. https://doi.org/10.1016/j.addr.2023.115116

Salem SS (2023) A mini review on green nanotechnology and its development in biological effects. Arch Microbiol 205(4):128. https://doi.org/10.1007/s00203-023-03467-2

Salem SS, Fouda A (2021) Green synthesis of metallic nanoparticles and their prospective biotechnological applications: an overview. Biol Trace Elem Res 199(1):344–370. https://doi.org/10.1007/s12011-020-02138-3

Salem SS, Hammad EN, Mohamed AA, El-Dougdoug W (2023) A comprehensive review of nanomaterials: types, synthesis, characterization, and applications (article). Biointerf Res Appl Chem 13(1). https://doi.org/10.33263/BRIAC131.041

Seil JT, Webster TJ (2008) Decreased astroglial cell adhesion and proliferation on zinc oxide nanoparticle polyurethane composites. Int J Nanomedicine 3(4):523–531. https://doi.org/10.2147/ijn.s4346

Selim S, Salem SS, Owda ME, Almuhayawi MS, Gattan HS, Alruhaili MH, Saddiq AA, Hussein S, Al-Sanea MM, Al Jaouni SK (2025) Biosynthesized zinc oxide nanoparticles: Multifunctional potential applications in anticancer, antibacterial, and *B. subtilis* DNA gyrase docking. Green Process. Synth. 1;14(1):20240218. https://doi.org/10.1515/gps-2024-0218

Sengupta S, Eavarone D, Capila I, Zhao G, Watson N, Kiziltepe T et al (2005) Temporal targeting of tumour cells and neovasculature with a nanoscale delivery system. Nature 436(7050): 568–572. https://doi.org/10.1038/nature03794

Sercombe L, Veerati T, Moheimani F, Wu SY, Sood AK, Hua S (2015) Advances and challenges of liposome assisted drug delivery (review). Front Pharmacol 6. https://doi.org/10.3389/fphar.2015.00286

Shahrtash SA, Ghnim ZS, Ghaheri M, Adabi J, Hassanzadeh MA, Yasamineh S et al (2024) Recent advances in the role of different nanoparticles in the various biosensors for the detection of the chikungunya virus. Mol Biotechnol. https://doi.org/10.1007/s12033-024-01052-6

Shi C, Xie H, Ma Y, Yang Z, Zhang J (2020) Nanoscale technologies in highly sensitive diagnosis of cardiovascular diseases (review). Front Bioeng Biotechnol 8. https://doi.org/10.3389/fbioe.2020.00531

Shim MS, Kim CS, Ahn Y-C, Chen Z, Kwon YJ (2010) Combined multimodal optical imaging and targeted gene silencing using stimuli-transforming nanotheragnostics. J Am Chem Soc 132(24): 8316–8324. https://doi.org/10.1021/ja100580y

Silva MM, Calado R, Marto J, Bettencourt A, Almeida AJ, Gonçalves LMD (2017) Chitosan nanoparticles as a mucoadhesive drug delivery system for ocular administration. Mar Drugs 15(12):370. https://www.mdpi.com/1660-3397/15/12/370

Silvestrini AVP, Morais MF, Debiasi BW, Praça FG, Bentley MVLB (2024) Nanotechnology strategies to address challenges in topical and cellular delivery of siRNAs in skin disease therapy. Adv Drug Deliv Rev 207:115198. https://doi.org/10.1016/j.addr.2024.115198

Soliman MKY, Salem SS (2025) Uncovering the potential of biofabricated *Ananas comosus* peel selenium nanoparticles for antibacterial antibiofilm suppression of virulence genes (can and LuxS) anticancer and antioxidant properties. BMC Biotechnol 25:51. https://doi.org/10.1186/s12896-025-00999-x

Sousa de Almeida M, Susnik E, Drasler B, Taladriz-Blanco P, Petri-Fink A, Rothen-Rutishauser B (2021) Understanding nanoparticle endocytosis to improve targeting strategies in nanomedicine (10.1039/D0CS01127D). Chem Soc Rev 50(9):5397–5434. https://doi.org/10.1039/D0CS01127D

Su H, Wang Y, Gu Y, Bowman L, Zhao J, Ding M (2018) Potential applications and human biosafety of nanomaterials used in nanomedicine. J Appl Toxicol 38(1):3–24. https://doi.org/10.1002/jat.3476

Supe S, Takudage P (2021) Methods for evaluating penetration of drug into the skin: a review. Skin Res Technol 27(3):299–308. https://doi.org/10.1111/srt.12968

Tauseef A, Hisam F, Hussain T, Caruso A, Hussain K, Châtel A et al (2023) Nanomicrobiology: emerging trends in microbial synthesis of nanomaterials and their applications. J Clust Sci 34(2): 639–664. https://doi.org/10.1007/s10876-022-02256-z

Tayeb AH, Amini E, Ghasemi S, Tajvidi M (2018) Cellulose nanomaterials—binding properties and applications: a review. Molecules 23(10):2684. https://www.mdpi.com/1420-3049/23/10/2684

Tiwari H, Rai N, Singh S, Gupta P, Verma A, Singh AK et al (2023) Recent advances in nanomaterials-based targeted drug delivery for preclinical cancer diagnosis and therapeutics. Bioengineering 10(7):760. https://www.mdpi.com/2306-5354/10/7/760

Unterweger H, Janko C, Schwarz M, Dézsi L, Urbanics R, Matuszak J et al (2017) Non-immunogenic dextran-coated superparamagnetic iron oxide nanoparticles: a biocompatible, size-tunable contrast agent for magnetic resonance imaging. Int J Nanomedicine 12 (null):5223–5238. https://doi.org/10.2147/IJN.S138108

Upadhyay K, Tamrakar RK, Thomas S, Kumar M (2023) Surface functionalized nanoparticles: a boon to biomedical science. Chem Biol Interact 380:110537. https://doi.org/10.1016/j.cbi.2023.110537

Veiga N, Diesendruck Y, Peer D (2023) Targeted nanomedicine: lessons learned and future directions. J Control Release 355:446–457. https://doi.org/10.1016/j.jconrel.2023.02.010

Vyas K, Rathod M, Patel MM (2023) Insight on nano drug delivery systems with targeted therapy in treatment of oral cancer. Nanomedicine 49:102662. https://doi.org/10.1016/j.nano.2023.102662

Wang T, Wang L, Li X, Hu X, Han Y, Luo Y et al (2017) Size-dependent regulation of intracellular trafficking of polystyrene nanoparticle-based drug-delivery systems. ACS Appl Mater Interfaces 9(22):18619–18625. https://doi.org/10.1021/acsami.7b05383

Woodruff MA, Hutmacher DW (2010) The return of a forgotten polymer—polycaprolactone in the 21st century. Prog Polym Sci 35(10):1217–1256. https://doi.org/10.1016/j.progpolymsci.2010.04.002

Wright CJ, Shah MK, Powell LC, Armstrong I (2010) Application of AFM from microbial cell to biofilm. Scanning 32(3):134–149. https://doi.org/10.1002/sca.20193

Xie S, Zhu L, Dong Z, Wang X, Wang Y, Li X et al (2011) Preparation, characterization and pharmacokinetics of enrofloxacin-loaded solid lipid nanoparticles: influences of fatty acids. Colloids Surf B: Biointerfaces 83(2):382–387. https://doi.org/10.1016/j.colsurfb.2010.12.014

Xie Z, Zhang G, Meng Y, Tong Y, Deng J, Li H et al (2024) A natural nano-platform: advances in drug delivery system with recombinant high-density lipoprotein. Chin Chem Lett 109584. https://doi.org/10.1016/j.cclet.2024.109584

Xue L, Thatte AS, Mai D, Haley RM, Gong N, Han X et al (2024) Responsive biomaterials: optimizing control of cancer immunotherapy. Nat Rev Mater 9(2):100–118. https://doi.org/10.1038/s41578-023-00617-2

Youshia J, Lamprecht A (2016) Size-dependent nanoparticulate drug delivery in inflammatory bowel diseases. Expert Opin Drug Deliv 13(2):281–294. https://doi.org/10.1517/17425247.2016.1114604

Yu B, Cong H, Peng Q, Gu C, Tang Q, Xu X et al (2018) Current status and future developments in preparation and application of nonspherical polymer particles. Adv Colloid Interf Sci 256:126–151. https://doi.org/10.1016/j.cis.2018.04.010

Yue O, Wang X, Hou M, Zheng M, Hao D, Bai Z et al (2023) Smart nanoengineered electronic-scaffolds based on triboelectric nanogenerators as tissue batteries for integrated cartilage therapy. Nano Energy 107:108158. https://doi.org/10.1016/j.nanoen.2022.108158

Yusuf A, Almotairy ARZ, Henidi H, Alshehri OY, Aldughaim MS (2023) Nanoparticles as drug delivery systems: a review of the implication of nanoparticles' physicochemical properties on responses in biological systems. Polymers 15(7):1596. https://www.mdpi.com/2073-4360/15/7/1596

Zhang D-Y, Shen X-Z, Wang J-Y, Dong L, Zheng Y-L, Wu L-L (2008) Preparation of chitosan-polyaspartic acid-5-fluorouracil nanoparticles and its anti-carcinoma effect on tumor growth in nude mice. World J Gastroenterol: WJG 14(22):3554

Zhang W, Mehta A, Tong Z, Esser L, Voelcker NH (2021) Development of polymeric nanoparticles for blood–brain barrier transfer—strategies and challenges. Adv Sci 8(10):2003937. https://doi.org/10.1002/advs.202003937

Zhang W, Taheri-Ledari R, Ganjali F, Mirmohammadi SS, Qazi FS, Saeidirad M et al (2023) Effects of morphology and size of nanoscale drug carriers on cellular uptake and internalization process: a review. RSC Adv 13(1):80–114. https://doi.org/10.1039/D2RA06888E

Zhao X, Li H, Lee RJ (2008) Targeted drug delivery via folate receptors. Expert Opin Drug Deliv 5(3):309–319. https://doi.org/10.1517/17425247.5.3.309

Zhao Z, Ukidve A, Krishnan V, Mitragotri S (2019) Effect of physicochemical and surface properties on in vivo fate of drug nanocarriers. Adv Drug Deliv Rev 143:3–21. https://doi.org/10.1016/j.addr.2019.01.002

Zheng Y, Oz Y, Gu Y, Ahamad N, Shariati K, Chevalier J et al (2024) Rational design of polymeric micelles for targeted therapeutic delivery. Nano Today 55:102147. https://doi.org/10.1016/j.nantod.2024.102147

Zhou Z, Zheng C, Li S, Zhou X, Liu Z, He Q et al (2013) AFM nanoindentation detection of the elastic modulus of tongue squamous carcinoma cells with different metastatic potentials. Nanomedicine 9(7):864–874. https://doi.org/10.1016/j.nano.2013.04.001

Chapter 16
Nano Sensors: Recent Developments in Environmental and Health Care Sectors

Sumod George and Dhannia P. Narayanan

Abstract Interdisciplinary alliances are extremely important for developing a safe and green environment as well as overcoming health and environmental challenges in the contemporary world. Recently nanotechnological principles are used in the rational design of nanoscale devices called nano sensors that measure physical quantities and convert these signals that can be detected and evaluated. Nano sensors thus enable the rapid detection of pathogens and pollutants and thus have application in health care sectors as well. Nano sensors can be prepared through various approaches like top-down lithography and bottom-up self-assembly supramolecular synthesis. Due to their sensitivity to pollutants nano sensors have found applications in food and environment sectors in analyzing pollutants and harmful metal ions. Nano sensors have also found crucial applications in medical fields and they include early detection of important diseases like cancer, infectious diseases caused by microorganisms and monitoring organ health post implementation. Other applications of nano sensors are molecular imprinting, selective detection of immune proteins, advances in diagnosis of pathogen like viruses etc. However, there are challenges like cytotoxic effects of nano sensors; technical challenges like precision in manufacturing, high cost of raw materials such as silicon, nanowires and nanotubes etc. With its application in many of the challenges associated with the environmental and health care sectors, nano sensors are an attractive area to both scientific and commercial world.

Keywords Environment · Gas sensors · Medical devices · Metal ions · Nano sensors · Nanotechnology · Virus detection

S. George (✉)
Department of Chemistry, Nirmala College, Muvattupuzha, Ernakulam, Kerala, India
e-mail: sumod@nirmalacollege.ac.in

D. P. Narayanan
Department of Biology, Mar Athanasius College, Ernakulam, Kerala, India
e-mail: dhanniapn@macollege.in

© The Author(s), under exclusive license to Springer Nature Switzerland AG 2025
J. M. Al-Khayri et al. (eds.), *Green Nanotechnology Applications for Ecosystem Sustainability*, Nanotechnology in Plant Sciences 4,
https://doi.org/10.1007/978-3-031-90766-1_16

16.1 Introduction

Nano sensors are minor devices that sense a physical, chemical, or biological stimulus and give a response in the form of a quantifiable signal (Abdel-Karim et al. 2020). They are fundamental for the invention of new molecular level devices and ensuing progress of nanotechnology. These nano devices have been receiving growing attention, because of the necessity to detect and quantify the physicochemical properties of toxins pollutants and contaminants at molecular level, in general. In addition, they also critical in detecting carcinogenic cells, monitoring organ health post transplantation and detection of pathogens like virus. The objectives of this chapter will be to highlight the recent developments and application of nano sensors in diverse fields and their future impact on our environment and life.

The prefix 'nano' means dwarf in Greek, i.e., something very minute and describes one thousand millionth of a meter (10^{-9}). The idea of nanoscience was introduced by Richard Feynman in 1959 (Feynman 1960) and was defined by Norio Taniguchi in 1974 (Taniguchi 1974). Nanoscience is the study of molecules and molecular assemblies on the nano scale and the technology that utilizes these understandings and converting it into devices for practical applications is called nanotechnology. Several inventions like: scanning tunneling microscope (STM), atomic force microscope (AFM), scanning probe microscope (SPM), which enables interventions at atomic scale, together with active research in nanoscience, in the last decades, has bought it to the forefront of cutting-edge technologies. A single unit of nanomaterial is sized between 1 and 100 nm. There are top-down and bottom-up approach for the synthesis of nanomaterials. Among the various devices developed using nanotechnology, nano sensors are interesting because they have wide application such as: virus detection, detection of harmful gases, monitoring agriculture lands and protecting food products, utilization of new materials as gas sensors and testing their applications etc. This book chapter will examine some of the recent developments and how nano sensors are resourcefully used for making life better on the planet earth. Nano sensors are made from nanomaterials because they have interesting unique properties like high surface-to-volume ratio, that qualifies them as a model for sensing devices. Interestingly, nano sensors are classified based on their structure, application and different sensing mechanisms. There are two types of structure based nano sensors: optical nano sensors and electrochemical nano sensors. Chemical sensors, biosensors, electrometers, and deployable sensors, all comes under the application nano sensors. Optical nano sensors are utilized for optical signal detection and is another attractive application of nano sensors which are categorized under the structural nano sensors. There is also a classification based on the source of energy used in nano sensors, i.e., active nano sensors which requires an energy source and passive nano sensors where no energy source is required for sensing e.g. thermocouple, piezoelectric sensor. A wider classification is given in Table 16.1 and Fig. 16.1. The ability of a nano sensor to detect a single or multiple analytes is associated with the recognition element in the nano sensor device. Antibodies and aptamers are the most commonly used recognition elements in the

Table 16.1 Classification of Nano sensors

Stimuli	Properties
Mechanical	Position, acceleration, stress, strain, force, pressure, mass density, viscosity, moment, torque, acoustic, wave amplitude, phase, polarization, velocity
Optical	Absorbance, reflectance, fluorescence, luminescence, refractive index, light scattering
Thermal	Temperature, flux, thermal conductivity, specific heat
Electrical	Charge, current, potential, dielectric constant, conductivity
Magnetic	Magnetic field, flux, permeability
Chemical	Components (identities, concentrations, states)
Biological	Biomass (identities, concentrations, states)

Adapted from the reference R. Abdel-Karim et al. 2020 *J. Electrochem. Soc.* **167** 037554 under the License (CC BY-NC-ND, http://creativecommons.org/licenses/by-nc-nd/4.0/)

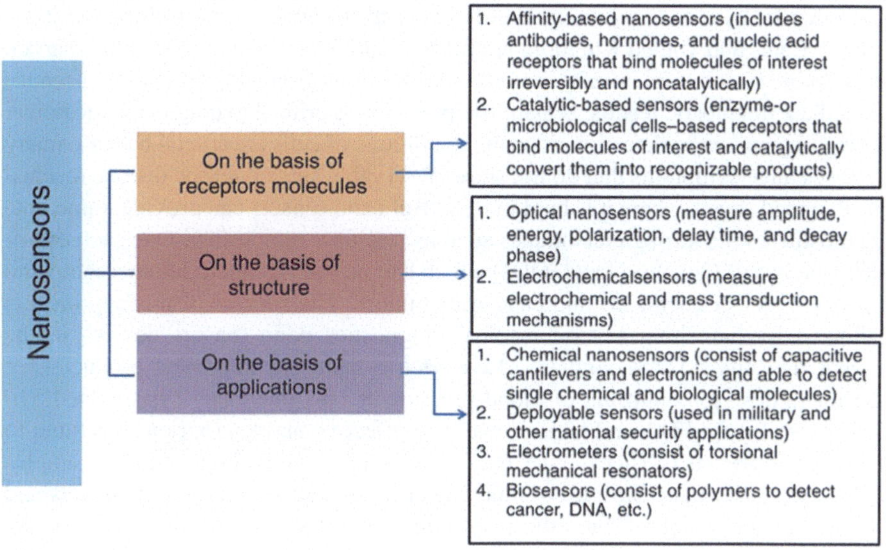

Fig. 16.1 Classification of Nano sensors. (Adopted from the reference R. Abdel-Karim et al. 2020 *J. Electrochem. Soc.* 167 037554 under the License (CC-BY-NC-ND, http://creativecommons.org/licenses/by-nc-nd/4.0/)

nano sensors. Enzymes and proteins are also used as recognition elements in nano sensors. Thus, nano sensors are advanced devices in the field of detection because they ensure fast and effective detection of microorganisms, toxins and additives compared to prevailing sensor devices in the market.

The first nano sensor was developed by Olaf Wolter in 1990 at IBM Sidelfigen, Germany for the successful batch processing of silicon Atomic Force Microscopy (AFM) cantilever probes. Subsequent decades witnessed the gradual emergence of nano sensors by different groups around the world. In 1999 researcher from Georgia Institute of Technology built a synthetic nano sensor and used it to measure the unknown mass of a particle. Later in 2001 a novel self-sensing and self-actuating probe was invented by Akiyama at SAMLAB, the Institute of Microtechnology (IMT), Switzerland. Ever since many nano sensors were invented and a variety of nanostructured materials were used to manufacture and they include: nanoscale wires (high sensitivity), carbon nanotubes (elevated surface area and high electron conductivity), metal oxide nano particles, thin films, polymeric and biomaterials. Today nano sensors are used in a variety of applications across a range of industries from biomedical to environmental devices to manufacturing and aerospace and defense. Sectors Some common applications include: as blood borne sensors in medical diagnosis, in plant biology to understand plant signaling and metabolism, to detect pesticides on the food surface and thus detect the carcinogens in food, to measure real-time conditions of soil and record pH, moisture and nutrients, real-time monitoring of metabolic activity of cancer cells. The global nano sensor market reached US$ 759.4 million in 2023 as per the *future market insights* (https://www.futuremarketinsights.com) the market is likely to hit US$ 1512.5 million by 2033. United States and United Kingdom are the two nations estimated to lead the market. Healthcare is the top priority among various nano sensor application because immediate diagnostic results, sensitivity, precision is critical to diagnostic medicine. Foretelling and early detection of health conditions of cells are crucial because many illnesses have symptoms that do not show observable signs until the disease reached an advanced stages where it is hard to treat. But nano sensors can provide immediate diagnostic results through techniques such as real-time monitoring. One such example is the use of Quantum Dots (QDs) which can be used as photo luminescent nano sensors to detect chemicals and cancerous tumors. The success of this approach is validated by launching several medical diagnostics nano sensors devices in the market. In June 2022 a company, Live Metric launched their new product Live one for intensive monitoring of blood pressure at 10-s intervals. It has gained 510 (k) approval from FDA and is projected cost-effective and appropriate substitute to the present BP cuff. This device uses nano sensor technology. Other examples include the onco-nano biosensor (June 2021) which can react to low pH levels and illuminate cancer cells. Thus, the paramount importance of Nano sensors in the contemporary environment is evident from the remarkable progress made in a short period (Li et al. 2020)

Nano sensors come under the nano-enabled products in the nanotechnology value chain starting from identifying a nanomaterial (nano particles, nanotubes, quantum dots etc.), nano intermediates (optical components, sensors, detectors) and nano-enabled products (Optical, electromagnetic, electrochemical sensors). The work is highly interdisciplinary requiring expertise in supramolecular chemistry, molecular recognition, medicinal chemistry, pharmacology further nanoscience mergers scientific branches from biology, chemistry, physics and engineering and opens new opportunities for interesting new applications in science and technology and particularly in medical sciences (Tarafdar et al. 2013).

16.2 Nano Sensor: Principles, Components and Nano Sensor Materials

The standard principle of electrochemical Nano sensors is a chemical reaction in which electrons are released, accepted or consume ions. This chemical reaction takes place between a controlled ligand and an analyte of interest that measurably effect the transduced signal, such as an electrical current or potential. Therefore, there are four important parts in a Nano sensor and they are analyte, sensor, transducer (a device which converts one form of energy to another) and a detector (Fig. 16.2). Nano sensors generally work by recording the change in electrical signals in the sensor material. For instance, some carbon-nanotube based Nano sensors that detect gaseous molecules work as follows. When a molecule of nitrogen dioxide(NO_2) is present it will remove an electron from the nanotube and as a result the conductivity of the nanotube reduces. On the other hand, if ammonia is present it will react with water vapor and donate an electron to the carbon nanotube, improving its conductivity. Further, by treating the nanotubes with various coating materials, they can be made sensitive to certain molecules and inert to certain others molecules.

Several materials are employed in building Nano sensors based on their physicochemical properties and they include metals, noble metals, metal oxides, carbon allotropes, polymer materials and biomaterials. Various nanomaterials and their important properties and mechanisms in Nano sensors are summarized in Table 16.2.

16.3 Nano Sensors for Virus Detection

Among the various microorganisms that cause significant threat to human health, those posed by viruses are most fatal. For example, corona pandemic caused by the Covid-19 by severe acute respiratory syndrome coronavirus 2(SARS-CoV-2) put everything nearly to the edge of collapse, all over the world (Vaughan 2020, 2021). Therefore, it is essential that early detection of virus infections is avital step towards the prevention of human miseries caused by these microorganisms. The important features of such detection tests are sensitivity, accuracy, cost effective, rapid shift time from sample collection for laboratory analysis to test time, and test assay in the absence of a particular equipment forms the bench mark of laboratory performance (Bardhan 2021). Many traditional methods already used for the detection of viruses include molecular methods, serological approaches and direct virus culture methods (Vainionpää 2008). Even though these conventional methods are brilliant in

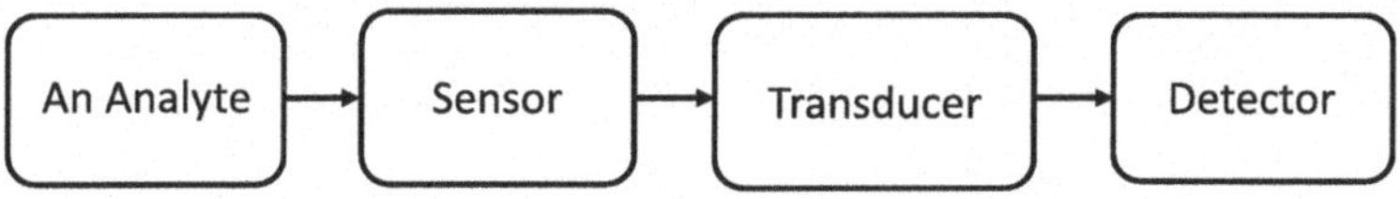

Fig. 16.2 Block diagram of Nano sensor components

Table 16.2 Nanomaterials applied for nanosensors

Sl No	Nano materials used in Nano sensors	Example	Mechanism	Advantages	Application	References
1	Metals and Noble metals	Au, Ag, Pt, Pd, Cu, Co	Arresting activities of bioreceptors, mediating electron transfer, enhancing bioreactions, amplifying mass change, and enhancing refractive index changes.	High sensitivity and selectivity, resistant to corrosion and oxidation, stable at high temperature	Environmental monitoring, food safety	Malekzad et al. (2017)
2	Metal oxides	SnO_2, ZnO, TiO_2	Metal oxides have change in electrical conductivity due to charge transfer between surface complexes, such as O−, O2−, H+, and OH−, and sensing ligands. This process requires an activation energy so that classical metal oxide sensors require high temperature (200 °C) for functioning	High surface area, low cost, stability, high electron communication, crystal morphology- diversity, ease of synthesis	Fabricating nano sensors, detection of NO_2, NMOS	Tyagi et al. (2022)
3	Carbon based nanomaterials	Carbon nanotubes (CNTs), graphene and nano/mesoporous carbon, fullerenes, Nanodiamond (ND)	ND has nitrogen vacancy (NV) defect that can be used to observe external perturbations such as magnetic or electrical fields. Fullerenes has fully conjugated π-electrons restricted in zero dimension with strong redox activity and good electron transfer kinetics and strong biomolecule	Good electric, thermal, mechanical and chemical high-quality properties for all allotropic forms of carbon. Fullerenes have unique cage like structure with distinguished properties	Has good selectivity suitable for gas sensing applications, detection of biomolecules	Bezzon et al. (2019)

			sensing properties, CNTs can adsorb molecules on to their surface through electronic interactions			
4	Polymer nanomaterials	Polyaniline, polypyrrole, poly(dopamine) (pDA)-modified indium tin oxide (ITO)	Nanocomposites of nano polymers with CNTs, metal nanoparticles, metal oxide nanoparticles can sense harmful oxidizing (NO_2, Cl_2) and reducing gases (NH_3, H_2S, alcohols)	Excellent biocompatibility, sensitivity, selectivity	Analysis of toxic gases like benzene, acetone, NO_2, Cl_2	Lee et al. (2016) and Abdel-Karim et al. (2020)
5	Bionanomaterial	L-phenylalanine ammonia-lyase enzyme	There is a biological recognition even in the transducing device where the analyte with the bioreceptor is converted to a appropriate signal output readable by the analyst	Immobilization of biomolecules, catalyzing the electrochemical reaction, improving the electron transfer	Helps in early diagnosis of neurodegenerative disorders and cancer.	Dhiman et al. (2021)

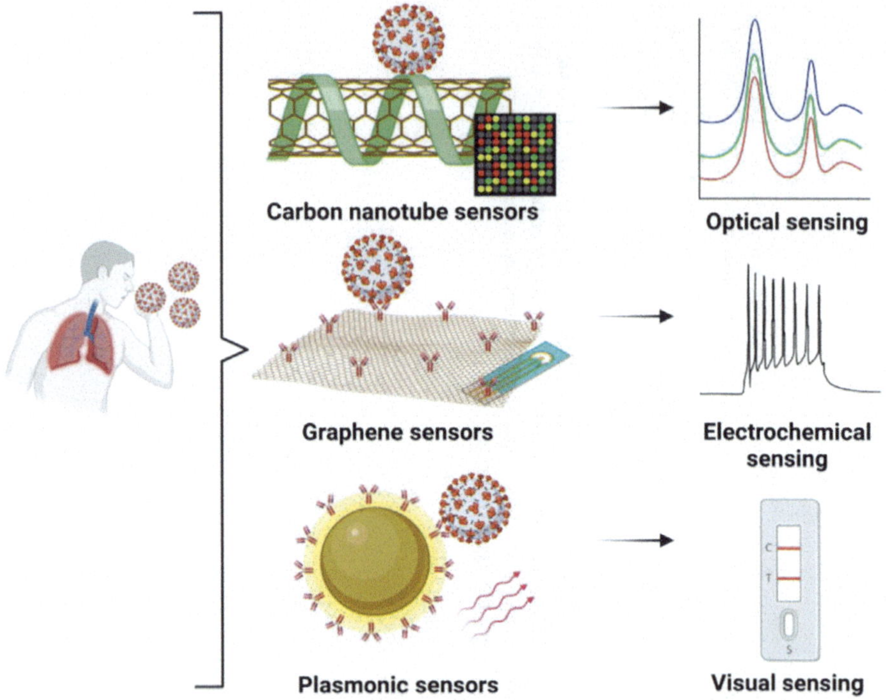

Fig. 16.3 Nano sensors based on nanoscale materials like carbon nanotubes, graphene or plasmonic nanomaterials and their application towards the detection of SARS-CoV-A in sample specimens from human blood samples: concepts based on optical, electrochemical or visual sensing. (Adopted from Bardhan (2021) open-access article distributed under the terms of the Creative Commons Attribution License (CC BY))

detecting viruses, yet they have drawbacks. Twenty-first century has seen the emergence of new technologies that can detect viruses by overcoming the disadvantages of the early traditional methods. Biosensors use nanotechnology to develop nano-biosensors that are promising in detecting virus (Fig. 16.3).

Plasmonic nano sensors were developed for almost 40 years for biosensing applications. The phenomenon of plasmonic resonance is based on the principle of the fluctuations in electric field induced by the flux (oscillation) of the free electrons of a metallic nanoparticle, obviously Au or Ag, which influences (a) the wavelength of incident light used (b) the dielectric medium (c) adjacent to other metal particles (plasmonic coupling), resulting in strong magnification of the electric field. Interestingly, plasmonic nanoparticles are strong absorbers and scatterers of light and in addition the optical response can be tuned by changing the size, shape and composition of nanoparticles, the unique wavelength and angle of the excitation of light source, configuration of the nanoparticles and its interaction with the target biomolecule analyte (Anker 2008). In the situation of the recent covid-19 pandemic, many research groups have worked on the idea of utilizing plasmonic resonance as a method for detecting SARS-CoV-2 viral pathogen. Li and co-workers recently reported a lateral flow immunoassay (LFIA) chip for the fast, rapid (<15 min) point-of-care diagnostic chip for the detection of the presence of IgM and IgG antibodies (Analyte for SARS-CoV-2 detection) against the SARS-CoV-2 virus in blood sample (Li et al. 2020). During analysis, the analyte (patient sample) flowing through the device, the SARS antibodies, if present undergo a binding to form a complex with the Au-COVID-19 antigen conjugates placed on the conjugate pad, subsequently results in a change in colour of the bound complex owing to localized surface plasmon resonance with the Au nanoparticles, and can be visually noticed as the appearance of a line on the diagnostic strip. This test is acute sensitive (87%) and specific (90%), require only 20 μl of blood samples and can return the test results in an average of15-minute interval. Li and team not only succeeded in producing rapid results using this nano sensor, but also achieved a lab-free detection of the SARS-CoV-2 virus (Li et al. 2020).

In order to mitigate and prevent the spread of virus pandemics the critical parameters include quick virus detection, ease of deployment of sample analyzers and quick turnaround time. Electrochemical nano sensors were utilized for producing rapid, portable and low-cost diagnostics for the detection of SARS-CoV-2 virus. Recently Mahari and coworkers (Mahari 2020) reported the ultrasensitive detection of the spike protein domain I of the SARS-CoV-2 virus using a novel biosensor, which is a printed circuit board-based electrochemical device. This device operates on the principle of detecting changes in electrical conductivity upon complexing of the sample analyte of interest (SARS-CoV-2) to its antibody, immobilized on the surface of a screen-printed carbon electrode. Remarkably the working range of this electrochemical nano sensor device was reported to be ~10fM-1 μM, under optimal conditions; which matched with that of expensive potentiostat-based immunosensor. In spiked saliva samples, this device had a limit of detection ~90fM, and can be used as a device for the instant detection of COVID-19 in less than 30s. This low-cost, portable instrument easily qualifies as a point-of-care diagnostics since it can be battery operated with no change in voltage, compared to expensive equipment such as potentiate which can be operated only in laboratory settings.

16.4 Nano Sensors for Gas Detection

Gas sensors were employed first in coal mines for the constant monitoring of hazardous gases like carbon monoxide, carbon dioxide and methane. In earlier days canary birds were bought down to the coal mine tunnels as a detector and the bird stops chirping and dies when it inhales poisonous gases. Industrial age witnessed the emergence of better precision equipment like Davy lamp (1815) for the detection of methane in underground mines. Modern era of gas detection started in 1926 when Dr. Oliver Johnson developed a catalytic combustion sensor in the Standard oil Company in California for the detection of combustible mixture in air to avert explosion in fuel storage tanks. In subsequent years gas sensors began appearing in chemical industries for harmful gas sensing and also for monitoring environmental pollution. Later gas sensors found application in hospitals for accurate and reliable monitoring of an aesthetic and oxygen during surgical events. Today it is used widely for monitoring air quality. In automobile industry gas sensors go beyond their earlier engine emissions controls applications and now it also ensures passenger comfort and safety (Lun and Xu 2022). Alarms for harmful gases like carbon monoxide and carbon dioxide levels are obtainable for office and domestic usage and mandatory in some jurisdiction. Further the development of high-performance, portable and miniaturized gas sensor devices is of great demand in the field of environmental monitoring, medical diagnosis and agriculture (Chakraborty et al. 2024).

Another interesting class of gas sensors are metal-oxide-semiconductor based (MOS) gas sensors. They are mostly used in sensors for (Dutta et al. 2023) for their attractive physicochemical properties as a gas sensing material. Among the various gas sensors MOS are attractive because they are inexpensive, ease of fabrication, instantaneous response, broad coverage of detection and sustainable under harsh conditions (Dey 2018), many of these properties being at the heart of the prerequisites for practical applications. One particular interesting factor about metal-oxide-semiconductor is that their operations are based on the semiconducting layer in their solid structure, which change their electrical conductivity proportional to the composition of gases and volatile organic compounds (VOCs) surrounding them and these semiconducting layers are based on metal oxides like WO_3, MoS_2, ZnO, SnO_2, TiO_2 and non-stoichiometric TiO_n. Therefore, the basic principles of MOS sensor are based on the reaction of the gas molecules and the semiconductor surface (Liu 2022). The physical and chemical properties of semiconductor materials change astonishingly when its size is reduced to nanoscale due to large surface area or quantum size effect. The large surface-to-volume ratio (surface area) of nanoparticle ensure high sensitivity. Nanostructured MOS has shown enormous potential in gas sensing applications because one of the ways to enhance the sensor performance is to use nano scale material. It is illuminating that the nanostructured MOS with exposed crystal facets has large difference in the gas sensing abilities which is attributed to their exceptional discrete surface properties. The facets with more crystal defects possess greater value of surface energy and have improved physicochemical properties (Zhu et al. 2019). For example, zinc oxide (ZnO) is a metal-oxide-semiconductor gas sensor material with improved gas sensing property. A study on the crystal plane's atomic structure diagram of metal oxide ZnO and calculating the densities of dangling bonds in every plane in ZnO, it was observed that the improvement in gas adsorption is attributed to the high density of dangling bonds and coordination of unsaturated oxygen. Upon calculating the density of dangling bonds in every plane, it was found that the crystal plane (0001) has higher dangling bond density (Gao and Zhang 2018). It is a known fact that high surface-to-volume ratio cannot promise an exceptional gas sensing properties (Xue et al. 2021) but factors like dangling bonds, active sites, surface atomic arrangement and surface states play a key role is determining the gas sensing characteristics among which surface atomic arrangement is critical. Dangling bond planes have abundant O_2

binding active sites and hence in the case of (0001) crystal plane, a good gas sensitivity could be attained.

Another interesting nanostructured MOS employed for gas detection is titanium oxide semiconductor compounds. There are stoichiometric (TiO_2), non-stoichiometric titanium oxides (TiO_{2-x}) as well as Magnéli phase (Ti_nO_{2n-1}, n = 4...10) used in MOS nano sensor designs (Ramanavicius and Ramanavicius 2020). Interestingly the n-type stoichiometric semiconductor TiO_2 exists in three phases and they include rutile (tetragonal, P42/mnm), anatase (tetragonal, I41/amd) and brookite (orthorhombic, Pbca) which can be interconverted by subjecting to thermal experimental conditions. Titanium anatase phase transforms into rutile phase at 800 °C, hence rutile is stable and anatase is the metastable phase. This is advantageous because annealing enables to fine tune the analytical properties of TiO_2 (Bai and Zhou 2014) which are sensitive towards detecting oxygen. Most TiO_2 nanomaterials are produced in anatase form. The remarkable gas sensing properties of titanium oxide semiconductors is ascribed to the relatively narrow band gap and more oxygen binding active site for the stoichiometric phases. An important example of titanium oxide is titanium dioxide (TiO_2) which crystallizes in the three above-described structural forms. A group of scientists from south Africa did a study on the synthesis of TiO_2 nanoparticles via hydrothermal method from aqueous sodium hydroxide and subsequently washed the material in distilled water and various concentrations of hydrochloric acid solution. This example of TiO_2 nanomaterial analyzed by X-ray diffraction and microscopy analysis revealed that the size of the TiO_2 particle could be controlled by varying concentration of HCl. In addition, they also observed that, at higher concentration of HCl there existed only a single phase i.e. TiO_2 anatase otherwise the material is having rutile as impurity. Thus, apart from synthesizing nanomaterials of desired size resulting in higher sensing response and selectivity to NO_2 gas at room temperature they could further elaborate in gaining polymorphic control on thethreeTiO_2phases (Tshabalala et al. 2016).

16.5 Bionanosensors and Their Role in Food and Agriculture

Among the various global challenges that humanity encounters today, agriculture and food safety occupy the top priority (Wang et al. 2022). But nanotechnology can overcome these challenges by producing advanced biosensors for monitoring agriculture lands, food products and next-generation diagnostic tools and techniques for agriculture. Nano biosensor is such an advanced technology and can play an important role in revolutionizing agriculture (Merkoçi 2021). Both nano sensors and nano biosensors represents cutting edge technologies that can help environmental monitoring including water, soil and many hazardous pesticides that would bring advantages in the food and agriculture areas that will protect our food industries to provide better living conditions (Huang 2021).

Nanotechnology has contributed several advancements in the area of agriculture. Agriculture sector has benefited by utilizing advanced nanomaterials for seed germination and micronutrient source. In addition, one of the interesting contributions is the development of nano biosensors that monitor soil quality parameters, plant diseases and stress, all of which are essential for precision agriculture (Yadav et al. 2023). Nano biosensor is a combination of nanomaterials with biosensors and are sensitive and specific, easy to operate with low power consumption, non-invasive and tailored for the monitoring of a plethora of environmental samples. Nano biosensors have various nanoparticles for target detection and they include metal nanoparticles (NPs) like Au, Ag, etc. metal oxides NPs (ZnO, TiO_2, MgO, etc.), carbon nanoparticles, magnetic nanoparticles, dendrimers etc. (Singh 2021). Interestingly due to its small size nano biosensors are portable and chemical nano sensors enable real-time monitoring of soil and plant physiological conditions such as soil fertility, pH, moisture content, the concentration of various agriculture minerals and their concentrations, toxic chemicals, prior detection of pests and pathogens. On the other hand, there are mechanical nanosensors that measure physical and chemical properties like pressure, mass, density of nanoscale areas (Han 2023). In addition, it is advantageous that nano sensors have a very low limit of detections (Sharma et al. 2021). An interesting example of a magnetic nanoparticle based nano sensor is discussed next.

A group of scientists from China reported a sensor that can detect organophosphates in pesticides to a very low concentration (Liang et al. 2013). A consequence of prolonged exposure to organophosphate is the inhibition of vital neurotransmitting enzymes like acetylcholinesterase that impart severe health dangers to human beings. In this study the researchers developed a new Iron oxide (Fe_3O_4) magnetic nanoparticle (MNP) peroxidase mimetic-based colorimetric method for the early detection of organophosphorus compounds and nerve agents. The detection material is composed of MNPs, and the enzymes acetylcholinesterase (AChE) and choline oxidase (CHO). The two enzymes AChE and CHO catalyze the formation of hydrogen peroxide (H_2O_2) in the presence of acetylcholine which in turn activates MNPs to catalyze the colorimetric substrates initiating a colour reaction. After treating with organophosphorus neurotoxins, the enzymatic activity of AChE was reduced and produced less H_2O_2 leading to diminished catalytic oxidation of colorimetric compounds over MNP peroxidase mimetics and subsequently reducing colour intensity. Thus, by utilizing the peroxidase-mimicking activity of MNPs, the developed colorimetric content has the potential to detect the neurotoxic organophosphate compounds in pesticides.

Nano sensors are also employed for monitoring the plant growth in a systematic approach. Recently a near-infrared (near-IR)fluorescent carbon nanotube sensor was employed, by a group of scientists from Singapore, to detect gibberellins (GAs), a class of plant hormones essential for plant growth (Boonyaves et al. 2023). Interestingly, the nano sensor successfully differentiates between GAs and other two related hormones: GA3 and GA4-diterpenoid phytohormones, which helps control and regulate the plant growth processes and development. Distinguishing different GAs is exceedingly challenging because of the similarities in their chemical structure making this approach remarkable. Thus, this strategy helps to monitor how crops are growing in real time and thus permitting timely intervention when necessary. The above-mentioned work has its roots in a concept developed by researchers from Massachusetts Institute of Technology (MIT) i.e., the corona phase molecular recognition (CoPhMoRe) platform (Gong 2022). With a new CoPhMoRe technique the Singapore-MIT team of researchers created nanoparticles that act like natural antibodies-they can recognize and interact onto specific molecules. After developing CoPhMoRe nano sensors for detecting hydrogen peroxide, synthetic auxins, arsenic and many other interesting compounds, the team's CoPhMoRe library screening has shown these nanoparticles are highly selective and bind specifically to GA3 and GA4, triggering significant fluorescence intensity variations(Boonyaves et al. 2023).

The GA detecting nano sensor comprise single walled carbon nanotubes (SWNTs) wrapped in polymers, containing CoPhMoRe corona phases that specifically bind to biologically active forms of GA, GA3 and GA4 causing near-IR fluorescence intensity changes, which are then measured to detect and quantify GA levels in living plants (Boonyaves et al. 2023).

16.6 Carbon Nanostructure-Based Nano Sensors

Carbon based nanostructures include diverse low-dimensional allotropes of carbon including: carbon black (CB), carbon nanotubes (CNTs), carbon fibers, carbon nanodiamonds, fullerene, graphene, graphene oxide and quantum dots (Fig. 16.4). Among the various allotropes CNTs and graphene have interesting physico-

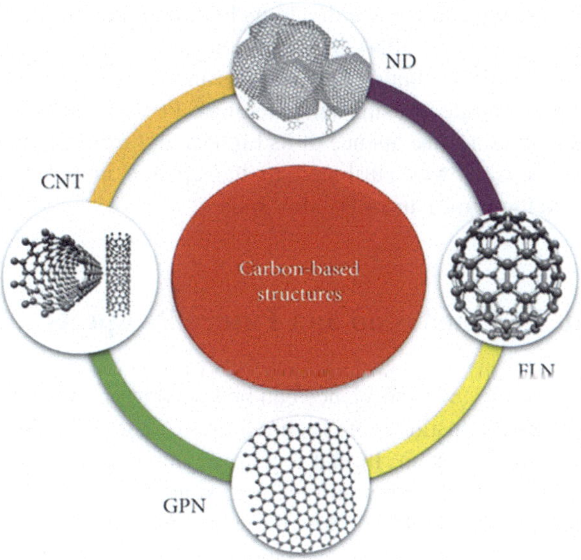

Fig. 16.4 Structural diversity of carbon allotropes which enables them with interesting physicochemical properties (*CN* Carbon Nanotubes, *ND* Nanodiamonds, *FLN* Fullerenes, *GPN* Graphene). (Adopted from the reference Bezzon et al. (2019). This is an open access article distributed under the Creative Commons Attribution License)

chemical properties therefore they are used to construct electrochemical sensors for food safety, sensing of harmful gases, optical sensors for heavy metal detection and devices for detection of pathogens (Yang 2022). Carbon nano tube based gas sensing utilizes a change in an electrical property during adsorption of gas molecules. Thus, after adsorption process, the adsorbed gas molecule transfers an electron to or take it from a CNT. This process subsequently alters the electrical properties of CNT and this change can be detected instantaneously. CNTs have an interesting structure that provides them unique physical properties. At atomic scale, carbon nanotubes are hexagonal sheets of graphite rolled into a nanoscale tube of single or multiple sheets and integrate many special properties of carbon bonds. This hollow architecture is appropriate for adsorption and desorption of gas molecules. In addition, CNTs are one-dimensional and have a cylindrical symmetry that gives them remarkable mechanical strength.

With vast knowledge on the physicochemical properties of CNTs and different combination with metal atoms open a new avenue for CNT based nano sensors. The employment of Li^+ and Pt metals in combination with CNTs are excellent examples for carbon nanostructure-based gas sensors. Chen et al. (2020) demonstrated a sensor fabricated with Li^+ doped CNTs producing a Li^+/CNT film could detect methane (CH_4) gas, a component of natural gas that can cause asphyxiation and detonations when exists in considerable concentration in closed atmosphere. When exposed to CH_4 gas, Li^+/CNT sensor adsorbs the methane gas and changes its conductivity through electrochemical reaction because of the gas adsorbed molecules. Li^+ reinforces the binding of CH_4 and induces dipole interactions between them. In addition, there is a change in hybridization from sp^2 to sp^3 due to the addition of CH_4 to the CNTs. The device is remarkable because of its sensitivity to CH_4 gas i.e. it has a sensitivity of 14.5% at 500 ppm of CH_4.

Although hydrogen gas is non-polluting, there are concerns about its large-scale production and storage due to its high flammability. A combination of CNTs coated with graphene (GPN) and platinum decked nano sensor has shown remarkable H_2 gas sensitivity (Jaidev 2018).CNTs and GPN promoted high surface area but did not certify H_2 gas sensing and therefore Pt was carefully selected and applied to the system as a consequence of its high catalytic affinity towards H_2. The mechanism of gas sensing was attributed to the change in resistance of the sensor film with switching from air to H_2 atmosphere.

16.7 Conclusion and Future Prospects

The history of nanoscience can be traced to an inspirational lecture given in 1959 in California Institute of Technology (Caltech) by Nobel prize winning physicist Richard Feynman. In that lecture Feynman introduced the concept of nanoscience to the scientific community and 15 years later, in 1974, Norio Taniguchi defined the term nanotechnology. This innovative idea caught the attention of several scientists around the world and they elaborated the concept of nanoscience by developing two approaches for synthesizing nanostructures which fall under two categories: top-down and bottom-up approaches. Top-down involves breaking down of bulk material, utilizing precision engineering and lithography technologies, to get nano-sized particles. While, bottom-up approach involves building of nanostructures from bottom using atoms and molecules where they self-assemble into precisely ordered nanostructures by chemical or physical interaction between them. Upon realizing the superior properties associated with nano sized material, subsequent decades saw a rapid progress in the field of nanoscience. During the early 1980s the emergence of nanoscience as an interesting subject became apparent with the invention of Scanning Tunneling Microscope (STM) which enables the design of novel structures from single atoms and molecules through their direct visualization. In addition, STM paved the way for the development of better microscopes like atomic force microscope (AFM) and scanning probe microscopes (SPM) that are essential to nanotechnology researchers today. In 1986, K. Eric Drexler published the first book on nanotechnology "Engines of Creation: The coming Era of Nanotechnology" and consequently saw the emergence of the theory of molecular engineering. The following decades saw a number of studies highlighting the potential of nanotechnologies in biomedicine for the diagnosis and therapy of human disease and this area being intensively researched and offering excellent results. Nano sensors were invented and utilized for the early detection of contagious virus rapidly, cost-effectively, quick turnaround time and provided lab-free analysis. During this period there were parallel developments where nanotechnology has been used to improve and sustain the environment by producing devices that could sense poisonous gases. New carbonaceous materials like single walled carbon nanotubes (SWCNTs), graphene was successfully tested for gas sensing applications. Gas sensors also found application in monitoring agriculture land for precision agriculture and protecting food products from pathogens and pesticides thus having a potent application in our society.

We feel that in a short span of time nanotechnology has gained an explosive growth from fundamental importance to advanced applications in healthcare sectors as diagnostic biosensors, to industries as protectors of harmful gases and safety equipment, to agriculture industries for monitoring the growth of crops. Nanoscience and its applications such as nano sensors, presented in this book chapter, will meet the contemporary challenges such as virus pandemics which affected almost every corner on the planet earth, food shortage and health and safety issues in industries etc. The present period and the coming decade will see the invention of advanced nano sensors capable of providing rapid test results through lab-free detection utilizing techniques such as plasmonic resonance as a method for detecting viral pathogens and thus reducing test result time. Nanosensors may also emerge as a new class of device for detecting pathogens and undesired poisonous gases in agriculture sectors. With many new nano device applications on the pipeline, many are optimistic that nano devices, including nano sensors, will have a revolutionary impact on our society. This may lead to improved health care services and accelerated crop production. The advancement made in physics, chemistry and material science not only provided new nanomaterials with unique properties, but a deeper understanding on nanoscience and could unveil new patterns in nanotechnology towards a better future.

Conflicts of Interest The authors have no conflicts of interest to declare.

References

Abdel-Karim R, Reda Y, Abdel-Fattah A (2020) Review nanostructured materials based nanosensors. J Electrochem Soc 167:037554–037565. https://doi.org/10.1149/1945-7111/ab67aa

Anker JN, Hall WP, Lyandres O, Shah NC, Zhao J, Van Duyne RP (2008) Biosensing with plasmonic nanosensors.Nature Materials 7(6):442–453. https://doi.org/10.1038/nmat2162

Bai J, Zhou B (2014) Titanium dioxide nanomaterials for sensor applications. Chem Rev 114: 10131–10176. https://doi.org/10.1021/cr400625j

Bardhan NM, Jansen P, Belcher AM (2021) Graphene, Carbon Nanotube and Plasmonic Nanosensors for Detection of Viral Pathogens: Opportunities for Rapid Testing in Pandemics like COVID□19. Frontiers in Nanotechnology 3:733126. https://doi.org/10.3389/fnano.2021.733126

Bezzon VDN, Montanheiro TLA, de Menezes BRC, Ribas RG, Righetti VAN, Rodrigues KF, Thim GP (2019) Carbon nanostructure-based sensors: a brief review on recent advances. Ad Mater Sci Eng Article ID 4293073. https://doi.org/10.1155/2019/4293073

Boonyaves K, Ang MC, Park M, Cui J, Khong DT, Singh GP, Koman VB, Gong X, Porter TK, Choi SW, Chung K, Chua NH, Urano D, Strano MS (2023) Near-infrared fluorescent carbon nanotube sensors for the plant hormone family gibberellins. Nano Lett 23:916–924. https://doi.org/10.1021/acs.nanolett.2c04128

Chakraborty U, Kaushik A, Chaudhary GR, Mishra YK (2024) Emerging nano-enabled gas sensor for environmental monitoring—perspectives and open challenges. Curr Opin Environ Sci Health 37:100532–100543. https://doi.org/10.1016/j.coesh.2024.100532

Chen Z, Zhang Z, Zhai X, Li Y, Lin L, Zhao H, Bian L, Li P, Yu L, Wu Y, Lin G (2020) Rapid and sensitive detection of anti-SARS-CoV-2 IgG, using lanthanide-doped nanoparticles-based lateral flow immunoassay. Anal Chem 92:7226–7231. https://doi.org/10.1021/acs.analchem.0c00784

Dey A (2018) Semiconductor metal oxide gas sensors: a review. Mater Sci Eng B 229:206–217. https://doi.org/10.1016/j.mesb.2017.12.036

Dhiman S, Gaba S, Varma A, Goel A (2021) Bio-nanosensors: synthesis and their substantial role in agriculture. In: Plant-microbes-engineered Nano-particles (PM-ENPs) nexus in agro-ecosystems. Springer, pp 165–172. https://doi.org/10.1007/978-3-030-66956-0_11

Dutta T, Noushin T, Tabassum S, Mishra SK (2023) Road map of semiconductor metal-oxide-based sensors: a review. Sensors 23:6849–6905

Feynman RP (1960). There's plenty of room at the bottom. Caltech Engineering and Science 23(5):22–36

Gong X, Cho S, Kuo S, Ogunlade B, Tso K, Salem DP, Strano MS (2022) Divalent metal cation optical sensing using single-walled carbon nanotube corona phase molecular recognition. Analytical Chemistry 94(47):16393–16401. https://doi.org/10.1021/acs.analchem.2c03648

Gao X, Zhang T (2018) An overview: facet-dependent metal oxide semiconductor gas sensor. Sensors Actuators B Chem 277:604–633. https://doi.org/10.1016/j.snb.2018.08.129

Han D, Hosamo H, Ying C, Nie RA (2023) Comprehensive review and analysis of nanosensors for structural health monitoring in bridge maintenance: innovations, challenges, and future perspectives. Appl Sci 13:11149–11171. https://doi.org/10.3390/app132011149

Huang X, Zhu Y, Kianfar E (2021) Nano biosensors: properties, applications and electrochemical techniques. J Mater Res Technol 12:1649–1672. https://doi.org/10.1016/j.jmrt.2021.03.048

Jaidev S, Baro M, Ramaprabhu S (2018) Room temperature hydrogen gas sensing properties of monodispersed platinum nanoparticles on graphene-like carbon-wrapped carbon nanotubes. International Journal of Hydrogen Energy 43(33):16421–16429. https://doi.org/10.1016/j.iijhydene.2018.06.178

Lee JY, Nguyen TL, Park JH, Kim BK (2016) Electrochemical detection of hydrazine using poly (dopamine)-modified electrodes. Sensors 16:647–657. https://doi.org/10.3390/s16050647

Li Z, Yi Y, Luo X, Xiong N, Liu Y, Li S et al (2020) Development and clinical application of a rapid IgM-IgG combined antibody test for SARS-CoV-2 infection diagnosis. J Med Virol 92:1518–1524. https://doi.org/10.1002/jmv.25727

Liang M, Fan K, Pan Y, Jiang H, Wang F, Yang D, Lu D, Feng J, Zhao J, Yang L, Yan X (2013) Fe_3O_4 magnetic nanoparticle peroxidase mimetic-based colorimetric assay for the rapid detection of organophosphorus pesticide and nerve agent. Anal Chem 85:308–312. https://doi.org/10.1021/ac302781r

Liu X, Zheng W, Kumar R, Kumar M, Zhang J (2022) Conducting polymer based nanostructures for gas sensors. Coord Chem Rev 462:214517. https://doi.org/10.1016/j.ccr.2022.214517

Lun D, Xu K (2022) Recent progress in gas sensor based on nanomaterials. Micromachines (Basel) 13:919–945. https://doi.org/10.3390/mi13060919

Mahari S, Roberts A, Shahdeo D, Gandhi S (2020) eCovSens - ultrasensitive novel in-house built printed circuit board -based electrochemical device for rapid detection of nCovid-19 antigen, a spike protein domain 1 of SARS-CoV-2. bioRxiv https://doi.org/10.1101/2020.04.24.059204

Malekzad H, ZangabadPS MH, Karimi M, Hamblin MR (2017) Noble metalnanoparticles in biosensors: recent studies and applications. Nanotechnol Rev 6(3):301–329. https://doi.org/10.1515/ntrev-2016-0014

Merkoçi A (2021) Smart nanobiosensors in agriculture. Nat Food 2:920–921. https://doi.org/10.1038/s43016-021-00426-2

Ramanavicius S, Ramanavicius A (2020) Insights in the application of stoichiometric and non-stoichiometric titanium oxides for the design of sensors for the determination of gases and VOCs (TiO_{2-x} and Ti_nO_{2n-1} vs TiO_2). Sensors 20:6833–6851

Sharma P, Pandey V, Sharma MMM, Patra A, Singh B, Mehta S, Husen A (2021) A review on biosensors and nano sensors application in agro ecosystems. Nanoscale Res Lett 16:1–24. https://doi.org/10.1186/s11671-021-03593-0

Singh RP (2021) Recent trends, prospects, and challenges of nanobiosensors in agriculture. In: Biosensors in agriculture: recent trends and future perspectives. Concepts and strategies in plant sciences. Springer, Cham, pp 3–31

Taniguchi, N. (1974). On the Basic Concept of 'Nano-Technology'. Proceedings of the International Conference on Production Engineering, Tokyo, Part II, Japan Society of Precision Engineering, pp. 18–23.

Tarafdar JC, Sharma S, Raliya R (2013) Nanotechnology: interdisciplinary science of applications. Afr J Biotechnol 12:219–226. https://doi.org/10.5897/AJB12.2481

Tshabalala ZP, Motaung DE, Mhlongo GH, Ntwaeaborwa OM (2016) Facile synthesis of improved room temperature gas sensing properties of TiO2 nanostructures: effect of acid treatment. Sensors Actuators B Chem 224:841–856. https://doi.org/10.1016/j.snb.2015.10.079

Tyagi S, Chaudhary M, Ambedkar AK, Sharma K, Gautam YK, Singh BP (2022) Metaloxide nanomaterial-based sensors for monitoring environmental NO2 and its impact on the plant ecosystem: a review. Sens Diagn 1:106–129. https://doi.org/10.1039/d1sd00034a

Vainionpää (2008) Diagnostic techniques: serological and molecular approaches. in: encyclopedia of virology, 3rd edition, pp. 29–37. Elsevier. https://doi.org/10.1016/B978-012374410-4.00585-9

Vaughan TG, Nadeau SA, Scire J, Stadler T (2020) Phylodynamic analyses of outbreaks in China, Italy, Washington State (USA), and the Diamond Princess. In: Virological.org, Jan 28, 2021 (preprint). (This work discusses multi-region outbreak dynamics using phylodynamics techniques)

Vaughan TG, Nadeau SA, Scire J, Huisman JS, Stadler T (2021) The origin and early spread of SARS-CoV-2 in Europe. Proceedings of the National Academy of Sciences 118(9): e2012008118. https://doi.org/10.1073/pnas.2012008118

Wang X, Luo Y, Huang K, Cheng N (2022) Biosensors for agriculture and food safety: recent advances and future perspectives. Adv Agrochem 1:3–6. https://doi.org/10.1016/j.aac.2022.08.002

Xue S, Cao S, Huang Z, Yang D, Zhang G (2021) Improving gas-sensing performance based on MOS nanomaterials: a review. Materials (Basel) 14:4263–4296. https://doi.org/10.3390/ma14154263

Yadav A, Yadav K, Ahmad R, Abd-Elsalam KA (2023) Emerging frontiers in nanotechnology for precision agriculture: advancements hurdles and prospects. Agrochemicals 2:220–256. https://doi.org/10.3390/agrochemicals2020016

Yang Z, Zhang X, Guo J (2022) Functionalized carbon-based electrochemical sensors for food and alcoholic beverage safety. A pplied Sciences 12(18):9082. https://doi.org/10.3390/app12189082

Zhu G, Yu X, Xie F, Feng W (2019) Ultraviolet light assisted heterogeneous Fenton degradation of tetracycline based on polyhedral Fe_3O_4 nanoparticles with exposed high-energy {110} facets. Appl Surf Sci 485:496–505. https://doi.org/10.1016/j.apsusc.2019.04.239

Chapter 17
Eco-Nanotoxicology: Emerging Concepts and Challenges for Sustainable Nanotechnological Applications

Asha Kumari Kumawat and Bhawana Pathak

Abstract Nanotechnology has emerged as a groundbreaking field with the potential to transform sectors ranging from medicine to energy and electronics. The use of natural and artificial nanoparticles in all aspects of life makes life easier. Nanoparticles have many benefits, but their uncontrolled use in commercial products and subsequent release into ecosystems raises concerns about their potential environmental and human health impacts. Eco-nanotoxicology is the study of how organisms and ecosystems are affected by man-made nanoparticles in the environment. Transforming nanotechnology into environmentally friendly technology through the adoption of bottom-up preparation methods from green chemistry helps mitigate related risks in industrial applications, toxic effects on ecosystem and aligns with the attributes of ideal technology. Recent trends in nanotoxicology involve integrating high-throughput screening methods to establish structure-activity relationships using various types of nanoparticles and in vitro screening tests along with computational analysis and omics-based approaches with a focus on developing sensitive toxicity tests for early detection of harmful effects. In view of this, the present paper takes an in-depth look at the key ideas and approaches in the growing subject of eco-nanotoxicology. It explores the interactions of nanomaterials with the environment, looking at their behavior, fate, bioavailability and toxicity in ecosystems. It will also include the importance of recognizing the potential hazards associated with nanoparticles to ensure their safe and long-term use in various industries contributing to the sustainable and harmonious coexistence of nanotechnology.

Keywords Engineered nanoparticles · Environmental impact · Human health impact · Nanomaterial · Nano-sensors · Nanotechnology · Nanotoxicology

---

A. K. Kumawat · B. Pathak (✉)
School of Environment and Sustainable Development, Central University of Gujarat, Gandhinagar, India
e-mail: bhawana.pathak@cug.ac.in

© The Author(s), under exclusive license to Springer Nature Switzerland AG 2025

J. M. Al-Khayri et al. (eds.), *Green Nanotechnology Applications for Ecosystem Sustainability*, Nanotechnology in Plant Sciences 4,
https://doi.org/10.1007/978-3-031-90766-1_17

17.1 Introduction

Nanotechnology has emerged as a groundbreaking field with the potential to transform sectors ranging from medicine to energy and electronics. Since its inception, nanotechnology has been seen as a field of research capable of exceeding the limit of diminishment in technology, as stated by Moore's law; yet, this diminishment has its own limitations in terms of nano. Nanoscale is a genuinely new scale, not merely another step towards diminishment (Joachim 2005). Materials at the nanoscale differ from bulk materials in their molecular structures and characteristics (Asha and Narain 2020). Nanoparticles have many benefits, but their uncontrolled use in commercial products and subsequent release into ecosystems raises concerns about their potential environmental impacts. Eco-nanotoxicology is a multidisciplinary discipline that investigates the potential environmental implications of nanomaterials, which are materials having nanoscale structures and characteristics (usually range from 1 to 100 nanometers). Nanomaterials have distinct and frequently increased physical and chemical characteristics when compared to bulk equivalents, making them useful in a variety of applications such as medicine, electronics, and materials research. The surface area-to-volume ratio and physic-chemical characteristics of nanomaterials have a major impact on biological activity (Vaish and Pathak 2023). Within the 1–100 nm range, a nanomaterial's size and shape affect its basic characteristics, which include melting temperature, color, and electrical conductivity (Rauscher et al. 2019). Nanomaterials have amazing potential advantages for both the environment and human health. Examples of these include bio robotics, nanoscale implants in medicine, nano remediation, waste management, medication delivery systems, bioimaging and tissue engineering, sensors for environmental monitoring, and nanoarrays (Dasari Shareena et al. 2018; Gupta et al. 2023).

Nanoparticles can be produced from natural and artificial activities, which helps to make all aspects of life easier. Figure 17.1 shows natural and artificial sources NPs as the production of nanoparticles is not limited to highly specialized target-oriented activities and laboratory experiments; a considerable amount of nanomaterials can also be obtained from natural sources, such as pollen grains, ultrafine particles in smoke, aerosols, dust particles, air pollutants, and many other naturally occurring exhaust mechanisms (Raut et al. 2024). Some natural nanoparticles are potentially harmful and were formed as a result of a variety of geological and biological processes. These nano products are novel to the environment and have some unfavorable effects that, when released into the environment to a certain extent or through exposure to nano-based products, can result in particle toxicity. Every product made using nanotechnology has some toxicological effects, varying depending on the factors that contribute to the toxicity of the nanoparticles (Egbuna et al. 2021).

Eco nanotoxicology, a multidisciplinary area seeks to understand the complex interactions between nanomaterials and ecosystems, concentrating on their impacts at many hierarchical levels ranging from molecular and cellular to organismal and

Fig. 17.1 Sources of nanoparticles. (Constructed by Asha Kumari Kumawat)

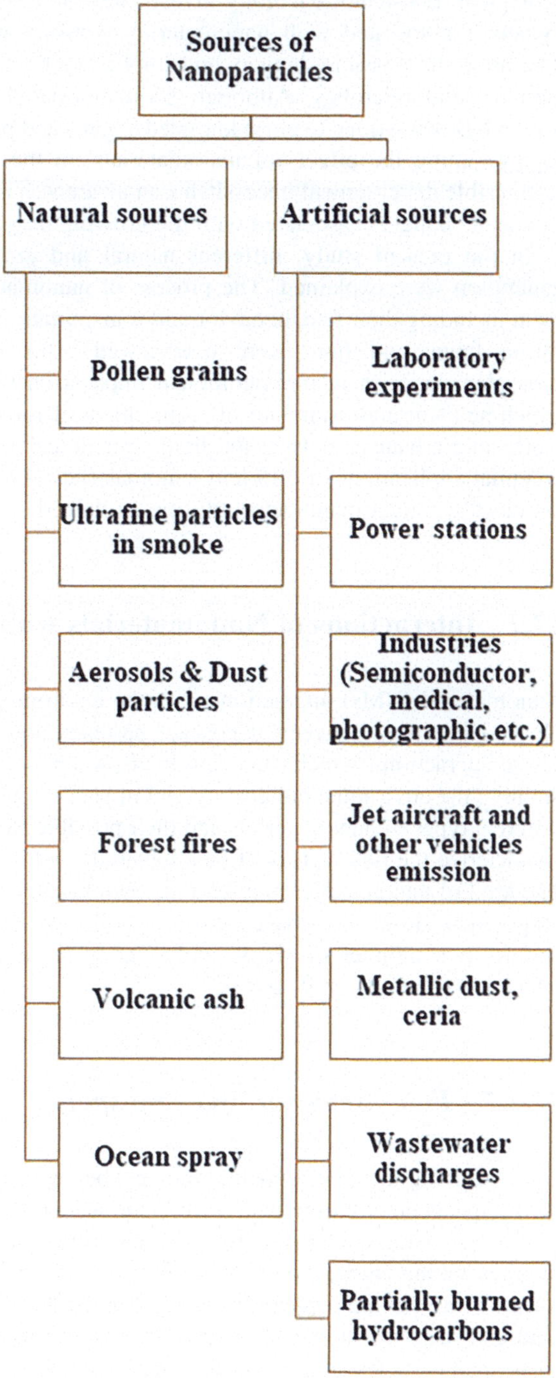

ecological. Eco-nanotoxicology investigates the biological, chemical, and physical dynamics associated with nanomaterial exposure in natural habitats in order to determine the possible dangers and implications for living species such as plants, animals, and microbes. Although eco-nanotoxicology is a relatively new area, research is being done to close knowledge gaps and provide standardized techniques for evaluating the effects of nanomaterials on the environment. Sustainable and responsible development necessitates an awareness and mitigation of possible environmental dangers associated with the growing usage of nanotechnology.

In the present study, different natural and artificial sources of nanoparticle generation were explained. The process of nanomaterial interaction with environment including their fate, behavior and transportation pathway with its bioavailability to fauna and flora were overviewed. This nanomaterial interaction with ecosystem leads to nanotoxicological impacts on human health and environment which helps in understanding of requirement of future research and developmental works and advancements in the field of nanotechnology. An in-depth analysis of potential applications of different nanomaterials in treating serious illness to different electric gadget manufacturing were explained.

17.2 Interactions of Nanomaterials with the Environment

Nanomaterial (NMs) interactions with the environment entail a dynamic and complicated interplay between nanoscale particles and other ecosystem components. These interactions, which can occur in air, water, soil, and live beings, are critical in defining the environmental destiny and impact of nanomaterials. Understanding the different types of nanomaterials and their possible routes of exposure is important to characterize the risk factors of nanomaterials and their destiny in the environment. The kind of nanomaterial may be unknown because to its manufacturing technique (vapor deposition gas phase, attrition, colloidal, and so on), which may expose people in a number of ways, including by ingesting, cutaneous absorption, or inhalation (El-Kady et al. 2023).

17.2.1 Fate, Behavior and Transport

Understanding the environmental fate and behavior of NMs is critical to understand their impacts on environmental health and is a critical component of environmental risk assessment. Understanding NM mobility, bioavailability, and ultimate fate requires taking into account a number of fate and transport processes. NM are released into the atmosphere by indirect or diffuse sources and eventually settle on land and surface waters (Ali et al. 2024). These include the following: sorption to suspended particles and the ensuing deposition to sediment; advection in and out of the system; diffusive transport; volatilization to air; transformation into other NMs or compounds; transportation, manufacture, and disposal of items at the end of their useful life; aggregation; sedimentation; dissolution; filtration; and NM emissions to air, water, and soil (Li et al. 2022a, b). For example, NMs are likely to be present in wastewater, sewage, sludge, waste incinerator residues, and landfill leachates (Bhardwaj et al. 2023). The distribution and behavior of these NMs in the environment are governed by their real characteristics as well as the ambient circumstances. Hartmann et al. (2014), have identified key transformation processes for NMs, determining their fate and behavior in the natural environment, are photochemical degradation, adsorption, oxidation, desorption, reduction, biotransformation, dissolution, agglomeration, precipitation, aggregation, speciation, complexation and sedimentation. These are all determined by their size, surface, qualities, and environmental conditions. Some engineered nanomaterials (ENMs) dissolve over time, resulting in exposure to both suspended NPs and dissolved ions (Abbas et al. 2024). Aggregation and other transformation processes might cause surprising behavior in NPs. Uncertainties about NP emissions, ENM coatings, interactions with NOM (natural organic matter), and ambient environmental properties make it difficult to draw broad conclusions about ENM fate and transport patterns (Vouitsis et al. 2023).

NMs undergo physical and chemical processes that affect their destiny and environmental concentrations in different compartments. In the processes of interaction with environment, when discharged into the atmosphere, NPs and their aggregates grow in size owing to condensation of organic and inorganic vapors on particle nuclei, resulting in dilution, coagulation, and deposition (Abbas et al. 2024; Razi et al. 2022). Nanoparticles were attached to aerosols, indicating that they might infiltrate soil and water through the atmosphere. The chemical characteristics of the water have a major impact on the fate and movement of NMs in it through water runoff (El-Kalliny et al. 2023). When these NPs contact with water, aggregation (formation of NM clusters in colloidal suspension), sedimentation (NMs are deposited in sediment compartments by gravity settling of aggregates or sorbed to NOM or other suspended particles), dissolution (release of dissolved ions from the nanoparticle, even inside an aggregation) and transformation processes takes place in water (El-Kalliny et al. 2023; Razi et al. 2022). Environmental changes to NPs will also have an impact on their toxicity and fate by changing their chemical and physical characteristics. Important transformation processes might involve phosphorous reactions, oxidation, and sulfidation. NMs in soil tend to aggregate, sorb to surfaces, sediment, and dissolve, similar to how they do in groundwater (estimated rates of aggregation, sedimentation, and dissolution). It will also be buried in sediment and deep sediment (Tourinho et al. 2012).

17.2.2 Bioavailability

Bioavailability of nanoparticles is described as a process in which NM metabolites play critical active roles in biomolecules, or affect biological activities or effectiveness as pathway modulators. Bioavailability is the extent and pace with which a substance is absorbed, transported, and made accessible to an organism. In the context of nanomaterials, their tiny size and huge surface area might impact their behavior and interactions with live beings and environmental matrices. The bioavailability of nanomaterials can have cascading effects on ecosystems, influencing the structure and function of populations and communities. Aggregation might also affect bioavailability. Aggregation can shift the primary exposure pathway from water to sediment or food by removing nanomaterials from the water column (Dube and Okuthe 2023). Experiments under actual exposure settings show that NM interactions with environmental matrices can reduce reactivity, bioavailability, and toxicity, implying a lower environmental risk in many situations (Lead et al. 2018).

17.3 Influence of Nanotoxicity on Human Health and Environment

Nanomaterials can elicit adverse effects on various organisms, from microbes to plants and animals, with potential consequences for ecosystem functioning. To provide some evaluation of the danger to the environment, toxicity assays include testing for mutagenicity, eco-toxicity, and mammalian toxicity such as acute and chronic tests, dermal toxicity, skin irritation tests and oral toxicity (Chauhan et al. 2023). Analyzing the physical-chemical characteristics of the materials that enable them to interact with and potentially harm biological systems is the first step in a logical approach to nanomaterial toxicity (Noga et al. 2023). In vitro cellular assay, which may be used to target various cell types at the portal of entry or systemic areas following nanomaterial absorption and dispersion, include another group of testing. Overall, the cellular *in vitro* data are more helpful in comprehending the bio-kinetics mechanism of ENMs.

Several studies have shown that nanoparticle toxicity is exceedingly complicated and multi-factorial, perhaps being governed by a number of physicochemical parameters, size and shape, as well as surface qualities including charge, area, and reactivity (Noga et al. 2023). Nowadays, a lot of nanomaterials and nano-based products are employed in biomedical applications, environmental cleanup, and health enhancement, which increases the likelihood that these nanoparticles will be released into the environment from the workplace and that people will be exposed to them for extended periods of time (Ghadam and Saboora 2024). The usage and release of nanomaterials or products derived from nanotechnology raises a number of health and environmental concerns, including those linked to the material's potential toxicity (Vouitsis et al. 2023).

17.3.1 Impact on Human Health

Numerous NPs at various doses have been linked to toxic effects in a range of living beings (Vouitsis et al. 2023). Several toxicity metrics are used in studies, such as half maximal effective concentration (EC50), median lethal dosage (LD50), median lethal concentration (LC50), minimum inhibitory concentration (MIC), least observed effect concentration (LOEC), and no observed effect concentration (NOEC). The specific toxic impact was not distinguished between acute and chronic toxicity for the purposes of this screening analysis. As a result, harmful impacts could include genetic changes, restricted development rates, impaired reproduction, and immediate death of individuals. Figure 17.2 shows the process of nanotoxicity influence on human health.

The physicochemical properties of NPs directly or indirectly affect the toxic effects of NPs on human health. It's crucial to remember that a nanoparticle's toxicity can vary greatly based on its unique properties and the exposure environment. Larger-sized particles typically cannot enter cells, tissues, or organs, but nanomaterials can pass through biological membranes. They can also enter the bloodstream by eating or breathing them in. The following are some possible harmful consequences of nanoparticles on human health.

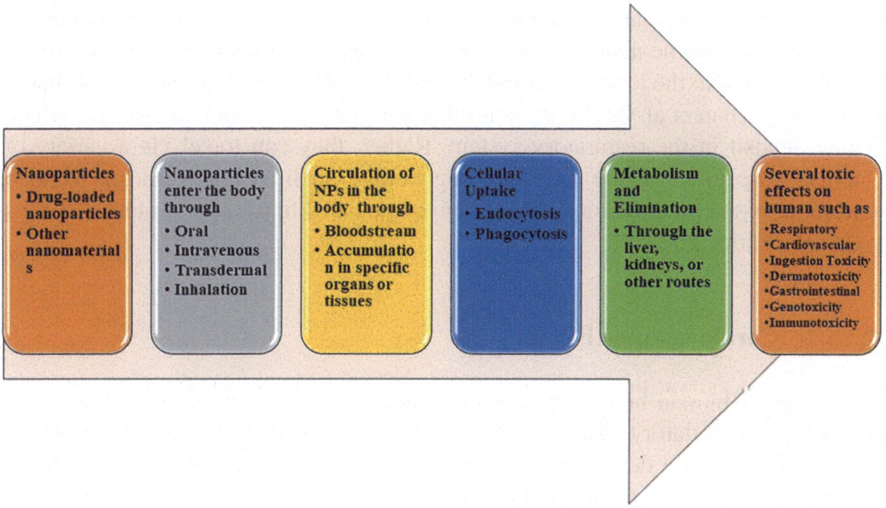

Fig. 17.2 Process of nanotoxicity influence on human health. (Constructed by Asha Kumari Kumawat)

17.3.1.1 Respiratory System Effects

Human inhalation of NPs results in a variety of respiratory problems. The burning of home fuels produces indoor air pollution, which is the eighth-largest contributing factor for the worldwide burden of disease. The internal surface area of a human lung is around 75–140 m^2, with approximately 3,00,106 alveoli (Ilyasu et al. 2024). Because of the lungs are directly exposed to the external environment, they serve as the primary point of passage for nanoparticles into the human circulatory system. Some serious effects on human respiratory system are explained below.

- **Inflammation:** Breathing in nanoparticles, particularly those with an ultrafine size, can lead to respiratory system inflammation. Inflammation damages cilia in the respiratory tract, affecting the passage of mucus (which ultimately traps pathogens and particles) through cilia. Inflammation also causes epithelial injury by disrupting the epithelial barrier between the outer layer of the lungs and the bloodstream, resulting in clotting, decreased lung function, and poor blood and oxygen circulation throughout the body (Sonwani et al. 2021). This inflammation can worsen pre-existing respiratory disorders like asthma or cause new ones like bronchitis (Taghavizadeh Yazdi et al. 2023).
- **Pulmonary Fibrosis:** Long-term exposure to certain nanoparticles has been linked to the development of this lung disease, which is typified by lung tissue scarring (Dong and Ma 2019). Pulmonary fibrosis and lung cancer appear with symptoms such as inflammation of the lung tissue and pneumonia (Sonwani et al. 2021). The primary causes of death for patients with chronic lung illnesses are different forms of pulmonary fibrosis (Kreuter et al. 2021).
- **Inhalation Toxicity:** Nanoparticles can be breathed and enter the respiratory system, causing inflammation, oxidative stress, and potentially damaging lung cells. This is particularly concerning in industrial contexts where workers may be exposed to airborne nanoparticles. Nanoparticles may translocate from the respiratory system to the lung interstitial, brain, liver, spleen, and, in pregnant women, the fetus (Khan et al. 2015). Particles that are inhaled may not always stay where they deposit in the respiratory system. Rather, they can travel via a variety of transit procedures in the different lung tissues, such as lung clearing. Human respiratory tract nanoparticles have an average half-life (t1/2) of around 700 days (Bhattacharjee 2023).

17.3.1.2 Cardiovascular Effects

NPs can enter human bodies by ingesting them to the gastro-intestinal tract, then moves to the circulatory system, or by inhaling (intranasal or intratracheal). These may result in notable deviations from the system's typical operation. Some of the first impacts of NPs on the cardiovascular system are elevated blood pressure, lowered heart rate, and changed vascular tone and dysfunction (Yu et al. 2016).

- **Blood Clotting and Thrombosis:** Increased blood clotting and thrombosis may result from nanoparticles interacting with blood components when inhaled or absorbed into the circulation. This may exacerbate heart-related conditions (Tran et al. 2022).

17.3.1.3 Ingestion Toxicity

If nanoparticles are swallowed, they can enter the digestive system and even reach the circulation. This can result in systemic dispersion and harm several organs. According to research, some nanoparticles might induce gastrointestinal discomfort and other side effects. The harmful consequences of injected nanoparticles are determined by their composition and charge. Hypersensitivity is a typical adverse effect of injecting nanoparticles intravenously, which happens in many patients and is most likely caused by complement activation (Moghimi et al. 2023).

17.3.1.4 Skin Exposure

Skin contact with some nanoparticles may result in irritation or allergic responses called dermatotoxicity. Furthermore, certain nanoparticles can pass through the skin and into the circulation, potentially producing systemic consequences. Consumer goods like ZnO and TiO_2 NPs in sunscreens can also expose skin to nanoparticles (NPs) (Lee et al. 2020). Certain evidences indicate that nanoparticles can enter the stratum corneum, the epidermis, and the dermis. They can also enter the skin by passing through or between epithelial cells, or via the appendages of the skin. One example of this would be the use of cosmetics containing nanoparticles. Both *in vitro* and *in vivo*, it has been shown that sunscreen containing sun-illuminated TiO_2 particles catalyzes DNA damage (Lee et al. 2020).

17.3.1.5 Gastrointestinal Effects

- **Inflammation and Irritation:** Ingested nanoparticles can cause irritation and inflammation in the gastrointestinal tract, leading to conditions such as gastritis or enteritis (Korani et al. 2015).

17.3.1.6 Genotoxicity

It has been demonstrated that some nanoparticles have genotoxic effects, which means they may harm cells' DNA (Yao et al. 2023). This raises questions regarding the possibility that nanoparticles might result in mutations and aid in the growth of cancer.

17.3.1.7 Oxidative Stress

- **Free Radical Generation:** By producing reactive oxygen species (ROS) in cells, nanoparticles have the ability to cause oxidative stress (Lin et al. 2023). Extended periods of exposure to oxidative stress have been linked to a number of illnesses and cellular component damage.

17.3.1.8 Neurotoxic Effects

- **Crossing Blood-Brain Barrier**: Some nanoparticles have the ability to cross the blood-brain barrier, potentially affecting the central nervous system which may have an effect on the central nervous system (Hasannejad-Asl et al. 2023). Concerns over long-term cognitive affects and neurological implications are raised by this. Neurotoxic effects include cognitive impairment, neuroinflammation, or damage to neuronal structures.

17.3.1.9 Reproductive and Developmental Effects

- **Impaired Fertility:** Certain nanoparticles have been associated with reproductive toxicity, including effects on sperm quality and female fertility (Xuan et al. 2023). NPs have an impact on the reproductive system through ovarian cell cytotoxicity, which disrupts the oogenesis process and ultimately leads to an excess of reactive oxygen species (ROS), apoptosis, and an imbalance in sex hormones in the body.
- **Developmental Toxicity:** Exposure to nanoparticles during pregnancy may pose risks to fetal development, leading to developmental abnormalities or birth defects (Iftikhar et al. 2023). Significant sperm DNA integrity loss was recorded as a result of TiO_2 nanoparticle's genotoxic effects, which are demonstrated when they penetrate the blood–testis barrier and induce inflammation and cytotoxicity (Santonastaso et al. 2019).

17.3.1.10 Immunotoxicity

When nanoparticles interact with the immune system, potentially detrimental immunological responses may result. This might involve allergic responses, inflammation, or compromised immune system performance (Bi et al. 2023).

17.3.2 *Impact on Environment*

Influence of nanoparticles on the environment is contingent upon several aspects, including their size, composition, surface charge, and the particular ecosystems they interact with. Although there are many useful uses for nanoparticles, there are worries about their possible environmental damage. The following are specific effects on the environment caused by nanoparticle toxicity. Figure 17.3 shows the process of nanotoxicity influence on environment.

17.3.2.1 Water Contamination

- **Aquatic Organisms:** Runoff or industrial discharges can introduce nanoparticles into aquatic systems. These nanoparticles may be exposed to aquatic species, including fish, algae, and invertebrates, which might have harmful consequences. However, dissolved silver ions in the environment are durable and extremely harmful to prokaryotes, as well as many freshwater and marine invertebrates and fish. For instance, it has been demonstrated that silver nanoparticles poisonous to aquatic life (Kang et al. 2023).

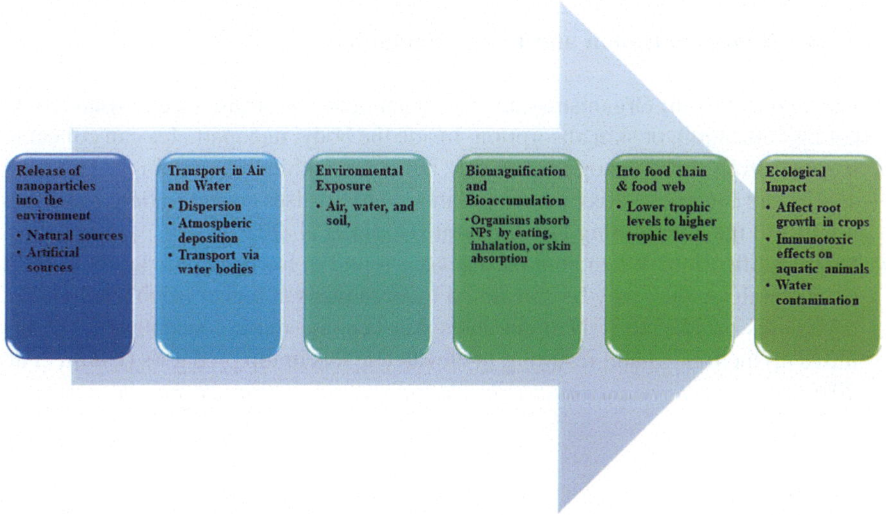

Fig. 17.3 Process of nanotoxicity influence on the environment. (Constructed by Asha Kumari Kumawat)

17.3.2.2 Soil Contamination

- **Plant Uptake:** Plants have the ability to absorb nanoparticles from soil, which might have an impact on food safety and agricultural productivity. Additionally, bioaccumulation in the food chain may result from this intake (Li et al. 2022a, b). Certain nanoparticles could potentially harm soil organisms. For instance, specific metal nanoparticles might release harmful ions, affecting the well-being of plants, microorganisms, and other creatures in the soil environment.
- **Impact on Microbial populations:** The health of the soil and the cycling of nutrients can be impacted by nanoparticles on soil microbial populations (Kundu et al. 2021). Metal and metal oxide nanoparticles can interact with microbial cell membranes. This contact causes damage to microbial cell membranes, decreasing viability. Nanoparticles generate reactive oxygen species (ROS), causing oxidative stress on microbial cells. High ROS levels harm proteins, lipids, and DNA, affecting microbial activity and survival (Kundu et al. 2021).

17.3.2.3 Airborne Dispersal

- **Atmospheric Transport:** Long distances can be covered by nanoparticles dispersed into the air, especially in industrial settings or during the usage of certain goods. This might lead to the deposition of nanoparticles in many settings, affecting ecosystems that are distant from the source (Bundschuh et al. 2018). Nanomaterials may be produced as powders or aerosols during production, exposing workers and neighboring ecosystems.

17.3.2.4 Biomagnification and Bioaccumulation

- **Bioaccumulation:** Organisms can absorb nanoparticles from the environment by eating, inhalation, or skin absorption. Inside the body, nanoparticles can collect in tissues. Over time, nanoparticles can build up within living things. Different species are capable of this bioaccumulation, which raises concentrations in tissues relative to the surrounding environment (Uddin et al. 2020).
- **Biomagnification:** When nanoparticles are found in lower trophic levels such as plants or algae, they may be consumed by organisms at higher trophic levels like herbivores or predators. Nanoparticles can become more concentrated as they move up the food chain, resulting in elevated levels in top predators (Uddin et al. 2020).

17.3.2.5 Disruption of Ecosystem Dynamics

- **Impact on Biodiversity:** Because certain creatures are toxic to nanoparticles, ecosystems may be disrupted by changes in species distribution and abundance, which might result in changes in biodiversity (Burketová et al. 2022).
- **Effects on pollinators (Bees):** Pollinators, including bees, may be negatively impacted by nanoparticles present in tainted pollen or used in some herbicides. This might lead to a disruption of the pollination processes that are essential for the growth of many plants (Hooven et al. 2019).

17.3.2.6 Microbial Ecology Effects

- **Microbial Community Disruption:** In soil and water, nanoparticles can alter the makeup and activity of microbial communities. There might be a domino effect from this disturbance on ecosystem functions including the cycling of nutrients. Technogenic activities contribute to the accumulation of nanoparticles in soil. These activities encompass intentional releases through soil and water remediation technologies, agricultural application (in the form of pesticides and fertilizers), and incidental releases through air, water, sewage, and sediment (Rajput et al. 2018).

17.3.2.7 Groundwater Contamination

- **Leaching:** Contaminated soils and waste disposal sites may release nanoparticles into groundwater, endangering aquatic ecosystems and human populations who depend on groundwater for drinking water (Kumar et al. 2023). Contaminated groundwater may seep into the soil due to groundwater-surface water interaction which could impact soil health and plant life.

17.3.2.8 Effects on Plants

While using nano pesticides and fertilizers offers many advantages for agriculture, there are drawbacks as well. Long-term usage of these substances can disrupt physiological processes and create deformities in plants, which lowers yield (Aqeel et al. 2022). Furthermore, through food chains and food webs, these nanomaterials can infiltrate plants and then make their way to people and other animals, where they may eventually cause dangerous illnesses. Plants may take up NMs from the soil, water, and environment. Figure 17.4 shows overview of nanoparticle uptake into plant parts through root causing some serious effects to plants such as damage to DNA (gene alteration), chloroplast disorganization, morphological barriers and effects on plant secondary metabolites.

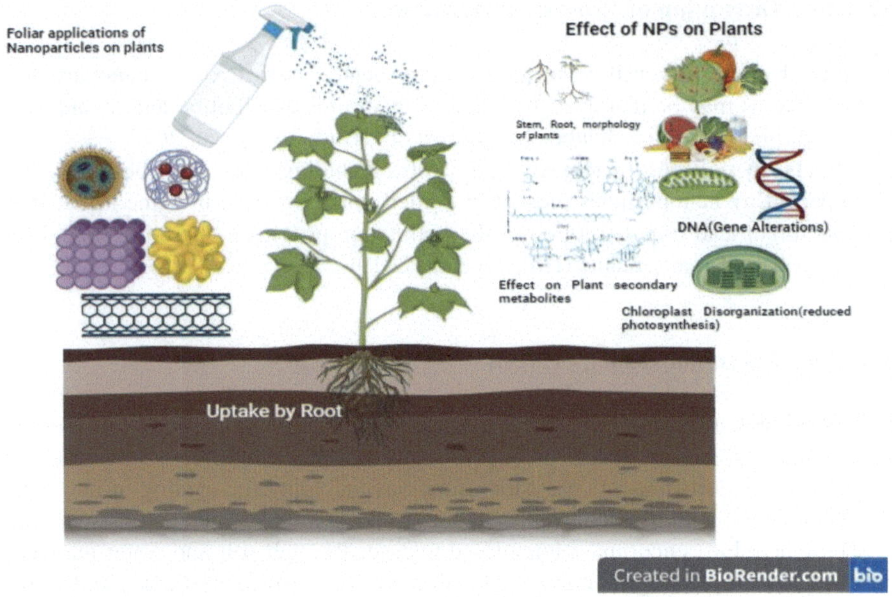

Fig. 17.4 Effects of Nanoparticles on plants. (Constructed by Asha Kumari Kumawat)

The environmental destiny and transport of nanoparticles, their interactions with various ecosystems, and the creation of sustainable and responsible nanotechnology practices is important to mitigate these challenges. The possible environmental dangers connected to the usage of nanoparticles are also being managed through the consideration of regulatory measures. Table 17.1 provides some direct effects of different NPs on humans and environment.

17.4 Emerging Sustainable Approaches in Nanotechnology

Nanotoxicology is a discipline that studies the possible negative effects of nanoparticles on biological systems. Nanotechnology is a fast-paced field experiencing constant progress and significant advancements. Recent trends in nanotoxicology involve integrating high-throughput screening methods to establish structure-activity relationships using various types of nanoparticles and *in vitro* screening tests along with computational analysis and omics-based approaches (Hristozov et al. 2024). There is also a focus on developing sensitive toxicity tests for early detection of harmful effects.

Recent developments in nanotoxicology include omics approaches, quantum nanotechnology, nanomedicine, nano-sensors, nanoelectronics, *in-vitro* models, improved characterization techniques, development of standardized testing protocols.

Table 17.1 Effects of some nanoparticles on human body and environment

Nanomaterial	Effects: On human health	Effects: On environment	References
Nickel nanoparticles (Ni-NPs)	Reproductive toxicity Exhibit carcinogenicity Skin allergies Lung fibrosis Cardiac toxicity	Toxic effects on aquatic organisms Groundwater-surface water contamination	Iftikhar et al. (2023) and Rizwan et al. (2024)
Titanium Oxide Nanoparticles (TiO_2 NPs)	DNA damage Embryonic toxicity in pregnant female Pulmonary inflammation	Embryonic toxicity in rats Immunotoxic effects on aquatic animals	Kuang et al. (2023), Kumah et al. (2023) and Lyashenko et al. (2023)
Silica nanoparticles (SiO_2-NPs)	Producing cardiotoxic effects Air–blood barrier in lungs	Induces oxidative stress and impairs reproductive performance in aquatic fauna Disturbance of plant metabolic activities	Al-Tabbal et al. (2024) and Rashidian et al. (2023)
ZrO_2 nanoparticles (NPs)	Embryonic toxicity in pregnant female	Embryonic toxicity in rats Plant DNA alteration	Barbasz and Dyba (2024), Lyashenko et al. (2023) and Mourya et al. (2024)
ZnO NPs	Affect respiratory track Respiratory immune toxicity	Effects on aquatic ecosystem Affect root growth in crops	Bordin et al. (2024), Mandal et al. (2024) and Sun et al. (2023)
Copper nanoparticles (cu-NPs)	Kidney, spleen, and liver damage	Induce oxidative stress, inflammation, and tumor lesions in rats. Plant chlorophyll damage	Naz et al. (2024) and Pesce et al. (2024)
Silver nanoparticles (AgNPs)	Liver and lungs dysfunction	Toxic effects on aquatic organisms Affect plant root growth	Jaswal and Gupta (2023) and Li and Cummins (2024)

17.4.1 Omics Approaches

Omics techniques entail the methodical examination of biological molecules including DNA, RNA, proteins, and metabolites to get a comprehensive comprehension of the structure and function of biological systems (Majumdar and Keller 2021). Various omics fields are dedicated to studying certain types of molecules. Recent developments in nanotoxicology involve utilizing omics approaches, such as metabolomics, microbiomics epigenomics proteomics, genomics and transcriptomics are used to identify molecular-level impacts of nanoparticles at low, non-lethal concentrations and the connections between nanoparticles and biological systems (Kahru and Mortimer 2021).

17.4.2 Quantum Nanotechnology

Quantum nanotechnology merges concepts from quantum mechanics and nanotechnology in an interdisciplinary approach. It entails manipulating and controlling quantum systems on a nanoscale to create innovative technologies and applications. Advancements in quantum computing and quantum communication with nanoscale quantum bits (qubits) helps in developing quantum sensors and detectors for use in areas like metrology and cryptography (Laucht et al. 2021). Some potential applications of quantum nanotechnology includes Quantum Dots, Quantum Computing, Quantum Sensing and Imaging, Quantum Communication, Quantum Materials, Quantum Sensors for Medicine, Quantum Dot Cellular Automata (QCA), and Quantum-enhanced Materials Design.

17.4.3 Nanomedicine

Nanomedicine utilizes nanotechnology in the medical field to diagnose, cure, and prevent illnesses. The focus is on designing and using nanoscale materials and technologies for medicinal applications. Recent developments are utilizing nanoparticles Therapeutics and Treatment Modalities, Diagnostic Tools, Biosensors, Regenerative Medicine and for targeted drug delivery to enhance the accuracy and efficiency of cancer therapies (Tiwari et al. 2023). Creating nanoscale imaging agents to improve disease diagnosis by detecting conditions earlier and with greater precision. Quantum dots and iron oxide nanoparticles can enhance fluorescence imaging and Magnetic Resonance Imaging (MRI) by improving contrast and sensitivity in diagnostic procedures. Nanomaterials are widely used in cancer because of their small size and high specificity. Nano-carriers including dendrimers, micelles, liposomes, and polymeric nanoparticles are employed to enhance the targeted delivery of anti-cancer medications to the tumor site. It is crucial to use quantum dots for real-time labelling and monitoring of cancer cells to assess the necessary dosage of therapy for treatment (Tiwari et al. 2023).

17.4.4 Nano-sensors

Nano-sensors are devices that use nanotechnology to determine and measure physical, chemical, or biological characteristics with exceptional sensitivity and precision. Nanoscale sensors are created using nanomaterials and concepts of nanotechnology to improve their performance. Nano-sensors have applications in many domains such as medical, environmental monitoring, food safety, and security. Nano-sensors have the potential to be utilized for early disease detection and continuous patient health tracking (Sahu et al. 2023). The Table 17.2 provides explanations of key nano-sensor devices and their applications.

17.4.5 Nanoelectronics

Conventional electronic parts like transistors and integrated circuits are reduced in size to the nanoscale by using materials and architectures with dimensions in the range of nanometers. Nanoelectronics is propelled by the necessity for electronic devices that are smaller, quicker, and more energy-efficient (Deleonibus 2021). Continuously striving to expand the boundaries of Moore's Law through the creation of nanoscale transistors and memory devices for example, exploring new materials such as 2D materials (e.g., graphene) developing transistors, photodetectors, and other electronic devices, Spintronic for non-volatile memory, magnetic sensors, and quantum computing, Nano-transistors for constructing nanoscale integrated circuits (ICs) used in computers, smartphones, and other electronic devices, Nanowires for nanoscale transistors, sensors, and interconnects, and Nano-circuits for development of advanced electronic devices, including memory devices and processors.

17.4.6 In Vitro Models

In vitro models are essential for understanding nanomaterial-biological system interactions. Scientists are increasingly utilizing advanced in vitro models to replicate the interactions between nanomaterials and various cell types (Ahmad et al. 2022). *In Vitro* Models include Cell Culture Models, Blood-Brain Barrier Model, Organ-on-a-Chip Models, Co-culture Model etc. These models strive to offer a more precise depiction of the potential toxicological impacts of nanomaterials.

17.4.7 Improved Characterization Techniques

Progress in imaging and analytical methods including Transmission Electron Microscopy (TEM), Scanning Electron Microscopy (SEM), X-ray Diffraction (XRD) and sophisticated spectroscopy have improved researchers' ability to characterize and comprehend the interactions between nanomaterials and biological organisms. The creation of state-of-the-art imaging technologies, including cryo-electron microscopy, which enable the visualization of structures and processes at the nanoscale (Li et al. 2020).

Table 17.2 Different nano-sensors and their applications

Type of non-sensor	Device type	Application	References
Optical Nano-sensors	Plasmonic Nano-sensors	Label-free detection of biomolecules, environmental monitoring, and medical diagnostics.	Matmin et al. (2024)
	Quantum dot Nano-sensors:	Used in biological imaging, drug delivery, and as fluorescent labels for biomolecule detection.	
Electrochemical Nano-sensors	Carbon nanotube-based sensors	Applied in glucose sensing for diabetes management, detection of specific biomolecules, and environmental monitoring.	Hassan (2022)
	Nano-electrode arrays	Used for high-sensitivity detection of neurotransmitters, DNA, and other biological molecules.	
Biological and chemical Nano sensors	Enzyme-based Nano sensors	Used for medical diagnostics, environmental monitoring, and detecting specific chemical species. Magnetic Nano-sensors	Kalyani and Kumawat (2020)
	DNA Nano sensors	Applied in genetic diagnostics, detection of specific DNA sequences, and studying DNA-protein interactions	
Thermal Nano sensors	Thermal resonance Nano sensors	Applied in ultrasensitive detection of small mass changes, suitable for detecting biomolecules	Behera et al. (2021)
	Thermoelectric Nano sensors	Used for temperature sensing in microelectronics, medical devices, and industrial applications	
Surface Acoustic Wave (SAW) Nano sensors	Nanomechanical SAW sensors	Applications include chemical and biological sensing, gas detection, and monitoring environmental pollutants	Peng et al. (2021)
Magnetic Nano-sensors	Magnetic nanoparticles	Applied in medical diagnostics for magnetic resonance imaging (MRI), drug delivery, and biomolecule detection	Slimani and Hannachi (2020)
	Magnetic Nano-sensors for biosensing	Used for detecting specific biological markers in medical diagnostics.	
Piezoelectric Nano-sensors	Nano-mechanical resonators	Used in medical diagnostics, environmental monitoring, and detecting specific chemical or biological substances	Salim et al. (2018)
Gas Nano sensors	Metal oxide Nano sensors	Applications in environmental monitoring for detecting gases such as CO2 and VOCs, and in industrial safety for detecting toxic gases	Zeng et al. (2021)
	Nanowire-based gas sensors	Used for detecting gases like methane and hydrogen for environmental and industrial safety.	

(continued)

Table 17.2 (continued)

Type of non-sensor	Device type	Application	References
Strain Nano sensors	Nanowire-based strain sensors	Applied in structural health monitoring, detecting tiny deformations in materials, and wearable technology for motion sensing	Raman and Sindhuja (2023)
Photonic Nano sensors	Photonic crystal Nano sensors	Used for label-free detection of biological molecules, environmental monitoring, and medical diagnostics	Altug et al. (2022)
Graphene-based Nano sensors	Graphene field-effect transistors (GFETs)	Applied in biosensing, environmental monitoring, and gas detection due to graphene's high surface area and conductivity	Huang et al. (2020)

17.4.8 Development of Standardized Testing Protocols

Standardization enables academics, regulators, and industry stakeholders to make well-informed choices on the safety of nanomaterials. There have been attempts to develop standardized testing protocols (Powers et al. 2007). This aids in comparing findings across various studies and guarantees a more uniform comprehension of the possible hazards linked to different nanomaterials. Organization for Economic Co-Operation and Development (OECD) has played a significant role in creating standards for chemical testing, which include cover nanomaterials. The OECD Test Guidelines (TGs) offer standardized techniques for different toxicity endpoints. OECD Test Guidelines 413, 432, and 487 have been modified to account for the unique characteristics of nanomaterials in assessing sub-chronic inhalation toxicity, in vitro skin corrosion, and in vitro micronucleus test, respectively. Similarly, United States National Institute for Occupational Safety and Health (NIOSH) is also engaged in researching and developing recommendations for evaluating the occupational safety and health impacts of nanomaterials. NIOSH has created procedures for evaluating the harmful effects of manmade nanomaterials, specifically concentrating on exposure by inhalation and testing in a controlled environment.

17.4.9 Clean and Green Energy Applications

Progress in nanomaterials for enhanced solar cells, such as quantum dots employed in solar cells to improve light absorption and energy conversion efficiency and perovskite solar cells including perovskite nanocrystals, are being researched to enhance the efficiency and stability of perovskite solar cells (Chen and Zhao 2020). Nanomaterials in supercapacitors like graphene and carbon nanotubes are being studied to improve battery performance by enhancing conductivity and charge storage capacity (Zhang et al. 2014). A new concept that is Nano-generators were introduced to harvest energy from mechanical vibrations, environmental movements, or other forms of ambient energy. Nanomaterial coatings like electrochromic nanoparticles are utilized in smart windows to control light and heat transfer, aiding in the development of energy-efficient building designs called smart windows and energy-efficient building materials (Nunes et al. 2019). Nanomaterials are also being used in power grids to enhance the efficiency and dependability of power networks, advancements in conductors and materials for energy storage.

17.4.10 Environmental Remediation

Utilizing nanomaterials like nano-catalysts for effective pollution control and wastewater treatment in environmental remediation. For example developing Metal-Organic Frameworks (MOFs), carbon nanotubes and graphene-based nanomaterial filtration systems for water purification because of their large surface area and exceptional ability to adsorb contaminants such as heavy metals and chemical substances (Hsu et al. 2023). Nanomaterial-coated filters, including Titanium Dioxide (TiO_2) and silver nanoparticles, are used in air purifiers to capture and deactivate airborne pollutants, such as particulate matter and Volatile Organic Compounds (VOCs). Combining semiconducting materials like zinc oxide or titanium dioxide with other nanoparticles, are designed for photocatalytic degradation of pollutants in the air (Roy et al. 2021). Nanoparticles such as clay-based nanomaterials and nanoscale zero-valent iron (nZVI) are employed in situ for soil remediation to capture and render contaminants including heavy metals and organic pollutants inactive. Apart from existing nanomaterials there are some emerging types of nanomaterials like MXene nano-sheets, a 2D nanomaterial, exhibits potential for water pollutant removal because of its large surface area and ion-exchange properties.

17.4.11 Food and Agriculture

The implementation of nanotechnology is to reduce food waste and extend expiration life in food packaging. Nano-pesticides are used to encapsulate and distribute insecticides more efficiently, decreasing environmental harm and enhancing precise pest management (Kaphle et al. 2018). Smart delivery systems are being used by farmers to accurately feed nutrients or agrichemicals to plants in order to optimize resource utilization and reduce environmental impact. Clay nanoparticles and silver nanoparticles are added to packaging materials to improve barrier qualities, prolong shelf life, and prevent microbiological development.

17.4.12 Nanorobotics

Advancements in the construction of nanoscale machines called DNA Origami Nano-robots that are capable of carrying out molecular-level duties; possible medical and industrial applications. Drug delivery nano-robots are also used as drug carriers for targeted delivery within the body (Sharma 2012). Rotaxane devices, which involve a ring threaded onto an axle, are being studied as synthetic nano-machines capable of controlled movement and mechanical operations on a very small scale.

17.4.13 Nano-ethics and Safety

As nanotechnology develops, it brings a number of moral, social, and security concerns that need to be carefully considered. Nano-ethics and safety regulates increasing focus on safety protocols, ethical considerations, and responsible research methods in the field of nanotechnology (Van de Voorde and Jeswani 2021). The nanomaterial is utilized in window coverings, cosmetics, textiles, meals, colors, medications, medical equipment, and more. Carbon, titanium, zinc, silica, and gold nanoparticles are the most frequent nanomaterials in consumer items (Vance et al. 2015). Nanoparticles are formed by three natural, human, and artificial ways, and environmental nanoparticles are more harmful than typical particles. Due to changes in nanoparticle features like regional surface influences solubility and chemical toxicity, manufactured nanoparticles may be beyond human and animal defenses. Lack of public knowledge of environmental toxicity and exposure, nanomaterial entry and transfer to humans, nanomaterial risks, and prevention can harm the environment and human societies. Nanoparticles swiftly penetrate into the skin, mucus, and respiratory tract, releasing undiscovered poisons that can harm humans. In addition to their benefits, nanoparticles may pose risks.

17.4.14 Nanotechnology in Space Exploration

An investigation into the potential applications of nanotechnology in sensors, propulsion systems, and spacecraft materials for forthcoming space missions. Nanosatellites and CubeSat have become increasingly popular in space exploration are used for various missions, including Earth observation, scientific research, and technology demonstration (Selva and Krejci 2012). Nanomaterials like carbon nanotubes and nanocomposites have made spaceship components lighter and stronger. These include heat protection, radiation shielding, and structural features. Nanomaterials improved spaceship propulsion, efficiency and energy density, advancing propulsion technology. Spacecraft surfaces are coated with nanotechnology for self-healing, temperature management, and environmental protection (Zhu et al. 2019).

17.5 Challenges

Nanotechnology presents several benefits, although it also poses drawbacks and obstacles in the advancement of nanotechnology and techniques using nanomaterials. Detecting and characterizing nanomaterials provide significant hurdles because of their extremely tiny size (Nizam et al. 2021). Distinguishing between the manmade nanoparticles and closely related natural materials proved challenging. Another barrier was the absence of accurate procedures that are both sensitive and selective for measuring size, number, and surface area exactly (Wang and Tao 2014). The interactions between engineered nanomaterials (ENMs) and complex environmental systems is difficult because of the wide variety of ecosystems, species, and environmental variables involved and predicting the fate and transport of nanomaterials in the environment is challenging. Many studies are being conducted to comprehend the health impacts of nanoparticles whether they are breathed, swallowed, or absorbed via the skin. Certain nanoparticles have been associated with harmful health impacts, prompting concerns over workplace safety and public exposure. Certain nanomaterials may display unexpected toxicity, with their dynamic properties being influenced by factors like as size, shape, and surface chemistry. This diversity complicates the full assessment of their safety. Furthermore, developing and implementing nanotechnology can be costly, potentially restricting its availability to specific industries and areas. It is a challenge to ensure fair access to nanotechnological advancements and its ethics.

17.6 Conclusion and Prospects

Eco-nanotoxicology shows the complex relationship between nanotechnology and environmental sustainability. Emerging concepts demonstrate the need for a sustainable nanotechnology paradigm change. Transforming nanotechnology into environmentally friendly technology through the adoption of bottom-up preparation methods from green chemistry helps mitigate related risks in industrial applications and aligns with the attributes of ideal technology. Also concerns about the negative impacts of nanomaterials on human health and the environment can be addressed by opting for eco-friendly synthesis techniques that operate at ambient temperature. The ecological effects of nanomaterials can be better understood using cutting-edge analytical and omics methods. Complex environmental interactions and the lack of standardized testing techniques necessitate concerted investigations. Ecologists, toxicologists, and material scientists must work together to solve these problems and comprehend consequences of nanoparticles on environment. Nanomaterials like nano-sensors, nano-lubricants, nano-metals, nano-coatings, nano-catalysts, nano-packaging, nanocomposites, metal-organic frameworks (MOFs), carbonaceous materials, nano-zeolites/silica, etc., can serve as valuable tools for ensuring environmental safety. Sustainable nanotechnological applications need green synthesis, biodegradable nanomaterials, and ethical concerns. The scientific community is at the forefront of eco-nanotoxicology, helping to responsibly integrate nanotechnology into varied applications.

Advancements in scientific methods and interdisciplinary collaboration are enhancing our understanding of the environmental impacts of nanomaterials in eco-nanotoxicology. Utilizing advanced analytical techniques such as high-resolution microscopy and omics can uncover eco-toxicological processes at the molecular level. Using standardized testing methodologies and eco-nanotoxicology prediction models will enhance risk assessments and streamline regulatory decision-making. Green nanotechnology is set to revolutionize the design and synthesis of nanomaterials, emphasizing sustainability and environmental considerations. Environmentally conscious nanomaterials and their uses will adhere to eco-friendly nanotechnology principles. Nanotechnology promises revolutionary advances in technology, medicine, energy, and materials research. Collaboration among researchers, policymakers, and the public, along with enhanced education, will encourage responsible nanotechnology and environmental sustainability practices and well-informed decision-making.

References

Abbas Q, Liu G, Ali MU et al (2024) Nanomaterials in environment: release, compartmental transformation, and nanotoxicity. In: Handbook of nanomaterials, vol 2. Elsevier, pp 745–789

Ahmad A, Imran M, Sharma N (2022) Precision nanotoxicology in drug development: current trends and challenges in safety and toxicity implications of customized multifunctional nanocarriers for drug-delivery applications. Pharmaceutics 14(11):2463

Ali S, Zahid A, Ahmed MU et al (2024) Properties of aerosol particles in the air and their effects on greenhouse gases dispersion. In: Advances and technology development in greenhouse gases: emission, capture and conversion. Elsevier, pp 71–102

Al-Tabbal J, Al-Harahsheh M, Al-Zou'by JY (2024) Silica nanoparticles as a waste product to alleviate the harmful effects of water stress in wheat. Int J Phytoremediation:1–17

Altug H, Oh SH, Maier SA et al (2022) Advances and applications of nanophotonic biosensors. Nat Nanotechnol 17(1):5–16

Aqeel U, Aftab T, Khan MMA et al (2022) A comprehensive review of impacts of diverse nanoparticles on growth, development and physiological adjustments in plants under changing environment. Chemosphere 291:132672

Asha AB, Narain R (2020) Nanomaterials properties. In: Polymer science and nanotechnology. Elsevier, pp 343–359

Barbasz AM, Dyba B (2024) Direct interaction of zirconia nanoparticles with human immune cells. Biophysica 4(1):83–91

Behera A, Pan J, Behera A (2021) Temperature nanosensors for smart manufacturing. In: Nanosensors for smart manufacturing. Elsevier, pp 249–272

Bhardwaj LK, Rath P, Choudhury M (2023) A comprehensive review on the classification, uses, sources of nanoparticles (NPs) and their toxicity on health. Aerosol Sci Eng 7(1):69–86

Bhattacharjee S (2023) Nanomaterials in imaging. In: Advances in medical imaging, detection, and diagnosis. Jenny Stanford Publishing, pp 683–723

Bi J, Mo C, Li S et al (2023) Immunotoxicity of metal and metal oxide nanoparticles: from toxic mechanisms to metabolism and outcomes. Biomater Sci 11(12):4151–4183

Bordin ER, Yamamoto FY, Mattoso Filho NP et al (2024) Ecotoxicity of doped zinc oxide nanoparticles: perspectives on environmental safety. Chemosphere 142185

Bundschuh M, Filser J, Lüderwald S et al (2018) Nanoparticles in the environment: where do we come from, where do we go to? Environ Sci Eur 30(1):1–17

Burketová L, Martinec J, Siegel J et al (2022) Noble metal nanoparticles in agriculture: impacts on plants, associated microorganisms, and biotechnological practices. Biotechnol Adv 58:107929

Chauhan S, Chakraborty P, Maharana C et al (2023) Impact of nanoparticles on human health and environment. In: Advances in nanotechnology for smart agriculture. CRC Press, pp 341–373

Chen Y, Zhao Y (2020) Incorporating quantum dots for high efficiency and stable perovskite photovoltaics. J Mater Chem A 8(47):25017–25027

Dasari Shareena TP, McShan D, Dasmahapatra AK et al (2018) A review on graphene-based nanomaterials in biomedical applications and risks in environment and health. Nano-Micro Lett 10:1–34

Deleonibus S (2021) The era of sustainable and energy efficient nanoelectronics and nanosystems. In: Convergence of more Moore, more than Moore, and beyond Moore: materials, devices, and nanosystems. Jenny Stanford Publishing, pp 3–76

Dong J, Ma Q (2019) Integration of inflammation, fibrosis, and cancer induced by carbon nanotubes. Nanotoxicology 13(9):1244–1274

Dube E, Okuthe GE (2023) Engineered nanoparticles in aquatic systems: toxicity and mechanism of toxicity in fish. Emerg Contam 9(2):100212

Egbuna C, Parmar VK, Jeevanandam J et al (2021) Toxicity of nanoparticles in biomedical application: nanotoxicology. J Toxicol 2021(1):9954443

El-Kady MM, Ansari I, Arora C et al (2023) Nanomaterials: a comprehensive review of applications, toxicity, impact, and fate to environment. J Mol Liq 370:121046

El-Kalliny AS, Abdel-Wahed MS, El-Zahhar AA et al (2023) Nanomaterials: a review of emerging contaminants with potential health or environmental impact. Discov Nano 18(1):68

Ghadam P, Saboora A (2024) Negative effects of nanonutrients on plants: nanotoxicity and genotoxicity issue. In: Nanofertilizer delivery, effects and application methods. Elsevier, pp 351–369

Gupta D, Boora A, Thakur A et al (2023) Green and sustainable synthesis of nanomaterials: recent advancements and limitations. Environ Res 231:116316

Hartmann NIB, Skjolding LM, Hansen SF et al (2014) Environmental fate and behaviour of nanomaterials: new knowledge on important transfomation processes. Danish Environmental Protection Agency

Hasannejad-Asl B, Pooresmaeil F, Choupani E et al (2023) Nanoparticles as powerful tools for crossing the blood-brain barrier. CNS Neurol Disord Drug Targets 22(1):18–26

Hassan RY (2022) Advances in electrochemical nano-biosensors for biomedical and environmental applications: from current work to future perspectives. Sensors 22(19):7539

Hooven LA, Chakrabarti P, Harper BJ et al (2019) Potential risk to pollinators from nanotechnology-based pesticides. Molecules 24(24):4458

Hristozov D, Badetti E, Bigini P et al (2024) Next generation risk assessment approaches for advanced nanomaterials: current status and future perspectives. NanoImpact 100523

Hsu CY, Rheima AM, Mohammed MS et al (2023) Application of carbon nanotubes and graphene-based nanoadsorbents in water treatment. BioNanoScience 13(4):1418–1436

Huang C, Hao Z, Qi T et al (2020) An integrated flexible and reusable graphene field effect transistor nanosensor for monitoring glucose. J Mater 6(2):308–314

Iftikhar M, Noureen A, Jabeen F et al (2023) Bioinspired engineered nickel nanoparticles with multifunctional attributes for reproductive toxicity. Chemosphere 311:136927

Ilyasu US, Aworunse OS, Olusanya CS et al (2024) Nanoparticles in air and their impact on air quality. In: Environmental nanotoxicology: combatting the minute contaminants. Springer, Cham, pp 183–202

Jaswal T, Gupta J (2023) A review on the toxicity of silver nanoparticles on human health. Mater Today Proc 81:859–863

Joachim C (2005) To be nano or not to be nano? Nat Mater 4(2):107–109

Kahru A, Mortimer M (2021) Advances in nanotoxicology: towards enhanced environmental and physiological relevance and molecular mechanisms. Nano 11(4):919

Kalyani VL, Kumawat P (2020) Recent advancement in nanosensors with special reference to biomedical applications. J Manage Eng Inform Technol 7(3):24–30

Kang J, Zhou N, Zhang YW et al (2023) Synthesis, multi-site transformation fate and biological toxicity of silver nanoparticles in aquatic environment: a review. Environ Technol Innov 32: 103295

Kaphle A, Navya PN, Umapathi A et al (2018) Nanomaterials for agriculture, food and environment: applications, toxicity and regulation. Environ Chem Lett 16:43–58

Khan SH, Fulekar MH, Pathak B (2015) Nanotoxicology-health and environmental impacts: a review. J Environ Nanotechnol 4(3):55–72

Korani M, Ghazizadeh E, Korani S et al (2015) Effects of silver nanoparticles on human health. Eur J Nanomed 7(1):51–62

Kreuter M, Mueller-Ladner U, Costabel U et al (2021) The diagnosis and treatment of pulmonary fibrosis. Dtsch Arztebl Int 118(9):152

Kuang Y, Guo H, Ouyang K et al (2023) Nano-TiO_2 aggravates immunotoxic effects of chronic ammonia stress in zebrafish (*Danio rerio*) intestine. Comp Biochem Physiol C Toxicol Pharmacol 266:109548

Kumah EA, Fopa RD, Harati S et al (2023) Human and environmental impacts of nanoparticles: a scoping review of the current literature. BMC Public Health 23(1):1059

Kumar V, Singh E, Singh S et al (2023) Micro-and nano-plastics (MNPs) as emerging pollutant in ground water: environmental impact, potential risks, limitations and way forward towards sustainable management. Chem Eng J 459:141568

Kundu D, Khan MF, Gogoi M et al (2021) Environmental impact and econanotoxicity of engineered nanomaterials. Nanotoxicol Nanoecotoxicol 1:287–312

Laucht A, Hohls F, Ubbelohde N et al (2021) Roadmap on quantum nanotechnologies. Nanotechnology 32(16):162003

Lead JR, Batley GE, Alvarez PJ et al (2018) Nanomaterials in the environment: behavior, fate, bioavailability, and effects—an updated review. Environ Toxicol Chem 37(8):2029–2063

Lee CC, Lin YH, Hou WC et al (2020) Exposure to ZnO/TiO_2 nanoparticles affects health outcomes in cosmetics salesclerks. Int J Environ Res Public Health 17(17):6088

Li Y, Cummins E (2024) Biotoxicity and exposure pathways of engineered silver nanoparticles released in the environment. In: Nanotoxicology for agricultural and environmental applications. Academic, pp 179–194

Li Y, Huang W, Li Y et al (2020) Opportunities for cryogenic electron microscopy in materials science and nanoscience. ACS Nano 14(8):9263–9276

Li J, Carlson BE, Yung YL et al (2022a) Scattering and absorbing aerosols in the climate system. Nat Rev Earth Environ 3(6):363–379

Li M, Zhang Y, Feng S et al (2022b) Bioaccumulation and biomagnification effects of nano-TiO_2 in the aquatic food chain. Ecotoxicology 31(6):1023–1034

Lin J, Li D, Li C, Zhuang Z, Chu C, Ostrikov KK & Wang P (2023) A review on reactive oxygen species (ROS)-inducing nanoparticles activated by uni-or multi-modal dynamic treatment for oncotherapy. Nanoscale, 15(28), 11813–11833

Lyashenko EN, Uzbekova LD, Polovinkina VV et al (2023) Study of the embryonic toxicity of TiO_2 and ZrO_2 nanoparticles. Micromachines 14(2):363

Majumdar S, Keller AA (2021) Omics to address the opportunities and challenges of nanotechnology in agriculture. Crit Rev Environ Sci Technol 51(22):2595–2636

Mandal AH, Ghosh S, Adhurjya D et al (2024) Exploring the impact of zinc oxide nanoparticles on fish and fish-food organisms: a review. Aquacult Rep 36:102038

Matmin J, Mohamad F, Wahab RA et al (2024) Advanced optical nanosensors. In: Handbook of nanosensors: materials and technological applications. Springer, Cham, pp 747–782

Moghimi SM, Haroon HB, Yaghmur A et al (2023) Perspectives on complement and phagocytic cell responses to nanoparticles: from fundamentals to adverse reactions. J Control Release 356: 115–129

Mourya D, Dubey K, Jha S et al (2024) In vitro effects of zirconia nanoparticles: uptake, genotoxicity, and mutagenicity in V-79 cells. Biol Trace Elem Res 202(3):927–940

Naz S, Raza N, Alhidary I et al (2024) Evaluation of copper nanoparticles on growth, organs histology and DNA damage in Japanese quails (Coturnix coturnix japonica). Toxin Rev 43(1): 127–136

Nizam NUM, Hanafiah MM, Woon KS (2021) A content review of life cycle assessment of nanomaterials: current practices, challenges, and future prospects. Nano 11(12):3324

Noga M, Milan J, Frydrych A et al (2023) Toxicological aspects, safety assessment, and green toxicology of silver nanoparticles (AgNPs)—critical review: state of the art. Int J Mol Sci 24(6): 5133

Nunes SC, Saraiva SM, Pereira RF et al (2019) Sustainable dual-mode smart windows for energy-efficient buildings. ACS Appl Energy Mater 2(3):1951–1960

Peng C, Chen M, Spicer JB et al (2021) Acoustics at the nanoscale (nanoacoustics): a comprehensive literature review. Part I: materials, devices and selected applications. Sensors Actuators A Phys 332:112719

Pesce S, Mamy L, Sanchez W et al (2024) The use of copper as plant protection product contributes to environmental contamination and resulting impacts on terrestrial and aquatic biodiversity and ecosystem functions. Environ Sci Pollut Res:1–17

Powers KW, Palazuelos M, Moudgil BM et al (2007) Characterization of the size, shape, and state of dispersion of nanoparticles for toxicological studies. Nanotoxicology 1(1):42–51

Rajput VD, Minkina T, Sushkova S et al (2018) Effect of nanoparticles on crops and soil microbial communities. J Soils Sediments 18:2179–2187

Raman S, Sindhuja M (2023) Advances in silicon nanowire applications in energy generation, storage, sensing, and electronics: a review. Nanotechnology 34(18):182001

Rashidian G, Mohammadi-Aloucheh R, Hosseinzadeh-Otaghvari F et al (2023) Long-term exposure to small-sized silica nanoparticles (SiO_2-NPs) induces oxidative stress and impairs reproductive performance in adult zebrafish (Danio rerio). Comp Biochem Physiol C Toxicol Pharmacol 273:109715

Rauscher H, Roebben G, Mech A et al (2019) An overview of concepts and terms used in the European Commission's definition of nanomaterial. Publications Office of the European Union

Raut SS, Singh R, Lekhak UM (2024) Naturally occurring nanoparticles (NONPs): a review. Next Sustain 3:100037

Razi M, Contreras-Mateus M, Hashlamoun KW et al (2022) Nanoparticles: preparation, stabilization, and control over particle size. In: Nanoparticles: an emerging technology for oil production and processing applications. Springer, Cham, pp 1–40

Rizwan M, Usman K, Alsafran M (2024) Ecological impacts and potential hazards of nickel on soil microbes, plants, and human health. Chemosphere 142028

Roy A, Sharma A, Yadav S et al (2021, 2021) Nanomaterials for remediation of environmental pollutants. Bioinorg Chem Appl

Sahu MK, Yadav R, Tiwari SP (2023) Recent advances in nanotechnology. Int J Nanomater Nanotechnol Nanomed 9(1):015–023

Salim M, Salim D, Chandran D et al (2018) Review of nano piezoelectric devices in biomedicine applications. J Intell Mater Syst Struct 29(10):2105–2121

Santonastaso M, Mottola F, Colacurci N et al (2019) In vitro genotoxic effects of titanium dioxide nanoparticles (n-TiO_2) in human sperm cells. Mol Reprod Dev 86:1369–1377. https://doi.org/10.1002/mrd.23134
Selva D, Krejci D (2012) A survey and assessment of the capabilities of Cubesats for Earth observation. Acta Astronaut 74:50–68
Sharma KR (2012) Nano-structuing of nano-robots for use in nano-medicine. Int J Eng Technol 2(2):116–134
Slimani Y, Hannachi E (2020) Magnetic nanosensors and their potential applications. In: Nanosensors for smart cities. Elsevier, pp 143–155
Sonwani S, Madaan S, Arora J et al (2021) Inhalation exposure to atmospheric nanoparticles and its associated impacts on human health: a review. Front Sustain Cities 3:690444
Sun Y, Qian Y, Geng S et al (2023) Joint effects of microplastics and ZnO nanoparticles on the life history parameters of rotifers and the ability of rotifers to eliminate harmful phaeocystis. Chemosphere 310:136939
Taghavizadeh Yazdi ME, Qayoomian M, Beigoli S et al (2023) Recent advances in nanoparticle applications in respiratory disorders: a review. Front Pharmacol 14:1059343
Tiwari H, Rai N, Singh S et al (2023) Recent advances in nanomaterials-based targeted drug delivery for preclinical cancer diagnosis and therapeutics. Bioengineering 10(7):760
Tourinho PS, Van Gestel CAM, Lofts S et al (2012) Metal-based nanoparticles in soil: fate, behavior, and effects on soil invertebrates. Environ Toxicol Chem 31(8):1679–1692
Tran HD, Moonshi SS, Xu ZP et al (2022) Influence of nanoparticles on the haemostatic balance: between thrombosis and haemorrhage. Biomater Sci 10(1):10–50
Uddin MN, Desai F, Asmatulu E (2020) Engineered nanomaterials in the environment: bioaccumulation, biomagnification and biotransformation. Environ Chem Lett 18(4): 1073–1083
Vaish S, Pathak B (2023) Mangrove synthesized bio-nanomaterial and its applications: a review. Environmental Nanotechnology, Monitoring & Management, p 100866
Van de Voorde M, Jeswani G (eds) (2021) Ethics in nanotechnology: emerging technologies aspects. Walter de Gruyter GmbH & Co KG
Vance ME, Kuiken T, Vejerano EP et al (2015) Nanotechnology in the real world: redeveloping the nanomaterial consumer products inventory. Beilstein J Nanotechnol 6(1):1769–1780
Vouitsis I, Portugal J, Kontses A et al (2023) Transport-related airborne nanoparticles: sources, different aerosol modes, and their toxicity. Atmos Environ 301:119698
Wang W, Tao N (2014) Detection, counting, and imaging of single nanoparticles. Anal Chem 86(1):2–14
Xuan L, Ju Z, Skonieczna M et al (2023) Nanoparticles-induced potential toxicity on human health: applications, toxicity mechanisms, and evaluation models. Med Comm 4(4):e327
Yao Y, Zhang T, Tang M (2023) The DNA damage potential of quantum dots: toxicity, mechanism and challenge. Environ Pollut 317:120676
Yu X, Hong F, Zhang YQ (2016) Bio-effect of nanoparticles in the cardiovascular system. J Biomed Mater Res Part A 104:2881–2897. https://doi.org/10.1002/jbm.a.35804
Zeng H, Zhang G, Nagashima K et al (2021) Metal–oxide nanowire molecular sensors and their promises. Chemosensors 9(2):41
Zhang L, Pu J, Wang L et al (2014) Frictional dependence of graphene and carbon nanotube in diamond-like carbon/ionic liquids hybrid films in vacuum. Carbon 80:734–745
Zhu Y, Cao K, Chen M et al (2019) Synthesis of UV-responsive self-healing microcapsules and their potential application in aerospace coatings. ACS Appl Mater Interfaces 11(36): 33314–33322

Chapter 18
Nanotechnology for Sustainable Future

Manu M. Joseph and Anu Mary Joseph

Abstract Nanotechnology deals with different material characteristics arising from quantum effects, modified electronic traits, and alterations in surface area. Nanotechnology deals with the fabrication of materials from the bottom up and the manipulation of existing structures from the top down, providing a unique perspective on the nature of matter. The merging of nanotechnology and environmental awareness has led to the emergence of 'green nanotechnology,' that combines transformative nanotech principles with a dedication to sustainable practices. This unique amalgamation gives priorities to environmentally friendly methodologies, the use of renewable materials, and contribute to the holistic development of our planet. The main objective of this approach is to harness the unique properties of nanomaterials while minimizing potential environmental risks, emphasizing the principles of green chemistry and the reduction of hazardous substances. In order to ensure the long-term sustainability of our fragile ecosystems, responsible management of resources, the conservation of biodiversity, and the preservation of ecosystem services are of vital importance. Human intervention can influence ecosystem health, and hence it is crucial to adopt sustainable practices while promoting awareness for responsible environmental safekeeping. This chapter aims to provide insights into the diverse applications of green nanotechnology in achieving ecosystem sustainability, highlighting innovations that address and mitigate environmental impacts. The central objective is to establish a harmonious balance between growing human needs and the preservation of natural ecosystems, offering a comprehensive perspective on integrating nanotechnology into environmental conservation efforts.

M. M. Joseph
Department of Life Sciences, School of Science, CHRIST University, Bengaluru, Karnataka, India
e-mail: manu.joseph@christuniversity.in

A. M. Joseph (✉)
Department of Chemistry, Newman College, Thodupuzha, Kerala, India
e-mail: dranumaryjoseph@newmancollege.ac.in

© The Author(s), under exclusive license to Springer Nature Switzerland AG 2025
J. M. Al-Khayri et al. (eds.), *Green Nanotechnology Applications for Ecosystem Sustainability*, Nanotechnology in Plant Sciences 4,
https://doi.org/10.1007/978-3-031-90766-1_18

Keywords Biogenic nanoparticles · Biomedicine · Bioremediation · Ecology · Green chemistry · Nanomedicine polysaccharides

18.1 Introduction

Nanotechnology is a fascinating field of science that involves manipulating materials at the nanoscale, typically between 1 and 100 nanometres, revealing unique properties due to quantum effects, surface area changes, and alterations in electronic and optical characteristics. Physicist Richard Feynman introduced the concept of nanotechnology in 1959, gaining recognition through Eric Drexler's 1986 book envisioning molecular nanotechnology (Bayda et al. 2019; Hulla et al. 2015). Operating at one billionth of a meter, nanoscience primarily explores atoms, molecules, and small structures, providing a distinctive perspective on matter. Quantum effects, such as tunneling and confinement, shape the physical and chemical behavior of materials. The bottom-up approach constructs structures from individual atoms and molecules through self-assembly, mimicking nature. On the other hand, top-down approach manipulates larger structures to achieve nanoscale features, exemplified by lithography techniques for precise control in creating patterns and structures (Johnson and Williams 2015; Smith and Brown 2019).

In the landscape of scientific exploration, convergence of nanotechnology and environmental consciousness has given rise to the unique concept of 'Green Nanotechnology.' This amalgamation integrates the transformative principles of nanotechnology with a commitment to sustainable practices, addressing environmental challenges and mitigating the ecological impact created by various technologies and human practices. Green nanotechnology emphasizes eco-friendly methodologies, sustainable materials, and applications promoting planetary health. Fundamentally, it seeks to harness the peculiar features of nanomaterials while minimizing potential environmental risks by giving priorities in green chemistry wherein renewable resources are employed by reducing the involvement of hazardous substances. This tactic reassures the adoption of environmentally friendly methods for nanomaterial synthesis, such as bio-inspired synthesis using plant extracts or microorganisms, to minimize the ecological footprint of nanoparticle production. A key aspect involves incorporating biodegradable and environmentally friendly materials, contributing to sustainable practices and reducing long-term ecological risks (Smith and Brown 2020; Johnson et al. 2018; Miller et al. 2016).

Characterized by a complex interplay of biotic and abiotic elements, ecosystems embody biodiversity, ecological processes, and the diverse services they offer. They encompass a variety of living organisms, from microscopic bacteria to towering trees, coexisting with the physical components of their surroundings. Biotic elements, including plants, animals, and microorganisms, intricately interact with abiotic factors such as soil, water, air, and sunlight, forming a complex network of relationships. Trophic levels define energy flow, with producers converting sunlight through photosynthesis and consumers occupying subsequent levels. Energy and nutrients circulate, shaping biodiversity and supporting life. Ecosystems provide crucial services for human well-being, including food, water, climate regulation, and disease prevention. Despite disturbances from natural events and human activities, they exhibit resilience, allowing recovery and adaptation. Concepts like ecological succession demonstrate their dynamic nature, endurance and evolution in the face of change (Smith and Brown 2023; Johnson et al. 2022a, b; Miller et al. 2024a, b). Holistic management of ecosystems is crucial, involving responsible resource management, biodiversity conservation, and preserving ecosystem services for present and future generations (Sidle et al. 2013). Sustainable practices aim for a harmonious balance, preventing degradation and fostering ecosystem resilience. Threats like habitat destruction, pollution, climate change, and human activities can disrupt this delicate equilibrium, impacting biodiversity and environmental stability. Conservation efforts, including protected areas and habitat restoration, play a vital role in promoting ecosystem sustainability. The focus is on balancing human needs with preserving natural ecosystems. Human actions, such as resource use, land management, and policy decisions, significantly influence ecosystem sustainability (Johnson; Miller et al. 2024a, b). This chapter explores how green nanotechnology can contribute to advancing ecosystem sustainability, highlighting innovations addressing environmental impacts.

18.2 Principles of Green Nanotechnology

Green nanotechnology represents a careful approach for integrating sustainability and environmental responsibility across the different domains of design, synthesis, and application of functional nanomaterials. In order to further explore this area, we have to adhere to several core principles, emphasizing the complex array if inter connecting networks between technology and ecological well-being.

18.2.1 Sustainable Synthesis Methods

An essential strategy involves substituting or reducing hazardous solvents with eco-friendly options like water, thereby minimizing toxicity and waste. Bio-inspired synthesis methods, employing plant extracts, microorganisms, or biomimetic approaches, provide sustainable alternatives to resource-intensive processes (Bahrulolum et al. 2021; Sundararajan et al. 2024; Niu et al. 2021). Microwave and ultrasound assisted synthesis enhance energy efficiency and reduce reaction times, aligning with the objective of lessening environmental impact. Catalysis is utilized to boost reaction efficiency, enabling milder conditions and lower energy consumption. Inorganic salt-assisted synthesis and green nanocomposites incorporate sustainable materials, promoting efficiency and diminishing reliance on non-renewable resources. Continuous flow synthesis methods cut waste, enhance scalability, and improve overall process efficiency. Green surface functionalization

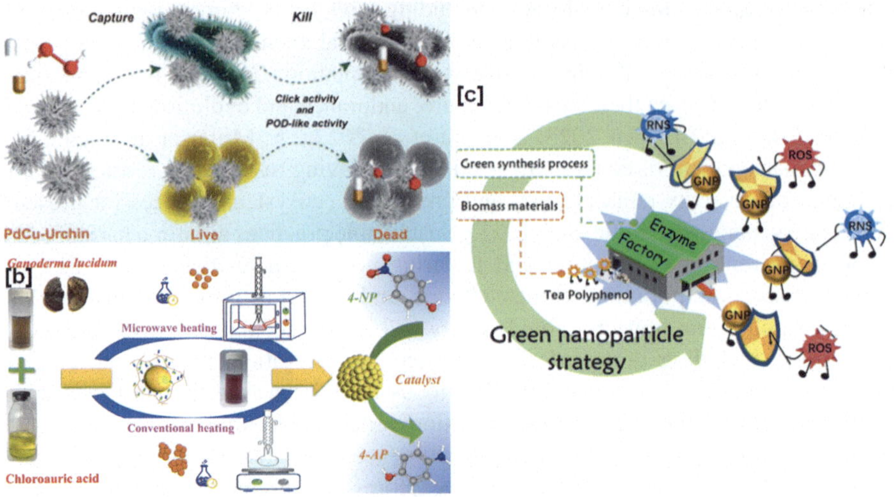

Fig. 18.1 Schematic illustration of (**a**) biomimetic PdCu-Urchin green nanoparticles with multibranched structures for superior bacteria adhesion. Enhanced interactions enable active molecules and reactive oxygen species to act effectively on bacteria, adapted with permission [16]. Copyright 2021, American Chemical Society. (**b**) Biosynthesis of gold nanoparticles (AuNPs) using *Ganoderma lucidum* extract in a microwave-assisted synthesis. This approach, yields smaller AuNPs in a shorter synthesis time and exhibit notable catalytic activity, efficiently reducing 4-nitrophenol to 4-aminophenol, adapted with permission [18]. Copyright 2021, American Chemical Society. (**c**) Schematic representation of green nanoparticles fabricated using tea polyphenol extracts. The resulting nanoparticles, with uniform spherical morphology, exhibited excellent stability along with impressive radical scavenging properties. (Adapted with permission [20]. Copyright 2021, American Chemical Society)

adopts environmentally friendly ligands, contributing to the biocompatibility of nanomaterials (Fig. 18.1). Prioritizing recyclability and reusability minimizes resource consumption and waste. Life cycle assessments evaluate the environmental impact of nanomaterial synthesis, ensuring a holistic understanding of sustainability from extraction to disposal (Nguyen et al. 2021; Liu et al. 2022; Yang et al. 2021).

Green nanoparticle synthesis involves employing natural resources, plant extracts, microorganisms, or other environmentally friendly methods for crafting nanomaterials. A widely embraced method is the utilization of plant extracts, serving as both reducing and stabilizing agents. Plant-derived phytochemicals like polyphenols and flavonoids play a pivotal role in metal ion reduction, leading to nanoparticle formation. Polysaccharides and proteins isolated from various plant sources were observed to be efficient materials in the green synthesis of functional nanoparticles spanning a spectrum of diverse biomedical applications. Microorganisms, spanning bacteria, fungi, and algae, are also harnessed for green synthesis, acting as both reducing and capping agents in an environmentally friendly process. This approach brings advantages such as biocompatibility, cost-effectiveness, and the absence of toxic by-products, establishing it as a preferred method for nanoparticle production (Fig. 18.2). Green-synthesized nanoparticles not only adhere to

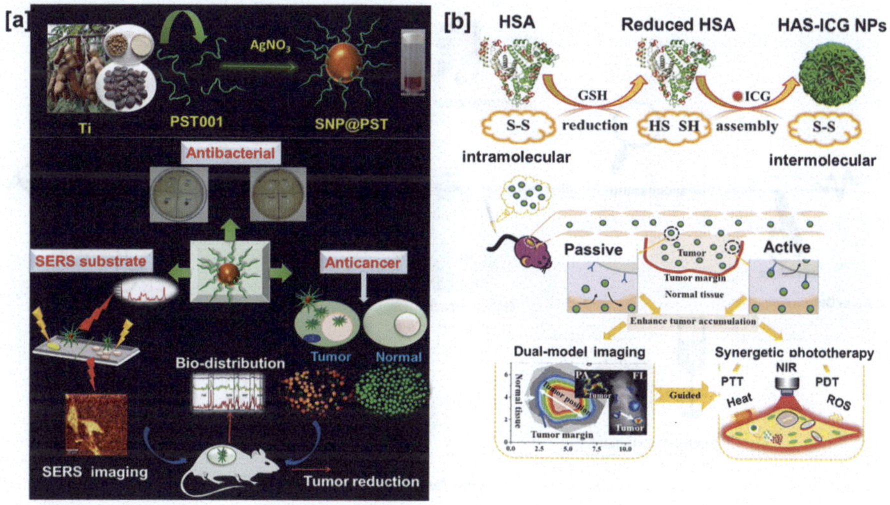

Fig. 18.2 Schematic illustration of (**a**) preparation of silver nanoparticles (SNP@PST) in a green chemistry approach using a naturally occurring galactoxyloglucan (PST001) isolated from the seed kernel of a popular tree *Tamarindus indica*. SNP@PST was evaluated for antibacterial and anticancer potential and later utilized as SERS substrate for tracing cellular internalization and biodistribution, adapted with permission [23]. Copyright 2017, American Chemical Society. (**b**) Schematic representation of a programmed assembly strategy for the synthesis of human serum albumin-indocyanine green nanoparticles (HSA-ICG NPs). The developed nanoparticles aim to enhance phototherapy treatment efficiency by improving drug accumulation and enabling precision-guided phototherapy, minimizing light-associated side effects. (Adapted with permission [25]. Copyright 2014, American Chemical Society)

sustainable practices but also pave the way for innovative applications across various fields Their versatility, biocompatibility, and reduced environmental impact contribute to their growing significance in cutting-edge research and technological advancements (Khan et al. 2022; Nasrollahzadeh et al. 2019; Joseph et al. 2017; Joseph et al. 2020; Sheng et al. 2014).

18.2.2 Biodegradability and Eco-friendly Materials

The biodegradability of nanoparticles produced through green methods is a crucial element in securing their environmental sustainability. Green synthesis methods, aim to minimize the use of toxic chemicals and foster the development of nanoparticles that can naturally decompose over time. Notably, nanoparticles synthesized through green methods possess inherent biocompatibility and reduced toxicity. Plant extracts rich in bioactive compounds, acting as both reducing and capping agents, are frequently employed in green synthesis. Examples include silver or gold nanoparticles, which exhibit enhanced biodegradability compared to

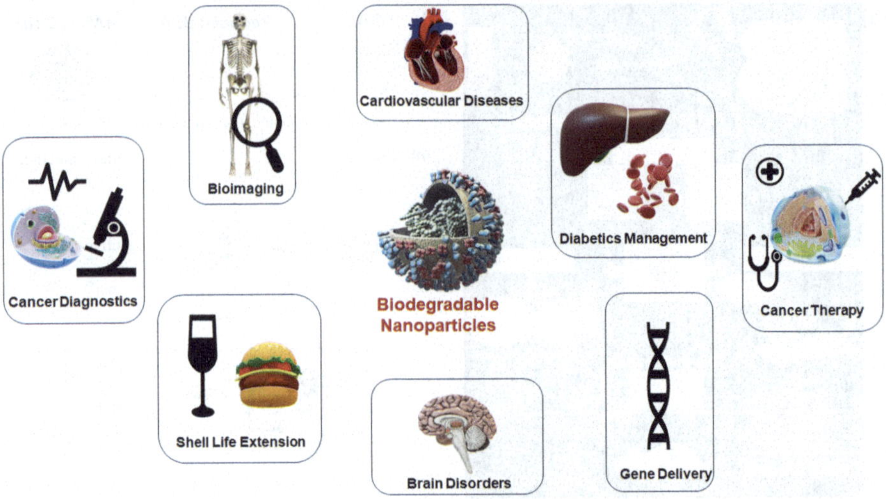

Fig. 18.3 Versatile applications of biodegradable nanoparticles. Biodegradable nanoparticles find diverse applications across various fields, serving as valuable tools in cancer diagnosis and therapy, management of diabetes, cardiovascular diseases, and brain disorders. Additionally, these nanoparticles are utilized as imaging agents, play a crucial role in gene delivery, and contribute to extending the shelf life of foods and beverages

conventionally synthesized counterparts. This reliance on natural compounds facilitates the creation of nanoparticles with minimal environmental persistence, aligning with green chemistry principles. The biodegradability of green-synthesized nanoparticles proves particularly advantageous in biomedical and environmental remediation applications. In biomedical uses, where nanoparticles are employed for drug delivery or imaging, their ability to biodegrade ensures minimized long-term impacts on both the human body and the environment (Fig. 18.3). In environmental remediation, deploying biodegradable nanoparticles contributes to sustainable pollution control solutions, as these particles can naturally degrade over time, diminishing potential ecological risks (Patil and Patel 2020; Kumari et al. 2010; Jeevanandam et al. 2022).

Considering the widespread use of nanoparticles, their fate in the environment is a crucial consideration. Green synthesis methods offer an eco-friendly approach by producing nanoparticles with enhanced biodegradability. However, thorough studies on the environmental fate and long-term impacts of these nanoparticles are essential to ensure responsible and sustainable deployment. As the intersection of nanotechnology and green chemistry progresses, the emphasis on biodegradability reflects a commitment to addressing environmental concerns and promoting the development of environmentally benign nanomaterials. Nanoparticles crafted through environmentally friendly methods, by utilizing natural sources like plant extracts, microorganisms, or biopolymers, not only reduces the reliance on hazardous chemicals but also enhances the biocompatibility and biodegradability of the resulting nanoparticles. Plant extracts, abundant in bioactive compounds, act as efficient reducing agents, while microorganisms function as nano-factories, contributing to nanoparticles with diminished toxicity (Saravanan et al. 2021; Bhardwaj et al. 2020).

18.2.3 Bio-inspired Nanoparticles

The bio-inspired fabrication of nanoparticles draws inspiration from natural processes to create nanomaterials with reduced environmental impact. This approach utilizes plant extracts, microorganisms, and biomimetic techniques as environmentally friendly alternatives. Plant-mediated synthesis of metallic nanoparticles leverages the reducing and stabilizing properties of phytochemicals present in various plants. For instance, silver nanoparticles synthesized using extracts from Aloe vera, neem, or green tea showcase antimicrobial properties, illustrating the potential of bio-inspired methods in biomedical applications. Microorganisms, especially bacteria and fungi, serve as living factories in nanoparticle synthesis, leveraging their enzymatic processes for metal ion reduction. *Bacillus subtilis*, for example, has been utilized in the synthesis of gold nanoparticles, highlighting the versatility of microbial-assisted green synthesis. The biomimetic approach mimics biological processes in the laboratory, using peptides, proteins, or other biomolecules to fabricate nanoparticles (Zhu and Pan 2014; Chen et al. 2020). An example is liposome-templated mineralization, where liposomes act as templates for calcium phosphate nanoparticle synthesis, replicating natural mineralization processes (Skrinda-Melne et al. 2023). The benefits of bio-inspired nanoparticle synthesis go beyond reduced environmental impact. The resulting nanoparticles often exhibit enhanced biocompatibility and functionality, making them suitable for medical applications. In drug delivery, bio-inspired silver or gold nanoparticles offer a platform for targeted and controlled release of therapeutic agents (Fig. 18.4). The application of bio-inspired green nanotechnology aligns with sustainability principles, providing a promising avenue for advanced nanomaterial development across medicine, catalysis, and environmental remediation (Dumontel et al. 2023; Ghose et al. 2023; 34. Koo et al. 2021).

18.3 Applications of Green Nanotechnology for Ecosystem Sustainability

Green nanotechnology focuses primarily on minimizing the environmental impact of nanomaterials, aiming to revolutionize industries while preserving ecosystems. Its applications extend beyond technological innovation, providing solutions aligned with ecosystem sustainability principles. In environmental stewardship, green nanotechnology emerges as a potent tool to address ecological challenges. By adopting environmentally friendly synthesis methods, reducing hazardous materials, and

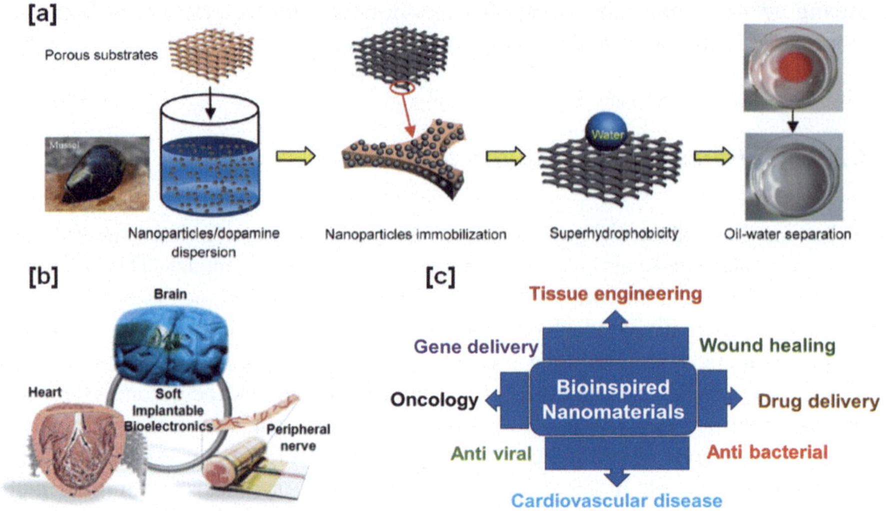

Fig. 18.4 Schematic illustration of (**a**) the immobilization of nanoparticles onto porous substrates and application for oil- water separation. Mussel-inspired direct immobilization of oxide nanoparticles on monolithic foams, spanning polymer to metals, in an aqueous dopamine solution. The resulting system is applied for efficient oil-water separation. [31]. Copyright 2014, American Chemical Society. (**b**) Scheme illustrating the key principles for the development of implantable bioelectronics with long-term stability in vivo. Multifunctional materials endowed with softness, biocompatibility, biodegradability, and self-healing capabilities are expanding the repertoire of materials applied to various electronic implants. (Adapted with permission [36]. Copyright 2021, American Chemical Society. (**c**) Divergent applications of bioinspired nanomaterials)

emphasizing biodegradable nanomaterials, it seeks to integrate technological advances with natural system balance. Applications span diverse fields, including environmental remediation, renewable energy, agriculture, and healthcare (Chincholika et al. 2023; Iavicoli et al. 2014). This holistic approach reflects a conscientious effort to leverage nanotechnology benefits while safeguarding the planet and interconnected ecosystems. Nanoparticles play a multifaceted role in ecosystem sustainability, presenting challenges and opportunities for environmental well-being in various natural systems and intentional applications (Table 18.1).

18.3.1 Environmental Contaminant and Challenges

Nanoparticles, whether originating from natural sources or human activities, can serve as environmental contaminants. When released into the environment through industrial processes or consumer products, engineered nanoparticles may pose risks to ecosystems due to their potential toxicity and persistence (Levard et al. 2012; Cazenave et al. 2019). It is essential to comprehend and address these challenges for overall ecosystem health. Nanoparticles are increasingly utilized in addressing

Table 18.1 Key applications and current advances in green nanotechnology for ecosystem sustainability

Key application	Description	Current advances
Pollution remediation	Use of nanomaterials for cleaning up pollutants in air, water, and soil.	Development of nanomaterials like nano adsorbents for efficient removal of heavy metals and organic pollutants.
Water purification	Nanotechnology-based systems for clean and safe water.	Advancement in nanofilters, nanomembranes, and photocatalysts for removing contaminants and pathogens.
Renewable energy	Enhancing energy generation and storage using nanotechnology.	Improvement in solar cells, nanobatteries, and supercapacitors for energy efficiency.
Carbon capture and storage	Nanotechnology for reducing greenhouse gases.	Creation of nanomaterials for efficient CO_2 capture and conversion into useful products.
Environmental monitoring	Nanosensors for detecting environmental pollutants.	Development of highly sensitive nanosensors for real-time monitoring of air and water quality.
Waste management	Conversion of waste materials into useful nanoproducts.	Advances in using waste-derived nanomaterials for applications like water purification and energy storage.
Soil remediation	Restoration of contaminated soil using nanomaterials.	Innovations in nanoclay and nanocomposites for immobilizing contaminants and restoring soil health.
Biodiversity conservation	Nanotechnology to protect and conserve species and habitats.	Use of nanotechnology for habitat restoration and protection of endangered species.
Sustainable packaging	Development of biodegradable and compostable packaging materials.	Creation of nanocomposites from renewable resources for packaging that reduces plastic waste.
Ecosystem restoration	Application of nanomaterials in ecological restoration projects.	Use of nanotechnology to enhance plant growth and restore damaged ecosystems, such as wetlands and forests.
Renewable resources utilization	Efficient use of renewable materials in nanotechnology applications.	Innovations in using cellulose nanofibers and other natural polymers in a wide range of applications.
Environmental risk assessment	Evaluation of the potential risks of nanomaterials on ecosystems.	Development of frameworks and models to assess and mitigate the ecological risks of nanotechnology.
Circular economy	Promoting resource efficiency and waste reduction through nanotechnology.	Integration of nanotechnology in recycling processes to recover valuable materials and minimize waste.
Sustainable urban development	Application of green nanotechnology in urban planning and construction.	Use of nanomaterials in energy-efficient buildings and green infrastructure projects.

environmental concerns. Carefully fabricated engineered nanoparticles, tailored for efficient removal of pollutants, heavy metals, and contaminants from air, water, and soil, take advantage of unique properties such as high surface area and reactivity.

Eco-friendly methods, like using plant extracts or microorganisms, are employed to synthesize green nanoparticles, significantly contributing to environmentally friendly contaminant remediation. In water purification, materials such as graphene oxide or iron nanoparticles establish notable adsorption capabilities, effectively eliminating pollutants and heavy metals, and also exhibiting antimicrobial activities for pathogen removal. In soil remediation, green nanoparticles assist in pollutant removal and the revitalization of contaminated land. Their environmentally friendly characteristics, coupled with effective remediation, position them as promising tools for mitigating environmental degradation and promoting sustainable practices in contaminant remediation (Singh et al. 2018; Ejaz et al. 2023).

18.3.2 Water Treatment

Green nanoparticles play a vital role in water purification, addressing challenges related to water contamination, and promoting the sustainability of ecosystems. Synthesized with environmentally friendly methods, effectively capture heavy metals, organic pollutants, and contaminants in water. Their increased surface area and reactivity make them efficient tools for water remediation, contributing to the preservation of aquatic ecosystems. It is important to note that silver nanoparticles, crafted through environmentally conscious methods, showcase powerful antimicrobial properties, assisting in sustainable water disinfection for ensuring access to clean drinking water. Moreover, iron nanoparticles synthesized in green processes have been utilized to tackle arsenic contamination in water, especially in regions facing issues with arsenic-contaminated groundwater. The environment friendly features of these nanoparticles alleviate concerns about pollution on aquatic life, aligning with sustainable water treatment practices (Song et al. 2023; Ebrahimbabaie et al. 2020).

Green nanoparticles also play a role in addressing emerging contaminants like pharmaceuticals and personal care products in water systems. It is worth to be notice that titanium dioxide nanoparticles, synthesized through environmentally friendly methods, exhibit photocatalytic properties capable of breaking down organic compounds under sunlight. This can promote rapid degradation of persistent pollutants, and the specificity of these nanoparticles in targeting contaminants minimizes the environmental impact of water treatment. Green nanoparticles also contribute to the development of nanocomposite materials for water purification. Nanostructured materials, like chitosan nanoparticles combined with other sustainable components, enhance pollutant and heavy metal removal, offering an eco-friendly alternative to conventional water treatment (Mumtaz et al. 2022; Ajiboye et al. 2021). The design of these nanocomposites aligns with broader goals of reducing the ecological footprint of water purification technologies. As global water scarcity and concerns about water quality persist, the role of green nanoparticles in water purification becomes increasingly crucial. By employing nanotechnology through eco-friendly synthesis, these nanoparticles provide a promising way forward for water management. Their applications, including adsorption, antimicrobial action, and photocatalysis, showcase the versatility of green nanoparticles in addressing various water contamination challenges, ultimately contributing to the long-term health and sustainability of our environment.

18.3.3 Soil Remediation

Soil contamination, resulting from harmful substances, raises substantial environmental concerns affecting ecosystems, human health, and agriculture. Industrial activities, improper waste disposal, agriculture, and accidental spills, introducing contaminants ranging from heavy metals to pesticides are all the major causative factors. The rise of nanotechnology in industries prompts worries about soil contamination from nanoparticles, demanding careful consideration due to potential adverse effects on soil and ecosystems (Horváth 2000; Rajput et al. 2020). Green nanoparticles offer an inventive and sustainable approach to addressing environmental contamination in soil remediation. Originating from nature or produced using eco-friendly methods, they represent a promising solution to soil pollution challenges. They exist in various forms, including plant-based, microbial-synthesized, and bio-inspired, each utilizing distinct mechanisms for effective remediation. Green nanoparticles employ varied mechanisms in soil remediation, with adsorption being crucial due to their high surface area facilitating contaminant binding. Neem-based nanoparticles, for example, effectively adsorb heavy metals like lead and chromium, diminishing their presence in soil. Furthermore, nanoparticles synthesized by microorganisms enhance microbial activity, promoting the biodegradation of organic pollutants; bacteria-generated silver nanoparticles illustrate this by breaking down pesticide residues.

Ion exchange is crucial, with clay-based nanoparticles immobilizing heavy metals through cation exchange, limiting their mobility. Certain green nanoparticles, acting as nano catalysts, accelerate chemical reactions, transforming contaminants into less harmful substances; iron oxide nanoparticles catalyze the degradation of chlorinated hydrocarbons in soil. Biosynthesized silver nanoparticles exhibit potent capabilities in degrading pesticides and herbicides, indicating the potential of microbial-synthesized nanoparticles against soil pollution. Aloe vera-derived nanomaterials are explored for their remediation potential, especially addressing organic pollutants, while tea leaf extract nanoparticles show effective adsorption for various contaminants. Despite promise, challenges exist in the application of green nanoparticles in soil remediation. Regulatory frameworks lack standardization, ecotoxicological concerns require careful consideration, and scalability and cost-effectiveness need further research for large-scale viability (Mazarji et al. 2023; Rajput et al. 2022). Ongoing collaboration, research, and integration with traditional methods are crucial for realizing the full benefits of green nanoparticle technologies in soil remediation.

18.3.4 Renewable Energy

Renewable energy plays a crucial role in the global transition toward sustainable and eco-friendly power supplies. Solar energy, harnessed through photovoltaic systems and solar thermal technologies, emerges as a versatile and abundant resource. Photovoltaic panels transform sunlight into electricity, offering an accessible option for both residential and large-scale applications. Concurrently, solar thermal systems focus sunlight to generate heat, subsequently converted into electricity. This dual approach diversifies energy portfolios, reducing reliance on conventional fossil fuels and mitigating associated environmental impacts. Wind energy is another key element, with wind turbines serving as symbols of sustainable power generation. By harnessing the kinetic energy of the wind, turbines convert it into mechanical power and then into electricity. Onshore and offshore wind farms capitalize on wind energy's reliability and scalability. The global push for wind power aligns with broader goals of reducing carbon emissions and establishing resilient energy infrastructures. Despite challenges like intermittency, advancements in energy storage and smart grid solutions enhance the reliability of renewable energy sources, marking a transformative era in the pursuit of clean and sustainable power (Sudarsan et al. 2023; Adu-Kankam and Camarinha-Matos 2022). A key area where green nanotechnology excels is in enhancing the efficiency of solar energy conversion. Nanomaterials, such as quantum dots and nanostructured thin films, possess unique optical and electronic properties that improve light absorption and electron transport in photovoltaic cells. This leads to more effective solar cells, capable of converting sunlight into electricity with increased efficiency (Guo 2011).

Nanomaterials such as graphene and nanostructured electrodes empower the creation of robust and high-performance energy storage devices. This advancement leads to enhanced charge/discharge rates and overall capacity, offering more sustainable solutions to manage the intermittent nature of renewable sources like solar and wind. In the field of energy harvesting, nanogenerators resulting from green nanotechnology can convert slight mechanical movements into electrical energy. This promising technology can produce self-powered sensors, reducing the reliance on external power sources. This approach aligns with the environmentally conscious ethos of green nanotechnology, ensuring a minimal ecological impact. Moreover, green nanotechnology plays a role in developing efficient and eco-friendly catalysts for energy-related reactions, benefiting sectors such as transportation and industry. Green nano catalysts demonstrate high catalytic activity, improving performance and lowering energy conversion costs. Sustainability remains a focal point in the manufacturing of nanomaterials within green nanotechnology. The adoption of bio-friendly synthesis methods serves to minimize the environmental footprint, supporting the broader objective of ecologically responsible nanotechnology application across its lifecycle (Rai et al. 2018; Wang and Wu 2012; Abd-Elrahman et al. 2022). In summary, green nanotechnology plays a transformative role in renewable energy, impacting various aspects. It improves solar cell efficiency, enhances energy storage, and contributes to eco-friendly catalysis, fostering the development of sustainable solutions essential for the widespread adoption and advancement of renewable energy technologies.

18.3.5 Agriculture and Crop Protection

Green nanotechnology brings palpable changes to agriculture, tackling issues in food production, environmental effects, and resource utilization. They are utilized in delivery systems for agrochemicals, where nanoparticles encapsulate pesticides, herbicides, and fertilizers for targeted release. This enhances effectiveness, reduces required quantities, and improves resource efficiency, diminishing environmental impact. Pesticide nanoencapsulation extends their release, enhancing persistence on crops and lessening the need for frequent reapplications. Green nanotechnology also plays a role in creating nano sensors for real-time monitoring of crop health and environmental conditions, detecting pathogens and nutrient levels. These nano sensors allow accurate interventions, supporting precision agriculture to optimize resource utilization and reduce environmental impact. For example, carbon-based nanomaterials such as graphene in nano sensors can identify plant diseases, providing an early-warning system to prevent extensive crop damage (Wang et al. 2021; Shang et al. 2019).

Nanomaterials play a pivotal role in enhancing nutrient delivery to plants through the use of nano-fertilizers, formulated with nanoscale carriers. These carriers protect nutrient loss, ensuring a higher proportion reaches the plant and improving crop nutritional content. For instance, silicon nanoparticles have been utilized to improve the delivery of nutrients like phosphorus, promoting sustainable agricultural practices. In pest management, green nanotechnology contributes eco-friendly alternatives to conventional pesticides, utilizing nanoparticles from natural sources like neem or chitosan with insecticidal properties. These nanomaterials offer safer and more sustainable options, minimizing harm to non-target organisms and reducing pesticide-resistant pests. Neem-based nanoparticles, for example, provide an effective and environmentally friendly means of crop protection. Nanoparticles, including those based on iron or carbon, remediate contaminated soils, enhancing fertility and ensuring the safety of agricultural produce. This aligns with the overarching goal of sustainable agriculture by mitigating pollutant impact on soil health (Garg et al. 2023; Aguirre-Becerra et al. 2022). Despite promising applications, ongoing considerations regarding potential environmental and health impacts emphasize the need to strike a balance between innovation and responsible development. As research progresses, green nanotechnology has the potential to reshape farming practices, making them more efficient, environmentally friendly, and resilient to emerging challenges in global food security.

18.3.6 Biodiversity Conservation

Biodiversity, encapsulating the abundance and variety of life on Earth, plays a vital role in establishing and maintaining sustainable ecosystems. The intricate interplay among diverse species, ecosystems, and genetic variations within species collectively contributes to the stability and resilience of natural systems. Essential for planetary health and diverse life, sustainable ecosystems are the major players in preserving biodiversity. Biodiversity decline heightens vulnerability to diseases, climate change, and invasive species. These ecosystems provide vital services like clean air, water, pollination, climate regulation, and nutrient cycling. Resilient ecosystems act as buffers, ensuring recovery and maintenance against disturbances in the biodiversity whether it is caused by natural causes or human interventions. Conservation efforts prevent species extinction by safeguarding habitats and addressing threats like habitat destruction, pollution, and climate change. Biogenic nanoparticles are crucial for biodiversity conservation, offering innovative solutions for monitoring and ecosystem management. Nano sensors excel in real-time environmental monitoring, assessing human activities' impact on biodiversity. In wildlife conservation, green nanoparticles contribute to non-invasive tracking and monitoring of animal species. Nano-sized radio transmitters or sensors attached to animals provide researchers with data on their movements, behavior, and physiological parameters, minimizing disturbance to wildlife while offering valuable insights for conservation efforts (Sharma et al. 2021; Koedrith et al. 2015; Wanekaya et al. 2008).

Green nanoparticles play a role in habitat restoration and remediation, aiding in the removal of contaminants from soil, water, or air to enhance habitat quality. Iron nanoparticles helps in cleaning aquatic ecosystems by precipitating heavy metals, aiding biodiversity recovery. Green nanoparticles play a role in sustainable agriculture by offering eco-friendly nano-fertilizers and nano pesticides, enhancing nutrient utilization, and reducing environmental harm. Sustainable farming practices driven by green nanotechnology are essential for preserving biodiversity. In conservation genetics, green nanoparticles supports non-invasive sampling and monitoring of genetic diversity. Nanoparticles efficiently extract and analyze DNA from environmental samples, providing genetic information without direct contact with organisms. This aids in understanding population dynamics, genetic health, and overall biodiversity patterns. Despite potential benefits, addressing risks and ethical considerations is crucial for responsibly deploying green nanoparticles in biodiversity conservation, ensuring adherence to conservation principles for lasting effectiveness (Thompson and Darwish 2019; Rahimizadeh et al. 2023).

18.4 Exploring Challenges in Green Nanotechnology for Sustainable Ecosystems

Green nanotechnology also raises little concerns about ecological impacts. Efficiently managing nanomaterial waste, understanding long-term effects on biodiversity, and responsible nanoparticle use in environmental applications are crucial challenges. Balancing benefits and risks are essential for a holistic coexistence between technology, ecosystem and living beings in the planet. In depth research investigations, ethics, and regulatory frameworks are imperative for responsible green nanotechnology advancement.

18.4.1 Toxicity and Ecotoxicity

Assessing the environmental impact of green nanotechnology, requires careful consideration in terms of toxicity it produce. Despite innovative solutions, concerns about potential adverse side effects on ecosystems and human health must be thoroughly investigated and carefully documented and calibrated. Nanoparticles, due to their small size and increased surface area, may have unique toxicological properties. Understanding the toxicity of green nanomaterials is crucial for their safe use. Say for an example, silver nanoparticles, widely used in antimicrobial coatings, have raised concerns due to potential toxicity to aquatic organisms, impacting plant and animal growth (Du et al. 2018). Ecotoxicity, is a crucial consideration for the sustainability of green nanotechnology. The introduction of nanoparticles into the environment sparks greater concerns about their prolonged impact on biodiversity, ecosystem operations and biotic components in the planet. In this point of view, carbon-based nanomaterials, such as nanotubes, exhibit potential ecotoxicity, impacting the growth of aquatic organisms. Understanding the interactions of nanoparticles with diverse organisms and ecosystems is vital to foresee and alleviate ecological consequences. For instance, titanium dioxide nanoparticles, extensively used in sunscreens, accumulate in soil and aquatic settings, prompting concerns about their long-term effects. Ongoing research seeks to comprehend nanomaterial behavior and formulate strategies to minimize environmental perseverance (Simonin and Richaume 2015; Minetto et al. 2014).

The engagement of biogenic nanoparticles with soil and water ecosystems introduces a new area of potential effects. For instance, nano-agrochemicals, which boost pesticide and fertilizer efficiency, might impact soil microbial communities and nutrient cycling. Grasping these interactions is essential for anticipating and controlling potential ecological disturbances while harnessing the advantages of green nanotechnology in agriculture sector. Current research focuses on safe-by-design strategies to tackle toxicity and ecotoxicity challenges, crafting nanomaterials with minimal ecological impact. This encompasses evaluating physicochemical properties, interactions with biological systems, and applying standardized testing methods for risk assessment (Ahmed et al. 2023; Hulkoti and Taranath 2014; Philip 2009). Globally, new regulatory frameworks are generated to steer the secure development and utilization of nanomaterials.

18.4.2 Ethical and Regulatory Frameworks

Ethical and regulatory aspects in green nanotechnology are essential for the sage use of nanomaterials. Addressing ethical worries tied to the environmental release of nanoparticles is essential as green nanotechnology advances, especially in applications like nano-enabled pesticides in agriculture, raising questions about impacts on non-target organisms and the broader ecosystem. Ensuring transparency in communicating risks and benefits to stakeholders is vital for ethical decision-making in adopting green nanotechnologies. Regulatory frameworks for nanotechnology face challenges due to the dynamic nature of nanomaterials and evolving risk understanding, leading to uncertainties in assessment and management. The use of nanomaterials in consumer products, such as nano-enabled cosmetics, poses challenges for agencies in establishing clear safety guidelines and labeling. The intricate nature of these particles demands a nuanced regulatory approach to address their unique characteristics and potential impacts on ecosystems (Dréno et al. 2019; Iavicoli et al. 2014).

The global regulatory landscape of green nanotechnology is complicated, with conflicting standards and nanomaterial definitions across countries. Organizations like the Organization for Economic Co-operation and Development (OECD) aim to create internationally recognized guidelines, addressing challenges and promoting responsible nanotechnology practices. Rigorous evaluation, especially in water treatment applications, ensures that any nanoparticle release into aquatic ecosystems avoids adverse effects on biodiversity. The regulatory challenge lies in implementing comprehensive risk assessments, considering nanomaterial life cycles and potential cumulative effects, to oversee the sustainable use of green nanotechnology (Resnik and Tinkle 2007a, b; Resnik 2012; Marchant and Lindor 2012). Addressing potential disparities in nanotechnology access and impacts, especially in environmental justice, is crucial. Ethical frameworks must ensure nanomaterials in pollution remediation do not disproportionately affect vulnerable communities, ensuring equitable distribution of benefits and risks. Collaboration among researchers, policymakers, industry stakeholders, and the public is essential for the ethical and sustainable integration of green nanotechnology.

18.5 Future Directions of Green Nanotechnology for Sustainable Ecosystems

In sustainable agriculture, the future of green nanotechnology revolves around precision farming and smart delivery systems. Nano-enabled precision agriculture could boost crop yield and reduce environmental impact by precisely delivering nutrients and pesticides. Nanoscale sensors integrated into soil or plants could provide real-time data on nutrient levels, enabling targeted and efficient resource use. Innovations in green nanotechnology like these have the potential to revolutionize farming practices, making them more sustainable and environmentally friendly (Abdelhamid and Mathew 2022; Goyal et al. 2023). Nanomaterials, such as graphene oxide or carbon nanotubes, exhibit potential for efficiently eliminating pollutants from water, offering sustainable solutions for clean water production. Additionally, nanomaterial-based membranes and filtration systems could play a pivotal role in addressing water scarcity issues. The development of these technologies exemplifies the future emphasis on sustainable water management through green nanotechnology. Energy storage stands out as a critical area where green nanotechnology can contribute to sustainability.

Progress in nanomaterials for batteries and supercapacitors could result in energy storage systems with increased efficiency and reduced environmental impact. Green nanotechnology explores eco-friendly electrodes and electrolytes for energy storage devices, ensuring alignment with sustainability goals throughout the entire life cycle of these technologies. In the realm of environmental monitoring, the future of green nanotechnology entails the creation of nano sensors for real-time detection of pollutants and contaminants. These sensors could be deployed in ecosystems to continuously monitor air and water quality, providing valuable data for timely interventions. The use of green nano sensors diminishes the environmental impact associated with monitoring activities, contributing to sustainable ecosystem management (Verma et al. 2023; Ren et al. 2020; Pandey et al. 2019; Han et al. 2023). The future trajectory of green nanotechnology underscores a commitment to circular economy principles. This entails the deliberate design of nanomaterials with consideration for end-of-life scenarios, emphasizing recycling and waste minimization. For instance, the advancement of biodegradable nanomaterials guarantees their natural breakdown after use, avoiding environmental harm. As these forward-looking initiatives unfold, addressing the ethical and regulatory aspects of green nanotechnology becomes imperative. Ensuring responsible innovation, transparent communication regarding potential risks, and establishing robust regulatory frameworks are essential for the sustainable integration of green nanotechnology across various sectors.

18.6 Conclusions and Prospects

Green nanotechnology emphasizes sustainable practices, utilizing nanomaterials while minimizing environmental risks. The chapter highlights eco-friendly methodologies, renewable materials, and applications for the proper ecosystem balance of the planetary health. Challenges like habitat destruction, pollution, and climate change require conservation and restoration efforts. Human actions impact ecosystems, necessitating sustainable practices and awareness promotion. Nanotechnology, manipulating structures from both ends, offers a distinctive perspective. Green nanotechnology seeks a balance between human needs and preserving natural ecosystems, promoting responsible technological integration into environmental conservation.

Green nanotechnology approach becomes very vital, especially when mounting challenges like habitat destruction, pollution, and climate change destabilize the balance of our planet's ecosystems. It also reviews some of the eco-friendly methodologies and renewable materials that ought to be used to maintain health as well as the overall well-being of our planet. Nanotechnology therefore heralds new ways in which environmental sustainability can be achieved through the manipulation of structures at the nanoscale to form materials and processes that not only become highly effective but also benign to any ensuing environment. For example, green nanotechnology can enable the development of biodegradable nanomaterials; therefore, it will have less ecological footprint from a given industrial process. It supports making energy-efficient technologies, from advanced solar cells to energy-storing batteries.

Therefore, in the future, green nanotechnology should be directed toward key areas in the interest of developing a potential impact on ecosystem sustainability.

Development of Biodegradable Nanomaterials: The development of self-degrading nanomaterials that leave behind non-polluting by-products has become essential. These materials will find applications in packaging, electronics, and medical devices without causing pollution and generating wastes.

Nanotechnology in remediation: Further development of nanomaterials is considered in the removal of various pollutants from water, air, and land; testing their nanostructure capacity for capturing and neutralizing many heavy metals, organic pollutants, and microplastics to restore health to contaminated ecosystems. In the context of renewable energy, green nanotechnology is explicitly posited to transform the performance of solar cells, upgrading the wind turbine and storage of the produced energy. Hence, the areas where future research is to be focused are those having potential for optimization in order that such technologies are made broadly accessible and cheap for general adoption of clean energy.

Agricultural Use: The applications of nanotechnology in agriculture could pose fresh hopes as sustainable solutions for the farming practice. Appropriate research should be carried out to develop nanosensors for precision farming, nanofertilizers that reduce nutrient runoff, and nanopesticides that are less detrimental to the environment.

Climate Change Mitigation: Nanotechnology can play a very critical role in the field of climate change mitigation through the development of advanced materials for carbon capture and storage technologies and through highly increased efficiency of renewable energy systems. New ways in which nanotechnology can be integrated into global climate strategies should be explored in the near future.

Education and Public Awareness: Education and public awareness regarding the opportunities and challenges offered by green nanotechnology are needed if it is going to achieve its full potential. A partnership of researchers and policy makers needs to create training programs to inform the public about sustainable nanotechnology and to promote responsible consumer behavior.

In other words, green nanotechnology is a promise to find a good equilibrium between the fulfillment of human needs and the protection of natural ecosystems. A sustainability perspective oriented toward the production and application of expanded potentials for green nanotechnology may be what defines the overall landscape for responsible technological incorporation into programs for environmental sustainability. Next-steps research will build upon these within green nanotechnology as it contributes to a more sustainable and durable planet for generations.

Acknowledgement Author (MMJ) thanks CHRIST University for the Seed Money Grant (CU-ORS-SM-24/04).

References

Abdelhamid HN, Mathew AP (2022) Cellulose-based nanomaterials advance biomedicine: a review. Int J Mol Sci 23:5405. https://doi.org/10.3390/ijms23105405

Abd-Elrahman NK et al (2022) Applications of nanomaterials in microbial fuel cells: a review. Molecules 27:7483. https://doi.org/10.3390/molecules27217483

Adu-Kankam KO, Camarinha-Matos LM (2022) Renewable energy communities or ecosystems: an analysis of selected cases. Heliyon 8:e12617. https://doi.org/10.1016/j.heliyon.2022.e12617

Aguirre-Becerra H et al (2022) Nanomaterials as an alternative to increase plant resistance to abiotic stresses. Front Plant Sci 13:1023636. https://doi.org/10.3389/fpls.2022.1023636

Ahmed T, Noman M, Gardea-Torresdey JL, White JC, Li B (2023) Dynamic interplay between nano-enabled agrochemicals and the plant-associated microbiome. Trends Plant Sci 28:1310–1325. https://doi.org/10.1016/j.tplants.2023.06.001

Ajiboye TO, Oyewo OA, Onwudiwe DC (2021) Simultaneous removal of organics and heavy metals from industrial wastewater: a review. Chemosphere 262:128379. https://doi.org/10.1016/j.chemosphere.2020.128379

Bahrulolum H et al (2021) Green synthesis of metal nanoparticles using microorganisms and their application in the agrifood sector. J Nanobiotechnol 19:86. https://doi.org/10.1186/s12951-021-00834-3

Bayda S, Adeel M, Tuccinardi T, Cordani M, Rizzolio F (2019) The history of nanoscience and nanotechnology: from chemical-physical applications to nanomedicine. Molecules 25:112. https://doi.org/10.3390/molecules25010112

Bhardwaj B, Singh P, Kumar A, Kumar S, Budhwar V (2020) Eco-friendly greener synthesis of nanoparticles. Adv Pharm Bull 10:566–576. https://doi.org/10.34172/apb.2020.067

Cazenave J, Ale A, Bacchetta C, Rossi AS (2019) Nanoparticles toxicity in fish models. Curr Pharm Des 25:3927–3942. https://doi.org/10.2174/1381612825666190912165413

Chen HY et al (2020) Hybrid cell membrane-coated nanoparticles: a multifunctional biomimetic platform for cancer diagnosis and therapy. Acta Biomater 112:1–13. https://doi.org/10.1016/j.actbio.2020.05.028

Chincholika P et al (2023) Green nanobiopolymers for ecological applications: a step towards a sustainable environment. RSC Adv 13:12411–12429. https://doi.org/10.1039/d2ra07707h

Dréno B, Alexis A, Chuberre B, Marinovich M (2019) Safety of titanium dioxide nanoparticles in cosmetics. J Eur Acad Dermatol Venereol 33(Suppl 7):34–46. https://doi.org/10.1111/jdv.15943

Du J et al (2018) A review on silver nanoparticles-induced ecotoxicity and the underlying toxicity mechanisms. Regul Toxicol Pharmacol 98:231–239. https://doi.org/10.1016/j.yrtph.2018.08.003

Dumontel B, Jiménez-Jiménez C, Vallet-Regí M, Manzano M (2023) Bioinspired extracellular vesicle-coated silica nanoparticles as selective delivery systems. Mater Today Bio 23:100850. https://doi.org/10.1016/j.mtbio.2023.100850

Ebrahimbabaie P, Meeinkuirt W, Pichtel J (2020) Phytoremediation of engineered nanoparticles using aquatic plants: mechanisms and practical feasibility. J Environ Sci 93:151–163. https://doi.org/10.1016/j.jes.2020.03.034

Ejaz M et al (2023) Nanotechnologies for environmental remediation and their ecotoxicological impacts. Environ Monit Assess 195:1368. https://doi.org/10.1007/s10661-023-11661-4

Garg D et al (2023) Nano-biofertilizer formulations for agriculture: a systematic review on recent advances and prospective applications. Bioengineering (Basel, Switzerland) 10:1010. https://doi.org/10.3390/bioengineering10091010

Ghose D, Swain S, Patra CN, Jena BR, Rao MEB (2023) Advancement and applications of platelet-inspired nanoparticles: a paradigm for cancer targeting. Curr Pharm Biotechnol 24:213–237. https://doi.org/10.2174/1389201023666220329111920

Goyal V et al (2023) Unlocking the potential of nano-enabled precision agriculture for efficient and sustainable farming. Plants (Basel, Switzerland) 12:3744. https://doi.org/10.3390/plants12213744

Guo KW (2011) Green nanotechnology of trends in future energy. Recent Pat Nanotechnol 5:76–88. https://doi.org/10.2174/187221011795909198

Han YY et al (2023) Green synthesis of carbon-supported ultrafine ZnS nanoparticles for superior lithium-ion batteries. Dalton Trans 52:16336–16344. https://doi.org/10.1039/d3dt02407e

Horváth A (2000) Soil pollution and soil limit values. Cent Eur J Public Health 8:53–58

Hulkoti NI, Taranath TC (2014) Biosynthesis of nanoparticles using microbes- a review. Colloids Surf B Biointerfaces 121:474–483. https://doi.org/10.1016/j.colsurfb.2014.05.027

Hulla JE, Sahu SC, Hayes AW (2015) Nanotechnology: history and future. Hum Exp Toxicol 34:1318–1321. https://doi.org/10.1177/0960327115603588

Iavicoli I, LesoV RW, Hodson LL, Hoover MD (2014) Opportunities and challenges of nanotechnology in the green economy. Environ Health 13:78. https://doi.org/10.1186/1476-069X-13-78

Jeevanandam J et al (2022) Green approaches for the synthesis of metal and metal oxide nanoparticles using microbial and plant extracts. Nanoscale 14:2534–2571. https://doi.org/10.1039/d1nr08144f

Johnson RS, Williams MA (2015) Nanoscale exploration and quantum effects. J Nanosci 8:45–58. https://doi.org/10.1234/js.2015.1234567890

Johnson RS, White M, Garcia L (2018) Eco-friendly methodologies in green nanotechnology. Environ Nanotechnol 22:112–128. https://doi.org/10.7890/en.2018.9876543210

Johnson RS, White M, Garcia L (2022a) Conservation strategies and restoration initiatives for ecosystem sustainability. Conserv Ecol 40:180–198. https://doi.org/10.7890/jce.2022.9876543210

Johnson RS, White M, Garcia L (2022b) Ecosystem services: a comprehensive review of human-well being. J Environ Sci 36:180–198. https://doi.org/10.7890/jes.2022.9876543210

Joseph MM et al (2017) Exploration of biogenic nano-chemobiotics fabricated by silver nanoparticle and galactoxyloglucan with an efficient biodistribution in solid tumor investigated by

SERS fingerprinting. ACS Appl Mater Interfaces 9:19578–19590. https://doi.org/10.1021/acsami.7b03191
Joseph MM, Hari N, Pillai RK, Nair AJ, Therakathinal TS (2020) Galactoxyloglucan endowed biogenic Nanoimmunobiotics arrests microbial growth and elicits antitumor immunity. ACS Appl Biol Mater 3:801–814. https://doi.org/10.1021/acsabm.9b00834
Khan F et al (2022) Green nanotechnology: plant-mediated nanoparticle synthesis and application. Nanomaterials (Basel, Switzerland) 12:673. https://doi.org/10.3390/nano12040673
Koedrith P et al (2015) Recent trends in rapid environmental monitoring of pathogens and toxicants: potential of nanoparticle-based biosensor and applications. Sci World J 510982. https://doi.org/10.1155/2015/510982
Koo JH, Song JK, Kim DH, Son D (2021) Soft implantable bioelectronics. ACS Mater Lett 3:1528–1540. https://doi.org/10.1021/acsmaterialslett.1c00438
Kumari A, Yadav SK, Yadav SC (2010) Biodegradable polymeric nanoparticles based drug delivery systems. Colloids Surf B Biointerfaces 75:1–18. https://doi.org/10.1016/j.colsurfb.2009.09.001
Levard C, Hotze EM, Lowry GV, Brown GE Jr (2012) Environmental transformations of silver nanoparticles: impact on stability and toxicity. Environ Sci Technol 46:6900–6914. https://doi.org/10.1021/es2037405
Liu X, Bai X, Wu W (2022) Ultrasound-assisted green synthesis of Ru supported on LDH-CNT composites as an efficient catalyst for N-ethylcarbazole hydrogenation. Ultrason Sonochem 91: 106227. https://doi.org/10.1016/j.ultsonch.2022.106227
Marchant GE, Lindor RA (2012) Prudent precaution in clinical trials of nanomedicines. J Law Med Ethics 40:831–840. https://doi.org/10.1111/j.1748-720X.2012.00711.x
Mazarji M et al (2023) Nanomaterials in biochar: review of their effectiveness in remediating heavy metal-contaminated soils. Sci Total Environ 880:163330. https://doi.org/10.1016/j.scitotenv.2023.163330
Miller LR, Davis K, Patel S (2016) Bio-inspired synthesis for sustainable nanoparticle production. Green Chem 35:45–58. https://doi.org/10.7890/jgc.2016.1234567890
Miller LR, Davis K, Patel S (2024a) Human actions and ecosystem sustainability: A policy perspective. Sustain Dev J 45:55–72. https://doi.org/10.7890/sdj.2024.1234567890
Miller LR, Davis K, Patel S (2024b) Resilience and adaptation in ecosystems: insights from ecological succession. Ecol Res 45:55–72. https://doi.org/10.7890/er.2024.1234567890
Minetto D, Libralato G, Volpi Ghirardini A (2014) Ecotoxicity of engineered TiO2 nanoparticles to saltwater organisms: an overview. Environ Int 66:18–27. https://doi.org/10.1016/j.envint.2014.01.012
Mumtaz N et al (2022) Nanoengineered metal-organic framework for adsorptive and photocatalytic mitigation of pharmaceuticals and pesticide from wastewater. Environ Pollut 308:119690. https://doi.org/10.1016/j.envpol.2022.119690
Nasrollahzadeh M, Mahmoudi-Gom Yek S, Motahharifar N, Ghafori Gorab M (2019) Recent developments in the plant-mediated green synthesis of ag-based nanoparticles for environmental and catalytic applications. Chem Rec 19:2436–2479. https://doi.org/10.1002/tcr.201800202
Nguyen VP, Le Trung HL, Nguyen TH, Hoang DQ, Tran TH (2021) Advancement of microwave-assisted biosynthesis for preparing au nanoparticles using Ganoderma lucidum extract and evaluation of their catalytic reduction of 4-nitrophenol. ACS Omega 6:32198–32207. https://doi.org/10.1021/acsomega.1c05033
Niu J, Zhao C, Liu C, Ren, J.&Qu, X. (2021) Bio-inspired bimetallic enzyme mimics as bio-orthogonal catalysts for enhanced bacterial capture and inhibition. Chem Mater 33:8052–8058. https://doi.org/10.1021/acs.chemmater.1c02469
Pandey RK, Chen L, Teraji S, Nakanishi H, Soh S (2019) Eco-friendly, direct deposition of metal nanoparticles on graphite for electrochemical energy conversion and storage. ACS Appl Mater Interfaces 11:36525–36534. https://doi.org/10.1021/acsami.9b09273
Patil V, Patel A (2020) Biodegradable nanoparticles: a recent approach and applications. Curr Drug Targets 21:1722–1732. https://doi.org/10.2174/1389450121666200916091659

Philip D (2009) Honey-mediated green synthesis of gold nanoparticles. Spectrochim Acta A Mol Biomol Spectrosc 73:650–653. https://doi.org/10.1016/j.saa.2009.03.007

Rahimizadeh K et al (2023) Nanoparticles-assisted aptamer biosensing for the detection of environmental pathogens. Environ Res 238:117123. https://doi.org/10.1016/j.envres.2023.117123

Rai PK et al (2018) Nanoparticle-plant interaction: implications in energy, environment, and agriculture. Environ Int 119:1–19. https://doi.org/10.1016/j.envint.2018.06.012. Pubmed:29909166

Rajput V et al (2020) ZnO and CuO nanoparticles: a threat to soil organisms, plants, and human health. Environ Geochem Health 42:147–158. https://doi.org/10.1007/s10653-019-00317-3

Rajput VD et al (2022) Nanotechnology in the restoration of polluted soil. Nanomaterials (Basel, Switzerland) 12:769. https://doi.org/10.3390/nano12050769

Ren D et al (2020) Recent environmental applications of and development prospects for immobilized laccase: a review. Biotechnol Genet Eng Rev 36:81–131. https://doi.org/10.1080/02648725.2020.1864187

Resnik DB (2012) Responsible conduct in nanomedicine research: environmental concerns beyond the common rule. J Law Med Ethics 40:848–855. https://doi.org/10.1111/j.1748-720X.2012.00713.x

Resnik DB, Tinkle SS (2007a) Ethical issues in clinical trials involving nanomedicine. Contemp Clin Trials 28:433–441. https://doi.org/10.1016/j.cct.2006.11.001

Resnik DB, Tinkle SS (2007b) Ethics in nanomedicine. Nanomedicine (London) 2:345–350. https://doi.org/10.2217/17435889.2.3.345

Saravanan A et al (2021) A review on biosynthesis of metal nanoparticles and its environmental applications. Chemosphere 264:128580. https://doi.org/10.1016/j.chemosphere.2020.128580

Shang Y et al (2019) Applications of nanotechnology in plant growth and crop protection: a review. Molecules 24:2558. https://doi.org/10.3390/molecules24142558

Sharma P et al (2021) A review on biosensors and nanosensors application in agroecosystems. Nanoscale Res Lett 16:136. https://doi.org/10.1186/s11671-021-03593-0

Sheng Z et al (2014) Smart human serum albumin-indocyanine green nanoparticles generated by programmed assembly for dual-modal imaging-guided cancer synergistic phototherapy. ACS Nano 8:12310–12322. https://doi.org/10.1021/nn5062386

Sidle RC, Benson WH, Carriger JF, Kamai T (2013) Broader perspective on ecosystem sustainability: consequences for decision making. Proc Natl Acad Sci USA 110:9201–9208. https://doi.org/10.1073/pnas.1302328110

Simonin M, Richaume A (2015) Impact of engineered nanoparticles on the activity, abundance, and diversity of soil microbial communities: a review. Environ Sci Pollut Res Int 22:13710–13723. https://doi.org/10.1007/s11356-015-4171-x

Singh J et al (2018) "Green" synthesis of metals and their oxide nanoparticles: applications for environmental remediation. J Nanobiotechnol 16:84. https://doi.org/10.1186/s12951-018-0408-4

Skrinda-Melne M, Locs J, Grava A, Dubnika A (2023) Calcium phosphates enhanced with liposomes—the future of bone regeneration and drug delivery. J Liposome Res 1–16. https://doi.org/10.1080/08982104.2023.2285973

Smith J, Brown PC (2019) Exploring nanoscale matter: an overview of quantum effects. J Nanosci 12:345–362. https://doi.org/10.1234/jns.2019.1234567890

Smith J, Brown PC (2020) Green nanotechnology: merging nanoscience with environmental sustainability. J Nanosci Sustain Dev 15:220–238. https://doi.org/10.1234/jnsd.2020.1234567890

Smith J, Brown PC (2023) Dynamics of biotic and abiotic interactions in ecosystems. Ecol Today 28:320–335. https://doi.org/10.1234/et.2023.1234567890

Song J, Xu Z, Chen Y, Guo J (2023) Nanoparticles, an emerging control method for harmful algal blooms: current technologies, challenges, and perspectives. Nanomaterials (Basel) 13:2384. https://doi.org/10.3390/nano13162384

Sudarsan A, Kurukkanari C, Bendi D (2023) A state-of-the-art review on readiness assessment tools in the adoption of renewable energy. Environ Sci Pollut Res Int 30:32214–32229. https://doi.org/10.1007/s11356-023-25520-9

Sundararajan N et al (2024) Mitigating global challenges: harnessing green synthesized nanomaterials for sustainable crop production systems. Global Chall (Hoboken NJ) 8: 2300187. https://doi.org/10.1002/gch2.202300187

Thompson LA, Darwish WS (2019) Environmental chemical contaminants in food: review of a global problem. J Toxicol 2019:2345283. https://doi.org/10.1155/2019/2345283

Verma N, Jujjavarapu SE, Mahapatra C, Mutra JKR (2023) Contemporary updates on bioremediation applications of graphene and its composites. Environ Sci Pollut Res Int 30:48854–48867. https://doi.org/10.1007/s11356-023-26225-9

Wanekaya AK, Chen W, Mulchandani A (2008) Recent biosensing developments in environmental security. J Environ Monit 10:703–712. https://doi.org/10.1039/b806830p

Wang ZL, Wu W (2012) Nanotechnology-enabled energy harvesting for self-powered micro–/nanosystems. Angew Chem Int Ed Eng English 51:11700–11721. https://doi.org/10.1002/anie.201201656

Wang CY, Yang J, Qin JC, Yang YW (2021) Eco-friendly nanoplatforms for crop quality control, protection, and nutrition. Adv Sci (Weinh) 8:2004525. https://doi.org/10.1002/advs.202004525

Yang P et al (2021) Green nanoparticle scavengers against oxidative stress. ACS Appl Mater Interfaces 13:39126–39134. https://doi.org/10.1021/acsami.1c12176

Zhu Q, Pan Q (2014) Mussel-inspired direct immobilization of nanoparticles and application for oil-water separation. ACS Nano 8:1402–1409. https://doi.org/10.1021/nn4052277

Index

© The Editor(s) (if applicable) and The Author(s), under exclusive license to Springer Nature Switzerland AG 2025

J. M. Al-Khayri et al. (eds.), *Green Nanotechnology Applications for Ecosystem Sustainability*, Nanotechnology in Plant Sciences 4,
https://doi.org/10.1007/978-3-031-90766-1

The manufacturer's authorised representative in the EU is Springer Nature Customer Service Centre GmbH, Europaplatz 3, 69115 Heidelberg, Germany. If you have any concerns regarding our products, please contact ProductSafety@springernature.com

Printed and bound by CPI Group (UK) Ltd, Croydon, CR0 4YY
30/09/2026
02225588-0017